FUNDAMENTALS, PROPERTIES, AND APPLICATIONS OF POLYMER NANOCOMPOSITES

This book is focused primarily on polymer nanocomposites, based on the author's research experience as well as open literature. This text is divided into three parts. Part I – Fundamentals, Processing, and Characterizations – consists of an introduction to nanotechnology, an overview of commonly used nanomaterials, their origins, manufacturing, properties, and current and potential applications as well as the basics of polymer matrices and composites, selection of resin matrix and nanomaterials, processing, and characterizing of multifunctional polymer nanocomposites. Part II – Multifunctional Properties and Applications – consists of seven chapters that discuss the important properties as well as current applications of these polymer nanocomposites. Part III – Concerns and Outlook – consists of two chapters that discuss the environmental and health impacts for nanomaterials and polymer nanocomposites. The environmental health and safety aspects of nanomaterials and polymer nanocomposites, risk assessment and safety standards, and fire toxicity of polymer nanocomposites are studied. In the final chapter, a brief overview of opportunities, trends, and challenges of polymer nanocomposites are included. Throughout the book, the theme is developed that polymer nanocomposites are a whole family of polymeric materials whose properties are capable of being tailored to meet specific applications. This volume serves as a general introduction to students and researchers just entering the field and to scholars from other subfields seeking information.

Dr. Joseph H. Koo is a senior research scientist/ research professor/ director of Polymer Nanocomposites Technology Laboratory of mechanical engineering at the University of Texas at Austin. His research focuses on polymer nanocomposites and multifunctional materials and structures for extreme environments, in four major areas: (a) Ablation research, (b) Flame retardant polymer research, (c) Conductive polymer research, and (d) Others: carbon/carbon and downhole drilling composites. He is the author of a textbook, *Polymer Nanocomposites: Processing, Characterization and Applications*, and more than 500 research publications. He is a Fellow in the Society for the Advancement of Material and Process Engineering (SAMPE), and an Associate Fellow in the American Institute of Aeronautics and Astronautics (AIAA).

Fundamentals, Properties, and Applications of Polymer Nanocomposites

JOSEPH H. KOO

The University of Texas at Austin

CAMBRIDGE
UNIVERSITY PRESS

CAMBRIDGE
UNIVERSITY PRESS

One Liberty Plaza, 20th Floor, New York, NY 10006, USA

Cambridge University Press is part of the University of Cambridge.

It furthers the University's mission by disseminating knowledge in the pursuit of education, learning and research at the highest international levels of excellence.

www.cambridge.org
Information on this title: www.cambridge.org/9781107029965

First published 2016

Printed in the United States of America by Sheridan Books, Inc.

A catalogue record for this publication is available from the British Library.

Library of Congress Cataloging-in-Publication Data
Names: Koo, Joseph H.
Title: Fundamentals, properties, and applications of polymer nanocomposites / Joseph H. Koo, University of Texas.
Description: Cambridge : Cambridge University Press, 2016. | Includes bibliographical references and index.
Identifiers: LCCN 2016017602 | ISBN 9781107029965 (hardback)
Subjects: LCSH: Nanocomposites (Materials) | Polymers.
Classification: LCC TA418.9.N35 K66 2016 | DDC 620.1/92–dc23
LC record available at https://lccn.loc.gov/2016017602

ISBN 978-1-107-02996-5 Hardback

To my wife, Penelope, and my daughter, Hilary, whose understanding, support, and encouragement made the completion of this book possible.

This book is also dedicated to my late graduate advisor, Professor Ken Kuo for his teachings.

Contents

Preface

Nanotechnology is the study and control of matter at dimensions of roughly 1 to 100 nanometers, where unique phenomena enable novel applications. Nanotechnology is also the design, fabrication, and application of nanostructures or nanomaterials. The excitement surrounding nanoscale science and technology gives us unique opportunities to develop revolutionary materials. Nanoscale science and technology is an emerging field that encompasses nearly every discipline of science and engineering. Nanophase and nanostructured materials, a new branch of materials science research, are attracting a great deal of attention because of their potential applications in areas, such as polymer nanocomposites, electronics, optics, catalysis, ceramics, magnetic data storage, and others.

The research on nanotechnology is evolving and expanding very rapidly, which makes it an impossible task for a volume of this size to cover all the aspects of the field. This book is focused primarily on polymer nanocomposites, based on the author's own research as well as others' research experiences. I have outlined a logical progression of what polymer nanocomposites are, how they are made, what their assets and deficiencies are, what they are designed for, and to what extent they are commercially exploited. Each specialized chapter is part of a sequence of developing polymer nanocomposites. I have been using my first book, *Polymer Nanocomposites: Processing, Characterization, and Application* (McGraw-Hill, 2006) since Spring 2007 to teach a semester course offered once a year at the University of Texas at Austin. As a result of my class, I developed this expanded volume and presented each chapter in a different manner than in my first book. One of the primary goals of this text is to aid the readers who wish to engage in the research and development of polymer nanocomposites in their professions.

As a redress, this text consists of fifteen chapters and is subdivided into three parts. Part I – Fundamentals, Processing, and Characterizations – consists of six chapters. Chapter 1 provides an introduction to nanotechnology. Chapter 2 consists of an overview of the most commonly used nanomaterials, their origins, manufacturing, properties, suppliers, and current and potential applications. The remaining chapters of Part I cover the basics of polymer matrices and composites, selection of resin matrix and nanomaterials, processing, and characterizing of multifunctional polymer nanocomposites.

Part II – Multifunctional Properties and Applications – consists of seven chapters that discuss the important properties as well as current applications of these polymer nanocomposites. Polymer nanocomposites can be thermoplastic-, thermoplastic elastomer-, or thermoset-based. The mechanical, thermal, flammability, ablation, electrical, optical, and other enhanced properties, such as tribological, permeability, and applications in the oilfield as well as selective applications of polymer nanocomposites, will be discussed.

Finally, Part III – Concerns and Outlook – consists of two chapters that discuss the environmental and health impacts for nanomaterials and polymer nanocomposites. The environmental health and safety aspects of nanomaterials and polymer nanocomposites, risk assessment and safety standards, and fire toxicity of polymer nanocomposites are discussed. In the final chapter, a brief overview of opportunities, trends, and challenges of polymer nanocomposites are included. Throughout the book, the theme is developed that polymer nanocomposites are a whole family of polymeric materials whose properties are capable of being tailored to meet specific applications. This volume serves as a general introduction to researchers just entering the field and to scholars from other subfields seeking information. This book is well suited as a textbook for a one-semester course tailored to upper-level undergraduate and entry-level graduate students, or for a professional short course.

During my preparation and writing, a great deal of help in terms of information and images was provided by a number of people and their organizations, whom I would like to thank, in no particular order: Dr. Richard A. Vaia of Air Force Research Laboratory/WPAFB; Max L. Lake, Patrick Lake, and Dr. Carla Leer Lake of Applied Sciences; Dr. Alexander Dole of Arkema; Joseph Ventura of Bayer MaterialScience; Thomas Pitstick of Carbon Nanotechnologies; Dr. Nate Hansen of Conductive Composites; Alan Hedgepeth and Dr. Stephan Sprenger of Evonik AG; Dr. Joseph D. Lichtenhan of Hybrid Plastics; Dr. Sanjay Mazumdar of Lucintel; Professor Brian Wardle of MIT; Professors Giovanni Camino and Alberto Fina of Polytechnic of Turin, Alessandria, Italy; Dr. Doug Hunter of Southern Clay Products; Eric Kuramoto of Thinky; and Professor Antonio Avila of Universidade Federal de Minas Gerais, Belo Horizonte, Brazil.

Many thanks are due to my friends Dr. Louis A. Pilato of Pilato Consulting/KAI, LLC and Dr. Maurizio Natali of University of Perugia/KAI, LLC for their help in collaborating with me on numerous research projects, tutorial class, and an online course that I offered at the University of Texas at Austin's Center for Lifelong Engineering Education. These activities are used as examples throughout this book. I would also like to thank Dr. Charles Y-C Lee and Dr. Jocelyn Harrison of Air Force Office of Scientific Research/NL, Dr. Shawn Phillips, Dr. Rusty Blanski, and Dr. Steven A. Svejda of Air Force Research Laboratory/Propulsion Directorate, Edwards AFB, CA for supporting our numerous research programs. I'm thankful to all my colleagues, especially to students who took my "Polymer Nanocomposites" class for the past nine years and provided excellent inputs to this volume, and to students in my research group who contribute to our research programs. Special thanks go to Kurt Schellhase (my current student) and Dr. Si Chon (Dickson) Lao (my former student) for obtaining permissions from different sources to use their data throughout this book. Many thanks go to Peter Gordon and Steven Elliot of Cambridge University Press for providing the opportunity to publish this book and particularly their staff for tolerance and assistance.

PART ONE

FUNDAMENTALS, PROCESSING, AND CHARACTERIZATION

1 Introduction to Nanotechnology

1.1 Definition of Nanotechnology

The Subcommittee on Nanoscale Science, Engineering, and Technology (NSET), Committee on Technology (CoT), National Science and Technology Council (NSTC) defines nanotechnology as [1]:

> Nanotechnology is the understanding and control of matter at dimensions between approximately 1 and 100 nanometers, where unique phenomena enable novel applications. Encompassing nanoscale science, engineering, and technology, nanotechnology involves imaging, measuring, modeling, and manipulating matter at this length scale.
>
> A nanometer is one-billionth of a meter. A sheet of paper is about 100,000 nanometers thick; a single gold atom is about a third of a nanometer in diameter. Dimensions between approximately 1 and 100 nanometers are known as the nanoscale. Unusual physical, chemical, and biological properties can emerge in materials at the nanoscale. These properties may differ in important ways from the properties of bulk materials and single atoms or molecules.

Mauro Ferrari, Professor of Molecular Medicine at the University of Texas Health and Science Center, of experimental therapeutics at the M.D. Anderson Cancer Center, and of Bioengineering at Rice University, provides the following definition [2]:

> At the nanoscale there is no difference between chemistry and physics, engineering, mathematics, biology or any subset thereof. An operational definition of nanotechnology involves three ingredients: (1) nanoscale sizes in the device or its crucial components, (2) the man-made nature, and (3) having properties that only arise because of the nanoscopic dimensions.

Professor Ferrari points out how the line between disciplines becomes blurred at such a small scale. This is why nanotechnology is characterized in such broad terms – it bridges the competencies of all the sciences.

Professor Robert Langer, one of thirteen institute professors at Massachusetts Institute of Technology, defines nanotechnology in simpler terms [2]:

> Nanotechnology is concerned with work at the atomic, molecular, and supramolecular levels in order to understand and create materials, devices and systems with fundamentally new properties and functions because of their small structure.

The common thread among these definitions is the mention of the molecular scale, or nanoscale. The importance of this molecular scale will be explored in this chapter.

The extent of U.S. government activity alone can be gauged from the fact that currently twenty federal agencies and departments participate in the National Nanotechnology Initiative (NNI website: http://www.nano.gov) in fiscal year (FY) 2016. The NNI, established in FY 2001, is a U.S. government research and development (R&D) initiative involving twenty federal departments, independent agencies, and commissions working together toward the shared and challenging vision of "a future in which the ability to understand and control matter at the nanoscale leads to a revolution in technology and industry that benefit society" [3]. The NNI is managed within the framework of the National Science and Technology Council (NSTC). The Nanoscale Science, Engineering, and Technology (NSET) Subcommittee of the NSTC Committee on Technology (CoT) coordinates planning, budgeting, program implementation, and review of the progress for the initiative. The National Nanotechnology Coordination Office (NNCO) acts as the primary point of contact for information on the NNI. The NNI website provides public outreach on behalf of the NNI and promotes access to and early application of the technologies, innovations, and expertise derived from NNI activities.

In 2014, federal agencies invested a total of \$1.57 billion in nanotechnology-related activities. The 2016 federal budget provides more than \$1.5 billion for NNI, affirming the administration's continuing commitment to a robust U.S. nanotechnology effort. Nearly half (43%) of this budget request is focused on applied R&D and support for the Nanotechnology Signature Initiatives (NSIs), reflecting an increased emphasis within the NNI on commercialization and technology transfer. The cumulative NNI investment since FY 2001, including 2016 request, now total more than \$22 billion. It has been more than one and a half decades of major investment and progress in nanoscience and nanotechnology, which was summarized by a "Special Report to the President and Congress on the Fourth Assessment of the National Nanotechnology Initiative" [4] and "The National Nanotechnology Initiative – Supplement to the President's Budget for Fiscal Year 2016" [1].

1.2 Brief History of Nanotechnology

Japanese scientist Norio Taniguchi first coined the term "nanotechnology" in 1974 in reference to small-scale semiconductor interactions, but unknowing uses of the technology and predictions of the concept came earlier [5]. One example of the early use of nanotechnology, however unknowingly, was by medieval artists making stained glass between 500 and 1450 AD. Little did they know that the metallic additives to molten glass used for coloration acted as nanoparticles, scattering light in different ways to produce different colors [5].

The idea of nanotechnology was first introduced to the public in 1959 by Richard Feynman in his esteemed lecture series entitled "There's Plenty of Room at the

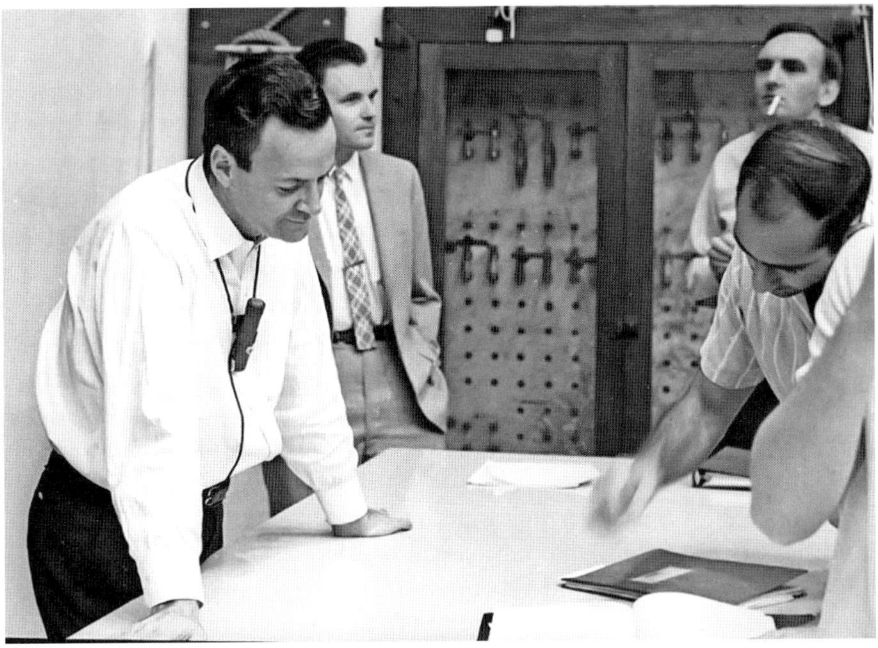

Figure 1.1. Richard Feynman at work (Courtesy of Caltech).

Bottom" [4–8]. In this talk Feynman describes a future where scientists control matter with atomic precision (Figure 1.1) [6]. He anticipates, in a seemingly prophetic fashion, the creation of microscopes that could resolve individual molecules, densely packed computer chips, and miniature robots that could aid in surgery and drug delivery [5, 6].

Part of Feynman's dream became realized in 1981 with the development of the Scanning Tunneling Microscope (STM) capable of observations with 0.1 nm resolution [7]. The STM enabled scientists to directly observe manipulations at the molecular scale and thereby served as a key fundamental tool for nanotechnology research.

Advances in microscopy have led to a number of other high-resolution visualization and characterization techniques, including transmission electron microscopy (TEM), scanning electron microscopy (SEM), atomic force microscopy (AFM), wide-angle X-ray diffraction (WAXD), and small-angle X-ray scattering (SAXS) [9]. These instruments will be described in Chapter 6.

Feynman's vision of nanoscale microchips came true in 2001 when IBM constructed the first logic gate using carbon nanotubes [7]. This sort of miniaturization in the realm of computing enabled by nanotechnology is what has given birth to smartphones, tablets, and ultrabooks we enjoy today. Other discoveries and applications anticipated by Feynman include the use of nanoscale devices in medicine. Readers should refer to the NNI website, www.nano.gov, under "Nanotechnology 101/Nanotechnology Timeline," which traces the development of nanotechnology from the first concept to the latest development.

1.3 What Is the Significance of *"Nanoscale Materials"*?

Nanoscale (nanoscopic scale) materials, which can be either stand-alone solids or subcomponents in other materials, are less than 100 nm in one or more dimensions. To put this dimension in perspective, a nanometer (nm) is one-billionth of a meter and one millionth of a millimeter, about four times the diameter of an atom. Truly grasping the scale of the nanometer is difficult for our macro-oriented brains, but real-world comparisons can help give us a sense. For example, the blink of an eye is to a year is what a nanometer is to a meter stick [6]. Figures 1.2 [7] and 1.3 show some interesting examples. Figure 1.2 shows some examples: a single-walled carbon nanotube is about 1 nm in diameter; DNA is about 2.5 nm in diameter; a nanoparticle is about 4 nm in diameter; a bacterium is about 2.5 μm in length; a single strand of hair is about 100 μm in diameter; a large raindrop is about 2.5 mm in diameter; an ant is about 4 mm long; an average house is about 10 m wide; and the Indianapolis Motor Speedway is 4 km per lap. Additional interesting materials for comparison are a human red blood cells (10,000 nm), a cell of bacteria *E. colis* (1,000 nm), a viral cell (100 nm), a polymer coil (40 nm), quantum rods (Q-rods) (30 nm in length) with an aspect ratio of 10:1, and a quantum dot (QD) (7 nm in diameter), all shown in Figure 1.3.

To illustrate the inherent value of manipulating matter on such a scale, Daniel Ratner, Professor of Bioengineering at the University of Washington, proposes a helpful thought experiment [7]. Suppose we have a 3 x 3 x 3 foot cube of pure gold. If we were to bisect this cube in each dimension, we would have eight smaller cubes. These new cubes would exhibit the same intrinsic properties as the original – each would still be heavy, shiny, and yellow, with the same chemical and structural properties. If we were to continue bisecting until we have cubes sized on the order of microns (10^{-6} of a meter), the inherent bulk properties of the material would still remain constant. And this is not specific to gold; the same holds true for steel, plastic, ice, or any pure solid. However, if we were to reach the nanoscale, quantum effects would begin to dominate, and the gold's properties, including its color, melting point, and intermolecular chemistry, would change. These quantum effects had been "averaged out of existence" in the bulk material [8]. At the nanoscale, the force of gravity gives way to van der Waal's forces, surface tension, and other quantum forces [10].

1.4 Why Is This *"Nanoscale"* So Special and Unique?

Nanomaterials are also known as *nanoscale structures, nanoscale materials, nanophase, or nanoparticles* in the literature. To distinguish nanomaterials from bulk, it is crucial to demonstrate the unique properties of nanomaterials and their prospective impacts in science and technology. Nanomaterials can be classified as one-dimensional nanoscale structures (platelet-type), two-dimensional nanoscale structures (fiber-type), and three-dimensional nanoscale structures (particulate-type). These three most important nanomaterials are illustrated schematically in Figure 1.4 and are described in detail in Chapter 2. Nanoscale structures have very

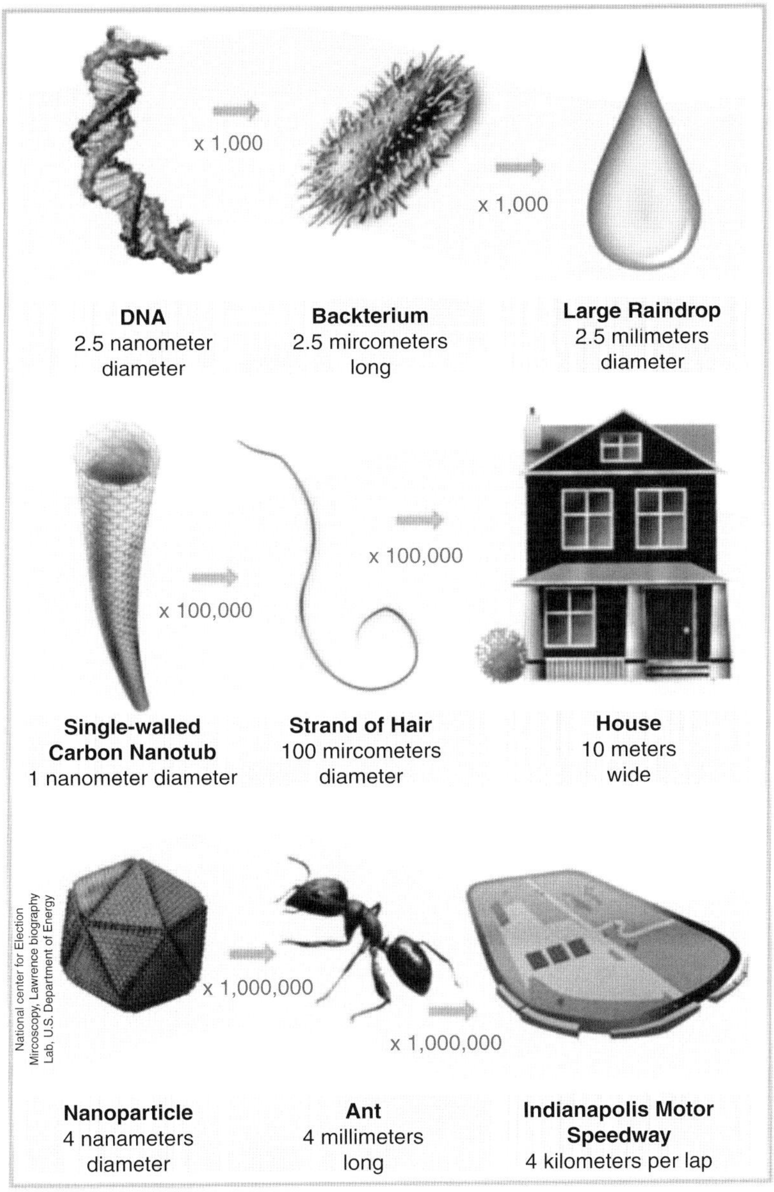

Figure 1.2. Visual examples of the size and scale of nanotechnology [5]. *Source:* www.nano.gov.

high surface-to-volume and aspect ratios, making them ideal for use in polymer nanocomposites.

Wave-like properties of electrons inside matter and atomic interactions are influenced by materials' variations in nanoscale length. By creating nanoscale structures it is possible to control the fundamental properties of materials, such as their melting temperature, magnetic properties, charge capacity, and even their color, without changing the nanoscale structures' chemical composition. This will lead to

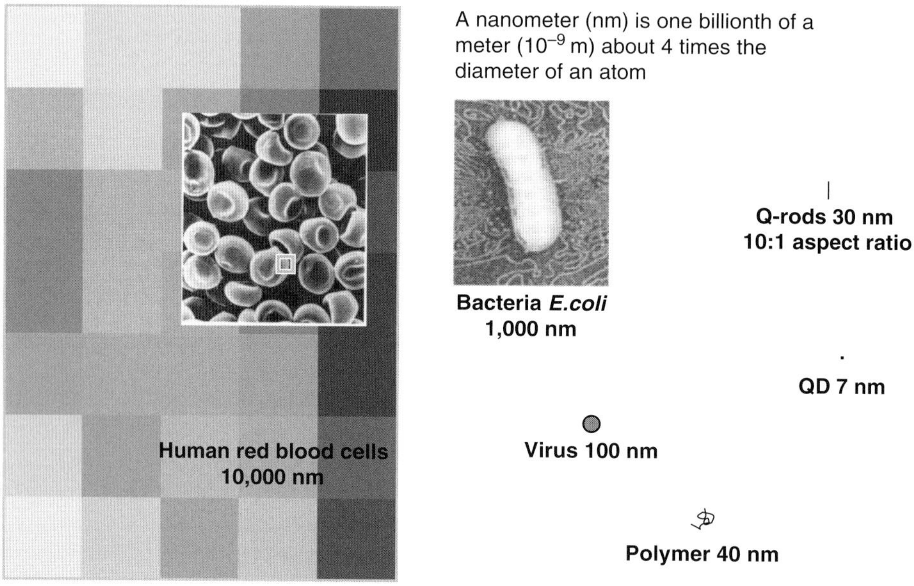

Figure 1.3. What "nano" really means? (Courtesy of R. Vaia).

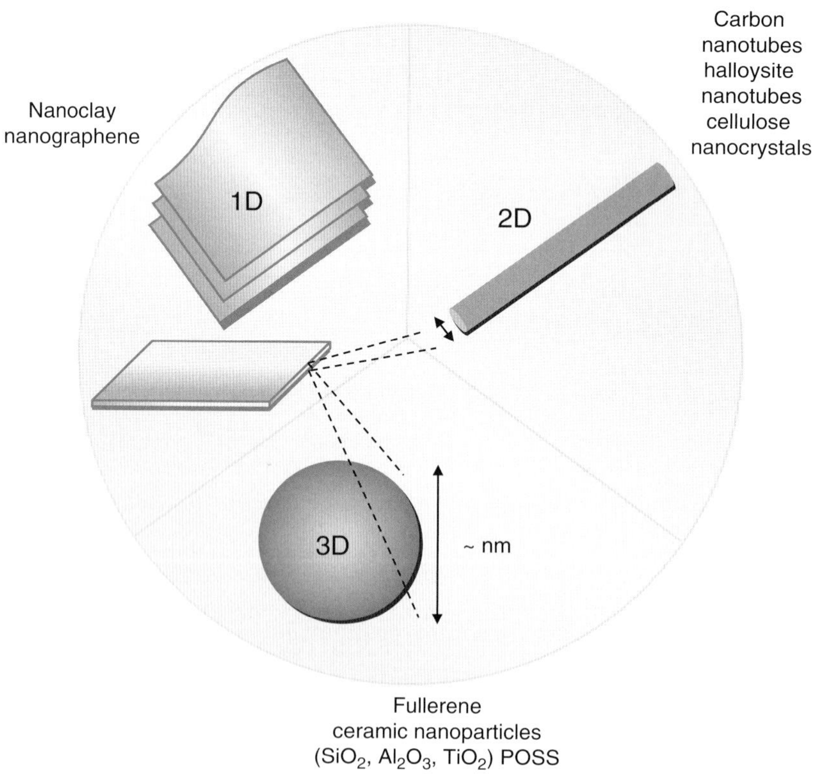

Figure 1.4. Schematic showing 1D, 2D, and 3D nanoscaled dimensions.

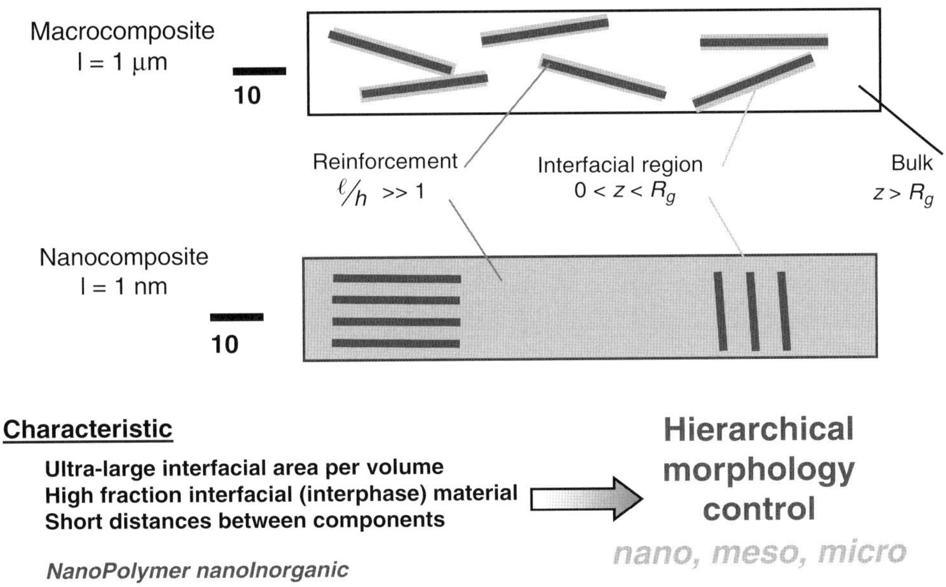

Figure 1.5. Nanoscale materials uniqueness (Courtesy of R. Vaia).

new, high-performance polymers and nanotechnologies that were impossible previously. Nanoscale structures, such as nanoparticles, have very high surface-to-volume and aspect ratios, making them ideal for use in polymer nanocomposites.

For the past five decades, we have been working with macrocomposites, such as filled polymers or fiber-reinforced polymer matrix composites, where the length scale of the polymer fillers or the fiber diameters is in micrometers, as shown in Figure 1.5. The reinforcement length scale is in micrometers, and the interface of fillers is close to the bulk polymer matrix. For the past two decades, we have been discovering nanocomposites, where the length scale of the reinforcement (nanoparticles) is on the nanometer scale (Figure 1.5). These nanocomposites have *ultra-large interfacial area per volume*, and the distances between the polymer and filler components are extremely short. Polymer coils and chains are 40 nm in diameter, and the nanomaterials are on the same order of magnitude as the polymer. As a result, molecular interaction between the polymer coils/chains and the nanomaterials will give polymer nanocomposites very unusual material properties that conventional polymers do not possess.

Before discussing the properties of nanomaterials, it may be advantageous to describe an example demonstrating the elementary consequences of the small size of nanoparticles [9]. The first and most important consequence of a small particle size is its huge surface area, and in order to obtain an impression of the importance of this geometric variable, the surface-over-volume ratio should be addressed. Assuming a particle is spherical, the surface a of one particle with diameter D is $a = \pi D^2$, and the corresponding volume v is $v = \pi D^3/6$. Therefore, the surface/volume ratio is

$$R = a/v = 6/D \tag{1.1}$$

This ratio is inversely proportional to the particle size, and as a consequence, the surface increases with decreasing particle size. The same is valid for the surface per mol A, a quantity that is extremely important in thermodynamic considerations.

$$A = na = M/(\rho\pi D^3/6)\,\pi D^2 = 6M/\rho D \tag{1.2}$$

In Equation (1.2), n is the number of particles per mol, M is the molecular weight, and ρ is the density of the material. Similar to the surface-over-volume ratio, the area per mol increases inversely in proportion to the particle diameter. Hence, huge values of surface area are achieved for particles that are only a few nanometers in diameter.

The unique properties and improved performance of nanomaterials are determined by their sizes, surface structures, and interparticle interactions. The role played by particle size is comparable to the role of the particle's chemical composition, adding another parameter for designing and controlling particle behavior. To fully understand the impacts of nanomaterials in nanoscale science and technology, one needs to learn why nanomaterials are so special!

The excitement surrounding nanoscale science and technology provides unique opportunities to develop revolutionary materials. Nanoscale science and technology is a relatively young field that encompasses nearly every discipline of science and engineering. Nanoscale structures are a new branch of materials research attracting a great deal of attention because of its potential applications in chemical catalysis, computing, imaging, material synthesis, medicine, printing, and many other fields.

1.5 How Polymer Nanocomposite Works?

The advent of academic, government, and industrial nanotechnology research has spurred significant growth opportunities for nanocomposites for the consumer, defense, aerospace, and health industries [12–15]. This is evident from the number of journal papers reporting on nanotechnology scientific research, and the number of patents issued by the U.S. Patent and Trademark Office (PTO) grew significantly in the past thirty years. It has been reported that from 2005 to 2006, the U.S. PTO issued 4,081 nano patents involving 9,491 inventors representing 34 countries [13]. This is a good indicator of how widespread the nanotechnology research activities have been across several disciplines. This book mainly focuses on a branch of nanotechnology research involving only the polymers, specifically polymer nanocomposites (PNCs).

The major objective of PNC research is to explore the different methods to achieve significant enhancement of properties over those of the traditional matrix polymers using only a small fraction of nanomaterials. The nanomaterials are defined as having at least one dimension measuring less than 50 nm [15]. These nanomaterials are expected to create new physical and chemical properties beyond what traditional filled polymeric material possess. It is well known from decades-long PNC research that some nanomaterials improve only one type of property, such as mechanical, thermal, flammability, or electrical, while a selected few nanomaterials can contribute to improving multifunctional properties.

The uniqueness of these novel PNCs will enable the circumvention of classic polymer performance by accessing new properties and exploiting synergies between PNCs that only occur when the nanoscale morphology and the fundamental physics associated with a property coincide. They represent a radical alternative to traditional filled polymers. In contract to traditional polymer systems where reinforcement is on the order of microns, PNCs are exemplified by discrete constituents on the order of a few nanometers. Polymer nanocomposites consist of improved mechanical, thermal, ablation, flammability, electrical, optical, permeability, charge dissipation, chemical resistance, magnetic, catalytic, and other properties (as discussed in detail in later chapters of this book). Polymer nanocomposite can differ completely from those of its component materials, thereby opening up unprecedented technologies. A common example of currently available PNCs are carbon nanotubes, which can help enhance electrical and thermal conductivities, increase stiffness, and enhance crack deflection and toughness. Other common nanomaterials used in polymer nanocomposites are TiO_2 and ZnO for optical properties, nitrides and carbides for hardness and wear resistance, nanoclay for flammability, and nanometals for color.

The challenges in the field of PNC research are threefold. First, the identification of the proper nanomaterial-polymer pairs is essential and critical. Otherwise, the desired properties cannot be realized. Second, the nanomaterials must be nanodispersed in the matrix polymers. Otherwise, anticipated properties cannot be achieved. Third, not all true polymer nanocomposites are cost-effective. The cost constraint largely governs the scope of current PNCs research and commercialization in industrial laboratories. Some novel, innovative academic PNC research will never be commercialized due to the prohibitive cost of nanomaterials, time-consuming processing and fabrication steps, and sometimes the lack of proper market. Of the three challenges presented in this paragraph, the first two challenges relate to academic research and the third challenge relates to industrial research concerning PNC preparation and commercialization.

In physical form, these nanomaterials (nanoparticles) are usually available as agglomerates. In the case of a platelet-type nanocomposite (see Chapter 2 for more detailed descriptions), a large number of individual platelets are held together as microscopic agglomerates by attractive forces originating in electrostatic, ionic, and van der Waals forces. The state of agglomeration of nanomaterials is usually dictated by the manufacturing methods, and the particle agglomerates serve as the starting material in the preparation of PNCs. The agglomerates, when combined with the matrix polymer in a solution or a melted state, are subjected to several dispersion forces, including shear, ultrasonic, and centrifugal, to disperse individual particles to the nanoscale in the matrix polymer (see Chapter 5 for detailed discussions). The transformation from initial microscopic particle agglomerates to individual nanoparticles by dispersion forces is among the challenges engineers and scientists have been confronting for the past two decades.

Uniform and individual dispersion of these nanoparticle agglomerates produces *ultra-large interfacial area per volume* between the nanoparticle and the matrix polymer. Polymer nanocomposites fundamentally differentiate from traditional filled polymers due to the immense internal interfacial area and the nanoscopic nature of the nanomaterials. As a result, the overall performance of PNCs cannot be understood by simple scaling rules that apply to traditional polymeric composites. New combinations of properties derived from the nanoscale structures of PNCs provide opportunities to circumvent traditional performance associated with conventional reinforced polymers, thus epitomizing the promise of PNCs.

Numerous examples can be found in the literature demonstrating substantial improvements in multifunctional properties of PNCs. Its value comes from providing value-added properties not present in the neat resin, without sacrificing the inherent processability and mechanical properties of the neat resin. The preparation of a blend or composite with multifunctionality requires a trade-off between desired performance, mechanical properties, cost, and processability. The term "nanocomposite," when used to describe properties comparisons, is intended to relate to traditional unfilled and filled polymers and not fiber-reinforced polymer matrix composites (PMCs). Polymer nanocomposites may provide matrix resins with "multifunctionality," but they should not be considered in the near and intermediate term as a potential replacement for current state-of-the-art carbon fiber–reinforced PMCs.

1.6 Strengths and Weaknesses of Nanoparticles

This is a relatively young field that has huge growth potential. In the last twenty to thirty years, extensive research has been conducted in understanding how nanoparticles affect the chemical nature of compounds and their overall effect on a material's properties. The most dramatic effect noticed is that when working on this level, the surface area of the nanoparticle is tremendous for its size. In a study conducted by the University of Rostock and the Fritz Haber Institute of the MPG in Germany [16], the surface area and diameter of gold catalyst nanoparticles were measured. Six samples of nanoparticles were synthesized, varying in shape: icosahedral, cuboctahedral, decahedral, truncated octahedron, and spherical. Given this small size and these various shapes, these nanoparticles have very large surface areas. The surface area is inversely proportional to the diameter of the nanoparticle. In this study, the smallest particles averaging 1.6 nm in diameter had a surface area of 173 m^2/g. If a hefty spoonful – about 2.5 grams – of this nanoparticle were flattened out, it could cover an entire basketball court. With a large surface area per volume, a nanoparticle is able to make numerous bonds with the surrounding molecules, far more than a particle on the micron scale. The ability to form so many bonds leads the nanoparticles to create an extremely solid molecular structure that, in turn, increases the mechanical properties of materials. Due to having a large surface area-to-volume ratio, fewer nanoparticles can be incorporated in materials that are able to outperform the same materials made by conventional mean using the micron scale.

Although nanoparticles have many strengths, there are some drawbacks. Although the cost of creating nanoparticles has declined within the last few years, it is still high. Mass-producing nanoparticles is a difficult task. Taking the production of nanoparticles in a laboratory to a high level of production for industrial use is a very challenging process. With this challenge come higher prices. Furthermore, even though nanoparticles can enhance materials, there is a limit. Because of the bonds the nanoparticles can create, when they are added to a liquid, the viscosity of the liquid increases. If too high a weight of nanoparticles is added to a mixture, the viscosity could increase to a level in which the material is unable to be utilized. This is especially important when working with a thermoset resin, which needs fluidity to be able to penetrate the fibers that it is designed to reinforce. Another problem area for nanoparticles has to do with the fact that sometimes a nanoparticle with very promising mechanical properties is created, but those properties do not transfer to a material to which this nanoparticle is introduced.

1.7 Safety of Nanoparticles

Due to the size of nanoparticles, it is important to study the effect they have on a human body. Nanoparticles combine the properties of solids, with the ability to move – a property of molecules [17]. Until recently, most items synthetically created by humans did not pose the threat that nanoparticles do. These materials were too large to penetrate the human body in the manner that a nanoparticle can. Nanoparticles are small enough to diffuse into biological system through lipid-based barrier that is effective against larger particles. With these small particles, reactive surfaces can enter a biological system and interfere with existing chemical transformation pathways. They can disrupt the body chemistry at the cellular level. The people most at risk are those that handle and process nanoparticles.

There are currently safety measures in place at different laboratories and companies for the treatment of nanoparticles, but there is no national standard. With an increase in use of nanoparticles, there will be an increase in the number of people that come in contact with the raw particles. There are regulations in place for the handling and transportation of these materials, but these regulations govern bulk materials, not nanoparticles. The NNI is the governing body in charge of monitoring nanotechnologies. It created the Environmental, Health, and Safety Research Strategy in 2011 to establish awareness and safety protocols.

Nanotechnology EHS (nanoEHS) activities have become a hallmark of the NNI, with R&D, policy, and regulation in this area extensively coordinated among federal agencies. NanoEHS activities, including relevant topics incorporated into the NSIs, account for more than 10% of the NNI budget request in 2016. In 2014, the NNI released the *Progress Review on the Coordinated Implementation of the National Nanotechnology Initiative 2011 Environmental, Health, and Safety Research Strategy* [18], which summarizes the extensive, ongoing collaborations and recent accomplishments in this area. A key component of the strategy is extensive communication and coordination with academic, industrial, and international nanoEHS communities. Examples of these efforts include the U.S. participation in the Innovative Nanoscience and Nanotechnology program [19]; the U.S.-EU Communities of Research [20]; collaboration with dedicated multidisciplinary academic centers, such as the NSF's Centers for Environmental Implications of Nanotechnology at the University of California, Los Angeles, and Duke University; and strong U.S. participation in the development of nanoEHS-related standards. With more research into the effects of nanoparticles on the health and the environment, the industry will most likely become more regulated in the future.

1.8 Overview of the Book

This book is aimed at the following readers:

- Advanced students and instructors in the fields of science and engineering.
- Professional scientists and engineers who may be trained in more traditional disciples but need to learn about this emerging technology.
- Policymakers and management looking for an understanding of the scientific challenges, prospective uses, and emerging markets for nanomaterials and polymer nanocomposites.

The overall goal is to capture the multidisciplinary and multifunctional flavor of nanomaterials and polymer nanocomposites while providing in-depth discussion of selected areas.

The book has been divided into three parts with fifteen chapters:

Part I, Fundamentals, Processing, and Characterization, consists of an introduction to nanotechnology (Chapter 1); an overview of different types of nanomaterials (Chapter 2); basics of polymer matrices and composites (Chapter 3); selecting the proper polymer matrix and nanomaterials for specific applications (Chapter 4); processing of the multifunctional polymer nanocomposites (Chapter 5); and structure and properties of these polymer nanocomposites (Chapter 6).

Part II, Multifunctional Properties and Applications, consists of mechanical properties of polymer nanocomposites (Chapter 7); thermal properties of polymer nanocomposites (Chapter 8); flammability properties of polymer nanocomposites (Chapter 9); ablation properties of polymer nanocomposites (Chapter 10); electrical properties of polymer nanocomposites (Chapter 11); optical properties of polymer nanocomposites (Chapter 12); and other enhanced properties of polymer nanocomposites (Chapter 13).

Part III, Concerns and Outlook, consists of nanotechnology environmental and health impacts (Chapter 14) and opportunities, trends, and challenges (Chapter 15).

The most commonly used nanomaterials in the market are discussed, including (a) *one nano-dimensional scaled materials*, such as montmorillonite (MMT) clays, nanographene platelets (NGPs), and layered double hydroxides (LDHs); (b) *two nano-dimensional scaled materials*, such as carbon nanofibers (CNFs), carbon nanotubes (CNTs): multi-walled carbon nanotubes (MWNTs), double-walled carbon nanotubes (DWNTs), and single-wall carbon nanotubes (SWNTs), halloysite nanotubes (HNTs), and nickel nanostrands (NiNS); and (c) *three nano-dimensional scaled materials*, such as polyhedral oligomeric silsesquioxane (POSS®), nano-silica (*n*-silica), nano-alumina (*n*-alumina), nano-titanium dioxide (n-TiO_2), nano-magnesium hydroxide (n-$Mg(OH)_2$), nano-silicon carbide (*n*-SiC), nano-silver (*n*-Ag), nano-zinc oxide (*n*-ZnO), and others, used to disperse into the families of polymers, such as thermoplastics, thermosets, and thermoplastic elastomers.

This volume is focused on research activities in areas of: (a) developing *processes* to disperse nanomaterials uniformly in the different types of polymers; (b) using wide-angle X-ray diffraction (WAXD), transmission electron microscopy (TEM), scanning transmission electron microscopy (STEM), scanning electron microscopy (SEM), and optical microscopy techniques to characterize nanomaterials and polymer nanocomposite *structures*; (c) studying the *processing-structure-property* relationships of these types of novel polymer nanocomposite materials; and (d) characterizing properties and evaluating the *performance* of these polymer nanocomposite materials using established laboratory devices and techniques based on ASTM, UL, and other established industry standards. The ASTM and UL test protocols were adapted to characterize the physical, mechanical, thermal, flammability, electrical, optical, tribological, permeability, and other properties of the thermoplastic-, thermoset-, and elastomer-based nanocomposites. Selected laboratory devices, such as a microscale combustion calorimeter (MCC), a cone calorimeter (CC), a simulated solid rocket motor (SSRM), oxyacetylene test bed (OTB), a subscale solid rocket motor, in situ ablation recession and thermal sensors, and char strength sensors were used to evaluate the materials performance in extreme environments.

In these studies, for solid thermosetting reactive prepolymers or thermoplastics with solid nanomaterials, four different processing methods were used to disperse nanomaterials in the polymers. These methods were: (a) solution intercalation, (b) melt intercalation, (c) three-roll milling, and (d) centrifugal mixing. For liquid thermosetting reactive prepolymers or thermoplastics with solid nanomaterials, four different processing were used, including (a) in situ polymerization, (b) emulsion polymerization, (c) high-shear mixing, and (d) ultrasonciation. The WAXD, TEM, SEM, and STEM analyses were used to characterize the degree of dispersion of these nanomaterials in the host polymer. These imaging techniques allowed us to screen formulations and distinguish compositions that exhibited either favorable or unfavorable nanodispersed nanoparticle/polymer blends. Furthermore, these analytical techniques facilitated and provided guidelines in the scale-up of favorable compositions.

1.9 Summary

In today's world, nanotechnology is all around us but remains invisible to our eyes because of its operating scale. From its origins in the vivid imagination of Richard Feynman, nanotechnology has evolved into a multidisciplinary field that is affecting positive change in many aspects of our lives. Thanks to smart nanoshells, our bodies are safer from cancer and other disease. Nanoscale semiconductors have enabled the dramatic miniaturization of computers, allowing them to now fit in the palm of our hand. And the energy we use to power these devices is becoming more environmentally friendly with advances in solar and wind energy capture thanks to nanotechnology. Today's airplanes are lighter and use less fuel, and spacecraft are becoming capable of increasingly more distant voyages.

The rapid growth of nanotechnlogy is due to its great potential in improving a vast number of properties in materials with just a small concentration of nanoparticles. Polymer is an important material for engineers, used in many applications that can benefit from its flexibility, light weight, and cost efficiency. Adding nanoparticles to the polymer matrix can greatly improve the bulk mechanical, thermal, flammability, ablation, and optical material properties as reported in the literature. Still, although the benefits of nanotechnology have been praised, it should be noted that there are still safety and health precautions that must be taken when handling these nanoparticles. Studies have reported potential harms to humans from certain nanoparticles, such as carbon if inhaled. As a result, industry, government, and academic organizations should emphasize nanosafety by providing guidelines to protect the health of people and the environment.

Nanotechnology plays important roles in the coming transformation of science, technology, and society marked by the convergence across technical fields and domains of human activity [21]. It is one of the fastest-growing technologies. However, since it is a relatively new technology, nanotechnology science and engineering raise many important questions, especially at the intersection of technology and society [22]. Discussed throughout this book is the uniqueness of polymer nanostructured material compared to other materials, such as micro-sized composites. This book also reviews studies that show different processes, ratios, and kinds of combinations of nanoparticles that perform at the highest levels of quality. With the discovery of this new technology, it is to our advantage to further our studies of it. Advancements in nanotechnology will create unique challenges and new opportunities in the future [23].

The potential impact of nanotechnology and nanomaterials on future products and services is tremendous. This emerging field has led to an enormous expansion in research and development related to nanomaterials, devices, and systems. It also has raised new questions of how these products will influence the environment, human health, business, education, and other areas. As with any emerging technology, a huge amount of disconnected information is emerging, often spread across multiple disciplines. The influence of these nanomaterials is truly multidisciplinary, and progress will depend on cross-communication between fundamental science and diverse applications. This work is an attempt to provide readers with a broader perspective by offering information on diverse nanomaterials and examples of how these materials can affect polymer properties. It also aims to help envision future products and possible applications.

An extensive list of references is included in this chapter for further reading; the majority of these publications are referenced throughout this book.

1.10 Questions and Problems

1. How difficult is it to obtain uniform dispersity of nanomaterial in a polymer?
2. What is the difference between polymer nanocomposites and a general polymer composite?
3. Instead of using polymer in a polymer nanocomposite, can a metal or ceramic be used to create nanocomposites?
4. What specific interactions are more apparent at the nanoscale that create the special properties?
5. What is the exfoliation process?
6. When nanoscale materials scale to nanostructured materials, do they retain the same properties or do they change because they are no longer at individual nano size?
7. Can a nanopolymer composite be classified as a composite?
8. For what applications can polymer nanocomposites be used?
9. What are typical ratios of nanoparticles to polymers in polymer nanocomposites?

REFERENCES FOR FUTURE READING

1. *"The National Nanotechnology Initiative – Supplement to the President's Budget for Fiscal Year 2016,"* National Science and Technology Council (NSTC), Committee on Technology (CoT), Subcommittee on Nanoscale Science, Engineering, and Technology (NSET), Washington, DC, March 2015, p. 7.
2. Nanotechnology (January 1, 2006). Retrieved December 15, 2014, from www.nature.com/nnano/journal/v1/n1/full/nnano.2006.77.html
3. *"National Nanotechnology Initiative Strategic Plan,"* National Science and Technology Council (NSTC), Committee on Technology (CoT), Subcommittee on Nanoscale Science, Engineering, and Technology (NSET), Washington, DC, February 2014, www.nano.gov/2014StrategicPlan, p. 5.
4. Executive Office of the President, President' Council of Advisors of Science and Technology (PCAST), Report to the President and Congress on the Fourth Assessment of the National Nanotechnology Initiative, April 2012.
5. A. Athavale (2011). Nanotechnology: Origin and history. *TechTalk@KPITCummins*, 4(3), 5–8. Retrieved December 15, 2014, from www.kpit.com/downloads/tech-talk/tech-talk-july-september-2011.pdf#page=3
6. R. P. Feynman, "There's Plenty of Room at the Bottom," Annual Meeting at American Physical Society at the California Institute of Technology, December 29, 1959. First published by *Caltech Engineering and Science* 23(5), February 1960, 22–36.
7. Size of the Nanoscale (n.d.). Retrieved December 15, 2014, from www.nano.gov/nanotech-101/what/nano-size
8. M. Ratner and D. Ratner (2003). *Nanotechnology: A gentle introduction to the next big idea.* Upper Saddle River, NJ: Prentice Hall, pp. 1–18.
9. J. H. Koo (2006). *Polymer Nanocomposites – Processing, Characterization, and Applications.* New York: McGraw-Hill.
10. What Is Nanotechnology? (n.d.). Retrieved December 15, 2014, from www.nano.gov/nanotech-101/what/definition
11. D. Vollath (2008). *Nanomaterials: An Introduction to Synthesis, Properties, and Applications.* Weinheim, Germany: Wiley-VCH Verlag GmbH, pp. 1–20.

12. X. Li, H. Chen, Y. Dang, Y. Lin, C. A. Larson, and M. C. Roco (2008). A longitudinal analysis of nanotechnology literature: 1976–2004. *Journal of Nanoparticle Research* 10, 3–22.
13. X. Li, D. Hu, Y. Dang, H. Chen, M. C. Roco, C. A. Larson, and J. Chan (2009). Nano mapper: An Internet knowledge mapping system for nanotechnology development. *Journal of Nanoparticle Research* 11, 529–552.
14. G. Giordano and H. Inmam (2010). Thinking small plays big. *Plastic Engineering* 66(1), 6–10.
15. J. Jancar, J. F. Douglas, F. W. Starr, S. K. Kumar, P. Cassagnau, A. J. Lesser, and S. S Sterstein (2010). Current issues in research on structure-property relationships in polymer nanocomposites. *Polymer* 51, 3321–3343.
16. M. C. Rocco (1998). Nanoparticle and nanotechnology research in the USA. *Journal of Aerosole Science* 29, 749–760.
17. A. Janz, A. Kockritz, L. Yao, and A. M. Adsorption (2010). Surface area: Fundamental calculations on the surface area determination of supported gold nanoparticles. *Langmuir* 26(9), 6783–6789.
18. "Progress Review on the Coordination Implementation of the National Nanotechnology Initiative 2011 Environmental, Health, and Safety Research Strategy," National Science and Technology Council (NSTC), Committee on Technology (CoT), Subcommittee on Nanoscale Science, Engineering, and Technology (NSET), Washington, DC, June 2014, www.nano.gov/2014 EHSProgressReview
19. The SIINN (Safe Implementation of Innovative Nanoscience and Nanotechnology) ERA-NET is a coordination activity funded by the European Commission within the 7th Framework Programme, www.siinn.eu. It promotes the safe and rapid transfer of European research results in nanoscience and nanotechnology (N&N) into industry applications.
20. The U.S.-EU organization that bridges nanoEHS research efforts; www.us-eu.org
21. W. J. Stark (2011). Nanoparticles in Biological Systems. *Angewante Chemie International Edition* 50, 1242–1250.
22. M. Roco and W. Bainbridge (2013). The new world of discovery, invention, and innovation: Convergence of knowledge, technology, and society. *Springer Science+Business Media Dordrescht* 15, 1946.
23. C. Tahan, R. Leung, G. Zenner, K. Ellison, W. Crone, and C. Miller (2006). Nanotechnology and society: A discussion-based undergraduate course. *American Journal of Physics* 74, 443.
24. A. Agarwal, S. R. Bakshi, and D. Lahiri (2011). *Carbon Nanotubes – Reinforced Metal Matrix Composites*. Boca Raton, FL: CRC Press.
25. P. M. Ajayan, L. S. Schadler, and P. V. Braun (2003). *Nanocomposite Science and Technology*. Weinheim, Germany: Wiley-VCH Verlag, GmbH.
26. W. Bauhofer and J. Z. Kovacs (2009). A review and analysis of electrical percolation in carbon nanotube polymer composites. *Composites Science and Technology* 69(10), 1486–1498.
27. S. Al-Malaika, A. Golovoy, and C. A. Wilkie (Eds.) (1999). *Chemistry and Technology of Polymer Additives*. Malden, MA: Blackwell Science.
28. G. W. Beall and C. E. Powell (2011). *Fundamentals of Polymer-Clay Nanocomposites*. Cambridge: Cambridge University Press.
29. S. N. Bhattacharya, R. K. Gupta, and M. R. Kamal (2008). *Polymeric Nanocomposites – Theory and Practice*. Cincinnati, OH: Hansen.
30. W. D. Callister, Jr. (2007). *Materials Science and Engineering – An Introduction*, 7th Edition. New York: Wiley & Sons.
31. G. Cao (2004). *Nanostructures and Nanomaterials – Synthesis, Properties & Applications*. London: Imperial College Press.

32. T. W. Chou, L. Gao, E. T., Thostenson, Z. Zhang, and B.-Y. Byun (2011). An assessment of the science and technology of carbon nanotube-based fibers and composites: A review. *Composites Science and Technology* 70(1), 1–19.
33. L. Dai (Ed.) (2006). *Carbon Nanotechnology – Recent Developments in Chemistry, Physics, Materials Science and Device Applications*. Amsterdam: Elsevier.
34. G. Duffa (2013). *Ablative Thermal Protection System Modeling*. Reston, VA: AIAA.
35. M. Endo, M. S. Strano, and P. M. Ajayan (2008). Potential applications of carbon nanotubes. *Carbon Nanotubes: Topics in Applied Physics* 111, 13–61.
36. H.-J. Fecht and M. Werner (Eds.) (2004). *The Nano-Micro Interface – Bridging the Micro and Nano Worlds*. Weinheim, Germany: Wiley-VCH Verlag, GmbH.
37. K. Friedrich, S. Fakirov, and Z. Zhang (Eds.) (2005). *Polymer Composites – From Nano- to Macro-Scale*. New York: Springer.
38. M. Galimberti (Ed.) (2011). *Rubber-Clay Nanocomposites*. Hoboken, NJ: Wiley.
39. K. E. Geckeler and E. Rosenberg (2006). *Functional Nanomaterials*. Stevenson Ranch, CA: American Scientific Publishers.
40. K. E. Geckeler and H. Nishide (Eds.) (2010). *Advanced Nanomaterials* Vols. 1 & 2. Weinheim, Germany: Wiley-VCH Verlag GmbH.
41. J.-F. Gerard (Ed.) (2001). *Fillers and Filled Polymers*. Weinheim, Germany: Wiley-VCH Verlag GmbH.
42. R. F. Gibson (2010). A review of recent research on mechanics of multifunctional composite materials and structures. *Composite Structures* 92(12), 2793–2810.
43. Y. Gogotsi (Ed.) (2006). *Nanotubes and Nanofibers*. Boca Raton, FL: CRC Press.
44. M. J. Green, N., Behabtu, M. Pasquali, and W. Adams (2009). Nanotubes as polymer. *Polymer* 50, 4979–4997.
45. Z. Guo and L. Tan (2009). *Fundamentals and Applications of Nanomaterials*. Norwood, MA: Artech House.
46. R. K. Gupta, E. Kennel, and K.-J. Kim (Eds.) (2010). *Polymer Nanocomposites Handbook*. Boca Raton, FL: CRC Press.
47. C. A. Harper (Ed.) (2002). *Handbook of Plastics, Elastomers, & Composites*, 4th edition. New York: McGraw-Hill.
48. G. L. Hornyak, H. F. Tibbals, J. Dutta, and J. J. Moore (2009). *Introduction to Nanoscience & Nanotechnology*. Boca Raton, FL: CRC Press.
49. T. R. Hull and B. K. Kandola (Eds.) (2009). *Fire Retardancy of Polymers: New Strategies and Mechanisms*. Cambridge: RSC Publishing.
50. S. Iijima (1991). Helical microtubules of graphitic carbon. *Nature* 354, 56–58.
51. B. Karn, T. Mascinangioli, W. Zhang, V. Colvin, and P. Alivisatos (Eds.) (2005). *Nanotechnology and the Environment – Applications and Implications*. ACS Symposium Series 890. Washington, DC: American Chemical Society.
52. D. W. van Krevelen and K. te Nigenhuis (2012). *Properties of Polymers*, 4th edition. Amsterdam: Elsevier.
53. R. Krishnamoorti and R. A. Vaia (Eds.) (2005). *Polymer Nanocomposites – Synthesis, Characterization, and Modeling*. Washington, DC: ACS Symposium Series 804.
54. R. N. Kostoff, J. A. Johnson, J. S. Murday, C. G. Y. Lau, and W. M. Tolles (2006). The structure and infrastructure of the global nanotechnology literature. *Journal of Nanoparticle Research* 8(3), 301–321.
55. R. N. Kostoff, J. S. Murday, C. G. Y. Lau, and W. M. Tolles (2006). The seminal literature of nanotechnology research. *Journal of Nanoparticle Research* 8(2), 193–213.
56. M. Le Bras, G. Camino, S. Bourbigot, and R. Delobel (Eds.) (1998). *Fire Retardancy of Polymers – The Use of Intumescence*. Cambridge: RSC Publishing.

57. M. Le Bras, C. A. Wilkie, S. Bourbigot, S. Duquesne, and C. Jama (Eds.) (2005). *Fire Retardancy of Polymers – New Applications of Mineral Fillers*. Cambridge, United Kingdom: RSC Publishing.
58. L. Liu, W. Ma, and Z. Zhang (2011). Macroscopic carbon nanotubes assemblies: Preparation, properties, and potential applications. *Small* 7(11), 1504–1520.
59. J. Marijnissen and L. Gradon (Eds.) (2010). *Nanoparticles in Medicine and Environment – Inhalation and Health Effects*. New York: Springer.
60. T. McNally and P. Potschke (2011). *Polymer-Carbon Nanotube Composites: Preparation, Properties, and Applications*. Cambridge: Woodhead Publishing.
61. V. Mittal (Ed.) (2010). *Optimization of Polymer Nanocomposites Properties*. Weinheim, Germany: Wiley-VCH Verlag GmbH.
62. V. Mittal (Ed.) (2010). *Polymer Nanotube Nanocomposites – Synthesis, Properties, and Application*. Salem, MA: Wiley & Scrivener Publishing.
63. V. Mittal (Ed.) (2011). *Thermally Stable and Flame Retardant Polymer Nanocomposites*. Cambridge: Cambridge University Press.
64. V. Mittal (Ed.) (2012). *Characterization Techniques for Polymer Nanocomposites*. Weinheim, Germany: Wiley-VCH Verlag GmbH.
65. W. Miziolek, S. P. Karna, J. M. Mauro, and R. A. Vaia (Eds.) (2005). *Defense Applications of Nanomaterials*, ACS Symposium Series 891. Washington, DC: American Chemical Society.
66. B. Morgan and C. A. Wilkie (Eds.) (2007). *Flame Retardant Polymer Nanocomposites*. Hoboken, NJ: Wiley.
67. S. M. Mukhopadhyay (Ed.) (2012). *Nanoscale Multifunctional Materials*. Hoboken, NJ: Wiley.
68. P. Mukhopadhyay and R. K. Gupta (Eds.) (2013). *Graphite, Graphene and Their Polymer Nanocomposites*. Boca Raton, FL: CRC Press.
69. J. K. Nelson (Ed.) (2010). *Dielectric Polymer Nanocomposites*. New York: Springer.
70. J. Paulo Davim (Ed.) (2013). *Tribology of Nanocomposites*. New York: Spring.
71. T. J. Pinnavaia and G. W. Beall (Eds.) (2000). *Polymer-Clay Nanocomposites*. New York: Wiley & Sons.
72. S. H. Priest (Ed.) (2012). *Nanotechnology and the Public – Risk Perception and Risk Communication*. Boca Raton, FL: CRC Press.
73. R. Pusch and R. N. Yong (2006). *Microstructure of Smectite Clays and Engineering Performance*. New York: Taylor & Francis.
74. M. Roco (2011). The long view of nanotechnology development: The NNI at 10 Years. *Journal of Nanoparticle Research* 13, 427–445.
75. J. M. Schnorr and T. M. Swager (2011). Emerging applications of carbon nanotubes. *Chemistry of Materials* 23(3), 646–657.
76. M. J. Schulz, A. D. Kelkar, and M. J. Sundaresan (Eds.) (2006). *Nanoengineering of Structural, Functional, and Smart Materials*. Boca Raton, FL: CRC Press.
77. S. Seal (Ed.) (2008). *Functional Nanostructures – Processing, Characterization, and Applications*. New York: Springer.
78. Z. Spitalskya, D. Tasisb, K. Papagelis, and C. Galiotis (2010). Carbon nanotube-polymer composites: Chemistry, processing, mechanical, and electrical properties. *Progress in Polymer Science* 35, 357–401.
79. J. A. Shatkin (2008). *Nanotechnology: Health and Environmental Risks*. Boca Raton, FL: CRC Press.
80. S. C. Tjong (2009). *Carbon Nanotube Reinforced Composites – Metal and Ceramic Matrices*. Weinheim, Germany: Wiley-VCH Verlag GmbH.
81. D. Tomanek and R. J. Enbody (2000). *Science and Application of Nanotubes*. New York: Kluwer Academic/Plenum Publishers.

82. R. A. Vaia (2002). Polymer nanocomposites open a new dimension for plastics and composites. *The AMPTIAC Newsletter* 6(1), 17–24.
83. M. Di Ventra, S. Evoy, and J. R. Heflin, Jr. (2004). *Introduction to Nanoscale Science and Technology*. Norwell, MA: Kluwer Academic Publishers.
84. T. A. Vilgis, G. Heinrich, and M. Klippel (2009). *Reinforcement of Polymer Nano-Composites – Theory, Experiments and Applications*. Cambridge: Cambridge University Press.
85. D. Vollath (2008). *Nanomaterials: An Introduction to Synthesis, Properties, and Applications*. Weinheim, Germany: Wiley-VCH Verlag GmbH.
86. L. Wang (Ed.) (2000). *Characterization of Nanophase Materials*. Weinheim, Germany: Wiley-VCH Verlag GmbH.
87. Z. L. Wang, Y. Liu, and Z. Zhang (Eds.) (2003). *Handbook of Nanophase and Nanostructured Materials, Vol. 4: Materials Systems and Applications (II)*. New York: Kluwer Academic/Plenum Publishers.
88. L. Wardle (2012). 3D Nano-engineering Composites and 2D Polymer Nanocomposites: Processing and Properties. In *ASC Series on Advances in Composite Materials, Nanocomposites*, Vol. 2, T.-W. Chou and C. T. Sun (Eds.), Lancaster, PA: DEStech Publishing.
89. L. Wardle, J. H. Koo, G. M. Odegard, and G. D. Seidel (2016). Advanced Nanoengineering Materials. In *Aerospace Materials and Applications*, B. Bhatt (Ed.), Reston, VA: AIAA.
90. N. Yao and Z. L. Wang (Eds.) (2005). *Handbook of Microscopy for Nanotechnology*. Boston, MA: Kluwer Academic Publishers.
91. J. Z. Zhang (2009). *Optical Properties and Spectroscopy of Nanomaterials*. Singapore: World Scientific Publishing.

2 An Overview of Nanomaterials

2.1 Introduction

The increasing availability of nanomaterials or nanoparticles offers the promise for researchers to develop novel polymer composite materials with new and improved properties. Nanomaterials lie at the intersection of material science, physics, chemistry, biology, medicine, and for many of the most interesting applications. With the increasing development and availability of nanomaterials or nanoparticles, new materials discover use in numerous applications, such as electronics, aerospace, military, pharmaceuticals, and medicine. One unique research area of particular interest is the development of polymer nanocomposites (PNCs). Polymer nanocomposites are created from a combination of inorganic nanomaterials and organic polymers, and they offer properties that are representative of both components. They are also known as hybrid materials. For traditional composite materials, they often exhibit localized heterogeneity that can lead to the loss of desired properties. When the micron-sized inorganic particles are incorporated into the polymer, the concept of true hybridization at the molecular level is not achieved. The nanomaterials make it possible to increase the interaction between the organic and inorganic phases by several orders of magnitude. In this chapter, numerous types of nanomaterials are introduced in terms of their origin, background, structure, surface treatment, manufacturing, fabrication, properties, features, potential applications, and suppliers.

Polymer nanocomposites are composites in which interphases dominate the composite properties because of the very small size of the nanomaterial component; at least one dimension of it is on the nanometer scale. This polymer can be thermoplastic, thermoset, or elastomer. The polymer-nanomaterial interface is a key determinant of PNC properties. The interaction of nano-sized particles with a polymer molecule is truly a molecular-level interaction. The chemical surface reactivity and bulk mixing dynamics would influence the incorporation of nanomaterials into the host polymer. Since their invention more than two decades ago, PNCs have become a new and important class of materials that not only offer superior performance as compared to conventional composites but also provide multifunctional properties

Table 2.1. *Characteristics of Nanoparticles to Polymers*

Improved Properties	Disadvantages
• Mechanical properties (tensile strength, stiffness, toughness) • Gas barrier • Synergistic flame retardant additive • Dimensional stability • Thermal expansion • Thermal conductivity • Electrical conductivity • Ablation resistance • Chemical resistance • Reinforcement • Other multifunctional properties	• Viscosity increase (limits processability) • Dispersion difficulties • Optical issues • Sedimentation • Black color when different carbon-containing nanomaterials are used

that extend their scope of application to new areas [1–10]. Polymer nanocomposites provide major enhancements in numerous properties, such as mechanical, thermal (stability and conductivity), flammability, ablation, electrical, optical, tribological, permeability, chemical resistance, and other areas. These multifunctional properties of PNCs are discussed in details in numerous references [1–10] and in subsequent chapters in this book. The properties and performance of PNCs can be affected by some important factors, such as the choice of:

- nanomaterials and their surface treatments;
- polymer matrix – that is, thermoplastic, thermoset, or elastomer;
- methods of preparation of polymer nanocomposites;
- structure and morphology of polymer nanocomposites.

It is a very complex matter to fully understand why the enhancement of properties occurs in polymer nanocomposites. To illustrate this point, several advantages and disadvantages are summarized in Table 2.1, when nanomaterials are incorporated into the polymer matrix to create a new PNC material.

2.2 Types of Nanomaterials/Nanoparticles

There are many different types of commercially available nanomaterials that can be incorporated into the polymer systems to form polymer nanocomposites. Depending on the application, the researcher must determine the type of nanomaterial to provide the desired effect. A brief discussion of the most commonly used nanomaterials in the literature is included in this chapter. In general, these nanomaterials or nanoparticles can be categorized based on their nanoscale dimensions, such as one-, two-, or three-dimensional nanoscale materials as shown in Figure 2.1 and further discussed later in the chapter.

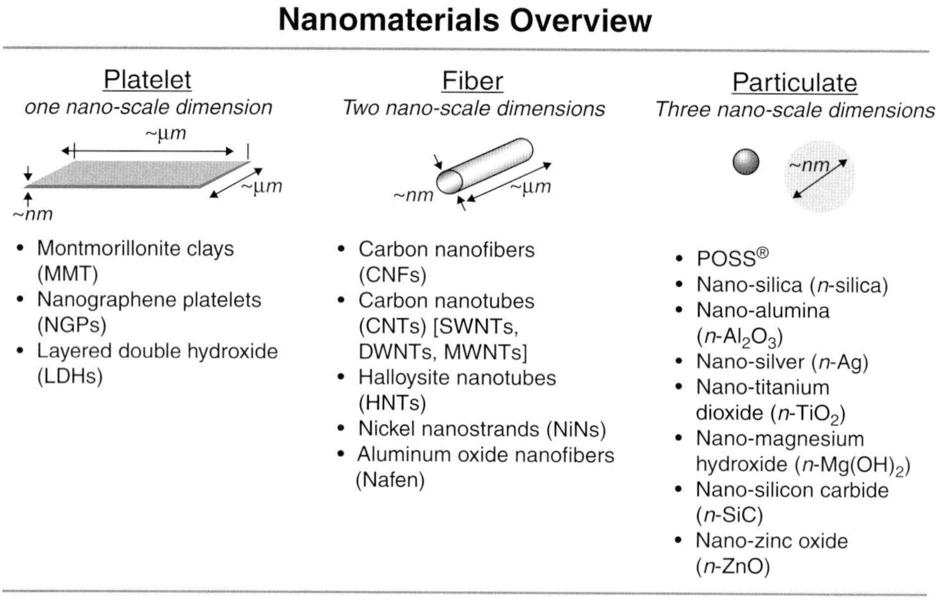

Figure 2.1. Overview of nanomaterials grouped in terms of their nanoscale dimensions.

One nano-scale dimension (lamellar) nanomaterials/nanoparticles:

- montmorillonite clays (MMT);
- nanographene platelets (NGPs);
- layered double hydroxides (LDHs).

Two nano-scale dimensions (fiber) nanomaterials/nanoparticles:

- carbon nanofibers (CNFs);
- carbon nanotubes (CNTs) (single-wall [SWNTs], double-wall [DWNTs], and multi-wall [MWNTs]);
- halloysite nanotubes (HNTs);
- nickel nanostrands (NiN_S).
- aluminum oxide nanofibers (Nafen)

Three nano-scale dimensions (particulates) nanomaterials/nanoparticles:

- polyhedral oligomeric silsesquioxanes (POSS®);
- nano-silica (*n*-silica);
- nano-alumina (*n*-alumina);
- nano-titanium dioxide (n-TiO_2);
- nano-magnesium hydroxide (n-$Mg(OH)_2$);
- nano-silicon carbide (*n*-SiC);
- nano-silver (*n*-Ag);
- nano-zinc oxide (*n*-ZnO);
- others (n-$BaSO_4$, n-$BaTiO_3$, n-$CaCO_3$, n-Fe_2O_3, n-Fe_3O_4, nano-diamond, and *n*-ZnS).

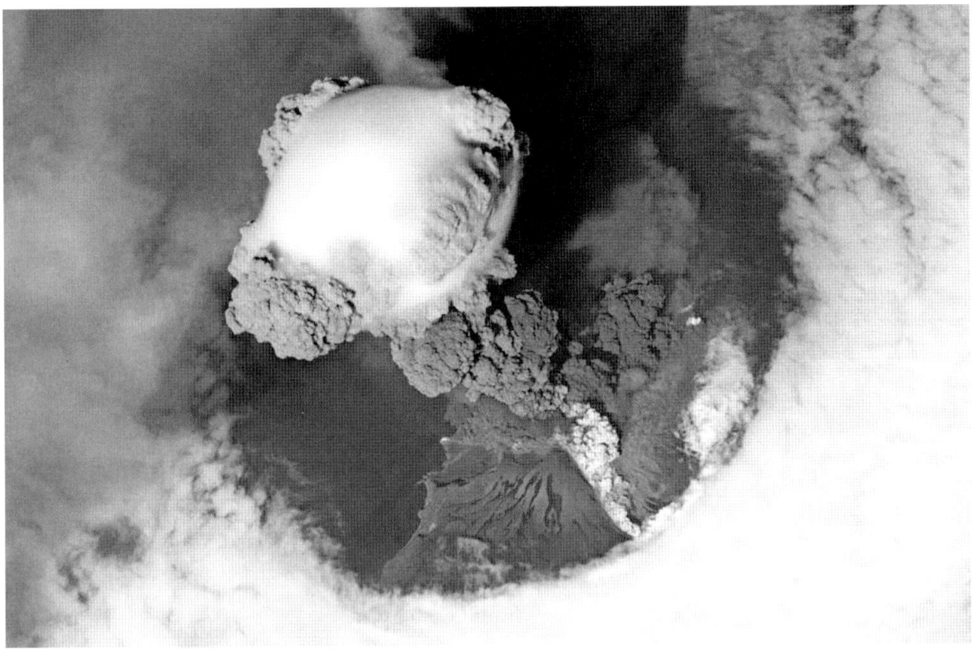

Figure 2.2. Bentonite comes from the deposition and alternation of volcanic ash from inland sea beds from 85 million years ago.

2.2.1 One Nanoscale Dimension in the Form of Lamellar

2.2.1.1 Montmorillonite (MMT) Clays

Origin. Nanoclay is one of the most widely studied nanomaterials in a variety of polymer matrices for various applications [1–3, 9, 11–13]. Bentonite (natural clay) is most commonly formed by the in situ alteration of volcanic ash, as shown in Figure 2.2. Another, less common method is the hydrothermal alteration of volcanic rocks. Bentonite contains montmorillonite (MMT) but can also contain glass, mixed-layer clays, illite, kaolinite, quartz, zeolite, and carbonates. Clay soil has particle sizes of less than 2 μm. The *expanding clays* are phyllosilicates, smectite, and montmorillonite, and the *nonexpanding clays* are talc, kaolin, and mica. The expanding clays are more versatile than the nonexpanding clays are. The starting material of Southern Clay Products' (SCP, Gonzales, Texas) commercial clay products comes from volcanic eruptions in the Pacific Ocean and the western United States during the Cretaceous period about 85–125 million years ago. During that time, magma – molten rock underneath the planet's crust – had forced its way to the surface and, under extreme pressure, exploded into the atmosphere in a series of global volcanic eruptions. Airborne ash from the eruption clouds subsequently fell to earth, forming deposits characterized by high-volume bedding of well - to moderately sorted ash. The parent ash was deposited under marine conditions – in other words, most of it landed in seas and oceans – and there was some minor accumulation in alkaline lakes, as illustrated in Figure 2.3 [11]. With time, the ash beds in the western United States' inland sea areas became what is now the clay-producing areas of Montana,

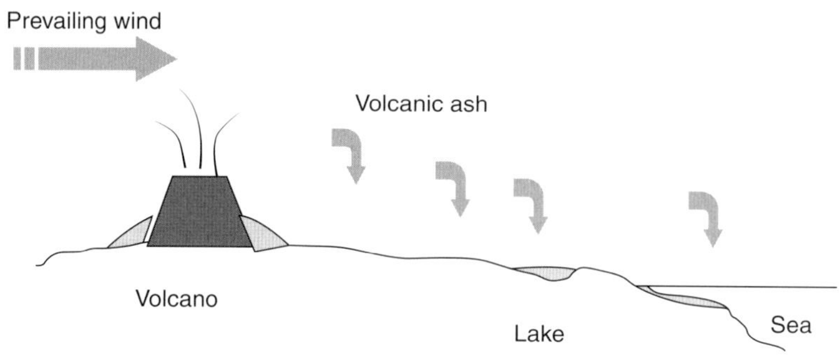

Figure 2.3. Formation of natural clays by the in situ alteration of volcanic ash.

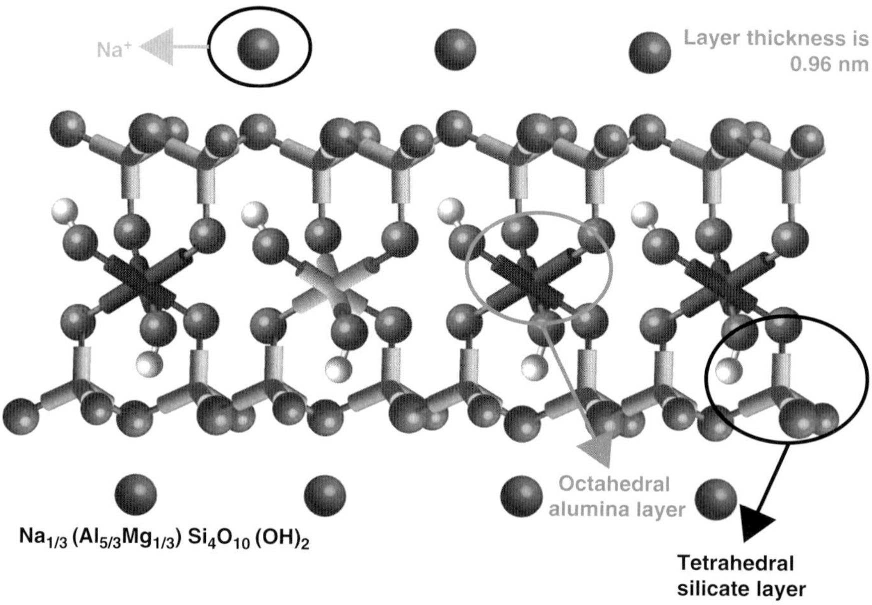

Figure 2.4. Chemical structures of montmorillonite nanoclays.

Wyoming, and the Dakotas. Opinions differ concerning the process and time of alternation of the ash to become clay. It is estimated that the resulting deposits in Wyoming alone contain more than 1 billion tons of clay [11, 12].

Structure. Silica is the dominant constituent of the montmorillonite (MMT) clays, with alumina being essential. The chemical structure of MMT clays is illustrated in Figure 2.4, showing its sheet structure consisting of layers containing the tetrahedral silicate layer and the octahedral alumina layer. The tetrahedral silicate layer consists of SiO_4 groups linked together to form a hexagonal network of the repeating units of composition Si_4O_{10}. The alumina layer consists of two sheets of closely packed oxygens or hydroxyls, between which octahedrally coordinated aluminum atoms are imbedded in such a position that they are equidistant from six oxygens or hydroxyls. The two tetrahedral layers sandwich the octahedral layer, sharing their apex oxygens with the latter. These three layers form one clay sheet that has a thickness of 0.96 nm.

The chemical formula of the MMT clay is $Na_{1/3}(Al_{5/3}Mg_{1/3})Si_4O_{10}(OH)_2$. In its natural state, Na^+ cation resides on the MMT clay surface.

Organic Surface Treatment. Because layered silicates are hydrophilic materials, they must be made organophilic (hydrophobic) to become compatible with the host polymers that are hydrophobic polymers. Without organic treatment, layer silicates will only disperse and phase-separate in the presence of very polar polymers. Organic treatment is typically accomplished via ion exchange between inorganic alkali cations on the clay surface with the desired organic cation. The organic treatment, at the interface between inorganic silicate and organic polymer, is a vital part of the polymer nanocomposite, and therefore it must be tailored to synthetic conditions. Synthetic methods for polymer nanocomposite preparation include solvent mixing, in situ polymerization, and melt compounding. One of the most "industry-friendly" methods of making polymer nanocomposite is the use of melt compounding. The polymer and organically treated clay are heated to the melting point of the polymer, and the two are mixed together using compounding equipment, such as an extruder or a mixing head. There are several types of clay surface treatments:

- quaternary ammonium salts based on textile antistatic agents;
- alkyl imidazoles, which provides improved thermal stability;
- coupling and tethering agents, reactive diluents that can be functional amino compounds;
- cation types containing phosphorous ionic compounds.

Achieving exfoliation of organo-montmorillonite clay in various polymer continuous phases is a function of the surface treatment of the MMT clays and the mixing efficiency of the dispersing apparatus or equipment. For hydrophilic nanoclay, such as SCP's Cloisite® Na^+, it is recommended to dissolve it either in water using high-shear mixing or to add Cloisite® Na^+ directly into a water-soluble, aqueous polymer solution. Three types of mixing equipment are recommended to disperse hydrophobic nanoclays, such as Southern Clay Products' Cloisite® Na^+, Ca^{++}, 3, 5, 10A, 11, 15, 20, 30B, 93, and 116 into different resin systems. The recommended equipment is as follows: high-shear mixer or three-roll mill for liquid resins, Brabender mixer for viscous resins, or twin-screw extruder for solid resins.

The preparation of polymer-clay nanocomposites is illustrated in Figure 2.5. Depending on the physical state of the polymer, one can incorporate the clay into the polymer by solution blending, melt blending, or in situ polymerization processes to form polymer-clay nanocomposites. Polymer-clay nanocomposites can be classified morphologically into (a) unmixed, (b) intercalated, and (c) exfoliated (delaminated) states, as shown in Figure 2.6. The most desirable morphological state for the polymer-clay nanocomposites is exfoliation, followed by intercalation. The processing challenge of nanoclay is to disperse the 8-μm particles into > 1 million platelets using the proper processing technique and conditions, as shown in Figure 2.7. The dispersion mechanism can be: (a) via a chemistry route, as in in situ polymerization; (b) via a processing route, as in extrusion; or (c) by combining the chemistry and the processing routes (the optimal way), as summarized in Figure 2.8.

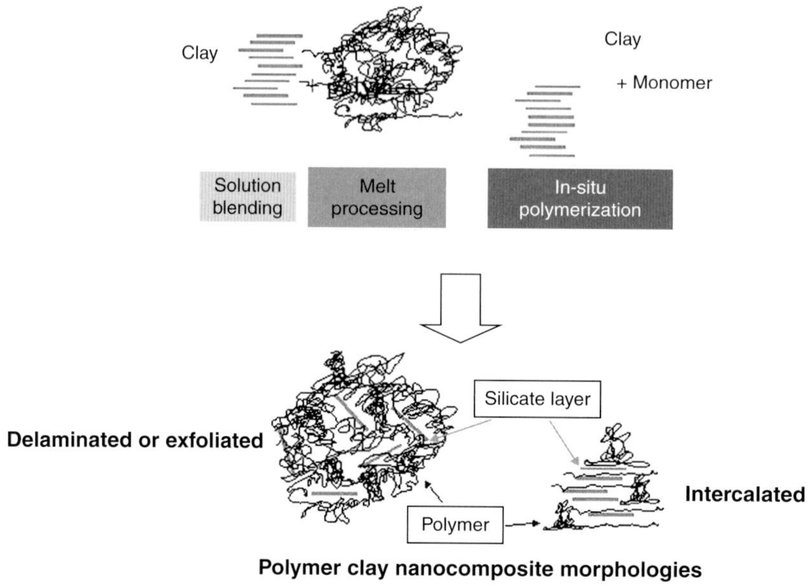

Figure 2.5. Schematic showing the preparation of polymer-clay nanocomposites (Courtesy of J. Gilman).

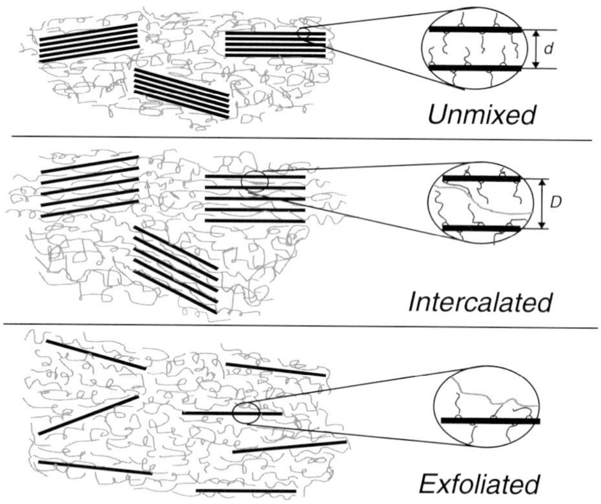

Figure 2.6. Schematic showing polymer-clay nanocomposite classifications.

Figures 2.9 and 2.10 further illustrate the concept of particles shearing apart and platelets peeling apart in cartoon (Figure 2.9) and in transmission electron micrographs (Figure 2.10). Transmission electron microscopy (TEM) is the most useful tool to determine the degree of dispersion of the nanomaterials in the polymer matrix. This will be discussed in more details in Chapter 5.

Organoclay Suppliers. There are several worldwide suppliers for organoclays. In the United States, the two major suppliers are Southern Clay Products (SCP) and Nanocor®. Southern Clay Products produces Cloisite®, Laponites®, and Nanofil®, types of clay materials. Cloisite and Nanofil materials are based on naturally occurring

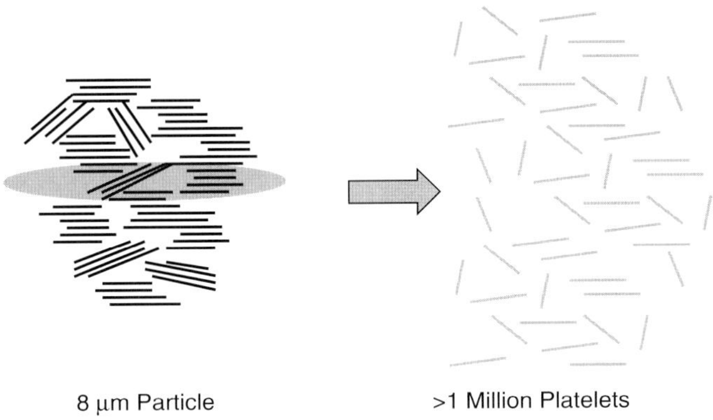

Figure 2.7. Processing challenge of nanoclay is turning 8 μm particles into > 1 million platelets (Courtesy of SCP).

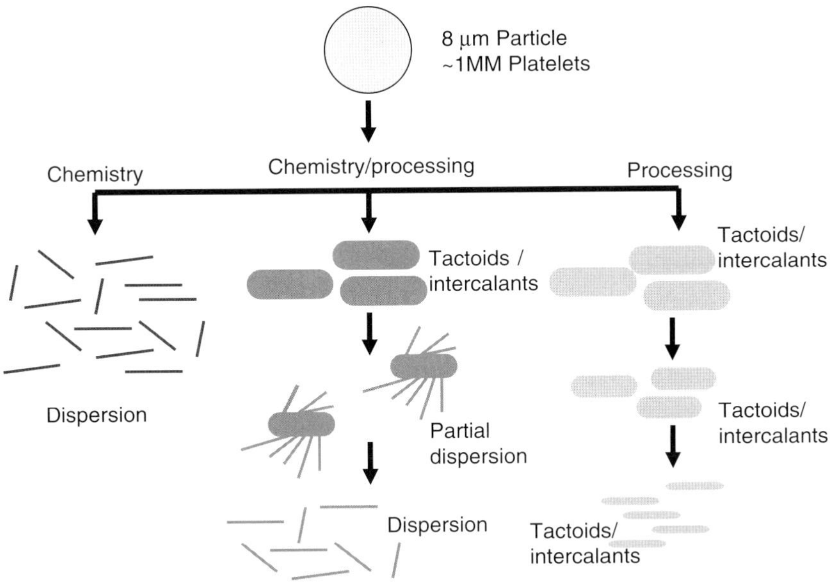

Figure 2.8. Different dispersion mechanisms of dispersing nanoclays (Courtesy of SCP).

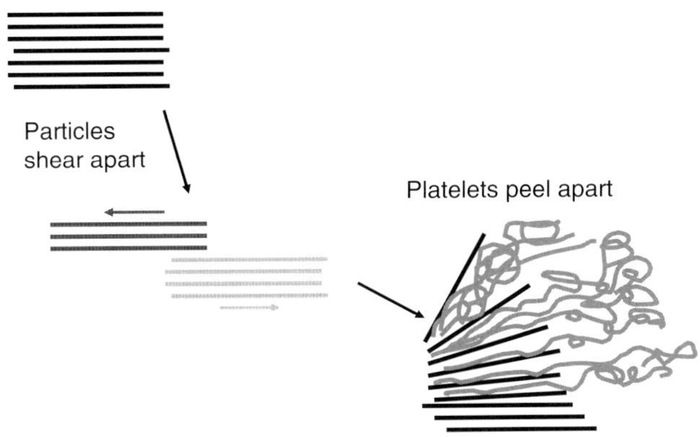

Figure 2.9. Cartoon showing particle shearing and platelets peeling apart (Courtesy of SCP).

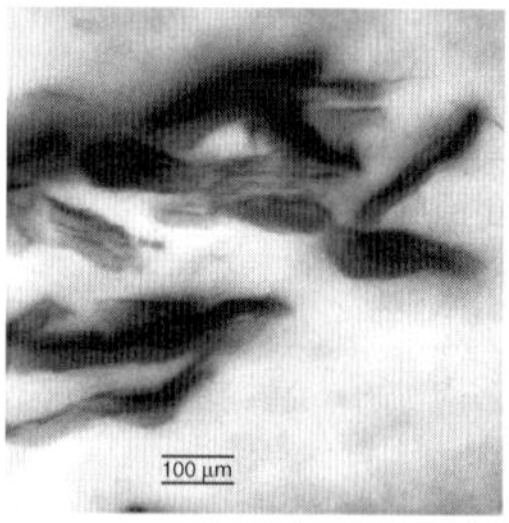

Particles shear apart

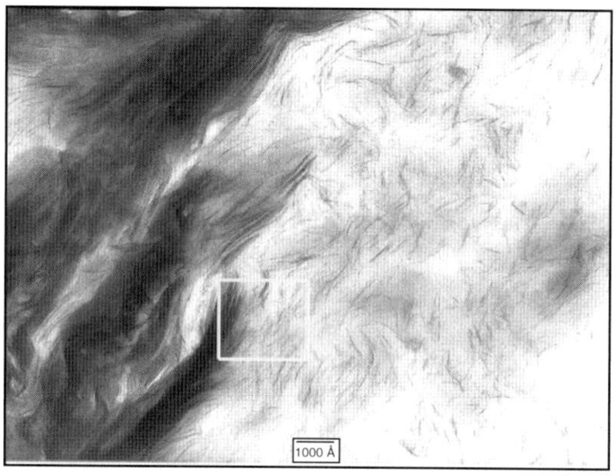

Platelets peel apart

Figure 2.10. TEM micrographs showing particle shearing and platelets peeling apart (Courtesy of SCP).

MMT, while Laponites are synthesized clays. Product information can be obtained on the company's website, www.nanoclay.com [12]. Nanocor, located in Chicago, IL, produces Nanomer® nanoclays, I.xxE (e.g., I.28E and I.30E), masterbatches/concentrates, and nanocomposites. Product information can be obtained on Nanocor's website, www.nanocor.com [13]. Other organoclay suppliers are Kunimine Industries Co., Ltd. (Japan, www.kunimine.co.jp), Laviosa Chimica Minerariain (Italy, www.laviosa.it), and several Chinese companies.

Cloisite Product Lines. Southern Clay Products manufactures and markets Cloisite additives [11, 12]. These montmorillonite organoclays are surface modified to allow complete dispersibility and miscibility with many different resin systems they were designed to improve. Cloisite MMT clays, such as Cloisite Na^{+}, Ca^{++}, 3, 5, 10A, 11, 15, 20, 30B, 93, and 116 are referenced in this section. Several of these MMT clays are mentioned throughout this book.

Cloisite Na^{+} is a natural bentonite. This polymer additive improves various polymer physical properties, such as reinforcement, heat deflection temperature (HDT), coefficient of linear thermal expansion (CLTE), and barrier properties. Typical properties, dry particle size, density, organic modified data of Cloisite Na^{+}, Ca^{++}, 5, 10A, 11, 15, 20, 30B, 93, and 116 are shown in Tables 2.2 and 2.3.

Cloisite Ca^{++} is a natural bentonite. It is designed to be used as an additive for plastics and rubber to improve various physical properties, such as reinforcement, coefficient of thermal expansion (CTE), synergistic flame retardant, and barrier.

Cloisite 3 has no data available on the SCP's website.

Cloisite 5 is a bis(hydrogenated tallow alkyl)dimethyl salt with bentonite. It is an additive for plastics and rubber to improve physical properties, such as reinforcement, CTE, synergistic flame retardant, and barrier.

Cloisite 10A is benzyl(hydrogenated tallow alkyl)dimethyl salt with bentonite. It is designed to be used as an additive for plastics and rubber to improve various physical properties, such as reinforcement, CTE, synergistic flame retardant, and barrier.

Table 2.2. *Typical Physical Properties of Cloisite Nanoclay*

MMT Clay	Moisture (%)	Dry Particle Size, d_{50} (μm)	Color	Packed Bulk Density (g/l)	Density (g/cc)	X Ray Results, d_{001} (nm)
Cloisite Na^+	4–9	<25	Off White	568	2.86	1.17
Cloisite Ca^{++}	4–9	<10	Off White	625	2.8	1.55
Cloisite 5	<3	<40	Off White	480	1.77	3.27
Cloisite 10A	<3	<10	Off White	265	1.9	1.9
Cloisite 11	<3	<40	Off White	265	2.0	1.84
Cloisite 15	<3	<10	Off White	165	1.66	3.63
Cloisite 20	<3	<10	Off White	175	1.77	3.16
Cloisite 30B	<3	<10	Off White	365	1.98	1.85
Cloisite 93	<3	<40	Off White	440	1.88	2.79
Cloisite 116	8–13	<15	Off White	340	2.8	1.25

Table 2.3. *Cloisite MMT Clay with Organic Modifier*

MMT Clay	Organic Modifier
Cloisite Na^+	Natural bentonite
Cloisite Ca^{++}	Natural bentonite
Cloisite 5	Bis(hydrogenated tallow alkyl)dimethyl salt
Cloisite 10A	Benzyl(hydrogenated tallow alkyl)dimethyl salt
Cloisite 11	Benzyl(hydrogenated tallow alkyl)dimethyl salt
Cloisite 15	Bis(hydrogenated tallow alkyl)dimethyl salt
Cloisite 20	Same as Cloisite 15
Cloisite 30B	Alkyl quaternary ammonium salt
Cloisite 93	Trialkyl ammonium salt
Cloisite 116	Natural bentonite

Cloisite 11 is benzyl(hydrogenated tallow alkyl)dimethyl salt with bentonite. It is designed to be used as an additive for plastics and rubber to improve various physical properties, such as reinforcement, CTE, synergistic flame retardant, and barrier.

Cloisite 15 is bis(hydrogenated tallow alkyl)dimethyl salt with bentonite. It is designed to be used as additive for plastics and rubber to improve various physical properties, such as reinforcement, CTE, synergistic flame retardant, and barrier.

Cloisite 20 is bis(hydrogenated tallow alkyl)dimethyl, salt with bentonite. It is designed to be used as an additive for plastics and rubber to improve various physical properties, such as reinforcement, CTE, synergistic flame retardant, and barrier.

Cloisite 30B is alkyl quaternary ammonium salt bentonite. It is designed to be used as an additive for plastics and rubber to improve various physical properties, such as reinforcement, CTE, synergistic flame retardant, and barrier.

Cloisite 93 is trialkyl ammonium bentonite salt with bentonite. It is designed to be used as an additive for plastics and rubber to improve various physical properties, such as reinforcement, CTE, synergistic flame retardant, and barrier.

Cloisite 116 is a natural bentonite. It is designed to be used as an additive for plastics and rubber to improve various physical properties, such as reinforcement, CTE, synergistic flame retardant, and barrier.

Nanocor Product Lines. Three general categories of products are available from Nanocor or its distributors: Nanomer nanoclays, masterbatches/concentrates, and nanocomposites.

Nanomer nanoclays are supplied as microfine powders under the "I" series designation, ready to be used for different resins, such as polyamides, epoxy, urethanes, EPDM, and engineering resins.

Masterbatches/concentrates in resin systems, such as polypropylene homopolymer, polypropylene copolymer, LLDPE, LDPE, HDPE, and EMA, are supplied in high-loading pellet forms. PNC nylon concentrate is designed for letdown into nylon 6, 6.6, and amorphous nylon.

Nanocomposites are supplied in pellet form, ready for injection molding, blow molding, and film casting by Nanocor's industrial partners. Imperm®, supplied by ColoMatrix Corporation, is an ultrahigh barrier nylon, providing protection for oxygen-sensitive products and CO_2 retention for carbonated soft drinks, waters, beers, and flavored alcoholic beverages. Durethan® KU2-2601, available from LANXESS Deutschland, is a nylon 6 nanocomposite for films and paper coating, designed for medium barrier applications requiring excellent clarity. Aegis™ NC is a nylon 6 nanocomposite available from Honeywell Polymers, designed for film and paper coating, and for engineering applications. Aegis™ OX is a nylon 6 nanocomposite available from Honeywell Polymers, combining nanocomposite barrier and oxygen scavenger in one system used for beer bottles. NanoTUFF™ is a nylon 6 nanocomposite available from Nylon Corporation of America for applications where high modulus and cold weather impact are required. NanoSEAL™ is formulated for fuel systems (tanks and hoses) to meet CARB fuel emissions standards, and is available from Nylon Corporation of America.

2.2.1.2 Nanographene Platelets (NGPs)

Origin. Graphene is the name given to a flat monolayer of carbon atoms tightly packed into a two-dimensional (2D) honeycomb lattice, and is a basic building block for graphitic materials for all other dimensionalities (Figure 2.11) as described by Geim and Novoselov [14]. It can be wrapped up into 0D fullerenes, rolled into 1D nanotubes, or stacked into 3D graphite. Theoretically, graphene (or 2D graphite) has been studied for sixty years, and is widely used for describing properties of various carbon-based materials [14]. A thin flake of this ordinary carbon, just one atom thick, lies behind the 2010 Nobel Prize in Physics awarded to Andre Geim and Konstantin Novoselov. Both of them originally studied and began their careers as physicists in Russia; presently they are professors at the University of Manchester, United Kingdom. They have shown that carbon in such a flat form has exceptional properties that originate from the remarkable world of quantum physics. Graphene is a form of carbon. As a material it is completely new, not only the thinnest ever but also the strongest. As a conductor of electricity it performs as well as copper.

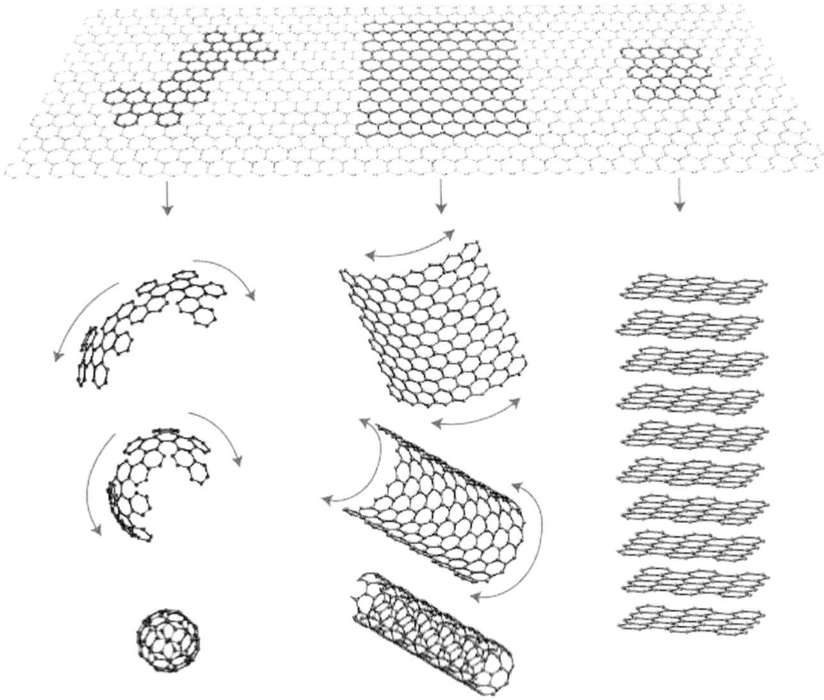

Figure 2.11. Mother of all graphitic forms. Graphene is a 2D building material for carbon materials of all other dimensionalities. It can be wrapped up into 0D buckyballs, rolled into 1D nanotubes, or stacked into 3D graphite [14].

As a conductor of heat it outperforms all other known materials. It is almost completely transparent, yet so dense that not even helium, the smallest gas atom, can pass through it. Geim and Novoselov extracted the graphene from a piece of graphite, such as found in ordinary pencils. Using regular adhesive tape, they managed to obtain a flake of carbon with a thickness of just one atom. This was done at a time when many believed it was impossible for such thin crystalline materials to be stable.

A nanographene platelet (NGP) is an emerging class of nanomaterials. Jang and Zhamu reported an excellent review of processing of NGPs and NGP nanocomposites [15], and their review forms the basis for this section. An NGP is a nanoscale platelet composed of one or more layers of graphene plane with a platelet thickness from about 0.34 nm to 100 nm. NGPs are predicted to have a range of unusual physical, electrical, chemical, and mechanical properties [15]. Although practical electronic device application for graphene is not envisioned to occur within the next 5 to 10 years, its application as nanofiller in a composite material is imminent. The availability of processable graphene sheets in large quantities is essential to the success in exploiting composite and other applications.

For more than six decades, scientists have presumed that a single-layer graphene sheet (one atom thick) could not exist in its free state based on the reasoning that its planar structure would be thermodynamically unstable. Several groups worldwide have recently succeeded in obtaining isolated graphene sheets [16–24]. Several

Table 2.4. *Comparison of Physical Properties of CNTs, CNFs, and NGPs [15]*

Property	Single-walled CNTs	Carbon nanofibers	NGPs
Specific gravity	0.8 g/cc	1.8 (AG)-2.1 (HT) g/cc; AG = as grown and HT = heat treated (graphitic)	1.8–2.2 g/cc
Elastic modulus	~1 TPA (axial direction)	0.4 (AG)–0.6 (HT) TPa	~1 TPa (in-plane)
Strength	50–500 GPa	2.7 (AG)–7.0 (HT) GPa	~100–400 GPa
Resistivity	5–50 μΩ cm	55 (HT)–1,000 (AG) μΩ cm	50 μΩ cm (in-plane)
Thermal conductivity	Up to 2,900 $Wm^{-1} K^{-1}$ (estimated)	20 (AG)–1,950 (HT) $Wm^{-1} K^{-1}$	5,300 $Wm^{-1} K^{-1}$ (in-plane), 6–30 $Wm^{-1} K^{-1}$ (*c*-axis)
Magnetic susceptibility	22×10^6 emu/g (radial); 0.5×10^6 emu/g (axial)	N/A	$22x10^6$ emu/g (perpendicular to plane); 0.5×10^6 emu/g (∥ to plane)
Thermal expansion	Negligible in the axial direction	$\sim 1 \times 10^{-6} K^{-1}$ (HT, axial)	$-1x10^{-6} K^{-1}$ (in-plane); $29 \times 10^{-6} K^{-1}$ (*c*-axis)
Thermal stability	>700°C (in air); 2,800°C (in vacuum)	450–650°C (in air)	450–650°C (in air)
Specific surface area	Typically 10–200 m^2/g; up to 1,300 m^2/g	10–60 m^2/g	Typically 100–1,000 m^2/g; up to >2,600 m^2/g

unique properties associated with these 2D crystals have been discovered [25–33]. In addition to single graphene sheets, double-layer or multiple-layer graphene sheets also exhibit unique and useful behavior. Single-layer and multiple-layer graphene sheets are collectively referred as nanographene platelets (NGPs).

Carbon nanotubes (CNTs) (discussed in subsection 2.2.2.2) and NGPs exhibit some similar behaviors but some vastly distinct properties. Selected physical and mechanical properties of NGPs, CNTs, and vapor-grown carbon fibers (VGCNFs) or carbon nanofibers (CNFs) (also discussed in a later section) are presented in Table 2.4.

Early and Recent Research on NGPs. In late 1980s, attempts to produce NGPs were made, although the significance of NGPs was not well recognized and they were known as thin graphite flakes. In 1988, Bunnell [34–36] developed entailed intercalating graphite with a strong acid to obtain a graphite intercalation compound (GIC), thermally exfoliating the GIC to obtain discrete layers by ultrasonization, mechanical shear forces, or freezing the layers into discrete flakes. Technically, the acid-treated graphite was actually graphite oxide (GO) rather than pristine graphite. Although most of the flakes presented were thicker than 100 nm [36], flakes as small as 20 nm were cited. Zaleski et al. [37] used air milling to further delaminate exfoliated graphite flakes to an average flake thickness of approximately 25 nm based on Jang and Zhamu's calculation [15]. Jang et al. [18, 19] succeeded in isolating single-layer and multi-layer graphene structures from partially graphitized polymeric carbon, which were obtained from a polymer or pitch precursor. Figure 2.12

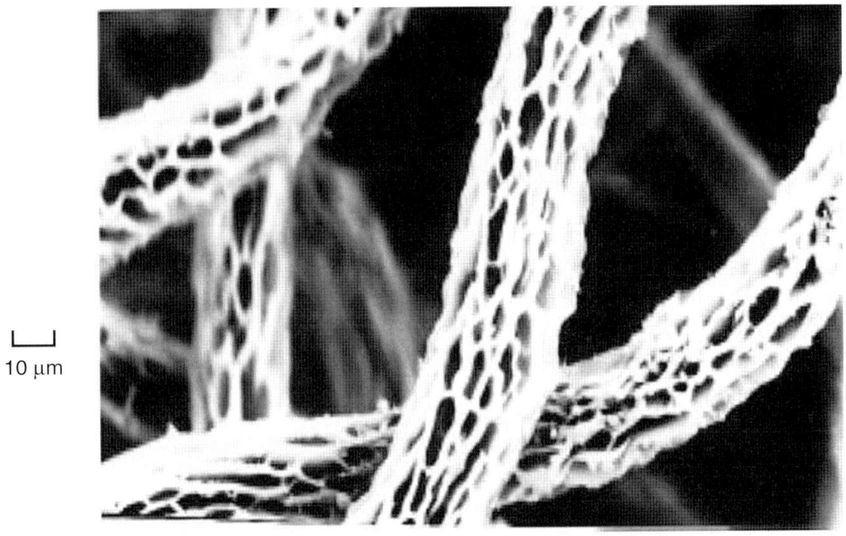

Figure 2.12. An SEM image of a partially exfoliated polymeric carbon fiber [15].

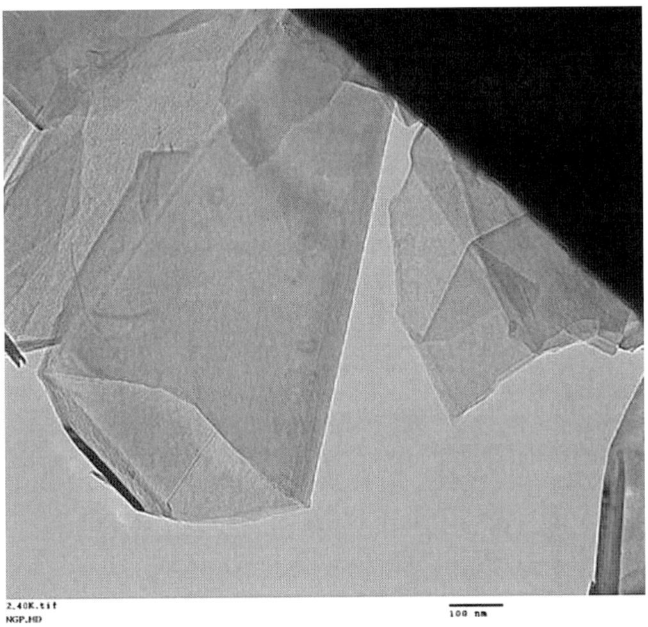

Figure 2.13. A TEM image of NGPs obtained from exfoliation and ball milling of a polymeric carbon [15].

shows polymeric carbon fibers obtained by intercalating and exfoliated carbonized polyacrylonitrile (PAN) fibers [15]. Graphene platelets extracted from these fibers using ball milling procedure are shown in Figure 2.13 [15]. Horiuchi et al. [38–42] have done some significant work on the preparation of nanoscale graphite oxide (GO) platelets, which they called carbon nanofilms. According to Horiuchi et al. [38], single-layer graphene was detected.

In 2004, Novoselov et al. [16, 17] prepared single-sheet graphene by removing graphene from a graphite sample one sheet at a time using a "scotch-tape" method.

Figure 2.14. An SEM image of exfoliated graphite (after sulfuric acid–nitric acid intercalation and thermal shock exposure at 1,050°C). Exfoliated graphite is characterized by having graphene sheets remaining somewhat interconnected with one another [15].

Although this method is not amenable to large-scale production of NGPs, their work did spur globally increasing interest in nanographene materials, mostly motivated that graphene could be useful for developing novel electronic device [24–32]. Small-scale production of ultra-thin graphene sheets on a substrate can be achieved by thermal decomposition–based epitaxial growth [29] and a laser desorption-ionization technique [43]. More promising techniques for mass-producing NGPs are likely those involving chemical or thermal exfoliation of intercalated graphite or oxidized graphite.

Chen et al. [44] exposed GO to 1,050°C for 15 s to obtain exfoliated graphite, which was then subjected to ultrasonic irradiation in a solution of water an alcohol. Li et al. [24] followed a similar approach. Jang et al. [45] thermally exfoliated GIC to produce exfoliated graphite (Figure 2.14) [15] and subjected exfoliated graphite to mechanical shear, such as ball milling, to obtain NGPs, which were mostly single- to five-layer structures (Figure 2.15) [15]. The specific surface area of the samples prepared by this method was found to be in a range of 300–1300 m^2/g, as measured by the BET method. Thermal exfoliation to produce nanostructured graphite was attempted by Petrik [46]. Drzal et al. [47] conducted thermal exfoliation of intercalated graphite using microwaves as a heat source.

McAllister et al. [21–23] used thermal exfoliation of GO to obtain exfoliated graphite oxide platelets, which were found to contain a high volume of single-layer graphene sheets, based on the BET method with nitrogen gas adsorption in the dry state and in an ethanol suspension with methylene blue dye as a probe. McAllister et al. [21] provided a good model to explain the exfoliation mechanisms and kinetics of GO.

Mack et al. [48, 49] developed a low-temperature process that involved intercalating graphite with potassium melt and contacting the resulting K-intercalated graphite with alcohol, producing violently exfoliated graphite containing many ultra-thin NGPs. This process must be carefully conducted in a vacuum or an extremely

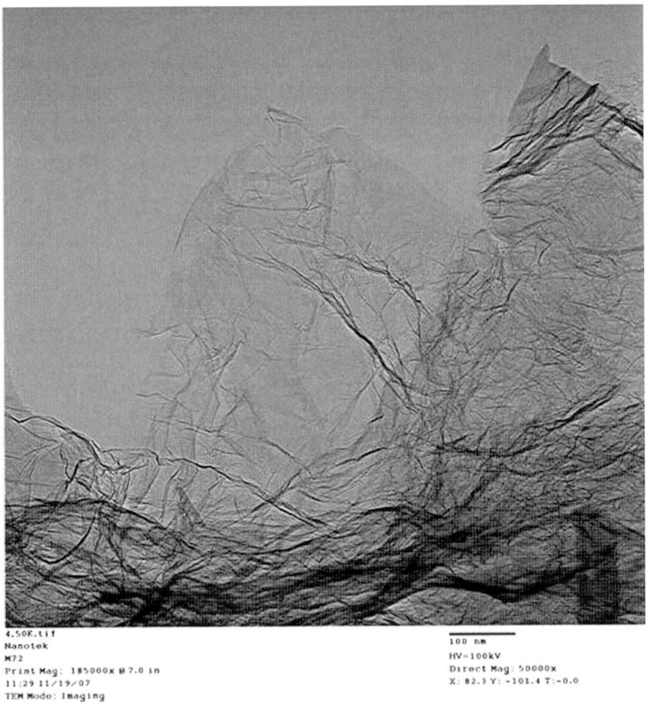

Figure 2.15. A TEM image of NGPs after ball milling of exfoliated graphite (which was obtained from sulfuric acid–nitric acid intercalation and thermal shock exposure at 1,050°C). Several graphene sheets collapsed to overlay one another after dispersion in acetone and then acetone removal during the TEM sample preparation procedure [15].

dry glove box environment, since pure alkali metals, such as potassium and sodium, are extremely sensitive to moisture and pose an explosion danger. This process is not easily amenable to the mass production of nanoscale platelets. One major advantage of this process is that it can produce nonoxidized graphene sheets, since no acid-oxidizer intercalation or a high temperature is involved.

Geim and Novoselov [14] recently presented an excellent progress article on "The Rise of Graphene." Lu et al. [50] wrote a recent review on "Graphene: Fundamentals and Functionalities." Mukhopadhyay and Gupta [51] also edited a new book on *Graphite, Graphene and their Polymer Nanocomposite*, which is a great resource for this subject matter.

Potential Applications of NGPs and NGP Nanocomposites. For scientific and engineering applications, anticipated features and benefits of NGP-based materials can be summarized as follows [15]:

- Nanographene exhibits many peculiar electronic, optical, magnetic, and chemical properties (others may not have yet been discovered) that enable many potential device applications [16, 17, 25–33].
- In addition to lower costs (compared to carbon nanotubes), another major advantage of graphene-based nanocomposites is their ability of forming a thin film, paper, or coating for electrostatic charge dissipation (ESD) and electromagnetic interference (EMI) applications. This occurs when the NGP loading exceeds the percolation threshold, so that NGPs form a network of electron transport path.

- Due to the ultra-high thermal conductivity of NGPs (four times more thermally conductive, yet four times lower in density compared to copper), a nanocomposite thin film, paper, or coating can be used as a thermal management layer in a densely packed microelectronic device.
- High loading of NGPs (5–75 wt%) is incorporated into a polymer or carbon matrix; the resulting nanocomposite possesses an exceptionally high electrical conductivity for fuel cell bipolar plate applications [52, 53].
- NGP nanocomposites have a good combination of mechanical, stiffness, strength, micro-cracking resistance, electrical and thermal conductivities, and barrier performance at a minimal filler concentration. The mechanical properties of NGPs have been used in making golf balls [54] and micro-composite containers for hydrogen storage [55].
- NGP composites (in the form of conductive paper/film/coating, structural adhesive, etc.) can be an integral part of protection strategies against lightning strikes for aircraft, telecommunication towers, and wind turbine blades.
- NGPs can be a component material for lithium ion battery electrodes. Self-assembled graphite oxide nanoplatelets and polyelectrolytes can be a cathode material that provides exceptionally high specific capacity [56].
- Ultra-thin graphene films, being optically transparent and electrically conductive, are a potential alternative for the metal oxide winder electrodes for solid-state dye-sensitized solar cells [57].
- For supercapacitor electrode applications, NGP-based materials possess these desirable features: (a) the dimensions of platelets can be tailored to obtain NGPs with lower thickness ~0.34 nm and length (width) range of ~100 nm–10 μm, yielding a specific area of 2,600 m^2/g; and (b) the surfaces of NGPs can be functionalized to achieve pseudo-capacitance-induced redox-like charge transfer.

Nanographene Platelet Suppliers. There are numerous NGP suppliers worldwide. Several major suppliers are included in this section. Koo et al.'s study [58] describes the methodology used to characterize the as-received nanographene platelets (NGP) from several nanomaterial manufacturers. The microstructures of these NGPs were first examined using scanning electron microscopy at several different magnifications. The thermal stability of these NGPs was evaluated using thermogravimetric analysis with at least three different heating rates, such as 5, 10, 20, and 40°C/min in argon or nitrogen atmosphere. The residual mass and rate of mass loss of these NGPs was compared. The kinetics parameters, such as the activation energy and pre-exponential factor, of these NGPs were then calculated using an isoconversion technique. Eleven NGPs were characterized using the above-described methods. The similarities and differences of these NPGs were compared. Based on these facile techniques, one can rank the morphological and thermal characteristics of these NGPs. The methodology serves as a unique screening technique to select the appropriate NPG that can be chosen for whatever specific application is desired. It is summarized in Table 2.5 [59–68].

Table 2.5. *Summary of Eleven Types of NGPs [58]*

	NGP	Source	Remarks
NGP-1	Grafmax HC 11-1Q	Nacional de Grafite, Sao Paulo, Brazil (www.grafite.com)	Provided by Prof. Antonio F. Avila, Universidade Federal de Minas Gerais, who used this product for his hybrid nanocomposites studies [59, 60], which showed thermal stability improvement for his epoxy system.
NGP-2	TG679	Graftech International, Parma, OH (www.graftech.com)	Dr. S. G. Miller used a functionalized TG679 in her thesis studies in epoxy resin [61], which showed mixed results in mechanical properties.
NGP-3	Russian NGP	Dr. Howard K. Schmidt, Chemical and Biomolecular Engineering Dept., Houston, TX	Dr. Schmidt obtained large quantities of NGP-3 from a Russian source; no technical information is available on this material [62].
NGP-4	XGnP™	XG Sciences, East Lansing, MI (www.xgsciences.com)	This is an earlier version of XGnP™ [63].
NGP-5	N006-010-P	Angstron Materials, Dayton, OH (www.angstronmaterials.com)	Thickness (Z) is ~10–20 nm, average X & Y is 14 μm, and specific surface area is 110 m^2/g [64].
NGP-6	0541DX	Skyspring Nanomaterials, Houston, TX (www.ssnano.com)	Thickness is 6–8 nm, average diameter is 15 μm, and specific surface area is 120–150 m^2/g [65].
NGP-7	Grade 2	Cheap Tubes, Brattleboro, VT (www.cheaptubesinc.com)	Thickness is <10 nm, average diameter is 15 μm, and specific surface area is 100 m^2/g [66].
NGP-8	Grade M	XG Sciences, East Lansing, MI (www.xgsciences.com)	Thickness is 6–8 nm, average diameter is 15 μm, and specific surface area is 120–150 m^2/g [63].
NGP-9	MEGD	Prof. Rod Ruoff, Dept. of Mechanical Engineering, The Univ. of Texas at Austin, Austin, TX (recently moved to Ulsan National Institute of Science and Technology (UNIST), Ulsan, S. Korea)	Abundant technical information can be found on Prof. Ruoff's website (bucky-central.me.utexas.edu) [67].
NGP-10	Grafmax HC 11	Nacional de Grafite, Sao Paulo, Brazil (www.grafite.com)	Nacional de Grafite provided this grade to us and has been supplying NGP to Prof. Avila's research [68]. D_{32} = 6.99 μm.
NGP-11	Grafmax HC 30	Nacional de Grafite, Sao Paulo, Brazil (www.grafite.com)	Nacional de Grafite provided this grade to us and has been supplying NGP to Prof. Avila's research [68]. D_{32} = 18.15 μm.

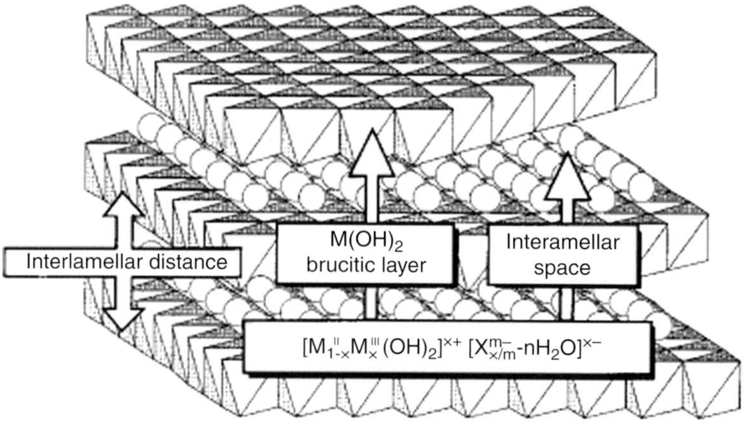

Figure 2.16. Layered crystal structure of layered double hydroxide [69] (Courtesy of ACS).

2.2.1.3 Layered Double Hydroxide (LDH)

Origin. The incorporation of polymers into layered double hydroxides (LDHs) to form polymer/LDH nanocomposites has been a subject of academic interest for more than 20 years according to Qiu and Qu's excellent review article on polymer/ LDHs nanocomposites [69]. These polymer/LDH nanocomposites often exhibit enhanced mechanical, thermal, optical, and electrical properties and flame retardancy. Among them, the thermal properties and flame retardancy are the most interesting and will be discussed in this section. Matusinovic and Wilkie [70] also published another review using LDHs as a new class flame retardant (FR) additive for polymers. These two review articles form the basis of the discussion of LDHs in this section. Layered double hydroxides are a different kind of layered crystalline filler for nanocomposites. LDHs combine the FR features of conventional metal hydroxide fillers (magnesium hydroxide and aluminum hydroxide) with those of layered silicate nanofillers (montmorillonite). LDHs are considered to be a new emerging class of nanofillers for FR polymer nanocomposites. Generally, their chemical structure is represented by the formula $[M^{2+}{}_{1-x}\ M^{3+}{}_{x}\cdot(OH)_2]^{x+}[A^{m-}]_{x/m}{}^{x-}$ $2H_2O$, where M^{2+} is a divalent metal ion (such as Mg^{2+} or Zn^{2+}), M^{3+} is a trivalent metal ion (such as Al^{3+} or Cr^{3+}), and A^{m-} is an anion with valency *m* (such as COC_3^{2-}, Cl-, or NO_3^-), and the value of x is equal to molar ratio of $M^{2+}/(M^{2+} + M^{3+})$ and is general in the range of 0.2 to 0.33 [69]. The typical structure of LDH materials is presented in Figure 2.16 [69]. The structure of LDHs is based on that of magnesium hydroxide, brucite, in which Mg^{2+} ions are arranged in sheets, each magnesium ion being octahedrally surrounded by six hydroxide groups, whereas each hydroxide spans three magnesium ions. Isomorphous substitution for some fraction of the divalent ions of trivalent ions of comparable size (e.g., Al^{3+}, Fe^{3+}) forms mixed metal layers – that is, $[M^{2+}{}_{1-x}\ M^{3+}{}_{x}\cdot(OH)_2]^{x+}$ – with a net positive charge. Electrical neutrality is maintained by the anion located in interlayer domains containing water molecules. These water molecules are connected to both the metal hydroxide layers and the interlay anions through extensive hydrogen bonding. The presence of anion and water molecules leads to enlargement of the basal spacing from 0.48 nm in brucite to about 0.78 nm in Mg-Al LDH. Because the interlayer ions are confined to the interlayer space by a relatively weak electrostatic force, they can be removed without destroying the layered structures of LDH. This anionic exchange capacity is a key feature of LDHs, which make them unique as far as inorganic materials are concerned.

An interesting property of LDHs is that the identity of both the divalent and trivalent metal ions and their ratio, as well as the identity of the interlayer anion, are adjustable resulting in varying charge densities and interlayer spacing. This makes a great variety of possible LDHs. The metals include lithium, magnesium, zinc, copper, cobalt, calcium, iron, aluminum, and a host of others. A wide variety of anions may also be used, including both inorganic and organic anions. There is evidence that both the identity of the metals and the anions are important to obtain good dispersion. The ratios of M^{2+} to M^{3+} may range from 2 to a maximum of 5 or 6, but the most common stoichiometries are Mg_2Al and Mg_3Al.

Properties and Applications. Due to their unique layered structure and highly tunable chemical composition based on different metal species and interlayer anions, LDHs have many interesting properties, such as unique anion-exchange ability, easy synthesis, high bond water content, "memory effect," nontoxicity, and biocompatibility. Based on these properties, LDHs are considered as very important layered crystals with potential applications in catalysis [71], controlled drugs release [72], gene therapy [73], improvement of heat stability and flame retardancy of polymer nanocomposites [69, 70], controlled release or adsorption of pesticides [74], and preparation of novel hybrid materials for specific applications, such as visible luminescence [75], UV/photo stabilization [76], magnetic nanoparticle synthesis [77], or wastewater treatment [78].

High charge density, small intergallery distance, and strong hydrophilicity prevent good dispersion of LDH at the nanometer level within the polymer matrix. Monomer and polymer molecules cannot easily penetrate between the LDH layers, nor can the layers be easily and homogeneously dispersed in a hydrophobic polymer matrix. By modifying LDH with organic anions, the intergallery distance is increased and the layers become more organophillic, so monomer or polymer molecules can more easily penetrate between the layers, making LDHs suitable for use as nanofillers. Even after organic modification, the charge density is still high and dispersion can be difficult. The identity of the metals and the anions controls dispersion. Montmorillonite (MMT) has been studied much more than the LDHs have been, and it is clear that the fire retardancy due to MMT is quite different than that from LDHs. For MMT, there is no reduction in the peak heat release rate (PHRR) unless the clay is well dispersed [70]. For an LDH-containing nanocomposite, there can be reduction even if the LDH is poorly dispersed. Also, the amount of MMT that must be used is around 3 wt%, while a much larger amount, about 10 wt%, is typically required for an LDH to lower the PHRR.

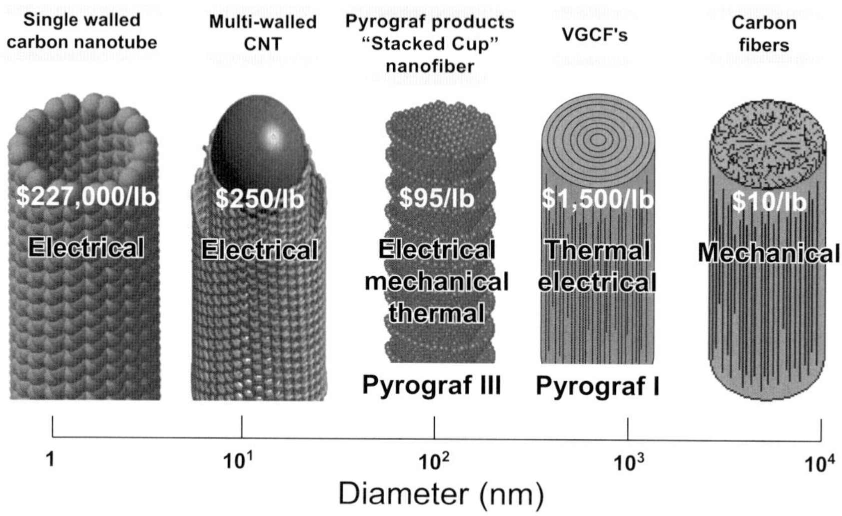

Figure 2.17. Sizes and costs from single-wall carbon nanotubes to conventional carbon fibers (Courtesy of ASI).

2.2.2 Two Nanoscale Dimensions in the Form of Fibers

2.2.2.1 Carbon Nanofibers (CNFs) – Vapor-Grown Carbon Fibers (VGCFs)

Origin. Carbon nanofibers (CNFs) are a form of vapor-grown carbon fiber that is a discontinuous graphitic filament produced in the gas phase from the pyrolysis of hydrocarbons [79–84]. With regard to properties of physical size, performance improvement, and product cost, CNFs complete a continuum bounded by carbon black, fullerenes, and single-wall to multiwall carbon nanotubes on one end and continuous carbon fiber on the other [80], as illustrated in Figure 2.17. Carbon nanofibers are able to combine many of the advantages of these other forms of carbon for reinforcement in commodity and high-performance engineering polymers. Carbon nanofibers have transport and mechanical properties that approach the theoretical values of single-crystal graphite, similar to the fullerenes, but they can be made in high volumes at low cost – ultimately lower than that of conventional carbon fibers. In equivalent production volumes, CNFs are projected to have a cost comparable to E-glass on a per-pound basis, yet they possess properties that far exceed those of glass and are equal to, or exceed, those of much more costly commercial carbon fiber. Maruyama and Alam published an excellent review of carbon nanotubes and nanofibers in composite materials [82].

Manufacturing. Carbon nanofibers are manufactured by Applied Sciences Inc./Pyrograf® Products (ASI) (Cedarville, Ohio; www.apsci.com) by pyrolytic decomposition of methane in the presence of iron-based catalyst particles at temperatures above 900°C. Figure 2.18 shows how Pyrograf carbon nanofibers are produced at ASI. Typically, about 100 mg of powdered catalyst is placed in a ceramic boat, which is positioned in a quartz tube located in a horizontal tube furnace. The catalyst is reduced in a dilute hydrogen-helium stream at 600°C and quickly brought to the

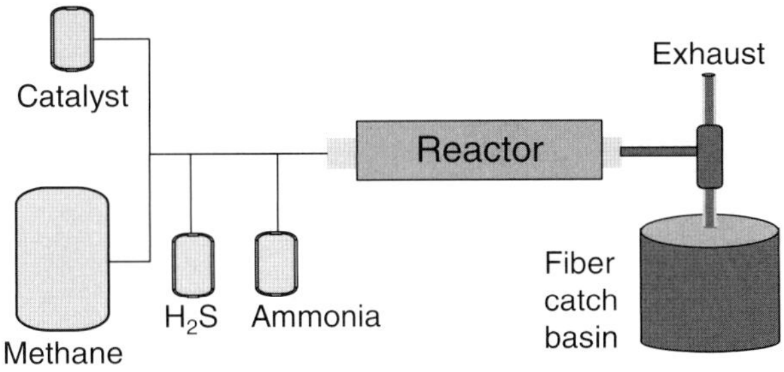

Figure 2.18. Schematic showing the production of raw Pyrograf® carbon nanofibers (Courtesy of ASI).

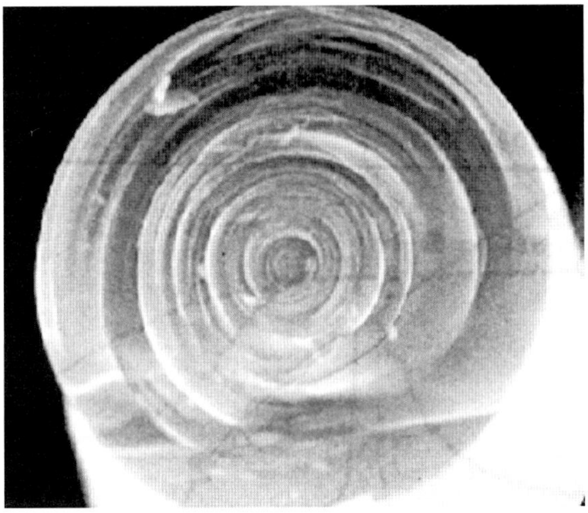

Figure 2.19. TEM micrographs of Pyrograf®-I VGCF (Courtesy of ASI).

desired reaction temperature. Following this step, a mixture of hydrocarbon, hydrogen, and inert gas is introduced into the system, and the reaction is allowed to proceed for about two hours. This approach will produce about 20 g of carbon fibers from the more active catalyst systems.

Properties. Pyrograf-I, an early product, is 10 micrometer in diameter and represents the first generation of products from Applied Sciences, as shown in an SEM micrograph in Figure 2.19. Pyrograf-III is a patented, very fine, highly graphitic CNF. Pyrograf-III is available in diameters ranging from 50 nm to 200 nm and lengths of 50 nm to 100 μm. Therefore, CNFs are much smaller than conventional continuous or milled carbon fibers (5–10 μm) but significantly larger than carbon nanotubes (1–10 nm). Compared to PAN and pitch-based carbon fiber, the morphology of CNFs is unique in that there are far fewer impurities in the filament, providing for graphitic and turbostatic graphite structures, and the graphene planes are more preferentially oriented around the axis of the fiber. Consequences of the circumferential orientation of high-purity graphene planes are a lack of cross-linking between the graphene layers and a relative lack of active sites on the fiber surface, making CNFs more resistant to oxidation and less reactive for bonding to matrix materials. Also, in contrast to carbon fiber derived from PAN or pitch precursors, CNFs are produced only in a discontinuous form, where the length of the fiber can be varied from about

Table 2.6. *Properties of CNF*

Property (Units)	As Grown	Heat Treated
Tensile strength (GPa)	2.7	7.0
Tensile modulus (GPa)	400	600
Ultimate strain (%)	1.5	0.5
Density (g/cc)	1.8	2.1
Electrical resistivity (μ' Ω-cm)	1000	55
Thermal conductivity (W/m-K)	20	1950

Table 2.7. *Thermoset Polyester/Pyrograf®-III Composite Properties [81]*

Fiber Content (wt%)	Tensile Strength (MPa)	Tensile Modulus (GPa)	Electrical Resistivity (ohm-cm)
17% PR-19	51.5	4.55	3.2
17% PR-19, OX	47.4	4.55	7.1
5% PR-19 in 10% ¼" glass	44.1	11.52	5.0
5% PR-19, OX in 10% ¼" glass	33.8	8.92	7.0

100 μm to several centimeters, and the diameter is on the order of 100 nm. As a result, CNFs possess an aspect ratio of about 1,000.

Carbon nanofibers exhibit exceptional mechanical and transport properties, thus demonstrating their excellent potential as an abstract component for engineering materials. Table 2.6 lists the properties of vapor-grown carbon fibers, both as grown and after a graphitizing heat treatment to 3,000°C. Note that, owing to the difficulty of direct measurements on the nanofibers, the values in Table 2.7 are measured on vapor-grown fibers that have been thickened to several microns in diameter. Such fibers consist almost exclusively of chemical vapor deposition (CVD) carbon, which is less graphitic and more defective than the catalytically grown carbon core that constitutes the CNF. Thus, the properties listed in the table represent an estimate for the properties of CNF.

One of the goals for the broad utility of CNFs is to provide mechanical reinforcement comparable to that achieved with continuous tow carbon fiber at a price that approaches that of glass fiber reinforcement, and with low-cost composite fabrication methods, such as injection molding. Theoretical models [81] suggest that reinforcement by discontinuous fibers, such as CNFs can closely approach that of continuous fibers, as long as the aspect ratio of the fibers is high and the alignment is good. Work is ongoing to improve the mechanical benefits of CNFs through fiber surface modification to provide physical or chemical bonding to the matrix. Such modifications have resulted in strength and modulus improvements of four to six times the values of neat resin; however, these values are still a modest fraction of what may be anticipated from idealized fiber-matrix interphase and alignment of the fibers within the matrix. The more immediate opportunities for use in structural composites lie in the prospect for modifying the properties of the matrix material.

Figure 2.20. An SEM micrograph showing Pyrograf®-III carbon nanofibers in bundles (Courtesy of ASI).

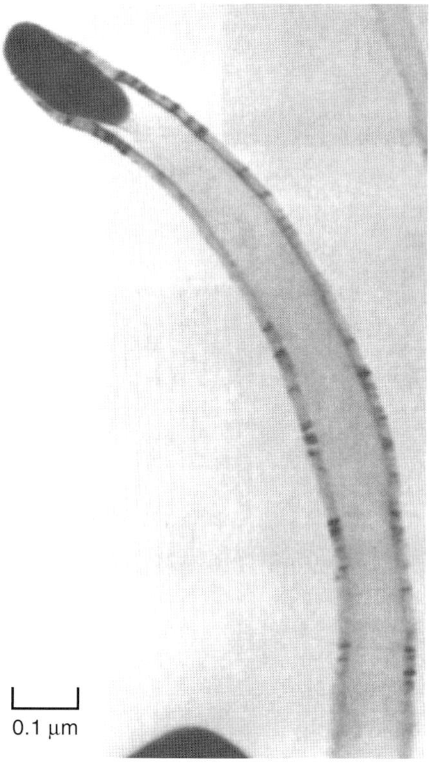

Figure 2.21. TEM micrographs showing single Pyrograf®-III carbon nanofibers (Courtesy of ASI).

For example, use of small-volume loadings of CNFs in epoxy may allow for improvement of interlaminar shear strength of PAN or pitch-based composites. The CNF additives to fiberglass composites could provide benefits to a suite of properties, including thermal and electrical conductivity, coefficient of thermal expansion, and mechanical properties, as suggested by the data in Table 2.4. Figures 2.20 and 2.21 show the SEM micrograph of a CNF bundle and a close-up of Pyrograf-III CNF, respectively. Figure 2.22 shows the TEM micrograph of an individual CNF with the hollow core in two distinct regions, catalytic and deposited.

Table 2.8. *CNF for Structural Applications for Better Strength and Modulus*

Generic Area	Military	Civilian
Synthetic Rubber	Tank tracks, aircraft tires, ship structures	Automotive tires, aircraft tires, ship structures, sporting goods
Thermoplastic Structures	Aircraft, satellite structures, ship structures, optical components	Auto body structures, sporting goods, optical components
Thermoset Structures	Aircraft, satellite structures, ship structures, optical components	Auto body structures, sporting goods, optical components
Carbon/Epoxy Structures	Airframes, especially z-direction enhancement	Sporting goods, automotive, aerospace structures

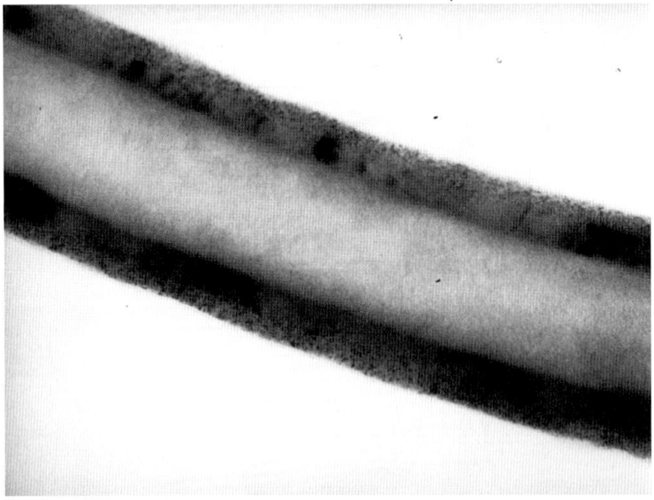

Figure 2.22. A TEM micrograph showing that carbon nanofibers have hollow core and two distinct regions – catalytic and deposited (Courtesy of ASI).

Carbon nanofiber nanocomposites can offer multifunctional performance for several potential aerospace applications, including:

- EMI shielding, electrostatic painting, antistatic;
- thermal conductivity of spacecraft, batteries, and electronics;
- improved mechanical properties in polymers (composite structure, injection-molded parts, tires);
- reduced coefficient of thermal expansion (CTE) for spacecraft resin systems.

Applied Sciences, Inc. has suggested that CNFs could be used for the following applications: (a) improving strength and modulus, as shown in Table 2.8; (b) lowering resistance, as shown in Table 2.9; (c) CTE control, as shown in Table 2.10; and (d) strain-compliant conductors, as shown in Table 2.11.

Glasgow et al. [83] demonstrated that the achievement of significant mechanical reinforcement in CNF composites requires high fiber loadings and is somewhat dependent on generating an appropriate interphase between the CNF and the

Table 2.9. *CNF for Applications for Lower Resistance*

Generic Area	Military	Civilian
Static Dissipation	Satellite charge control, fuel lines, hoses, tubing, ordnance	Fuel hoses and tubing for automotive, electronics assembly
Electrostatic Paint Spray	Aircraft, ground vehicles, ships, other structures	Auto body components, aircraft, other vehicles
Electromagnetic Interference Mitigation	Plane crash reduction, avionics, RF, secure facilities	High-speed computing, communications
Lightning Strike Mitigation	Aircraft (especially composite airframe), ships, radomes, ordnance bunkers	Aircraft, ground structures
Compliant Contacts	Solar cells, electronics SEM	Electronics

Table 2.10. *CNF for Applications for CTE Control*

Generic Area	Military	Civilian
Optics	Low-cost injection moldable mirrors, laser	Low-cost injection moldable optics
Structures	Satellite structures, especially where stability in orbit is important, aircraft, ground vehicles, ships	Auto body components (match aluminum or steel) and for aerospace and ships
Electronics	MCT devices, electronic boards, high-power electronics	Electronic boards, computers, cooling components

Table 2.11. *CNF for Applications for Strain Compliant Conductors*

Generic Area	Military	Civilian
Compliant Contacts	Solar cell contacts for spacecraft, electronic boards	Electronic boards, solar cell contacts, high-temperature lead free-solder alternative
Thermal Control Devices		Temperature actuators, temperature measurement devices

matrix. Novel surface treatments under development have yielded good improvements in the tensile modulus and strength of CNF-reinforced polypropylene. Adding surface functional groups, particularly oxygen groups, has also demonstrated benefits for interphase development. Carboxyl and phenolic groups contributing to a total surface oxygen concentration in the range of 5 to 20 atom percent have been added to CNF used to fabricate epoxy polymer matrix composites, providing improved flexural strength and modulus. The effect of similarly functionalized CNF on bismaleimide (BMI) polymer matrix composites also shows promise. Data for propylene, epoxy, and BMI/CNF reinforced composites indicate that higher fiber volume loadings will find a role in structural composite markets as price and availability improve.

Recently, Tibbets et al. [84] did a comprehensive review on the fabrication and properties of VGCNF/polymer composites. This review summarized the wide variety of composite properties achieved with VGCNF composites. When compounded with thermoplastics or thermosets, VGCNF can more than triple the polymer's tensile strength and modulus. Compressive strength is generally improved by an even larger margin. Preliminary research provides some indication that a practical method may be found to improve the orientation of VGCNF to achieve even greater improvements. The most beneficial application of VGCNF is to add conductivity to polymer matrices. Resistivities of below 0.15 Ω cm may be achieved with a fiber loading near 15 wt%, and a percolation threshold of below 1 wt% is possible. Furthermore, a tenfold improvement in thermal conductivity has been demonstrated in epoxy composites.

CNFs Suppliers. Carbon nanofibers are manufactured by Applied Sciences Inc./Pyrograf® Products (ASI) (Cedarville, Ohio; www.apsci.com).

2.2.2.2 Carbon Nanotubes (CNTs)

Origin. There are numerous excellent reviews and publications on the synthesis and the physical properties of carbon nanotubes (CNTs) [85–89]. In this section, only a brief introduction of the manufacturing, properties, and classification of carbon nanotubes is presented. Carbon nanotubes have attracted much attention in the last decade because of their unique potential uses to improve structural, electrical, and mechanical properties [90]. Carbon nanotubes have high Young's modulus and tensile strength, and they can be metallic, semiconducting, or semimetallic, depending on the helicity and diameter [90]; some typical material properties of CNTs, CNFs, and NGPs are compared in Table 2.4.

Manufacturing. Carbon nanotubes can be prepared by arc evaporation [91], laser ablation [92], pyrolysis [93], PECVD [94], and electrochemical methods [95, 96]. Carbon nanotubes were first synthesized in 1991 by Sumio Iijima in the carbon cathode by arc discharge [97]. However, the experimental discovery of single-wall carbon nanotubes came in 1993 [98, 99]. In 1996, a much more efficient synthesis route was developed, involving laser vaporization of graphite to prepare arrays on the order of single-wall carbon nanotubes [100]. The process offered major new opportunities for quantitative experimental studies of carbon nanotubes.

Properties. Properties of carbon nanotubes have been studied extensively. Carbon nanotubes are excellent candidates for stiff and robust structures, because the carbon-carbon bond in the graphite is one of the strongest in nature. Transmission electron microscopy data revealed that carbon nanotubes are flexible and do not break upon bending [101]. Thermal conductivity of carbon nanotubes can be extremely high, and the thermal conductivity of individual carbon nanotubes was found to be much higher than that of graphite and bulk nanotubes [102]. The mechanical (Table 2.4) and transport (Table 2.4) properties of CNTs are compared to other conductive materials. Overall, carbon nanotubes show a unique combination of stiffness, strength, and tenacity compared to other fibers that usually lack one of more of

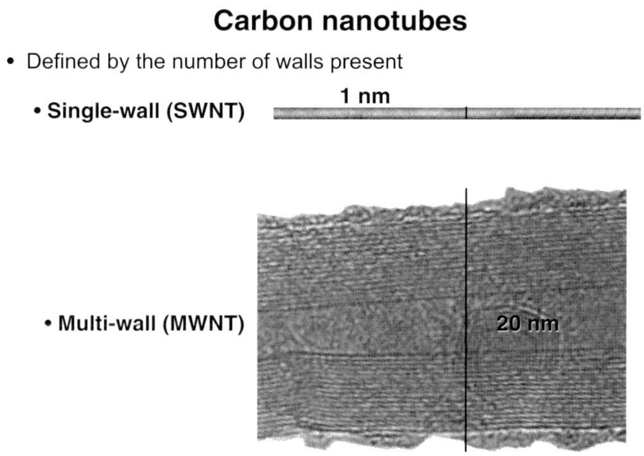

Figure 2.23. Definition of single- and multi-wall carbon nanotubes (Courtesy of CNI).

these properties (Table 2.4). Thermal and electrical conductivities are also very high and are comparable to other conductive materials (Table 2.4). Carbon nanotubes have a wide spectrum of potential applications. Examples include use in catalysis, storage of hydrogen and other gases, biological cell electrodes, quantum resistors, nanoscale electronic and mechanical devices, electron field emission tips, scanning probe tips, flow sensors, and nanocomposites [103].

Classifications. Carbon nanotubes have many structures, different in length types of helicity, and the number of walls or layers. Carbon nanotubes can be categorized by their structures as single-wall nanotubes (SWNT), multi-wall nanotubes (MWNT), and the newly established double-wall nanotubes (DWNT), as illustrated in Figure 2.23. By definition, SWNT are single-walled carbon nanotubes about 1 nm in diameter with micrometer-scale lengths; MWNT are multi-walled carbon nanotubes with an inner diameter of about 2–10 nm, an outer diameter of 20–70 nm, and a length of about 50 microns; and DWNT are double-walled carbon nanotubes with outer diameters of less than 2 nm and length ranging from several hundred nanometers to several microns.

Single-wall Carbon Nanotubes. Single-wall carbon nanotubes (SWNT) are tubes of graphite that are normally capped at the ends. They have a single cylindrical wall. The structure of SWNT can be visualized as a layer of graphite, a single atom thick, called graphene, which is rolled into a seamless cylinder. Most SWNT have a diameter close to 1 nm; the tube length can be many thousands of times that. Single-wall carbon nanotubes are more pliable yet harder to make than multi-wall carbon nanotubes are. They can be twisted and bent into small circles or around sharp-angle bends without breaking.

Buckytubes are tubular fullerenes, polymers that are part of the fullerene family of carbon molecules discovered by Dr. Richard E. Smalley and colleagues in 1985. Buckytubes comprise single-wall carbon nanotubes and nested (endohedral or endootopic) SWNTs (i.e., one, two, or more tubular fullerenes nested inside another tubular fullerene, as depicted in Figures 2.24 and 2.25).

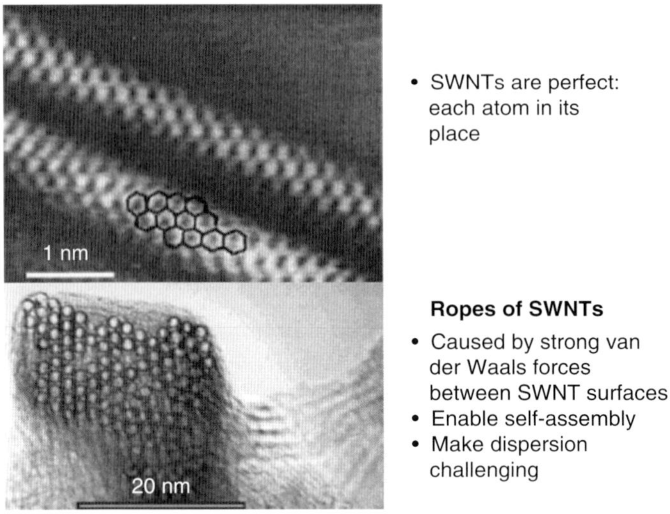

Figure 2.24. TEM micrographs showing SWNTs and ropes of SWNTs (Courtesy of CNI).

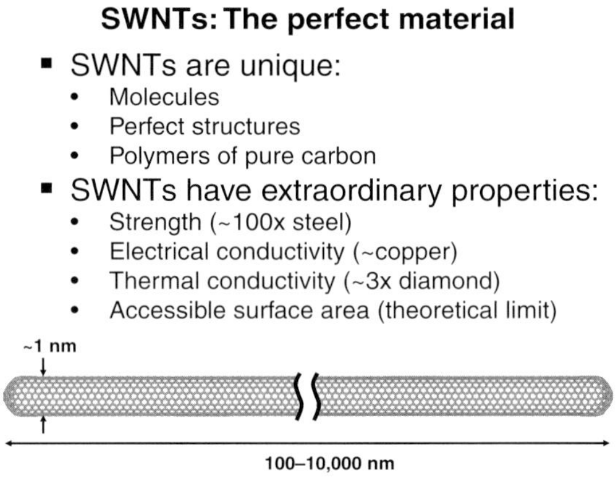

Figure 2.25. SWNTs are the perfect materials with unique and extraordinary properties (Courtesy of CNI).

Buckytubes are carbon nanotubes that are graphene (layers of graphite) rolled up into seamless tubes. Graphene consists of a hexagonal structure similar to chicken wire. Rolling up graphene into seamless tubes can be accomplished in various ways. For example, carbon-carbon bonds (like the wires in chicken wire) can be parallel or perpendicular to the tube axis, resulting in a tube where the hexagons circle the tube like a belt, but oriented differently. Alternatively, the carbon-carbon bond need not be either parallel or perpendicular, in which case the hexagons will spiral around the tube with a pitch, depending on how the tube is wrapped. Buckytubes have extraordinary electrical conductivity, thermal conductivity, and mechanical properties. They are the best electron field emitter. As polymers of pure carbon, they can be reacted and manipulated using the chemistry of carbon. This characteristic provides the opportunity to modify the structure and to optimize solubility and dispersion. Buckytubes (Figure 2.26) are molecularly perfect, which means that they are free of property-degrading flaws in the nanotube structure. Their material properties can

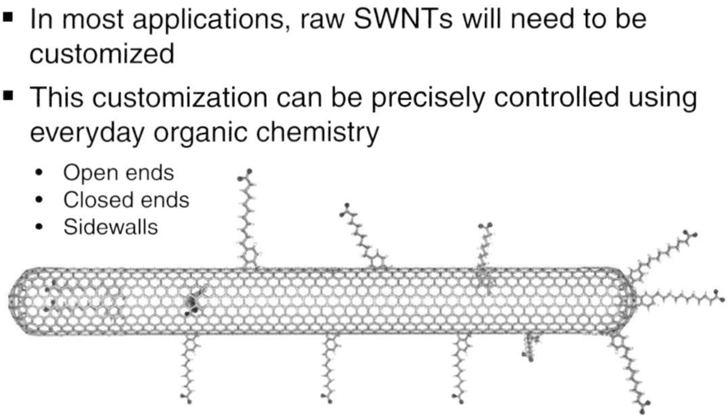

Figure 2.26. SWNTs can be customized and controlled using organic chemistry (Courtesy of CNI).

therefore approach closely the very high levels intrinsic to them. Buckytubes have extraordinary properties: their strength is about 100 times that of steel, their electrical conductivity is close to that of copper, their thermal conductivity is about 3 times their diameter, and they have tremendous accessible surface area. These extraordinary characteristics give buckytubes potential in numerous applications. In most applications, raw buckytubes will need to be customized. The customization can be precisely controlled using organic chemistry in open ends, closed ends, and sidewalls, as illustrated in Figure 2.26.

Single-wall carbon nanotubes offer incredible opportunities in electrical properties, mechanical properties, thermal properties, and field emission, as shown below:

- *Electrical properties*. Electrically conductive composites for electrostatic dissipation, shield, and conductive sealants; energy storage for super capacitors and fuel cells; electronic materials and devices for conductive inks and adhesives; electronic packaging; device and microcircuit components.
- *Mechanical properties*. High-performance composites; coatings for wear-resistance and low friction; high performance fibers; reinforced ceramic composites.
- *Thermal properties*. Thermally conductive polymer composites; thermally conductive paints and coatings.
- *Field emission*. Flat-panel displays; electron device cathodes and lighting.

Multi-wall Carbon Nanotubes. Multi-wall carbon nanotubes (MWNTs) can appear either in the form of a coaxial assembly of single-wall carbon nanotube similar to a coaxial cable, or as a single sheet of graphite rolled into the shape of a scroll. MWNTs have an interior diameter of 2–10 nm, an exterior diameter of 20–75 nm, and a length of 50 microns, as shown in Figure 2.27 [104]. They are produced by chemical vapor deposition (CVD) synthesis of xylene-ferrocene at a temperature of 725°C with high purity > 95 percent (Figure 2.28). Multi-wall carbon nanotubes are easier to produce in high volume than single-wall carbon nanotubes are. MWNTs are nanoscale

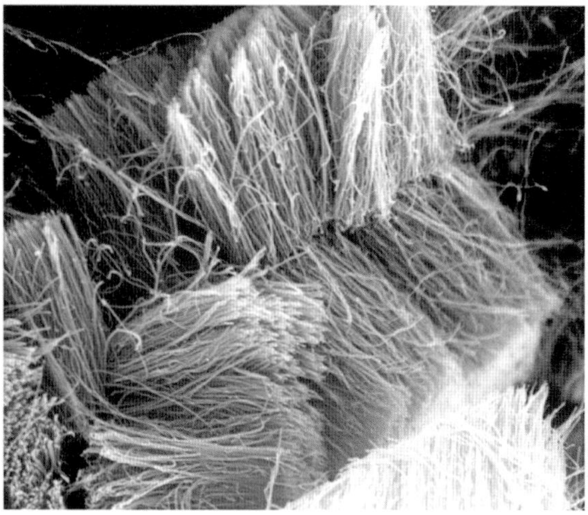

Figure 2.27. MWNTs have dimensions of 2–10 nm ID, 20–75 OD, and 50 microns long.

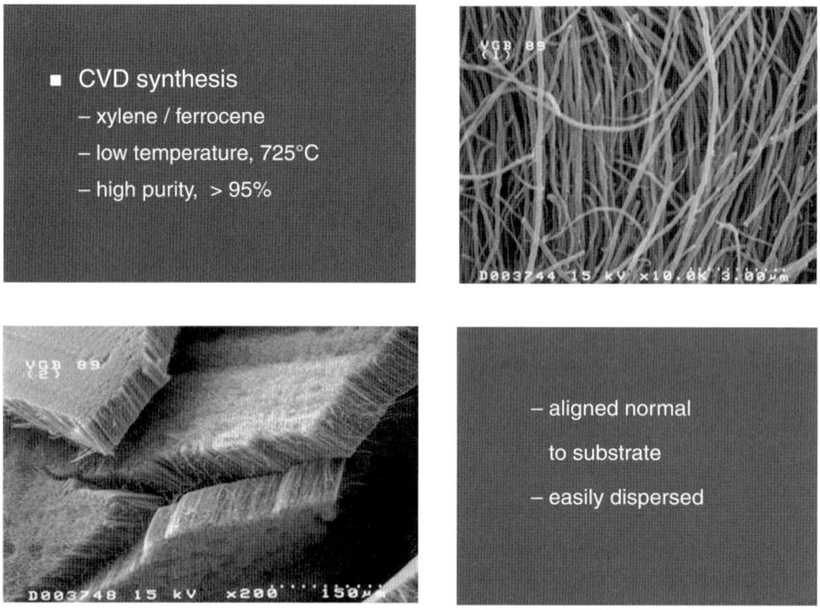

Figure 2.28. MWNTs are produced using CVD synthesis.

carbon fibers with a high degree of graphitization. Technically they are neither fullerenes nor molecular. The structure of MWNTs is less well understood because of its greater complexity and variety. Regions of structural imperfection may diminish its desirable material properties. The challenge in producing single-wall carbon nanotubes on a large scale as compared to multi-wall carbon nanotubes is reflected in the manufacturing prices of SWNTs, which currently remain significantly higher than that of MWNTs. They are very good materials, with vastly superior properties over graphite and carbon black (Figures 2.29 and 2.30) [104]. Single-wall carbon nanotubes have a performance of up to ten times better, and are outstanding for every specific application.

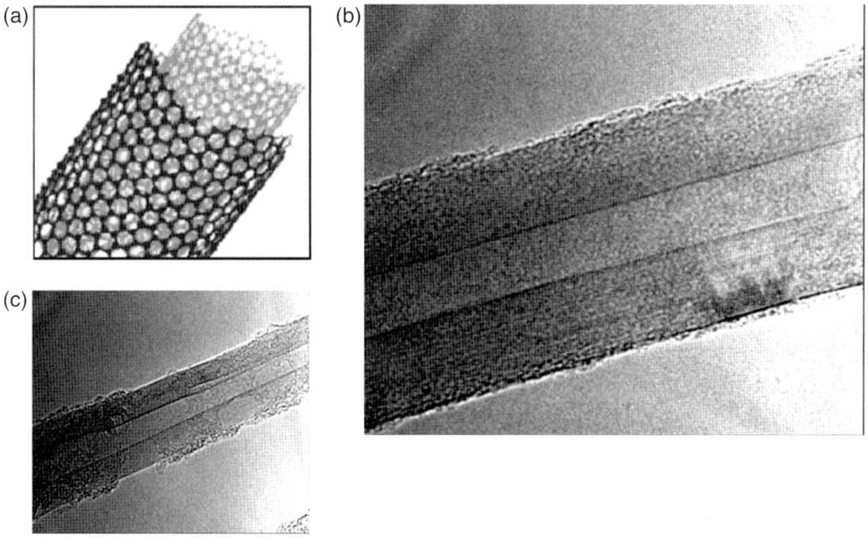

Figure 2.29. Microstructures of MWNTs in cartoon and TEM micrographs.

Double-wall Carbon Nanotubes. Double-wall carbon nanotubes (DWNTs) are an important sub-segment of multi-wall carbon nanotubes. DWNTs combine similar morphology and other material properties of single-wall carbon nanotubes while significantly improving their resistance to chemicals. This property is especially important when functionality is required to add new properties to DWNTs. Since double-wall carbon nanotubes are a synthetic blend of single-wall and multi-wall carbon nanotubes, which exhibit the electrical and thermal stability of a MWNT and the flexibility of a SWNT. Since the SWNT are developed for specific applications, they are more susceptible to breakage if they have been functionalized. When structural imperfections are created, they can modify their overall mechanical and electrical properties. With the DWNTs, only the outer wall is modified, thereby preserving their intrinsic properties. Research has shown that DWNTs have better thermal and chemical stability than SWNTs. As a result, a double-wall carbon nanotube can be used for gas sensors and dielectrics, as well as for demanding applications, such as field emission displays, polymer nanocomposites, and nanosensors.

Carbon Nanotubes Suppliers. There are numerous carbon nanotubes suppliers worldwide, and a few major ones are included in this section.

Nanocyl s.a. is a leading global manufacturer of specialty and industrial carbon nanotubes technologies, located in Sambreville, Belgium (www.nanocyl.com) [105]. It was founded as a spin-off from the Universities of Namur and Liege by Professor Nagy and Professor Pirard with the support of private investors in 2002. Nanocyl uses the catalytic carbon vapor deposition (CCVD) method for large-scale production of carbon nanotubes. Nanocyl focuses its technology and knowledge on the practical application of carbon nanotubes to new and existing materials, resulting in the improved performance of polymers, metals, composites, and biomaterials. Currently Nanocyl produces SWNTs, MWNTs, and DWNTs as well as masterbatches of

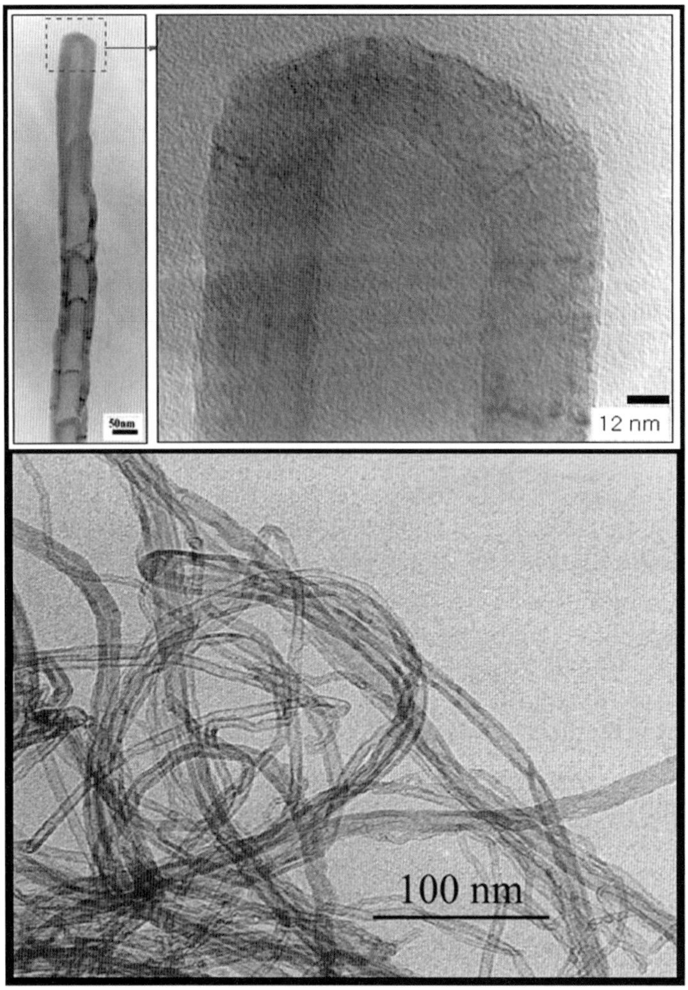

Figure 2.30. TEM micrographs showing MWNTs in bundles, as individual nanotube, and in high magnification (Courtesy of CNI).

polymer nanocomposites. Nanocyl manufactures products, such as PLASTICYL™, EPOCYL™, AQUACYL™, THERMOCYL™, BIOTYL™, PREGCYL™, and STATICYL™ that facilitate the integration of Nanocyl™ NC7000 in their customers' applications. The company also offers other high-quality carbon nanotube research grades and technologies.

Nanocyl NC7000 series are thin multi-wall carbon nanotubes manufactured via CCVD process. They are available in powder form, and each nanotube has an average diameter about 9.5 nm and an average length of 1.5 μm, 90% carbon purity, and 10% metal oxide. This MWNT product has also been pre-dispersed in other products, such as masterbatches (PLASTICYL™), epoxy resins (EPOCYL™), silicones (THERMOCYL™, BIOTYL™, and STATICYL™), and aqueous dispersions (AQUACYL™).

Bayer MaterialScience is marketing the nano-sized materials under the trade name Baytubes® (www.baytubes.com) [106]. Bayer MaterialScience set up a second pilot production plant with an annual capacity of 30 tons, expanding the capacity to 60 tons in 2007, which made Bayer MaterialScience one of the top three carbon nanotubes producers worldwide. Baytubes® differ from other natural or engineered nanotubes in many ways. They are of high purity (>95% standard grade; >99% can be achieved in high purity variations) and show no other residue than from the catalyst and catalyst carrier. In particular, TEM analysis indicates the absence of carbon soot in the product. They show high bulk density consistent with the large agglomerate structure seen even with the naked eye, without using electron microscopy. The entangled nature of the Baytubes® can only be observed with electron microscopy. The size of the agglomerates (>100μm) in Baytubes® facilitates the handling at the workplace by reducing the potential of exposure via inhalation.

Baytubes® are extremely resilient. They might only have a quarter of the mass of steel, but they are five times stronger when subjected to mechanical loads. They also conduct electric current as effectively as copper does. The tiny tubes can be used to produce extremely robust plastics that are constructed in the same way as reinforced concrete is. Baytubes® are already used in epoxy systems, thermoplastic systems, and coating systems. Final products are typically high-tech sport equipment, injection-molded and extruded parts, as well as conductive coatings. An excellent example is the current use of Baytubes® in skis, ski poles, hockey sticks, baseball bats, and surfboards. Baytubes® help improve the mechanical strength of some of those articles as well as reduce weight of some others. Baytubes® can be used to make rotor blades for wind turbine plants lighter, longer, and thus more efficient. In the automotive industry, the use of Baytubes® in plastics can avoid the need for expensive production steps. These examples will open the door to more demanding applications in energy and aircraft industry.

Baytubes® are agglomerates of multi-wall carbon nanotubes with low outer diameter, narrow diameter distribution, and an ultra-high aspect ratio (length-to-diameter ratio). Baytubes® show excellent tensile strength and E-modulus, as well as exceptional thermal and electrical conductivity. Baytubes® are produced via a high-yield catalytic process based on chemical vapor deposition. The process yields easy-to-handle agglomerates with high bulk density. The optimized process results in a high degree of purity (low concentration of residual catalyst and absence of free amorphous carbon). Baytubes C 150P is carbon purity > 95 wt%, an outer mean diameter of ~ 13 nm, an inner mean diameter of ~ 4 nm, and length is > 1 um. It has a bulk density of 130–150 kg/m^3. Figure 2.31 shows some representative SEM micrographs of Baytube C150P MWNT in progressive magnifications from (a) to (d) [106].

Arkema. In 2003, Arkema launched a research project to study CNTs and their applications. In 2006, Arkema started up at its Lacq facility the first pilot laboratory capable of producing some 20 tons of CNTs annually, sold under the trademark Graphistrength® (www.graphistrength.com) [107].

Arkema has developed in its laboratories and in collaboration with universities a continuous process for the synthesis of multi-wall carbon nanotubes for the

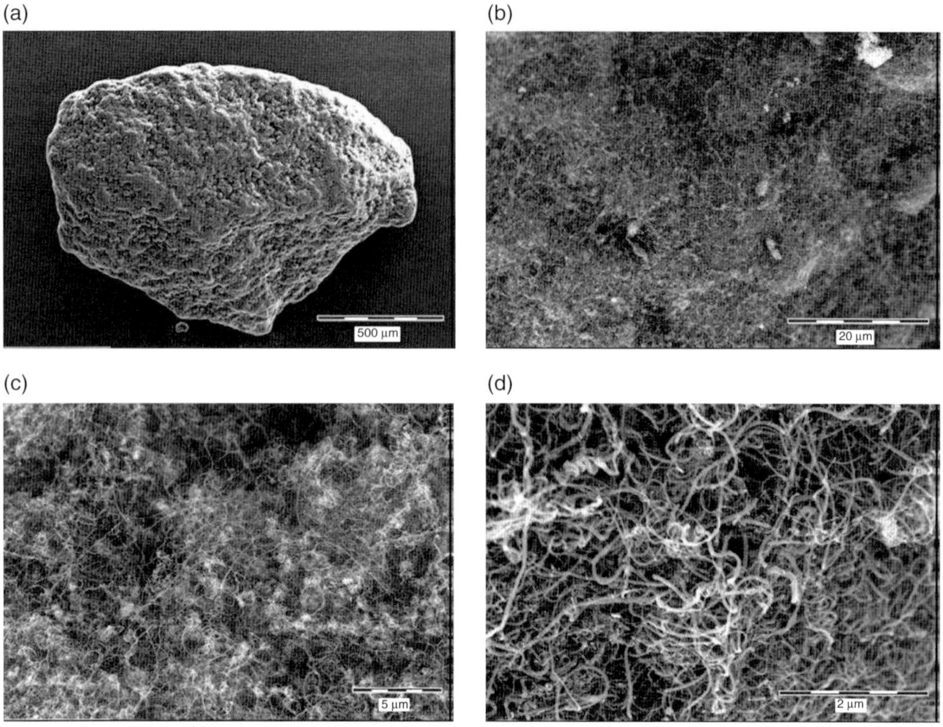

Figure 2.31. TEM micrographs of typical Baytube C150P MWNT in progressive magnifications from (a) to (d) (Courtesy of Bayer MaterialScience).

large-scale manufacture of products with good quality, consistency, and perfect reproducibility. Arkema uses a continuous synthesis process based on catalytic chemical vapor deposition (CCVD) in a fluidized bed–type reactor. Nanotubes grow on iron particles. This produces a black powder with characteristics that are perfectly controlled and stable over time.

Graphistrength® C100 MWNT is a black powder with an apparent density of 50–150 kg/m^3, mean agglomerate size of 200–500 µm, C content > 90wt%, mean number of walls of 5 to 15, an outer mean diameter of 10–15 nm, and lengths about 0.1–10µm. Arkema in particular developed a range of innovative masterbatches that are easy to process within various thermoplastic, elastomer, and thermoset matrices. These high-tech masterbatches help optimize the application properties of end-products.

Arkema is currently testing on a pilot scale for the use of agricultural ethanol as a source of carbon for the production of its Graphistrength® multi-wall carbon nanotubes. Future production will be based on this renewable raw material.

2.2.2.3 Halloysite Nanotubes (HNTs)

Origin. Halloysite nanotubes (HNTs) are types of naturally occurring clay mineral with nanotubular structures, and they have attracted a lot of attention recently. Du et al. [108], did an excellent review on newly emerging applications of halloysite nanotubes. The authors [108] discussed due to the unique characteristics, such as nanosized lumens, high L/D ratio, low hydroxyl group density on the surface of HNTs, exciting applications have been discovered for these cheap and abundantly deposited clays. Holloysite nanotubes have been widely used in controlled or sustained release, in nanoreactors or nanotemplates, and for the elimination of contaminants or pollutants. HNT-polymer nanocomposites exhibit improved properties, such as superior mechanical, flame retardant, thermal stability, and reduced CTE properties. Liu et al. [109] provided an overview of the recent advances in HNT-polymer nanocomposite research, paying particular attention to interfacial interactions of the HNT-polymer nanocomposites. The authors [109] summarized the characteristics of HNT-polymer nanocomposites and predict the development of the potential applications in high-performance composites for aircraft and automobile industries, environment protection, and biomaterials. Yuan et al. [110], reviewed the recent research advances on the properties and applications of HNTs, paying attention to the structures and morphology of HNTs and their changes, the formation of tubular structures, the physicochemical properties, and surface chemical modifications, and the HNT-based advanced materials, and related applications. Zhang et al. [111] reviewed applications and interfaces of halloysite-polymer nanocomposites. Zhang et al. [111] reviewed the recent development in functionalized HNTs with meatal nanoparticles and the diverse application of the resulting materials. The interfacial characteristics between HNTs and nanoparticles ere emphasized by combining HNTs' advantages like nature, biocompatibility, nontoxicity, which allow them to be expanded to applications, such as drug delivery system, multi-phase nanocomposites in the near future. The above four review papers form the basis of this section.

HNTs were first reported by Berthier in 1826 [112] as a dioctahedral 1:1 clay mineral of the kaolin group, and are found deposited in soils worldwide [108-119]. Many countries, such as China [116], France [117], Belgium [118], and New Zealand, [119] have discovered HNT deposits on their respective territory.

Based on the state of hydration, HNTs are generally classified into two groups: hydrated HNTs with a crystalline structure of 10 Å d_{001} spacing, and dehydrated HNTs with 7 Å d_{001} spacing. Figure 2.32 shows a typical crystalline structure of HNTs. The presence or history of interlayer water in HNTs is one of the most important features to distinguish HNTs from kaolinite [114]. The d_{001} spacing of HNTs changes from 10 to 7 Å through an irreversible dehydration process. HNTs have a similar chemical composition to kaolinite, and the ideal chemical formula can be expressed as $Al_2Si_2O_5(OH)_4.nH_2O$, where n equals 2 and 0, representing the hydrated and dehydrated states, respectively [120, 121].

Due to the different crystallization conditions and geological occurrence, HNTs adopt different morphologies, such as tubular, spheroidal, and plate-like particles. The tubular structure is the most common and useful [114]. Typically, the inner and outer diameters of HNTs are 1–30 nm and 30–50 nm, respectively, and their lengths are 100–2,000 nm [122-124]. Some typical morphologies of HNTs are shown in Figure 2.33 [123] in SEM of the entire view and Figure 2.34 [123] in TEM of an HNT. HNTs contain two types of hydroxyl groups, inner and outer, which are situated

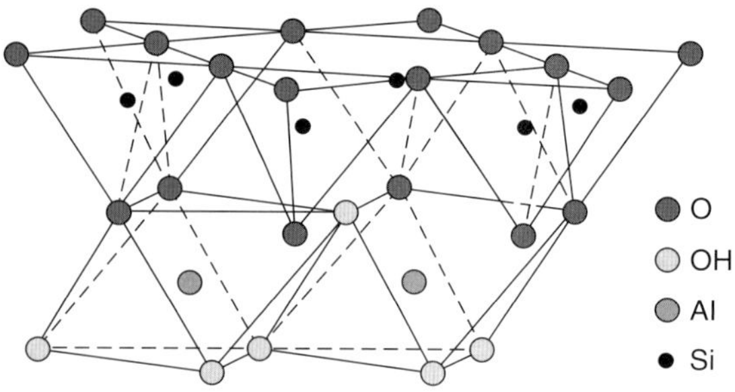

Figure 2.32. Crystalline structure of HNTs.

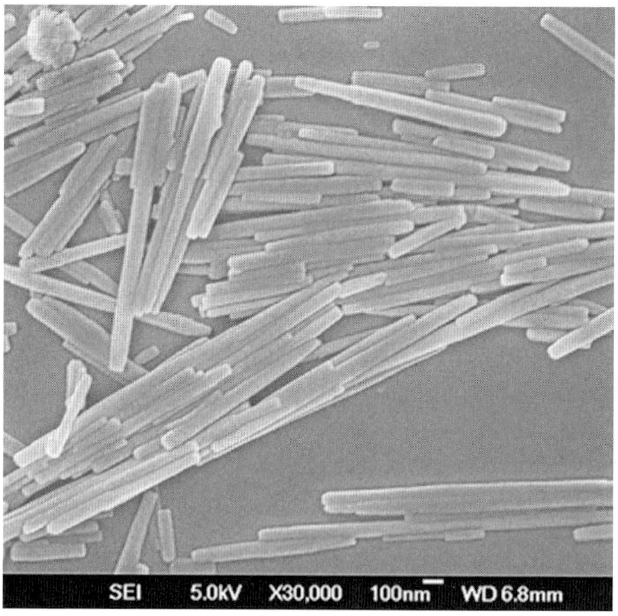

Figure 2.33. Morphology of HNTs as shown by TEM [123].

between layers on the surface of the HTNs, respectively. Due to the multi-layer structure, most of the hydroxyl groups are inner groups and only a few hydroxyl groups are located on the surface of the HNTs. The surface of HNTs is mainly composed of O-Si-O groups, and the siloxane surface can be confirmed from Fourier transform infrared spectra. As a result, when compared with other silicates, such as kaolinite and montmorillonite, the density of surface hydroxyl groups of HNTs is smaller.

Properties. HNTs have been widely used in controlled or sustained release, in nanoreators or nanotemplates, and for the elimination of contaminants or pollutants [108]. Due to their characteristics, such as nanoscale lumens, high aspect ratio, and low hydroxyl group density on the surface, HNTs have been the focus for numerous researchers. Traditionally, due to the high aspect ratio and superior high-temperature resistance, HNTs were used to produce high-quality ceramics, such as thin-walled porcelain or crucible products, which have been the main applications of HNTs [125, 126]. Recently, scientists and engineers have discovered and developed a wide range of exciting new applications for this unique, cheap, and abundantly available naturally occurring clay with nanoscale lumens. Due to HNTs characteristics, it

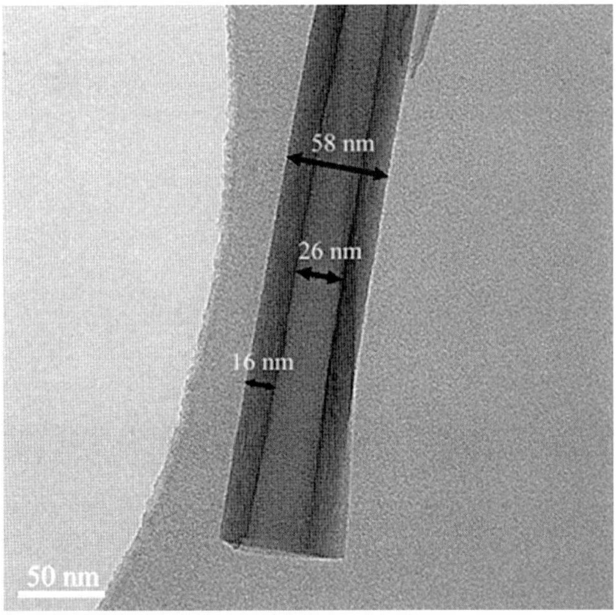

Figure 2.34. Morphology of HNTs as shown by SEM 123].

is expected that HNTs will prove promising as reinforcing filler for polymer materials. Recently, an increasing number of studies have focused on the fabrication of polymer-HNT nanocomposites and their properties [108-111].

Fabrication of Polymer-HNT Nanocomposites. For *thermoplastics*, melt blending is an industrially friendly method to fabricate their composites with inorganics. Due to the ease of dispersion of HNTs, they can be dispersed relatively uniformly in thermoplastics by direct melt blending, especially for polymers with high polarity, such as polyamides. Some polymer-HNT nanocomposites, such as polyamide 6 (PA6), PP, and poly(butylene terephthalate), have been successfully manufactured via the melt-blending process without any surface modification of HNTs [108-111]. Compatibility between polyolefins and inorganics is challenging due to the great polarity discrepancy and the chemical inertness of polyolefins. To increase the compatibility between PP and HNTs, Du et al. proposed a two-step method of grafting a PP chain onto the HNTs surface, which is shown in Figure 2.35 [108, 127]. Their work effectively increases the mechanical performance of the PP-HNT nanocomposite, and this approach may be used for other polymers. Other methods, such as hydrogen bonding and charge transfer mechanisms, have also been proposed for the fabrications of PP-HNT nanocomposites [108]. Due to their fewer surface hydroxyl groups, HNTs will expect to disperse better than other silicates, such as montmorillonite and kaolinite. HNTs can be dispersed uniformly by direct melt blending in polyamides with high polarity (Figure 2.36) [108]. Du et al. use hydrogen bonding approach to disperse HNTs in PP. The morphology of the PP-HNT nanocomposites is shown in reference [127].

For *rubber/clay nanocomposites*, a good dispersion of the clay and strong interfacial interactions are two crucial factors in determining the performance of the

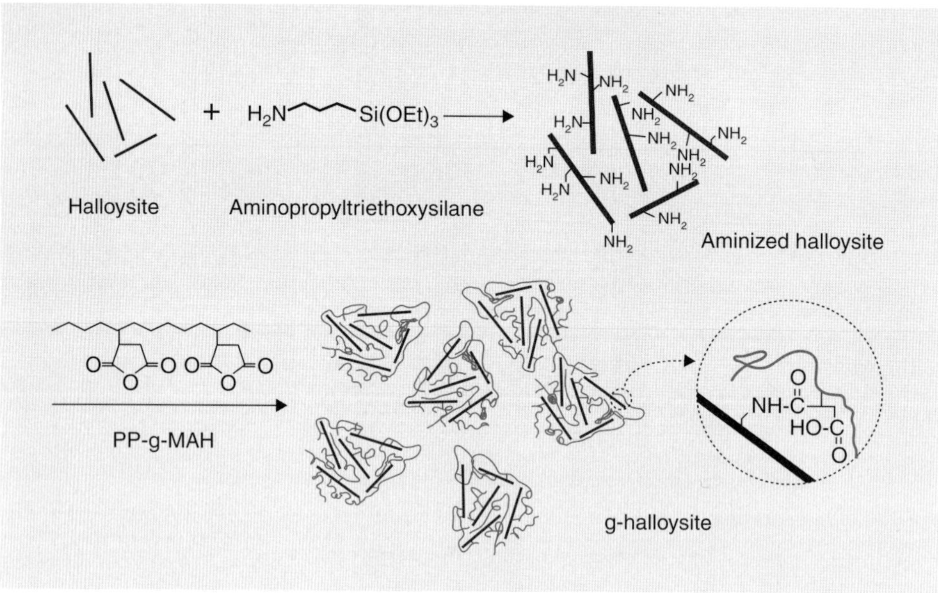

Figure 2.35. Surface grafting of HNTs [108].

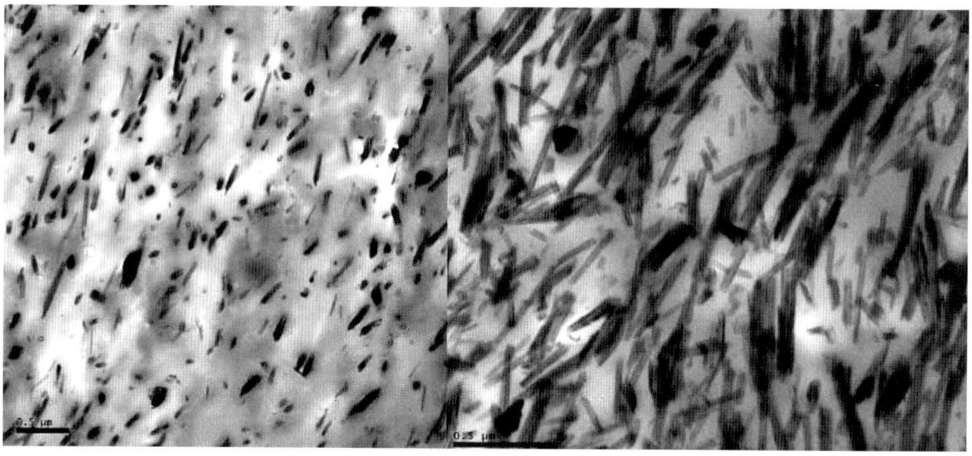

Figure 2.36. TEM images of PA6/HNT (100/10) nanocomposite (left) and SBR/HNT/MAA (100/40/2) nanocomposite (right) [108].

nanocomposites [128]. Guo et al. utilized methacrylic acid (MMA) to improve the performance of styrene-butadiene rubber SBR-HNT nanocomposites by direct blending [129]. Carboxylated butadiene-styrene rubber (xSBR) is a copolymer of styrene, butadiene, and a small amount of acrylic acid. Due to the presence of carboxyl groups, it is expected that HNTs will have good compatibility with xSBR. HNTs and xSBR have been used to prepare nanocomposites with strong interfacial interaction via hydrogen bonding [130]. To obtain good dispersion of HNTs in xSBR, a co-coagulation process of xSBR latex and HNTs aqueous solution was used (Figure 2.37) [108]. Hybrids containing up to 65 wt% HNTs were fabricated in an

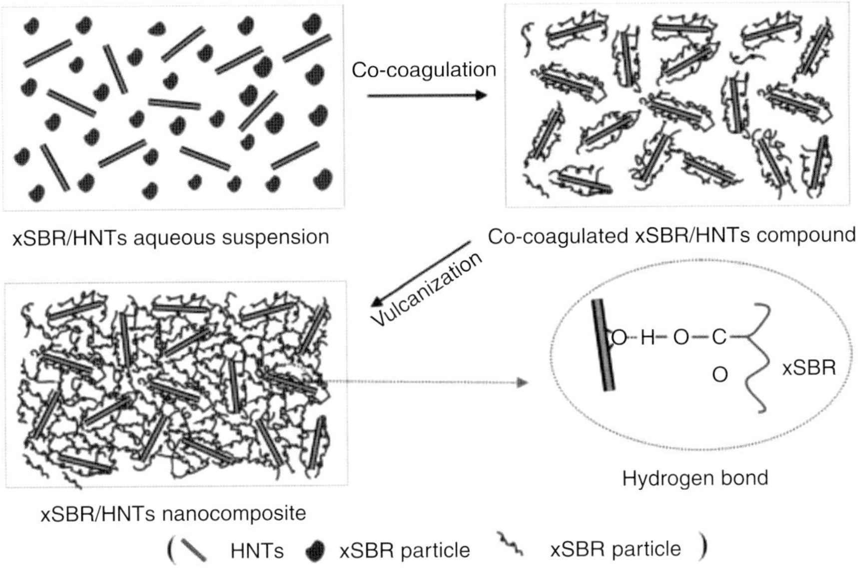

Figure 2.37. Preparation process of xSBR/HNT nanocomposites [108].

SBR matrix with the in situ formation of zinc diborbate by Guo et al. [131]. Apart from the mechanical strength and heat resistance comparable to general engineering plastics, the hybrids exhibits every good transparency, which is attributed to the excellent dispersion of nanotubular clay, as shown in Figure 2.36 [108].

For *thermosetting resins*, which are traditionally processed by casting or potting, it is rather difficult to achieve a uniform dispersion of HNTs. Liu et al. [124, 132] fabricated epoxy/HNT nanocomposites with good dispersion of HNTs via co-curing with cyanate ester resin and surface modification of HNTs.

Features. Incorporating HNTs into polymers will provide the following properties enhancements:

- *Mechanical performance.* Mechanical properties, especially the modulus of PP, can be effectively improved using HNTs. The flexural properties, tensile strength, and impact strength all are noticeably improved with the addition of HNTs [108, 109].
- *Thermal stability and flame retardancy.* Apart from acting as reinforcements, HNTs have also been demonstrated to be effective flame retardant additives for PP, as reported by Du et al. [133]. The thermal stability and flame retardant effects of HNTs on PP are credited to heat and mass transport and the presence of iron in the HNTs. This suggests that HNTs are promising as non-halogen flame-retardant fillers for other polymers [108-110].
- *Crystallization behavior.* Nano-sized inorganic inclusions generally influence the crystallization process of a polymer, such as PP nanocomposites [134, 135]. Ning et al. [136] and Du et al. [137] found that well-dispersed HNTs in a PP matrix can serve as nucleation agents, resulting in an enhancement of overall crystallization rate. These results suggest that, similar to other nano-sized inorganics, such as

silica and montmorillonite, HNTs serve as nucleation agents and facilitate the crystallization of the nanocomposites [108, 109].

- *Reduced coefficient of thermal expansion.* Introduction of interfacial reactions in thermosetting systems may effectively reduce the thermal expansion of the cured resin. The coefficient of thermal expansion (CTE) of epoxy-HNT hybrids with low HNT concentration is substantially lower than that of the neat cured resin [131]. It is believed that interfacial reactions take place during the curing of the hybrids between the aluminols and silanols and the cyanate ester. Consequently, the covalently linked interface formed is responsible for the improved performance and morphological characteristics of the hybrids [108-110].

In conclusion, due to such characteristics as nano-sized lumens, high L/D ratio, and low hydroxyl group density on the surface, more exciting applications have been discovered for these unique, cheap, and abundant clays. So far, HNTs have been widely used in controlled or sustained release, in nanoreactors or nanotemplates, and for the elimination of contaminants or pollutants. Recently, research has focused on polymer-HNT composites. These investigations suggest that the nanocomposites exhibit markedly improved properties, such as superior mechanical performance, higher thermal stability, better flame retardancy, and reduced CTE. HNTs possess promising prospects in the preparation of new structural and functional materials [108-111].

HNTs Suppliers. There are only a few HNTs nanotubes suppliers worldwide. In the United States, the main supplier is NaturalNano (Rochester, New York; www.naturalnano.com) and Applied Minerals Inc. (New York, New York; www.appliedminerals.com).

2.2.2.4 Nickel Nanostrands (NiNs)

Origin. Nickel nanostrands™ were developed under Phase I of an Air Force Small Business Innovation Research (SBIR) program. Nickel nanostrands are strands of sub-micron-diameter nickel particles linked in chains and are microns to millimeter in length. Figures 2.38 shows the different magnifications of the nickel nanostrands [138]. They are analogous to carbon nanofibers or multi-wall carbon nanotubes but provide the additional properties of nickel, significantly expanding the variety of options available for developing tomorrow's nanostructured technologies. They bring the additional electromagnetic, chemical, catalytical, and metallurgical properties associated with nickel to the nanostructured designer's toolbox. These new materials are manufactured by Conductive Composites (Heber City, Utah; www.conductivecomposites.com) [139]. Conductive Composites focuses on creating materials and technologies that enable electrical conductivity and electromagnetic capabilities in composites and plastics.

Properties. These nickel nanostrands have already demonstrated their utility in creating conductive resins, paints, adhesives, and thermoplastics for a wide range of conductive polymer and conductive composites applications. Conductive Composites created paints with a sheet resistance less than 1 ohm per square and adhesives and thermoplastics with conductivities of 40 S/cm and 150 S/cm, respectively.

Figure 2.38. Images depicting nickel nanostrands in different magnifications). (Courtesy of Conductive Composites)

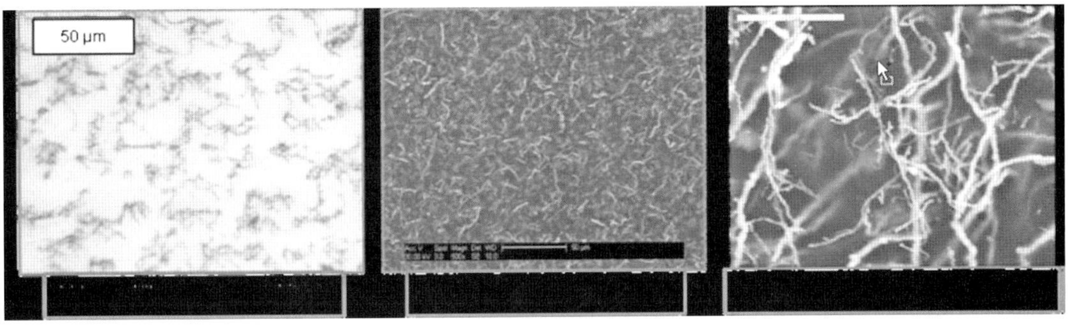

Figure 2.39. Nickel nanostrands of 0.5 vol% in water 500x (left), 5.0 vol% in epoxy 500x (center), and 10 vol% in urethane 2500x (right)). (Courtesy of Conductive Composites)

Nickel nanostrands can also further enhance the conductivity of fiber- or particle-reinforced composites, thus providing a three-dimensional conductive lattice in otherwise insulating polymer resin matrix. A loading of only 2% by volume (2 vol%) in standard carbon prepreg will double the conductivity of a carbon fiber-reinforced composite. Nickel nanostrands were dispersed in different median, such as 0.5 vol% in water 500x (right), 5.0 vol% in epoxy 500x (center), and 10 vol% in urethane 2500x (right) (Figure 2.39) [139]. This is important for the Air Force and industry, because infiltrated carbon composites have proven highly useful in lightning strike protection for aircraft and other composite structures. When nickel nanostrands are added to elastomers, the resulting composite exhibits remarkable changes in conductivity with respect to tensile or compressive strain. Since nanostrands are a magnetic material, they may be magnetically aligned, while the carrier is still in the liquid phase, yielding other unique applications, such as magnetically oriented inks or magnetically aligned conductive fibers. The unique microstructure and chemistry of nickel nanostrands could also lead to important advancements in filtering, catalysis, energy storage, and nanometallurgy. Though nickel nanostrands are similar to carbon nanofibers in terms of size and shape, their anticipated applications are quite different. The proprietary chemistry by which they are produced will no doubt result in a volume production pricing comparable to other nickel commodity powders.

Features and Applications. Nickel nanostrands can be engineered to meet the diameter and length specifications in many fields of sub-micron and nanostructure design. Diameters ranging from 50 nm up to 2 microns have been produced, with aspect ratio generally in the 50:1 to 500:1 range. Currently, Conductive Composites is in the process of producing nanostrands in the range of 10–30 nm in diameter. Further development could lead to advancements in nanotechnology directly benefiting the Air Force, aerospace community, and other industries. In general, nickel nanostrands have the following features:

- Highly branched, three-dimensional nanostructures that offer the ultimate balance of conductivity, electromagnetic, and mechanical properties.
- Can be used as a high-performance additive or as a standalone material.
- Proven performance advantages over carbon- and silver-based technologies.
- Supplied in an as-manufactured continuous form or as a 3D powder.
- Easily dispersed in fluid systems using low shear techniques.
- Mixing and dispersion guide is available.

Present and potential applications and markets of nickel nanostrands are summarized as follows:

• Paints	• Filtering
• Thermoplastics	• Catalysis
• Elastomers	• Thermosets
• Adhesives	• Coatings
• Prepregs	• Coatings
• Energy storage	

Currently, Conductive Composites offers three different grades of nickel nanostrands: nickel nanostrands, premium grade; nickel nanostrands; and fine nickel nanostrands. Table 2.12 summarizes some of their physical properties [138, 139].

Nickel Nanostrands Suppliers and Products. Conductive Composites is the sole manufacturer of nickel nanostrands and other technologies, given the company the ability to create a wide variety of highly electrically conductive fibers, nonwovens, plastics, resins, adhesives, polymer coatings, composites, and other systems. A summary of conductive material products is as follows:

- *Coated fiber (NiFiber).* Lightweight structural fibers achieve an ultimate combination of strength and conductivity.
- *Coated nonwoven sheets (NiShield).* Conductive lightweight layers with exceptional shielding and conductivity properties in a continuous roll format.
- *Nanostructured powders (Nanostrands).* Unique branched geometry and excellent percolation properties lead to highly conductive real-world properties.

All the materials on this list are available in commercial volume, and Conduct Composites' production lines run high-volume continuous-throughput manufacturing

Table 2.12. *Typical Properties of Nickel Nanostrands*

Property	Values
Bulk density	0.090–0.250 g/cc
Specific Surface Area (BET)	2–5 m^2/g
Purity	99.999%
Density	8.91 g/cc

in ISO 9001 compliant facilities. Conductive Composites' production processes are clean and "green" with no process exhaust effluents and only closed-loop fluid systems.

2.2.2.5 Aluminum Oxide Nanofibers

ANF Technology Ltd, a Surrey, UK-registered company with a production facility in Tallinn, Estonia, and a representative office in California, produces aluminum oxide nanofibers branded as Nafen™ (www.nafen.eu) [140]. It consists of very long co-aligned fibers 10–20 nm in diameter built of perfect alumina monocrystals in gamma/chi phase. Figure 2.40 shows several SEM and TEM micrographs of the Nafen aluminum oxide nanofibers as grown [140]. Table 2.13 shows some typical physical and mechanical properties of Nafen [140].

Nafen is the only type of superior-grade nanofibers recently available on the market in industrial volumes at a reasonable price. ANF has teamed up with some of the world's leading technological research institutions and centers of excellence to uncover the potential benefits of Nafen. Testing is under way at Fraunhofer Institute of Polymeric Materials and Composites (PYCO, Germany), Moscow Institute of Steels and Alloys (MISiS), University of South Carolina, The University of Texas at Austin, University of Cambridge, and other industrial and research institutes. Initial results achieved provide strong evidence of significant performance enhancement from infusing Nafen into major polymer systems, such as polycyanurate thermoset, epoxy, polyester acrylate, polyether, and PVB for a variety of applications.

Several advantages are noted in comparing Nafen with other nanofibers:

- inexpensive
- easy to disperse
- transparent, no color issues
- low loading levels below 5 wt%
- high aspect ratio > 10
- compatible with carbon and glass fibers
- dielectric
- flame retardant
- thermally conductive

Features and Applications. When 0.1 wt% Nafen was incorporated in DER 330, a low-viscosity epoxy system, using ultrasonication technique and hardened with polyethylene polyamine (PEPA), significant improvement in fracture energy and flexural strength was achieved (Figure 2.41) [140].

Table 2.13. *Typical Physical and Mechanical Properties of Nafen Aluminum Oxide Nanofibers*

Fiber diameter, nm	10–20
Fiber length, mm	1–150
Fiber tensile strength, GPa	12
Fiber tensile modulus, GPa	400
Specific surface area (BET), m^2/g	150
Phase	gamma/chi
Specific gravity, g/cm^3	3.98
Bulk density, g/cm^3	0.1 to 0.3

Courtesy of ANF Technology.

Figure 2.40. SEM and TEM micrographs of the Nafen™ aluminum oxide nanofibers as grown (Courtesy of ANF Technology).

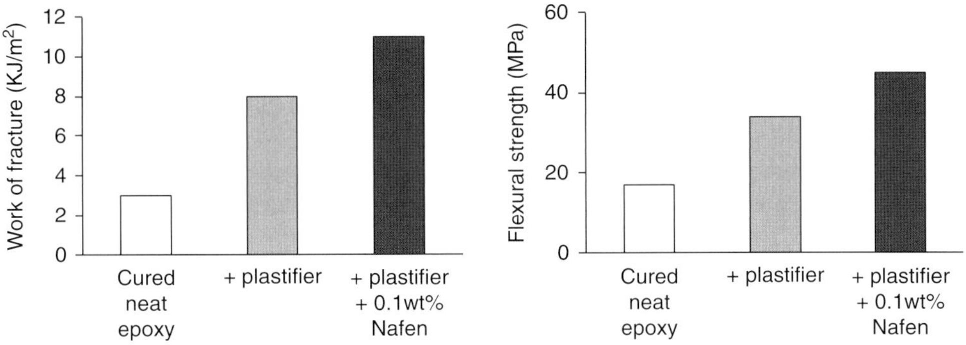

Figure 2.41. Facture energy and flexural strength improvement of epoxy resin with 0.1 wt% Nafen (Courtesy of ANF Technology).

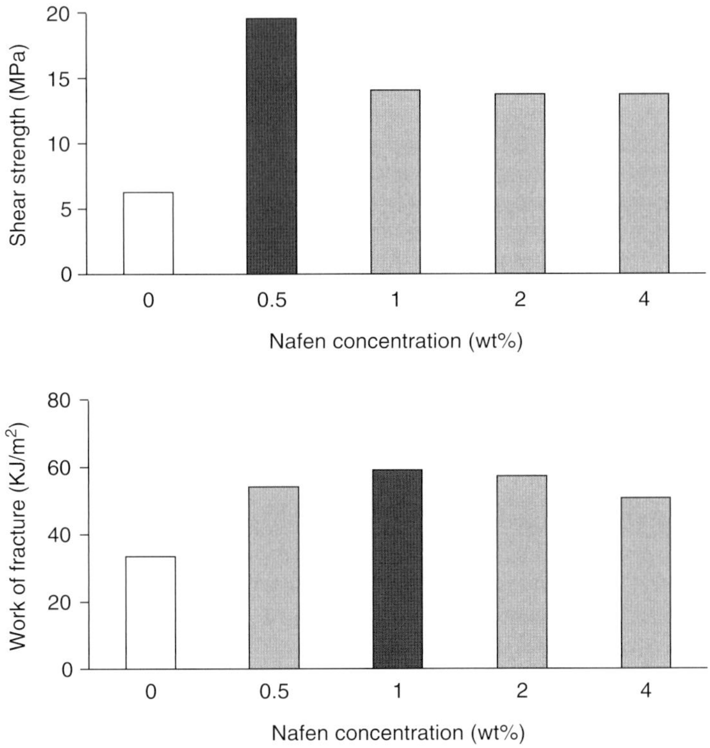

Figure 2.42. Fracture energy and shear strength improvement when 0.5–4 wt% of Nafen was added to epoxy resin (Courtesy of ANF Technology).

When 0.5–4 wt% Nafen was incorporated in DER 330 using ultrasonication technique in ethanol and mechanically mixed and hardened with diamino diphenyl sulfone (DDS), significant improvement in fracture energy and shear strength was achieved (Figure 2.42) [140].

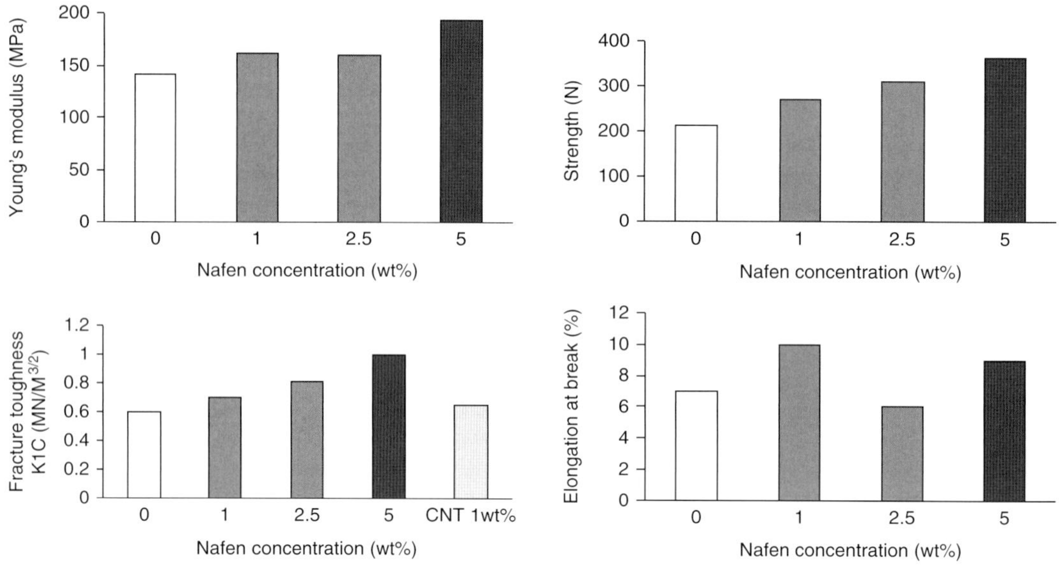

Figure 2.43. Young's modulus, tensile strength, fracture toughness, and elongation at break were improved with the addition of Nafen nanofibers in polycycyanurate resin (Courtesy of ANF Technology).

When 1–5 wt% of Nafen was added to polycyanurate by direct mixing with a three-roll mill, improvement in Young's modulus, tensile strength, elongation at break, and fracture toughness was achieved (Figure 2.43) [140].

Nafen was dispersed in ethanol using ultrsonication and mechanically mixed with PVB powder to produce transparent film that possesses increased toughness and ultimare strength while still remaining transparent (Figure 2.44) [140]. PVB can be used for triplex safety glass (Figure 2.45) [140].

When 0.5 wt% Nafen was pre-dispersed in glycidyl methacrylate (GMA) at 10 wt% and mechanically mixed with the polyether, hardness (6X) and Young's modulus were both increased while elasticity was enhanced, ensuring no cracking at bending points (Figure 2.46) [140]. Polyether-Nafen coating can be used as anticorrosion coatings, topcoat materials, potting compounds, plywood patch, wire and cable coatings, and thermal break (Figure 2.47) [140].

Aluminum Oxide Nanofibers Suppliers and Products. ANF is the sole manufacturer of Nafen™. ANF also develops Nafen-based dispersion and compounds for polymeric products. Currently, Nafen has shown promises in epoxy for fiber-reinforced composites, polycyanurate high-temperature applications, polyester acrylate UV-curable transparent coatings, polyether protective coatings, PVB transparent films, and other systems. A summary of Nafen products is as follows:

- *Nafen Aluminum Oxide nanofibers.* Controlled diameter of 7–40 nm; narrow size distribution; lengths up to 15 cm; uniaxial aligned or randomly oriented fiber blocks; industrial production volumes; chemically and thermally stable; 400^+

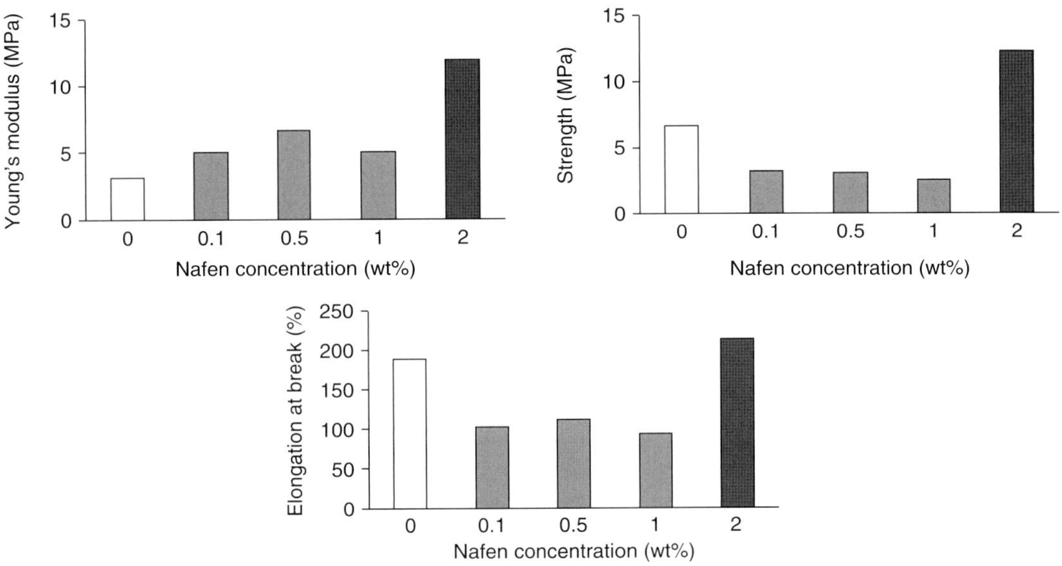

Figure 2.44. Improved mechanical properties of PVB transparent films when 0.1–2 wt% of Nafen was added, still retaining transparency (Courtesy of ANF Technology).

Figure 2.45. Nafen in PVB transparent film for safety glass application (Courtesy of ANF Technology).

GPa Young's modulus; well wetted and dispersed in water, alcohol, and other liquids; resistant to radiation; high specific surface area for catalysis or functionalizing the surface.

- *Nafen Aluminum Oxide nano powders.* Controlled diameter of 7–40 nm; industrial production volumes; chemically and thermally stable; well wetted and dispersed in water, alcohol, and other liquids; resistant to radiation; high specific surface area for catalysis or functionalizing the surface.

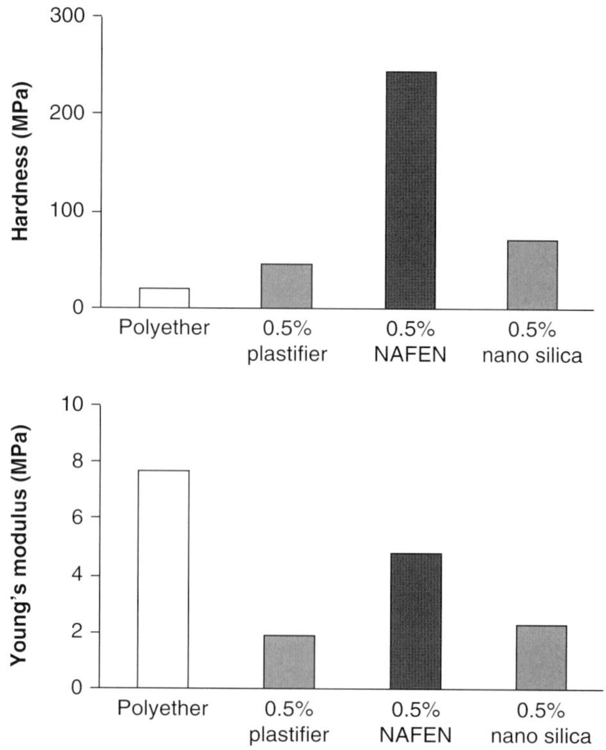

Figure 2.46. Small amount of Nafen was added to polyether for protective coatings (Courtesy of ANF Technology).

Figure 2.47. Polyether-Nafen coating can be used as anticorrosion metal coatings (Courtesy of ANF Technology).

2.2.3 Three Nanoscale Dimensions in the Form of Particulates

2.2.3.1 Polyhedral Oligomeric Silsesquioxanes (POSS®)

Origin. A departure from the use of one-dimensional and two-dimensional nanomaterials is the concept involving the creation of three-dimensional inorganic-organic structural materials. One approach to developing better materials is to create inorganic-organic composite materials in which inorganic building blocks are incorporated into organic polymers. In the early 1990s, a scientific team under the direction of Dr. Joseph Lichtenhan at Edwards Air Force Base in California invented the first new class of chemical feedstock to be developed in 50 years – Polyhedral Oligomeric Silsesquioxanes or POSS®. In 1998, Dr. Lichtenhan and his team spun this POSS technology out of the Air Force Research Laboratory and founded Hybrid Plastics®, taking the first step in evolving from a small R&D facility to a full-scale manufacturing company. In 2004, the company took its next major step and relocated from California to Hattiesburg, Mississippi (www.hybridplastics.com) [141].

Polyhedral oligomeric silsesquioxanes (POSS) nanostructured materials, representing a merger between chemical and filler technologies, can be used as multifunctional polymer additives, acting simultaneously as molecular-level reinforcements, processing aids, and flame retardants. POSS nanoscopic chemicals provide unique opportunities to create revolutionary material combinations through a melding of the desirable properties of ceramics and polymers at the 1 nm length scale. These new combinations enable the circumvention of classic material performance trade-offs by exploiting the synergy and properties that only occur between materials at the nanoscale. They release no VOCs and therefore produce no odor or air pollution, while offering easy incorporation into existing manufacturing protocols. These elegant chemicals also push the technology into new applications, such as soluble yet biologically inert delivery vehicles for pharmaceuticals and medical applications such as biocompatible prosthetics.

Properties. POSS technology is derived from a continually evolving class of compounds closely related to silicones through both composition and a shared system of nomenclature. POSS chemical technology has two unique features:

1. The chemical composition is a hybrid, intermediate ($RSiO_{1.5}$) between that of silica (SiO_2) and silicone (R_2SiO).
2. POSS molecules are physically large with respect to polymer dimensions and nearly equivalent in size to most polymer segments and coils.

POSS molecules can be thought of as the smallest particles of silica possible. However, unlike silica or modified clays, each POSS molecule contains covalently bonded reactive functionalities suitable for polymerization or grafting POSS monomers to polymer chains. Each POSS molecule contains nonreactive organic functionalities for solubility and compatibility of the POSS segments with the various polymer systems.

The chemical diversity of POSS technology is very broad, and a large number of POSS monomers and polymers are currently available or undergoing development. POSS chemical technology is easy to use, with monomers available in both liquid and solid form and soluble in most common solvents. Hence, POSS technology can be used in the same manner as common organics, in either monomer or polymer (resin) form. POSS chemical feedstock can be added to nearly all polymer types (glassy, elastomeric, rubbery, semi-crystalline, and crystalline) and compositions.

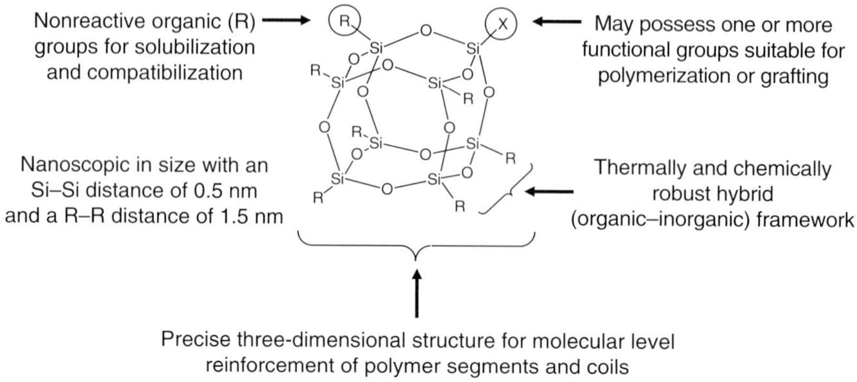

Figure 2.48. Anatomy of a POSS® molecule (Courtesy of Hybrid Plastics).

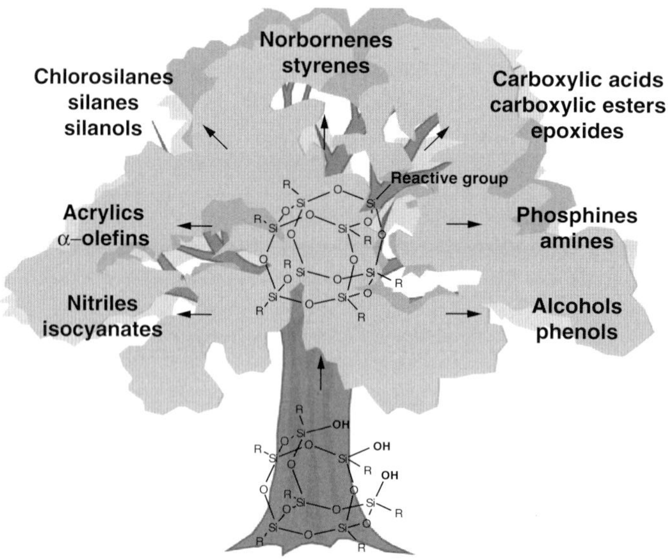

Figure 2.49. The POSS® chemical tree (Courtesy of Hybrid Plastics).

Enhancements in the physical properties of polymers incorporating POSS segments result from POSS's ability to control the motions of the chains while still maintaining the processability and mechanical properties of the base resin. This is a direct result of POSS's nanoscopic size and its relationship to polymer dimensions.

The anatomy of a POSS nanostructured chemical is illustrated in Figure 2.48 [141]. It is a precise, three-dimensional silicone-oxygen cage structure for molecular-level reinforcing of polymer segments and coils. Hybrid Plastics currently offers a large variety of nanostructured chemicals, as shown in Figure 2.49 [141]. The key purpose of POSS technology (Figure 2.50) is to create hybrid materials that are tough, lightweight, and as easy to process as polymers, yet possessing the characteristics of high-use temperature and oxidation resistance, like ceramics [141]. A list of property enhancements via POSS® is listed in Figure 2.51 [141].

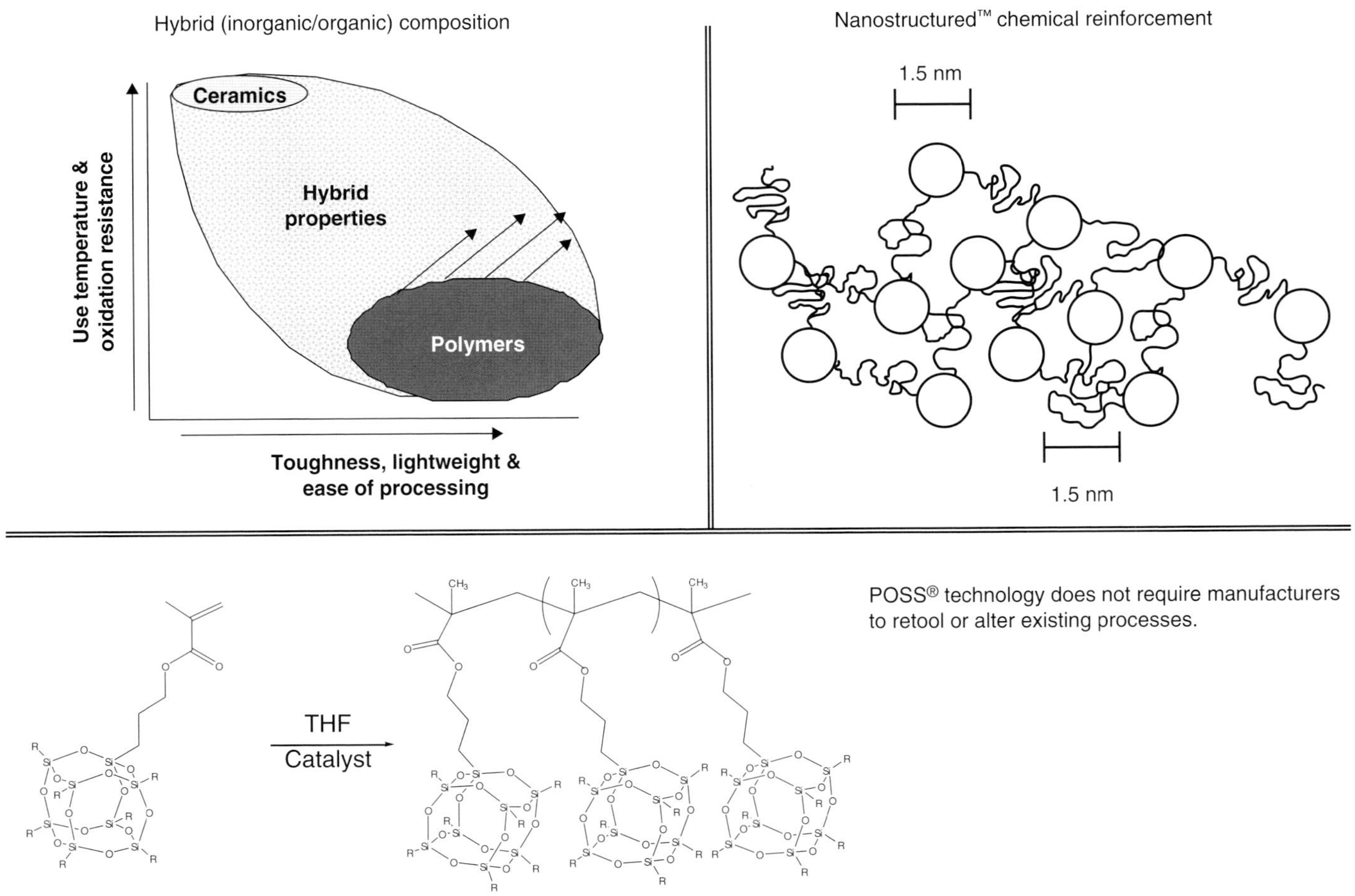

Figure 2.50. Key aspects of POSS® technology (Courtesy of Hybrid Plastics).

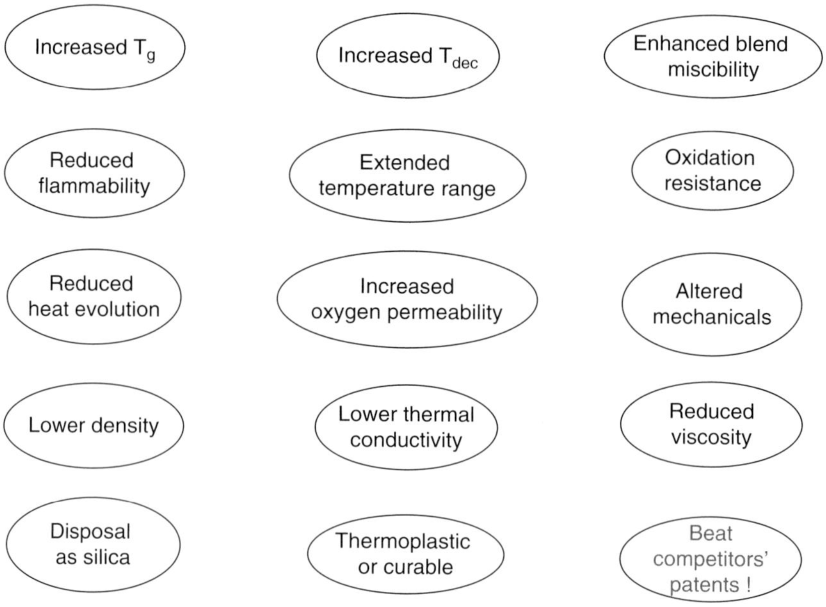

Figure 2.51. Property enhancements of polymers via POSS® (Courtesy of Hybrid Plastics).

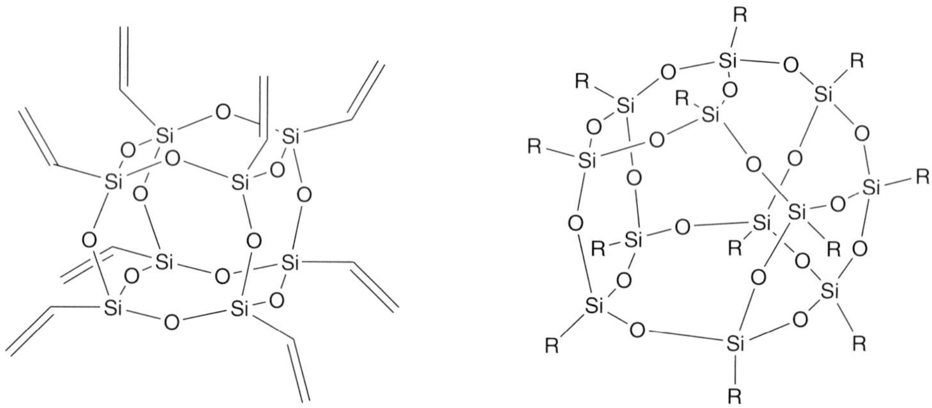

Figure 2.52. The chemical structures of $Vinyl_8T_8$ and $Phenyl_{12}T_{12}$ (Courtesy of Hybrid Plastics).

Features. Polyhedral oligomeric silsesquioxane materials are one type of hybrid inorganic-organic material of the form $(RSiO_{1/2})_n$ or R_nT_n, where organic substituents are attached to a silicon-oxygen cage [142]. The most common POSS® cage is the T_8 (e.g., $Vinyl_8T_8$ in Figure 2.52), although other cages with well-defined geometries include n = 6, 10, 12, 14, 16, and 18 [143]. By incorporating these Si-O cages into organic polymers, properties superior to the organic material alone are realized, offering exciting possibilities for the development of new materials [144–148]. Vinyl

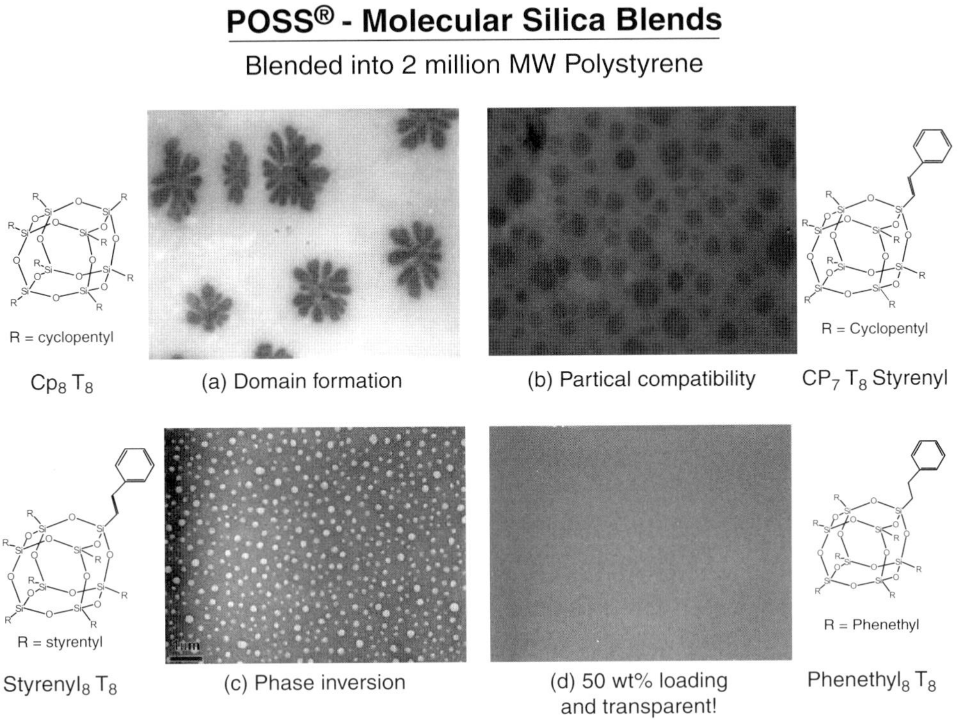

Figure 2.53. TEM micrographs showing the change of morphologies of polystyrene by changing the R group of the POSS® materials (Courtesy of A. Lee).

and phenyl-based POSS materials have been used in composites that have shown excellent fire resistance performance [149].

Another example of POSS®-molecular silica blends is shown in Figure 2.53, where each of four types of POSS has a different R group that was blended into 2 million molecular weight (MW) polystyrene. The morphology of the polystyrene-POSS nanocomposites was changed significantly, as shown in the TEM images in Figure 2.53a–d. Figure 2.53a illustrates R = cyclopentyl (Cp_8T_8) in the POSS material when blended with the polystyrene, where a snowflake domain has formed. Figure 2.53b shows R = cyclopentyl (CP_7T_8Styrenyl) in the POSS material when blended with the polystyrene, where a partial compatibility has formed. Figure 2.53c demonstrates R = styrenyl (Styrenl$_8$T$_8$) of the POSS® material when blended with polystyrene, where a phase inversion occurred. Figure 2.53d illustrates R= Phenethyl (Phenethyl$_8$T$_8$) in the POSS material when blended with 50 wt% loading of POSS® into the polystyrene – a transparent material. This example shows that the POSS material can be customized to control the morphology and properties of the resulting nanocomposite blends. A valuable reference on the “applications of POSS” was recently published using POSS in plastics, composite materials, space materials, dental materials, tissue engineering, drug delivery, lithography, fuel cells, batteries, lubricants, liquid crystal, LED, sensor, photovoltaic, and biomedical devices [150].

POSS Suppliers and Products. Hybrid Plastics manufactures POSS® Nanostructured® chemicals, flow aids, dispersion aids, thermoset resins, thermoplastic masterbatch, and others. The full list of products is as follows:

- *R&D chemicals.* A variety of R&D quantities of POSS chemicals can be purchased from Hybrid Plastics' online catalog.
- *Bulk chemicals.* Large production quantities of a few of Hybrid Plastics' most popular POSS chemicals.
- *Thermosets.* Formulated thermosetting resins for coatings, adhesives, and composites.
- *Thermoplastics.* Masterbatch formulations of common thermoplastics for improved flow, reduced surface energy, better dispersion, and more.
- *Electronics.* POSS chemicals and formulated materials for use in electronics applications.
- *POSS & nanopowders.* POSS chemicals with a metal in one corner for catalysis and predispersed pigments/fillers.

2.2.3.2 Nano-silica (*n*-silica)

Origin. Dispersed nano-silica (also known as silicon dioxide) ranges from 5 nm to 70 nm in size [151] and can be spherical or irregular in shape. Nano-silica can be produced by high-temperature hydrolysis [152–153] or sol-gel (macrosurfaced nano-silica) [154] methods. Both methods are briefly discussed in this section.

In 1941, Degussa (now known as Evonik) patented a high-temperature hydrolysis of metallic oxides to produce extremely fine–particle oxides. The method was converted into large-scale production in the 1950s and has become the process for the preparation of nanoparticles based on silicon dioxide, aluminum oxide, and titanium dioxide. Under TEM analyses, the primary particles of the three oxides show cubic forms with rounded corners. All materials exhibit no internal surface. Silicon dioxide was marketed under the trademark of AEROSIL®. The raw materials are chlorosilanes that are hydrolyzed in an oxygen-hydrogen flame. The resultant silicon dioxide occurs in aerosol and is subsequently separated from the gaseous phase. A special process is used to remove residual hydrochloric acid that is still absorbed in the material.

Properties. AEROSIL is highly dispersed, amorphous, very pure silica produced by high-temperature hydrolysis of silicon tetrachloride in an oxygen-hydrogen gas flame [151–153]. It is a white, fluffy powder consisting of spherically shaped primary particles. AEROSIL OX 50 has the largest average primary particle size of 40 nm; AEROSIL 300 and 380 have the smallest size of 7 nm. The primary particles are spherical and free of pores. The primary particles interact in the flame to develop aggregates, which join together reversibly to form agglomerates. Figure 2.54 shows a TEM micrograph of AEROSIL 300 in which the primary particles, aggregates, and agglomerates can be clearly seen [153]. The average diameters of the primary particles are 7–40 nm, according to the AEROSIL® grade. The specific surface areas range between 50 m^2/g and 380 m^2/g. In contrast to precipitated silicas, AEROSIL

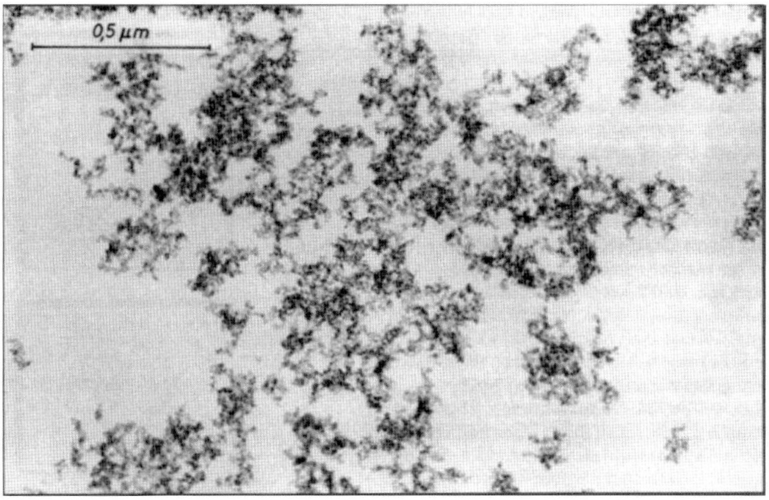

Figure 2.54. TEM micrographs of AEROSIL® 300 showing particle in aggregates (Courtesy of Evonik).

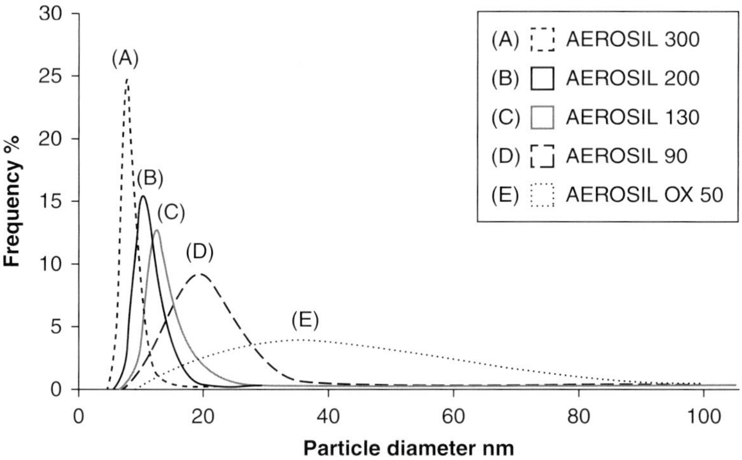

Figure 2.55. Particle size distributions of various grades of AEROSIL® nanosilicas (Courtesy of Evonik).

does not have a clearly defined agglomerate size. Particle size distribution becomes wider as the average primary particle size increases and the tendency to form agglomerates is reduced (Figure 2.55) [153]. The size of the surface area is easily illustrated if one considers that approximately 20 g of AEROSIL 200 (average primary particle size 12 nm) has the same surface area as a football field. AEROSIL consists entirely of amorphous silicon dioxide. It starts to sinter and turn into glass above 1,200°C. Crystallization only occurs after heat treatment. AEROSIL is nearly insoluble in water. It is also insoluble in acids. It does, however, dissolve in strong alkaline media to form silicates.

Applications. Siloxane and silanol groups are situated on the surface of the AEROSIL particles. The latter is responsible for the hydrophilic behavior of the

Table 2.14. *Material Properties of AEROSIL® R805 and R202 [155]*

PROPERTIES	AEROSIL® R 805	AEROSIL® R 202
Specific surface area (BET) (m^2/g)	150 ± 25	100 ± 20
Carbon content (%)	4.5–6.5	3.5–5.0
Average primary particle size (nm)	12	14
Tapped density standard material (g/l)	50	50
Moisture 2 hours at 105°C	<0.5	<0.5
Ignition loss 2 hours at 1,000°C based on material dried for 2 hours at 105°C	5.0–7.0	4.0–6.0
pH value in 4% dispersion	3.5–5.5	4.0–6.0
SiO_2 – content based on ignited material (%)	>99.8	>99.8

untreated AEROSIL. These silanol groups also determine the interaction of the AEROSIL with solids, liquids, and gases. One gram of AEROSIL 200 with a specific surface area of 200 m^2/g contains approximately 1 mmol of silanol groups. The surface of AEROSIL particles can be easily chemically modified by reacting the silanol groups with various silanes and siloxanes, resulting in hydrophobic AEROSIL particles. For example, AEROSIL R972 and AEROSIL R812 are made hydrophobic with dimethylsily or trimethylsily groups, which are stable against hydrolysis. Hydrophilic and hydrophobic grades of AEROSIL have proved successful for use in numerous applications, such as a reinforcing filler, a thickening and thixotropic agent, an anti-settling agent, and a free-flow aid. Hydrophobic types of AEROSIL are characterized by the following properties:

- extremely low moisture absorption;
- ability to impart hydrophobicity to other systems;
- easier to disperse in organic media;
- effective rheology control in complex liquid.

AEROSIL is a sub-micrometer amorphous silica that has been used successfully for decades as a thickening agent and thixotropic in liquid system. It is also used to adjust the rheological properties of epoxy resins. It also serves as a rheological aid to provide effective, stable thickening and thixotropy of epoxy systems [151–153]. AEROSIL fumed silica for rheology control is widely used in silicone rubber, coatings, plastics, printing inks, adhesives, lubricants, creams, ointments, and toothpaste. Nelson and coworkers used this nanoparticle with polystyrene in their recent study demonstrating significant improvement in reduced flammability [155]. Properties of AEROSIL R805 and R202 are listed in Table 2.14 [153].

AEROSIL has a proven track record in the following traditional applications:

- *In plastics and adhesives:* coating polymers; sealant polymers; film; gel coats and laminating resins; casting resins; cable compounds; adhesive; silicone rubber; fluorinated rubber
- *In coatings:* solvent-based coatings (two-pack systems); solvent-based coatings (air drying systems); water-based coatings (clear coats and pigmented systems); UV-curing coatings; powder coatings; unsaturated polyester

- *In printing inks:* letterpress printing; flexographic printing; offset printing; gravure printing
- *As an external additive for toner:* negative toner (black); positive toner (black); color toner (negative)
- *In pharmaceuticals and cosmetics:* antiperspirants; aerosols, sprays; coated tablets; creams, lotions; fragrances; powders; hair formulations, shampoos; lipstick; nail polish; ointments, pastes; mascara, makeup; suppositories; tablets, capsules; toothpastes, dental powder

The introduction of most nanoparticles into low-viscosity, reactive resin systems, such as epoxies, vinyl esters, cyanate esters, bismaleimides (BMI), and phenolics results in a rapid increase in viscosity as the nanoparticle content is increased to 5–7 wt%. This leads to limited use of the nanomodified reactive resin system in many low- to medium-viscosity demanding processing applications, such as vacuum-assisted resin transfer molding (VARTM), impregnation of fabric, filament winding, and other processing methods. Although there are instances that indicate 5–7 wt% nanoparticles is optimal, reduction in viscosity through the use of solvent and/or increased temperature allows reasonable dispersion of the nanoparticles into the resin matrix system. Questionable uniformity of nanoparticle dispersion using high-shear mixing conditions followed by removal of solvent is an additional issue, as is "scale-up."

As mentioned earlier, there are processing problems (viscosity increase) associated with the introduction of 5–7 wt% nanoparticles into low-viscosity resin systems (epoxies, vinyl esters, etc.). A novel technique of preparing macrosurfaced silica appears to avoid the increased viscosity problem when 5–7 wt% nanoparticles are introduced.

Macrosurfaced Nanosilica. The use of Hanse Chemie Technology [154, 156, 157] of macrosurface-treated spherical nanosilica will allow the introduction of ≥ 10 percent nanosilica with minimal increase in viscosity in the nanomodified reactive resin system and still provide multifunctionality to the resulting cured resin system. Macrosurfaced nanosilica is currently manufactured by Evonik Nanoresins GmbH (Geesthacht, Germany; www.nanoresins.ag). Nanoresins' proprietary synthesis of silica nanocomposites is shown in Figure 2.56 [154]. A silicate solution is transformed into silica sol using modified sol-gel process. The silicas were surface-modified to be compatible with selective polymer matrixes to form transparent polymer nanocomposites with low-viscosity and no sedimentation. Evonik Nanoresins is able to create homogeneous dispersions of amorphous silicon dioxide (SiO_2) in standard, chemically unchanged organic monomers, prepolymers, and oligomers using this technique. The silica phase consists of discrete nanospheres (20 nm) with an extremely narrow size distribution, as shown in Figure 2.57 [154]. Owing to agglomerate-free colloidal dispersion of the nanosilica particles, the nanocomposites are highly transparent, with low viscosity, and do not show sedimentation even with SiO_2 loading up to 60%, as shown in a TEM micrograph in Figure 2.58 [154]. This nanosilica can also be used as raw materials for coatings or varnishes and as additives for a wide variety of other applications.

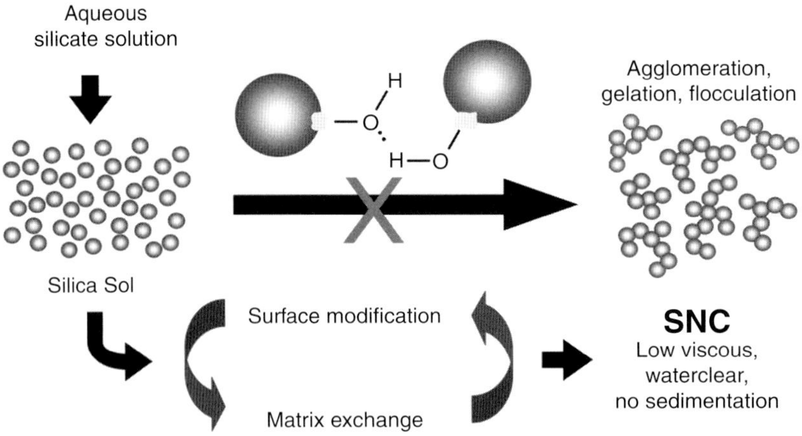

Figure 2.56. Hanse Chemie's route to produce nanosilicas (Courtesy of Hanse Chemie).

Figure 2.57. Narrow particle size distribution is a characteristic of Hanse Chemie's nanosilicas (Courtesy of Hanse Chemie).

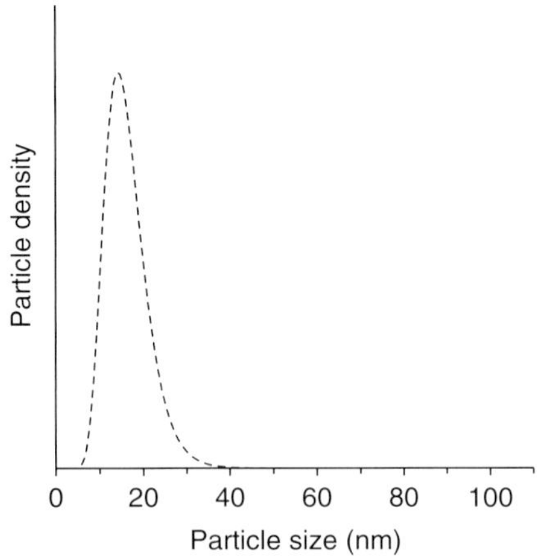

Figure 2.58. TEM micrographs showing Hanse Chemie's homogenous dispersion of nanosilica in polymer (Courtesy of Hanse Chemie).

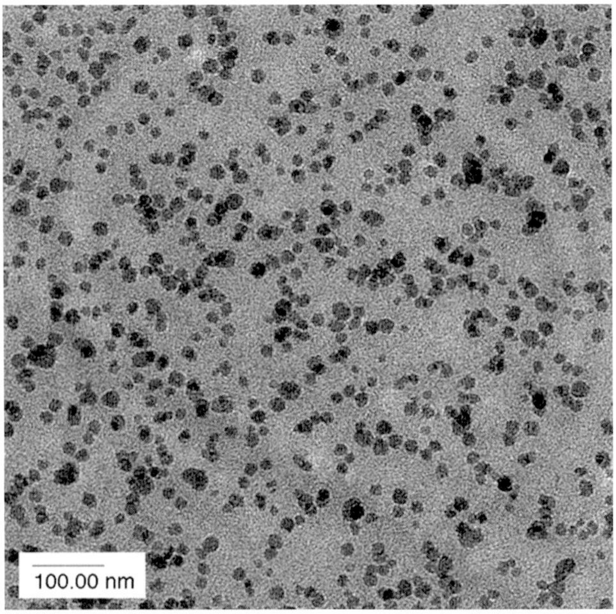

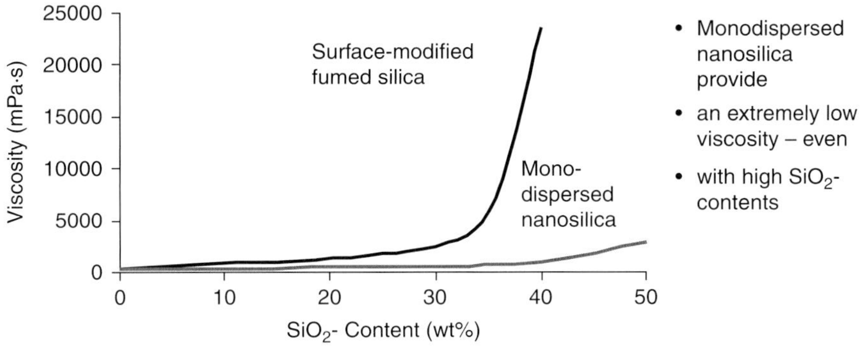

Figure 2.59. Viscosities of fumed silica and monodispersed nanosilica in acrylate ester (Courtesy of Hanse Chemie).

It is the versatility that makes this sol-gel process so unique: the nanoparticles can be incorporated into a wide range of different polymer materials. Nanosilica thus produced are very well suited for further processing due to their low viscosity. They achieve specific improvement in the mechanical and thermal properties without the disadvantages usually associated with inorganic fillers. Thus the toughness and hardness of a material can be increased without any loss in optical clarity.

Properties. Nanoresins claims that these nanosilica particles can easily penetrate "closed meshed fabrics" and are therefore well suited for reinforcement of composites, especially VARTM. This characteristic would similarly apply to uniform impregnation of fabric in prepreg preparation. Furthermore, it is possible that with a further increase in nanosilica content to 15–20%, with a modest increase in viscosity and reasonable economics for the nanosilica, that a flame-resistant epoxy polymer matrix composites (PMCs) may emerge without jeopardizing the expected benefits of nanomodification, such as, among others, improved heat strength and higher T_g, improved toughness. Nanoresins currently manufactures three nanocomposites products under the trade names of Nanocryl®, Nanopox®, and Nanopol®.

Nanocryl is created via low-viscosity dispersions of nanosilica in unsaturated (meth-)acrylates. It increases the surface hardness of coatings and their scratch and abrasion resistance without any loss in transparency and elasticity. It also improves other properties, such as gas permeability, cure shrinkage, and fracture toughness. Whenever acrylates are used, Nanocyrl optimizes the qualities of coatings, adhesives, electronics, and high performance. Figure 2.59 illustrates that there is no viscosity buildup of nanosilica in acrylate ester up to 50 wt%. Nanocryl may be used in all applications that require radiation and peroxide curing [154]. The stability of the dispersions is affected by incompatible additives (gelation and flocculation).

Nanopox is made by dispersing nanosilicas in epoxy resins. Nanopox optimizes the quality of epoxy resins. With the nanosilicas only 20 nm in size, Nanopox penetrates even the tiniest structures, such as carbon fibers, thus increasing the material's fractural toughness and impact resistance. Combine Nanopox with standard epoxy resins and obtain the hardness and stiffness of much more expensive, highly cross-lined epoxies, along with toughness. It can be used for applications, such as coatings, structural adhesive, structural composites, casting resins (e.g., plotting, impregnation, and encapsulation), electronics, and general purpose.

Nanopol is a product family of surfaced-modified nanosilicas dispersed in solvent carriers. Nanopol is a raw material that substantially improves the properties and application of solvent-based coating systems. They allow formulation of low-viscosity, sedimentation-free system even at high filler contents. The monodispersed nanosilicas provide optimum coating formulations that perform like silica-filled products while avoiding the disadvantages of conventional inorganic fillers. It offers outstanding scratch resistance and abrasion resistance without sacrificing other desired properties, such as transparency, gloss, and flexibility for coatings on wood, plastics, or metal substrates. Its major application is for coatings.

Applications. A variety of applications can be considered with the diversity of Nanoresins nanosilica products ranging from epoxies, acrylates, polyurethanes, and silicone elastomers for coatings to fiber-reinforced composites, structural adhesives, and castings. The Nanoresins nanoparticles and reactive resin modifiers provide the following general features:

Processing:

- Commercially available, stable dispersion for reactive systems, such as epoxies, acrylates, urethanes, and silicones
- Ease of processing, low viscosity, providing excellent wet-out of woven fabric with reduced heat of reaction and reduced cure shrinkage
- Cure rate unaffected
- Combine with typical fillers

Nanocomposites Products:

- Enhanced scratch resistance, abrasion resistance without impairing transparency in UV-curing clear coatings
- Enhanced barrier properties to water vapor, gases, and solvents in UV-curing coatings
- Improved impact strength and elastic modulus of fiber-reinforced composites
- Increased toughness and tensile shear strength in epoxy-resin based and other structural adhesives
- Reduced thermal yellowing, reduced thermal expansion and inner tension
- Depending on Nanoresins product, enhanced thermal conductivity or improved electrical insulation properties

2.2.3.3 Nano-alumina (n-alumina)

Origin. In 1941, Degussa (now known as Evonik) developed and patented a process where silicon tetrachloride was vaporized and hydrolyzed in an oxyhydrogen flame. AEROSIL nanosilicas were produced using this process. The basis of the AEROSIL process is the hydrolysis of the gaseous metallic chlorides under the influence of water that develops during the oxygen-hydrogen reaction, and at the temperature

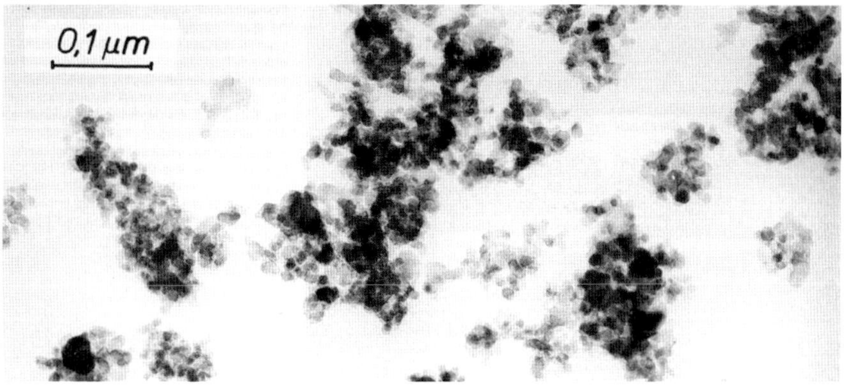

Figure 2.60. TEM micrographs of Aluminum oxide C (Courtesy of Evonik).

characteristics for such a reaction. The formation of the highly dispersed oxides takes place schematically, according to the following equations [152]:

Aluminum oxide C

$$4\ AlCl_3 + 6\ H_2 + 3\ O_2 \rightarrow 2\ Al_2O_3 + 12\ HCl$$

Nano-titanium dioxide P 25

$$TiCl_4 + 2\ H_2 + O_2 \rightarrow TiO_2 + 4\ HCl$$

Under TEM analyses, the primary particles of the three oxides show cubic forms, with the corners rounded off. Similar to AEROSIL particles, they have no internal surface.

Properties. Aluminum oxide C has an average primary particle size of about 13 nm and a specific surface of about 100 m^2/g. The aluminum oxides produced through precipitation of aluminum hydroxide from aluminate solution followed by calcinations consist of particles in the order of magnitude of 10 micrometers. High-surface aluminum oxide gels, in contrast to Aluminum oxide C, have a high proportion of internal surface. Figure 2.60 shows TEM micrographs of Aluminum oxide C [152]. The specific surface is also a function of the specific weight, and for "heavy" oxides these must be evaluated correspondingly higher. In Figure 2.61, the BET surfaces are represented in comparison with the specific weights of four nanoparticles [152].

The compacted apparent density of Aluminum oxide C is about 80 g/L. For use in powdered coatings or as a free-flow aid in powders that tend to form lumps, a less compact material with a compacted apparent density of about 50 g/L is offered as Aluminum oxide CS. Through dispersion of Aluminum oxide C in water, followed by drying and grinding, a compacted apparent density of about 500 g/L can be attained. Figure 2.62 shows 10 g each of commercial Aluminum oxide C (compacted apparent density about 80 g/L) and a variant compacted via an aqueous dispersion (compacted apparent density about 500 g/L) [152]. In combination with AEROSIL, Aluminum oxide C has proven to be effective in the thickening and the

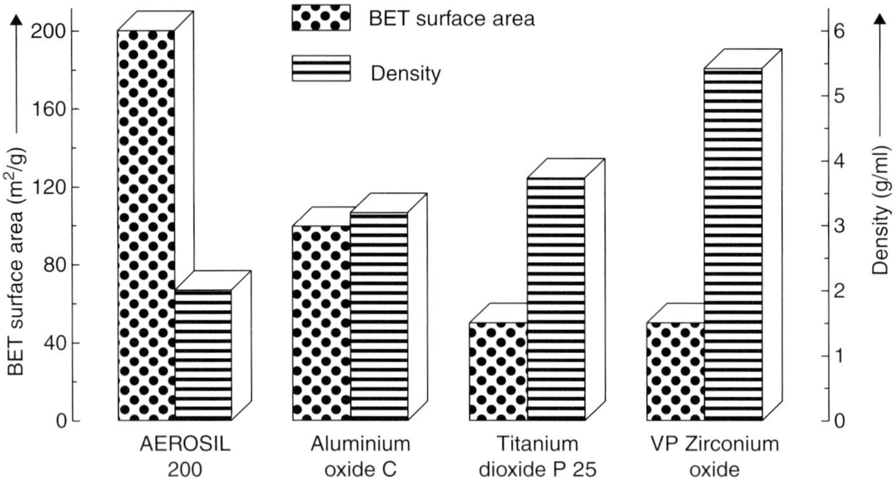

Figure 2.61. Comparison of BET surface areas and densities of AEROSIL® 200, Al_2O_3 C, TiO_2 P 25, and ZrO (Courtesy of Evonik).

Figure 2.62. Comparison of 10g of commercial Aluminum oxide C (compacted apparent density about 80 g/l) (left) and 10 g of Aluminum oxide C compacted by the wet process (compacted apparent density about 550 g/l) (right) (Courtesy of Evonik).

formation of thixotropizing of polar liquids. For the thickening and the formation of thixotropizing of water, AEROSIL COK 84 – a mixture of 84% AEROSIL 200 and 16% Aluminum oxide C – has been established as optimal.

A partial listing of potential Aluminum oxide C applications includes: production of aluminum nitride, inkjet papers, thermo-transfer printing, cable insulation, high-voltage insulators, transparent coatings, acrylate suspensions containing pigments, solder-resistant masks, photoresists for the production of ICs, coating of steel, heat insulation mixtures, powdered coating, and hair shampoo. Aluminum oxide nanoparticles have been used to form a nanocomposite for improved stereolithography [158].

Table 2.15. *Chemical Purity of Disperal® and Dispal® Alumina Powders [159]*

Impurity	ppm (typical)
Na_2O	20
Fe_2O_3	100
SiO_2	120

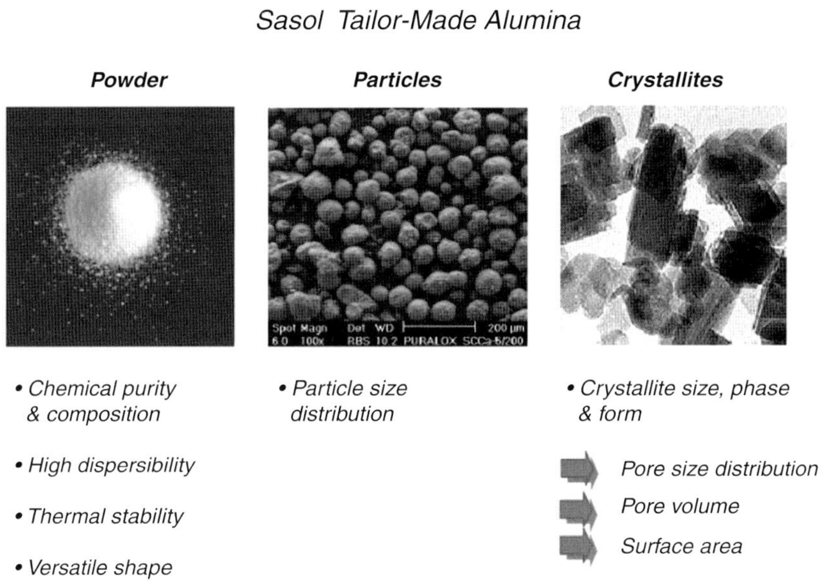

Figure 2.63. Sasol's tailor-made alumina in chemical purity, particle size, and crystallite size, phase, and form (Courtesy of Sasol).

Nano-alumina. Sasol pioneered processes utilizing alkoxide chemistry to convert primary aluminum metal into synthetic boehmite alumina of exceptional purity. DISPERAL® and DISPAL® are the trade names for the high-purity, highly dispersible, boehmite alumina powders and sols/dispersions manufactured by Sasol in Brunsbuttel, Germany and in Lake Charles, Louisiana (www.sasolalumina.com). These alumina, which are nano-sized in the dispersed phase, exhibit a unique combination of purity, consistency, and dispersibility that makes them excellent materials for use in colloidal applications. These Sasol nano-alumina (n-alumina) have traditionally been used in applications, such as sol-gel ceramics, catalysis, refractory materials, rheology control, and surface fractionizing. Other, more recently developed uses include surface coating and paint detackyfication. Sasol's processes yield alumina with significantly lower levels of common impurities, such as iron, sodium, and silica (Table 2.15) [159]. Due to their extensive experience in their business, Sasol can tailor-make alumina per specific client requests (Figure 2.63) [159]. Sasol is able to produce highly dispersible alumina with a wide range of physical

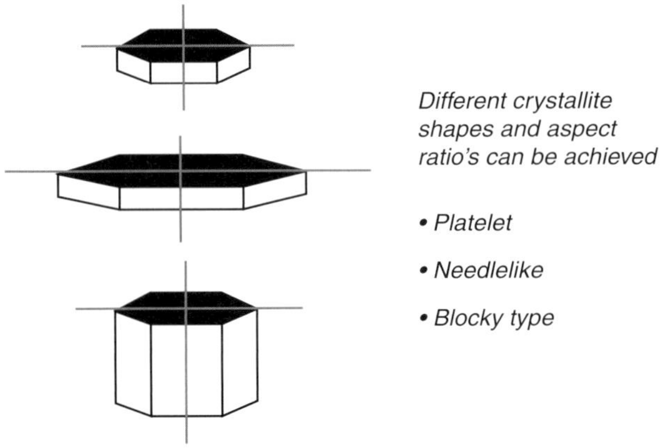

Figure 2.64. Different crystalline size and shape of Sasol *n*-alumina (Courtesy of Sasol).

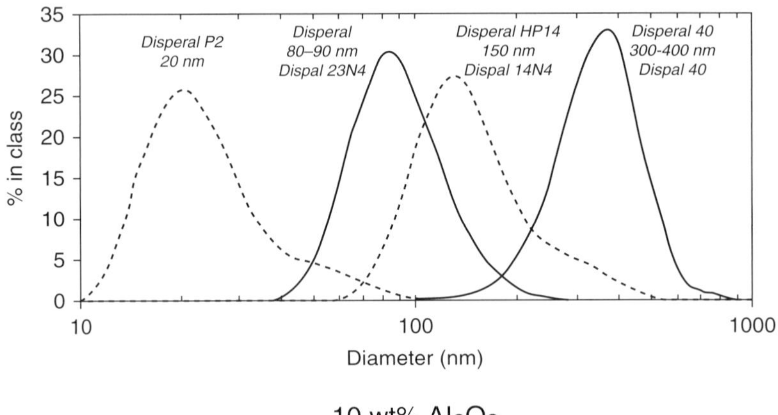

Figure 2.65. Different particle size distributions of dispersed Sasol 10 wt% alumina (Courtesy of Sasol).

properties, such as different crystalline size and shape: platelet, needlelike, and blocky type (Figure 2.64) [159]. They can be in dispersed particle size (Figure 2.65) ranging from 15 nm to 600 nm and resulting in translucent to opaque dispersions (Figure 2.66) [160].

Features. The unique product characteristics of Sasol dispersible alumina can lead to many advantages in different applications. Some key features are:

- *The alumina are synthetic.* These high-purity materials are produced under careful control to yield products of excellent quality.
- *Both powders and pre-dispersed alumina are available.* One can choose the form most appropriate for one's processing needs.
- *The powders are highly dispersible.* Low-viscosity nanoparticle sols can be prepared at room temperature in 10–30 min.

(a)
(b)

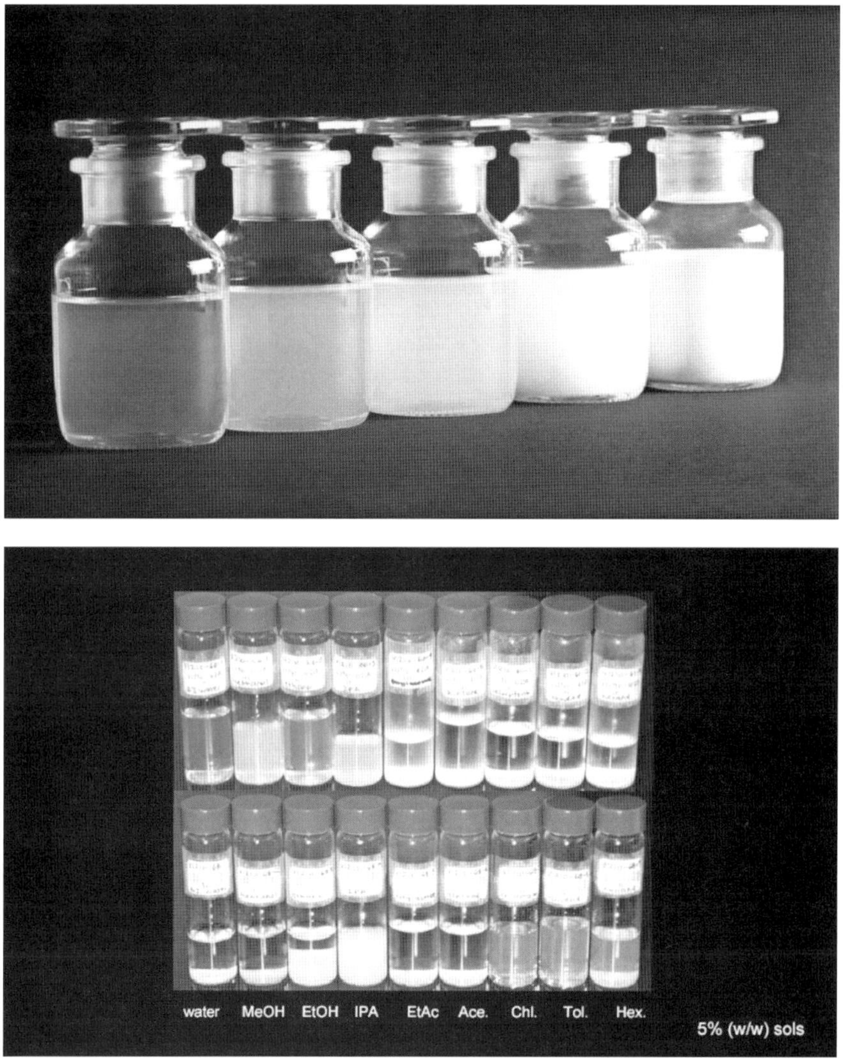

Figure 2.66. Dispersibility of *n*-alumina is clearly shown in (a) translucent to opaque alumina solutions, each with a concentration of 10 wt% Al_2O_3; and (b) in how well the 5 wt% of *n*-alumina dissolves in different types of solvents (Courtesy of Sasol).

- *The alumina are versatile.* They can be used under a variety of conditions, including low or high ph and low or high shear.
- *Sasol also offers alumina that are dispersible in organic media*, such as alcohols, dimethylformamide (DMF), and others.

Nano-alumina can be synthesized by sol-gel technique [159, 160], hydrothermal synthesis, precipitation, and combustion method [159, 160]. Like other nanoparticles, *n*-alumina aggregates to micron-sized particles and can be broken down by mechanical and chemical methods. Figure 2.67 shows a schematic representation of what occurs to the powder during dispersions using chemical attack and mechanical energy

Figure 2.67. SEM micrographs of as-received *n*-alumina (Courtesy of Sasol).

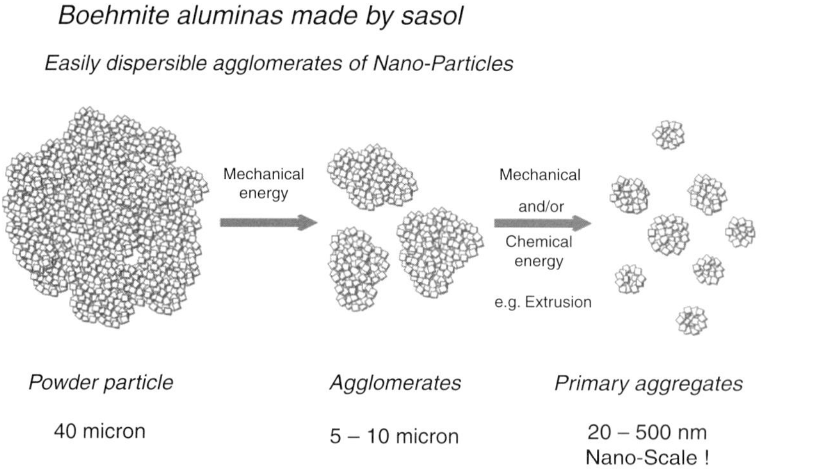

Figure 2.68. Schematics showing how *n*-alumina are dispersed by chemical and mechanical means (Courtesy of Sasol).

[160]. Figure 2.68 shows the SEM micrograph of an as-received *n*-alumina [160]. Figure 2.69 shows the TEM micrograph of a dispersible alumina at magnification of 180,000 [160]. *N*-alumina may improve hardness, thermal conductivity and stability, wear resistance, strength, and stiffness, among other characteristics.

2.2.3.4 Nano-titanium Dioxide (*n*-TiO_2)

Origin. *N*-titanium dioxide P 25 has an average primary particle size of about 21 nm and a specific surface of about 50 m^2/g [152]. It is therefore considerably more finely divided than the titanium dioxides that are produced on a major industrial scale according to the sulphate or chloride process for pigment applications, and that have

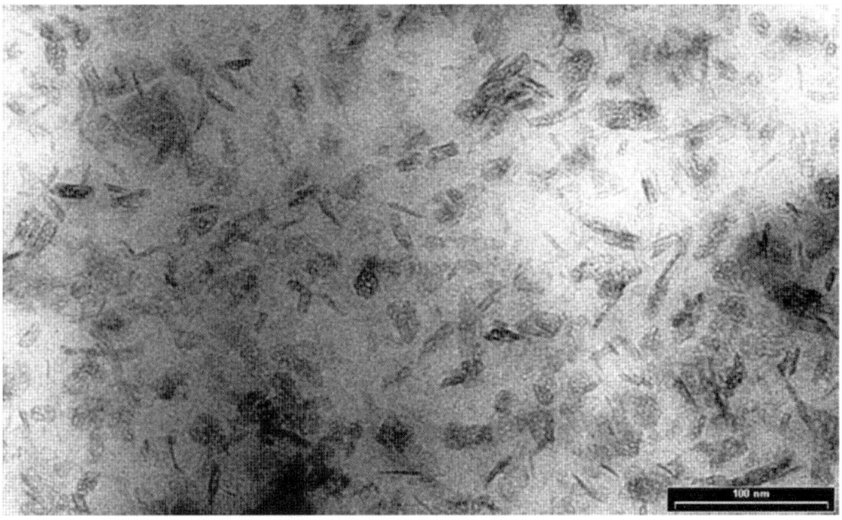

Figure 2.69. TEM micrographs of a dispersed alumina in PA6 polymer (Courtesy of Sasol).

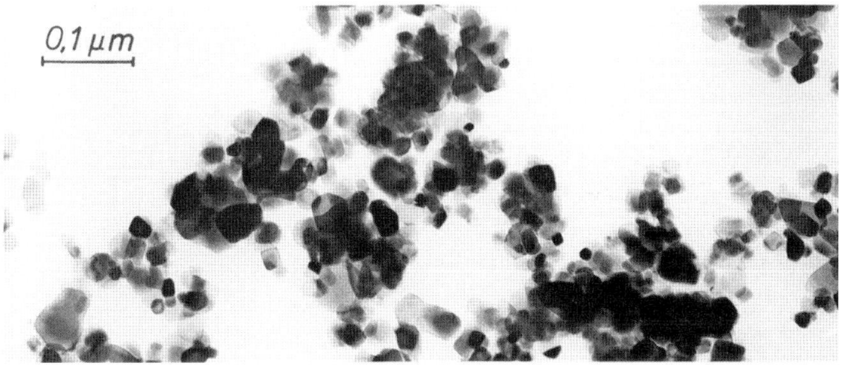

Figure 2.70. TEM micrographs of Titanium dioxide P 25 (Courtesy of Evonik).

a particle diameter of about 0.3 µm. Their surfaces consequently lie in the order of magnitude of 10 m^2/g. Figure 2.70 shows a TEM micrograph of *n*-TiO_2 P 25, and Figure 2.71 shows the compacted apparent density of a commercial titanium dioxide pigment and of *n*-TiO_2 P 25 [152].

Properties. *N*-titanium dioxide P 25 has been shown to be an excellent additive to improve heat stability of room-temperature vulcanized (RTV) silicone adhesive/sealant. *n*-TiO_2 P 25 can offer long-term heat stability at high temperatures. It also has the added benefit of flammability retardancy to RTV silicone using only a small amount of *n*-TiO_2 P 25. It also can offer semi-reinforcing qualities, which larger-particle-size heat stabilizers cannot. For optimum properties, a minimum loading level of three parts is recommended.

Potential *n*-TiO_2 P 25 applications consist of high-voltage insulation material, IC substrate boards, toners, florescent tubes, battery separators, and polystyrene.

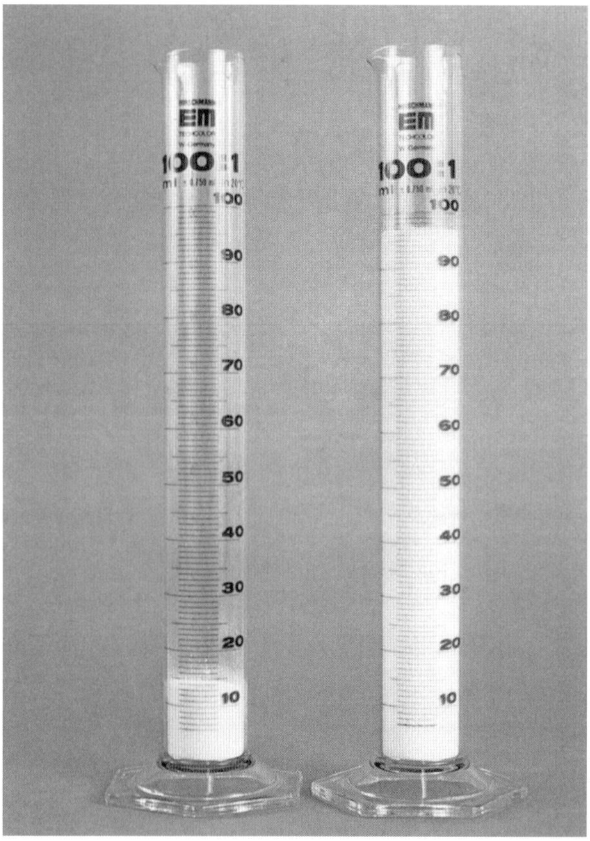

Figure 2.71. Comparison of the compacted apparent densities of a commercial TiO_2 and pigment (left) and of TiO_2 P 25 (right) (Courtesy of Evonik).

2.2.3.5 Nano-magnesium Hydroxide (n-$Mg(OH)_2$)

Origin. Nano-magnesium hydroxide (n-$Mg(OH)_2$) is used industrially as a non-hazardous alkali to neutralize acidic wastewater. It also takes part in the block method of building artificial reefs. Solid $Mg(OH)_2$ also has smoke-suppressing and fire retardant properties. This is due to the endothermic decomposition it undergoes at 332°C:

$$Mg(OH)_2 \rightarrow MgO + H_2O$$

The heat absorbed by the reaction acts a retardant by delaying ignition of the associated substance. The water released dilutes any combustible gases and inhibits oxygen from aiding the combustion. Common usage of $Mg(OH)_2$ as a fire retardant additive includes plastics [161], roofing, and coatings. It is also used for folk remedy. Figure 2.72 shows an ESEM image and Figure 2.73 shows SEM images of $Mg(OH)_2$ nanoparticles. *N*-magnesium hydroxide has platelet structure and requires exfoliation, otherwise the mechanical properties of the resulting polymer nanocomposites would be reduced [161].

Global Market and Applications. Nano-magnesium hydroxide has various applications ranging from flame retardants, to health care, to functional fillers. Almost 70% of the n-$Mg(OH)_2$ is presently used as flame retardant by cable and wire industries

Figure 2.72. An ESEM image of n-$Mg(OH)_2$ [161].

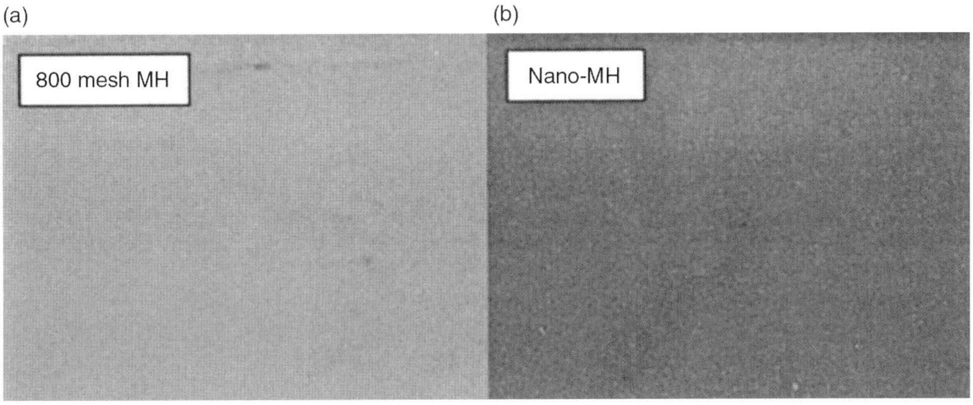

Figure 2.73. The appearance of composites filled with 70 wt % $Mg(OH)_2$ [161].

[162]. However, it also finds limited applications in other industries, such as textile because of its ease of dispersion; production of chemicals, especially polymers; building materials; and in health care because of its antibacterial properties. The emerging application areas in which research efforts are under way are in weaponry, aerospace, defense, and specialty cables.

About 77% of the global consumption is in East Asia and North America. Despite its technical superiority, the cost and low awareness of this product had restricted its consumption. The price is affected by both the complexity of the manufacturing process and the material's size and purity. The hub for the manufacturers of n-$Mg(OH)_2$ is located in China and the United States, accounting for 80% of the global capacity. There is large underutilization of existing production capacity due to low demand. The Indian market is still in the process of development due to lack of awareness. It is expected that the market for *n*-$Mg(OH)_2$ will reach 646 metric tons in 2015 despite its high cost and lack of awareness [162].

Magnesium Hydroxide Nanoparticle Suppliers. There are several U.S. suppliers of magnesium hydroxide nanopowder as follows:

US Research Nanomaterials, Inc. is located in Houston, Texas (www.us-nano.com). Their $Mg(OH)_2$ nanoparticles product is a white powder with 99% purity, average diameter of 10 nm, bulk density of ~0.23 g/cc, true density of 2.36 g/cc, and specific surface area of 90 m^2/g.

American Elements is located in Los Angeles, California (www.americanelements.com). They have four $Mg(OH)_2$ nanoparticles products with 99%, 99.9%, 99.99%, and 99.999% purity. It is a white flaky powder with an MW of 58.83, true density of 2.34 g/cc, bulk density of ~0.23 g/cc, melting point of 350°C, average diameter of 10 nm, and specific surface area of 90 m^2/g.

Nanostructured & Amorphous Materials, Inc. is located in Houston, Texas (www.nanoamor.com). Their $Mg(OH)_2$ nanoparticles product is 99% pure. It is a white powder with true density of 2.36 g/cc, bulk density of ~ 0.21 g/cc, average particle diameter of 15 nm, and specific surface area of 80 m^2/g.

There are also suppliers from China, such as Haicheng Xinnuoxier Mining Industry Co., Ltd that sells *n*-magnesium hydroxide particles in industrial quantities.

2.2.3.6 Nano-silicon Carbides (*n*-SiC)

Origin. Non-oxide ceramic, such as silicon carbide, boron carbide, and silicon nitride are attractive structural materials for high-temperature applications because of their low density, very high hardness, and excellent thermal and chemical stability [163–166]. Despite these advantages, they are susceptible to fast fracture during mechanical loading due to their inherently brittle nature. Silicon carbide nanoparticles possess high purity, narrow range particle size distribution, and larger specific surface area. They exhibit chemical stability, high thermal conductivity, smaller thermal expansion coefficient, and better abrasion resistance. The micro hardness of SiC nanoparticles is 2840 ~ 3320 kg/mm^2 and hardness between that of corundum and diamond. Their mechanical strength is higher than that of corundum. Figures 2.74 [164] and 2.75 [165] show the TEM and HRTEM images of a typical *n*-SiC powders, respectively.

Applications. Silicon carbide nanoparticles applications are numerous. They can be used in the following applications:

A reinforcing phase for silver alloys by adding *n*-SiC particles as an effective reinforcing phase for silver alloy to improve material properties. With 1 wt% of *n*-SiC particles added to the silver alloy, tensile strength, thermal conductivity, and intensity and abrasion resistance are all greatly improved.

Strengthening copper compound materials. Copper matrix compound material of high intensity and high conductivity has been widely applied to electronic parts

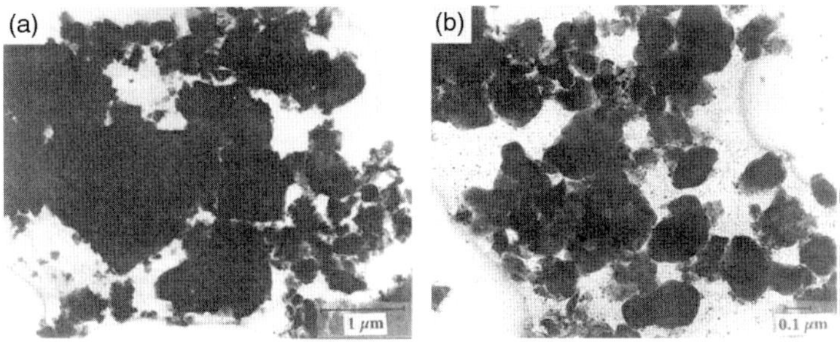

Figure 2.74. SEM micrographs of (a) SiC(C) and (b) SiC(F) powders [164].

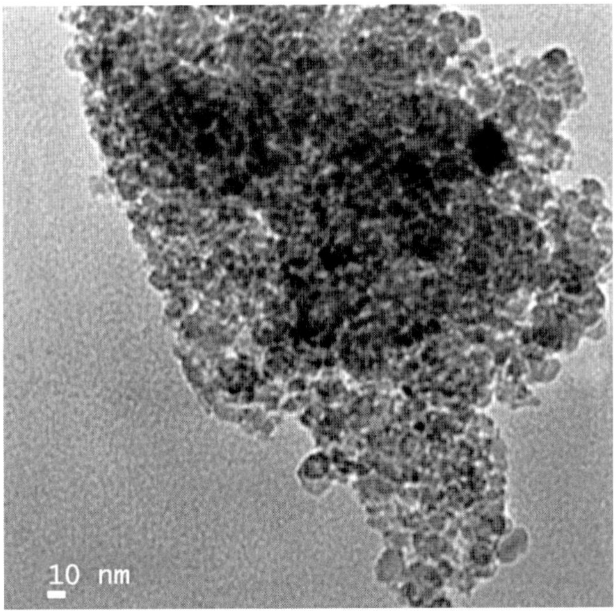

Figure 2.75. High-resolution TEM micrograph of SiC powders. Reproduced from [163] under the terms of a Creative Commons CC-BY 3.0 license.

and equipment. The high intensity and high conductivity of copper alloys are seemingly contradictory properties. Usually, mechanical properties are improved at the expense of thermal conductivity. The stable *n*-SiC dispersion-strengthened copper matrix material is an ideal solution to this problem. Adding a small amount of n-SiC to the matrix reinforces the copper alloy.

Rubber tires and rubber products. By adding a small amount of *n*-SiC particles to rubber tires, improvements of abrasion resistance of the tires by 15–30% have been reported without alternating original rubber formula of degrading the performance of the tires. SiC nanopowder can also be applied to other rubber products with the requirements of abrasion resistance, heat dissipation, and heat resistant, such as rubber roller and fuser film sleeve for printers.

High-performance engineering polymers. When *n*-SiC particles are treated by coupling agent and added at about 10 wt% to engineering polymers, such as PI, PEEK, PTFE, improvements have been reported in abrasion, thermal conductivity, insulation, tensile, impact resistant, and high-temperature resistance of these polymers.

Ceramic matrix composites. Ceramic matrix composite is prepared by compounding multiple nanoparticles with ceramic as matrix. Types of ceramic matrix include silicon nitride, silicon carbon, and others. These advanced ceramics are advantageous in high-temperature resistance, are high intensity, rigid, lightweight, and corrosion resistant, but quite brittle. Combining nanoparticles and ceramic matrix into a compound is an effective approach to improving the reliability of ceramics and producing nano-reinforced ceramic matrix composites with excellent tenacity. Ceramic matrix composites have been applied to tools, sliding components, engine parts, and energy device. They can be used as structural components where wear resistance, high-temperature resistance, and corrosion resistance have been sorted in machinery, chemical, energy aviation, space, and defense industries.

Radar and infrared stealth materials. Nano-ceramic powder is a new type of ceramic infrared absorbing agent, mainly including silicon carbide and silicon nitride nanopowders. They can absorb a larger range of wave band while exhibiting high absorption strength. The compound of silicon carbon nanoparticles and nano-magnetic absorbing agent (i.e., magnetic metal powder) will have its absorption effect improved greatly.

Coating materials. Incorporating *n*-SiC into coating materials can improve abrasion resistance, corrosion resistance, and oxidative stability properties. It has been shown that incorporating nanoparticles of silicon carbide, zirconium carbide, titanium carbide, titanium nitride, or boron carbide into compound coating of metal can provide superior abrasion resistance and self-lubrication. The achieved abrasion resistance is 100 times higher than steel bearing, friction coefficient being 0.06 to 0.1; in addition, *n*-SiC also provides the coating material with high-temperature stability and abrasion resistance. These coatings have been applied to specific components of liquid rockets to extend the service life of these parts.

Nano-silicon Carbide Suppliers. There are several U.S. suppliers of nano-silicon carbide.

US Research Nanomaterials, Inc. is located in Houston, Texas (www.us-nano.com). Their silicon carbide nanopowder (SiC, beta) is a grayish-white powder with 99+% purity manufactured by plasma CVD process with a cubic morphology, average diameter of <80 nm, bulk density of 0.05 g/cc, true density of 3.216 g/cc, specific surface area of 25–50 m^2/g, and zeta potential of -27.8 mV.

American Elements is located in Los Angeles, California (www.americanelements.com). They produce four SiC nanoparticles with 99%, 99.9%, 99.99%, and 99.999% purity. These are spherical high-surface particles with a MW of 40.1, true density of 3.22 g/cc, melting point of 2,730°C, diameter range of 10–150 nm, and specific surface area range of 10–75 m^2/g.

Nanostructured & Amorphous Materials, Inc. is located in Houston, Texas, (www.nanoamor.com). Their SiC beta nanoparticles is 97.5% purity manufactured by plasma CVD process. It is a grayish-white powder with a spherical morphology, true density of 3.22 g/cc, bulk density of 0.068 g/cc, and a particle diameter range of 45–55 nm.

NanoAmor is located in Houston, Texas (www.nanoamor-europe.com). Their SiC beta nanoparticles are 97+% purity with a melting point of 2,700°C. It is a black powder with a nearly spherical morphology, true density of 3.22 g/cc, bulk density of ~0.051 g/cc, an average particle diameter of 10 nm and a maximum particle diameter of 20 nm, and a specific surface area of 150–200 m^2/g.

There are also manufacturers/suppliers from China, such as Linyi Jinmeng Carborundum Co., Ltd. (www.lyjmsic.en.alibaba.com) and NaBond Technologies Co., Ltd. (www.nabond.com) that sell *n*-silicon carbide particles in industrial quantities.

2.2.3.7 Nano-silver (*n*-Ag)

Origin. Silver nanoparticles have unique optical, electrical, and thermal properties and are being incorporated into products that range from photovoltaic to biological and chemical sensors. Examples include conductive inks, pastes, and fillers that utilize silver nanoparticles for their high electrical conductivity, stability, and low sintering temperatures. Additional applications include molecular diagnostics and photonic devices that take advantage of the novel optical properties of these nanomaterials. An increasingly common application is the use of silver nanoparticles for antimicrobial coatings, and many textiles, keyboards, wound dressings, and biomedical devices now contain silver nanoparticles that continuously release a low level of silver ions to provide protection against bacteria. Understanding how the size, shape, surface, and aggregation state of the silver nanoparticles change after integration into a target application is critical for optimizing performance. Silver nanoparticle manufacturer offers precisely manufactured monodisperse silver nanoparticles that are free from agglomeration, making them ideal for research, development, and use in a variety of innovative applications. Each batch of nanoparticles is extensively characterized using transmission electron microscopy (TEM) images, dynamic light scattering (for particle size analysis), Zeta potential measurements, and UV/Visible spectral analysis to ensure consistent materials in every batch [167].

Properties and Applications. Silver nanoparticles (*n*-Ag) exhibit antimicrobial, conductive, and optical properties making them suitable for a wide variety of application, such as coatings and ink, polymers and nanocomposites, printed electronics, medical devices, and molecular diagnostics [167].

Medical Applications. Silver nanoparticles are widely incorporated into wound dressings, and are used as an antiseptic and disinfectant in medical applications and in consumer goods. They have a high surface area per unit mass and release a continuous level of silver ions into their environment. The silver ions are bioactive and have broad-spectrum antimicrobial properties against a wide range of bacteria. By controlling the size, shape, surface, and agglomeration state of the *n*-Ag particles, specific silver ion release profiles can be developed for a given application.

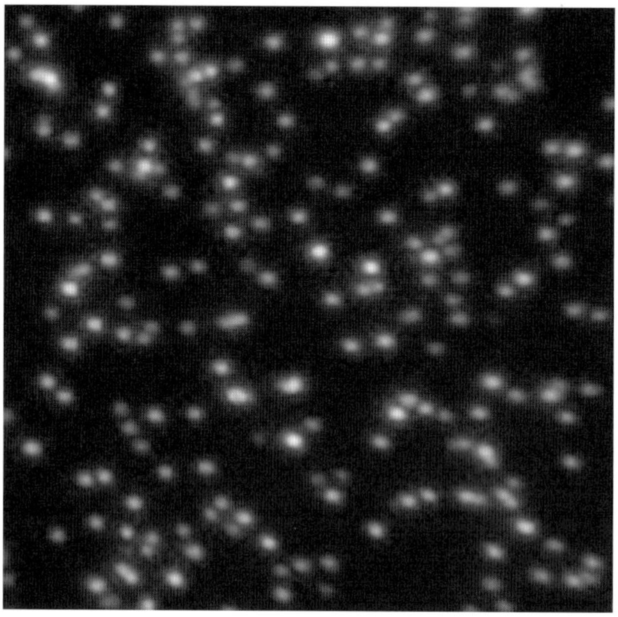

Figure 2.76. Dark field microscopy image of 60 nm *n*-Ag particles (Courtesy of nanoComposix).

Conductive Composites. Incorporation of *n*-Ag particles into polymers, composites, and adhesives increases the electrical and thermal conductivity of the material. Silver pastes and epoxies are widely utilized in the electronics industries. Silver nanoparticle-based inks are used to print flexible electronics and have the advantage in the fact that the melting point of the small *n*-Ag particles in the ink is reduced by hundreds of degrees compared to bulk silver. When sintered, these *n*-Ag particle-based inks have excellent conductivity.

Plasmonics. Silver nanoparticles have unique optical properties because they support surface plasmons. At specific wavelengths of light the surface plasmons are driven into resonance and strongly absorb or scatter incident light. This effect is so strong that it allows for individual nanoparticles as small as 20 nm in diameter to be imaged using a conventional dark field microscope. This strong coupling of metal nanostructures with light is the basis for the new field of plasmonics. Applications of plasmonic silver nanoparticles include biomedical labels, sensors, and detectors. Figure 2.76 shows the dark field microscopy image of nanoComposix's 60 nm silver nanoparticles [167]. It is also the basis for analysis techniques, such as Surface Enhanced Raman Spectroscopy (SERS) and Surface Enhanced Fluorescent Spectroscopy.

Photovoltaics. There is increasing interest in utilizing the large scattering and absorption cross sections of plasmonic silver nanoparticles for solar applications. Since the *n*-Ag particles act as efficient optical antennas, very high efficiencies can be obtained when the nanoparticles are incorporated into collectors. Figure 2.77 shows that silver nanosphere optical properties can be tuned by precisely controlling the nanoparticle size [167]. Here, nanoComposix's NanoXact silver nanospheres with sizes ranging from 20 nm to 80 nm are shown.

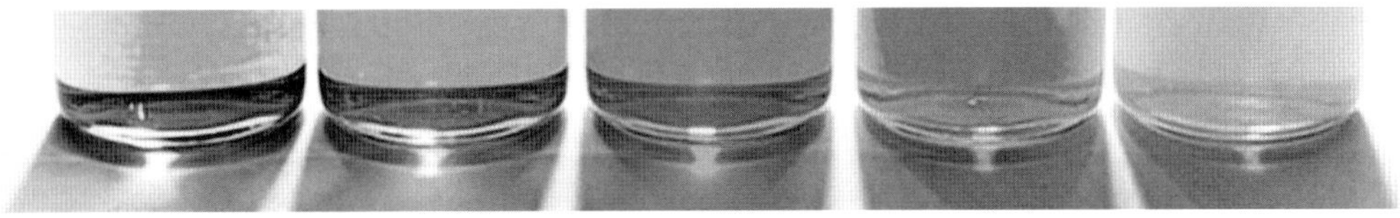

Figure 2.77. Silver nanosphere optical properties can be tuned by precisely controlling the nanoparticle size. Here, NanoXact silver nanospheres with sizes ranging from 20 nm to 80 nm are shown
(Courtesy of nanoComposix).

Nano-silver Particle Suppliers. There are several U.S. suppliers of nano-silver particles:

nanoComposix, Inc. is located in San Diego, California (www.nanocomposix.com). They specialize in silver nanoparticles and have several types of products: (1) precisely engineered silver nanospheres, (2) silver plates with tuned wavelength, (3) re-dispersible powders in customer's choice of solvent, (4) silica-coated spheres with increased versatility, (5) silver nanowires with high aspect ratio, and (6) custom nanoparticles with a particle size, material, or surface coating.

American Elements is located in Los Angeles, California (www.americanelements.com). They have five silver nanoparticles with 99%, 99.5%, 99.9%, 99.95%, and 99.99% purity. They are silver high surface particles with spherical morphology, cubic crystal phase, MW of 107.87, true density of 10.49 g/cc, bulk density of 0.312 g/cc, melting point of 962°C, boiling point of 2,162°C, diameter range of 80–100 nm, and specific surface area of 5.37 m^2/g.

Nanogap subnmparticles, Inc. is located in San Francisco, California (www.nanogap-usa.com). They have three types of silver products: (1) Atomic Quantum Clusters (AQC) are stable clusters of between 2 and 150 atoms, with average particle sizes in the 0.5–2 nm range; (2) NanoParticles are manufactured by different processes than those used to produce AQC, and are composed of larger particles, typically in the size range of 5 nm to 50 nm; and (c) NanoFibers are a high aspect ratio fibrous product supplied as a dispersion (containing a small amount of polymer to avoid aggregation).

NovaCentrix, LLC is located in Austin, Texas (www.nanocentrix.com). NovaCentrix manufactures silver nanopowder for both the conductive ink market and biological uses. It has an average particle size of 25 nm (using TEM image analysis) and a specific surface area of 23 m^2/g. X-ray diffraction data for silver nanopowder shows the material to be highly crystalline. Silver nanopowder is used in electrically conductive inks for inject and flexographic printing.

2.2.3.8 Nano-zinc oxide (*n*-ZnO)

Origin. Nano-zinc oxide powders and dispersions are inorganic zinc oxide nanoparticles with antibacterial, antifungal, anticorrosive, catalytic, and UV filtering properties [168, 169]. Particles are available in size range of 10 nm to 200 nm shown by Fan and Lu in their review article of *n*-ZnO [169]. They are also available as a nanofluid.

Nanofluids are generally defined as nanoparticles suspended in a solution either using surfactant or surface charge technology.

Applications. A variety of research is underway in the following areas:

Nano electronic and photonic materials. MEMS, NEMS, and piezoelectrics.

Bio nano materials. Bio markers, bio diagnostics, and bio sensors.

Medical. Sunscreen agent used in cosmetics, antibacterial, antiaging, UV protection, and dental cements.

High-tech. High-temperature lubricant in gas turbine engines, polymer nanocomposites, composites, flame retardant additives, and textiles.

Energy. Electrodes for solar cells, fuel cell layers, and pigments for paints.

Environmental. Environmental remediation, gas sensors, photocatalytic decontamination, attenuation of UV light, demilitarization of chemical and biological warfare agents.

Nano-zinc Oxide Particle Suppliers. There are several U.S. suppliers of nano-zinc oxide:

US Research Nanomaterials, Inc. is located in Houston, Texas (www.us-nano.com). Their ZnO nanopowder is a milky white powder with 99+% purity in single crystal phase with a nearly spherical morphology, average particle diameter of 10–30 nm, true density of 5.61 g/cc, and specific surface area of 20–60 m^2/g.

American Elements is located in Los Angeles, California (www.americanelements.com). They have three zinc oxide nanopowders with 99.9% (powder), 99.9% (dispersion), and 99.99% purity. The *n*-ZnO are spherical, white powders, with a MW of 81.37, true density of 5.6 g/cc, melting point of 1,975°C, boiling point of 2,360°C, diameter range of 10–200 nm, and exact mass of 79.92 g/mol.

Sciventions is located in Toronto, Canada (www.sciventions.com). Their ZnO nanoparticles are sold in a suspension in concentration of 1.5 mg/mL. It is a colorless, transparent suspension with UV absorption of 288 nm, pH of 7–8, and particle size of 1–10 nm (90%).

2.2.3.9 Others Nanoparticles

There are many other nanomaterials, such as barium sulfate ($BaSO_4$) [170–172], barium titanate ($BaTiO_3$) [173–175], calcium carbonate ($CaCO_3$) [176–178], Fe_3O_4 [179–181], Fe_2O_3 [182–184], nanodiamond [185–187], zinc sulfide (ZnS) [188–191], and others that have been used to produce polymer nanocomposites with enhanced material properties.

2.3 Summary

In this chapter we discussed several commonly used nanomaterials in one-, two-, and three-dimensional nanoscale materials needed to form polymer nanocomposites. The origin, properties, features, applications, and suppliers of these nanomaterials were presented, providing the readers with the fundamentals and properties of these nanomaterials and the knowledge to select the proper nanoparticles for their applications.

2.4 Questions and Problems

1. Give examples of other nanomaterials that are not discussed in this chapter and quote their references.
2. Why is organic treatment so important for montmorillonite nanoclays? Name and describe the dispersion mechanisms and equipment for nanoclay within the polymer. What are some advantages of Southern Clay Products' *Cloisite* Na^+ nanoclay?
3. For the nanoclay dispersion categories, is there a set standard for industries using this material? For example, when is an unmixed polymer-clay nanocomposite acceptable versus when an intercalated polymer-clay nanocomposite is acceptable?
4. What is the intended major goal of carbon nanofiber in the industry? Describe some of the multifunctional performance that carbon nanofibers can offer as well as the potential use for each of the factors listed.
5. Why have carbon nanotubes attracted so much attention for research? Why does carbon nanotube have such a stiff and robust structure? Why do materials like buckytubes and small-diameter carbon nanotubes possess such amazing properties?
6. Are there specific industries using specific carbon nanotubes (SWNT, MWNT, or DWNT) for their applications?
7. What are some special features of polymer-HNT nanocomposite properties enhancements?
8. What are some special features of nickel nanostrands?
9. What are the two unique structural factors that POSS possess and how do these factors enhance POSS? What is the key purpose of POSS technology and what are some of the property enhancements that POSS can offer?
10. How does nano-titanium dioxide compare to other flame retardant and improved heat stability polymer nanocomposites?
11. How does nano-titanium influence sunscreen? For instance, does nano-titanium actually reflect UV rays? If so, does the amount of nano-titanium influence the SPF rating of different sunscreens?
12. What is the one important characteristic of nano-silica that stands out? Are there any problems that occur with the introduction of these nanoparticles into polymers?

REFERENCES

1. T. J. Pinnavaia and G. W. Beall (Eds.) (2000). *Polymer-Clay Nanocomposites*. New York: John Wiley & Sons.
2. R. Krishnamoorti and R. A. Vaia (Eds.) (2001). *Polymer Nanocomposites: Synthesis, Characterization, and Modeling*. ACS Symposium Series 804, Washington, DC: American Chemistry Society.
3. J. H. Koo (2006). *Polymer Nanocomposites: Properties, Characterization, and Applications*. New York: McGraw-Hill.
4. A. B. Morgan and C. A. Wilkie (Eds.) (2007). *Flame Retardant Polymer Nanocomposites*. Hoboken, NJ: Wiley.

5. R. A. Gupta, E. Kennel, and K. J. Kim (Eds.) (2010). *Polymer Nanocomposites Handbook*. Boca Raton, FL: CRC Press.
6. V. Mittal (Ed.) (2010). *Polymer Nanotube Nanocomposites: Synthesis, Properties, and Applications*. Hoboken, NJ: Wiley.
7. V. Mittal (Ed.) (2010). *Optimization of Polymer Nanocomposites Properties*. Weinheim, Germany: Wiley-VCH.
8. V. Mittal (Ed.) (2011). *Thermally Stable and Flame Retardant Polymer Nanocomposites*. Cambridge: Cambridge University Press.
9. G. W. Beall and C. E. Powell (2011). *Fundamentals of Polymer-Clay Nanocomposites*. Cambridge: Cambridge University Press.
10. V. Mittal (Ed.) (2012). *Characterization Techniques for Polymer Nanocomposites*. Weinheim, Germany: Wiley-VCH.
11. B. Briell (2000). Nanoclay – Counting on Consistency, presented at *Nanocomposite 2000*, Southern Clay Products, Gonzales, TX.
12. Southern Clay Products, Gonzales, TX (www.nanoclay.com).
13. Nanocor, Chicago, IL (www.nanocor.com).
14. A. K. Geim and K. S. Novoselov (2007). The rise of graphene. *Nature Materials* (6), 183–191.
15. B. Z. Jang and A. Zhamu (2000). Processing of nanographene platelets (NGPs) and NGP nanocomposites: A review. *Journal of Materials Science* 43, 5092–5101.
16. K. S. Novoselov, A. K. Geim, S. V. Morozov, et al. (2004). Electric field effect in atomically thin carbon films. *Science* 306, 666–669.
17. K. S. Novoselov, D. Jiang, F. Schedin, et al. (2005). Two-dimensional atomic crystals. *Proceedings of the National Academy of Sciences USA* 102(30), 10451–10452.
18. B. Z. Jang and W. C. Huang (2006). US Patent 7,071,258 (July 4).
19. B. Z. Jang (2006). US Patent 11/442,903 (June 20); a divisional of 10/274,473 (October 22, 2002).
20. W. Schwalm, M. Schwalm, and B. Z. Jang (2004). Local Density of States for Nanoscale Graphene Fragments. American Physical Society, Paper No. C1.157, Montreal, Canada, March 2004.
21. M. J. McAllister, J. L. Li, D. H. Adamson, et al. (2007). Single sheet functionalized graphene by oxidation and thermal expansion of graphite. *Chemistry of Materials* 19(18), 4396–4404.
22. J. L. Li, K. N. Kudin, M. J. McAllister, et al. (2006). Oxygen-driven unzipping of graphitic materials. *Physical Review Letters* 96(17), 176101–176104.
23. H. C. Schniepp, J. L. Li, M. J. McAllister, et al. (2006). Functionalized single graphene sheets derived from splitting graphite oxide. *Journal of Physical Chemistry B* 110(17), 8535–8539.
24. X. Li, X. Wang, L. Zhang, S. Lee, et al. (2008). Chemically derived, ultrasmooth graphene nanoribbon semiconductors. *Science* 319(5867), 1229–1232.
25. K. S. Novoselov, A. K. Geim, S. V. Morozov, et al. (2005). Two-dimensional gas of massless Dirac fermions in graphene. *Nature* 438(7065), 197–200.
26. Y. Zhang, and T. Ando (2002). Hall conductivity of a two-dimensional graphite system. *Physical Review Letters* B 65(24), 245420–245431.
27. Y. Zhang, Y. W. Tan, H. L. Stormer, et al. (2005). Experimental observation of the quantum Hall Effect and Berry's phase in graphene. *Nature* 438, 201–204.
28. Y. Zhang, J. P. Small, M. E. Amori, et al. (2005). Electric field modulation of Galvanomagnetic properties of mesoscopic graphite. *Physical Review Letters* 94(17), 176803(4pp). doi:10.1103/PhysRevLett.94.176803.
29. C. Berger, Z. Song, T. Li, et al. (2004). Ultrathin epitaxial graphite: 2D Electron gas properties and a route toward graphene-based nanoelectronics. *The Journal of Physical Chemistry B* 108(52), 19912–19916. doi: 10.1021/jp040650f.

30. T. Enoki and Y. Kobayashi (2005). Magnetic nanographite: an approach to molecular magnetism. *Journal of Materials Chemistry* 15, 3999–4002. doi: 10.1039/b500274p.
31. H. B. Heersche, P. Jarillo-Herrero, J. B. Oostinga, et al. (2007). Bipolar supercurrent in graphene. *Nature Letter* 446, 56–59. doi: 10.1038/nature05555.
32. Y. W. Soon, M. L. Cohen, and S. G. Louie (2006). Half-metallic graphene nanoribbons. *Nature Letter* 444, 347–349. doi: 10.1038/nature05180.
33. J. C. Meyer, A. K. Geim, M. I. Katsnelson, et al. (2007). The structure of suspended graphene sheets. *Nature Letter* 446, 60–63. doi: 10.1038/nature 05545.
34. L. R. Bunnell, Sr. (1991). US Patent 987(4):175.
35. L. R. Bunnell, Sr. (1991). US Patent 019(5):446.
36. L. R. Bunnell, Sr. (1993). US Patent 186(5):919.
37. P. L. Zaleski, D. J. Derwin, R. J. Girkant, et al. (2001). US Patent 287(6):694.
38. S. Horiuchi, T. Gotou, M. Fujiwara, et al. (2004). Single graphene sheet detected in a carbon nanofilm. *Applied Physics Letter* 84, 2403–2405.
39. S. Horiuchi, T. Gotou, M. Fujiwara, et al. (2003). Carbon nanofilm with a new structure and property. *Japan Journal of Applied Physics* 42(Part 2), L1073–L1076. doi:10.1143/JJAP.42.L1073.
40. M. Hirata and S. Horiuchi (2003). US Patent 596(6):396.
41. M. Hirata, T. Gotou, and M. Ohba (2005). Thin-film particles of graphite oxide. 2: Preliminary studies for internal micro fabrication of single particle and carbonaceous electronic circuits. *Carbon* 43, 503–510. doi: 10.1016/j.carbon.2004.10.009.
42. M. Hirata, T. Gotou, S. Horiuchi, et al. (2004). Thin-film particles of graphite oxide 1: High-yield synthesis and flexibility of the particles.*Carbon* 42, 2929–2937. doi: 10.1016/j.carbon.2004.07.003.
43. J. D. Udy (2006). US Patent Application No. 11/243,285 (October 4); Pub. No. 2006/0269740 (November 30).
44. G. H. Chen, W. Weng, C. Wu, et al. (2004). Preparation and characterization of graphite nanosheets from ultrasonic powder technique. *Carbon* 42, 753–759. doi:10.1016/j.carbon.2003. 12.074.
45. B. Z. Jang, S. C. Wong, and Y. Bai (2005). US Patent Appl. No. 10/858,814 (June 3, 2004); Pub. No. US 2005/0271574 (December 8).
46. V. I. Petrik (2006). US Patent Appl. No. 11/007,614 (December 7, 2004); Pub. No. US 2006/0121279 (June 8).
47. L. T. Drzal and H. Fukushima (2006). US Patent Appl. No. 11/363,336 (February 27); 11/361,255 (February 24); 10/659,577 (September 10, 2003).
48. J. J. Mack, L. M. Viculis, R. B. Kaner, et al. (2005). US Patent 872(6):330.
49. L. M. Viculis, J. J. Mack, O. M. Mayer, et al. (2005). Intercalation and exfoliation routes to graphite nanoplatelets. *Journal of Material Chemistry* 15, 974–978. doi: 10.1039/B413029D.
50. W. Lu, P. Soukiassian, and J. Boecki (2012). Graphene: Fundamentals and functionalialities. *MRS Bulletin* (December), 37.
51. P. Muhopadhyay and R. K. Gupta (Eds.) (2013). *Graphite, Graphene and Their Polymer Nanocomposites*. Boca Raton, FL: CRC Press.
52. B. Z. Jang, A. Zhamu, and L. Song (2006). US Patent Application No. 11/324,370 (January 4).
53. L. Song, J. Guo, A. Zhamu, et al. (2006). US Patent Application No. 11/328,880 (January 11).
54. M. J. Sullivan and D. A. Ladd (2006). US Patent 7,156,756 (January 2, 2007) and No. 7,025,696 (April 11).
55. B. Z. Jang (2007). US Patent 186(7):474.

56. T. Szabo, A. Szeri, and I. Dekany (2005). Composite graphitic nanolayers prepared by self-assembly between finely dispersed graphite oxide and a cationic polymer. *Carbon* 43, 87–94. doi: 10.1016/j.carbon.2004.08.025.
57. X. Wang, L. Zhi, and K. Mullen (2008). Transparent, conductive graphene electrodes for dyesensitized solar cells. *Nano Letters* 8(1), 323–327. doi: 10.1021/nl072838r.
58. J. H. Koo, D. Pinero, A. Hao, S. C. Lao, B. Johnson, et al. (2013). Methodology for assessment of the morphological and thermal characteristics of nanographene platelets, AIAA-2013-1584. Presented at the 54th AIAA/ASME/ASCE/AHS/ASC, SDM, Boston, MA, April 8–11.
59. A. F. Ávila of Universidade Federal de Minas Gerais, Department of Mechanical Engineering, Belo Horizonte, Brazil (aavila@netuno.lcc.ufmg.br).
60. A. F. Ávila (2009). Composite Laminates Performance Enhancement by Nanoparticles Dispersion: An Investigation on Hybrid Nanocomposite. In *Composites Performance and Trends*, F. Columbus (Ed.). Hauppauge, NY: Nova Science Publishers.
61. S. G. Miller (2008). Effects of Nanoparticle and Matrix Interface on Nanocomposite Properties. Ph.D. dissertation, University of Akron, Akron, OH.
62. H. K. Schmidt of Rice University, Chemical and Biomolecular Engineering Dept., Houston, TX (hks@rice.edu).
63. XG Sciences, Inc. at East Lansing, MI (www.xgsciences.com).
64. Angstron Materials, LLC at Dayton, OH (www.angstronmaterials.com).
65. Skyspring Nanomaterials, Inc., Houston, TX (www.ssnano.com).
66. Cheap Tubes, Inc., Brattleboron, VT (www.cheaptubesinc.com).
67. R. Ruoff of Dept. of Mechanical Engineering, The University of Texas at Austin (r.ruoff@mail.utexas.edu). Abundant technical information can be found in Professor Ruoff's website: www.bucky-central.me.utexas.edu. Professor Ruoff has moved to Ulsan National Institute of Science and Technology (UNIST), Ulsan, S. Korea.
68. Nacional de Grafite, Sao Paulo, Brazil (www.grafite.com).
69. L. Qiu and B. Qu (2011). Polymer/Layered Double Hydroxide Flame Retardant Nanocomposites. In *Thermally Stable and Flame Retardant Polymer Nanocomposites*, V. Mittal (Ed.). Cambridge: Cambridge University Press, pp. 332–359.
70. Z. Matusinovic, and C. A. Wilkie (2012). Fire retardancy and morphology of layered double hydroxide nanocomposites: A review. *Journal of Material Chemistry* 22, 18701–18704.
71. B. M. Choudary, B. Bharathi, C. V. Reddy, M. L. Kantam, and K. V. Raghavan (2001). The first example of catalytic n-oxidation of tertiary amines by tungstate-exchanged mg-al layered double hydroxide in water: a green protocol. *Chemical Communications* 18, 1736–1737.
72. J. H. Choy, S. Y. Kwak, Y. J. Jeong, and J. S. Park (2000). Inorganic layered double hydroxides as nonviral vectors. *Angewandte Chemie International Edition* 39, 4042–4045.
73. L. Desigaux, M. Ben Belkacem, P. Richard, J. Cellier, P. Leone, et al. (2006). Self-assembly and characterization of layered double hydroxide/DNA hybrids. *Nano Letters* 6, 199–204.
74. M. Lakraimi, A. Legrouri, A. Barroug, A. de Roy, and J. P. Besse (1999). Removal of pesticides from water by anionic clays. *Journal de Chimie Physique et de Physico-Chimie* 96, 470–478.
75. D. Yan, J. Lu, M. Wei, J. Ma, D. G. Evans, and X. Duan (2009). A combined study based on experiment and molecular dynamics: Perylene tetracarboxylate intercalated in a layered double hydroxide matrix. *Physical Chemistry Chemical Physics* 11, 920–929.

76. Y. Tian, G. Wang, F. Li, and D. G. Evans (2007). Synthesis and thermo-optical stability of methyl red-intercalated Ni-Fe layered double hydroxide material. *Materials Letters* 61, 1662–1666.
77. A. V. Lukashin, A. A. Vertegel, A. A. Eliseev, M. P. Nikiforov, P. Gornert, and Y. D. Tretyakov (2003). Chemical design of magnetic nanocomposites based on layered double hydroxides. *Journal of Nanoparticle Research* 5, 455–464.
78. D. Mohan and C. U. Pittman (2007). Arsenic removal from water/wastewater using adsorbents: A critical review. *Journal of Hazardous Materials* 142, 1–53.
79. G. G. Tibbetts (1984). Why are carbon filaments tubular? *Journal of Crystal Growth* 66, 632–638.
80. M. L. Lake and J.-M. Ting, (1999). Vapor Grown Carbon Fiber Composites. In *Carbon Materials for Advanced Technologies*, T. D. Burchell (Ed.). Oxford: Pergamon, pp. 139-167.
81. G. G. Tibbetts, J. C., Finegan, J. J. McHugh, J.-M. Ting, D. G. Glasgow, and M. L. Lake (2000). Applications Research on Vapor-Grown Carbon Fibers. In *Science and Application of Nanotubes*, E. Tomanek and R. J. Enbody (Eds.). New York: Kluwer Academic/Plenum Publishers, pp. 35-51.
82. B. Maruyama and K. Alam (2002). Carbon nanotubes and nanofibers in composite materials. *SAMPE Journal* 38(3), 59–70.
83. D. G. Glasgow, G. G. Tibbetts, M. J. Matuszewski, K. R. Walters, and M. L. Lake (2004). Surface Treatment of Carbon Nanofibers for Improved Composite Mechanical Properties. *Proc. SAMPE 2004 Int'l Symposium*, SAMPE, Covina, CA.
84. G. G. Tibbetts, M. L. Lake, K. L. Strong, and B. P. Rice (2007). A review of the fabrication and properties of vapor-grown carbon nanofiber/polymer composites. *Computer Science and Technology* 67(7–8), 1709–1718.
85. M. Terrones (2003). Science and technology of the twenty-first century: Synthesis, properties, and applications of carbon nanotubes. *Annual Review of Materials Research* 33, 419–501. doi: 10.1146/annurev.matsci.33.012802.100255.
86. P. J. F. Harris (1999). *Carbon Nanotubes and Related Structures, New Materials for the Twenty-First Century*. Cambridge: Cambridge University Press.
87. K. Tanaka, T. Yamabe, and K. Fukui (1999). *The Science and Technology of Carbon Nanotubes*. Amsterdam: Elsevier.
88. R. Saito, G. Dresselhaus, and M. S. Dresselhaus (1998). *Physical Properties of Carbon Nanotubes*. London: Imperial College Press.
89. L. Dai (Ed.) (2006). *Carbon Nanotechnology*. Amsterdam: Elsevier.
90. M. S. Dresselhaus, G. Dresselhaus, and P. C. Eklund (1996). *Science of Fullerenes and Carbon Nanotubes*. San Diego, CA: Academic Press.
91. T. W. Ebbesen (1994). Carbon Nanotubes. *Annual Review of Materials Science* 24, 235–264. doi: 10.1146/annurev.ms.24.080194.001315.
92. T. Guo, P. Nikolaev, A. Thess, D. T. Colbert, and R. E. Smalley (1995). Catalytic growth of single-walled nanotubes by laser vaporization. *Journal Physics Letters* 243, 49–54. doi: 10.1016/0009-2614(95)00825-O.
93. M. Endo, K. Takeuchi, S. Igarashi, K. Kobori, M. Shiraishi, and H. W. Kroto (1993). The production and structure of pyrolytic carbon nanotubes. *Journal Physics and Chemistry of Solids* 54, 1841–1848.
94. O. Groning, O. M. Kuttel, C. Emmenegger, P. Groning, and L. Schlapbach (1999). Field Emission Properties of Carbon Nanotubes. *Journal of Vacuum Science & Technology B: Microelectronics and Nanometer Structures* B18, 665–678. doi: 10.1116/1.591258.
95. W. K. Hsu, J. P. Hare, M. Terrones, H. W. Kroto, D. R. M. Walton, and P. J. F. Harris (1995). Condensed-phase nanotubes. *Nature* 377, 687.
96. W. K. Hsu, M. Terrones, J. P. Hare, H. Terrones, H. W. Kroto, and D. R. W. Walton (1996). Electrolytic formation of carbon nanostructures. *Chemical Physics Letters* 262, 161–166.

97. S. Iijima (1991). Helical microtubules of graphitic carbon. *Nature* 354, 56–58. doi: 10.1038/354056a0.
98. S. Iijima and T. Ichihashi (1993). Single-shell carbon nanotubes of 1-nm diameter. *Nature* 363, 603–605. doi: 10.1038/363603a0.
99. D. S. Bethune, C. H. Kiang, M. S. de Vries, G. Gorman, R. Savoy, et al. (1993). The discovery of single-wall carbon nanotubes at IBM. *Nature* 363, 605–607.
100. A. Thess, R. Lee, P. Nikolaev, H. Dai, P. Petit, et al. (1996). Crystalline ropes of metallic carbon nanotubes. *Science* 273, 483–487. doi: 10.1126/science.273.5274.483.
101. P. M. Ajayan, O. Stephan, C. Colliex, and D. Trauth (1994). Aligned carbon nanotube arrays formed by cutting a polymer resin-nanotube composite. *Science* 265, 1212–1214. doi: 10.1126/science.265.5176.1212.
102. P. Kim, L. Shi, A. Majumdar, and P. L. McEuen (2001). Thermal transport measurements of individual multiwalled nanotubes. *Physical Review Letters* 87(21), 215502(4pp).
103. G. Cao (2004). *Nanostructures & Nanomaterials: Synthesis, Properties & Applications*. London: Imperial College Press.
104. K. Smith (2005). Carbon Nanotechnologies, Inc., Houston, TX, personal communication, June.
105. Nanocyl, Sambreville, Belgium, Nanocyl™ SWNT, DWNT, MWNT (www.nanocly.com).
106. Bayer MaterialScience, Leverkusen, Germany, Baytubes® MWNT (www.baytubes.com).
107. Arkema, Lacq, France, Graphistrength® MWNT (www.graphistength.com).
108. M. Du, B. Guo, and D. Jia (2010). Newly emerging applications of halloysite nanotubes: a review. *Polymer International* 59, 574–582. doi:10.1002/pi.275.
109. M. Liu, Z. Jia, D, Jia, and C. Zhou (2014). Recent advance in research on halloysite nanotubes-polymer nanocomposite. *Progress in Polymer Science* 39(8), 1498-1525. doi: 10.1016/j.progpolymsci.2014.04.004.
110. P. Yuan, D. Tan, and F. Annabi-Bergaya (2015). Properties and applications of halloysite nanotubes: recent research advances and future prospects. *Applied Clay Science* 112-113, 75-93. doi: 10.1016/j.clay.2015.05.001.
111. Y. Zhang, A. Tang, H. Yang, and J. Quyang (2016). Applications and interfaces of halloysite nanocomposites. *Applied Clay Science* 119, 8–17. doi: 10.1016/j.clay.2015.06.034.
112. P. Berthier (1826). Analyse de l'halloysite. *Annales de Chimie et de Physique* 32, 332–325.
113. M. I. Prudencio, M. A. S. Braga, H. Paquet, J. C. Waerenborgh, L. C. J. Pereira, and M. A. Gouveia (2002). Clay mineral as severages in weathered basalt profiles from central and southern Portugal: Climate. *Catena* 49(1), 77–89. doi: 10.1016/S0341-8162(02)00018-8.
114. E. Joussein, S. Petit, J. Churchman, B. Theng, D. Righi, and B. Delvaux (2005). Halloysite clay minerals-a review. *Clay Minerals* 40, 383-426.
115. S. Nakagaki and F. Wypych (2007). Nanofibrous and nanotubular supports for the immobilization of metalloporphyrins as oxidation catalysts. *Journal of Colloid and Interface Science* 315(1), 142–157.
116. I. R. Wilson (2004). Kaolin and halloysite deposits of China. *Clay Minerals* 39(1), 1–15.
117. A. Perruchot, C. Dupuis, E. Brouard, D. Nicaise, and R. Ertus (1997). L'halloysite Kartstique: Comparaison des Gisements Types de Wallonie (Belgique) et du Perigord (France). *Clay Minerals* 32(2), 271–287. doi: 10.1180/claymin.1997.032.2.08.

118. J. T. Kloprogge and R. L. Frost (1999). Raman microprobe spectroscopy of hydrated halloysite from Neogene Cryptokarst from Southern Belgium. *Journal of Raman Spectroscopy* 30, 1079–1085.
119. G. J. Churchman and B. K. G. Theng (2002). Clay research in Australia and New Zealand. *Applied Clay Science* 20(4-5), 153–156.
120. C. Q. Kautz and P. C. Ryan (2003). The 10 Å and 7 Å halloysite transition in a tropical soil sequence, Costa Rica. *Clays and Clay Minerals* 51(3), 252–263. doi: 10.1346/CCMN.2003.0510302.
121. S. Hillier and P. C. Ryan (2002). Identification of halloysite (7 Å) by ethylene glycol solvation: the 'MacEwan effect.' *Clay Minerals* 37, 487–496. doi: 10.1180/0009855023730047.
122. M. L. Du, B. C. Guo, X. J. Cai, Z. X., Jia, M. X. Liu, and D. M. Jia (2008). Morphology and properties of halloysite nanotubes reinforced polypropylene nanocomposites. *e-Polymers* 130, 1–14.
123. Y. P. Ye, H. B. Chen, J. S. Wu, and L. Ye (2007). High strength epoxy nanocomposites with natural nanotubes. *Polymer* 48, 6426–6433.
124. M. X. Liu, B. C. Guo, M. L. Du, X. J. Cai, and D. M. Jia (2007). Properties of halloysite nanotube-epoxy resin hybrids and the interfacial reactions in the systems. *Nanotechnology* 18, 455703 (9pp).
125. T. Imai, Y. Naitoh, T. Yamamoto, and M. Ohyanagi (2006). Translucent nano mullite based ceramic fabricated by spark plasma. *Journal of the Ceramic Society of Japan* 114, 138–140.
126. *Am Ceram Soc Bull* (2007): 86, pp. A19.
127. M. L. Du, B. C. Guo, M. X. Liu, and D. M. Jia (2006). Preparation and characterization of polypropylene grafted halloysite and their compatibility effect of polypropylene/halloysite. *Polymer Journal*. 38, 1198–1204. doi: 10.1295/polyj.PJ2006038.
128. J. Ma, P. Xiang, Y. W. Mai, and L. Q. Zhang (2004). A novel approach to high performance elastomer by using clay. *Macromolecular Rapid Communication* 25, 1692–1696.
129. B. C. Guo, Y. D. Lei, F. Chen, X. L. Liu, M. L. Du, and D. M. Jia (2008). Styrene-butadiene rubber/halloysite nanotubes nanocomposites modified by methacrylic acid. *Applied Surface Science*: 255, 2715-2722. doi: 10.1016/j.apsusc.2008.07.188.
130. M. L. Du, B. C. Guo, Y. D. Lei, M. X. Liu, and D. M. Jia (2008). Carboxylated butadiene-styrene rubber/halloysite nanotube nanocomposites: interfacial interaction and performance. *Polymer* 49(22), 4871–4876.
131. B. C. Guo, F. Chen, Y. D. Lei, W. Y. Zhou, and D. M. Jia (2010). Tubular clay composites with high strength and transparency. *Journal of Macromolecular Science B: Physics* 49, 111–121.
132. M. X. Liu, B. C. Guo, M. L. Du, Y. D. Lei, and D. M. Jia (2008). Natural inorganic nanotubes reinforced epoxy resin nanocomposites. *Journal of Polymer Research* 15, 205–212.
133. M. L. Du, B. C. Guo, and D. M. Jia (2006). Thermal stability and flame retardant effects of halloysite nanotubes on poly(propylene). *European Polymer Journal* 42, 1362–1369. doi: 10.1016/j.eurpolymj.2005.12.006.
134. T. Labour, C. Gauthier, R. Seguela, G. Vigier, Y. Bomal, and G. Orange (2001). Influence of the β crystalline phase of the mechanical properties of unfilled and $CaCO_3$-filled polypropylene. I. Structural and mechanical characterization. *Polymer* 42, 7127–7135.
135. P. Tordjeman, C. Robert, G. Marin, and P. Gerard (2001). The effect of α, β crystalline structure on the mechanical properties of polypropylene. *European Physics Journal E* 4, 459–465.

136. N. Y. Ning, Q. J. Yin, F. Luo, Q. Zhang, R. Du, and Q. Fu (2007). Crystallization behavior and mechanical properties of polypropylene/halloysite composites. *Polymer* 48, 7374–7384. doi: 10.1016/j.polymer.2007.10.005.
137. M. L. Du, B. C. Guo, J. J. Wan, Q. L. Zou, and D. M. Jia (2010). Effects of halloysite nanotubes on kinetics and activation energy of non-isotherm crystallization of polypropylene. *Journal of Polymer Research* 17, 109–118.
138. Nanostrand User Guide, Conductive Composites, Heber City, Utah (www.conductivecomposites.com).
139. Conductive Composites, Huber City, Utah, (www.conductivecomposites.com).
140. ANF Technology Ltd, Warlingham, Surrey, United Kingdom (www.nafen.eu).
141. Hybrid Plastics, Inc., Hattiesburg, Mississippi (www.hybridplastics.com).
142. M. G. Voronkov and V. I. Vavrent'yev (1982). Polyhedral oligosilsesquioxanes and their homo derivatives. *Topics in Current Chemistry* 102, 199–236.
143. P. A. Agaskar, W. G. Klemperer (1995). The higher hydridospherosiloxanes: synthesis and structures of $H_nSi_nO_{1.5n}$ (n=12, 14, 16, 18). *Inorganica Chimica Acta* 229, 355–364.
144. R. H. Baney, M. Itoh, A. Sakakibara, and T. Suzuki (1995). Silsesquiosanes. *Chemical Reviews* 95(5), 1409–1430.
145. J. D. Lichtenhan (1995). Polyhedral oligomeric silsesquioxanes: Building blocks for silsesquioxane-based polymers and hybrid materials. *Comments on Inorganic Chemistry* 17(2), 115–130. doi: 10.1080/02603599508035785.
146. J. D. Lichtenhan, (1996). In *Polymeric Materials Encyclopaedia*, J. C. Salamore (Ed.). Boca Raton, FL: CRC Press, pp. 7769–7778.
147. G. Z. Li, L. C. Wang, H. L. Ni, and C. U. Pittman Jr.(2001). Polyhedral oligomeric silsesquioxane (POSS) polymers and copolymers: A review. *Journal of Inorganic and Organometallic Polymers* 11(3), 123–154. doi: 10.1023/A: 1015287910502
148. S. H. Phillips, T. S. Haddad, and S. J. Tomczak (2004). Developments in nanoscience: polyhedral oligomeric silsesquioxane (POSS)-polymers. *Current Opinion in Solid State & Materials Science* 8, 21–29. doi: 10.1016/j.cossms.2004.03.002.
149. U. Sorathia, and I. Perez (2004). Improving fire performance characteristics of composite materials for naval applications. *Polymeric Materials: Science & Engineering* 91, 292–296.
150. C. Hartman-Thompson (Ed.) (2011). *Applications of Polyhedral Oligomeric Silsesquioxanes*. New York: Springer.
151. Technical Bulletin AEROSIL® No. 27, Degussa AG, D-63403 Hanau-Wolfgang, Germany, October 2001.
152. Technical Bulletin AEROSIL® No. 56, Degussa AG, D-63403 Hanau-Wolfgang, Germany, October 1990.
153. Technical Bulletin AEROSIL® Fumed Silica, Degussa AG, D-63403 Hanau-Wolfgang, Germany, September 2002.
154. S. Sprenger and M. Pyrlik (2004). Nanoparticles in Composites and Adhesives: Synergy with Elastomers. *Proceedings of the 11th International Conference on Composites/Nano Engineering*, Hilton Head Island, SC, August.
155. F. Yang, R. Yngard, and G. L. Nelson (2005). Flammability of polymer-clay and polymer-silica nanocomposites. *Journal of Fire Sciences* 23, 209–226.
156. U.S. Patent Application, 20040147029 (July 29, 2004).
157. J. Cinquin, S. Bechtel, K. Schmidtke, and T. Meer (2004). Polymer Nano-Composites of Aeronautic Applications: From Dream to Reality? *Proceedings of the 11th International Conference on Composites/Nano Engineering*, Hilton Head Island, SC, August.
158. A. D. Pool and H. T. Hahn (2003). A Nanocomposite for Improved Stereolithography. *Proceedings of the 2003 SAMPE ISSE*, SAMPE, Covina, CA.

159. Inorganic Specialty Chemicals-Alumina Nano-particles, Sasol NA, Houston, TX.
160. Disperal®/Dispal®-High purity dispersible alumina. Technical datasheet, Sasol NA, Germany.
161. H. Huang, M. Tian, L. Liu, W. Liang, and L. Zhang (2006). Effect of particle size of flame retardancy of Mg $(OH)_2$-filled ethylene vinyl acetate copolymer composites. *Journal of Applied Polymer Science* 100, 4461–4469.
162. T. Chen and A. Isarov (2007). New Magnesium Hydroxides Enabling Low-Smoke Cable Compounds. 56th IWCS Conference, Pittsburg, PA, November.
163. V. Yong and H. T. Hahn (2004). Kevlar/Vinyl Ester Composites with SiC Nanoparticles. *Proceedings of the 2004 SAMPE ISSE*, SAMPE, Covina, CA.
164. Y. Sakka, D. D. Bidinger, and I. A. Aksay (1995). Processing of silicon carbide-mullite-alumina nanocomposites. *Journal of the American Ceramics Society* 78(21), 479–486.
165. P. Padhi and K. Sachikanta (2011). A Novel Route for Development of Bulk Al/SiC Metal Matrix Nano Composites. Department of Mechanical Engineering, Konark Institute of Science & Technology, Bhubaneswar, India & Central Tool Room of Training Center, Bhubaneswar, India.
166. A. Kassiba et al. (2007). Some fundamental and applicative properties of [polymer/nano-SiC] hybrid nanocomposites. *Journal of Physics: Conference Series Volume 79*, doi:10.1088/1742–6596/79/1/012002.
167. S. J. Oldenburg (2005). Silver Nanoparticles: Properties and Applications. San Diego, CA: nanoComposix.
168. Z. L. Wang (2004). Zinc oxide nanostructures: Growth, properties and applications. *Journal of Physics: Condensed Matter* 16, R829–R858.
169. Z. Fan and J. G. Lu (2005). Zinc oxide nanostructures: Synthesis and properties. *Journal of Nanoscience and Nanotechnology* 5(10), 1–13.
170. A. Ricker, P. Liu-Snyder, and T. J. Webster (2008). The influence of nano MgO and $BaSO_4$ particle size additives on properties of PMMA bone cement. *International Journal of Nanomedicine* 3(1), 125–132.
171. G. Aninwene, D. Stout, Z. Yang, and T. J. Webster (2013). Nano-$BaSO_4$: a novel antimicrobial additive to pellethane. *International Journal of Nanomedicine* 8, 1197–1205.
172. G. Aninwene, D. A. Stout, Z. Yang, and T. J. Webster (2013). Nano $BaSO_4$: A Novel Means to Create Antimicrobial Radiopaque Thermoplastics. *Proceedings of the 2013 AlChE Annual Meeting*, November 3–8, San Francisco, CA.
173. C. V. Chanmal and J. P. Jog (2008). Dielectric relaxations in PVDF/$BaTiO_3$ nanocomposites. *Express Polymer Letters* 2(3), 294–301.
174. H. Beltran, N. Maso, E. Cordoncillo, and A. R. West (2007). Nanocomposite ceramics based on La-doped $BaTi_2O_3$ and $BaTiO_3$ with high temperature-independent permittivity and low dielectric loss. *Journal of Electroceramics* 18(3–4), 277–282.
175. K. C. Singh and C. Jiten (2013). Production of $BaTiO_2$, nanocrystalline powders by high energy milling and piezoelectric properties of corresponding ceramics. *Key Engineering Materials* 547, 133–138.
176. A. Chatterjee and S. Mishra (2013). Rheological, thermal, and mechanical properties of nano-calcium carbonate ($CaCO_3$)/Poly(methyl methacrylate) (PMMC) core-shell nanoparticles reinforced polypropylene (PP) composites. *Macromolecular Research* 21(5), 474–483.
177. K. Shelesh-Nezhad, H. Orang, and M. Motallebi (2012). The Effects of Adding Nano-Calcium Carbonate Particles on the Mechanical and Shrinkage Characteristics and Molding Process Consistency of PP/nano-$CaCO_3$ Nanocomposites. In *Polypropylene*, F. Dogan (Ed.), pp. 357–368,

ISBN: 978-953-51-0636-4, InTech (www.intechopen.com). doi: 10.5772/35272. Available from: http://www.intechopen.com/books/polypropylene/the-effects-of-adding-nano-calcium-carbonate-particles-on-the-mechanical-and-shrinkage-character.

178. T. Sato and J. J. Beaudoin (2011). Effect of nano-$CaCO_3$ on hydration of cement containing supplementary cementitious materials. *Advances in Cement Research* 23(1), 33–43.
179. C. Hu, Z. Mou, G. Lu, N. Chen, Z. Dong, et al. (2013). 3D graphene-Fe_3O_4 nanocomposites with high-performance microwave absorption. *Physical Chemistry Chemical Physics* 15, 13038–13043.
180. H. Gu, Y. Huang, X. Zhang, Q. Wang, J. Zhu, et al. (2012). Magnetoresistive polyaniline-magnetite nanocomposites with negative dielectric properties. *Polymer* 53, 801–809.
181. K. Kalantari, M. B. Ahmad, K. Shemeli, and R. K. Khandanlou (2013). Synthesis of talc/Fe_3O_4 magnetic nanocomposites using chemical co-precipitation method. *International Journal of Nanomedicine* 8, 1817–1823.
182. A. Mahapatra, B. G. Mishra, and G. Hota (2013). Electrospun Fe_2O_3-Al_2O_3 nanocomposite fibers as efficient absorbent for removal of heavy metal ions from aqueous solution. *Journal of Hazard Materials* 258–259, 116–123.
183. D. Ortega, J. S. Garitaonandia, C. Barrera-Solano, M. Ramirez-del-Solar, E. Blanco, and M. Dominguez (2006). γ-Fe_2O_3/SiO_2 nanocomposites for magneto-optical applications: Nanostructural and magnetic properties. *Journal of Non-Crystalline Solids* 352, 2801–2810.
184. L. Menon, S. Patibandla, K. Bhargava Ram, S. I. Shkuratov, D. Aurongzeb, et al. (2004). Ignition studies of Al/Fe_2O_3 energetic nanocomposites. *Applied Physics Letters* 84(23), 4735–4737.
185. S. V. Kidalov, F. M. Shakhov, and A. Y. Vul (2007). Thermal conductivity of nanocomposites based on diamonds and nanodiamonds. *Diamond and Related Materials* 16(12), 2063–2066.
186. V. N. Mochalin, O. Shenderova, D. Ho, and Y. Gogotsi (2012). The properties and applications of nanodiamonds. *Nature Nanotechnology* 7, 11-23. doi: 10.1038/nnao.2011.209.
187. I. Neitzel (2012). Nanodiamond-Polymer Composites, Ph.D. dissertation, Drexel University, Dept. of Materials Engineering, Philadelphia, PA.
188. D. Pugh-Thomas, B. M. Walsh, and M. C. Gupta (2011). CdSe (ZnS) nanocomposite luminescent high temperature sensor. *Nanotechnology* 22(18), 185503 (7pp). doi:10.1088/0957-4484/22/18/185503.
189. S. Pan and Z. Liu (2012). ZnS-Graphene nanocomposites: Synthesis, characterization and optical properties. *Journal of Solid Chemistry* 191, 51–56.
190. S. Ummartyotin, N. Bunnak, J. Juntaro, M. Sain, and H. Manuspiya (2012). Hybrid organic-inorganic of ZnS embedded PVP nanocomposite film for photoluminescent application. *Computes Rendus Physique* 13(9–10), 994–1000.
191. B. N. Patil and S. A. Acharya (2013). Preparation of ZnS-graphene nanocomposites and its photocatalytic behavior for dye degradation. *Advanced Materials Letters* (May 12). doi: 10.5185/amlett.2013.fdm.16.

3 Basics of Polymer Matrices and Composites

3.1 Classification of Materials

Solid materials are often classified as metals, ceramics, and polymers. This scheme is based primarily on chemical makeup and atomic structures, and most materials fall into one distinct grouping or another, although there are some intermediates. In addition, there are the composites, combinations of two or more of the basic material classes. Polymers are organic compounds that are chemically based on carbon, hydrogen, and other nonmetallic elements (e.g., O, N, and Si). They have very large molecular structures, often chain-like in nature, with a backbone of carbon atoms. Some of the common polymers are polyethylene (PE), nylon, poly(vinyl chloride) (PVC), polycarbonate (PC), polystyrene (PS), silicone rubber, epoxy, and phenolic. Polymers are different from the other materials in numerous ways but generally possess lower densities, thermal conductivities, and moduli. The lower densities of polymeric materials offer an advantage for applications where weight is a concern. The addition of thermally and/or electrically conductive fillers allows polymers to be used for insulative or conductive applications. As a result, polymers may find applications in electromagnetic interference (EMI) shielding and antistatic protection applications. Polymers can further be classified into three basic polymeric categories: *thermoplastics*, *thermosets*, and *elastomers*.

3.2 Types of Polymers

Thermoplastics, consisting of individual long-chain molecules, can be reprocessed; products can be graduated and fed back into the appropriate machine. *Thermosets* contain an infinite three-dimensional network, which is only created when the product is in its final form, and cannot be broken down by reheating. *Elastomers (rubbers)* contain looser three-dimensional networks, where the chains are free to change their shapes. Neither thermosets nor elastomers can be reprocessed. Some polymers, such as polyurethanes, can be produced in both thermoplastic and thermoset variants.

Table 3.1. *U.S. Resin Production, Sales, and Captive Use in 2013 and 2014 (millions of pounds, dry weight basis)*[a]

Resin	Production			Total Sales & Captive Use		
	2014	2013	% Chg	2014	2013	% Chg
Epoxy[b]	561	499	12.4	532	484	9.9
Other Thermosets[c]	15,253	14,476	5.4	15,201	14,433	5.3
Total Thermosets	**15,814**	**14,975**	**5.6**	**15,733**	**14,917**	**5.5**
LDPE[b,d]	7,112	6,919	2.8	6,967	6,940	0.4
LLDPE[b,d]	13,856	13,853	0.0	13,731	13,833	−0.7
HDPE[b,d]	17,514	17,899	−2.2	17,599	17,810	−1.2
PP[b,e]	16,446	16,427	0.1	16,356	16,396	−0.2
PS[b,e]	4,455	4,515	−1.3	4,437	4,581	−3.1
EPS[b,d]	953	890	7.1	945	899	5.1
Nylon[b,e]	1,299	1,238	4.9	1,301	1,242	4.8
PVC[d]	15,038	15,373	−2.2	14,994	15,255	−1.7
Other Thermoplastics[f]	15,625	15,433	1.2	16,913	16,787	0.8
Total Thermoplastics	**92,298**	**92,547**	**−0.3**	**93,243**	**93,743**	**−0.5**
GRAND TOTAL PLASTICS	108,112	107,522	0.5	108,976	108,660	0.3

[a] Except phenolic resins, which are reported on a gross weight basis.
[b] Sales and captive use data include imports.
[c] Includes: polyurethanes (TDI, MDI, and polyols), phenotic, urea, melamine, unsaturated polyester, and other thermosets.
[d] Canadian production and sales data included.
[e] Canadian and Mexican production and sales data included.
[f] Includes: PET, ABS, engineering resins, SB latex, and other thermoplastics.

Source: American Chemistry Council, www.americanchemistry.com. Reprinted with permission.

3.2.1 Thermoplastics

The relative importance of thermoplastics can be judged from their U.S. annual production, sales, and captive use data (Table 3.1) [1]. The 2014 total thermosets production was 15,814 million lb. (14.62% of the total resins) vs. total thermoplastics production of 92,298 million lb. (85.38% of the total resins). The production rate increase of thermoplastics and thermoset from 2013 to 2014 was –0.3% and 5.6%, respectively. The first five items in the table are regarded as *commodity thermoplastics*. Many manufacturers compete to supply these polymers. Prices change quite rapidly, in response to the price of crude oil, so the table indicates relative prices. The remaining thermoplastics in Table 3.1 are called *engineering thermoplastics* because of their superior mechanical properties. They are produced on a smaller scale and have prices about twice that of the commodity thermoplastics. Finally, there are specialty plastics that only sell a few thousand tons annually. An example is polytetrafluoro ethylene (PTEE), which has unique low-friction properties.

Thermoplastics can be divided into amorphous and semi-crystalline solids. The amorphous polymers are glassy at temperatures lower than T_g (glass transition temperature) and rubbery liquids at higher temperatures. Semi-crystalline thermoplastics have an amorphous phase and a crystalline phase with a melting temperature,

Table 3.2. *Polymer Densities, Melting and Transition Temperatures of Common Thermoplastics*

Abbreviations	Polymer	Density (kg m^{-3})	T_g(°C)	T_m(°C)	Even if bent through 90°
Semi-crystalline plastics					
P4MP	Poly (4-methyl-pentene-l)	830	25	238	Semi-brittle
PP	Polypropylene	900–910	−10	170	Whitens
LDPE	Low-density polyethylene	920–925	−120	120	Ductile
MDPE	Medium-density polyethylene	935–945	−120	130	Ductile
HDPE	High-density polyethylene	955–965	−120	140	Ductile
PA 6	Polyamide 6	1,120–1,150	50	228	Ductile
PA 66	Polyamide 6,6	1,130–1,160	57	265	Ductile
PET	Polyethylene terephthalate	1,336–1,340	80	260	Ductile
POM	Polyoxymethylene (Acetal)	1,410	−85	170	Semi-brittle
PVDC	Polyvinylidene chloride	1,750	−18	205	
PTFE	Polytetrafluoro ethylene	2,200	−73	332	Ductile
Glassy plastics					
PS	Polystyrene	1,050	100		Brittle
SAN	Styrene acrylonitrile copolymer	1,080	100		
ABS	Acrylonitrile butadiene styrene copolymer	990–1,100	100		Whitens
PC	Polycarbonate	1,200	145		Ductile
PVCu	Polyvinyl chloride unplasticised	1,410	80		Ductile
PMMA	Polymethyl methacrylate	1,190	105		Brittle

T_m, crystal melting temperature; T_g, glass transition temperature.

T_m. The densities, glass transition, and crystal melting temperatures of some commonly used thermoplastics are listed in Table 3.2 [2].

3.2.2 Thermosets

The basic thermosetting resins all possess a commonality is that they will, upon exposure at elevated temperature from ambient to upward of 232°C, undergo an irreversible chemical reaction often referred to as *polymerization* or cure [3]. Each family has its own set of individual chemical characteristics based on its molecular component

Table 3.3. *Comparisons of Thermoplastic and Thermosetting Resin Characteristics*

Thermoplastic Resin	Thermosetting Resin
▪ High MW solid	▪ Low MW liquid or solid
▪ Stable material	▪ Low-medium viscosity, requires cure
▪ Re-processable, recyclable	▪ Cross-linked, non-processable
▪ Amorphous or crystalline	▪ Liquid or solid
▪ Linear or branched polymer	▪ Low MW oligomers
▪ Liquid solvent resistance	▪ Excellent environmental/solvent resistance
▪ Short process cycle	▪ Long process cycle
▪ Neat up to 30% filler	▪ Long or short fiber reinforced
▪ Injection/Compression/Extrusion	▪ RTM/FW/SMC/Prepreg/Pultrusion
▪ Limited structural components	▪ Many structural components
▪ Neat resin + nanoparticles	▪ Neat or fiber reinforced + nanoparticles
▪ Commodity – high-performance areas for automotive, appliance housings, toys	▪ Commodity – advanced materials for construction, marine, aircraft, aerospace

and its ability to either homopolymerize, copolymerize, or both. This transformation process represents the line of demarcation separating the thermosets from the thermoplastic polymers. Crystalline thermoplastic polymers are capable of a degree of crystalline cross-linking, but there is little, if any, of the chemical cross-linking that occurs during the thermosetting reaction. The important beneficial factor lies in the inherent enhancement of thermoset resins in their physical, electrical, thermal, and chemical properties due to their ability to maintain and retain these enhanced properties when exposed to severe environments. The cross-linking reaction, which occurs during the fabrication of thermosets, also provides good adhesion to other materials. As a result, epoxy and polyester resins are used for fiber-reinforced composites, and phenolics are used for bonding fibers to brake pads, and sand for metal casting. Table 3.3 shows some comparison of thermoplastic and thermoset resin characteristics [4].

A *thermosetting matrix* is defined as a composite matrix capable of curing at some temperature from ambient to several hundred degrees and cannot be reshaped by reheating. In general, thermosets contain two or more ingredients, a resin matrix with a curing agent that causes the matrix to polymerize (cure) at room temperature. Thermosets can also be thermally cured at elevated temperature with a long curing cycle without a curing agent. Some commercially available resins (matrices) are polyester and vinyl ester, polyurea, epoxy, phenolic, bismaleimide, polyimide, cyanate ester, and phenyl triazine.

Polyester and vinyl esters. Polyester matrices have had the longest period of use, with wide application in many large structural applications. They will cure at room temperature with a catalyst (peroxide), which produces an exothermic reaction. The resultant polymer is nonpolar and very water resistant, making it an excellent choice for marine constructions. The isopolyester resins are the most water-resistant polymers in this polymer group. They have been chosen as the prime matrix material for use for a fleet of U.S. Navy minesweepers.

Epoxy. The most widely used matrices for advanced composites are the epoxy resins, even though they are more costly and do not have the high-temperature

Table 3.4. *Epoxy Resin Selection Factors [3]*

Advantages
- Adhesion to fibers and to resins
- No by-products formed during cure
- Low shrinkage during cure
- Solvent and chemical resistance
- High or low strength and flexibility
- Resistance to creep and fatigue
- Good electrical properties
- Solid or liquid resins in uncured state
- Wide range of curative options

Disadvantages
- Resins and curatives somewhat toxic in uncured form
- Absorb moisture – heat distortion point lowered by moisture and change in dimensions and physical properties due to moisture absorption
- Limited to about 200°C for upper temperature use (dry)
- Difficult to combine toughness and high temperature resistance
- High thermal coefficient of expansion
- High degree of smoke liberation in a fire and flammable
- May be sensitive to ultraviolet light of degradation
- Slow curing

capability of the bismaleimides or polyimides. The selection factors of epoxy are listed in Table 3.4 [3]. When epoxy is reinforced with fibers, it forms polymer matrix composites (PMCs). The properties of continuous and aligned glass-, carbon-, and aramid-fiber reinforced epoxy composites are included in Table 3.5 [5]. Thus, a comparison of the mechanical characteristics of these three composites may be made in both longitudinal and transverse directions.

Bismaleimides (BMIs). The BMI resins have found their niche in high-temperature aircraft design applications where temperature requirements are in the 177°C (350°F) range. BMI is the primary product and is based on the reaction product from methylene dianiline (MDA) and maleic anhydride: bis (4 maleimidophynyl) methane (MDA BMI). Variations of tis polymer with compounded additive to improve impregnation are now on the market and can be used to impregnate suitable reinforcement to result in high-temperature mechanical properties.

Polyimides (PIs). The PI resins are the highest temperature polymers in the general advanced composite, with a long-term upper temperature limit of 232 to 316°C (450 to 600°F).

Polyureas. Polyureas involve the combination of novel MDA polymers and either amine or imino-functional polyether polyols. The resin systems can be reinforced with milled glass fibers, flaked glass, wollastanite, or treated mica, depending on the compound requirements as to processability or final product.

Cyanate ester (CE) and phenolic triazine (PT). The CE resins have shown superior dielectric properties, and much lower moisture absorption than any other structural resin for composites. The physical properties of CE resins are comparable to

Table 3.5. *Properties of Continuous and Aligned Glass-, Carbon-, and Aramid-Fiber Reinforced Epoxy Matrix Composites in Longitudinal and Transverse Directions (Fiber Volume Fraction is 0.60 in All Cases) [5]*

Property	Glass (E-glass)	Carbon (High Strength)	Aramid (Kevlar 49)
Specific gravity	2.1	1.6	1.4
Tensile modulus			
Longitudinal [GPa (10^6 psi)]	45 (6.5)	145 (21)	76 (11)
Transverse [GPa (10^6 psi)]	12 (1.8)	10 (1.5)	5.5 (0.8)
Tensile strength			
Longitudinal [GPa (10^6 psi)]	1020 (15)	1240 (180)	1380 (200)
Transverse [GPa (10^6 psi)]	40 (5.8)	41 (6)	30 (4.3)
Ultimate tensile strain			
Longitudinal	2.3	0.9	1.8
Transverse	0.4	0.4	0.5

those of a representative BMI resin. The PT resins also possess superior elevated temperature properties, along with excellent properties at cryogenic temperatures. They are available in several viscosities, ranging from a viscous liquid to powder, which facilitates their applications that use liquid resins, such as filament winding (FW) and resin transfer molding (RTM).

3.2.3 Elastomers

Another important group of polymers is the group that is elastic or rubberlike, known as *elastomers*. This group of materials including thermoplastic elastomers (TPEs), melt processable rubbers (MPRs), thermoplastic vulcanizate (TPV), synthetic rubbers, and natural rubbers (NR) [6]. In applications, such as conveyor belts and pressure hoses, thick layers of either steel wire or polymeric fiber reinforcement take the main mechanical loads. These layers, with elastomer interlayers, allow flexibility in bending, whereas the reinforcement limits the in-plane stretching of the product. The applications are dominated by natural rubber (NR) and styrene butadiene copolymer rubber (SBR). Other rubbers have specialized properties: butyl rubbers have low air permeability, nitrile rubbers have good oil resistance, while silicone rubbers have high and low temperature resistance. The rubbery behavior of the amorphous phase in semi-crystalline thermoplastics is an important property.

3.2.4 Fillers

The term "fillers" refers to solid additives that are incorporated into the polymer matrix. Fillers are generally inorganic materials and can be classified according to their effect on the mechanical properties of the resulting mixture. Inert or extender fillers are added mainly to reduce the cost of the compound, while reinforcing

Table 3.6. *Forms of Various Fillers*

Spherical	Lamellar	Fibrous
Sand/quartz powder	Mica	Glass fibers
Silica	Talc	Asbestos
Glass spheres	Graphite	Wollastonite
Calcium carbonate	Kaolin	Carbon fibers
Carbon black		Whiskers
Metallic oxides		Cellulose
		Synthetic fibers

fillers are added to improve certain mechanical properties, such as modulus or tensile strength. Although termed *inert*, inert fillers can affect other properties of the compound besides cost. They may increase the density of the compound, reduce the shrinkage, increase the hardness, and increase the heat deflection temperature. Reinforcing fillers will increase the tensile, compressive, and shear strengths; increase the heat deflection temperature; reduce shrinkage; increase the modulus; and improve the creep behavior. Reinforcing fillers improve properties via several mechanisms. In some cases, a chemical bond is formed between the filler and the polymer; in other cases, the volume occupied by the filler affects the properties of the thermoplastic. As a result, the surface properties and the interaction between the filler and the thermoplastic are of great importance. The particle shape, particle size, distribution of sizes, and the surface chemistry of the particle affect the polymer properties. In general, the smaller the particle, the greater the improvement of the mechanical property, such as tensile strength. Larger particle size may have adverse effects. Particle shape can also influence the properties. Plate-like particles or fibrous particles may be oriented during processing. This may result in properties that are anisotropic. The surface chemistry of the particle is important to promote interaction with the polymer and to allow for good interfacial adhesion. It is important that the polymer wet the particle surface and have good interfacial bonding so as to obtain the best property enhancement.

Examples of inert or extender fillers include china clay (kaolin), talc, and calcium carbonate. Calcium carbonate is important filler with a particle size of about one micron. It is a natural product from sedimentary rocks and is separated into chalk, limestone, and marble. Calcium carbonate may improve interaction with the thermoplastic. Glass spheres are also used as thermoplastic fillers. They may be either solid or hollow, depending on the particular application. Talc is a type of filler with a lamellar particle shape. It is a natural, hydrated magnesium silicate with good slip properties. Kaolin and mica are also natural materials with lamellar structures. Other fillers include wollastonite, silica, barium sulfate, and metal powders. Carbon black is used as a filler primarily in the rubber industry, but it also finds application in thermoplastics for conductivity, UV protection, and as a pigment. Fillers in fiber form are often used in thermoplastics. Types of fibers include cotton, wood flour, fiberglass, and carbon. Table 3.6 shows the fillers and their forms [7].

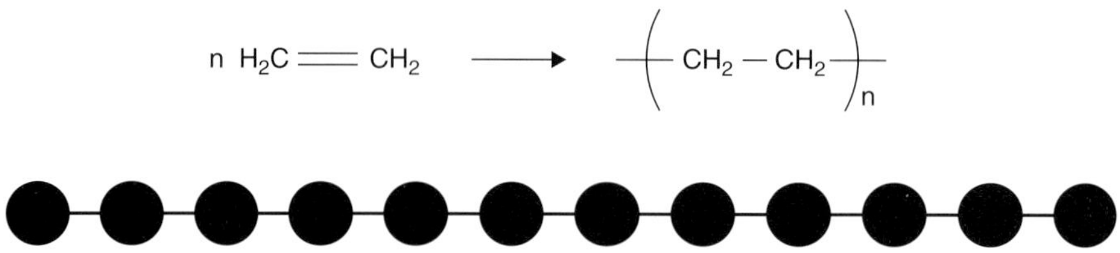

Figure 3.1. Polymerization.

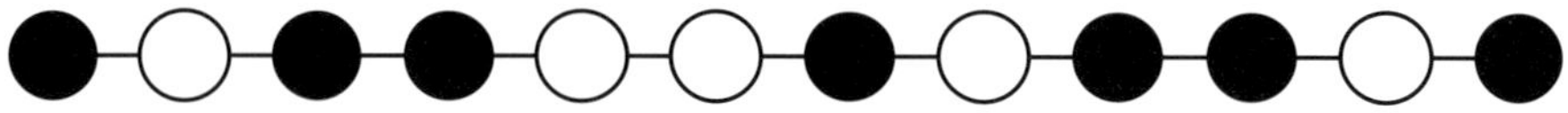

Figure 3.2. Copolymer structure.

3.2.5 Reinforcements

Fiberglass. Fiberglass possess high tensile strength and strain to failure, but the real benefits of its use relate to its heat and fire resistance, chemical resistance, moisture resistance, and very good thermal and electrical values.

Graphite. Graphite fibers have the widest variety of strength and moduli and also the greatest number of suppliers. These fibers start out as organic fiber, rayon, polyacrylonitrile, or pitch called the precursor. The precursor is stretched, oxidized, carbonized, and graphitized. The relative amount of exposure to temperatures from 2,500 to 3,000°C will then determine the graphitization level of the fiber. A higher degree of graphitization will usually result in a stiffer (high modulus) fiber with greater electrical and thermal conductivities.

Aramid. The organic fiber Kevlar® 49, an aramid, essentially revolutionized pressure vessel technology because of its great tensile strength and consistency coupled with low density, resulting in very weight-effective designs for rocket motors.

Boron. Boron fibers, the first fibers to be used in the production of aircraft, are produced as individual monofilaments on a tungsten or carbon substrate by pyrolytic reduction of boron trichloride (BCL) in a sealed glass chamber.

3.3 Polymer Structure and Synthesis

A polymer is prepared by stringing together a series of low-molecular-weight species, such as ethylene, into an extremely long chain, such as polyethylene, similar to one would string together a series of beads to make a necklace (Figure 3.1) [7]. The chemical characteristics of these starting species will determine the final properties of the polyethylene. When two different low-molecular-weight species are polymerized, the resulting polymer is termed copolymer, such as ethylene vinylacetate depicted in Figure 3.2 [7]. Polymers can be separated into thermoplastics and thermosets as stated in the previous section. A thermoplastic material is a high-molecular-weight polymer that is not cross-linked. It can exist in either a linear or a branched structure (Figure 3.3) [7]. Upon heating, thermoplastics soften and melt, which allows them to be shaped using polymer-processing equipment. A thermoset has all of the chains tied together with covalent bonds in a three-dimensional

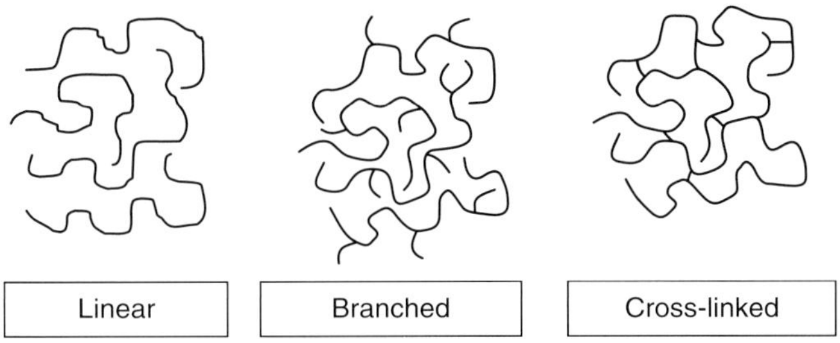

Figure 3.3. Linear, branched, and cross-linked polymer structures.

network (cross-linked) as depicted in Figure 3.3. Thermoset materials will not flow once it is cross-linked, but a thermoplastic material can be reprocessed simply by heating it to the appropriate temperature. The different types of structures are shown in Figure 3.3. The properties of different polymers can vary widely. Properties can be varied for each individual polymer material as well, simply by varying the microstructure of the material.

There are two primary polymerization approaches: step-reaction polymerization and chain-reaction polymerization [7]. In step-reaction (also referred to as *condensation polymerization*), reaction occurs between two poly-functional monomer, often liberating a small molecule, such as water. As the reaction proceeds, high-molecular-weight species are produced as longer and longer groups react together. For example, two monomers can react to from a dimer, then react with another monomer to form a trimer. The reaction can be described as n-mer + m-mer → (n + m) mer, where n and m refer to the number of monomer units for each reactant. Molecular weight of the polymer builds up gradually with time, and high conversions are usually required to produce high-molecular-weight polymers. Polymers synthesized by this method typically have atoms other than carbon in the backbone, such as polyesters and polyamides.

Chain-reaction polymerization (also referred to as *addition polymerizations*) require an initiator for polymerization to occur. Initiation can occur by a free radical or an anionic or cationic species, which opens the double bond of a vinyl monomer and the reaction proceeds as shown in Figure 3.1. Chain-reaction polymers typically contain only carbon in their backbone and include such polymers as polystyrene and polyvinyl chloride.

3.4 Processing of Polymers

Processing is the technology of converting raw polymer, or compounds containing raw polymer, to parts of desired shape. Considering the wide variety of polymer types and even wider variety of components made from them, a complete description and analysis of the processing techniques would be impossible in this section. So here is an outline of the common processing techniques, introducing relevant terminology and considering how the different techniques are used in the fundamentals previously discussed.

3.4.1 Basic Processing Operations

Polymer-processing operations can be classified into five broad categories: extrusion, molding, spinning, calendaring, and coating. Of these techniques, extrusion is perhaps the most widely used. Applications of extrusion include the continuous production of plastic pipe, sheet, and rods. Molding is normally a batch process, principally in the form of injection – and compression molding used to make components as diverse as the cap to a ballpoint pen, fiber-spinning textiles, coating, calendaring of plastic sheet, deposition of organic coatings on plastic sheet, molding, stamping, and compounding.

3.4.2 Extrusion

Thermoplastic parts with a uniform cross-sectional area are formed by extrusion. This includes items, such as pipe, hose and tubing, gaskets, wire and cable insulation, and sheeting. Polymer pellets are conveyed down an electrically heated barrel by a rotating screw. It melts as it proceeds down the barrel and is forced through a die that provides its final shape. Vented extruders incorporate a section in which a vacuum is applied to the melt to remove volatiles, such as traces of unreacted monomer, moisture, solvent from the polymerization process, or degradation products.

Screws are optimized for the particular polymer being extruded. Basically, a screw consists of three sections: melting, compression, and metering. The function of the melting sections is to convey the solid pellets forward from the hopper and convey them into molten polymer. Its analysis involves a combination of fluid and solid mechanics and heat transfer. The compression section, in which the depth of the screw flight decreases, is designed to compact and mix the molten polymer to provide a more or less homogenous melt to the metering section, the function of which is to pump the molten polymer out through the die. This last section is well understood. Analysis of the metering section is an interesting application of the rheological principles and the die design. The determination of the die cross section needed to produce a desired product is still mostly a trial-and-error process. Viscoelastic polymer melts well upon emerging from the die, and the degree of die swell cannot be reliably predicted.

In addition to die swell, as the extrusion rate is increased, the extrudate begins to exhibit roughness and then an irregular, severely distorted profile comes up. The phenomenon is known as melt fracture. This is not yet totally understood, but melt fracture can be minimized by increasing die length, smoothly tapering the entrance to the die, and raising the die temperature.

Extruders are normally specific by screw diameter and length-to-diameter ratio. Typical L/D ratio seems to grow each year, with values now in the 20-1 to 36-1 range. Single-screw extruders depend for their pumping ability on the drag flow of material between the rotating screw and stationary barrel. They are not particularly good mixing devices. Counter-rotating twin-screw extruders are true positive-displacement pumps, capable of generating the high pressures needed in certain profile extrusion applications. Co-rotating twin-screw extruders, though not positive-displacement pumps, with proper screw design, can give excellent mixing and a narrow residence time distribution. They are used extensively in polymer compounding (mixing) operations, and to a certain extent in continuous polymerization reactors.

The heaters on extruders are needed mainly for start-up. In steady-state operation, most of the heat for plasticizing the polymer is supplied by the drive motor through viscous energy dissipation. In fact, cooling through the barrel walls and/or screw center is sometimes necessary.

3.4.3 Calendaring

Polymer sheet (> 25.4 micron (0.01 inch) thickness) may be produced by extrusion through thin flat dies or by calendaring. Basically, a calendar consists of a series of rotating heated rolls between which the compounded polymer is squeezed into sheet form, the thickness of the sheet being determined by the clearance between the rolls [9]. Commercial calendars may be very large (rolls up to 0.914 m (3 ft.) in diameter by 2.44 m (8 ft.) in length) and represent a big capital outlay, but they are capable of tremendous production rates (up to 91.44 m/min (100 yards/min)). In addition to forming unsupported sheeting, the polymer may be laminated to a layer of fabric between two rolls to give a supported sheeting. The final rolls may also be embossed to impart a pattern to the sheet. Shower curtains, vinyl upholstery materials, and vinyl floor titles are produced by calendaring.

3.4.4 Sheet Forming

Thermoplastic sheet is converted to a wide variety of finished parts by sheet forming. Although the details vary considerably, these processes all involve heating the sheet above its softening temperature and forcing it to conform to a mold [9]. In vacuum forming, the heat-plasticized sheet is drawn against the mold surface by the application of a vacuum from beneath the surface. Similarly, a positive pressure may be used to force the sheet against the mold surface; where very deep draws are required, mechanical assists are used. In drape forming, the plasticized sheet is draped over a male mold, perhaps with a vacuum assist. One of the great advantages of sheet forming is that relatively inexpensive molds are required, since no high pressures are involved and epoxy molds are often used. Among the many items made by sheet forming are drinking cups, meat trays, cigarette packs, aircraft canopies, advertising displays, and lightbulbs.

3.4.5 Stamping

The high production rates are possible with a stamping process, and work is currently under way to adapt the production process to forming thin-walled (2.54 cm (0.1 inch)) parts from glass-reinforced thermoplastic sheet. The sheet is heated in an infrared oven and then stamped in modified metal-forming equipment, with a dwell time of about 8 sec as the material cools in the mold. Because of the thin sections, cooling is rapid, and the cycle times as low as 15 to 20 sec can be achieved with automated feeding and part removal systems. So far, this process has been limited to rather simple shapes and shallow draws, but because of its economic potential, it should see more widespread use in the near future. Parts such as crankcase oil pan, battery trays, and bucket seats are made using this process.

3.4.6 Casting

The raw materials for many thermosetting polymers are available as low molecular weight liquids that are converted to solid or rubbery materials by cross-linking. These materials are easily cast to shape at atmospheric pressure in inexpensive molds, such as "carved wood" furniture panels from polyester resins. A mold is made by pouring curing silicone rubber over the original at room temperature. When cured, the rubber mold is stripped from the original and used to cast the polyester copies from a liquid unsaturated polyester. A free-radical initiator causes cross-linking to the final solid object, which reproduces the detail of the carved original at a fraction of the cost.

Such furniture component provides a good example of how the processing method of choice often depends on the production volume. These parts can also be injection molded from high-impact polystyrene. If the panels are large, they require a large and very expensive molding machine. Parts can be injection-molded with a cycle time of less than a minute and very little manual labor. When the capital investment is amortized over a large number of panels, it becomes a small contribution to the total panel cost. On the other hand, the long cycle times and extensive manual labor required for the casting process make it prohibitively expensive for large production runs, but it is the preferred process for small volumes.

3.4.7 Solution Casting

Plastic sheet or film may be produced by dissolving the polymer in an appropriate solvent, spreading the viscous solution on a polished surface, and evaporating the solvent. In the manufacture of photographic film base, the solution is spread with a "knife" on a slowly rotating wheel about 0.6096 m (2 ft.) wide and 6.096 m (20 ft.) in diameter. As the wheel revolves, hot air evaporates the solvent, and the dried film is stripped from the wheel before the casting point is again reached.

3.4.8 Blow Molding

Plastic bottles are nearly all made by blow molding. In one form of this process, extrusion blow molding [9], a hollow cylindrical tube or *parison* is extruded downward. The parison is then clamped between halves of a water-cooled mold. The mold pinches off the bottom of the parison and forms the thread in the neck of the bottle. Compressed air expands the parison against the inner mold surfaces, and when the part has cooled sufficiently, the mold opens, the part is ejected, and the mold is returned to grab another parison. Extrusion blow molding has passed well beyond the bottle production stage. It is now being used to produce large parts, such as drums and gasoline tanks for various vehicles. In such metal replacement applications, blow-molded containers offer light weight and great design flexibility.

3.4.9 Reinforced Thermoset Molding

Many polymers do not possess enough mechanical strength for structural applications. When these polymers reinforced with high modulus fibers, the polymeric composites have high strength-to-weight ratios and can be fabricated in a wide variety of complex shapes. Glass and other fibers are used extensively to reinforce thermosetting resins, such as polyesters and epoxies. Fiberglass-reinforced plastics (FRPs) are now used for practically all boats under 12.19 m (40 ft.) in length as well as for truck cabs, low-production-volume automobile bodies, structural panels, and aircraft components, among others. Many of these parts are fabricated by hand layup process. The mold surface is often first sprayed with a pigmented but nonreinforced gel coat of the liquid resin to provide a smooth surface finish. The gel coat is followed by successive layers of fiberglass, either in the form of woven cloth or random matting, impregnated with liquid resin, which is then cured to give the finished product. The molds are relatively inexpensive because no pressure is required. The major drawback of this process is the expense of the manual labor involved.

A special gun chops continuous fibers into pieces of approximately 2.54 cm (1 inch) long each. The chopped fibers are combined with a stream of liquid resin and sprayed directly onto the mold surface. Although the random chopped fibers do not reinforce as well as woven cloth or random mat does, the labor savings are substantial. This process is used to fabricate sinks, bathtubs, and recreational vehicle bodies, among other things.

In fiberglass or graphite fishing rods, vaulting poles, and golf club shafts, the fibers are arranged along the long axis to resist the bending stress. Filament winding extends this principle to more complex structures. Continuous filaments of the reinforcing fiber are impregnated with liquid resin and then wound on a rotating mandrel. The winding pattern is designed to resist most efficiently the anticipated stress distribution. This technique is used to produce rocket motors and missile structures, gun barrels in the defense industry, and tanks and pipes for the chemical process industries.

Pultrusion is used to produce continuous lengths of objects with a constant cross section, such as structural beams. Continuous fibers (rovings) and/or mat are impregnated by passing them through a bed of liquid resin and pulled slowly through a heat die of the desired cross section. The resin cures to a solid in the die.

3.4.10 Fiber Spinning

Polymers used for synthetic fibers are similar to those used as plastics, but for fibers, the processing operation must produce an essentially infinite length-to-diameter ratio. This is accomplished by forcing the plasticized polymer through a spinnerette, a plate with multiple holes used to produce the individual fibers, which are then twisted together to form a thread for subsequent weaving operations. The cross section of the spinneret holes has a lot to do with the fiber cross section, which in turn greatly influences the fiber properties. The three basic types of spinning operations differ mainly in the method of plasticizing and deplasticizing the polymer.

Melt spinning is basically an extrusion process. The polymer is plasticized by melting and is then pumped through the spinnerette. The fibers are usually solidified by a cross-current blast of air as they proceed to the drawing rolls. The drawing step stretches the fibers, orienting the molecules in the direction of stretch and inducing high degree of crystallinity, which provides good fiber properties. Nylons are commonly melt spun. In dry spinning, a solution of the polymer is forced through the spinnerette. As the fibers proceed downward to the drawing rolls, a countercurrent stream of warm air evaporates the solvent. In this process, the cross section of the fiber is determined not only by the shape of the spinnerette holes but also by the complex nature of the diffusion-controlled solvent evaporation process, because there is considerable shrinkage as the solvent evaporates. The acrylic fibers, mainly polyacrylonitrile, are produced by dry spinning.

3.5 Mechanical Behavior of Polymers

3.5.1 Stress-Strain Behavior

The mechanical properties of polymers are modulus of elasticity, as well as yield and tensile strengths. For many polymeric materials, the simple stress-strain test is employed for the characterization of some of the mechanical parameters [10]. The mechanical characteristics of polymers are highly sensitive to the rate of deformation (strain rate), temperature, and chemical nature of the environment (the presence of water, oxygen, organic solvents, etc.).

Typically three different types of stress-strain behavior are found for polymeric materials, as illustrated in Figure 3.4 [11]. Curve *A* illustrates the stress-strain behavior of a brittle polymer; it fractures while deforming elastically. Curve *B* is similar to that of many metallic materials; the initial deformation is elastic, which is followed by yielding and a region of plastic deformation. The deformation displayed by curve *C* is totally elastic; this rubber-like elasticity (large recoverable strains produced at low stress levels) is displayed by elastomers.

Modulus of elasticity (termed tensile modulus) and ductility in percent elongation are determined for polymers using the stress-strain curve (Figure 3.4). For plastic polymers (curve *B*, Figure 3.4), the yield point is taken as a maximum on the curve, which occurs just beyond the termination of the linear-elastic region

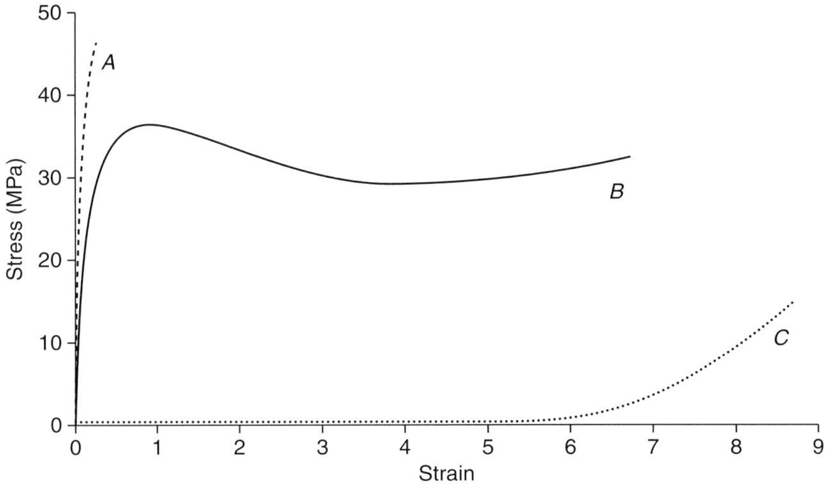

Figure 3.4. The stress-strain behavior for brittle (curve *A*), plastic (curve *B*), and highly elastic (elastomeric) (curve *C*) polymers.

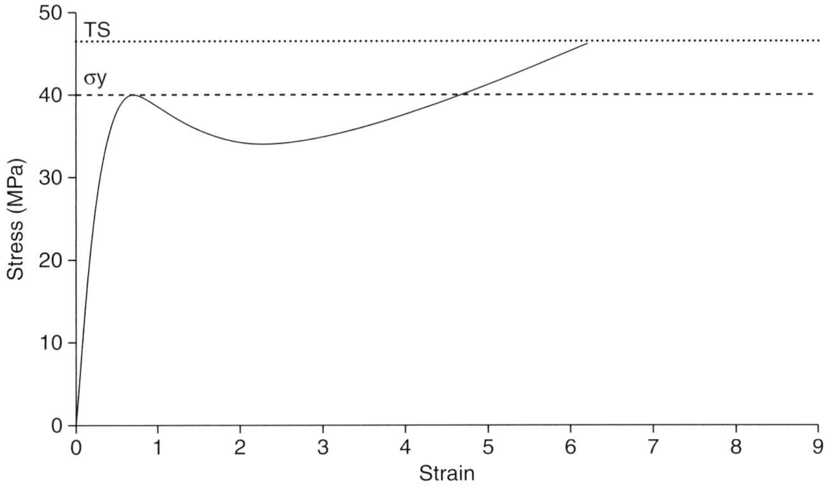

Figure 3.5. Schematic stress-strain curve for a plastic polymer showing how yield and tensile strengths are determined.

(Figure 3.5) [11]. The stress at this maximum is the tensile yield strength (σ_y). Tensile strength (*TS*) corresponds to the stress at which fracture occurs (Figure 3.5); *TS* may be greater or less than σ_y. Table 3.7 shows some mechanical properties for several polymeric materials [11].

3.5.2 Strength

There are different kinds of strength. There is *tensile* strength, which is important for a material when stretched or under tension. A polymer has *compression* strength, exhibited under compression. A polymer has *flexural* strength, exhibited under

Table 3.7. *Room-Temperature Mechanical Properties of Some Common Polymers*

Material	Specific Gravity	Tensile Modulus [GPa (ksi)]	Tensile Strength [MPa (ksi)]	Yield Strength [MPa (ksi)]	Elongation at Break (%)
Polyethylene (low density)	0.917–0.932	0.17–0.28 (25–41)	8.3–31.4 (1.2–4.55)	9.0–14.5 (1.3–2.1)	100–650
Polyethylene (high density)	0.952–0.965	1.06–1.09 (155–158)	22.1–31.0 (3.2–4.5)	26.2–33.1 (3.8–4.8)	10–1200 40–80
Poly(vinyl chloride)	1.30–1.58	2.4–4.1 (350–600)	40.7–51.7 (5.9–7.5)	40.7–14.8 (5.9–6.5)	
Polytetrafluoroethylene	2.14–2.20	0.40–0.55 (58–80)	20.7–34.5 (3.0–5.0)	–	200–400
Polypropylene	0.90–0.91	1.14–1.55 (165–225)	31–41.4 (4.5–6.0)	31.0–37.2 (4.5–5.4)	100–600
Polystyrene	1.04–1.05	2.28–3.28 (330–475)	35.9–51.7 (5.2–7.5)	–	1.2–2.5
Poly(methyl methacrylate)	1.17–1.20	2.24–3.24 (325–470)	48.3–72.4 (7.0–10.5)	53.8–73.1 (7.8–10.6)	2.0–5.5
Phenol-formaldehyde	1.24–1.32	2.76–4.83 (400–700)	34.5–62.1 (5.0–9.0)	–	1.5–2.0
Nylon 6,6	1.13–1.15	1.58–3.80 (230–550)	75.9–94.5 (11.0–13.7)	44.8–82.8 (6.5–12)	15–300
Polyester (PET)	1.29–1.40	2.8–4.1 (400–600)	48.3–72.4 (7.0–10.5)	59.3 (8.6)	30–300
Polycarbonate	1.20	2.38 (345)	62.8–72.4 (9.1–10.5)	62.1 (9.0)	110–150

Source: Modern Plastics Encyclopedia '96. Copyright 1995, The McGraw-Hill Companies. Reprinted with permission.

bending. A polymer sample's *torsional* strength is tested by twisting it. Finally, a polymer sample has *impact* strength, tested when hit with a sharp, heavy, or otherwise damaging object.

3.5.3 Elongation

Elongation is a type of deformation that occurs when a polymer sample is stretched. It is usually measured in percent (%), calculated by the length of the polymer sample after it is stretched (L) divided by the original length of the sample (L_o) and multiplied by 100.

$$L/L_o \times 100 = \%\ \text{elongation}$$

Two important elongations that are measured are ultimate elongation and elastic elongation.

3.5.4 Fracture

If a glassy polymer is stressed very rapidly or is stressed at a temperature that is much lower than its glass transition temperature, it tends to become brittle and break, or fracture. An amorphous polymer tends to draw down in a homogenous manner if stressed at a temperature above its glass transition temperature, and it shows large strains before fracturing. At intermediate temperatures and low rates of deformation, the polymer can yield before fracturing or fracture in a ductile manner by neck formation. At low strain (i.e., <1%), the deformation of most polymers is elastic, which means that the deformation is homogenous and full recovery can occur over a finite time. At high strains, the deformation of glass polymers occurs by either crazing – characteristic of brittle polymers – or by a process called shear bonding, which is the dominant mechanism for ductile polymers. Such deformations are not reversible unless the polymer is heated above its glass transition temperature.

3.5.5 Modulus

Modulus is measured to know how well a material resists deformation. There are different types of modulus. The Young's modulus is the material resistance to tensile deformation in a uniaxial tensile deformation, and is calculated by the slope of the stress-strain curve in the initial linear region. Shear modulus is the material resistance to shear deformation. The flexural modulus is the material resistance to bending.

3.5.6 Toughness

Toughness is a measure of the energy a polymer sample can absorb before it breaks. Strength tells how much force is needed to break the sample, and toughness tells how much energy is needed to break the sample. Just because a polymer sample is strong does not mean it is also tough.

3.5.7 Crazing

When some polymers, such as polystyrene, are deformed to a certain level, the critical strain (ε_o), manifesting as small cracks, develops in a direction perpendicular to the principal direction of deformation. These crazes reflect light and result in visual cloudiness or whitening of the polymer sample.

A craze is a unique morphological feature of polymers and is morphologically different from a true neck. A craze, which can be from a nm to a few m in thickness, consists of polymer microfibrils (0.6 to 30 nm in diameter) stretched in the direction of tensile deformation. The microfibrils are surrounded by void shape, which can represent as much as 90% of the total volume of the craze.

Although the reasons for craze initiation are still uncertain and several different theories have been proposed, it is recognized that crazes constitute the defects from which brittle cracks initiate. The mechanical feature of a polymer sample that has begun crazing is initiated by the breakdown of the fibillar microstructure to form additional voids that grow slowly until some critical size has been reached. Beyond this point, the craze will rapidly propagate as a crack. As the crack propagates, crazes are formed at the crack tip and act to retard its advance.

3.6 Aging of Polymers

When exposed to elevated temperatures, moisture, oxygen, and ultraviolet rays for extended periods of time, polymers are known to exhibit changes in their molecular structure that can have profound effects on the long-term durability of the polymer and the associated composite material. This process is known as *aging* [12]. Aging is classified into three different types: physical, chemical, and hydrothermal. *Physical aging* is the slow contraction of the molecular network that occurs over time when exposed to sub-T_g temperatures. This densification of the material leads to embrittlement and loss of toughness. *Chemical aging* is the oxidation and continued curing mechanisms that act on the polymer structure during extreme environmental exposure. This aging mechanism can lead to embrittlement, loss of toughness, and discoloration of the polymer. Hydrothermal aging introduces changes into the polymer structure due to the presence of moisture. As water molecules diffuse into the polymer from the structural surface, the water can change the bulk polymer density, can reduce the mechanical integrity of the material, and can chemically react with the polymer to alter the bulk polymer properties. All three aging mechanisms are generally detrimental to the durability of a polymer. Many polymers are formulated to specifically resist the effects of one or more aging mechanisms.

3.7 Polymers for Aerospace Composites

Polymers used as the matrix component of composites in aerospace structures must be exceptionally durable to resist the cyclic mechanical loads associated with aircraft operation. Furthermore, many aerospace structures must also be tolerant of high temperatures, moisture levels, thermal cycling, and UV radiation. For most aerospace vehicles, cost is not as big a consideration as in other industries, so higher-performing polymers that may be relatively expensive are often chosen over less-expensive polymers with lower-performing qualities.

Thermosetting polymers are typically used in aerospace applications where high temperatures are expected. Examples of thermosetting composites include the skin of supersonic aircraft and some reentry vessels. The most common thermosets used in aerospace vehicles are epoxies, bismaleimides (BMI), cyanate esters (CE), phenolics, and polyimides (PI) [13]. Epoxy resins are the most commonly used matrix material for composites, but their use in aerospace structures is limited by relatively low service temperatures and susceptibility to brittleness and moisture attack. BMIs are similar to epoxies except for improved temperature performance with a relatively high performance-to-cost ratio. CE resins are more easily processed than BMIs are and offer excellent strength, moisture absorption, and electrical properties. Similar to epoxy and BMIs, CE resins can be brittle if not toughened properly. Phenolic resins are low-cost, flame-resistant, and low-smoke resins that are primarily used in interior aircraft panels and ablative and rocket nozzle applications. Thermosetting PI resins have good adhesion and heat and chemical resistance as well as superior mechanical properties. The processing of some PI resins can be difficult due to the need for using highly corrosive chemicals.

Thermoplastic polymers are a more cost-effective alternative for thermosets when extremely high temperature resistance requirements are not needed. Thermoplastic also have a much higher impact strength than do thermosets. Applications of these materials in the aerospace industry include aircraft interiors, wing ribs and panels, and land gear doors. Common thermoplastic polymers used in aerospace structures include poly(ether ketone) (PEEK), polyetherimide (PEI), poly(p-phenylene sulfide) (PPS), and thermoplastic polyimides. PEEK has good resistance to moisture attack, corrosion, abrasion, and chemical and radiation exposure. PEEK has low smoke emission and excellent stiffness. PEI polymers have high heat resistance, strength, and modulus, but can degrade with exposure to aggressive fluids, such as hydraulic fluid. PPS has excellent resistance to oils, fuels, solvents, anti-icing agents, and acids/bases. It also has excellent hardness, dimensional stability, and excellent fire resistance. Thermoplastic polyimides, like their thermosetting counterparts, offer excellent performance at elevated temperatures.

3.8 Polymer Foams

Polymer foams consist of polymer resin that contains large pockets of air (cells). The presence of the cells creates a lattice network of polymer, and the overall bulk density is much lower than that of pure resin. There are two basic types of polymer foams: open-cell and closed-cell [14]. In open-cell foams air can pass freely between adjacent cells in the material. In a closed-cell foam the cells are completely enclosed, which prevents movement of air between cells. Open-cell foams tend to be flexible (e.g., foam materials used in seat cushions) while closed-cell foams tend to be rigid (e.g., Styrofoam). Because of the rigid nature of closed-cell foams, they are often used as the core material in sandwich composites used in aerospace structures. Polymethacrylimide (PMI) is a polymer that is commonly used for this purpose.

3.9 Shape-Memory Polymers

Shape-memory polymers (SMPs) are polymers that can be deformed from one shape to another from an external stimulus, most commonly a temperature variation above or below the SMP *transformation temperature*. Other external stimuli include magnetic fields, electric fields, irradiation, and pH levels [15]. Researchers in the aerospace industry are particularly interested in SMPs for use in morphing aerodynamic structures in which the surface areas of control surfaces can be quickly adjusted to enable optimal performance and efficiency under a wide range of flight conditions. SMPs have also been proposed as the base material in deployable solar arrays for space vehicles.

SMPs have been formulated based on thermoplastic and thermosetting chemistries for a wide range of polymers including polyurethane, epoxy, polyolefin, poly(vinyl chloride), and poly(ether esters). SMPs can be categorized into four classes depending on their chemical architecture and the origin of their transformation temperature: chemically cross-linked glassy thermosets, chemically cross-linked semi-crystalline elastomers, physically cross-linked amorphous thermoplastics, and physically cross-linked semi-crystalline block copolymers. The shape memory effect has been attributed to the existence of a frozen-in network configuration when the polymer is cooled to its glassy state. That is, upon deforming the polymer in its rubbery state (above T_g), the conformational rearrangement of the macromolecules along the deformation axis results in a network with reduced entropy. The reduced entropy is stored as elastic energy upon cooling to the glassy state where the deformed network configuration is frozen in. When the SMP is heated back to the rubbery state and is not mechanically constrained, the elastic energy is released as a mechanical restoring force, and the material will attempt to recover its original shape.

The main advantage of SMPs with respect to shape memory alloys is reduced density, easier processing, lower manufacturing cost, and large attainable strains. Specifically, strains of up to 700% can be obtained compared to less than 10% for shape memory alloys. Some of the drawbacks of SMPs are their relatively low modulus (up to 10 MPa), their long response time, and their relatively low fatigue toughness.

3.10 Summary

This chapter provides a brief introduction of the three different categories of polymers: *thermoplastics*, *thermosets*, and *elastomers*, with examples of each.

3.11 Questions and Problems

1. Briefly explain what thermoplastic, thermoset, and elastomer are, as well as their characteristics.
2. Discuss some representative applications of thermoplastics, thermosets, and elastomers in the industry.
3. What are some advantages and disadvantages of thermoplastics and thermosets? Why do thermosets retain their enhanced properties under severe environmental conditions?
4. What are nanomodified polymer matrix composites? Provide examples.
5. Is it possible to have a copolymer resin matrix? Provide examples.
6. How does the density of the polymer affect its properties?
7. Define glass transition temperature and heat deflection temperature and list the differences between the two.
8. What are the mechanical property comparisons between thermoplastic resins and thermoset resins?
9. Why thermoplastic elastomers are classified as elastomers and not thermoplastics?
10. Do thermosets have a glass transition temperature?
11. Is there a big difference in properties between the woven and discontinuous reinforcement options?

12. Because nanomaterials are normally used to reinforce a material, which makes it stiffer, is it possible for a nanomaterial to enhance the flexibility of a polymer?
13. As we all know, thermoplastics are solid, and the thermosets can be solid and liquid. Why?
14. At the temperature of T_g, what are the changes in the structures of thermoplastics, thermosets, and elastomers, respectively?

REFERENCES

1. American Chemical Council Plastics Statistics Data (2015). (www.americanchemistry.com).
2. N. J. Mills (2005). *Plastics – Microstructure and Engineering Application, 3rd Edition*. Oxford: Elsevier.
3. R. E. Wright (2002). Thermosets, Reinforced Plastics, and Composites. In *Handbook of Plastics, Elastomers, and Composites, 4th Edition*. New York: C. A. Harper, Ed., McGraw-Hill, pp. 109–188.
4. J. H. Koo (2006). *Polymer Nanocomposites: Processing, Characterization, and Applications*. New York: McGraw-Hill.
5. R. F. Floral and S. T. Peters (1989). Composite Structures and Technologies. Tutorial notes.
6. J. M. Margolis (2002). Elastomeric Materials and Processes. In *Handbook of Plastics, Elastomers, and Composites, 4th Edition*. New York: C. A. Harper, Ed., McGraw-Hill, pp. 189–228.
7. A-M. M. Baker and J. Mead (2002). Thermoplastics. In *Handbook of Plastics, Elastomers, and Composites, 4th Edition*. New York: C. A. Harper, Ed., McGraw-Hill, pp. 1–108.
8. W. D. Callister (2007). *Materials Science and Engineering – An Introduction, 7th Edition*. New York: John Wiley & Sons, pp. 567.
9. V. K. Ahluwalia and A. Mishra (2008). *Polymer Science – a Textbook*. Boca Raton, FL: CRC Press.
10. ASTM Standard D 638. "Standard Test Method for Tensile Properties of Plastics."
11. W. D. Callister (2007). *Materials Science and Engineering – an Introduction, 7th Edition*. New York: John Wiley & Sons, pp. 525–543.
12. G. M. Odegard and A. Bandyopadhyay (2011). Physical aging of epoxy polymers and their composites. *Journal of Polymer Science Part B: Polymer Physics* 49, 1695–1716.
13. J. R. Fried (2008). *Polymers in Aerospace Applications*. Shrewsbury, UK: Smithers Rapra Technology.
14. N. Mills (2007). *Polymer Foams Handbook*. New York: Butterworth-Heinemann, p. 2.
15. I. A. Rousseau (2008). Challenges of shape memory polymers: A review of the progress towards overcoming SMP's limitations. *Polymer Engineering and Science* 48, 2075–2089.

4 Selecting Resin Matrix and Nanomaterials for Applications

4.1 Characteristics of Polymer Nanocomposites

There are two main challenges to developing polymer nanocomposites after the desired nanomaterials have been selected for the polymer of interest. First, the choice of nanomaterials requires an interfacial interaction and/or compatibility with the polymer matrix. Second, the proper processing technique should be selected to uniformly disperse and distribute the nanomaterials or nanomaterial aggregates within the polymer matrix. In Chapter 2, the different types of nanomaterials were briefly described. In Chapter 3, the three different kinds of polymers were also discussed. In this chapter, we will discuss the selection of resin matrix and nanomaterials for any specific applications.

A range of thermoplastic nanocomposites, thermoset nanocomposites, and elastomer nanocomposites were surveyed in this chapter. Selective polymer nanocomposites for different applications will be used to demonstrate the technical approach. Several of the functions of these materials are listed below:

- *Mechanical*: increased modulus, strength, toughness, elongation (in some cases); detailed discussions are in Chapter 7
- *Thermal*: increased thermal resistance, enhanced thermal stability, higher glass transient temperature (T_g), increased heat deflection temperature (HDT), reduced coefficient of thermal expansion (CTE); detailed discussions are in Chapter 8
- *Flammability*: low peak heat release rates, reduced CO, reduced smoke, passed UL 94 V0; detailed discussions are in Chapter 9
- *Ablation*: low erosion, enhanced charred strength, better insulation; detailed discussions are in Chapter 10
- *Electrical*: improved electrical conductivity, lower resistivity (depends on the nanomaterials); detailed discussions are in Chapter 11
- *Optical*: transparency provided in selective systems; detailed discussion are in Chapter 12
- *Barrier*: reduced oxygen and moisture transmission; detailed discussions are in Chapter 13

Table 4.1. *Comparisons of Thermoplastic and Thermosetting Resin Characteristics*

Thermoplastic Resin	Thermosetting Resin
▪ High MW solid	▪ Low MW liquid or solid
▪ Stable material	▪ Low to medium viscosity, requires cure
▪ Reprocessable, recyclable	▪ Cross-linked, nonprocessable
▪ Amorphous or crystalline	▪ Liquid or solid
▪ Linear or branched polymer	▪ Low MW oligomers
▪ Liquid solvent resistance	▪ Excellent environmental and solvent resistance
▪ Short process cycle	▪ Long process cycle
▪ Neat up to 30% filler	▪ Long or short fiber reinforced
▪ Injection/compression/extrusion	▪ Resin transfer molding (RTM)/filament winding (FW)/sheet molding compound (SMC) prepreg/pultrusion
▪ Limited structural components	▪ Many structural components
▪ Neat resin + nanoparticles	▪ Neat or fiber reinforced + nanoparticles
▪ Commodity: high-performance areas for automotive, appliance housings, toys	▪ Commodity: advanced materials for construction, marine, aircraft, aerospace

- *Tribological*: reduced scratch resistance, better abrasion resistance; detailed discussions are in Chapter 13
- *Chemical*: improved solvent resistance, improved moisture resistance; detailed discussions are in Chapter 13
- Others: reduced shrinkage.

Selected nylon 6 nanocomposites, epoxy nanocomposites, and thermoplastic olefin (TPO) nanocomposites are in this chapter as typical examples for thermoplastic-, thermoset-, and elastomer-based nanocomposites. Physical, mechanical, and thermal properties as well as the performance of these polymer nanocomposite materials are discussed. In most cases, polymer nanocomposites exhibit multifunctionality.

4.2 Different Types of Polymer Nanocomposites

In general, polymers can be classified into the three basic families of resin, namely *thermoplastics, thermosets*, and *elastomers*. Chapter 3 provided a brief discussion of these resin systems. Table 4.1 lists several characteristics of the thermosetting and thermoplastic resin systems.

4.2.1 Thermoplastic-Based Nanocomposites

Materials are often classified as metals, ceramics, or polymers. Polymers differ from the other materials in a variety of ways, but generally they exhibit lower densities and moduli. The lower densities of polymeric materials offer an advantage in applications where lighter weight is desired. The addition of thermally and/or electrically conducting fillers allows the polymer formulator the opportunity to develop materials from insulating to conducting characteristics.

Thermoplastics are used in a vast array of products. In the automotive area they are used for interior parts and in under-the-hood applications. The packaging applications area is a large area for thermoplastics, from carbonated beverage bottles to plastic wrap. Application requirements vary widely; fortuitously, plastic materials can be formulated to meet these different business opportunities. It remains for the designers to select from the array of thermoplastic materials that are available to meet the particular needs of their applications.

Many thermoplastics have been nano-modified into polymer nanocomposites by incorporating nanomaterials into the polymers. Several of these polymer nanocomposite are in details in Chapters 7 through 13. A typical example of this class of thermoplastic-based nanocomposites is nylon 6 nanocomposites, covered in the following subsection.

4.2.1.1 Nylon 6 Nanocomposites

Mica-type silicates, such as montmorillonite, hectorite, and saponite are attractive nanoclays [1–3] functioning as reinforcing fillers for polymers because of their high aspect ratio and unique intercalation and exfoliation characteristics. The incorporation of organoclays into polymer matrices has been known for many decades. In 1950, Carter et al. [4] developed organoclays that were surface-treated with organic bases to reinforce latex-based elastomers. In 1963, Nahin et al. [5] incorporated organoclay into a thermoplastic polyolefin matrix. Organoclay-modified composites were obtained with strong solvent resistance and high tensile strength by irradiation-induced crosslinking. In 1976, Fujiwara and Sakomoto [6] of Unichika patented the first organoclay hybrid nanocomposite. Ten years later, a Toyota research team disclosed improved methods to produce nylon 6-clay nanocomposites using in situ polymerization similar to the Unichika process [7–10]. They reported these nylon 6-clay nanocomposites exhibited increased solvent resistance, decreased CTE, reduced permeability, and enhanced flame retardant characteristics. A very comprehensive review of polymer/layered silicate nanocomposites was published by Ray and Okamoto of Toyota Technical Institute in 2003 [11]. These materials were commercialized in Japan by Ube and Unitika. Gianellis and Vaia et al. [12–15] reported that nanocomposites can be obtained by direct polymer intercalation, where polymer chains diffuse into the space between the clay galleries. They also suggested that this approach can be combined with conventional polymer extrusion to decrease the time to form these hybrids by shearing clay platelets and leading to sample uniformity.

Dennis and Paul et al. [16] demonstrated the degree of exfoliation and dispersion of layered silicate nanocomposites formed from nylon (polyamide) 6 by melt compounding. This polymer processing is affected by both the chemistry of the clay surface and the type of extruder and its screw design. Increasing the mean residence time in the extruder generally improves the exfoliation and dispersion. It appears that an optimum extent of back mixing is necessary as judged by the broadness of the residence time distribution and optimal shear intensity. However, excessive shear intensity or back mixing also causes poor exfoliation and dispersion. Shear intensity is required

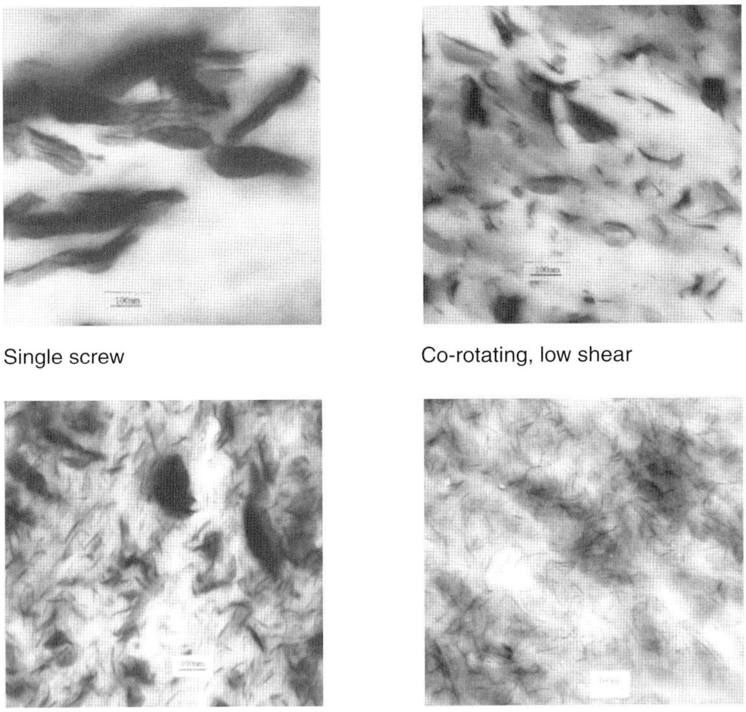

Figure 4.1. Transmission electron micrographs for elective nylon 6-Closiste 15A nanocomposites using different extruders and screw configurations [16].

to start the dispersion process, by shearing particles apart into tactoids or intercalants. Residence time in a low-shearing or mildly shearing environment is required to allow polymer to enter the clay galleries and peel the platelets apart (Chapter 2). The non-intermeshing twin-screw extruder used in the study yielded the best exfoliation and dispersion. Excellent exfoliation and dispersion can be achieved with both co-rotating and counter-rotating, intermeshing types of extruders when a fully optimized screw configuration is used. Figure 4.1 shows TEM micrographs of selective nylon 6-clay nanocomposites using different extruder and screw configurations [16]. It is recognized that the extruder processing conditions are important variables that must be optimized to affect a high degree of exfoliation and dispersion.

Cho and Paul [17] prepared nylon 6-organoclay nanocomposites by melt compounding using a typical co-rotating twin-screw extruder and compared them with nylon 6 composites containing glass fibers and untreated clay. Transmission electron microscopy and WAXD indicate that the organoclay was well dispersed into the nylon 6 matrix and the mechanical properties of these materials compare well with nylon 6-clay nanocomposites formed by in situ polymerization and melt processing. Melt processing of these nylon 6-clay nanocomposites in a single-screw extruder failed to give similar levels of dispersion or exfoliation.

A high degree of exfoliation by melt processing requires an adequate residence time in the extruder, as well as sufficient shear. Although highly exfoliated nylon 6-clay

Table 4.2. *Mechanical Properties of Nylon 6 Composites [17]*

Nylon 6 Composites	Mineral Content (%)	Izod Impact Strength (J/m)	Modulus (GPa)	Yield Strength (MPa)	Elongation at Break (%)[a]
Nylon 6	0	38 ± 4	2.66 ± 0.2	64.2 ± 0.8	200 ± 30
N6/glass fiber	5	53 ± 8	3.26 ± 0.1	72.6 ± 0.8	18 ± 1.3
N6/montmorillonite	5	40 ± 2	3.01 ± 0.1	75.4 ± 0.3	22 ± 6.0
N6/organoclay	3.16	38 ± 3	3.66 ± 0.1	83.4 ± 0.7	126 ± 25
N6/organoclay/glass fiber	8	44 ± 3	4.82 ± 0.1	95.0 ± 0.9	8 ± 0.5

[a] Crosshead speed 0.51 cm/min.

nanocomposites show continuous improvement in strength and modulus relative to the neat nylon 6 matrix as more organoclay is added, there is a loss of ductility beyond a certain organoclay loading. The type of organo-treatment and the speed of testing seem to have significant effects on ductility. Adding additional organoclay to nylon 6 significantly increases the ductile-to-brittle transition temperature. The mechanical properties of the organoclay nanocomposites showed greater values than glass fiber composites (Table 4.2) [17].

Fornes and Paul et al. [18] examined the effect of varying molecular weight of nylon 6 and nanoclay. Nylon 6 (low, or LMW; medium, or MMW; and high, or HMW) was prepared by a co-rotating twin-screw extruder. Wide-angle X-ray diffraction and TEM results collectively reveal a mixed structure for LMW-based nanocomposites having regions of intercalated and exfoliated clay platelets, whereas the MMW and HMW composites revealed highly exfoliated structures. Qualitative TEM observations were supported by quantitative analyses of high-magnification TEM images. The average number of platelets per stack was shown to decrease with increasing nylon 6 molecular weights. The TEM particle density increased with increasing molecular weight, revealing clay platelet exfoliated for the nanocomposites in the order of HMW > MMW > LMW composites (Figure 4.2) [18]. Tensile properties revealed superior performance for the higher molecular weight nylon 6 composites, with highest properties for HMW. Additionally, the HMW-based nanocomposites had the highest moduli, yield strengths, and elongation at break (Figure 4.3) [18]. The notched Izod impact strength was relatively independent of clay concentration for the HMW and MMW composites, but gradually decreased for the LMW composites [18]. The melt viscosity, and consequently the shear stress, increased with increasing nylon 6 molecular weight. The higher shear stress is believed to have been the major contributor to exceptional exfoliation of clay platelets in the higher molecular weight nylon 6-matrix.

Figure 4.4 schematically illustrates the various roles of shear stress during the melt compounding of nanocomposites [18]. Initially, the stress should assist in rupturing large organoclay particles into dispersed stacks of silicate tactoids (Figure 4.4a). As the extrudate travels in the extruder, transfer of the stress from the molten polymer to the silicate tactoids is proposed, shearing tactoids into smaller stacks of silicate platelets (Figure 4.4b). Ultimately, individual platelets peel apart through

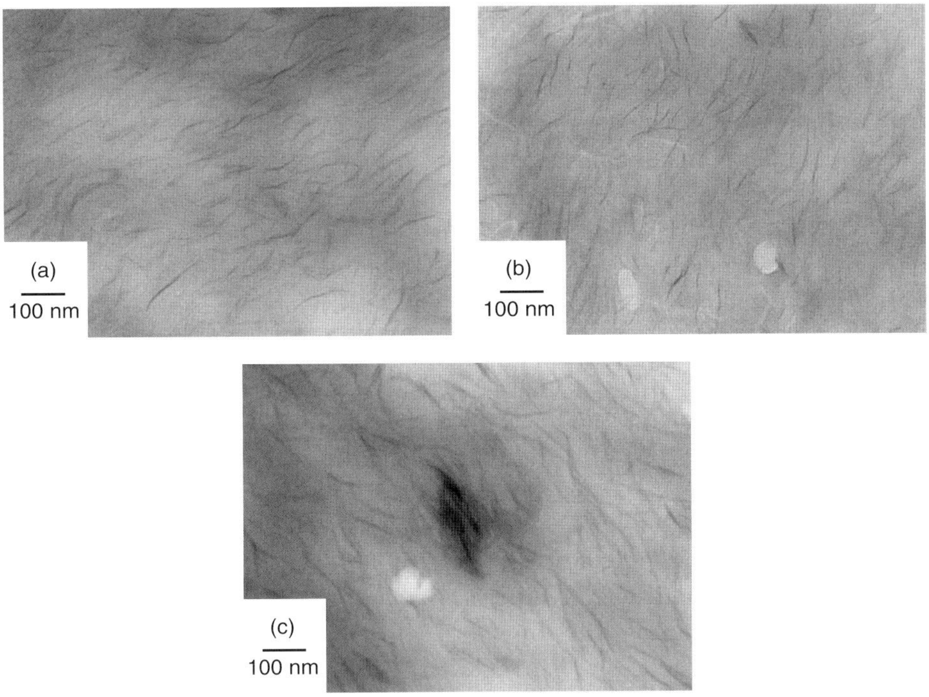

Figure 4.2. Transmission electron micrographs of melt-compounded nanocomposites containing ~3.0 wt% montmorillonite based on (a) HMW, (b) MMW, and (c) LMW nylon 6 [18].

a combination of shear and diffusion of polymer chains in the organoclay gallery (Figure 4.4c).

The Paul Research Group [19] also studied the *structure-property* relationships for nanocomposites formed by melt processing a series of organically modified montmorillonite clays with high and low molecular weight nylon. The structure of the alkyl ammonium on the clay was systemically varied to determine how specific groups affect nylon 6 nanocomposite morphology and physical properties. As demonstrated by WAXD, galleries of the organoclays expand in a systematic manner to accommodate the molecular size and the amount of amine surfactant exchanged for the inorganic cation of the native montmorillonite. The density of the organic material in the galleries is in the range expected for organic liquids and solids. Both the modulus and the yield strength of the nanocomposites appear to decrease as the original organoclay *d*-spacing increases, in contrast to some earlier suggestions in the literature.

Three distinct surfactant structural effects have been identified that lead to a high amount of exfoliation, high stiffness, and increased yield strengths for nanocomposites based on high molecular weight polyamide. They are (a) one long alkyl substitute group on the ammonium ion rather than two, (b) methyl groups rather than 2-hydroxy-ethyl groups on the amine, and (c) an equivalent amount of amine surfactant on the clay (to avoid an excess amount). Similar trends, but lower extends of exfoliation, were seen for a lower molecular weight grade of nylon 6. The authors proposed that these effects are due to the amount of exposed silicate surface. Alkyl

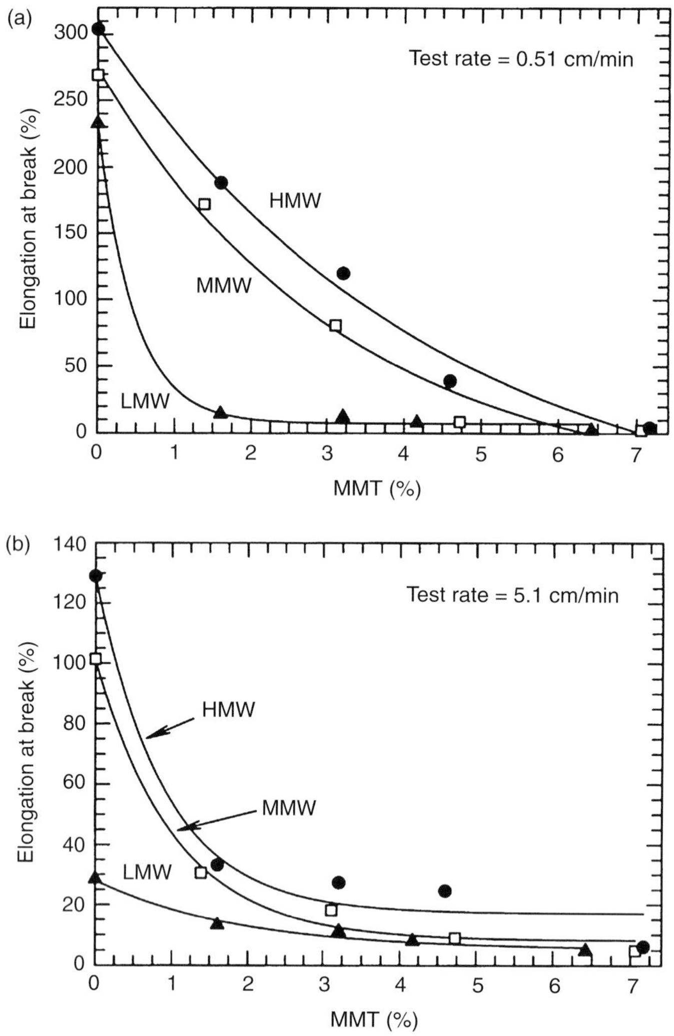

Figure 4.3. Effect of montmorillonite content on elongation at break for LMW-, MMW-, and HMW-based composites at a crosshead speed of (a) 0.51 cm/min and (b) 5.1 cm/min [18].

ammonium ions, which cover a larger percentage of the silicate surface, shield desirable polar polyamide–polar clay interactions and ultimately lead to lesser platelet exfoliation. However, this finding does not imply that unmodified clay – that is, sodium montmorillonite – would be optimum. Organic surface modification of the clay is still required to overcome the cohesive forces between neighboring platelets, so that polymer intercalation and exfoliation can occur during melt processing. It is possible that these observations may "only apply" to nylon 6 polymer. Thus, the nature of polymer-organoclay thermodynamic interactions may be different for other polymers and may have different *structure-property* effects. Continued studies in this area are encouraged.

The molecular weight of the polyamide 6 has a significant effect on nanocomposite morphology and mechanical properties. The higher molecular weight

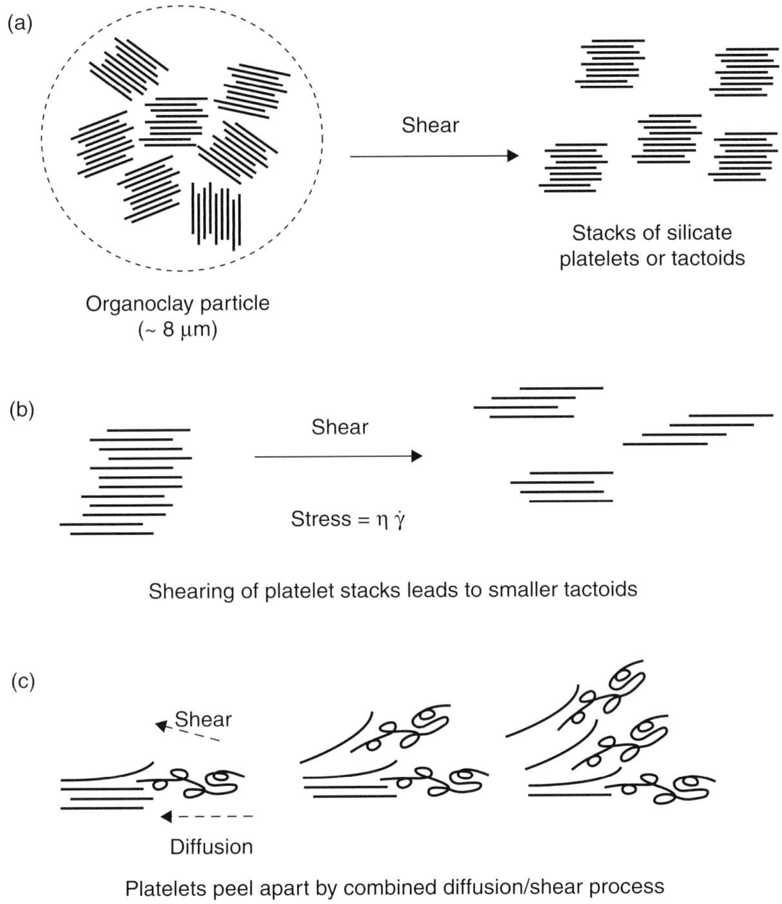

Figure 4.4. Stepwise mechanism of clay platelet exfoliation in the melt compounding of nanocomposites: (a) organoclay particle breakup, (b) clay tactoid breakup, and (c) platelet exfoliation [18].

polyamide consistently leads to better exfoliation and greater reinforcement than lower molecular weight polyamides. In summary, with proper selection of the nylon 6, organoclay modifier, and processing conditions, one can produce high-performance nylon 6 nanocomposites by melt processing. These features contributed to the development of commercial nylon type products by Toyota, Ube, Unitika, Honeywell, and others.

4.2.2 Thermoset-Based Nanocomposites

Thermosetting type resins consist of solid, semisolid, or liquid organic reactive intermediate material that cures or cross-links into a high molecular weight product, with no observable melting point. The basic characteristic of these intermediate thermosetting resins is that they will, upon exposure to elevated temperature from ambient to above 232°C (450°F), undergo an irreversible chemical reaction often referred to as *polymerization*, or *cure*. Each family member has its own set of individual chemical characteristics based on its molecular structure and its ability to either homopolymerize, copolymerize, or both. This transformation process separates the thermosets from the thermoplastics. The important beneficial factor lies in the inherent enhancement of the physical, thermal, chemical, and electrical properties of thermoset resins, because of that chemical cross-linking polymerization reaction, which contributes to their ability to maintain and retain these enhanced properties when exposed to severe environmental conditions.

An intermediate reactive thermosetting species is defined as a liquid, semisolid, or solid composition capable of curing of some defined temperature that can be ambient to several hundred degrees and that cannot be reshaped by subsequent reheating. In general, these intermediate reactive thermosetting compositions contain two or more ingredients: a reactive resinous matrix with a curing agent that causes the intermediate material to polymerize (cure) at room temperature, or a low molecular weight resinous matrix and curing agent that, when subjected to elevated temperatures, will commence to polymerization and cure.

Several thermosetting resins have been nano-modified into polymer nanocomposites by introducing the nanomaterial into the thermosetting resin. These intermediate thermosetting nanocomposites can be impregnated into fiber reinforcements, such as glass, silica, quartz, carbon/graphite, aramid, poly(p-phenylene-2-6-benzobisoxazole) (PBO), polyethylene, boron, or ceramic and upon curing lead to laminates or nano-modified polymer matrix composites (PMCs). Several of these thermoset-based nanocomposites are described in details in Chapters 7 through 13. An epoxy-based nanocomposite as a typical example for this class of material.

4.2.2.1 Epoxy Nanocomposites

Epoxy resins are widely used in commercial and military applications due to their high mechanical/adhesion characteristics, solvent and chemical resistance, combined with the versatility cure over a range of temperatures without the emission of volatile by-products. The properties of epoxy-based organic-inorganic composites can be fine-tuned by an appropriate choice of the structures of both epoxy prepolymer and hardener and of the type and amount of inorganic filler or fiber.

The use of inorganic nanoparticles can be particularly interesting to the common processing techniques used for conventional epoxy-based polymer matrix composites (PMCs). Micrometer-sized inorganic particles are currently widely used for the reinforcement of epoxy matrices to lower shrinkage on curing, thermal expansion coefficients, improve thermal conductivity, and meet mechanical requirements. The final properties of the PMCs are affected by several factors, such as intrinsic characteristics of each component, the contact, the shape and dimension of fillers, and the nature of the interface. A strong interface between matrix and filler is needed to achieve high performance. It requires that the load applied on the PMCs is mainly transferred to the fillers via the interface. To enhance these properties, the use of submicron particles can lead to a significant improvement of the mechanical

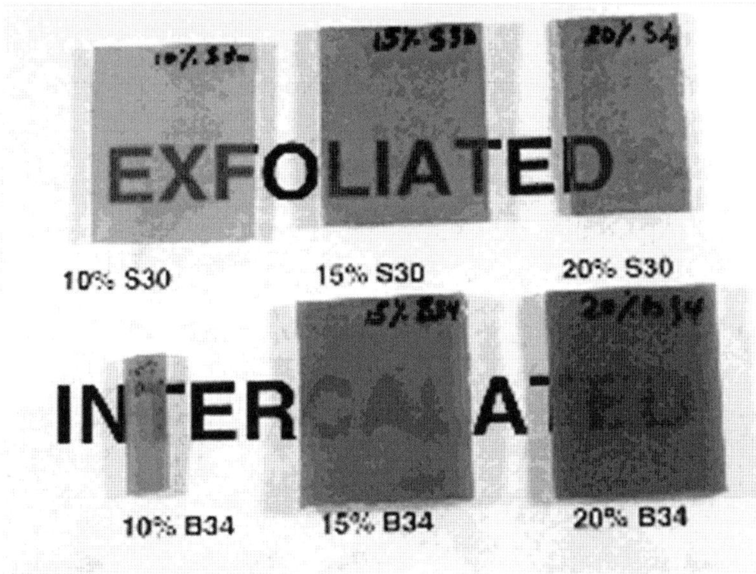

Figure 4.5. Optical image of 5-mm-thick plaques of Epon 828/D400 containing 10%, 15%, and 20% S30A and 10%, 15%, and 30% B34 [21].

properties of the PMCs. In the last decade, tremendous amount of research has been conducted in the preparation of submicron inorganic particles, leading to the possibility of preparing PMCs reinforced with nanofillers [20]. The enhanced modulus, decreased coefficient of thermal expansion, and potential rigid-phase toughening afford thermoset nanocomposites opportunities in PMCs. Epoxy resin reinforced with nanoparticles represents one of the most studied systems. Details of epoxy nanocomposites studies are included in this section.

Brown, Curliss, and Vaia [21] studied the role of various quaternary ammonium surfactants in the epoxy/diamine nanocomposite formation and processing techniques, which are necessary in ultimately fabricating PMCs with a nanocomposite matrix resin. The use of a hydroxyl-substituted quaternary ammonium modifier provides flexibility to combine both catalytic functionality (increases the intragallery reaction rate) and enhanced miscibility toward both components (resin and clay). Balancing these two factors is necessary to maintain rheological properties that are compatible with PMCs.

Initially, the authors examined the effect of various processing steps on nanocomposite formation. The generalized processing procedure to fabricate the Epon 828 layered silicate nanocomposites is shown in [21]. For all the three organically modified layered silicates (OLS) of this study, intercalation rather than exfoliation of Epon 828 was observed. This study showed that autoclave processing (evacuated bag with 100 psi hydrostatic pressure) was useful in producing bubble-free uniform plaques, especially for systems containing greater than 10 wt% OLS. Figure 4.5 shows optical micrographs of the optically transparent plaques of autoclaved nanocomposites containing up to 20 wt% S30A in Epon 828/D400 resin, and turbid plaques containing up to 20 wt% B34 in Epon 828/D400 resin [21].

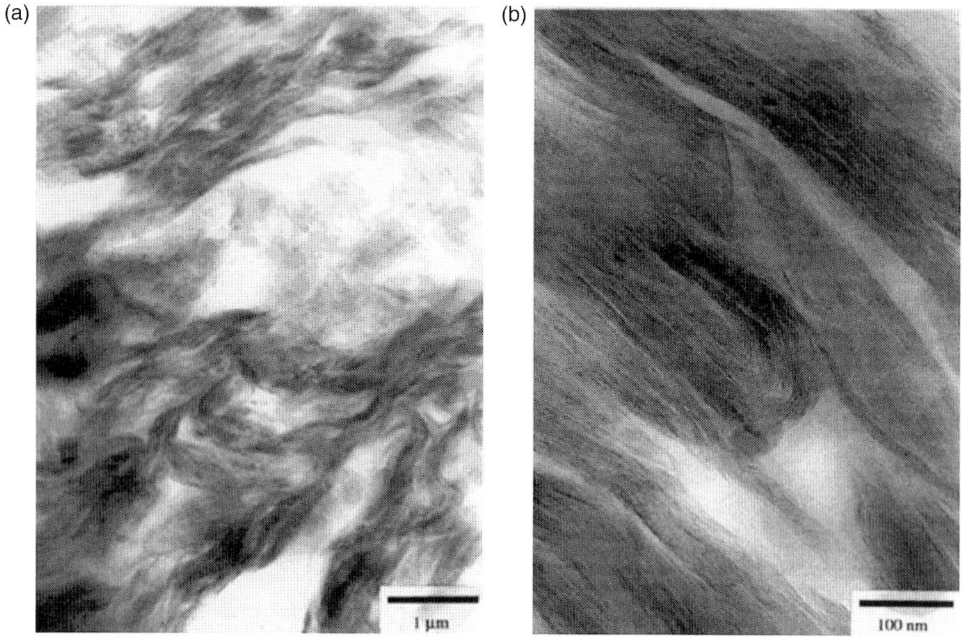

Figure 4.6. Bright-field TEM micrographs of 10% B34 in Epon 828 cured with D2000: (a) low magnification of large aggregate and (b) high magnification within aggregate [21].

Detailed TEM verified the intercalated structure of B34-containing resin (Figures 4.6a and 4.6b) [21]. Individual layers, oriented perpendicular to the sample surface, appear as dark lines, with lateral size around 500 nm. Primary particles consisting of groups of parallel layers are larger, with dimensions from 0.5 to 1.5 μm. Although the S30A nanocomposites were suggested as exfoliated by X-ray diffraction and the plaques were optically clear, a mixture of partially intercalated as well as exfoliated layers exists.

The temperature dependence of the dynamic shear moduli and tan (δ) from torsional bar measurements (0.1% strain 1 Hz) for unmodified Epon 828/D2000 were compared with nanocomposites containing 10 wt% OLS, which showed small differences at temperature range from 0 to 50°C [21]. The temperature dependency of the dynamic shear moduli and tan (δ) at various concentrations of intercalated B34 and partially exfoliated S30A in Epon 828/D2000 were also compared [21]. Table 4.3 summarizes the glass transition temperature and dynamic shear moduli at various temperature [21].

Incorporation of the layered silicate into the epoxy resulted in flammability [22] and ablation resistance [23]. When exposed to an open flame, unmodified Epon 828/D2000 burns, leaving oily, tacky residue. Resin containing intercalated B34 burns but produces a rigid graphitic char. Increased exfoliation of the layers, as in the S30A-containing resin, resulted in a self-extinguishing behavior upon removal of the flame. The resulting char exhibited a highly uniform microcellular microstructure consisting of 10- to 20-μm diameter cells separated by 1- to 2-μm struts of silica-alumina-oxycarbide (Figure 4.7) as determined by microprobe system [21].

Table 4.3. *Dynamic Mechanical Summary for OLS/Epon 828/D2000 [21]*

	Wt% OLS		G (± 25%)			
		T_g[a]	−140°C[b]	−60°C[b]	−25°C[c]	0°C[c]
828/D2000		−24	1.6	1.1	31	2.4[d]
B34	2.5	−18	1.7	1.3	35	1.2[d]
	5.0	−18	1.8	1.3	52	2.5[d]
	10	−18	1.9	1.4	80	3.1
S30A	2.5	−18	1.7	1.3	90	2.5[d]
	5.0	−22	1.8	1.3	50	1.0[d]
	10	−15	2.3	1.7	120	8.0
	15	−22	2.4	1.9	270	23
B24	10	−22	1.4	1.1	28	2.9[d]

[a]Plus or minus 3°C. [b]In GPa. [c]In MPa. [d]Lowest resolution for transducer ~1–2 MPa.

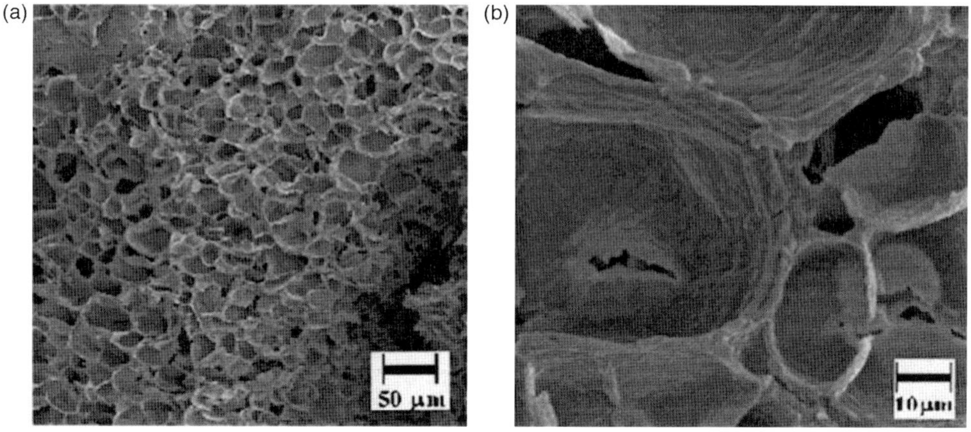

Figure 4.7. An SEM of char produced by burning 10% S30A/Epon 828/D2000 in air [21].

This study suggests that surface modifiers should include a catalyst to enhance reactivity within the OLS crystallites. A critical balance must be maintained for use with PMCs. If miscibility between the OLS and thermoset mixture is great, viscosity increases to an unacceptably large level, and potential gelation of the system will occur though impingement of exfoliated sheets. If miscibility toward both components is insufficient, a non-stoichiometric mixture of reagents will be presented in the interlayer, and enhanced interlayer reactivity will not produce the desired network formation. Thus, optimal methodology for multi-component thermosets would entail a choice of an OLS modifications that (a) partially compatibilizes the mixture, leading to a disordered intercalate containing both components and separation of the OLS aggregates into individual crystallites (slight viscosity increase); and (b) catalyzes interlayer reactivity to enable rapid layer separation before matrix gelation. Overall, this study further refines the criteria for organic modifier necessary to fabricate blends of nanoscopic inorganic phases in high-performance resins.

Most of the recent focus on epoxy systems has been on difunctional resins, predominantly using the diglycidyl ether of bisphenol-A (DGEBA). The more highly cross-linked systems are difficult to toughen, and a smaller increase than those achieved for flexible and low glass transition epoxy systems may still prove to be beneficial in comparison to using higher concentrations of more commonly used fillers. Although DGEBA is widely used, most aerospace materials and other high-performance applications demand epoxies of higher functionality because of the required higher modulus and glass transition temperature.

Becker, Varley, and Simon [24] investigated the possibilities of improving the mechanical properties of multifunctional epoxy resins through dispersion of octadecyl ammonium ion-modified layered silicates within the polymer matrix. They used three different resins in their studies: (a) difunctional diglycidyl ether of bisphenol-A (DGEBA), (b) trifunctional triglycidyl *p*-amino phenol (TGAP), and (c) tetrafunctional tetraglycidyldiamino diphenylmethane (TGDDM) [24, 25]. All resins were cured with diethyltoluene diamine (DETDA). The morphology of the final, cured materials was probed by wide-angle X-ray scattering (WAXS), optical, and atomic force microscopy. The α- and ß-relaxation temperatures of the cured systems were determined using dynamic mechanical thermal analysis. It was found that the presence of organoclay steadily decreased both transition temperatures with increasing filler concentration. The effect of different concentration of the alky ammonium-modified layered silicate on the toughness and stiffness of the different epoxy resins was analyzed. All resin systems have shown improvement in both toughness and stiffness of the materials through the incorporation of layered silicates, despite the fact that these two properties cannot be simultaneously achieved.

The clay used in the studies of Becker et al. [24, 25] is a commercially available octadecyl ammonium ion-modified montmorillonite layered silicate (Nanomer I.30E) from Nanocor Inc. The three epoxy resins used are DGEBA, TGAP, and TGDDM. The hardener is a mixture of two DETDA. Structures of these materials are shown in Table 4.4 [24].

A wide range of epoxy nanocomposites with varying amounts of organoclay (0, 2.5%, 5%, 7.5%, and 10%) by weight were prepared. Wide-angle X-ray scattering traces of the clay concentrations show that the organoclay, with an initial *d*-spacing of 23 Å, is exfoliated in the DGEBA-based system (Figure 4.8) [24]. Only the 10% concentration shows a distinct peak correlating to a *d*-spacing of 48 Å. In contrast to the DGEBA-based system, resins of high functionality show distinctive peaks even at a lower organoclay concentration, indicating that these systems have a lower degree of exfoliated layered silicate. Wide-angle X-ray scattering traces are shown in Figure 4.9 for TGAP and Figure 4.10 for TGDDM [24].

Before investigating the microstructure of the nanocomposites, optical microscopy was applied to image the bulk surface. Optical micrographs for the DGEGA system containing 5% of organoclay show large particulate phases [24]. This finding indicates that even for DGEBA-based system, not all clay particles are fully dispersed in the polymer phase. Both optical microscopy and AFM images show that at least part of the silicate content remains in tactoids or stackings of layers, rather than forming a homogenous morphology through the whole material.

Table 4.4. *Epoxy Resins and Hardener as Used for Nanocomposite Synthesis [24]*

Substance	Formula
DGEBA	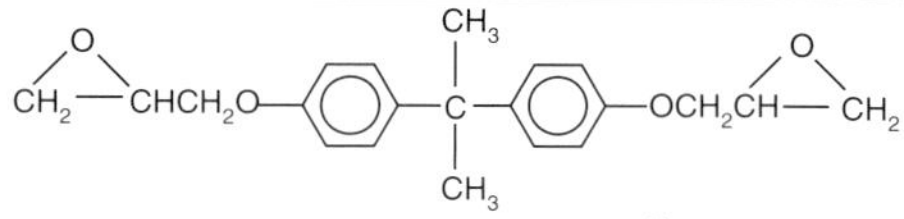
TGAP	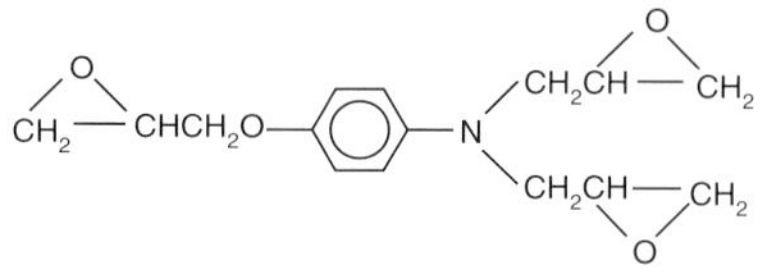
TGDDM	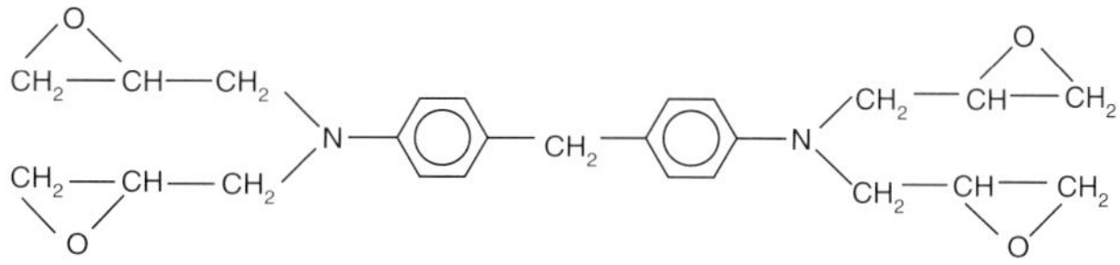
DETDA	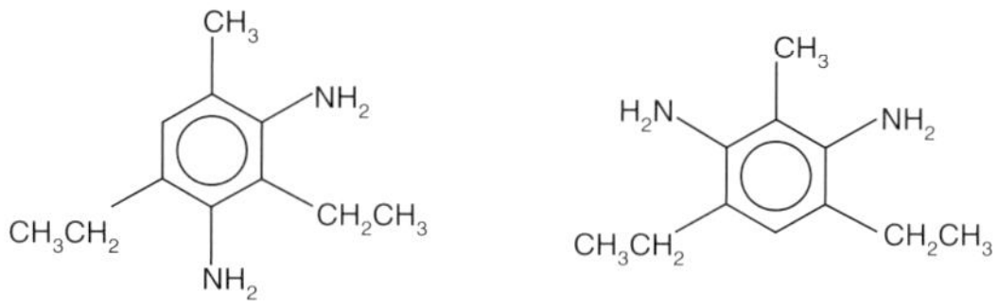

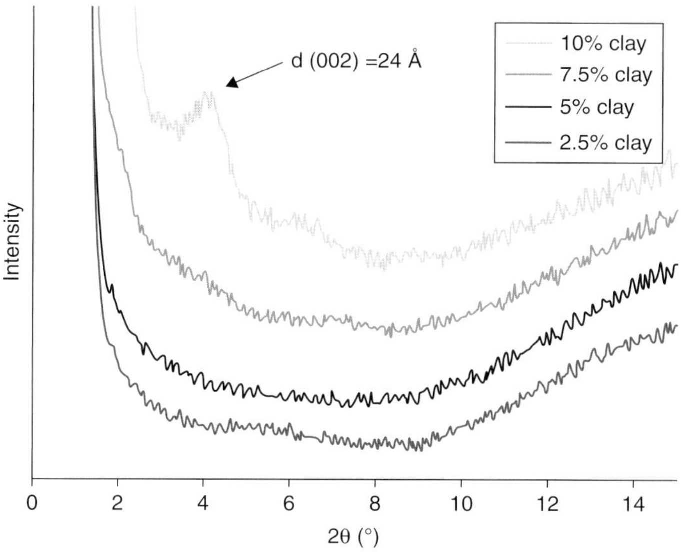

Figure 4.8. Wide-angle X-ray scattering of DETDA-cured DGEBA nanocomposites containing 0 to 10% organoclay [24].

Dynamic mechanical thermal analysis was applied in two steps, from –100°C to 50°C and from 50°C to 300°C, to investigate the influence of the organoclay on the α- and ß-transition. The value of glass transition temperature (T_g) decreased steadily with increasing organoclay concentrations. The reduction in T_g was found to be in the order of 15°C for TGAP- and DGEBA-based system and 20°C for the TGDDM-based system at an organoclay content of 10%. In summary, it appears that in all

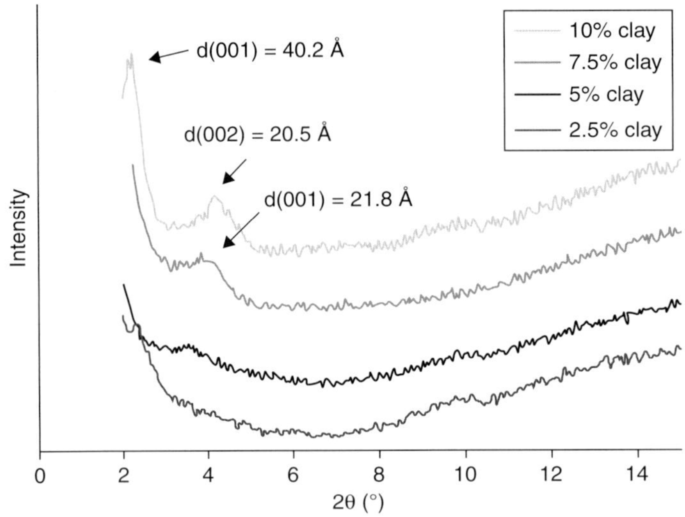

Figure 4.9. Wide-angle X-ray of DETDA-cured TGAP nanocomposites containing 0 to 10% organoclay [24].

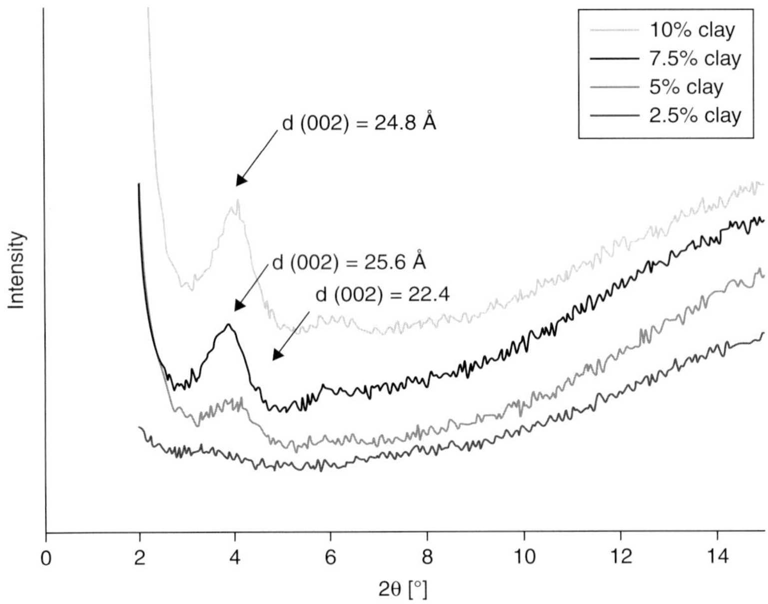

Figure 4.10. Wide-angle X-ray scattering of DETDA-cured TGDDM nanocomposites containing 0 to 10% organoclay [24].

systems, incorporation of clay leads to a decrease in both glass transition and secondary relaxation temperatures, possibility due to a decrease in effective crosslink density.

Differential scanning calorimetry data of peak temperatures and percentage residual cure are summarized in Table 4.5 [24]. Differential scanning calorimetry traces for the DBEBA-based systems confirmed full cure, while the TGAP- and TGDDM-based nanocomposites showed small peaks around 250°C, indicating

Table 4.5. *DSC Results of Postcured Epoxy-Clay Nanocomposite Systems [24]*

Clay Concentration (%)	TGAP		TGDDM	
	Peak (°C)	% Residual Cure	Peak (°C)	% Residual Cure
0	266.4	5.1	253.7	4.3
2.5	270.2	2.9	237.4	0.7
5	251.0	3.0	248.5	3.7
7.7	253.7	2.8	233.0	1.1
10	255.7	1.6	229.5	0.6

further reaction. Although the amount of reaction was very small, the filled resin was more fully cured than the neat systems were.

Density results of the various nanocomposites after curing were determined by gas displacement pycnometry. Theoretical densities according to the rule-of-mixture were calculated based on an organoclay density of 1.68 g/cm^3. All densities are within a very small range of 1.19 to 1.23 g/cm^3. The densities are generally slightly below predicted values according to the rule-of–mixture. The lower density is possibly related to poorer packing of the polymer molecules near the silicate layers.

Modulus was determined using the three-point bend test. All resin systems show a monotonic increase in modulus with increasing organoclay concentrations. A normalized modulus was calculated by dividing the actual nanocomposite modulus value through the modulus of the unfilled system. The increase in modulus for all resin systems is in a range of 20% for an organoclay concentration of 10 wt% is demonstrated [24]. The improved stiffness can be directly ascribed to the reinforcement of exfoliated, high-aspect-ratio platelets. In the work by Lan and Pinnavaia [26], it was found that a Jeffamine D2000-cured DGEBA nanocomposite increased the modulus by approximately 500% at an organoclay concentration of 10 wt%. The DGEBA system in this study shows only slightly greater relative improvements in modulus than the TGAP or TGDDM resins, respectively.

The fracture toughness, as quantified by stress intensity factor, K_{IC}, was determined using the compact tension test. Most toughening techniques show a loss in stiffness; however, in this study both toughness and stiffness were improved through the organoclay incorporation. The stress intensity factor indicates that DGEBA and TGDDM show an increase in toughness in a similar range, whereas improvement in the TGAP system is significantly lower.

Improvement in stiffness of the high-functionality epoxy resins is comparable with those achieved for the difunctional (DGEBA) resin system. Although it is often found that improvement in modulus compromises toughness of the material, both toughness and stiffness were improved through the incorporation of organoclay. The improved modulus and toughness make nanocomposite strategy an attractive alternative to the commonly used micrometer-sized fillers.

Becker et al. [25] extended the above research to study the relationship between cure temperature, morphological properties, and mechanical properties of the di-, tri-, and tetrafunctional, high-performance, epoxy layered-silicate nanocomposites. Wide-angle X-ray scattering was used to monitor the different stages of organoclay exfoliation kinetics. It was found that some degree of conversion was required to obtain significant intercalation. The polymer nanocomposite morphology was probed using TEM, WAXD, and positron annihilation lifetime spectroscopy. Comparison of the nanocomposite formation based on three different epoxy resins with different structures and functionalities has shown that the structure and chemistry of the epoxy resin, its mobility, and reactivity are key factors controlling exfoliation, and therefore the morphology of the cured nanocomposite. Two different effects have to be considered: improved dispersion and better exfoliation of the silicate platelets in the polymer matrix, and changes in the nature of the reaction and thus the network formation. The difunctional DGEBA resin gave better exfoliation than the high-functionality resins. It was observed for all resins that higher cure temperatures led to improved intercalation and exfoliation. Tactoids were observed in all three resin systems. This is attributed to better catalysis of the intragallery reaction by organo-ions, which reside within the galleries. The structural mobility (diffusion) as indicated by the viscosity of the resin seemed less relevant.

High cure temperatures were found to improve clay exfoliation and simultaneously increased toughness and modulus in the case of DGEBA- and TGAP-based materials. The mechanical properties remained relatively unaffected by the cure temperature. It is assumed that the effect of cure temperature on the reaction chemistry and the effect on cross-link density also have a major impact on mechanical properties in addition to the changes in organoclay dispersion. More dramatic changes in the nanocomposite morphology may be required to significantly change the mechanical properties. Because the processing temperatures window is limited by side reactions and thermal degradation, variation of cure temperature alone may not be sufficient to form a fully dispersions, "true" nanocomposite. Free volume properties did not vary significantly between resins or with curing temperature and generally followed the rule-of-mixtures, although there was a suggestion that the presence of clay leads to increased free volume. This finding was consistent with decreased T_g upon the addition of layered silicate and decreased cross-link density in interfacial regions of clay and epoxy matrix.

Wang et al. [27] developed a "slurry-compounding" process for the preparation of epoxy-clay nanocomposites. The microstructures of the nanocomposites were characterized by optical and transmission electron microscopy. It was found that clay was highly exfoliated and uniformly dispersed in the resulting nanocomposite. Characterization of mechanical and fracture behaviors revealed that Young's modulus increases monotonically with increasing clay concentration while the fracture toughness shows a maximum at 2.5 wt% of clay. The micro-deformation and fracture mechanism were studied throughout TEM and SEM analyses of the microstructures of the arrested crack tips and damage zone. The initiation and development of the micro-cracks are the dominant micro-deformation and fracture mechanisms in these nanocomposites. The formation of a large number of micro-cracks and the

Table 4.6. *Experimental and Model Elastic Modulus of Epoxy-Silica Nanocomposites [27]*

Composite	SiO_2 Content (wt%)	E (GPa) Experimental	Standard Deviation (Experimental) (GPa)	E (GPa) Two-Phase Model
Unfilled Epoxy	0	2.000	0.100	2.000
Epoxy/SiO_2 from Ethanol	1	2.135	0.149	2.018
"	3	2.272	0.181	2.062
"	5	2.370	0.123	2.099
Epoxy/SiO_2 from Methanol	1	2.428	0.065	2.019
"	3	2.506	0.147	2.061
"	5	2.537	0.202	2.104

increase in the fracture surface area due to crack deflection are the major toughening mechanisms.

Silica nanoparticles having different sizes were characterized by Bondioli et al. [28] using the sol-gel process. Epoxy-silica nanocomposites were prepared with nano-silicas (reaction of TEOS with methanol or ethanol) ranging from 1 wt% to 5 wt%. Scanning electron microscopy analysis and tensile tests were carried out on these epoxy-silica nanocomposites. The results indicated the absence of particle aggregation and enhanced elastic modulus. Mechanical properties were also modeled using a finite element code (OOF [29]) able to construct a numerical model based on microstructural analyses of the material (Table 4.6) [28]. Table 4.6 summarizes the mechanical properties (elastic moduli experimentally determined and modeled) for the nanocomposites [28]. A more reliable three-phase model was prepared by considering the presence of an interphase layer surrounding the particles with the intermediate elastic properties between the epoxy and the inclusion, and a characteristic size proportional to the particle radius. The composite elastic moduli were calculated from the stress-strain curves. The results were recorded (Table 4.6) as a function of particle concentration. The elastic modulus increases with higher amount of reinforcement, but the values predicted by the computational model are considerably lower. It is worth noting that analytical equations, such as the Halpin-Tsai rule [30] or the Lewis-Nielsen equation [31], give estimates that are in good agreement with the numerical results, and thus differ from the experimental data. This means that analytical and numerical approaches commonly used for particulate composite are not suitable when a nanophase is involved.

A more refined model was used by the authors by taking into account of the interphase between the matrix and the particle as a third constituent phase. It was assumed that such an interphase had properties in between the polymeric matrix and the inorganic filler. It was hypothesized that the dimension of the interphase's extension was proportional to the particle size. It was set equal to half of the average diameter. Nanocomposites with nano-silicas were modeled, and the respective elastic moduli were evaluated with the same procedure previously described. The results are plotted in Figure 4.11 [28].

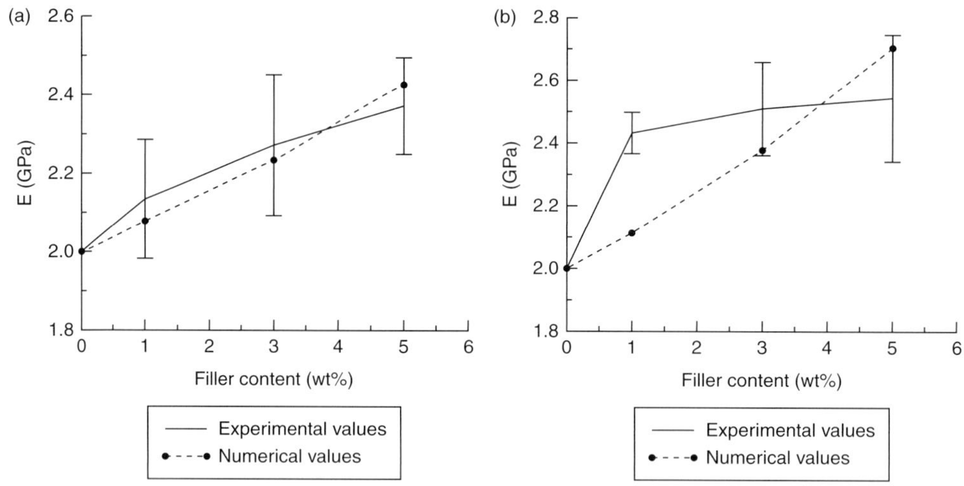

Figure 4.11. Elastic modulus as a function of filler content (three-phase model): (a) nanosilica from methanol and (b) nanosilica from ethanol [28].

The numerical results are in good agreement with the experimental data, especially if the error bars are taken into consideration. These results demonstrate that such a model is a reliable tool for the prediction of elastic properties of epoxy-silica nanocomposites. Future development of this work should aim at the experimental characterization of the nanocomposite interphase to support the hypotheses of this research. The characteristic size and stiffness should be determined. This could be achieved by means of AFM techniques [32] or nanoindentation equipment [33].

4.2.3 Elastomer-Based Nanocomposites

Another important group of polymers is the elastic or rubberlike ones, known as elastomers. These elastomeric materials can be block copolymers or multiphase systems containing soft segments (low T_g) and hard segments (high T_g, possible crystalline). They are processable under thermoplastic conditions. A few elastomers have been nano-modified into elastomer nanocomposites, which are described in detail in Chapters 6 through 13. For completeness, some rubber materials (synthetic, isoprene, natural, etc.) that are nano-modified are included. A typical example of thermoplastic olefin nanocomposites will be included in this section.

4.2.3.1 Thermoplastic Olefin (TPO) Nanocomposites

Most of the early work on the formation of nanocomposites focused on their preparation via in situ polymerization of caprolactam (Toyota group and others) in the presence of expanded silicate material. Recently, researchers have been investigating melt processing as the preferred method for the preparation of thermoplastic nanocomposites, such as nylon [15–18], polystyrene [11, 34], and polypropylene [35, 36]. These researchers have primarily been interested in the effect of processing conditions on the physical and mechanical property enhancement of the resulting PNCs.

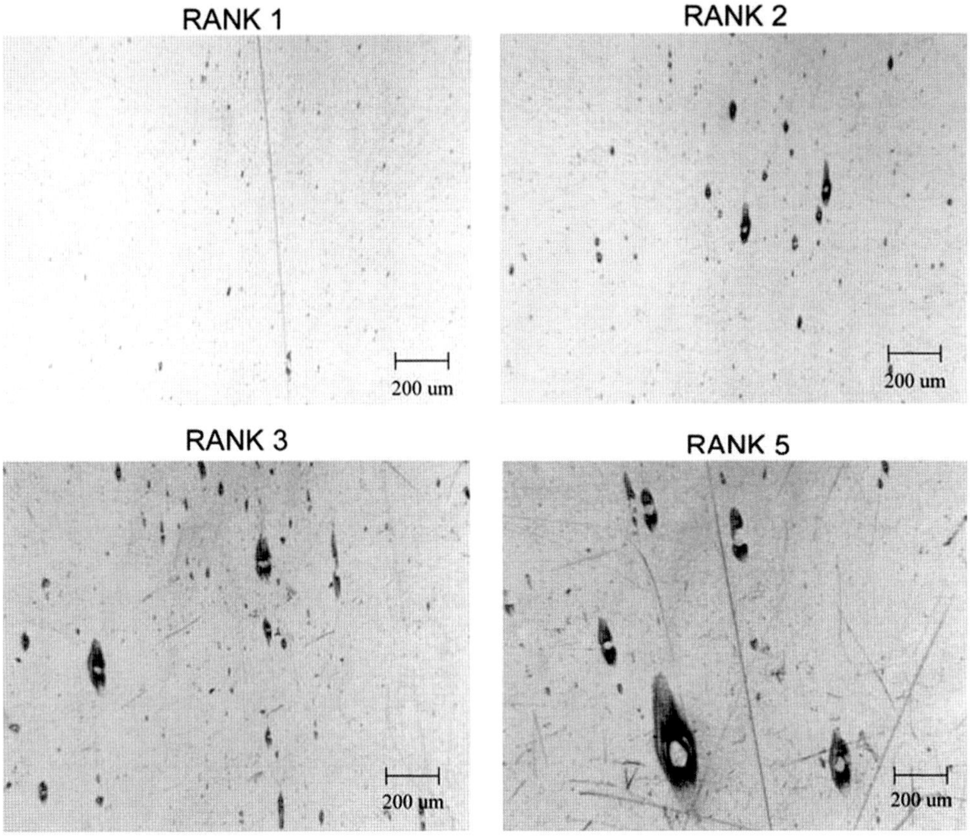

Figure 4.12. Micrographs showing representative surfaces with various ranks [37].

Fasulo et al. [37] conducted extrusion trials designed to determine the role of processing conditions with regards to the dispersion of the clay-based filler systems in thermoplastic olefin (TPO) resin systems. The TPO resins were supplied by Basell Polyolefins, Inc. These resins were based on a blend of polypropylene homopolymer, impact-modified polypropylene, and ethylene-propylene based elastomers. The organically modified MMT nanoclays were provided by Southern Clay Products, Inc. These fillers were refined from Wyoming bentonite clay and modified using an ammonium-based surfactant.

The authors examined more than 300 varying nanocomposite formulations and evaluated their surface appearance. The surface evaluation consisted of visual ranking, controls, and microscopy. Surface imperfections of the injection-molded samples were examined visually and using scanning electron microscopy (SEM). Energy-dispersive X-ray spectroscopy (EDS) analysis was utilized to examine the surfaces and cross-sections of these samples. Figure 4.12 shows representative surface with a numerical ranking [37].

Property evaluation included physical and mechanical properties. The physical properties all appeared to be relatively independent of the processing conditions. The only property that appeared to correlate with the amount of agglomerated

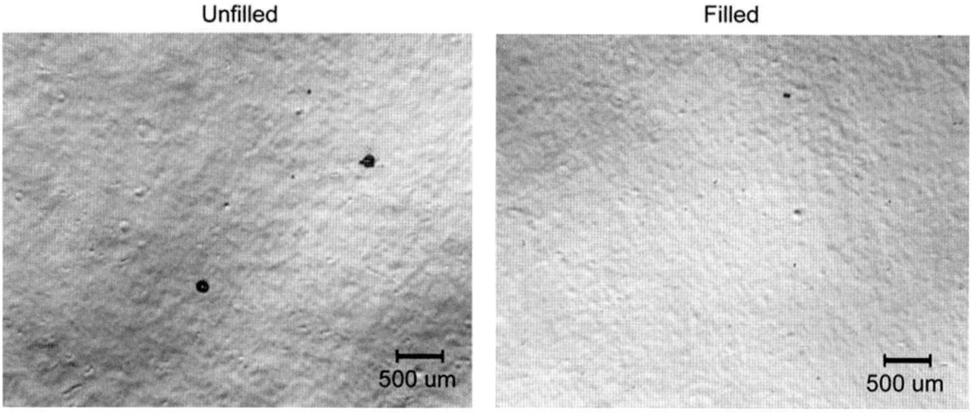

Figure 4.13. Micrographs of a coated unfilled formulation and a coated nanocomposite formulation [37].

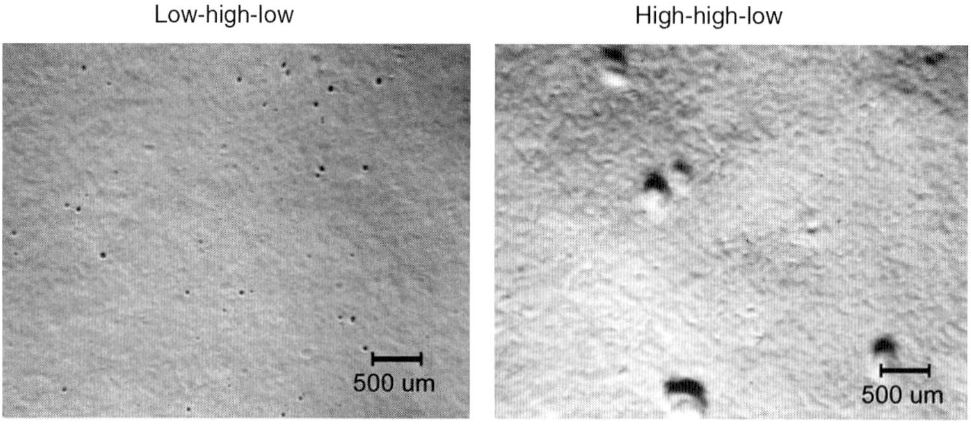

Figure 4.14. Micrographs of coated nanocomposite formulations processed under the worst conditions [37].

material is the number of brittle failures from multiaxial instrumented impact, and even this result was minimal. Figure 4.13 shows two micrographs, the control formulation, and a nanocomposite prepared using the same processing conditions [37]. Both materials exhibited with few imperfections. The imperfections visible on the control formulation can be attributed to either the coating process or poor elastomer dispersion.

Figure 4.14 shows micrographs of the painted surfaces from two of the worst-ranked processing conditions [37]. These imperfections are more numerous and larger than those observed on the control panel and can be attributed only to the surface imperfections observed on the substrate. For the low-high-low processing conditions, the imperfections are small but are visible in bright light. For the high-high-low processing conditions, the imperfections are large and visible in any light. These results provided a reasonable estimate of the size of surface imperfections required for the imperfection to be visible after the painting operation.

This work has shown that certain processing conditions maximize the clay agglomeration whereas others minimize it. A balance of processing parameters is required. The clay can agglomerate when the processing temperature is higher because there is an opportunity to degrade the intercalant or surfactant that exists between the clay sheets prior to the wetting of the filler by the molten resin. Without the surfactant, the surface tension of the unmodified clay sheets leads to agglomeration. When the feed rate is high, there is great chance of forming a mass of clay that can then experience increased pressure as it is processed in the extruder and creates agglomeration. A low screw rotation speed imparts less energy to the clay sheets, leading to slower reduction in the breakdown in the height of the clay stacks, and ultimately reduces the exfoliation of the filler material.

4.3 Summary

Several examples of thermoplastic-, thermoset-, and elastomer-based nanocomposites are presented in this chapter. These examples demonstrate that physical, mechanical, and other material properties will be enhanced with the selection of a nanomaterial compatible to the polymer. Suitable processing and curing conditions are essential to create polymer nanocomposites for a specific application. Characterization tools are useful to understand the polymer nanocomposites' behavior, properties, and performance.

Scientists and engineers can develop formulations for a range of polymer nanocomposites to fit their requirements and applications. Customer-designed polymer nanocomposites can be prepared to fit one's needs and fulfill one's requirements. The challenge is to select the baseline polymers with the proper nanomaterials to solve specific problems. Processing and morphological characterization are the keys to understanding the fundamentals of this new class of materials.

4.4 Questions and Problems

1. Why do polymer nanocomposites exhibit such multifuncitonality? Provide examples using thermoplastic, thermoset, and elastomer nanocomposites.
2. Do different types of processing methods for polymer nanocomposites create different dispersions of nanoparticles within the polymer? In other words, is a specific processing method more reliable and creates a better dispersion than do other methods?
3. Is it possible to perform more than one nanocomposite processing method on a material?
4. Is it easier to exfoliate with lower wt% of nanoparticles in the composite rather than with higher wt%? Why (or why not)?
5. Why can't exfoliation or intercalation occur if MMT is modified with a surface agent such as benzyl dimethyloctadecylammonium cation, or if too much water is added?

6. How do you classify whether a polymer nanocomposite is intercalated or exfoliated? In other words, what magnification visually expresses these classifications and to what extent can you decipher between the two? How reliable is determining between exfoliation and intercalation?
7. What "under-the-hood" applications are thermoplastics used in?
8. What are the applications for thermoset-based nanocomposites?
9. What are the differences between thermoplastic- and elastomer-based nanocomposites?
10. What are the applications for elastomer-based nanocomposites?
11. Do amorphous polymers usually have better impact resistance than do more crystalline polymers?
12. Is there a significant relationship between electrical conductivity and polymer crystallinity?
13. Is it harder to disperse nanoparticles/fillers into branched thermoplastics?
14. Are elastomers' properties highly dependent on whether they are thermoplastic or thermosets?
15. How does the incorporation of the nanoparticle alter the thermosetting resin? Does it change the curing at all?

REFERENCES

1. T. J. Pinnavaia (1983). Intercalated clay catalysts. *Science* 220, 365–371.
2. V. Mehrotra and E. P. Giannelis (1990). Conducting molecular multilayers: intercalation of conjugated polymers in layered media. *Materials Research Society Symposium Proceedings* 171, 39–44.
3. E. P. Giannelis (1992). A new strategy for synthesizing polymer-ceramic nanocomposites. *Journal of Minerals* 44, 28–30.
4. L. W. Carter, J. G. Hendricks, and D. S. Bolley (1950). United States Patent No. 2531396 (assigned to National Lead Co.).
5. P. G. Nahin and P. S. Backlund (1963). United States Patent No. 3084117 (assigned to Union Oil Co.).
6. S. Fujiwara and T. Sakamoto (1976). Japanese Kokai Patent Application No. 109998 (assigned to Unichika K.K., Japan).
7. Y. Fukushima and S. Inagaki (1987). Synthesis of an intercalated compound of montmorillonite and 6-polyamide. *Journal of Inclusion Phenomena* 5, 473–482.
8. A. Okada, Y. Fukushima, M. Kawasumi, S. Inagaki, A. Usuki, S. Sugiyama, T. Kuraunch, and O. Kamigaito (1988). United States Patent No. 4739007 (assigned to Toyota Motor Co., Japan).
9. M. Kawasumi, M. Kohzaki, Y. Kojima, A. Okada, and O. Kamigaito (1989). United States Patent No. 4810734 (assigned to Toyota Motor Co., Japan).
10. A. Usuki, Y. Kojima, M. Kawasumi, A. Okada, Y. Fukusima, et al. (1993). Synthesis of nylon 6-clay hybrid. *Journal of Material Research* 8, 1179–1184.
11. S. S. Ray and M. Okamoto (2003). Polymer/layered silicate nanocomposites: A review from preparation to processing. *Progress in Polymer Science* 28, 1539–1641.
12. R. A. Vaia, H. Ishii, and E. P. Giannelis (1993). Synthesis and properties of two-dimensional nanostructures by direct intercalation of polymer melts in layered silicates. *Chemistry of Materials* 5, 1694–1696.

13. R. A. Vaia, K. D. Jandt, E. J. Kramer, and E. P. Giannelis (1995). Kinetics of polymer melt intercalation. *Macromolecules* 28, 8080–8085.
14. R. A. Vaia and E. P. Giannelis (1997). Lattice model of polymer melt intercalation in organically-modified layered silicates. *Macromolecules* 30, 7990–7999.
15. R. A. Vaia and E. P. Giannelis (1997). Polymer melt intercalation in organically-modified layered silicates: Model predictions and experiment. *Macromolecules* 30, 8000–8009.
16. H. R. Dennis, D. L. Hunter, D. Chang, S. Kim, et al. (2001). Effect of melting processing conditions on the extent of exfoliation in organoclay-based nanocomposites. *Polymer* 42, 9513–9522.
17. J. W. Cho and D. R. Paul (2001). Nylon 6 nanocomposites by melt compounding. *Polymer* 42, 1083–1094.
18. T. D. Fornes, P. J. Yoon, H Keskula, and D. R. Paul (2001). Nylon 6 nanocomposites: The effect of matrix molecular weight. *Polymer* 42, 9929–9940.
19. T. D. Fornes, P. J. Yoon, D. L. Hunter, H. Keskula, and D. R. Paul (2002). Effect of organoclay structure on nylon 6 nanocomposite morphology and properties. *Polymer* 43, 5915–5933.
20. M. I. Baraton (2002). *Synthesis, Fictionalization and Surface Treatment of Nanoparticles.* Los Angeles, CA: American Science Publishers.
21. J. M. Brown, D. Curliss, and R. A. Vaia (2000). Thermoset-layered silicate nanocomposites: Quaternary ammonium montmorillonite with primary diamine cured epoxies. *Chemistry of Materials* 12, 3376–3384.
22. J. W. Gilman (1999). Flammability and thermal stability studies of polymer layered-silicate (clay) nanocomposites. *Applied Clay Science* 15, 31–59.
23. R. A. Vaia, G. Price, P. N. Ruth, H. T. Nguyen, and J. Lichtenhan (1999). Polymer/layered silicate nanocomposites as high performance ablative materials. *Applied Clay Science* 15, 67–92.
24. O. Becker, R. Varley, and G. Simon (2002). Morphology, thermal relaxations and mechanical properties of layered silicate nanocomposites based upon high-functionality epoxy resins. *Polymer* 43, 4365–4373.
25. O. Becker, Y-B. Cheng, R. J. Varley, and G. P. Simon (2003). Layered Silicate nanocomposites based on various high-functionality epoxy resins: The influence of cure temperature on morphology, mechanical properties, and free volume. *Macromolecules* 36, 1616–1625.
26. T. Lan and J. T. Pinnavaia, (1994). Clay-reinforced epoxy nanocomposites. *Chemistry of Materials* 6, 2216–2219.
27. K. Wang, L. Chen, J. Wu, M. L. Toh, C. He, and A. F. Yee (2005). Epoxy nanocomposites with highly exfoliated clay: Mechanical properties and fracture mechanisms. *Macromolecules* 38, 788–800.
28. F. Bondioli, V. Cannillo, E. Fabbri, and M. Messori (2005). Epoxy-silica nanocomposites: Preparation, experimental characterization, and modeling. *Journal of Applied Polymer Science* 97, 2382–2386.
29. W. C. Carter, S. A. Langer, and E. R. Fuller (1998). OOF Manual: Version 1.0. www.ctcms.nist.gov/oof
30. E. J. Barbero (1998). *Introduction to Composite Materials Design (Solution Manual)*. New York: Taylor & Francis.
31. T. B. Lewis and L. E. Nielsen (1970). Dynamic mechanical properties of particulate-filled composites. *Journal of Applied Polymer Science* 14, 1449–1471.
32. S.-L. Gao and E. Mader (2002). Characterisation of interphase nanoscale property variations in glass fibre reinforced polypropylene and epoxy resin composites. *Composites Part A* 33, 559–576.
33. H. Wang, Y. Bai, S. Liu, J. Wu, and C. P. Wong (2002). Combined effects of silica filler and its interface in epoxy resin. *Acta Materialia* 50, 4369–4377.

34. B. Hoffmann, C. Dietrich, R. Thomann, C. Friedrich, and R. Mulhaupt (2000). Morphology and rheology of polystyrene nanocomposites based upon organo-clay. *Macromolecular Rapid Communications* 21, 57–61.
35. M. Kawasumi, N. Hasegawa, M. Kato, A. Usuki, and A. Okada (1997). Preparation and mechanical properties of polypropylene-clay hybrids. *Macromolecules* 30, 6333–6338.
36. P. Reichert, H. Nitz, S. Klinke, R. Brahdsch, R. Thomann, and R. Mulhaupt (2000). Poly(propylene)/organoclay nanocomposite formation: Influence of compatibilizer functionality and organoclay modification. *Macromolecular Materials Engineering* 275, 8–17.
37. P. D. Fasulo, W. R. Rodgers, R. A. Ottaviani, and D. L. Hunter (2004). Extrusion processing of TPO nanocomposites. *Polymer Engineering and Science* 44, 1036–1045.

5 Processing of Multifunctional Polymer Nanocomposites

5.1 Synthesis Methods

After the selection of a particular polymer matrix and the appropriate nanoparticles for a specific application, the next challenge is to determine the proper synthesis method to create the desired polymer nanocomposite. Figure 5.1 shows the processing challenge of transforming 8μm particles into > 1 million platelets.

In general, for solid thermosetting reactive prepolymers or solid thermoplastic polymers with solid nanoparticles, the following processing methods are recommended:

- Solution intercalation
- Melt intercalation
- Three-roll milling
- Centrifugal

For liquid thermosetting reactive prepolymers or thermoplastic polymers in solvent with solid nanoparticles, the following processing methods are recommended:

- In situ polymerization
- Emulsion polymerization
- High-shear mixing
- Ultrasonication

Examination of Figures 5.2 and 5.3 identifies the "real" challenge that confronts the investigator as conditions are considered to suitably disperse the chosen polymer into the aggregated or layered nanoparticle(s). Solution intercalation, mesophase mediated emulsion or suspension, in situ polymerization, and melt processing are considered convenient methods to disperse layered silicates into nanocomposites in an intercalated or exfoliated state (Figure 5.3). As mentioned later in this chapter, optimum mechanical properties of the resulting nanocomposite are obtained when the layered silicate is fully exfoliated with polymer inserted into the silicate galleries. Intercalation of layered silicate with polymer is somewhat beneficial in polymer properties enhancement but not as optimal as exfoliated systems are [1–10]. An excellent review by Alexandre and Dubois lists recent developments in synthesis,

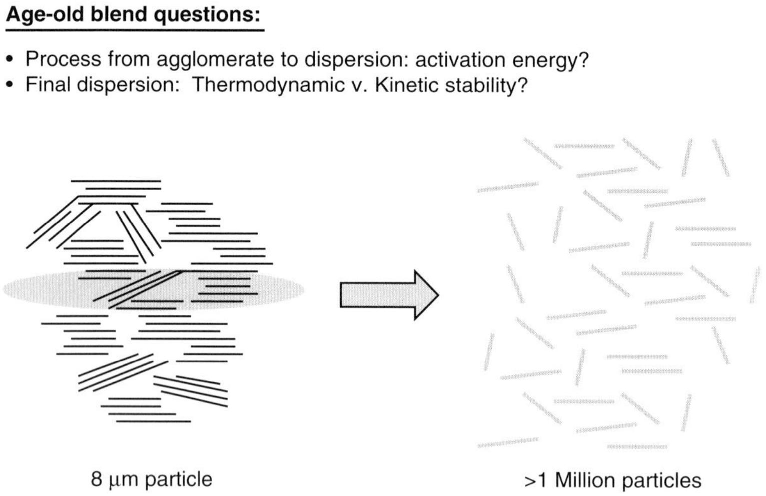

Figure 5.1. Processing challenge of layered silicate (Courtesy of SCP).

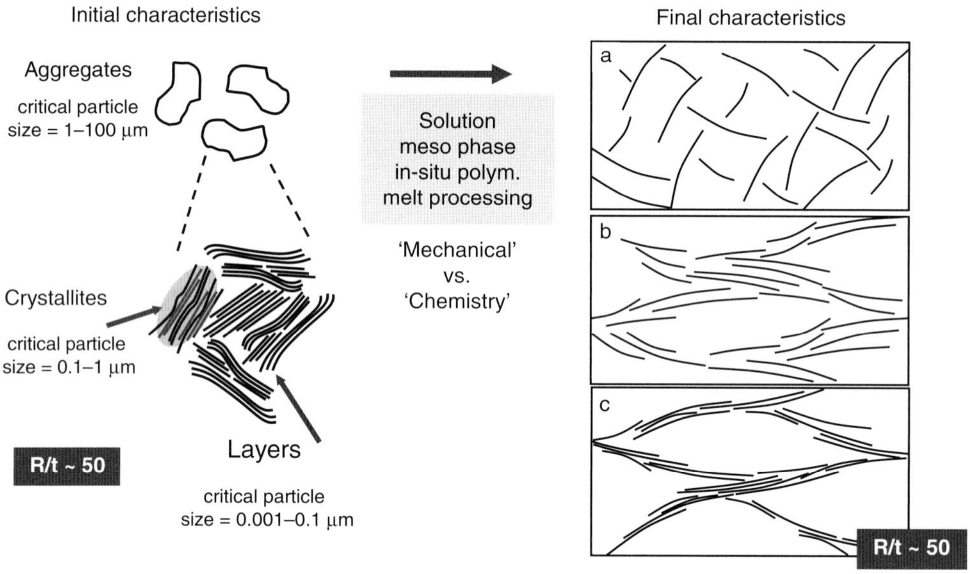

Figure 5.2. Dispersion of layered silicates using different processing methods (Courtesy of R. Vaia).

properties, and future applications of polymer-layered silicate nanocomposites [8]. Different approaches and a wide range of polymer matrices (i.e., thermoplastics, thermosets, and elastomers) were considered. Parallel to process selection is the use of WAXD and TEM (Chapter 6) to characterize the uniformity, or lack thereof, of the dispersion. TEM is vital in appreciation of "viewing" an intercalated or exfoliated nanocomposite system.

Two types of structures are obtained: intercalated nanocomposites, where the polymer chains are sandwiched between silicate layers; and exfoliated nanocomposites, where the separated, individual silicate layers are more or less uniformly dispersed in the polymer matrix. This new family of materials exhibits enhanced

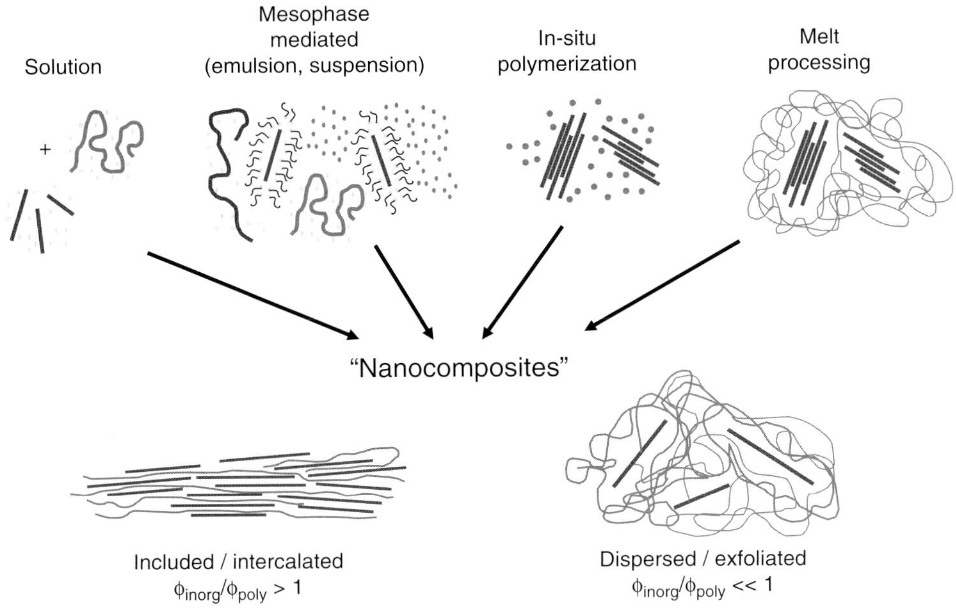

Figure 5.3. Various synthesis methods to disperse layered silicate into nanocomposites (Courtesy of R. Vaia).

properties, such as increased Young's modulus and storage modulus, increased thermal stability and gas barrier properties, and good flame retardancy at a low filler level, usually about 5 wt%.

Sanchez et al. surveyed the applications of hybrid organic-inorganic nanocomposites in their recent review [9]. This new generation of hybrid materials is opening many promising applications in several diverse areas: optics, electronics, ionics, mechanics, energy, environment, biology, medicine, functional smart coatings, fuel and solar cells, catalysts, and sensors, among others. These hybrid materials are described in Chapter 15 under new trends and opportunities.

The following sections show how to use these seven synthesis approaches and different types of nanoparticles to form polymer nanocomposites.

5.2 Solution Intercalation

The layered silicate is exfoliated into single layers using a solvent in which the polymer (or a prepolymer, in the case of insoluble polymers, such as polyimide) is soluble. It is well known that such layered silicates, due to the weak forces that stack the layers together, can be easily dispersed in an adequate solvent. The polymer then absorbs onto the delaminated sheets, and when the solvent is evaporated (or the mixture precipitated), the sheet reassembles, sandwiching the polymer to form an ordered, multilayered structure.

The layered silicate is swollen using a solvent, with which the polymer is also soluble. By swelling the layers in the nanoparticle the solvent allows the polymer access to the space in between the silicates' layers and thus intercalates them. The solvent is then removed and the intercalated silicate and polymer structure remains intact.

5.2.1 Solution Intercalation from Polymers in Solution

This technique has been widely used with water-soluble polymers to produce intercalated nanocomposites [11, 12] based on polyvinyl alcohol (PVOH) [13, 14], polyethylene oxide (PEO) [14–18], polyvinyl pyrrolidone (PVPyr) [19], or polyacrylic acid (PAA) [18]. When polymeric aqueous solutions are added to dispersions of fully delaminated sodium layered silicate, the strong interaction between the water-soluble macromolecules and the silicate layers often triggers the reaggregation of the layers as it occurs for PVPyr [18] or PEO [14]. In the presence of PVOH, the silicate layers are colloidally dispersed, resulting in a colloidal distribution of nanoparticles in PVOH [14].

Polymer intercalation using this technique can also be performed in organic solvents. Polyethylene oxide has been successfully intercalated in sodium MMT and sodium hectorite by dispersion in acetonitrile [20], allowing stoichiometric incorporation of one or two polymer chains between the silicate layers and increasing the intersheet spacing from 0.98 to 1.36 and 1.71 nm, respectively. Study of the chain conformation using two-dimensional, double-quantum Nuclear Magnetic Resonance (NMR) on ^{13}C-enriched PEO intercalated in sodium hectorite [21] reveals that the conformation of the -OC-CO- bonds of PEO is 90 ± 5% gauche, including constraints on the chain conformation in the interlayer.

5.2.2 Solution Intercalation from Prepolymers in Solution

The Toyota Research group was the first to use the solution intercalation technique to produce polyimide (PI) nanocomposites [22]. The polyimide-montmorillonite (MMT) nanocomposite has been synthesized by mixing in dimethylacetamide, a modified montmorillonite with the polyimide precursor, that is, a polyamic acid obtained from the step polymerization of 4,4'-diaminodiphenyl ether with pyromellitic dianhydride. The organic-modified montmorillonite was prepared by previous intercalation with dodecylammonium hydrochloride. After elimination of the solvent, an organoclay-filled polyamic acid film was recovered and thermally treated up to 300°C to trigger the imidization reaction and to produce the polyimide nanocomposite. The WAXD patterns (Chapter 6) of these filled PI films do not show any diffraction-peak-typical structure of an intercalated morphology. This led the authors to conclude that an exfoliated structure was formed and explained the excellent gas barrier properties of the resultant films. Enhanced properties and excellent gas barrier performance are attributable to the exfoliated nanocomposite structure. This experiment has been extended to other layered silicates (hectorite, saponite, and synthetic mica) with different aspect ratios, as shown in Table 5.1 [22].

X-ray diffraction (see Chapter 6 where absence of a peak for layered silicates is strong evidence for exfoliation) data of the resulting polyimide-based

Table 5.1. *Nature, CEC, and Length of the Layered Silicates Used in the Synthesis of Polyimide*

Layered Silicate	CEC (meq/100 g)	Length of Dispersed Particles (Angstroms)[a]
Hectorite Na^+	55	460
Saponite Na^+	100	1650
Montmorillonite Na^+	119	2180
Synthetic Mica Na^+	119	12,300

[a] Longerparticle dimension as determined by TEM observation.

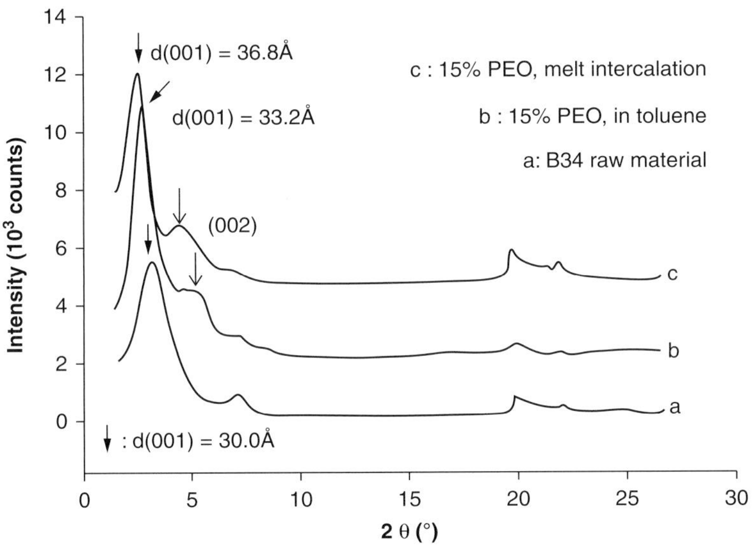

Figure 5.4. WAXD patterns for the PEO/B34 system: (a) pure B34, (b) 15% PEO via solution intercalation in toluene, and (c) 15% PEO via melt interaction.

nanocomposites show no noticeable peak, indicating a possible exfoliated structure for the montmorillonite and the synthetic mica. For both hectorite and saponite, a board peak centered on a value of 15 Å is observed, indicating that for these layered silicates, polymer intercalation occurs, probably mixed with some exfoliation. For saponite, the measured interlayer spacing is even smaller than the value measured for the starting organically modified clay (18 Å), suggesting that the organic cation may have been expelled from the clay interlayer during the imidization reaction.

Shen et al. [23] used solution intercalation and melt intercalation to create a polymer nanocomposite using poly-ethylene oxide (PEO) and two nanomaterials: Na^+-montmorillonite (MMT) and organo-modified bentonite (B34). They used X-ray diffraction (WAXD) and Fourier transform infrared (FTIR) analyses to characterize their polymer nanocomposites. The researchers did not perform TEM analysis of their samples, so it is impossible to definitively say which processing method resulted in better dispersion. Figure 5.4 shows the WAXD patterns they obtained for three of their samples [23]. It shows that neither the melt intercalation nor the solution intercalation in toluene dispersed the B34 in the PEO.

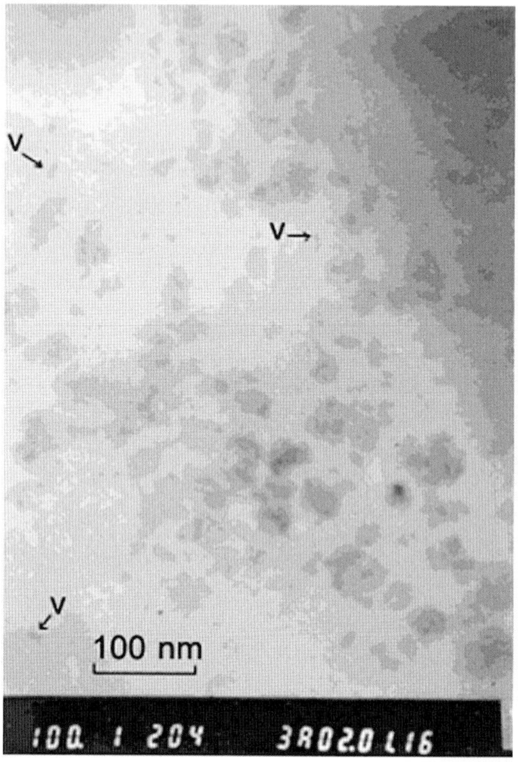

Figure 5.5. TEM micrograph of PE-g-MA/MgAl-LDH nanocomposite (Courtesy of ACS).

Chen and Qu [24] used solution intercalation to exfoliate polyethylene-grafted-maleic anhydride (PE-g-MA) into organo-modified MgAl layered double hydroxide (OMgAl-LDH) under reflux in xylene. They found that they were able to create exfoliated nanocomposites using solution intercalation. Figure 5.5 shows a TEM image of the nanocomposite they synthesized [24] (courtesy of ACS).

In summary, solution intercalation is a good processing technique because of its relative simplicity. It does not require large amounts of expensive equipment to perform. It is also a good technique for producing thin films of polymer nanocomposites with oriented clay layers. However, because it does require the use of solvents, many of which are hard to deal with, it is not considered an environmentally friendly method. For this same reason, it is not easily scalable because dealing with large amounts of organic solvents is cost prohibitive on a commercial scale.

5.3 Melt Intercalation

The layered silicate is mixed with the solid polymer matrix in the molten state. Under these conditions, and if the layer surfaces are sufficiently compatible with the selected polymer, the polymer can be inserted into the interlayer space and form either an intercalated or an exfoliated nanocomposite. This melt intercalation method does not rely on a solvent to disperse the nanomaterials, but rather uses heat and kinetic energy to break apart the nanomaterial into individual nanoparticles

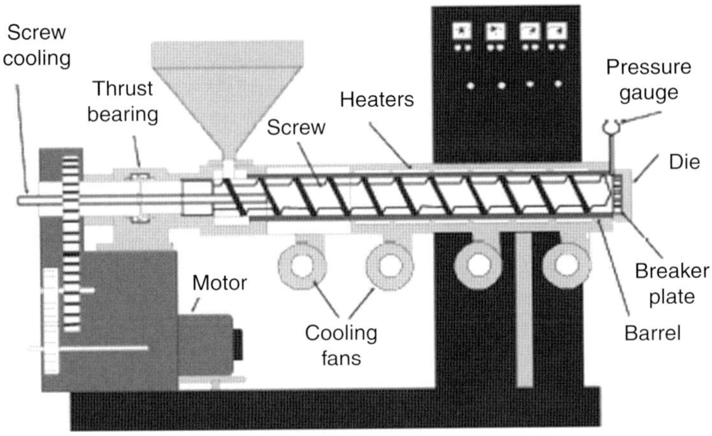

Figure 5.6. Single-screw extruder.

while allowing the polymer to insert itself between the nanoparticles. In addition to this method being compatible with a number of thermoplastic polymers, melt intercalation has the added benefit of being environmentally friendly. This is due to the fact that melt intercalation does not require a solvent like solution intercalation does, and these solvents used in solution intercalation can be harmful to humans and environmentally damaging. This leads to added costs of solution intercalation, especially when done on a large scale, to properly dispose of the solvents used in the process.

As mentioned earlier, melt intercalation in addition to heat uses kinetic energy to disperse the nanomaterials within the polymer matrix. One (single-screw extruders) or two (twin-screw extruders) screws with varying configurations usually provide this kinetic energy. In a single-screw extruder, the screw has a deep-channeled feed zone that transports solids from the feed throat and compresses them; a transition or compression zone where the channels become shallower and the polymer is melted; and a shallow-channeled metering zone where the melt is conveyed to the adapter and die (Figure 5.6). Single-screw extruders are widely used for extruding film, sheet, and other applications that use pre-compounded pellets.

Several different types of twin-screw extruders are used for different applications. Parallel twin-screw extruders have two screws that rotate in the same direction (co-rotating) or opposite direction (counter-rotating) with non-intermeshing or fully intermeshing flights. A non-intermeshing counter rotating screw is shown in Figure 5.7 and an intermeshing co-rotating twin screw is shown in Figure 5.8. The sizes of twin-screw extruders can vary considerably depending on how much polymer they need to process and under what kind of shear rates. Figure 5.9 shows an example of a schematic of twin-screw extruder.

While single-screw extruders are typically flood-fed, with the feed filling the screw channel, twin-screw extruders and extruder equipment are typically starve-fed, with only a partially filled screw channel. Because the screw channel is not completely filled, downstream feed ports or vents can be easily added.

Figure 5.7. Non-intermeshing counter-rotating twin screw.

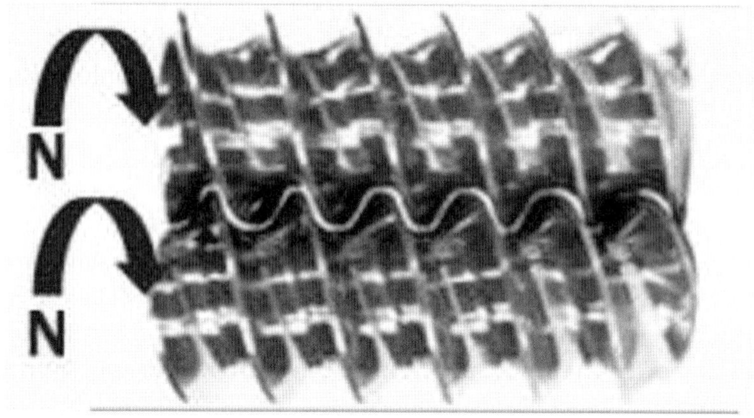

Figure 5.8. Intermeshing co-rotating twin screw.

Compared to single-screw extruders, twin-screw extruders are better at feeding and conveying, especially of powdered material or large-volume fillers, and typically create more dispersive mixing.

5.3.1 Thermoplastic Nanocomposites

Vaia and Giannelis [25] have studied polystyrene (PS) as the matrix for dispersing different types of clays. Li-fluorohectorite [cation exchange (CEC) = 150 meq/100 g], saponite (100 meq/100 g), and sodium montmorillonite (80 meq/100 g) were modified accordingly, using various ammonium cations: dioctadecyldimethylammonium, octadecyltrimethylammonium, and a series of primary alkylammonium cations with carbon chains of 6, 9–16, and 18 carbon atoms. The nanocomposites were synthesized

Table 5.2. *Characteristics of Polystyrene Melt Intercalation within Octadecylammonium Modified Clays*[a]

Entry	Clay (CEC in meq/100g)[a]	Ammonium Cation	Initial Gallery Height (nm)	Final Gallery Height (nm)	Net Change (nm)
1	M (80)	PODA[b]	0.75	0.75	0
2	S (100)	PODA[b]	0.83	0.83	0
3	F (120)	PODA[b]	1.33	2.16	0.83
4	F (120)	QODA[c]	1.57	2.69	1.12
5	M (80)	DODMDA[d]	1.43	2.25	0.82
6	S (100)	DODMDA[d]	1.50	2.35	0.85
7	F (120)	DODMDA[d]	2.85	2.85	0

[a] M = montmorillonite, S = saponite, F = fluorohectorite
[b] PODA: primary octadecylammonium
[c] QODA: quaternized octadecylammonium
[d] DODMDA: dioctadecyldimethylammonium

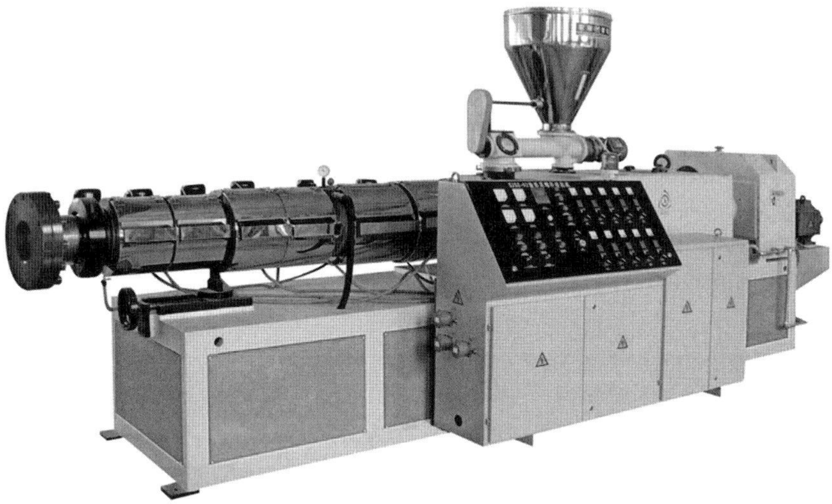

Figure 5.9. Twin-screw extruder.

by statically annealing (without any mixing or shearing) a pelletized intimate mix of the modified silicate in PS (M_w = 30,000, M_w/M_n = 1.06) under vacuum at 170°C, a temperature well above the PS glass transition, as shown in Table 5.2 [25].

Comparison of the first three entries indicates that, for a given alkyl surfactant, increasing the cation exchange from 80 to 120 meq/100 g allows for PS intercalation to occur. At a low CEC and for single alkyl chain buildup cations (entries 1 and 2), no intercalation is observed. Under these conditions, the aliphatic alkyl chain of the organic cation adopts a pseudo-monolayer arrangement characterized by low gallery height. Intercalation can take place at lower CECs when the clay surface possesses an ammonium cation bearing two long alkyl chains (entries 5 and 6). However, excessive packing of the chains all along the layer surface (high CEC and two long alkyl chains per cationic head; entry 7) leads to the formation of a non-intercalated structure, as predicted by the theory introduced by Vaia and Giannelis [26]. The authors

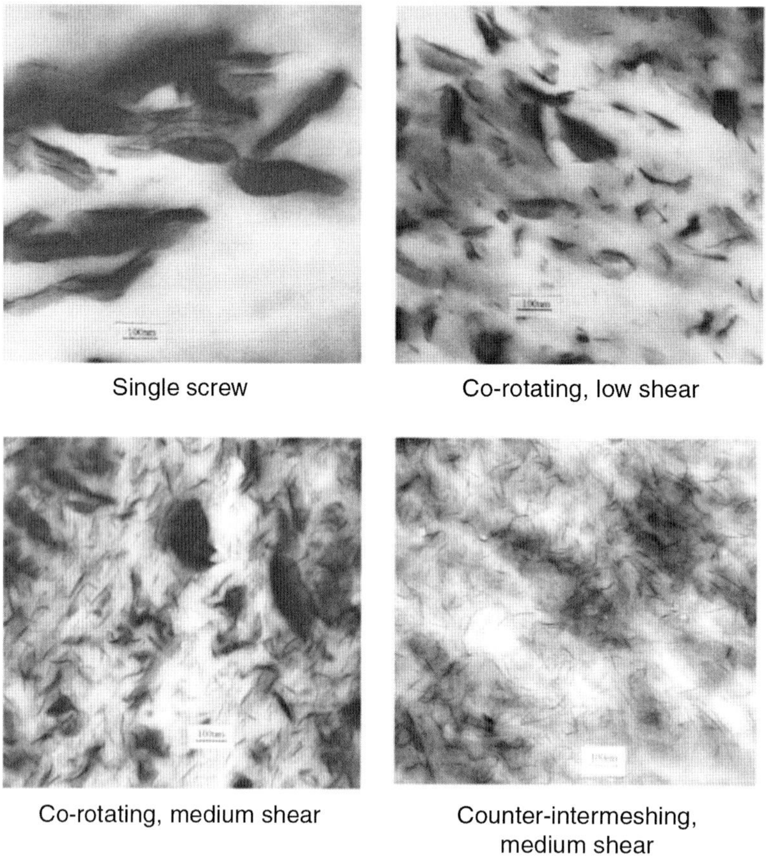

Figure 5.10. TEM images of polymer nanocomposites (polyamide 6 and Cloisite 15A) using various extruders.

also found that polymer intercalation depends on the length of the exchange ammonium cation as well as on the annealing temperature.

Dennis et al. [27] did a study to see how different extruder setups would affect the processing of a polymer nanocomposite. These researchers studied the polyamide 6 polymer (from Honeywell) in conjunction with both Cloisite 15A and Cloisite 30B (from Southern Clay Products). To perform their experiment they used four different extruders in conjunction with multiple-screw designs. For this polymer nanocomposite system, they found that the non-intermeshing twin-screw extruder performed best in terms of delamination and dispersion. However, the researchers noted that it is very likely that co-rotating and counter-rotating twin-screw extruders can provide excellent results if a fully optimized screw configuration is used.

Figure 5.10 shows TEM images of some of the processed polymer nanocomposites that these researchers produced [27]. It is clearly evident that the counter-intermeshing, medium-shear image shows the best dispersion. Table 5.3 shows the mechanical properties of all of the different test samples [27]. In this table, it is evident that the counter-rotating non-intermeshing extruder does provide very good results. It is also interesting to note that the two high-shear samples had very

Table 5.3. *Montmorillonite Feed Ratios, PMMA Contents in Nonextracted and Extracted Composites, Average Molecular Weights, and Polydispersities of Extracted PMMAs*

Sample	Feed Ratio of PMMA/ clay (g/g)	PMMA Content (wt%)[a] Nonextracted	PMMA Content (wt%)[a] Extracted[b]	$M_n \times 10^{-3}$ (g/mol)	$M_w \times 10^{-3}$ (g/mol)	M_w/M_n
PMMA	100/0	–	–	23	160	6.6
PMMA10	100/10	87.4	58.7	44	250	5.8
PMMA20	100/20	79.3	49.6	60	200	3.4
PMMA30	100/30	60.4	33.4	63	150	2.4
PMMA40	100/40	58.6	22.8	82	390	4.8
PMMA50	100/50	46.1	18.4	38	290	7.6

[a] As determined by TGA
[b] Composite recovered after Soxhlet extraction in toluene for 5 days

high percent elongation at break results – even higher than, or at least as high as, the neat polymer.

5.3.2 Elastomer Nanocomposites

Burnside and Giannelis [28] have described the two-step preparation of silicon-rubber-based nanocomposites. First, silanol-terminated polydimethylsiloxane (PDMS, *Mw* = 18,000) was melt blended at room temperature with dimethylditallowammonium-exchanged montmorillonite. Then the silanol end groups were cross-linked with tetrathylorthosilicate (TEOS) in the presence of tin bis(2-ethylhexanoate) as a catalyst at room temperature. To obtain exfoliated nanocomposites (as determined by featureless XRD patterns – see Chapter 6), several conditions were required, such as mixing the modified clay and PDMS under sonication (for better mixing) and with the addition of a small amount of water extending/cross-linking TEOS end groups. The nature of both the silicone matrix and the clay modifier play an important role. Neither exfoliation nor intercalation can occur if MMT is modified with a different surfacing agent, such as benzyl dimethyloctadecylammonium cation, or if too much water is added. On the other hand, only intercalation was observed when a PDMS-polydiphenylsiloxane random copolymer containing 14–18 mol% diphenylsiloxane units was used. The results stress the key importance of getting the right match between the matrix and the organoclay to optimize the layered silicate exfoliation.

Li and Shimizu [29] used melt intercalation to fabricate a MWCNT-thermoplastic elastomer nanocomposite. In their research, the authors mixed poly[styrene-b-(ethylene-co-butylene)-b-styrene] triblock copolymer (SEBS) from Asahi Kasei Corporation (Japan) with high-purity MWCNT's from Nikkiso Co., Ltd. (Japan). The researchers' aim was to create an elastic, conductive nanocomposite. They found that they were able to disperse the MWCNTs uniformly throughout the SEBS matrix using high-shear melt intercalation. They used a HSE3000 mini high-shear extruder (Imoto, Japan) at a speed of 1,000 rpm.

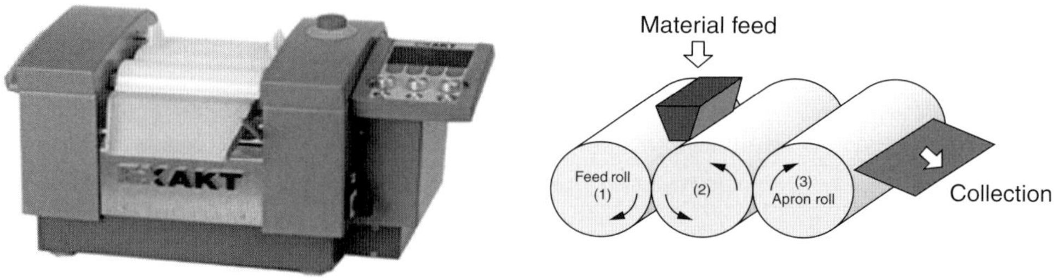

Figure 5.11. Three-roll milling equipment and schematic.

Elastomer nanocomposites and thermoplastic elastomer nanocomposites that possess enhanced ablation and fire retardancy characteristics were developed by the Air Force Research Laboratory/Polymer Working Group at Edwards AFB, CA [30–35]. The TPSiV X1180 thermoplastic elastomer (TPE) is a polyamide-based, vulcanized silicone thermoplastic resin manufactured by Dow Corning, which was selected as a potential replacement material as insulative materials in rocket propulsion systems. Three formulations – TPSiV X1180:Ph12 T12 POSS in weight ratio of 80:10, TPSiV X1180: Cloisite 30B in weight ratio of 92.5: 7.5, and TPSiV X1180: PR-24-PS CNF in weight ratio of 85: 15 – were processed by twin-screw extrusion. None of these nanoparticles was compatible with the resin system as observed by TEM analysis. The above blends with the baseline Kevlar-filled EPDM rubber were tested for ablation performance using a subscale solid rocket motor. All three TPEs were outperformed by the baseline material. These observations clearly demonstrated that if the nanoparticle are poorly dispersed in the polymer matrix, no thermal and ablation performance improvement can be expected [35].

Overall, melt intercalation is a very powerful technique for processing polymer nanocomposites. This technique is environmentally friendly, scalable, and effective at dispersing a variety of nanomaterials in polymer matrices.

5.4 Three-Roll Milling

Three-roll milling is considered as low-shear mixing for incorporating solid nanoparticles into a liquid polymer, as compared to high-shear mixing, which is discussed in Section 5.6. This process uses multiple cylinders rotating in different directions to mix nanomaterials with a given polymer. Figure 5.11 shows what a three-roll mill looks like as well as how the various cylinders rotate in relation to each other [36]. The roll mill, like melt intercalation, uses shearing to disperse nanomaterials throughout a polymer matrix. Roll milling is an effective processing technique for some polymer nanocomposites. Like melt intercalation, it is environmentally friendly in that it uses no solvents. Roll mills also have the advantage of being relatively low-shear devices, so this processing technique does not damage CNTs, which is a large benefit.

Polymer nanocomposites were developed by incorporating different POSS® nanostructured chemicals in Derakane™ Momentum™ 441-400 epoxy vinyl ester (EVE) resin [36]. The various POSS® polymers and monomers used in this study

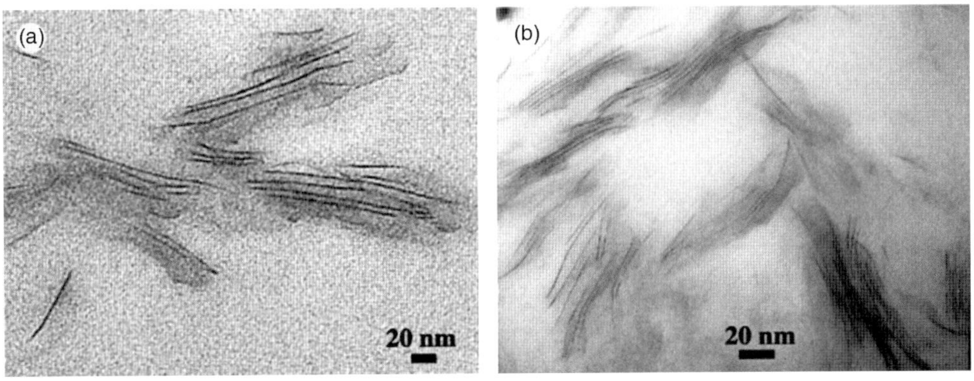

Figure 5.12. TEM images of clay nanocomposites: (a) 5 wt% and (b) 10 wt%.

were polyphenyl silsesquioxane uncured (phenyl resin), polyvinyl silsesquioxane uncured (vinyl resin), polyvinyl silsesquioxane fully cured (Firequench®), and vinyl-POSS®-cage mixture (vinyl POSS®). These formulations were prepared by blending POSS® (15 wt%) into the EVE resin using three-roll milling, low-shear mixing, and solvent-assisted blending synthesis methods. Three-roll milling is the preferred method for this study. It is very fast, cost-effective, and results in the optimum POSS® dispersion in the EVE resin. The E-glass-reinforced EVE-POSS® nanocomposites exhibit better flammability properties than pure E-glass-reinforced Derakane™ Momentum™ 441-400 resin, with reduced smoke, a reduced heat release rate, and an increased ignition time using cone calorimetry. The EVE-POSS® nanocomposites were reported to have equivalent or better mechanical, thermal, cure, and processing characteristics. The properties were also found to be comparable or better than the halogenated EVE resin (Derakane™ Momentum™ 510A-40). The nano-reinforced polymers acquired fire retardancy via two major mechanisms: (a) reduced volatilization of fuel (organic monomer/polymer) and (b) the formation of thermo-oxidative stable and nonpermeable surface chars.

A diglycidyl ether of bisphenol A (DGEBA) epoxy resin cured with methyl tetrahydrophdralic anhydride hardener was blended with Nanocor I.28E using three-roll milling [37]. Good dispersion was reported by the authors using this mixing technique.

Rosca et al. [38] used roll milling to disperse multi-wall carbon nanotubes in epoxy. These researchers found that for higher aspect ratio MWCNTs (NLIG, NL15L, NL15S), shear alignment occurred. The shearing forces from the roll milling equipment forced the MWCNTs into lower energy configuration. This did not happen for all MWCNTs in their experiment, which the authors hypothesized was either because in order for this to occur the MWCNTs need to be of a certain length or because the viscosity of some mixtures was so high that the MWCNTs could not move within it.

Yasmin et al. [39] used roll milling to disperse clay nanoparticles (Cloisite 30 B) in an epoxy matrix. They found that the three-roll mill yielded good dispersion of clay particles within a short period of time. Figure 5.12 shows TEM images of their samples. It does appear they have broken down most large agglomerates, but most of the clay platelets are still in tactoids [39].

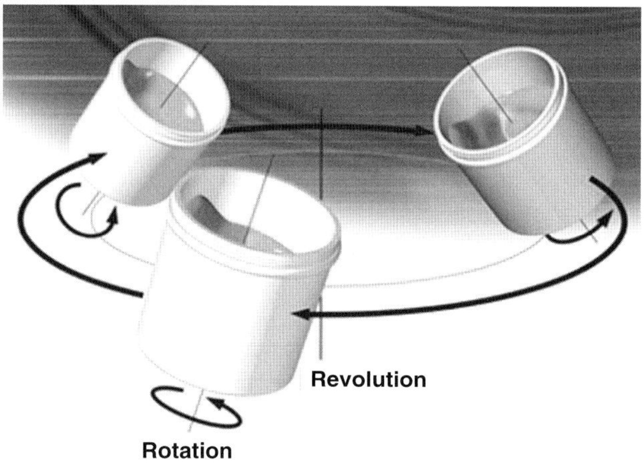

Figure 5.13. Schematics showing the principle of the centrifugal processing (Courtesy of Thinky USA).

Seyhan et al. [40] used roll milling to disperse MWCNTs and DWCNTs within a styrene-free resin (POLIYA 420) and a vinyl-ester-epoxy resin (POLIPOL 701). They noted that they chose the roll milling technique to disperse the CNTs due to the fact that roll milling does not damage or rupture CNTs like other processing techniques. They also note that the three-roll mill they used, much like the equipment shown in Figure 5.11, produces almost entirely shear loads with very little compressive loads.

5.5 Centrifugal Processing

Centrifugal processing is used primarily in powder-to-powder processing. It can also be used to mix nanoparticles with aqueous solutions. Materials are deposited into the container directly, and the rotation axis has the angle of 45 degrees to revolution. Defoaming by 400G acceleration is generated from 2,000 rpm of revolution. Mixing by centrifugal force from rotation can be applied to materials simultaneously. As a result, this equipment manufactured by Thinky Corporation (Tokyo, Japan; www.thinky.com) can be used for blending and degassing at the same time. Figure 5.13 shows the principles of this centrifugal processing technique. Figure 5.14 shows the Thinky ARV-310 mixer with different types of adapters for a variety of applications. When using standard 300 ml container and vacuum state, it has a maximum mixing capacity of 200 ml and a maximum mixing weight of 310 g (gross), the revolution speed of 2,000 rpm (max), and the rotation speed of 1,000 rpm (max). Both mixing and defoaming abilities have proven to be superior to other mixing techniques.

As an example, powder-to-powder blending using a rotation and revolution Thinky® ARE-310 mixer (centrifugal processing) were used to combine PA11 polymer and MWCNTs via centrifugal processing [41]. The PA11 powder was mixed gradually in three steps of 30 g, 30 g, and 39 g to 1 g of MWCNT (based on the desired final loading of MWCNTs, where the total mixture is 100 g in this example), each time rotating at 2,000 rpm for 10 seconds, resulting in PA11-1wt% MWCNT

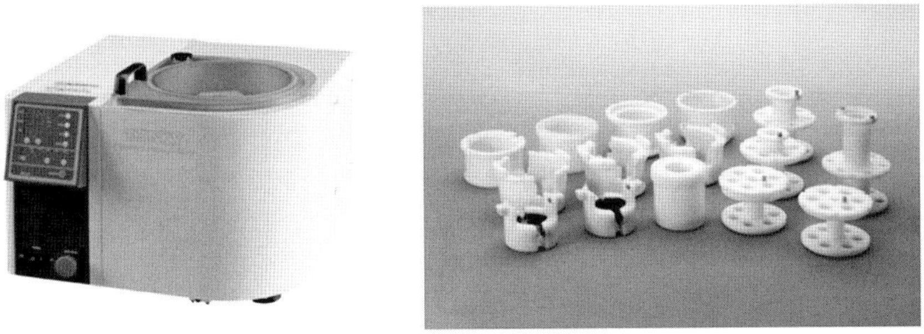

(a) (b)

Figure 5.14. Thinky (a) model ARV-310 mixer and (b) different adapters to be used with ARV-310 mixer for different type of experiments (Courtesy of Thinky USA).

mixtures. Substantial improvements in electrical conductivity were observed with increasing loading of MWCNT.

Lu et al. [42] processed and fabricated poly(3,4-ethylenedioxythiophene): poly(4-styrene sulfonate) (PEDOT:PSS)/nickel nanostrands nanocomposites using the centrifugal mixing method. The electrical properties of the nanocomposites were studied, and it was found that the addition of nanostrands into the intrinsically conductive polymer system with shorter mixing time did not improve the electrical conductivity, since there was no conductive network of nanofillers formed. With longer mixing time, better dispersion of the nanostrands was achieved and a highly significant, two-orders-of-magnitude improvement in electrical conductivity was observed with the 10% nanostrands loading. Additional processing parameters (mixing speed) of the centrifugal mixing method will be explored by Lu et al. to obtain the optimum conditions for the processing of the nanocomposites.

Hansen et al. [43] used a desired volume fraction of nickel nanostrands added to a polymer sample, then "wet" the nickel nanostrands by manual mixing methods. Dispersion was achieved by using a Thinky AR-25 centrifugal double planetary mixer. A mixing duration of 20 s at 2,000 rpm was used for baseline samples. This polymer-nanostrand composite was then used as a conductive coating. Coating samples were screened through a –50 woven stainless mesh before spray application. Polymer-nanostrand systems typically are visibly percolated at volume fractions of nanostrands above 0.5%, and imaging of sample with this loading level is challenging. However, the electrical conductivity of nanocomposite at these levels is very close to the percolation threshold, and results can be highly sensitive with respect to differences in conductor volume fraction. Test specimens of nanostrand-polymer mixtures were sprayed on an electrically insulating Kapton® polyimide film substrate and tested for electrical resistivity. Figure 5.15 shows the electrical resistivity of nanostrands in water-based acrylic/urethane polymer films [43].

In general, centrifugal processing is a very effective powder-to-powder and powder-to-liquid mixer and also an environmentally friendly method. Thinky Corporation has a total product line from R&D to production as shown in Figure 5.16 and a mixer for larger volumes (Figure 5.17).

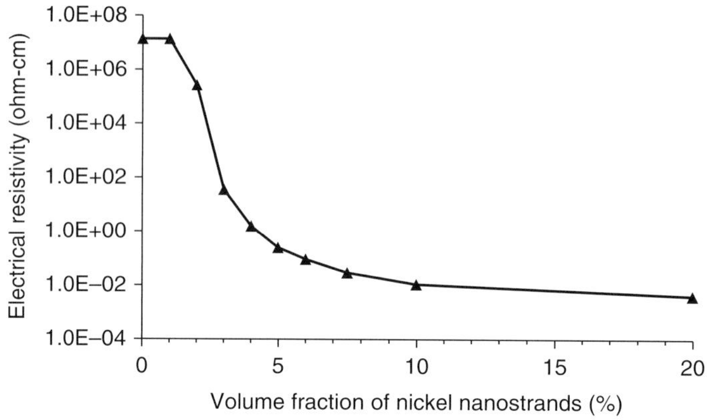

Figure 5.15. Electrical resistivity of nanostrands in water-based acrylic/urethane polymer films.

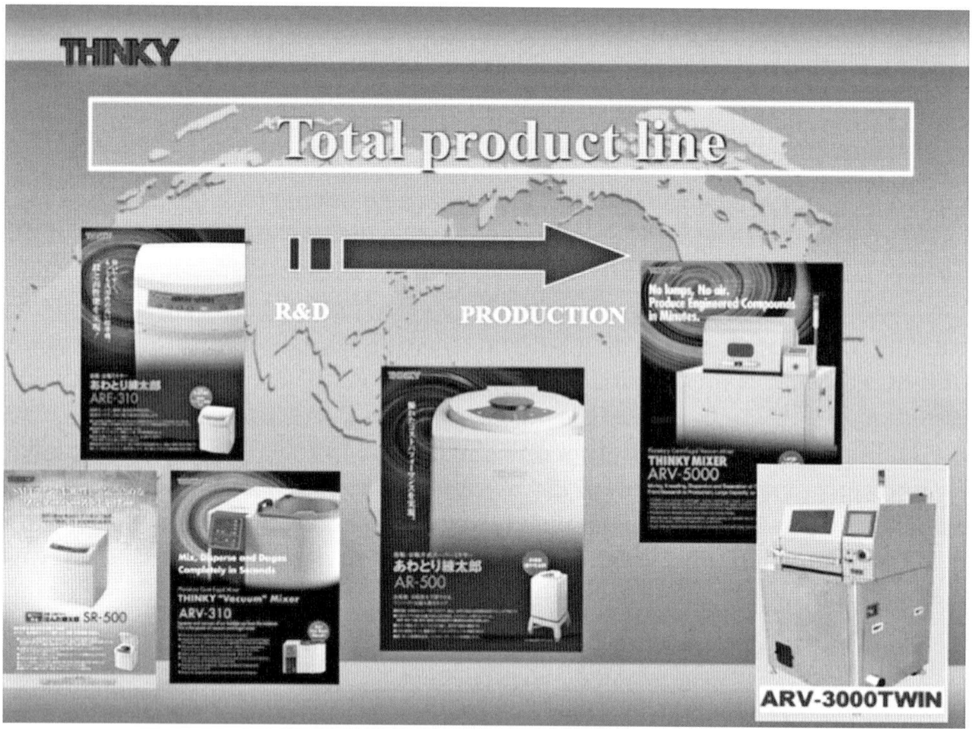

Figure 5.16. Total product line of Thinky mixers (Courtesy of Thinky USA).

5.6 In Situ Polymerization

The layered silicate is swollen within the liquid monomer (or a monomer solution) so that the polymer formation can occur between the intercalated sheets. Polymerization can be initiated by different polymerization methods, such as heat or radiation, diffusion of a suitable initiator, or an organic initiator or catalyst fixed through cationic exchange inside the interlayer spaces before the swelling step by the monomer. The in situ polymerization technique initially mixes a monomer with the nanoparticle

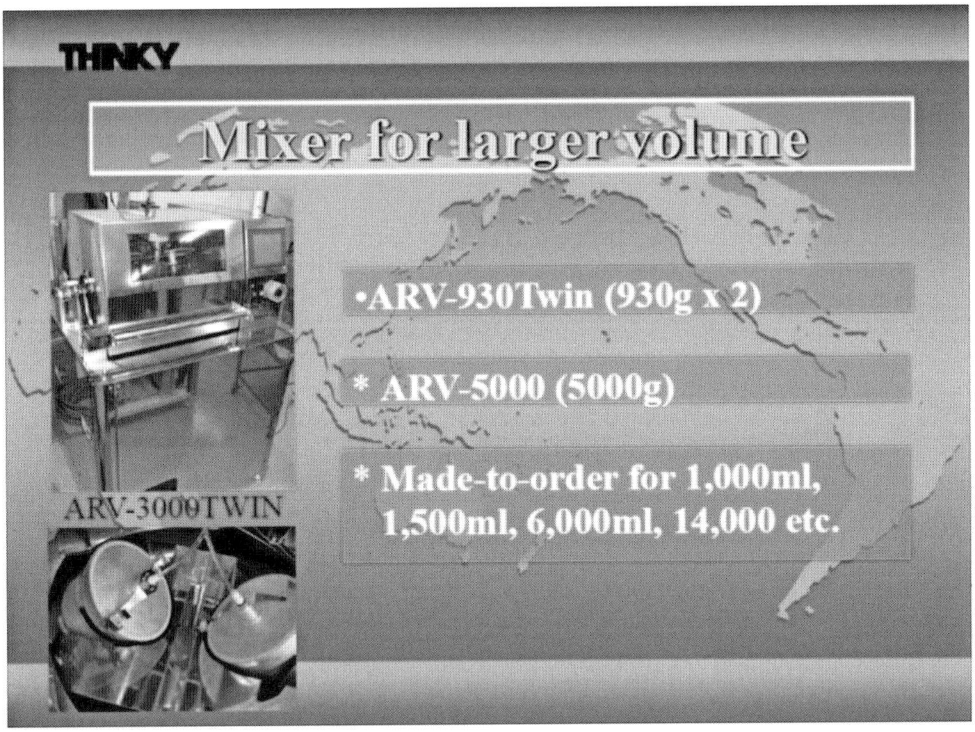

Figure 5.17. Thinky mixer for larger production volumes (Courtesy of Thinky USA).

being used. The addition of the monomer swells the layered silicate, creating small openings between the layers of the nanomaterial. The monomer is then polymerized through a variety of methods. During the polymerization, the growing polymer uses the small openings to grow into the nanomaterial, thus intercalating it. As this progresses and more polymer grows in between layers of the nanomaterial, the nanomaterial becomes more and more dispersed. The polymerization reaction can be driven by a variety of factors, including heat, radiation, or catalysts.

5.6.1 Thermoplastic Nanocomposites

Many "interlamellar" polymerization reactions were studied by various research teams in the 1960s and 1970s using layered silicate [12, 44, 45], but it is with the work initiated by the Toyota research team [46, 47] about 20 years ago that the study of polymer-layered silicate nanocomposites became recognized. The Toyota team studied the Na-montmorillonite organically modified by protonated α,ω-amino acid ($^{+}H_3N$–$(CH_2)_{n-1}$–COOH, with n=2, 3, 4, 5, 6, 8, 11, 12, 18) to be swollen by the ε-caprolactam monomer (melting temperature is 70°C) at 100°C. Subsequently, the nanomodified caprolactam was transformed by ring opening to nylon 6–based nanocomposites [48–50]. A clear difference occurs in the swelling behavior between the montmorillonite with relatively short ($n < 11$) and longer alkyl chains, as depicted in Table 5.4, indicating that a larger amount of monomer can be intercalated for longer alkyl chains.

Table 5.4. *Basal Spacing on Organomodified Montmorillonite in the Presence of ε-caprolactam (ε-CLa) at 100°C*

$^{+}H_3N–(CH_2)_{n-1}–COOH$ (n)	Interlayer Spacing of the Modified Clay (Å)	Interlayer Spacing when Swollen by (ε-CLa) at 100°C (Å)
2	12.7	14.1
3	13.1	19.7
4	13.2	19.9
5	13.2	20.4
6	13.2	23.4
8	13.4	26.4
11	17.4	35.7
12	17.2	38.7
18	28.2	71.2

Montmorillonite modified by 12-aminolauric acid (n = 12) was selected during the intercalative ring-opening polymerization of ε–caprolactam [49]. The modified montmorillonite (12-Mont) was mixed with the monomer containing a small amount of 6-aminocaproic acid as a polymerization accelerator. The mixture was heated first at 100°C for 30 minutes, then at 250°C for 6 hours. The cooled and solidified product was crushed, washed with water at 80°C, and then dried. Depending on the amount of 12-Mont introduced, either exfoliated (for less than 15 wt%) or delaminated structure (from 15 to 70 wt%) was obtained, as evidenced by X-ray diffraction (WAXD) and transmission electron microscopy (TEM) measurements (Chapter 6). Comparison of the titrated amounts of COOH and NH_2 end groups present in the synthesized nanocomposites with given values, such as the cation exchange (CEC) of the montmorillonite used (119 meq/100 g), led to the conclusion that the COOH end groups present along the 12-Mont surface are responsible for the initiation of polymerization. Nearly all of the $^{+}NH_3$ end groups present in the matrix should be interacting with the montmorillonite anions. Finally, the ratio of bonded to nonbonded polymer chains increased with the amount of incorporated MMT (from 32.3% of bonded chains for 1.5 wt% MMT to 92.3% of bonded chains for 59.6 wt% clay).

Maneshi et al. [50] used in situ polymerization to produce polyethylene-clay nanocomposites. They used Cloisite 93A clay (Southern Clay Products) as their nanomaterial. To facilitate the polymerization reaction, the researchers used bis(cyclopentadienyl)zirconium dichloride (Cp_2ZrCl_2, Aldrich). They noted that the larger the amount of Zr catalyst used in the reactor, the higher polymerization activities occurred and better clay exfoliation was visible.

Namazi et al. [51] used in situ polymerization to create starch-g-polycaprolactam (Starch-g-PCL). They used Cloisite 15A (Southern Clay Products) along with potato starch from Merck. These researchers were mainly interested in comparing the results of creating this polymer nanocomposite through in situ polymerization and solution intercalation. They found that the best diffusion and intercalation of the polymer occurred through the solution intercalation method.

5.6.2 Thermoset Nanocomposites

In situ polymerization has also been explored to create thermoset-based nanocomposites. This method has been extensively used for the production of both intercalated and exfoliated epoxy-based nanocomposites. Messersmith and Giannelis [52] have analyzed the effects of different curing agents and conditions in the formation of epoxy nanocomposites based on the diglycidyl ether of bisphenol A (DGEBA) and a montmorillonite modified by bis(2-hydroxyethyl)methyl hydrogenated tallow alkyl ammonium cation. They observed increases in viscosity at relatively low-shear rates, and the transition of opaque to semitransparent suspension indicated that the modified MMT clay dispersed readily in DGEBA when sonicated for a short period of time. The increase in viscosity was attributed to the formation of a "house-of-cards" structure, in which edge-to-edge and edge-to-face interactions between dispersed layers form percolation structures. WAXD patterns of uncured clay/DGEBA also indicate that intercalation occurred during mixing and that this intercalation improves as the temperature is increased from room temperature to 90°C.

5.6.3 Rubber-Modified Epoxy Nanocomposites

Lan and Pinnavaia [53] examined the formation of nanocomposites with a rubber-epoxy matrix obtained from DGEBA derivative (Epon 828) and cured with a polyether diamine (Jeffamine D2000) to obtain low glass transition temperature materials. Two montmorillonites modified by the protonated n-octylamine and the protonated n-octadecylamine, respectively, were used in this study. It was shown that depending on the alkyl chains length of the modified clays, an intercalated and partially exfoliated (*n*-octyl) or a totally exfoliated (*n*-octadecyl) nanocomposite can be obtained.

These authors have also studied other parameters, such as the nature of alkylammonium cations present in the gallery and the effect of the cation-exchange capacity of the clay [54] when Epon 828 was cured with *m*-phenylene diamine. It was demonstrated that when mixed with the epoxide and any length of the protonated primary alkylamine (with 8, 10, 12, 16, or 18 carbons), the modified clays adopt a structure where the carbon chains are fully extended and oriented perpendicularly to the silicate plane incorporating a maximum of one monolayer of epoxide molecules.

In summary, in situ polymerization is an effective method of creating certain polymer nanocomposites. It has the advantages of being able to be scaled up for large volumes of production. However, it does not always provide the necessary amount of dispersion of nanoparticles due to their being no shearing force forcing the tightly bound layers apart.

5.7 Emulsion Polymerization

In a manner analogous to the solution intercalation technique (Section 5.2), the layered silicate is dispersed in the aqueous phase, and the polymer nanocomposites are formed. Emulsion polymerization process is also somewhat of a mix of in situ

Table 5.5. *Summary of Mechanical Properties of Polymer Nanocomposites [57]*

Extruder and Screw Type	XRD Basal Spacing (A)	XRD Area Under Curve	TEM Platelets or Intercalates per 6.25 cm^2 (at 130, 500X)	Extruder Mean Residence Time (s)	Normalized Variance (σ^2)
Single screw extruder					
30B	30.9	120	13	141	0.0049
15A	32.2	82.5	4	141	0.0049
Twin screw extruder					
Co-rotating intermeshing					
Low shear 15A	36.2	382	7	67	0.090
Medium shear 15A	37.7	146	16	153	0.113
Counter-rotating intermeshing					
Low shear 10A	34.4	263	8	47	
Medium shear 15A	38.0	106	14	102	0.0557
Medium shear 30B	No Peak	No Peak	35	102	0.0557
High shear 15A	37.9	164	10	117	0.049
Counter-rotating non-intermeshing					
Low shear 15A	34.7	581	11	108	0.0895
Medium shear 15A	No Peak	No Peak	27	162	0.0653
High shear 15A	37.9	277	20	136	0.0579

polymerization (Section 5.6) and solution intercalation. In emulsion polymerization, a monomer is added to an emulsion, which contains the nanoparticles. The nanoparticles are swollen by the emulsion, which allows the monomer, when an appropriate initiator is added, to grow into a polymer. The growing polymer is then able to intercalate itself throughout the nanoparticle layers that are present.

The emulsion polymerization technique has been used to study the intercalation of water-insoluble polymers within Na^+-MMT, which is well known to readily delaminate in water [55–57]. Polymethyl methacrylate (PMMA) was first tested using this technique [55]. Emulsion polymerization was carried out in water in the presence of various amounts of MMT clay. Distilled methyl methacrylate monomer (MMA) was dispersed in the aqueous phase with the aid of sodium lauryl sulfate as a surfactant. Polymerization was conducted at 70°C for 12 hours, using potassium persulfate as the free-radical initiator. The resultant latex was then coagulated with an aluminium sulfate solution, filtered, and dried under reduced pressure. The obtained composites were extracted with hot toluene for 5 days (Soxhlet extraction). The amounts of intercalated polymer were determined for both extracted and nonextracted materials, and are given in Table 5.5, with the molecular weights and polydispersities of the extracted polymers.

These results demonstrate that part of the PMMA chain remains immobilized on and/or inside the layered silicates and is not extracted. This finding is further confirmed by Fourier Transform Infrared Spectroscopy (FTIR) of the extracted

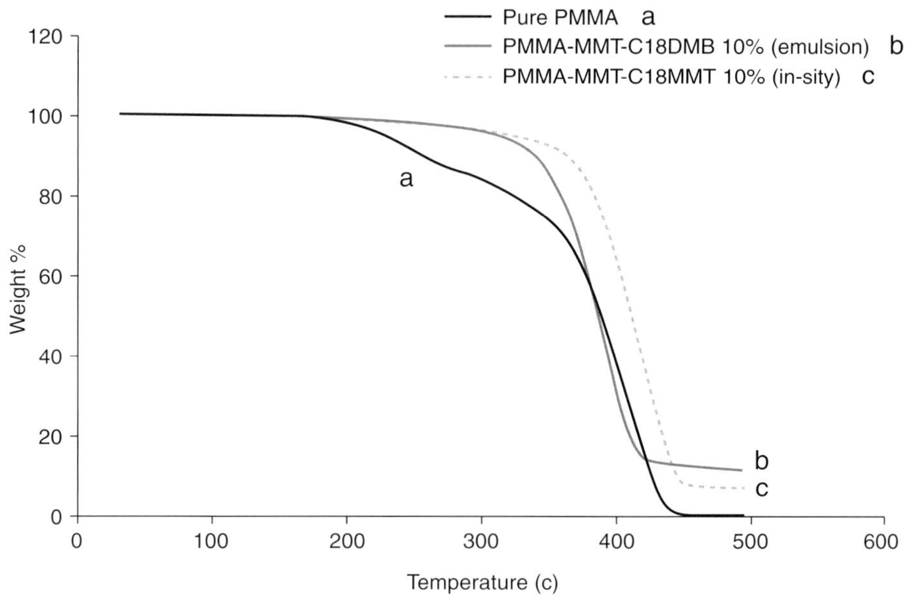

Figure 5.18. TGA analysis of PMMA/clay nanocomposite.

composite, which shows the absorption bands typical of PMMA chains. It can be observed that the relative content of clay does not substantially modify the PMMA molecular weights (M_w), the value of which is quite comparable to M_w of PMMA polymerized in the absence of clay (entry 1 in Table 5.5). The presence of layered silicates does not seem to perturb the free-radical polymerization. Intercalation is evidenced by X-ray diffraction (WAXD) (Chapter 6), where an increase of about 5.5 Å is observed for PMMA 10, 20, and 30. This increase correlates relatively well with the thickness of the polymer chain in its extended form. Differential scanning calorimetry (DSC) data obtained for the extracted nanocomposites did not show any glass transition, in accordance with what is usually observed for intercalated polymers. Ion-dipole interactions are believed to be the driving force for the immobilization of the organic polymer chains that are flat on the layered silicate surface.

Zhang et al. [58] used emulsion polymerization to create a polystyrene-SiO_2 microsphere polymer nanocomposite. In their experiment, the researchers actually dispersed the SiO_2 ultrasonically before adding them to the oil phase created by dissolving AIBN in styrene. To complete the polymerization reaction they agitated the mixture (50 rpm) and kept it at 78°C for 24 hours.

Meneghetti and Qutubuddin [59] studied the properties of poly(methyl methacrylate) (PMMA) and a montmorillonite clay nanocomposite. Figure 5.18 shows the results from their TGA analysis of the nanocomposites created using the two different methods [59]. Both nanocomposites begin breaking down at higher temperatures; however, it appears the PMMA/clay nanocomposite created using in situ polymerization was more effective to enhance thermal properties than the polymer nanocomposite created using emulsion polymerization.

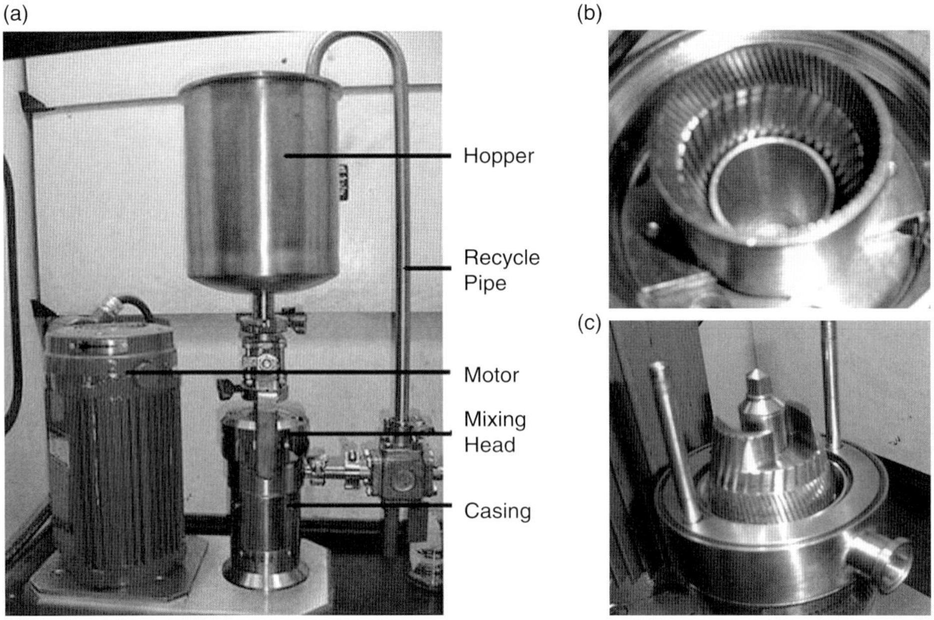

Figure 5.19. IKA colloid mill: (a) high-shear mixer overall setup, (b) mixing head stator top (from underside up), and (c) mixing head rotor bottom on pump housing and collector plate.

5.8 High-Shear Mixing

The solid or liquid nanoparticles are mixed with the liquid polymer matrix using high-shear equipment (e.g., an IKA® mixer, IKA, Wilmington, NC) as shown in Figure 5.19. Under these conditions and if the surface-treated nanoparticles are compatible with the selected polymer, high-shear mixing will disrupt the nanoparticle aggregates and disperse the polymer matrix into the nanoparticle layers. An intercalated or an exfoliated nanocomposite will result. This technique may or may not (especially in the case of water-based polymer) require solvent.

In this process, the liquid polymer is mixing with the nanomaterial and then is run through a high-shear mixer to disperse and exfoliate the nanomaterial. The mixer does this by shearing the nanomaterial into platelets, which the polymer can then incorporate into its matrix. An example of an industrial-scale high-shear mixer is the one manufactured by Ross (Hauppauge, New York; www.mixers.com). Two rotor configurations that physically shear the polymer-nanoparticle mixer are: (a) Quad slot mixing head and (b) MegaShear mixing head. The mixing process is done in a loop so that the polymer-nanoparticle liquid can be processed as many times as is needed.

Mulhaupt et al. [60] used a mixture of two epoxy resins, tetraglycidyl 4,4'-diaminodiphenyl methane (TGDDM) and bisphenol-A-diglycidylether (BADGE), cured with 4,4'-diaminodiphenyl sulfone (DDS), as the matrix material for high-performance hybrid nanocomposites containing organophilic layered silicate (OLS, synthetic fluorohectorite) and functional six-arm star poly(propylene oxide-block-ethylene) (PPO) as a toughening agent. The hybrid nanocomposites were composed of OLS as well as separated liquid PPO spheres in the epoxy matrix. Synthetic fluorohectorite (Somasif MX-100) was supplied by Unicoo Japan Ltd. (Tokyo, Japan) and octadecylamine (ODA) was supplied by Fluka (Buchs, Switzerland). The liquid PPO was supplied by Bayer AG (Leverkusen, Germany).

To prepare the epoxy hybrid nanocomposites, mixtures of equal amounts of BADGE (Aradite GY 250) and TGDDM (Araldite MY 720) were used as epoxy components. BADGE (130.5 g) and TGDDM (130.5 g) were mixed together with 20 g of the modified PPO with setearate end groups for tailored polarity as well as phenol end groups for covalent bonding to epoxy resin at 80°C and 8 hPa pressure over a period of 30 minutes with a Molteni Labmax high-shear mixer (Rheinfelder, Switzerland). Somasif/ODA (10 g) was then added to the liquid prepolymers and dispersed therein using an IKA Ultra-Turrax T25 basic high-performance disperser (IKA-Werke, Staufen, Germany) for 6 minutes. The organophilic layered silicate was swollen in the resin for 3 hours at 90°C under reduced pressure with the Labmax mixer. Then 109 g DDS (Hardener HT 976) was dissolved in the reaction mixture at 135°C over a period of 25 minutes of mixing at 8 hPa.

The epoxy resin was then poured into a mold and cured at 140°C for 3 hours and at 220°C for 7 hours in a vented oven to produce epoxy hybrid nanocomposites containing 5 wt% of PPO liquid polymer and 2.5 wt% of Somasif/ODA OLS. Various other hybrid nanocomposites with different PPO and OLS were produced by this method. Controls containing either liquid PPO toughening agent or surface-treated clay (OLS) were also prepared for comparison.

Various analytical procedures, such as gas permeation chromatography (GPC), differential scanning calorimetry (DSC), transmission electron microscopy (TEM), dynamic mechanical analyzer (DMA), linear coefficient of thermal expansion (CTE), tensile properties, and fracture toughness were determined. To evaluate the effect of morphology on the fracture mechanical response, the surface of broken compact-tension specimens was examined using scanning electron microscopy (SEM).

The hybrid nanocomposites possessed high glass transition temperature of 220°C. The T_g values were elevated compared to those of the neat resin (212°C) by adding the modified liquid PPO. The CTE of the composite material was increased by adding liquid polymer toughening agent. The stiffness was only slightly increased, and the strength of the hybrid composites was improved by 20%, mainly because of the addition of the layered silicates. The K_{Ic} was decreased by 15% to values of 1.15 MPa $m^{-1/2}$, compared to 1.35 MPa $m^{-1/2}$ for the neat resin. Hence, the extensive matrix yielding of the neat resin with the generation of shear ribs was the most effective failure mode. The predominant failure mode of the hybrid nanocomposites initiated by the silicate particles, crack bifurcation and branching, was not effective. Crack pining by the liquid PPO spheres could be resolved only when the modified polymer was used as a single additive and afforded the lowest K_{Ic} values.

Koo and coworkers have been successfully using high-shear mixing techniques to incorporate layered silicates, nanosilicas, carbon nanofibers, and POSS® to form polymer nanostructured materials in several of their research programs [61–67]. Layered silicates, carbon nanofiber, and POSS® were incorporated separately into resole phenolic (Hexion SC-1008) using high-shear, nonsparking paint-mixing equipment to form different types of polymer nanocomposites [61–67]. Neat resin castings (without fiber reinforcement) were made for WAXD and TEM analyses (Chapter 6). The polymer nanocomposites were then impregnated with carbon fabric to form polymer matrix composites. These nanocomposite rocket ablative materials (NRAMs) were tested under a simulated solid rocket motor environment. Ablation and heat transfer characteristics of these NRAMs are discussed in more detail in Chapter 10.

Koo's research team [65–67] also incorporated layered silicates, carbon nanofibers, and POSS® into resole phenolic (Hitco 134A) and cyanate ester (Lonza PT-15 and PT-30) using high-shear mixing equipment to form polymer nanocomposites. These polymer nanocomposites were used to create a new type of nano-modified carbon/carbon composite (NCCC). Wide-angle X-ray diffraction (WAXD) and TEM analyses (Chapter 6) were used to determine the degree of dispersion. Six resin/nanoparticle materials were selected to produce prepregs. These prepregs were fabricated into composites for carbonization and densification to produce C/C composites. Heat aging was conducted for 24 hours for the six candidates and the baseline Hitco CC139 materials. Mechanical properties, such as the tensile, compressive, and flexural strengths and moduli were determined and compared with the baseline CCC [65, 66]. Thermo-oxidative studies of these NCCCs are discussed in more detail in Chapter 8.

The incorporation of layered silicates, nanosilicas, and carbon nanofibers into a high-temperature, damage-tolerant epoxy resin system (CYCOM® 977-3, from Cytec Engineering Materials) was reported by the Koo research group [67]. Wide-angle X-ray diffraction and TEM were used to determine the degree of dispersion. Dynamic mechanical thermal analysis (DMTA) was used to determine the T_g and complex modulus of the polymer nanocomposites. The TEM analyses indicated the layered silicates, nanosilicas, and CNFs dispersed very well in the epoxy resin system (Chapter 6). High T_g and complex modulus values from DMTA for the nano-modified materials are presented as evidence for nanophase presence in the epoxy system as compared to lower T_g and complex moduli for the epoxy resin control. The DMTA data of the neat epoxy nanosilica nanocomposite (2 wt% Aerosil® R202) show the high T_g (258°C) and the highest complex modulus (964 MPa). The epoxy nanocomposites for carbon-fiber-reinforced polymer matrix composites are discussed in more detail in Chapter 8.

Wang et al. [68] used a high-shear mixer to help fabricate a nano-$CaCO_3$/polypropylene composite. The mixer they used had a top speed of 6,000 rpm and was outfitted with special rotors for nanocomposites. The researchers did not include the name of the company that manufactured the high-shear mixer. In the experiment, the researchers only used the high-shear mixer to break down the $CaCO_3$ into nano-sized particles before using melt intercalation to actually produce the polymer nanocomposite. The researchers found that their final polymer nanocomposites were much better dispersed and had better mechanical properties when the $CaCO_3$ was first processed by the high-shear mixer.

Zunjarrao et al. [69] used high-shear mixing and ultrasonication to create a composite out of onium ion surface modified montmorillonite (MMT) clay and epoxy resin (DEGBF). They operated their high-shear mixer at 15,000 rpm for 30 minutes

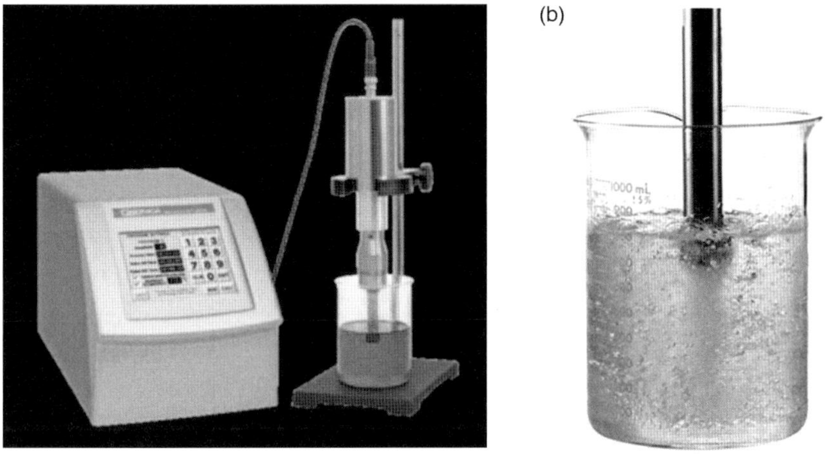

Figure 5.20. (a) A typical probe sonicator, Q700 manufactured by QSconica and (b) probe sonicator creating cavitation field in a liquid (Courtesy of QSonica).

and kept the mixture at 65°C using an ice bath. They then used a vacuum chamber for 12 hours to degas the composite. The researchers found that the high-shear mixing samples showed a higher degree of dispersion than those created using ultrasonication. However, the researchers found at low clay loadings, both processing techniques yielded exfoliated clay particles.

In summary, high-shear mixing is an effective processing technique of polymer nanocomposites. It is very effective in exfoliating nanoparticles within polymers due to the extremely high shear rates involved. It is relatively scalable, but the number of times you can cycle a polymer-nanomaterial mixture through is limited due to viscosity increase of the mixture.

5.9 Ultrasonic Mixing

The ultrasonic mixing is used to disperse nano-sized particles into liquids, such as water, oil, solvents, or resins. The application of ultrasonic mixing on nanomaterials has multiple effects. The most obvious is the dispersing of nanomaterials in liquids to break up particle agglomerates. Another process is using ultrasound during particle synthesis or precipitation. Nanomaterials – nanoclays, carbon nanotubes, graphenes or metal oxides – tend to be agglomerated when mixed into a liquid. Effective means of de-agglomeration and dispersing are needed to overcome the bonding forces after wetting the powder. The ultrasonic breakup of the agglomerated structures in aqueous and non-aqueous suspensions will utilize the full potential of nanomaterials.

The ultrasonic electronic generator transforms AC line powder to a 20kHz signal that drives a piezoelectric convertor/transducer (Figure 5.20a) (Qsonica, LLC, Newtown, CT; www.sonicator.com,). This electrical signature is converted by the transducer to a mechanical vibration due to the characteristics of the internal piezoelectric crystals. The vibration is amplified and transmitted down the length of the horn/probe where the tip longitudinally expands and contracts. The distance the tip

Figure 5.21. Solubility of MWNCTs in aqueous solution (Courtesy of QSonica).

travels is dependent on the amplitude selected by the user through the amplitude control knob. In liquid, the rapid vibration of the tip causes cavitation, the formation and violent collapse of the microscopic bubbles (Figure 5.20b). The collapse of thousands of cavitation bubbles releases tremendous energy in the cavitation field. Objects and surfaces within the cavitation field are processed. The probe tip diameter dictates the amount of samples that can be effectively processed. Smaller tip diameter delivers high-intensity sonication, but the energy is focused within a small, concentrated area. Large tip diameters can process larger volumes but offer lower intensity. The choice of a generator and horn/probe types is matched to the volume, viscosity, and other parameters of the particular application.

The key when considering a sonicator is the sample volume range. The sample volume indicates which size probe is required for the experiment. Probe sonicators are more powerful and effective than ultrasonic cleaner baths are. A cleaner bath takes hours to accomplish what a probe sonicator can do in minutes. It is clear that MWCNTs are not completely soluble in water by using bath sonicator for eight hours; there is considerable sedimentation of MWNCTs at the bottom of a small bottle (Figure 5.21a). Upon operating 20KHz applied by a probe sonicator, the MWCNTs are entirely dispersed in aqueous solution, forming a homogeneous-free solution (Figure 5.21b). Remarkably, there is no sedimentation observed even after four months of sitting at room temperature (Figure 5.21c). The concentration of MWCNTs is 2500 mg/L and the MWCNTs/SDS ratio is 1:10. Figure 5.21d shows MWCNTs of Figure 5.21c were diluted to 25 mg/L with deionized water.

The dispersion of nano-zirconium dioxide (ZrO_2) and its effect on the mechanical properties were investigated by Halder et al. [70]. The authors fabricated ZrO_2-epoxy nanocomposites via mechanical mixing (MM) and ultrasonic dual mode mixing (UDMM) methods. The mechanical mixing of ZrO_2 nanoparticles in epoxy resin was employed using glass rod stirring, and the ultrasonic dual mode mixing was employed by ultrasonic vibration along with magnetic stirring to produce ZrO_2-epoxy

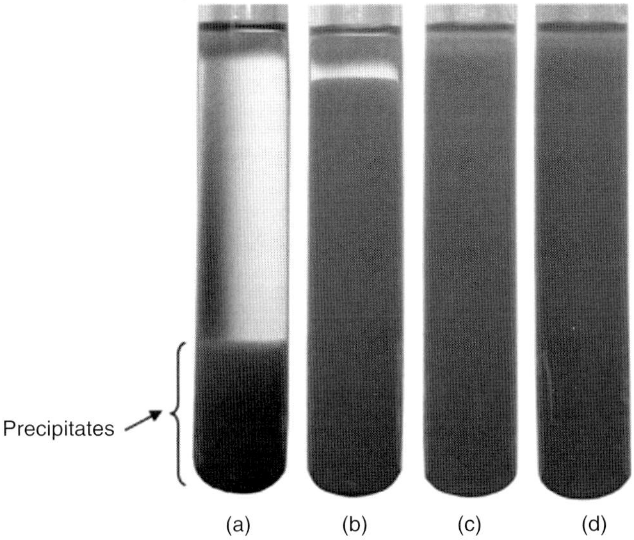

Figure 5.22. Suspension state of 3 wt% organoclay/castor oil dispersions: (a) after 2-h mechanical agitation, (b) after 15-min sonication, (c) after 30-min sonication, and (d) after 60-min sonication.

nanocomposite. Micrographs obtained using a field emission scanning electron microscope (FESEM) revealed an improved dispersion quality of ZrO_2 nanoparticles, especially for the UDMM method. A series of FESEM images of ZrO_2-epoxy nanocomposite processed by UDMM in which nanoparticles of 5 wt%, 10 wt%, and 20 wt% revealed good dispersions of ZrO_2 in the epoxy resin at low, high, and very high magnifications are shown in [70]. The improvement in dispersion was reflected in much improved tensile and fracture properties of the nanocomposite.

Heidarian et al. [71] investigated the effect of duration of ultrasonication process on structural characteristics and barrier properties of solvent-free castor oil–based polyurethane (PU)/organically modified montmorillonite (OMMT) nanocomposites. A series of PU/OMMT composites were synthesized by in situ polymerization technique through an ultrasonication-assisted process at various processing durations. Effect of ultrasonication duration on de-agglomeration of clay stacks in castor oil dispersions was evaluated by optical microscopy, sedimentation test, and viscosity measurement. Figure 5.22 shows the suspension state of castor oil/ OMMT mixtures, prepared after 2-hour mechanical agitation and 15-, 30-, and 60-minute sonication [71]. Suspensions were placed at 50°C for 2 weeks after preparation. Wide-angle X-ray diffraction and Fourier transform infrared spectroscopy were employed to investigate the effect of processing time on degree of delamination of clay platelets and interfacial strength between clay layers and PU matrix. The surface morphology of the nanocomposites was analyzed by atomic force microscopy (AFM). The topographic morphology of the surface of molded specimens was scanned by AFM. Figure 5.23 reveals the representative AFM pictures [71]. The results showed that by increasing the ultrasonication time up to 60 minutes, the size of clay agglomerates decreased and the interlayer spacing of clay platelets increased. To evaluate the

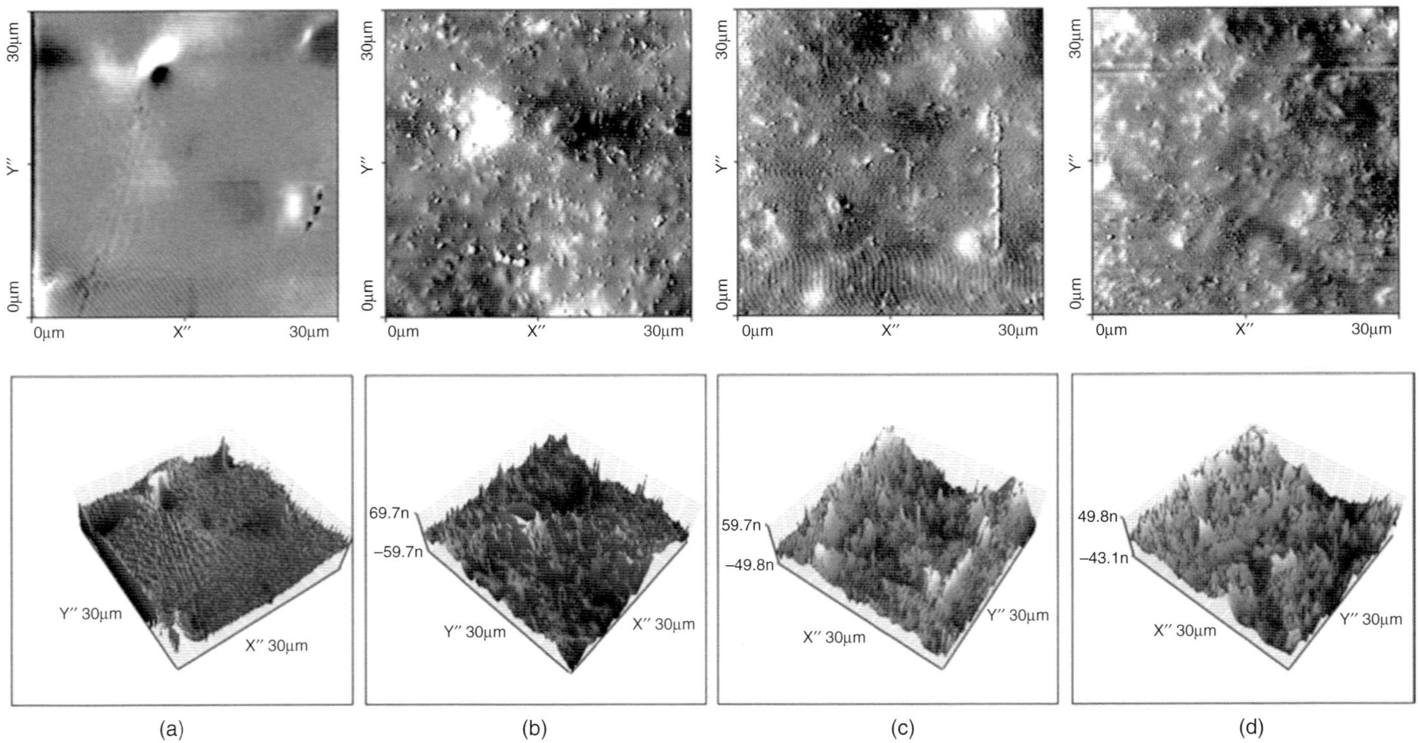

Figure 5.23. Typical topographic images of (a) pristine PU surface and the surfaces of 3 wt% PU/OMMT composites premixed by (b) 15-min sonication, (c) 30-min sonication, and (d) 60-min sonication.

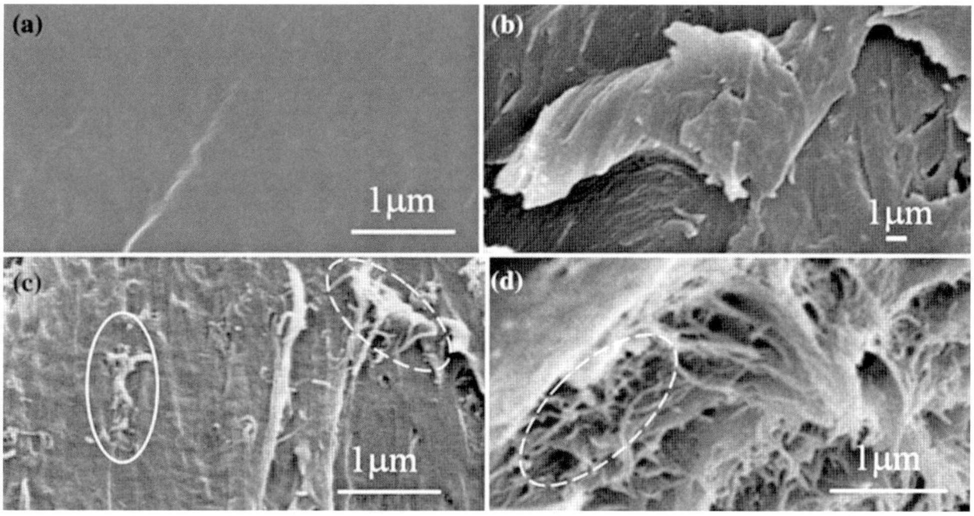

Figure 5.24. SEM images of (a) PS, (b) PS/MWNT 1 wt%, (c) PS/MWNT 2 wt%, and (d) PS/MWNT 4 wt%.

effect of ultrasonication duration on transport properties of the PU/OMMT composites, diffusion coefficient and permeability were determined through water uptake test. Electrochemical impedance spectroscopy was carried out to analyze the barrier properties and to evaluate the corrosion performance of these composite coatings on carbon steel panels. It was found that by increasing sonication time, the barrier property of nanocomposites against diffusion of water molecules improved, which is due to further separation of clay platelets, enhancement of the traveling pathways for water molecules, and improvement of interactions between the two components.

Polystyrene (PS) and multi-walled carbon nanotube (MWNT) nanocomposites were synthesized by Zhang et al. [72] via an in situ bulk polymerization by employing an ultrasonicator without adding an initiator. The authors found the ultrasonication technique produced well-dispersed MWNT in the PS matrix. Morphology of the as-synthesized PS/MWNT nanocomposite was investigated by both scanning electron microscopy and transmission electron microscopy. Figure 5.24 shows several SEM images of PS (a) and PS/MWNT nanocomposites (b, c, and d) [72]. Figure 5.25 shows photographs of PS/MWNT nanocomposites at 20°C without ultrasonication (a, d), at 80°C without ultrasonication (b, e), and at 20°C with ultrasonication (c, f) [72]. Electrical conductivity of the PS/MWNT nanocomposite film fabricated by a solvent casting method was also examined to be enhanced with MWNT content, while average molecular weights of the synthesized PS in the PS/MWNT nanocomposites analyzed by a gel permeation chromatography increased and then saturated at 2 wt% MWNT. Rheological properties of MWNT containing PS were enhanced because of improved dispersion of the MWNT through an interaction between MWNT and PS.

Ultrasonic processors and flow cells for de-agglomeration and dispersion from laboratory to production level are manufactured by Hielscher Ultrasonics GmbH (Teltow, Germany; www.hielscher.com). Hielscher offers a broad range of ultrasonic devices and accessories for dispersion of nanomaterials. Compact laboratory devices

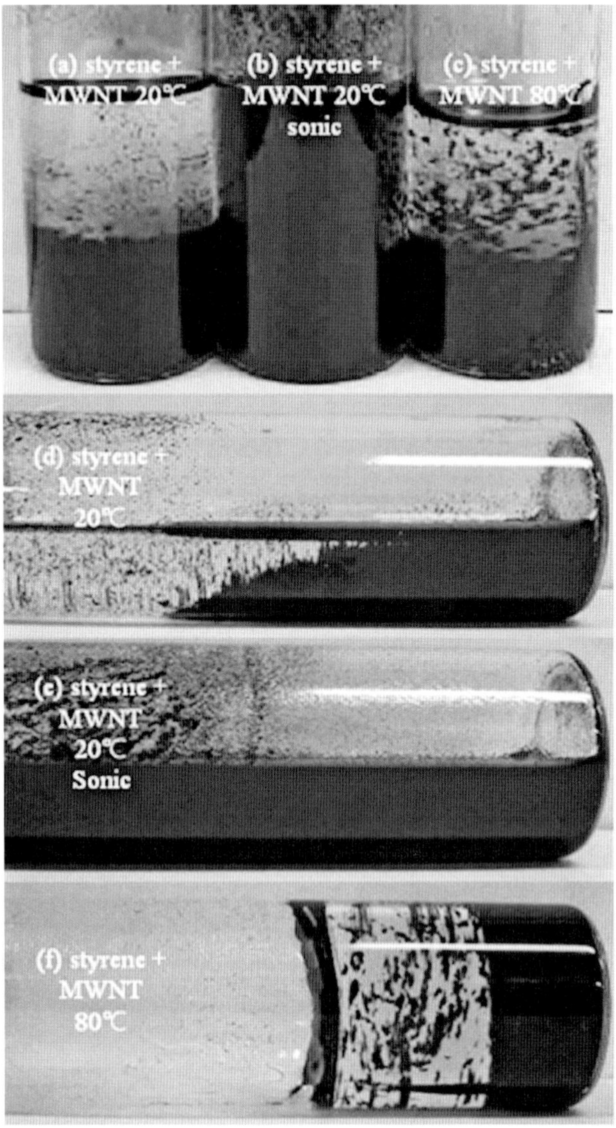

Figure 5.25. PS/MWNT nanocomposites at 20°C without ultrasonication (a, d), at 80°C without ultrasonication (b, e), and at 20°C with ultrasonication (c, f).

of up to 400 watts power are mainly used for sample preparation or initial feasibility studies. Larger devices, such as 500-, 1,000-, and 2,000-watt ultrasonic processors with flow cell and various booster horns and sonotrodes can process larger volumes. Ultrasonic processors of 2, 4, 10, and 16kW and larger clusters can process production volumes at almost any level. The industrial systems can be retrofitted to work inline.

5.10 Summary

In this chapter, a brief overview of seven popular synthesis approaches to incorporating different types of nanoparticles into a variety of polymer matrices was presented. The fundamentals behind each technique were examined as well as their advantages and disadvantages. Selected literature examples were used to demonstrate these synthesis methods employed successfully by researchers. Selective examples will be described in more details in Chapters 7 to 13, and will emphasize the properties and performance of these polymer nanostructured materials.

5.11 Questions and Problems

1. Besides the processing techniques described in the chapter, identify other common processing techniques used to formulate polymer nanocomposites and discuss the advantages and disadvantages of these processing techniques. Present your answers in a format similar to the one used in the chapter.
2. Why a longer curing cycle usually means a better final polymer?
3. What are common catalysts used in the curing cycle for thermosets? Provide several examples.
4. How may the density of the polymer affect the dispersion and processing techniques?
5. In Figure 5.2, describe the methods that were used to produce the final characteristics as shown in a, b, and c.
6. Why did Toyota Research Group (Section 5.2.2) select polyamide nanocomposites for their applications?
7. What is solution intercalation? How can one tell if the nanoparticles exfoliated if X-ray diffraction is used to evaluate intercalation within the nanocomposite?
8. What is melt intercalation? Provide some examples. What are some advantages of melt intercalation method compared to solution intercalation?
9. What are some nanoparticles that have been dispersed via three-roll milling, emulsion polymerization, and in situ polymerization? Which of these three methods seem more versatile?
10. Describe high-shear mixing and the nanoparticles that have been successfully dispersed using this method. Is there any limitation to this method?
11. Is the solution intercalation from polymers in solution widely used with some non-water-soluble polymers? If so, how does it work?
12. Can we use the solution intercalation method to prepare the solvent-soluble polymer? Why or why not?
13. Which test method do we normally use to characterize the polymer prepared from the solution intercalation from polymers in solution?
14. What are the property differences (e.g., mechanical, electrical, etc.) among the polymers of thermoset-based nanocomposites, thermoplastic-based nanocomposites, and elastomer-based nanocomposites?
15. What are some of the processing methods for dispersing nanoparticles in polymers? How can we tell whether or not the nanoparticles are properly dispersed?
16. Provide three examples of water-soluble polymers used in solution intercalation.
17. Describe the melt intercalation process and how surface modification affects the nanoparticles dispersion.
18. Is it expected to find a glass transition temperature (T_g) when observing the DSC data of an intercalated polymer?

19. Why are solution intercalation, mesophase mediated, in situ polymerization, and melt processing considered convenient processing techniques?
20. What are the water-soluble polymers that are used with the solution intercalation technique?
21. For solution intercalation from prepolymers in solution, how is the solvent eliminated?

REFERENCES

1. E. Tomanek and R. J. Enbody (Eds.) (2000). *Science and Application of Nanotubes*. New York: Kluwer Academic/Plenum Publishers.
2. T. J. Pinnavaia and G. W. Beall (Eds.) (2000). *Polymer-Clay Nanocomposites*. New York: John Wiley & Sons.
3. R. Krishnamoorti and R. A. Vaia (Eds.) (2001). *Polymer Nanocomposites: Synthesis, Characterization, and Modeling*. ACS Symposium Series 804. Washington, DC: ACS.
4. Z. L. Wang, Y. Liu, and Z. Zhang (Eds.) (2003). *Handbook of Nanophase and Nanostructured Materials, Vol. 4: Materials Systems and Applications (II)*. New York: Kluwer Academic/Plenum Publishers.
5. G. Cao (2004). *Nanostructures & Nanomaterials Synthesis, Properties & Applications*. London: Imperial College Press.
6. M. Di Ventra, S. Evoy, and J. R. Heflin Jr. (Eds.) (2004). *Introduction to Nanoscale Science and Technology*. New York: Kluwer Academic Publishers.
7. M. J. Schulz, A. D. Kelkar, and M. J. Sundaresan (Eds.) (2006). *Nanoengineering of Structural, Functional, and Smart Materials*. Boca Raton, FL: CRC.
8. M. Alexandre and P. Dubois (2000). Polymer-layered silicate nanocomposites: preparation, properties, and uses of a new class of materials. *Material Science Engineering* R28, 1–63.
9. C. Sanchez, B. Julian, P. Belleville, and M. Popall (2005). Applications of hybrid organic-inorganic nanocomposites. *Journal of Material Chemistry* 15, 3559–3592.
10. M. Lerner and C. Oriakhi (1997). *Handbook of Nanophase Materials*. New York: Mekker Decker.
11. B. Lagaly (1999). Introduction from clay mineral-powder interactions to clay mineral-polymer nanocomposites. *Applied Clay Science* 15, 1–9.
12. D. J. Greenland (1963). Adsorption of polyvinylalcohols by montmorillonite. *Journal of Colloidal Science* 18, 647–664.
13. N. Ogata, S. Kawakage, and T. Orgihara (1997). Poly (vinyl alcohol)-clay and poly (ethylene oxide)-clay blend prepared using water as solvent. *Journal of Applied Polymer Science* 66, 573–581.
14. R. L. Parfitt and D. L. Greenland (1970). Absorption of poly(ethylene glycols) on montmorillonites. *Clay Mineral* 8, 305–323.
15. X. Zhao, K. Urano, and S. Ogasawara (1989). Adsorption of polyethylene glycol from aqueous solutions on montmorillonite clays. *Colloidal Polymer Science* 267, 899–906.
16. E. Ruiz-Hitzky, P. Aranda, B. Casal, and J. C. Galvan (1995). Nanocomposite materials with controlled ion mobility. *Advanced Materials* 7, 180–184.
17. J. Billingham, C. Breen, and J. Yarwood (1997). Adsorption of polyamide, polyacrylic acid and polyethylene glycol on montmorillonite: An in situ study using ATR-FTIR. *Vibrational Spectroscopy* 14, 19–34.
18. R. Levy and C. W. Francis (1975). Interlayer adsorption of polyvinylpyrrolidone on montmorillonite. *Journal of Colloid Interface Science* 50, 442–450.

19. J. Wu and M. M. Lerner (1993). Structural, thermal, and electrically characterization of layered nanocomposites derived from sodium-montmorillonite and polyethers. *Chemistry of Materials* 5, 835–838.
20. D. J. Harris, T. J. Bonagamba, and K. Schmidt-Rohr (1999). Conformation of poly(ethylene oxide) intercalated in clay and MoS_2 studied by two-dimensional double-quantum NMR. *Macromolecules* 32, 6718–6724.
21. K. Yano, A. Usuki, A. Okada, T. Kurauchi, and O. Kamigaito (1993). Synthesis and properties of polyimide-clay hybrid. *Journal of Polymer Science Part A: Polymer Chemistry* 31, 2493–2498.
22. K. Yano, A. Usuki, and A. Okada (1997). Synthesis and properties of polyimide-clay hybrid films. *Journal of Polymer Science Part A: Polymer Chemistry* 35, 2289–2294.
23. Z., Shen, G. P. Simon, and Y. B. Cheng (2002). Comparison of solution intercalation and melt intercalation of polymer-clay nanocomposites. *Polymer* 43(15), 4251–4260.
24. W. Chen and B. Qu (2003). Structural characteristics and thermal properties of Pe-G-Ma/Mgal-Ldh exfoliation nanocomposites synthesized by solution intercalation. *Chemistry of Materials* 15(16), 3208–3213.
25. R. A. Vaia and E. P. Giannelis (1997). Polymer melt intercalation in organically-modified layered silicates: Model predictions and experiment. *Macromolecules* 30, 8000–8009.
26. R. A. Vaia, H. Ishii, and E. P. Giannelis (1993). Synthesis and properties of two-dimensional nanostructures by direct intercalation of polymer melts in layered silicates. *Chemistry of Materials* 5, 1694–1696.
27. H. R. Dennis, D. L. Hunter, D. Chang, S. Kim, J. L. White, et al. (2001). Effect of melt processing conditions on the extent of exfoliation in organoclay-based nanocomposites. *Polymer* 42(23), 9513–9522.
28. S. D. Burnside and E. P. Giannelis (1995). Synthesis and properties of new poly(dimethylsiloxane) nanocomposites. *Chemistry of Materials* 7, 1597–1600.
29. Y. Li and H. Shimizu (2009). Toward a stretchable, elastic, and electrically conductive nanocomposite: Morphology and properties of poly[styrene-b-(ethylene-co-butylene)-b-styrene]/multiwalled carbon nanotube composites fabricated by high-shear processing. *Macromolecules* 42(7), 2587–2593.
30. R. Blanski, J. H. Koo, P. Ruth, N. Nguyen, C. Pittman, and S. Phillips (2004). Polymer Nanostructured Materials for Solid Rocket Motor Insulation – Ablation Performance. *Proceedings of the 52nd JANNAF Propulsion Meeting*, CPIA, Columbia, MD.
31. J. H. Koo, D. Marchant, G. Wissler, P. Ruth, S. Barker, et al. (2004). Polymer Nanostructured Materials for Solid Rocket Motor Insulation – Processing, Microstructure, and Mechanical Properties. *Proceedings of the 52nd JANNAF Propulsion Meeting*, CPIA, Columbia, MD.
32. P. Ruth, R. Blanski, and J. H. Koo (2004). Preparation of Polymer Nanocomposites for Solid Rocket Motor Insulation. *Proceedings of the 52nd JANNAF Propulsion Meeting*, CPIA, Columbia, MD.
33. R. Blanski, J. H. Koo, P. Ruth, N. Nguyen, C. Pittman, and S. Phillips (2004). Ablation Characteristics of Nanostructured Materials for Solid Rocket Motor Insulation. *Proceedings of the National Space & Missile Materials Symposium*, Seattle, WA, June 21–25.
34. J. H. Koo, D. Marchant, G. Wissler, P. Ruth, S. Barker, et al. (2004). Processing and Characterization of Nanostructured Materials for Solid Rocket Motors. *Proceedings of the National Space & Missile Materials Symposium*, Seattle, WA, June 21–25.

35. D. Marchant, J. H. Koo, R. Blanski, E. H. Weber, P. Ruth, et al. (2004). Flammability and Thermophysical Characterization of Thermoplastic Elastomer Nanocomposites. *ACS National Meeting, Fire & Polymers Symposium*, Philadelphia, PA, August 22–26.
36. S. K. Gupta, J. J. Schwab, A. Lee, B. X. Gu, and B. S. Hsiao (2002). POSS® Reinforced Fire Retarding EVE Resins. *Proceedings of the SAMPE 2002 ISSE*, SAMPE, Covina, CA, pp. 1517–1526.
37. A. Yasmin, J. L. Abot, and I. M. Daniel (2003). Processing of clay/epoxy nanocomposites with a three-roll mill machine. *Materials Research Society Symposium Proceedings*, 740, 75–80.
38. I. D. Rosca and S. V. Hoa (2009). Highly conductive multiwall carbon nanotube and epoxy composites produced by three-roll milling. *Carbon* 47(8), 1958–1968.
39. A. Yasmin, J. L. Abot, and I. M. Daniel (2003). Processing of clay/epoxy nanocomposites by shear mixing. *Scripta Materialia* 49(1), 81–86.
40. A. T. Seyhan, M. Tanoğlu, and K. Schulte (2009). Tensile mechanical behavior and fracture toughness of MWCNT and DWCNT modified vinyl-ester/polyester hybrid nanocomposites produced by 3-roll milling. *Materials Science and Engineering: A* 523(1), 85–92.
41. M. Yuan, B. Johnson, J. H. Koo, and D. Bourell (2014). Polyamide 11-MWNT nanocomposites: Thermal and electrical conductivity measurements. *Journal of Composite Materials* 48(15), 1833-1841.
42. C. Lu, M. Krifa, and J. H. Koo (2013). Conductive Poly(3,4 ethylenedioxythiophene):poly(4-styrene sulfonate) (PEDOT:PSS)/Nickel Nanostrands Nanocomposites. *Proceedings of the SAMPE 2013 ISSE*, SAMPE, Covina, CA, May.
43. N. Hansen, D. O. Adams, and D. T. Fullwood (2012). Quantitative methods for correlating dispersion and electrical conductivity in conductor-polymer nanostrand composites. *Composites: Part A* 43, 1939–1946.
44. G. Lagaly (1999). Introduction: From clay mineral-polymer interactions to clay mineral-polymer nanocomposites. *Applied Clay Science* 15, 1–9.
45. M. P. Eastman, E. Bain, T. L. Porter, K. Manygoats, R. Whitehorse, et al. (1999). The formation of poly(methyl-methacylate) on transition metal-exchanged hectorite. *Applied Clay Science* 15, 173–185.
46. Y. Fukushima, A. Okada, M. Kawasumi, T. Kurauchi, and O. Kamigaito (1988). Swelling behavior of montmorillonite by poly-6-amide. *Clay Mineral* 23, 27–34.
47. A. Usuki, Y. Kojima, M. Kawasumi, A. Okada, Y. Fukushima, et al. (1993). Synthesis of nylon-6-clay hybrid. *Journal of Material Research* 8, 1179–1183.
48. A. Usuki, M., Kawasumi, Y. Kojima, A. Okada, T. Kurauchi, and O. Kamigaito (1993). Swelling behavior of montmorillonite cation exchanged for ω-amino acid by ε-caprolatam. *Journal of Material Research* 8, 1174–1178.
49. Y. Kojima, A. Usuki, M. Kawasumi, A. Okada, T. Kurauchi, and O. Kamigaito (1993). Synthesis of nylon-6-clay hybrid by montmorillonite intercalated with ε-caprolatam. *Journal of Polymer Science Part A: Polymer Chemistry* 31, 983–986.
50. A. Maneshi, J. B. P. Soares, and L. C. Simon. (2011). An efficient in situ polymerization method for the production of polyethylene/clay nanocomposites: Effect of polymerization conditions on particle morphology. *Macromolecular Chemistry and Physics* 212(18), 2017–2028.
51. H. Namazi, M. Mosadegh, and A. Dadkhah (2009). New intercalated layer silicate nanocomposites based on synthesized starch-g-pcl prepared via solution intercalation and in situ polymerization methods: A comparative study. *Carbohydrate Polymers* 75(4), 665–669.

52. P. B. Messersmith and E. P. Giannelis (1994). Synthesis and characterization of layered silicate-epoxy nanocomposites. *Chemistry of Materials* 6, 1719–1725.
53. T. Lan and T. J. Pinnavaia (1994). Clay-reinforced epoxy nanocomposites. *Chemistry of Materials* 6, 2216–2219.
54. T. Lan, P. D. Kaviratna, and T. J. Pinnavaia (1995). Mechanism of clay tactoid exfoliation in epoxy-clay nanocomposites. *Chemistry of Materials* 7, 2144–2150.
55. D. C. Lee and L. W. Jang (1996). Preparation and characterization of PMMA-clay hybrid composite by emulsion polymerization. *Journal of Applied Polymer Science* 62, 1117–1122.
56. D. C. Lee and L. W. Jang (1998). Characterization of epoxy-clay hybrid composite prepared by emulsion polymerization. *Journal of Applied Polymer Science* 68, 1997–2005.
57. M. W. Noh and D. C. Lee (1999). Synthesis and characterization of PS-clay nanocomposite by emulsion polymerization. *Polymer Bulletin* 42, 619–626.
58. K. Zhang, W. Wu, H. Meng, K. Guo, and J. F. Chen (2009). Pickering emulsion polymerization: Preparation of polystyrene/nano-SiO_2 composite microspheres with core-shell structure. *Powder Technology* 190(3), 393–400.
59. P. Meneghetti and S. Qutubuddin (2006). Synthesis, thermal properties and applications of polymer-clay nanocomposites. *Thermochimica Acta* 442(1), 74–77.
60. J. Frohlich, R. Thomann, O. Gryshchuk, J. Karger-Kocsis, and R. Mulhaupt (2004). High-performance epoxy hybrid nanocomposites containing organophilic layered silicates and compatibilized liquid rubber. *Journal of Applied Polymer Science* 92, 3088–3096.
61. J. H. Koo, H. Stretz, A. Bray, W. Wootan, S. Mulich, et al. (2002). Phenolic-Clay Nanocomposites for Rocket Propulsion System. *Proceedings of the SAMPE 2002 ISSE*, SAMPE, Covina, CA.
62. J. H. Koo, H. Stretz, A. Bray, J. Weispfenning, Z. P. Luo, and W. Wootan (2003). Nanocomposites Rocket Ablative Materials: Processing, Characterization, and Performance. *Proceedings of the SAMPE 2003 ISSE*, SAMPE, Covina, CA, pp. 1156–1170.
63. J. H. Koo, W. K. Chow, H. Stretz, A. C-K. Cheng, A. Bray, and J. Weispfenning, (2003). Flammability Properties of Polymer Nanostructured Materials. *Proceedings of the SAMPE 2003 ISSE*, SAMPE, Covina, CA, pp. 954–964.
64. J. H. Koo, H. Stretz, J. Weispfenning, Z. P. Luo, and W. Wootan (2004). Nanocomposite Rocket Ablative Materials: Subscale Ablation Test. *Proceedings of the SAMPE 2004 ISSE*, SAMPE, Covina, CA.
65. J. H. Koo, C. U. Pittman, Jr., K. Liang, H. Cho, L. A. Pilato, et al. (2003). Nanomodified Carbon/Carbon Composites for Intermediate Temperature: Processing and Characterization. *Proceedings of the SAMPE 2003 ISTC*, SAMPE, Covina, CA.
66. J. H. Koo, L. A. Pilato, P. Winzek, S. Shivakumar, C. U. Pittman, Jr., and Z. P. Luo (2004). Thermo-Oxidative Studies of Nanomodified Carbon/Carbon Composites. *Proceedings of the SAMPE 2004 ISSE*, SAMPE, Covina, CA.
67. J. H. Koo, L. A. Pilato, G. Wissler, A. Lee, A. Abusafieh, and J. Weispfenning, (2005). Epoxy Nanocomposites for Carbon Fiber Reinforced Polymer Matrix Composites. *Proceedings of the SAMPE 2005 ISSE*, SAMPE, Covina, CA.
68. G. Wang, X. Y. Chen, R. Huang, and L. Zhang (2002). Nano-$CaCO_3$/polypropylene composites made with ultra-high-speed mixer. *Journal of Materials Science Letters* 21(13), 985–986.
69. S. C. Zunjarrao, R. Sriraman, and R. P. Singh (2006). Effect of processing parameters and clay volume fraction on the mechanical properties of epoxy-clay nanocomposites. *Journal of Materials Science* 41(8), 2219–2228.

70. S. Halder, P. K. Ghosh, and M. S. Goyat (2012). Influence of ultrasonic dual mode mixing on morphology and mechanical properties of ZrO_2-epoxy nanocomposite. *High Performance Polymers* 24(4), 331–341.
71. M. Heidarian and M. R. Shishesaz (2012). Study on effect of duration of the ultrasonication process on solvent-free polyurethane/organoclay nanocomposite coatings: Structural characteristics and barrier performance analysis. *Journal of Applied Polymer Science* 126, 2035–2048.
72. K. Zhang, J. Y. Lim, H. J. Choi, J. H. Lee, and W. J. Chio (2013). Ultrasonically prepared polystyrene/multi-walled carbon nanotube nanocomposites. *Journal of Materials Science* 48, 3088–3096.

6 Structures and Properties Characterization

6.1 Global Characterization Methods

Three are three key steps in the development of polymer nanocomposites: (a) materials preparation, (b) property characterization, and (c) material performance or device fabrication. The materials preparation involved in the processing of nanoparticles with polymer matrix into a polymer nanocomposite was discussed in Chapter 5. The next challenge is to determine the degree or quality of dispersion of the nanoparticles in the polymer matrix. Characterization involves two main processes: structure or morphological analysis and property measurements. Structure analysis is carried out using a variety of microscopy and spectroscopy techniques, while property characterization is rather diverse and depends on the individual application.

Due to the high selectivity of the size and structure of the polymer nanocomposites (PNCs), their physical properties can be quite diverse. It is known that the physical properties of nanostructures depend strongly on their size and shape. The properties measured from a large quantity of nanomaterials could be an average of the overall properties, so that the unique characteristics of individual nanostructure could be embedded. An essential task as one develops capability in the preparation of nanomaterials is property characterization of an individual nanostructure with a well-defined atomic structure.

Characterizing the properties of individual nanoparticles presents a challenge to many existing testing and measuring techniques because of the following constraints. First, the size (diameter and length) is rather small, prohibiting the application of well-established techniques. Second, the small size of nanostructures makes their manipulation rather difficult, and specialized techniques are needed for identifying and analyzing individual nanostructures. Finally, new methods and analytical tools must be developed to quantify the properties of individual nanostructures. In this chapter, several commonly used structural characterization techniques are briefly discussed, along with examples demonstrating the techniques. More detailed characterization of nanophase materials can be found elsewhere [1–5]. Property characterization for the individual PNCs is also briefly discussed. It is also necessary to employ more than one characterization technique in order to accurately characterize the nanomaterials and polymer nanocomposites. The following section presents an overview of some commonly used material characterization techniques. It is divided into two general groups: the first group is used for morphology characterization and the second group is used for material properties characterization.

Morphology Characterization:

- Optical microscopy
- Wide-angle X-ray diffraction (WAXD)
- Scanning electron microscopy (SEM)
- Transmission electron microscopy (TEM)
- Energy-dispersive X-ray spectroscopy (EDS or EDX)
- Small angle X-ray scattering (SAXS)
- Scanning tunneling microscopy (STM)
- Atomic force microscopy (AFM)
- Raman spectroscopy
- X-ray photoelectron spectroscopy (XPS)
- Others

Global Methods for Polymer Nanocomposite Properties Characterization:

- Mechanical properties (tensile, flexural, and compressive strength and modulus, creep, shear, and impact) – Chapter 7
- Thermal and rheology properties (thermogravimetric analysis (TGA), differential scanning calorimetry (DSC), dynamic mechanical thermal analysis (DMTA), and thermal conductivity) – Chapter 8
- Flammability properties (cone calorimetry, mass loss calorimetry, microscale combustion calorimetry, limiting oxygen index, UL 94, and Steiner tunnel test) – Chapter 9
- Ablation properties (solid rocket motor simulation (SSRM), subscale solid rocket motor (Char motor), oxyacetylene test bed (OTB), in situ ablation recession and thermal sensor, and char strength sensor) – Chapter 10
- Electrical properties – Chapter 11
- Optical properties – Chapter 12
- Other properties – Chapter 13

The following sections (6.2 to 6.9) discuss the techniques that are commonly used to characterize the morphology of the microstructures and nanostructures of polymer nanocomposites.

6.2 Optical Microscopy

Optical microscopy is the most inexpensive imaging technique to inspect the dispersion of nanomaterials in a polymer nanocomposite. If the nanomaterials are not uniformly dispersed in a specific resin matrix, aggregates of nanomaterials can easily be seen using this optical technique before moving to the high-power imaging instruments, such as SEM, STM, and TEM.

6.3 X-Ray Diffraction

Wide-angle X-ray diffraction (WAXD) is the most commonly used technique to characterize the degree of nanodisperions of MMT organoclay in a specific polymer. Wide-angle X-ray diffraction measures the spacing between the ordered crystalline layers of the organoclay. By using Bragg's law,

$$\sin\theta = n\lambda/2d \tag{6.1}$$

where d is the spacing between atomic planes in the crystalline phase and λ is the X-ray wavelength. The intensity of the diffracted X-ray is measured as a function of the diffraction angle 2θ and the specimen's orientation. This diffraction pattern is used to identify the specimen's crystalline phases and to measure its structural properties. WAXD is nondestructive and does not require elaborate sample preparation, which partly explains the wide usage of this technique in materials characterization. Spacing change (increase or decrease) information can be used to determine the type of polymer nanocomposite formed, such as:

- Immiscible (no *d*-spacing change)
- Decomposed/deintercalated (*d*-spacing decrease)
- Intercalated (*d*-spacing increase)
- Exfoliated (*d*-spacing outside of wide-angle X-ray diffraction, or expansive and disordered as to give a signal)

However, WAXD can be affected by many parameters:

- Sampling (powder vs. solids, alignment of clay platelets)
- Experimental parameters (slit width, count time, angle step rate)
- Layered silicate order (disordered/amorphous materials exhibit no signal by WAXD)
- WAXD measures *d*-spacing, not overall (global) clay dispersion in the sample

To reveal the capabilities of a modern X-ray characterization of nanoparticles and the fundamentals of WAXD, the reader should refer to other sources [3, 5]. Figure 6.1 displays the WAXD data showing the intensity of various wt% (5, 10, and 15) of SCP's Cloisite® 30B MMT organoclay dispersing in a resole phenolic (Hexion's SC-1008). The fact that no peaks were shown around 2.5 for all three wt% indicates that all three formulations were possibly in an exfoliated state. Unfortunately, in this study case, WAXD provided the wrong diagnostics, which is evident when TEM analysis was done on the same three cases. This is shown later in Section 6.4.2 and in more detail in Chapter 10.

6.4 Electron Microscopy and Spectroscopy

6.4.1 Scanning Electron Microscopy (SEM)

A scanning electron microscope scans an electron beam across the sample surface and collects scattered electrons for imaging. Because the image is formed using

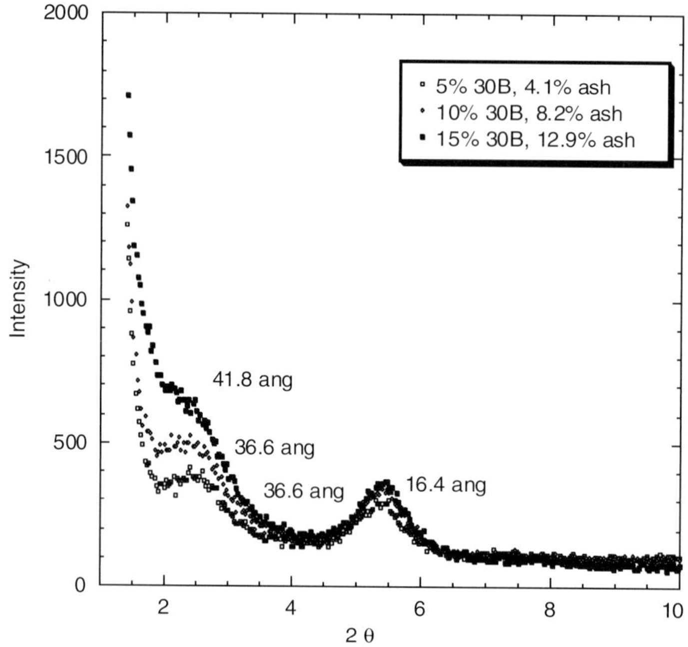

Figure 6.1. WAXD data showing the intensity of 5%, 10%, and 15% of Cloisite 30B MMT nanoclay in resole phenolic.

backscattered signals instead of forward-transmitted signals, the electron beam energy does not need to be high (<40 keV) and the sample also does not require electronic transparency. It has only a conductive requirement to prevent charging. A schematic representation SEM is shown in Figure 6.2 [6]. A high-powered electron beam originating from an electron gun near the top of the column is accelerated down a column through sets of collimators and magnetic focusing lenses. Detectors are place near and around a sample. Many kinds of 2° signals are collected and assimilated. If the SEM has EDX capability, in situ elemental analysis is also conducted. Figure 6.3 shows an FEI Quanta 650 SEM, capable of performing chemical analysis by using a Bruker EDX system. The resolution of this instrument is less than 1 nm. Magnification is on the order of 2 million. The resolution is 0.6 nm at 5 kV and 1.0 nm at 1 kV. Magnification capability ranges from on the order 25 to 2.0 million. Accelerating voltage ranges from 0.5 kV to approximately 30 kV.

At least four different kinds of signals are scattered backward that can potentially be used: secondary electrons, backscattered electrons, characteristic X-ray, and Auger electrons. The most popular SEM imaging method is secondary electron imaging (SEI). SEI can produce good resolution down to 1–5 nm with a large depth of field; this feature help reveal a lot of information about surface topography. On the other hand, backscattered electrons are often scattered by samples elastically and thus are strongly related to atomic number (Z) near the surface as well as crystallographic orientation of the surface. Therefore, backscattered electrons are often used to identify chemical composition together with characteristic X-rays. Backscattered electrons are also used to image the distribution of elements with

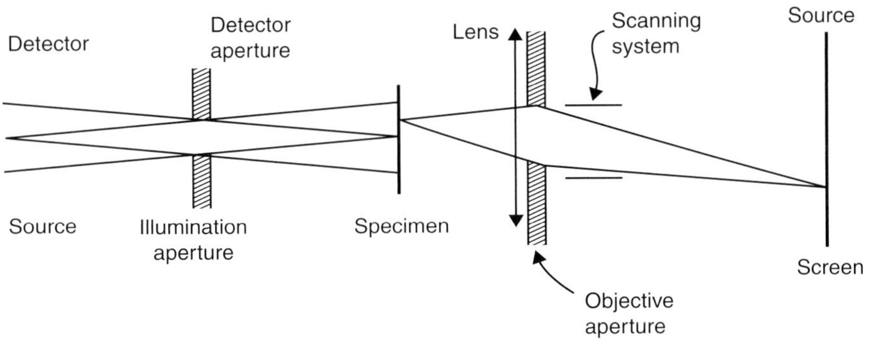

Figure 6.2. A schematic representation of a scanning electron microscope (SEM).

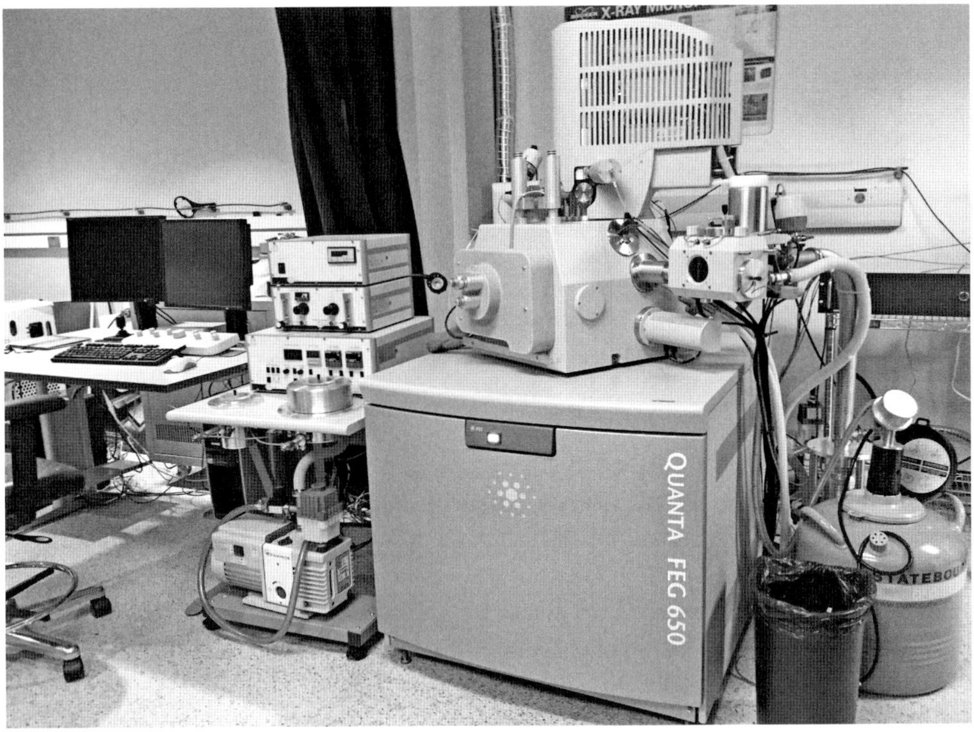

Figure 6.3. An FEI Quanta 650 SEM.

significant difference in weight. Characteristic X-rays have also been widely used in SEM to identify materials chemical composition. This is called energy-dispersive X-ray spectroscopy (EDX) (this technique is discussed later in this chapter).

6.4.2 Transmission Electron Microscopy (TEM)

One of the characteristics of polymer nanostructured materials is the extremely small particle size of the added nanomaterials. Although some structural features can be revealed by X-ray and neutron diffraction, direct imaging of individual nanomaterials is only possible using transmission electron microscopy (TEM) and

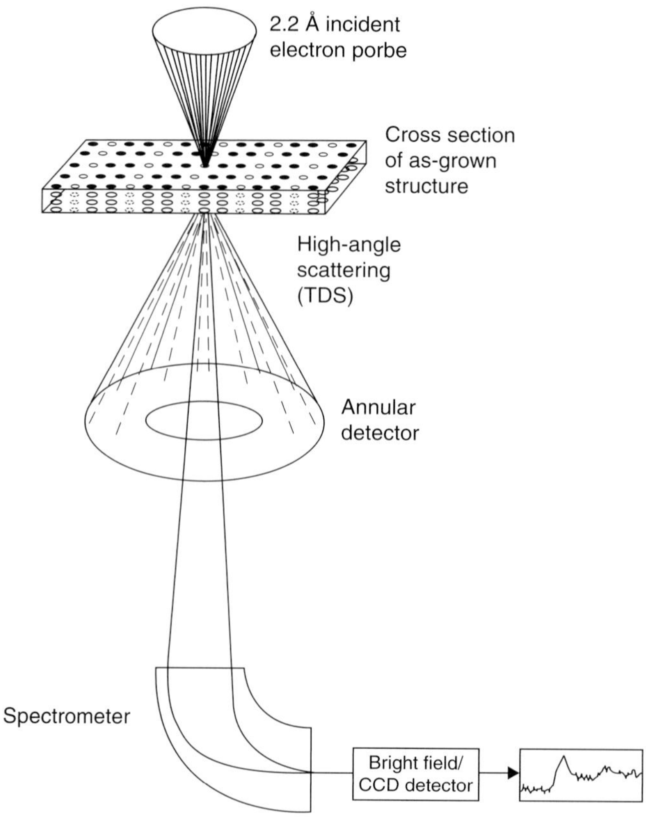

Figure 6.4. A schematic representation of a transmission electron microscope (TEM).

scanning probe microscopy. Transmission electron microscopy is unique because it can provide a real space image on the atom distribution in the nanocrystal and on its surface [1, 5, 7]. Today's TEM is a versatile tool that provides not only atomic resolution lattice images but also chemical information at a spatial resolution of 1 nm or better, allowing direct identification of the chemistry of a single nanocrystal. With a finely focused electron probe, the structural characteristics of a single nanomaterial can be fully characterized. To reveal the capabilities of a modern TEM and the fundamentals of TEM and its applications in characterization of nanophase materials, the reader should refer to other sources [1–9].

A schematic representation of a transmission electron microscope (TEM) is shown in Figure 6.4 [8]. TEM has resolving powder in the range of 0.05 nm. TEMs are also equipped with in situ EDX and electron diffraction capability. Figure 6.5 shows an FEI Tecnai Transmission Electron Microscope (TEM). The TEM is equipped with a 20–120kV kV FEG (field emission gun). The chemical analysis system is the energy-dispersive X-ray spectroscopy (EDS) and electron energy-loss spectroscopy (EELS); these can be used in a complimentary manner to quantify the chemical composition of the specimen. Electron energy-loss spectroscopy can also provide information about the electronic structure of the specimen. The element in the electron gun is ZrO/W and its point-image resolution capability is 0.17 nm. Bean spot

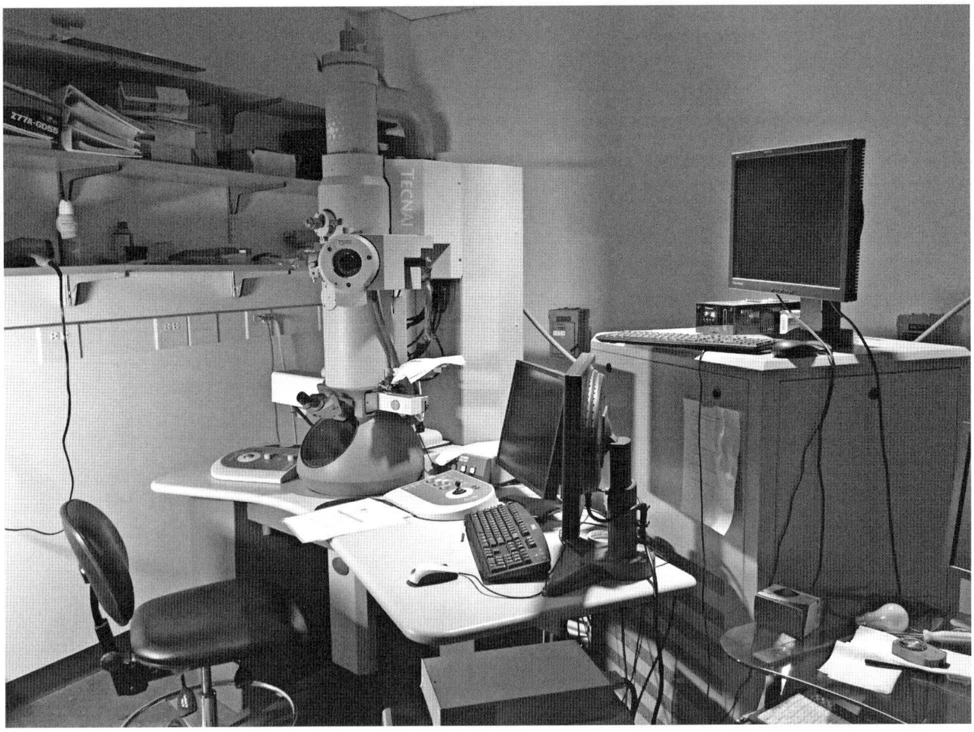

Figure 6.5. An FEI Tecnai Transmission Electron Microscope (TEM).

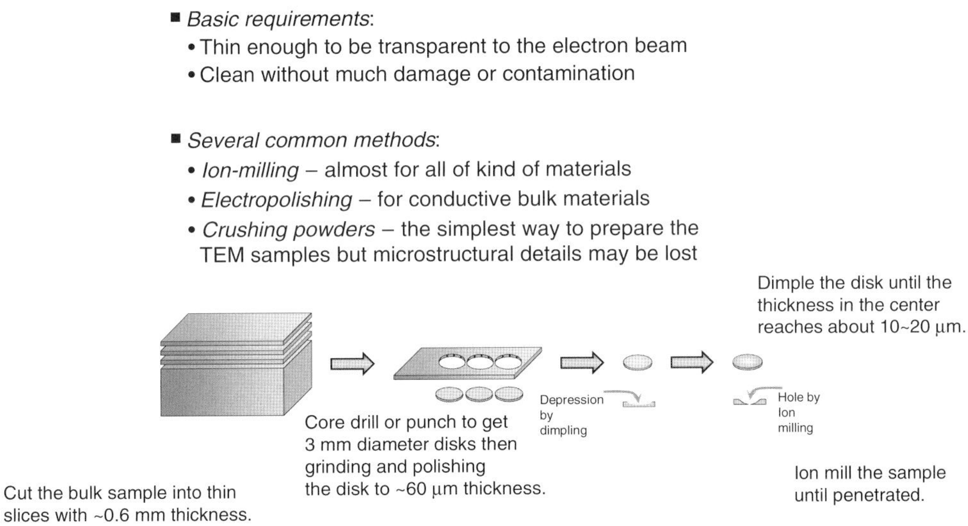

Figure 6.6. TEM sample preparation process by ion-milling.

size can be focused down to 0.4 nm in diameter. Magnification ranges from on the order of 100 to 1.5 million.

TEM sample preparation is of utmost importance in obtaining TEM with good resolution. Figure 6.6 illustrates the sequence of TEM sample preparation that the author followed in his study. The basic requirements are that the specimen has to

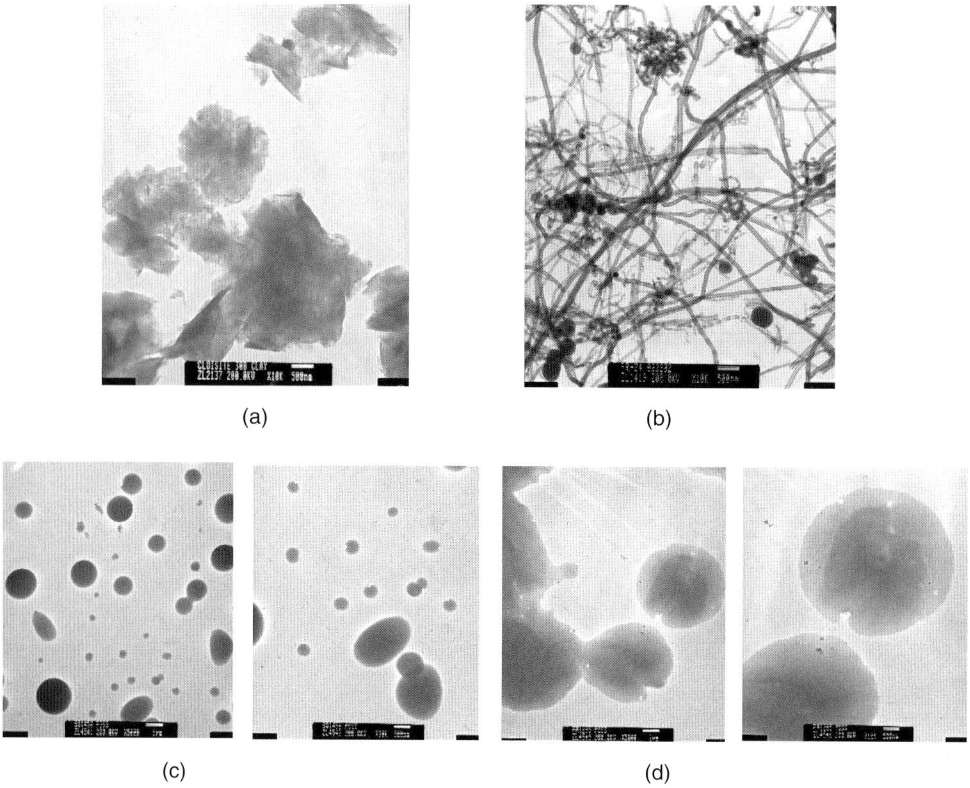

Figure 6.7. TEM micrographs of nanoparticles in IPA solvent: (a) Cloisite® 30B, (b) PR-24-PS CNFs, (c) TriSilanolPhenyl-POSS SO1458, and (d) Poly(vinyl)silsesquioxane uncured-PM1285.

be thin enough to be transparent to the electron beam, and must be clean, without much damage or contamination. There are several common methods for TEM specimen preparation:

- *Ion-milling*: for nearly all kinds of materials
- *Electropolishing*: for conductive bulk materials
- *Crushing powders*: the simplest way to prepare for TEM specimens, but microstructural details may be lost

The bulk sample is first cut into thin slices about 0.6 mm thick. The thin sample slices are then either core-drilled or punched to obtain a disk 3-mm in diameter with a thickness of ~ 60 μm. The disk is then dimpled until the thickness in the center reaches about 10–20 μm, and the specimen is then ion-milled until it is penetrated.

Figure 6.7 shows TEM micrographs of three types of nanoparticles in IPA solvent as a 99.5% solution in IPA [10]. The unit bar is 500 nm in all three micrographs. Cloisite® 30B in Figure 6.7a shows intercalated nanoclay. Figure 6.7b shows PR-24-PS CNFs entangled in bundles. Figure 6.7c shows Trisilanolphenyl-POSS®-SO1458 ($C_{48}H_{38}O_{12}Si_7$), and Figure 6.7d shows polyvinyl silsesquioxane uncured-PM1285 $(C_2H_3O_{1.5}Si)n$ in micron sizes.

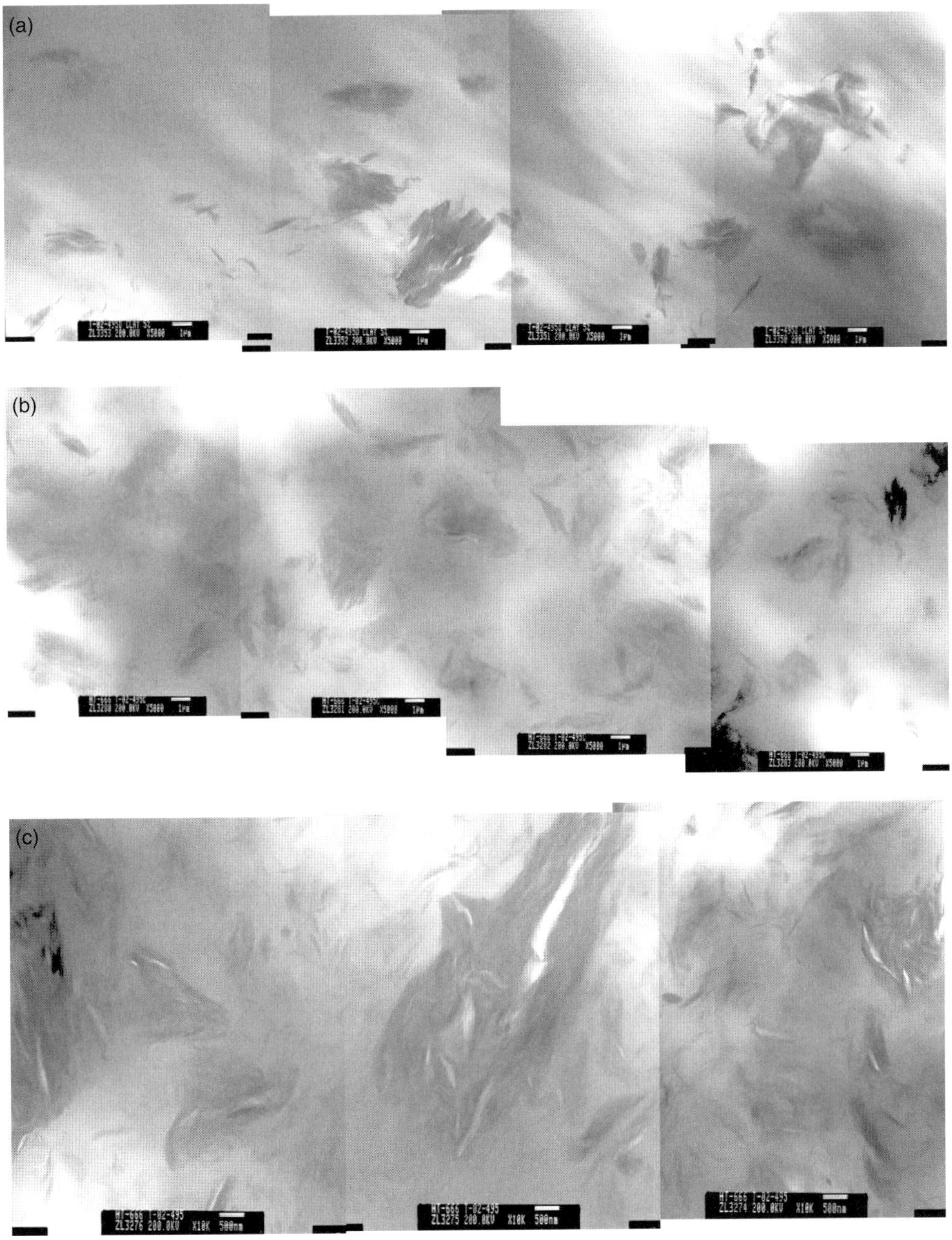

Figure 6.8. TEM Images of (a) 5 wt% Cloisite® 30B in 95 wt% SC-1008 (Scale Bar is 1 μm), (b) 10 wt% Cloisite® 30B in 90 wt% SC-1008 (Scale Bar is 1 μm), and (c) 15wt% Cloisite® 30B in 85 wt% SC-1008 (Scale bar is 500 nm).

Figure 6.8 shows the TEM images of (a) the 5 wt% Cloisite 30B:95 wt% SC-1008 (Figure 6.8a), (b) 10 wt% Cloisite 30B:90 wt% SC-1008 (Figure 6.8b), and 15 wt% Cloisite 30B:85 wt% SC-1008 (Figure 6.8c). The TEM images indicated intercalation and exfoliated of the nanoclay before committing a 20-lb run at Cytec Engineered Materials of these nanoclay-resin mixtures for the preparation of prepregs. These

TEM results verified the WAXD results as shown in Figure 6.1 were incorrect. This proved to be a very cost-effective and efficient technique for screening different formulations on a small scale.

Transmission electron microscopy allows the observation of the overall organoclay dispersion in the polymer nanocomposite sample. Clay dispersion and structure observed under the microscope can determine the nature of a clay nanocomposite as follows:

- *Immiscible*: usually large clay tactoids, undispersed clay particles
- *Intercalated*: clay layers in ordered stacks can be observed
- *Exfoliated*: single clay layers can be observed

Global microscale dispersion, as well as nanoscale dispersion and structure, can be determined by TEM.

Several limitations and drawbacks of the transmission electron microscopy procedure include the following:

1. Extensive sample preparation is time consuming and limits the number of samples that can be imaged and analyzed.
2. In addition to the small physical size of samples, the field of view is relatively small as well. For example, the section under analysis may not be representative of the sample as a whole.
3. Sample structure and morphology have been known to change drastically under exposure to electrons with extremely high energy, and damage to biological samples in particular can occur.
4. TEM cannot measure *d*-spacing of clay; therefore, it cannot easily determine the difference between intercalated clay nanocomposite and well-dispersed, immiscible nanocomposite.
5. TEM is a high-vacuum instrument that is costly to operate and maintain, certainly playing a role in the high barrier entry required to conduct nanoscience research.

However, the advantages of the TEM technique are numerous. Any type of sample, whether electrically insulating, semiconducting, or conducting, is able to be imaged by TEM. The incredible resolution capability allows for atomic-level inspection. Transmission electron microscopy measures overall "characteristics" of clay, such as uniformity, tactoids or non-uniformity of the clay within the material. It is recommended that TEM measurements should be combined with WAXD data for thorough analyses of the nanomaterials.

6.4.3 Energy-Dispersive X-Ray Spectroscopy (EDS or EDX)

X-rays emitted from atoms are characteristics of the elements, and the intensity distribution of the X-rays represents the thickness-projected atom densities in the specimen. This is known as energy-dispersive X-ray spectroscopy (EDS), which has played an important role in microanalysis, particularly for heavier elements. EDS is

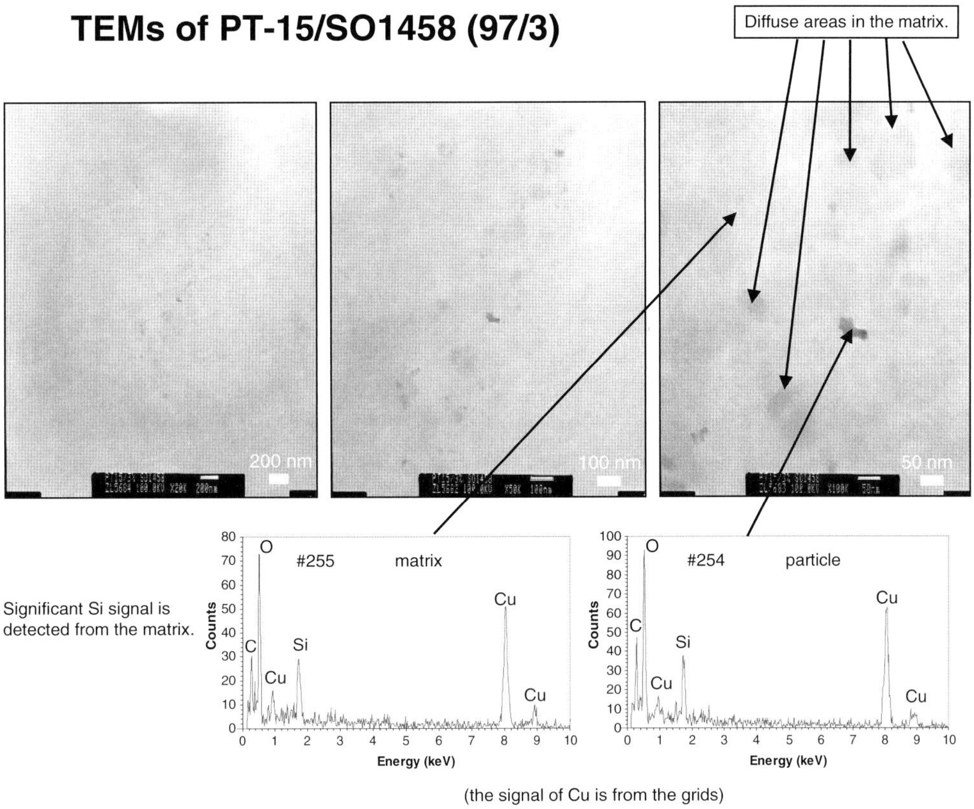

Figure 6.9. TEM micrograph of PT-15/SO1458 POSS in progressive magnification, EDS shows Si is dispersed very well in the PT-15 resin system.

a key tool for identifying the chemical composition of a specimen. A modern TEM is capable of producing a fine electron probe smaller than 2 nm, allowing direct identification of the local composition of an individual element. Examples will be given in later chapters. When POSS® materials were incorporated into the cyanate ester resin, EDS was used to demonstrate that molecular dispersion was achieved by the PT-15 cyanate ester-Trisilanolphenyl-POSS® nanocomposite. Figure 6.9 shows the TEM micrographs of PT-15/SO1458 POSS® in progressive magnification. EDS shows Si is dispersed very well in the PT-15 resin system. Si counts were observed in the particle and matrix areas [10]. Thus molecular dispersion of POSS particle in the PT-15 CE resin is evident.

6.5 Small-Angle X-Ray Scattering (SAXS)

Small-angle X-ray scattering (SAXS) is an analytical X-ray application technique for the structural characterization of solid and fluid materials in the nanometer range. In SAXS experiments, the sample is irradiated by a well-defined, monochromatic X-ray beam, as shown in Figure 6.10. When a nonhomogeneous medium, for example proteins in water, is irradiated, structural information about the scattering

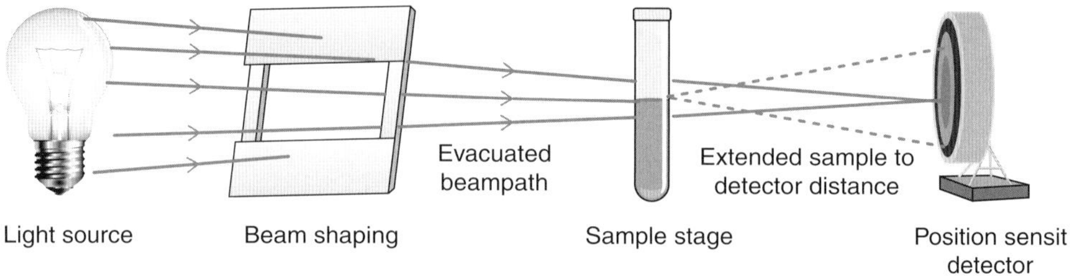

Figure 6.10. Schematic of small-angle X-ray scattering (SAXS).

particles can be derived from the intensity distribution of the scattered beam at very low scattering angles.

With SAXS it is possible to study both monodisperse and polydisperse systems. In the case of monodisperse systems, one can determine size, shape, and internal structure of the particles. For polydisperse systems, a size distribution can be calculated under the assumption that all particles have the same shape. Small-angle X-ray scattering is applied to investigate structural details in the 0.5–50 nm size range in materials, such as:

- Particle sizing of suspended nanopowders
- Materials in life science and biotechnology (proteins, viruses, DNA complexes)
- Polymer films and fibers
- Catalyst surface per volume
- Microemulsions
- Liquid crystals

Small-angle X-ray scattering can provide the following information:

- Lamella repeat distance
- Radius of gyration
- Particle size and shape
- Large-scale structure and long-range order

Readers should refer to existing texts describing the SAXS technique [4, 11, 12].

6.6 Scanning Probe Microscopy (SPM)

Scanning probe microscopes were developed as an important technique to characterize the surface properties of materials, especially nanomaterials. Scanning probe microscopes are capable to create detailed three-dimensional images of specimen surfaces with atomic resolution [5]. Two basic types of scanning probe microscopes are used in nanoscience. In 1981, Binnig et al. developed the scanning tunneling microscopy (STM). In the STM technique, the magnitude of the tunneling current between the probe tip and the atoms of a substrate surface is monitored. STM samples must be electrically conducting. Atomic force microscopy (AFM) and its derivatives compromise a second class of SPMs. In AFM, the size of the force between a probe tip and the atoms of substrate surface is monitored. AFM was also developed by Binning et al. in 1985 specially for insulating materials. AFM is able to measure forces on the order of 1 μN and less. There are many similarities between STM and AFM. Both are equipped with a probe tip fastened to a cantilever, a scanning (motion) mechanism, and a detector system. Four achievements that contributed to the development of scanning probe methods are summarized by Horyak et al. [5]:

1. Physical cushioning mechanisms, motion-damping eddy currents induced by magnets and copper plates, and nitrogen-gas-regulated suspension systems serve to isolate the SPM from external vibration sources. Protective enclosures shield the microscope from room drafts. If extremely high-resolution study is required, cryogenic temperatures eliminate atomic and molecular movement.
2. Computerized feedback system control of piezoelectric devices with nanometer-level precision guides the probe during its descent to the surface – less than a nanometer above where van der Waals attraction exert force on the probe tip (in AFM) or tunneling current is able to flow (in STM).
3. Computerized control of piezoelectric scanners moves the probe tip across the sample surface with nanometer-level precision.
4. The fabrication of the sharper probes allows for better resolution of surface features. For example, carbon nanotube probes with sharpness of less than 1 nm have shown promise in the past few years.

6.6.1 Scanning Tunneling Microscopy (STM)

The fundamental principle of STM is quantum tunneling, as electrons can tunnel through a finite energy barrier if the physical distance of the barrier is small enough. Electron (or quantum) tunneling is attained when a particle (an electron) with lower kinetic energy is able to exist on the other side of an energy barrier with higher potential energy, thus disobeying a fundamental law of classical mechanics. Tunneling is the penetration of an electron into a classically forbidden region [13]. Electrons exhibit wave behavior and their position is represented by a wave (probability) function. The wave function represents a finite probability of finding an electron on the other side of the potential barrier. Since the electron does not possess enough kinetic energy to overcome the potential barrier, the only way the electron can appear on the other side is by tunneling through the barrier. The principle of quantum tunneling was also used to explain exponential radioactive decay rates. In quantum mechanics, a finite probability is allocated to tunneling, and hence nuclear decay is allowed.

STM relies on an electronic signal to relay information about a sample – the strength of a tunneling current potential that exists between the probe tip and the substrate surface. Small changes in the distance between the probe tip and the substrate surface translate into large changes in the tunneling current. By this phenomenon, atomic scale resolution by STM is possible in the x, y, and z directions. The density of states (DoS) of solid-state materials can be also mapped by the technique called scanning tunneling spectroscopy (STS). Chemical reactions induced and oriented by the STM probe are also made available by means of this technique.

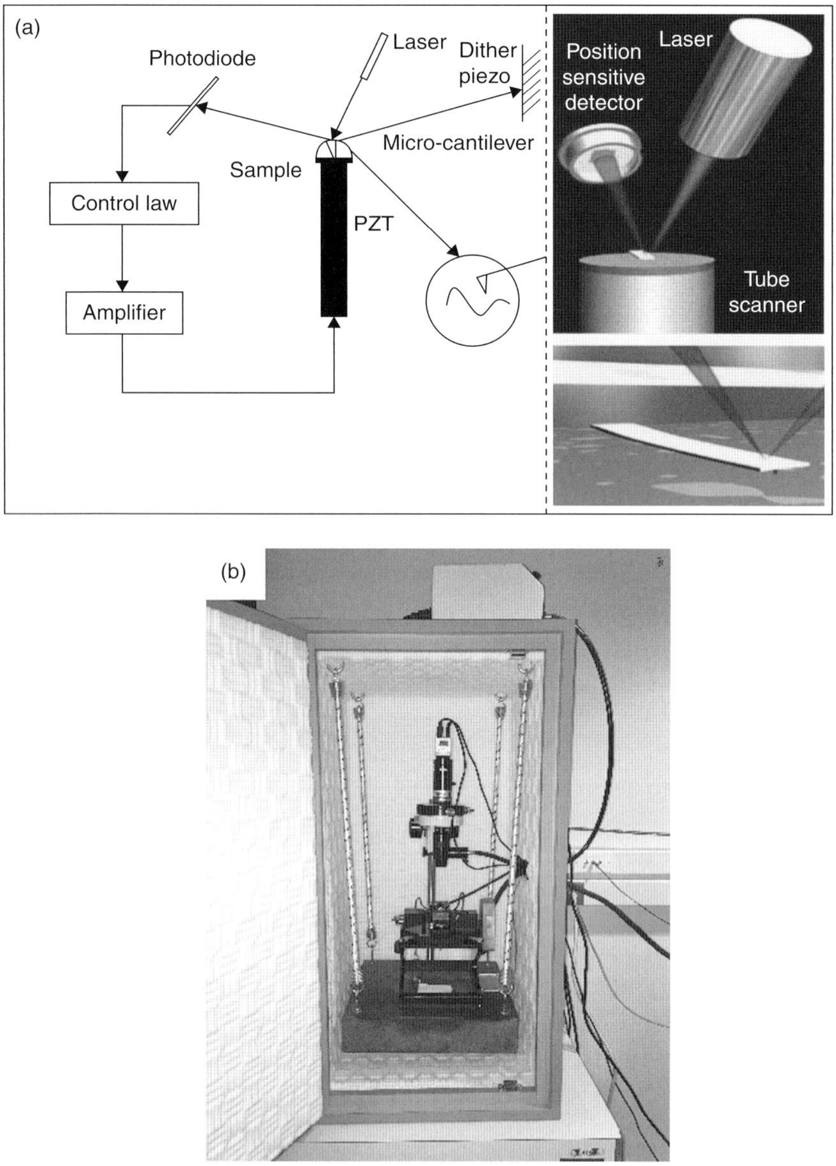

Figure 6.11. (a) A schematic of an AFM system. (b) An Agilent 5500 Atomic Force Microscope.

6.6.2 Atomic Force Microscopy (AFM)

Atomic force microscopy relies on the mechanical deflection of a cantilever to replay information about the contour of a sample surface. An atomically sharpened probe tip (20–50 nm or less) descends perpendicularly from the distal end of a cantilever. The tip-to-sample is fixed by means of a feedback mechanism that maintains constant force between them (i.e., constant height). A schematic of an AFM system is shown in Figure 6.11a [14]. A focused beam is reflected off the back of the cantilever equipped with a sharpened probe and into a photodiode detector. Any deflection

Table 6.1. *Atomic Force Microscopy Primary Modes*

AFM mode	Configuration	Capability
Contact Mode	The probe-cantilever assembly applies a constant force to the surface; the force constant is <1 $N \cdot M^{-1}$. The deflection of the cantilever is due to topographical changes characteristic of the surface. The mechanical deflection is translated into an optical signal by reflection of an aligned laser light from the top surface of the cantilever into an aligned-calibrated dual photodetector collector. Due to intimate contact with the surface, large lateral forces can influence the probe. Surface contamination, electrostatic forces, and heterogeneous surfaces are able to impact the action of the probe. AFM is capable of imaging nanomaterials in air, vacuum, and liquids.	Samples with hard surfaces are appropriate for contact mode AFM analysis. AFM is capable of imaging insulator, semiconductor, and conductor surfaces. Qualitative information includes 3D visualization and material sensing by phase contrast. Quantitative information includes topographic mapping, particle and pore size, particle and pore morphology, surface roughness and texture, particle count, size distribution, and volume-mass distribution.
Tapping Mode	Strong repulsive regime Oscillating probe (>100 kHz) Intermittent contact with surface Lateral forces reduced significantly Amplitude and phase imaging	Three-dimensional topography Use for soft samples or weakly bound to surface Biological materials DNA Carbon nanotubes
Noncontact Mode	Weak-attractive regime Oscillating probe Can be applied with water layer	Soft surfaces

of the beam is translated into a topographical feature. Rastering of the tip over the surface produces a topographical image. The photodiode is split into two compartments (or vice versa) and is able to detect differences in beam position to the level of a nanometer. An Agilent 5500 Atomic Force Microscope is shown in Figure 6.11b.

The AFM technique relies on a balance between attractive van der Waals and repulsive electrostatic forces between the probe tip and the surface. The net force is a function of the distance between the two. The force is also a function of the dielectric constant ε of the medium. Hand in hand with atomic-order resolution, the theoretical magnification potential of the AFM technique is on the order of 10^9 [5]. The versatility of AFM is demonstrated further by its ability to image large objects that are tens of micron or more in size. There are three major modes of operation of AFM: contact, tapping, and noncontact (Table 6.1), as summarized by Hornyak et al. [5].

Different cantilever-probe tip combinations are available for different types of AFM technique. Cantilevers are fabricated from single crystal silicon. The length of AFM contact and tapping mode cantilevers is between 100 to 500 μm; the width is between 25 to 40 μm, and the thickness is between 1 to 10 μm. The resonant frequency of these cantilevers falls between 10 kHz and 300 kHz and the spring constant between 0.1 $N \cdot m^{-1}$ and 50 $N \cdot m^{-1}$. The probe tip diameter ranges from a few nanometers for finer resolution, to 20 nm for an average AFM, to as large as 50 nm. Noncontract AFM mode cantilevers are made of highly doped single crystal silicon.

Nakajima et al. [15] investigated the surface mechanical properties including Young's modulus, adhesion, viscoelasticity, and energy dissipation during probe-sample interactions by employing AFM nanomechanical mapping. The ability of nanomechanical mapping to identify the composition and evaluate the mechanical properties at nanoscale will open a new way to study surface properties and explore the *microstructure-properties* relationship in a large range of polymeric and biological materials.

6.7 Raman Spectroscopy

Far-infrared, infrared (IR), near-infrared, Raman spectroscopy, and surface-enhanced Raman Spectroscopy (SERS) are extremely relevant methods used to characterize nanomaterials. These methods provide information about vibrational energy; IR methods rely on asymmetrical (dipolar) vibrations and Raman methods on symmetrical (polarizable) vibrations. In this way, the two methods complement each other.

Raman and Krishnan in 1928 discovered that light scattered off certain molecules changed the polarization state of the molecules. The Raman Effect is a scattering phenomenon that links the vibrational frequencies of the molecule to the energy difference between the incident and scattered light. However, only molecules with symmetric vibrational (polarization) modes and transitions were amenable to Raman spectroscopic analysis. The SERS process depends on nanomaterial facets or particles that are able to enable signal intensity by multiple orders of magnitude – an example of a spectroscopic method enabled by nanotechnology.

Two types of photon scattering exist: elastic scattering and inelastic scattering. Elastic scattering, in which there is no energy loss, is called Rayleigh (Rayleigh-Debye) or Mie scattering. Such scattering occurs when particle dimensions are on the order of or smaller than the wavelength of the incident radiation. Remarkably, in that case, scattering is a true nanoscale phenomenon. If particle size does not conform to this limit (e.g., larger materials), then reflection and refraction also occur. Rayleigh scattering (elastic) involves no change in the polarizability of the molecule. However, a small portion of the incident photons is scattered at frequencies that are different form the incident light. This is called inelastic scattering, and the result is exhibited in the form of vibrational, rotational, or electronic energy changes in the molecule. Another form of Raman spectroscopy called resonant Raman spectroscopy (RRS) occurs when there is resonance between the incident radiation and an electronic transition of the molecule.

SERS is a technique that exhibits incredible sensitivity and has great importance to nanoscience. Substrates in the SERS technique are noble metals like Au and Ag that happen to be in nanoparticle form. The primary mechanism of SERS relies on the surface plasmon resonance of nanometal particles. An electromagnetic enhancement on the order of 10^6 of the vibrational signal from an adsorbed species is imparted to the spectrum. Analysis of SWNTs by SERS is a diagnostic of their presence. Single-molecule detection is also possible with this method.

6.8 X-Ray Photoelectron Spectroscopy (XPS)

An X-ray photoelectron spectroscope (XPS) is considered a modern tool for carbonaceous materials characterization used to determine the chemical composition, impurity presence, and nature of chemical bonds. The most important application of XPS in nanocomposites based on polymer-CNT is to identify the chemical composition of the modified carbon nanotubes surface in order of confirm whether or not surface modification has occurred. The XPS analysis performed on the CNT surface was focused on measuring the binding energy of photoelectrons ejected when CNTs are irradiated with different agents pointed out by XPS analysis. Garea et al. [16] investigated nanocomposites synthesis of nanocomposites based on POSS and polymer matrix using Raman spectroscopy and X-ray photoelectron spectroscopy.

6.9 Other Techniques

A variety of other morphological characterization techniques can be found in [1–16].

The following sections (6.10 to 6.16) discuss the global methods commonly used to characterize the bulk material properties, such as mechanical, thermal, flammability, ablation, electrical, optical, and others, of the polymer nanocomposites. These material properties are discussed in detail in Chapters 7 to 13.

6.10 Mechanical Properties

When nanomaterials are added to neat polymers to enhance specific properties, the most important task is to maintain the mechanical properties of these polymer nanocomposites. The mechanical properties, among all the properties of these polymer nanocomposites, are often the most important properties because virtually all service conditions are the majority of end-use applications involve some degree of mechanical loading. The basic understanding of stress-strain behavior of polymers is of utmost important to design engineers. A typical stress-strain diagram is shown in Figure 6.12. For a better understanding of the stress-strain curve, a few basic terms associated with the stress-strain diagram are included.

Stress. The force applied to produce deformation in a unit area of a test specimen. Stress is a ratio of applied load to the original cross-sectional area expressed in kg/m^2 (lb/in^2).

Strain. The ratio of the elongation to the gauge length of the test specimen, or simply stated, change in length per unit of the original length ($\Delta l/l$). It is expressed in a dimensionless ratio.

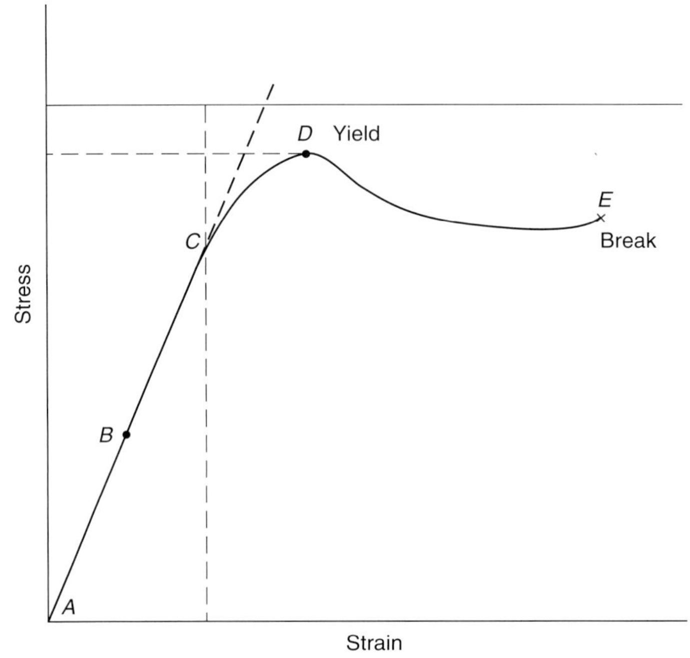

Figure 6.12. A typical stress-strain curve.

Elongation. The increase in length of a test specimen produced by a tensile load.

Yield Point. The first point on the stress-strain curve at which an increase in strain occurs without the increase in stress.

Yield Strength. The stress at which a material exhibits a specified limiting deviation from the proportionality of stress to strain. This stress will be at the yield point.

Proportional Limit. The greatest stress at which a material is capable of sustaining the applied load without any deviation from proportionally of stress to strain (Hooke's Law), expressed in kg/m2 (lb/in^2).

Modulus of Elasticity. The ratio of stress to corresponding strain below the proportion limit of a material. It is expressed in F/A, usually in kg/m^2 (lb/in^2). This is also known as Young's modulus. A modulus is a measure of the material's stiffness.

Ultimate Strength. The maximum unit stress a material will withstand when subjected to an applied load in compressing, tension, or shear, expressed in kg/m^2 (lb/in^2).

Some commonly used mechanical property tests as listed below [17]:

- Tensile Properties Test (ASTM D 638, ISO 527-1)
- Flexural Properties Test (ASTM D 790, ISO 178)
- Compressive Properties Test (ASTM D 695, ISO 75-1 and 75-2)
- Creep Properties Test
- Impact Properties Test (Izod and Charpy impact tests)
- Shear Properties Test (ASTM D 732)
- Others tests, such as abrasion, fatigue resistance, and hardness tests.

6.11 Thermal Properties

The thermal properties of polymer nanocomposites are equally as important as the mechanical properties are. Unlike metals, polymers are extremely sensitive to changes in temperature. Four commonly used analyses are: thermogravimetric analysis, differential scanning calorimetry, dynamic mechanical and thermal analysis, and thermal conductivity instruments, all of which are discussed in this section.

6.11.1 Thermogravimetric Analysis (TGA)

Thermogravimetric analysis (TGA) is a technique in which the mass of a substance is monitored as a function of temperature or time as the sample specimen is subjected to a controlled temperature program in a controlled atmosphere. TGA is a technique in which, upon heating a material, its weight increases or decreases. A TGA consists of a sample pan that is supported by a precision balance. The pan resides in a furnace and is heated or cooled during the experiment. The mass of the sample is monitored during the experiment. A sample purge gas controls the sample environment. This gas may be inert, or it may be a reactive gas that flows over the sample and exits through an exhaust. TGA can determine the thermal stability of the polymer nanocomposites compared to the base polymer.

Lao et al. studied the thermal and flammability properties of polyamide 12 (PA12) with different wt% loadings of Cloisite 30B nanoclay (NC2) and PR-19-PS carbon nanofibers (CNF) [18]. The authors performed TGA on neat and nanoparticle-reinforced PA12 (Figure 6.13) under nitrogen using scan rates of 10°C/min and 20°C/min [18]. The test could not be performed under air since the intumescent flame retardant (FR) additives would expand at high temperature under an air environment, resulting in erratic results. Since lower and higher scan rates showed similar trend of curve shifts with the lower scan rate exhibiting a better separation of curves, only the TGA data obtained using the scan rate of 10°C/min was shown. From this raw data the decomposition temperatures for both 10% and 50% mass loss, $T_{10\%}$ and $T_{50\%}$, were inferred and summarized [18]. The $T_{10\%}$ and $T_{50\%}$ of the PA12/NC2 blends were higher than those of neat PA12, while those for the PA12/CNF blends were about the same as those of neat PA12 (Figure 6.13).

6.11.2 Differential Scanning Calorimetry (DSC)

Vaia et al. [19] reported that the modification chains interacted in the filler interlayers exist in states with varying degrees of order. In general, as the interlayer packing density or the chain length decreases (or the temperature increases), the intercalated chains adopt a more disordered, liquid-like structure resulting from an increase in the gauche/trans conformer ratio. In order to quantify such phase behavior transitions and chain dynamics for organic monolayers immobilized on the surface of montmorillonites, DSC analysis is performed. Figure 6.14 shows an example of determination of thermal transitions in the C18-4C18-modified filler using DSC

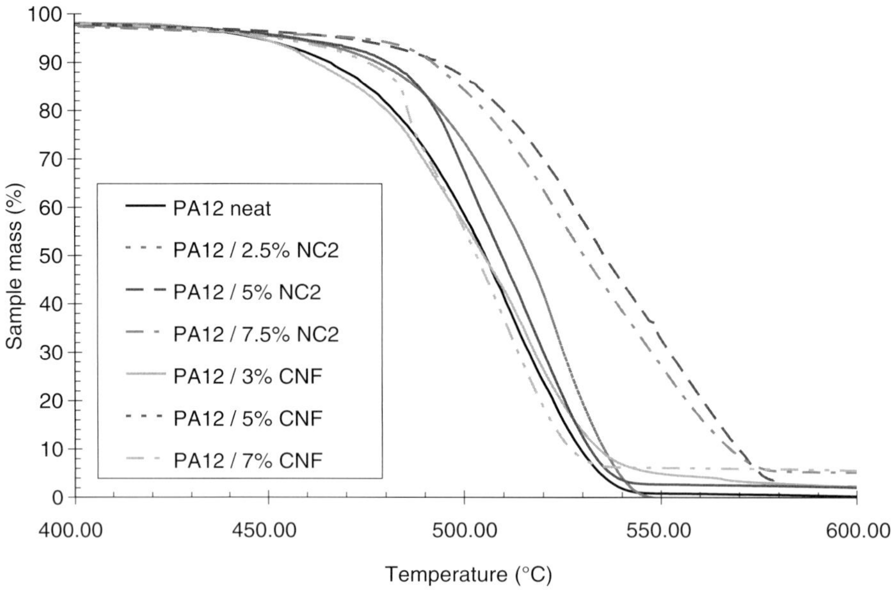

Figure 6.13. Sample mass (%) of neat and nanoparticle-filled PA12 as a function of temperature (TGA, scan rate 10 °C/min).

reported by Osman et al. [20]. Two heat cycles and one cooling cycle were used. C18-modified filler did not show any transition indicating that the modifications chains did not pack in an ordered fashion possibly owing to incomplete coverage of the filler surface as well as more area available per cation on the filler surface as compared to the cross-sectional area of the C18 ammonium molecules. 2C18-, 3C18-, and 4C18-modified fillers exhibited phase transition temperatures of 57°C, 51°C, and 37°C, respectively. The transitions signals also became sharper as the chain density in the ammonium modification molecules was enhanced.

6.11.3 Dynamic Mechanical Thermal Analysis (DMTA)

The introduction of nanomaterials into a polymeric matrix causes difficulties in processing. The mobility of the nanomaterials is reflected by viscosity, viscoelasticity, moduli, and other characteristics. Therefore, understanding the rheological properties of such nanocomposites is of significant importance for optimizing the process parameters that lead to controlled microstructure, which enables the design of the properties of the polymer nanocomposites [21, 22]. Rheology has been extensively used in the study of nanocomposite system in conjunction with basic characterization techniques, such as XRD, SEM, and TEM [23]. A large number of rheological studies on polymer nanocomposites have been published in the past 20 years. Rheology offers an effective way to assess the state of nanomaterial dispersion and to detect the presence of interconnecting microstructure of the nanocomposites in molten state directly. Song and Jin [24] reported the characterization of rheological properties of polymer nanocomposites. The authors provided the fundamental rheological theory for studying polymer nanocomposites.

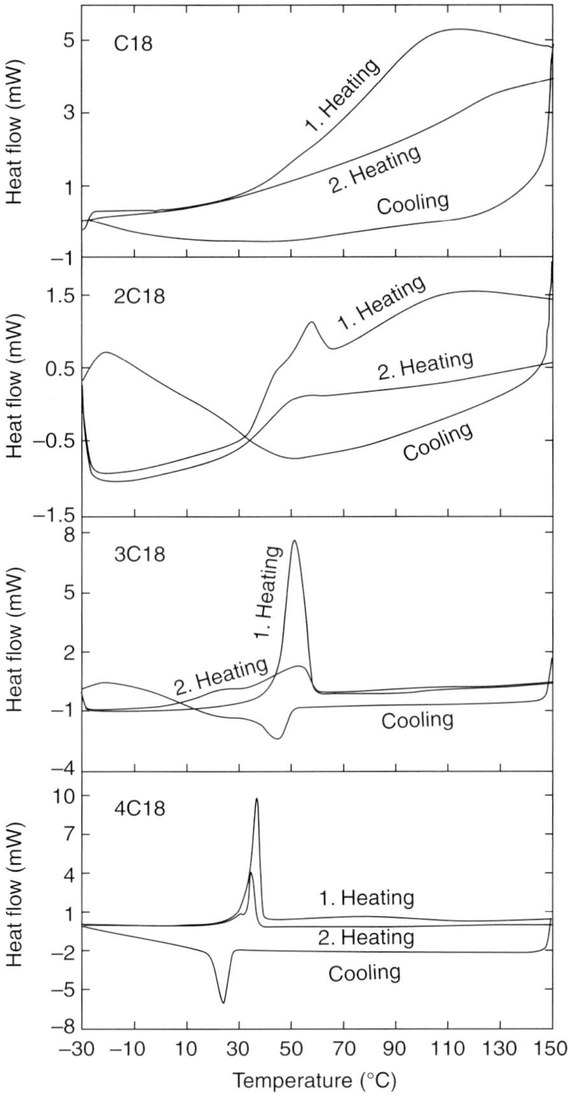

Figure 6.14. DSC thermograms of C18-4C18 modified clay material [13].

Krishnamoorti et al. [25] described the steady shear rheological behavior of intercalated or exfoliated nanocomposites based on poly(dimethyldiphenyl siloxane) (PDS) and layered silicate. Figure 6.15 shows steady shear viscosity behavior of a series of exfoliated nanocomposites [25]. The viscosity systemically increases with silicate loading but still obeys the Newtonian behavior in the lower shear rate region, even at the highest silicate loadings.

6.11.4 Thermal Conductivity

The thermal conductivity property is commonly characterized according to ASTM E1461 [26]. A Netzsch LFA 457 Microflash is normally used, as shown in Figure 6.16, for thermal diffusivity (α) measurements. Thermal diffusivities of samples 8×8 mm^2 is

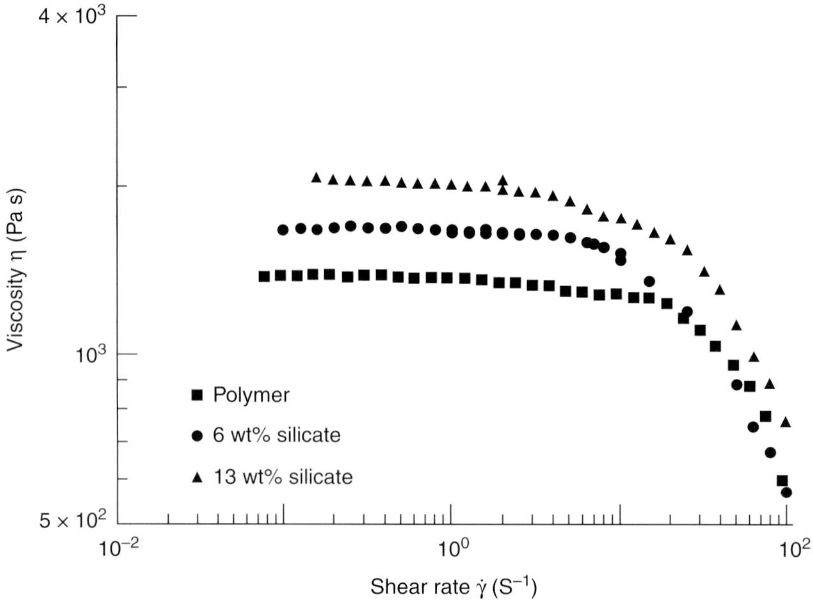

Figure 6.15. The steady shear viscosity behavior of a series of exfoliated nanocomposites of poly(dimethylsiloxane) with a dimethyl ditallow-montmorillonite with varying amounts of silicate loadings at T = 28°C.

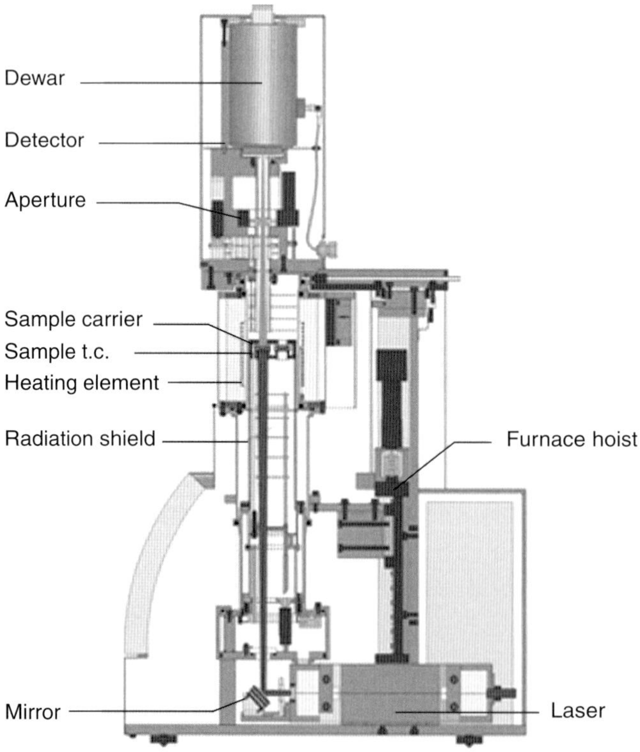

Figure 6.16. Netzsch LFA 457 Microflash for thermal diffusivity measurements.

measured at a temperature range from 30°C to 175°C. Two from each set of NGP and n-Al loadings of the samples are simultaneously tested with a reference sample of either graphite or Pyrex. After diffusivity results are obtained, the specific heat (C_p) of each sample is calculated through the use of LFA software. This is accomplished by using the comparative method of contrasting the change in temperature curves used to calculate the thermal diffusivity of the test sample with those of the reference samples. The reference samples possess a known specific heat, while the test samples' specific heat is determined. Gathering the diffusivity results along with the measured density (ρ) and Cp, using Equation 6.1, the thermal conductivity (K) of the samples will be determined:

$$K = \alpha \cdot \rho \cdot C_p \tag{6.2}$$

6.11.5 Other Thermal Properties

Besides the aforementioned thermal properties, other thermal properties are also characterized [27], such as (a) heat deflection temperature (HDT) measured by ASTM D 648, ISO 75-1 & 75-2, (b) vicat softening temperature measured by ASTM D 1525, ISO 306, (c) torsion pendulum test using ASTM D 2236, (d) long-term heat-resistance test using ASTM D 794, (e) UL temperature index, (f) creep modulus/creep rupture tests, (g) coefficient of linear thermal expansion, and (h) brittleness temperature measured by ASTM D 746, ISO 974.

6.12 Flammability Properties

There are numerous techniques to evaluate the thermal, flammability, and smoke obscuration of the polymer nanocomposites, depending on the type of information desired. A brief discussion of the techniques is not intended to be exhaustive or authoritative, but only as an aide to the reader in identifying the test methods used worldwide; the reader should refer to the original ASTM, UL, and other organizations for more detailed information. The reader is referred to Troitzsch [28], Shah [29], and ASTM fire and flammability standards [30] for an overview for all flammability and fire test methods.

6.12.1 Cone Calorimetry (CC)

The potential fire hazard of any material depends on many factors, including ignitability, rate of surface flame spread, heat release rate (peak, average, and total), mass loss rate, smoke evolution, and evolution of toxic gases (e.g., CO). The Cone Calorimetry or Cone Calorimeter (CC) has been regarded as the most significant bench scale instrument for evaluating reaction-to-fire properties of materials. Cone Calorimeter data address most of these factors, either directly (e.g., heat release rate HRR]) or indirectly (e.g., time to ignition and HRR as functions of external heat flux as indicators of flame spread potential). The data obtained from the CC test are available in engineering units that are conducive to extrapolation to other fire sizes

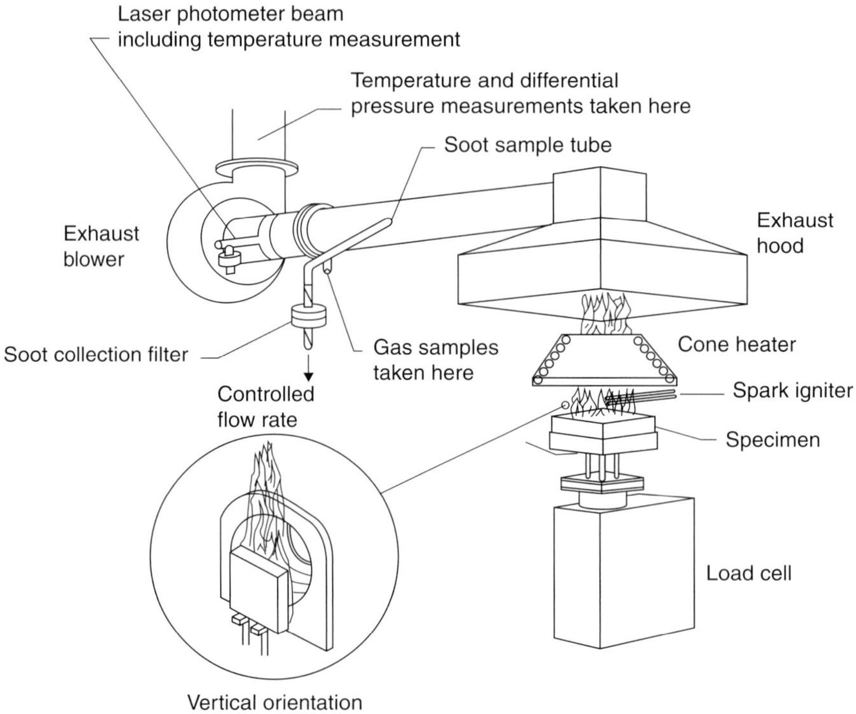

Figure 6.17. Schematic view of the Cone Calorimeter.

and scenarios. This section briefly discusses how the CC operates. Readers should refer to Babrauskas's excellent description of the CC in [31], the ASTM E 1354 [32], or the ISO 5660 [33] for more details on the Cone Calorimeter.

A conceptual view of the CC is shown in Figure 6.17. Figure 6.18 shows a commercial Cone Calorimeter instrument. Figure 6.19 shows a test specimen ignited under the Cone Calorimeter during testing. The CC is a fire test instrument based on the principle of oxygen consumption calorimetry. This empirical principle is based on the observation that, generally, the net heat of combustion of any organic materials is directly related to the amount of oxygen required for combustion. Approximately 13.1 MJ of heat are released per kilogram of oxygen consumed. At the core of the instrument is a radiant electrical heater in the shape of a truncated cone (hence the name). This heating element irradiates a flat horizontal sample 100 mm × 100 mm and up to 50 mm thick, placed beneath it, at a preset heat flux of up to 100 kW/m^2. The sample is placed on a load cell for continuous monitoring of its mass as it burns. Ignition is provided by an intermittent spark igniter located 13 mm above the sample.

The gas stream containing the combined combustion products is captured through an exhaust duct system consisting of a high-temperature centrifugal fan, a hood, and an orifice-plate flow meter. The typical air flow rate is 0.024 m^3/s. Oxygen concentration in the exhaust stream is measured with an oxygen analyzer capable of an accuracy of 50 ppm, and the heat release rate is determined by comparing the oxygen concentration with the value obtained when no sample is burning.

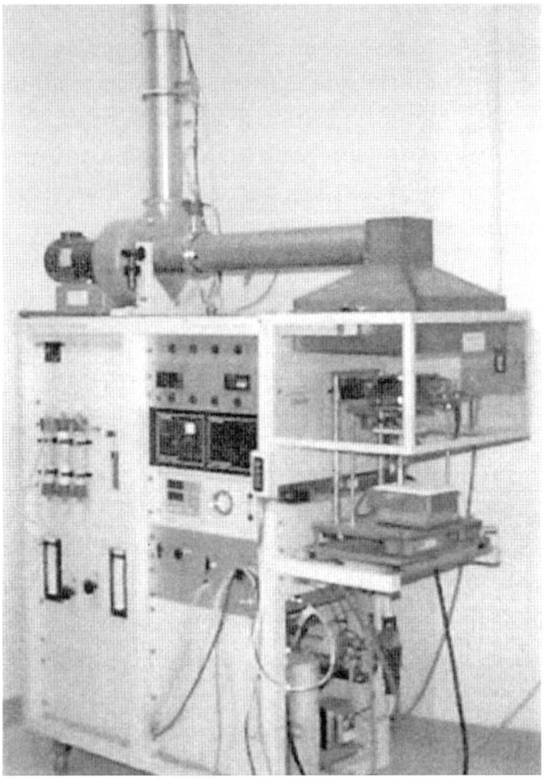

Figure 6.18. A commercial Cone Calorimeter.

Figure 6.19. Close-up showing test specimen igniting during Cone Calorimeter testing.

Smoke obscuration measurement is made in the exhaust duct by a helium-neon laser, with silicon photodiodes as the main beam and reference detectors and appropriate electronics to derive the extinction coefficient and to set the zero reading. Locations are also provided in the exhaust duct for additional sampling probes, to determine concentrations of other combustion products, such as carbon dioxide and carbon monoxide. All data are collected with a PC that records data continuously at fixed intervals of a few seconds during the test.

The CC is used to determine the following principal flammability properties:

- Heat release rate per unit area (kW/m^2)
- Cumulative heat released (kW)
- Effective heat of combustion (MJ/kg)
- Time to ignition (s)
- Mass loss rate (kg/s)
- Total mass loss (kg)
- Smoke density (m^2/kg)

Ignitability is defined as the propensity to ignition as measured by the time in seconds to sustain flaming (the existence of flame on or over most of the specimen surface for periods of at least 4 seconds) at a specified heat flux. The ignitability is thus a measure of how quick the material will ignite in a fire situation.

Heat release is defined as the heat generated in a fire due to various chemical reactions occurring within a given weight or volume of material. The major contributors are those reactions in which carbon monoxide and carbon dioxide are generated and oxygen is consumed. Heat release data provide a relative fire hazard assessment for materials. Materials with low heat release per unit weight or volume will do less damage to the surroundings than will materials with high heat release rate. The heat release rate, especially the peak amount, is the primary characteristic determining the size, growth, and suppression requirements of a fire environment. The rate of heat release is determined by measurement of the oxygen consumption as determined by the oxygen concentration and the flow rate in the exhaust product stream.

Effective heat of combustion, mass remaining, smoke ratio, carbon dioxide, and carbon monoxide are the additional Cone Calorimeter data that are obtained during the experiment.

6.12.2 Mass Loss Calorimetry (MLC)

The Mass Loss Calorimetry or Mass Loss Calorimeter (MLC) per ASTM E2102 [34] or ISO 13927 is used to determine mass loss rate and ignitability at any heat flux condition defined within the Cone Calorimeter fire model of ISO 5660 or ASTM E1354. This instrument as shown in Figure 6.20 is manufactured by Fire Testing Technology Ltd. (FTT), West Sussex, UK, and is a complete assembly that is also suitable to replace the cone heater/load cell assembly inside a Cone Calorimeter. The MLC as a stand-alone instrument serves as a quality control tool. The instrument is supplied as a two-part component system: the cone heater assembly (with conical heater,

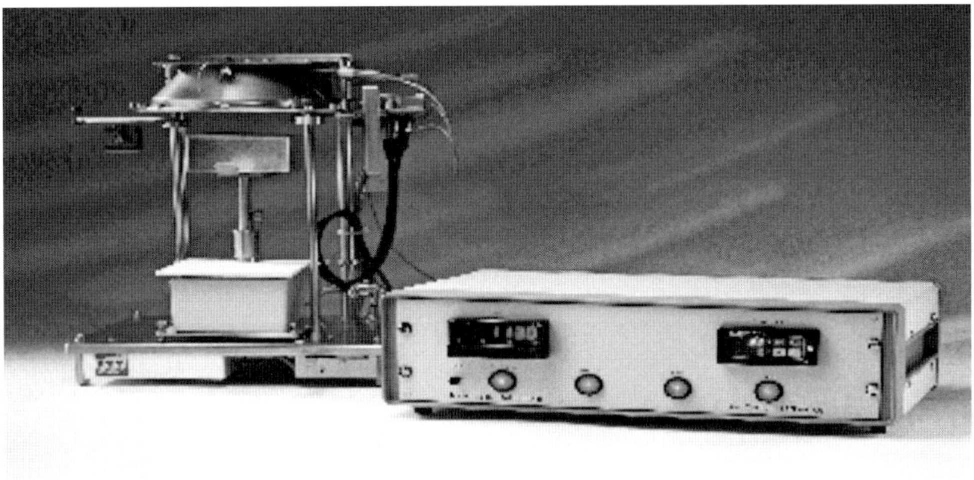

Figure 6.20. FTT Mass Loss Calorimeter (MLC) setup with cone heat assembly (left) and control unit (right).

shutter mechanism, spark ignition, load cell) and the control as shown in Figure 6.20. It can also be provided as an enclosed system, in order to measure combustion products and toxicity yields or to use in low-oxygen conditions. Heat release rate instrument can be used to determine the effect of combustion as a ratio of heat release rate and mass loss rate. Thus, if the effective heat of combustion is known, mass loss rate data can be used to calculate approximate values of heat release rate. The MLC is an inexpensive way of assessing fire properties of materials. Heat release is calculated using thermopile in the MLC.

6.12.3 Microscale Combustion Calorimetry (MCC)

The microscale combustion calorimetry was developed by Federal Aviation Administration (FAA) to offer the industry a research tool to assist FAA in its mandate to dramatically improve fire safety of aircraft materials. It is a standardized ASTM D 7309 test method [35] (Figure 6.21). The tester is becoming a useful research tool due to its ability to obtain meaningful test data with a sample size in the range of 0.5–50 mg. Potential measurements include heat of combustion, ignition temperature, heat release, heat release capacity, and flame resistance.

6.12.4 Oxygen Index (Limiting Oxygen Index – LOI)

Limiting Oxygen Index (LOI) is another test commonly used in the R&D laboratory; it is also known as the Oxygen Index (OI). It is included in some national or international standards, such as ASTM D2863-97 [36] or BS ISO 4589-2 (Figure 6.22). The specimen size and shape is not as strictly specified as the UL 94 test is, but usually bars of about 100 × 65 × 3 mm are used in testing rigid plastics. The specimen is positioned in a candle-like position vertically at the exit of a glass chimney and clamped at the bottom.

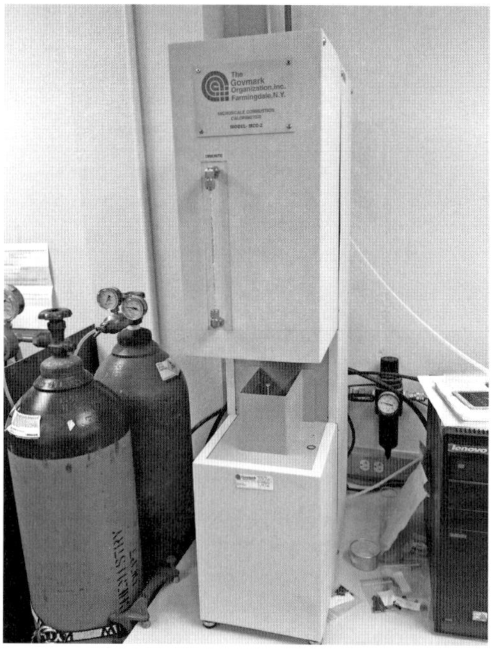

Figure 6.21. Microscale combustion calorimetry (MCC).

Figure 6.22. Limit oxygen index (LOI) chamber.

The chimney is continuously fed with a controlled mixture of nitrogen and oxygen. The flame of a small igniting burner is applied to the top of the specimen until the entire top surface is ignited. If the specimen does not ignite or extinguishes in less than 30 seconds, the concentration of oxygen is then increased, in steps with repeated trial ignitions, until the specimen shows stable candle-like combustion for more than 3 minutes after

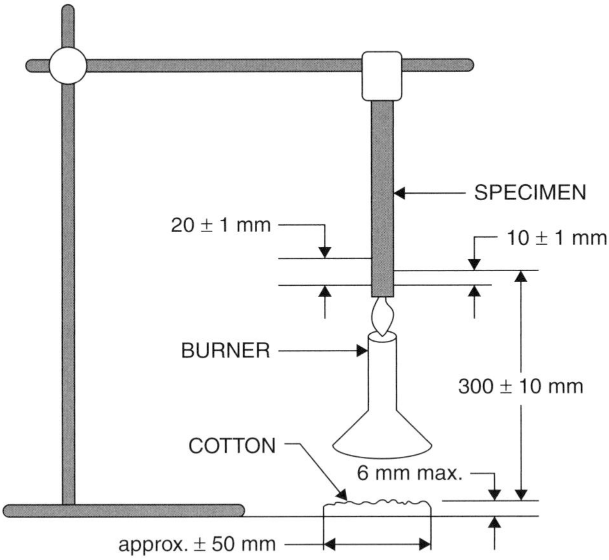

Figure 6.23. UL 94 vertical burn test for V-0, V-1, and V-2 rating.

removal of the ignition source; if more than 5 cm length of the sample is consumed, a new specimen should be put in place and tested at a slightly lower oxygen concentration. By iteration of these steps, finally the LOI value is determined; this is the highest concentration of oxygen at which the tested sample self-extinguished is less than 3 minutes and less than 5 cm of the material is consumed. The LOI test does not represent any real fire scenario, but it is quite reproducible and fairly insensitive to sample thickness. It is often used as a screening, research statistical design tool because it gives a numerical value typically to two significant figures, instead of a mere classification as given by the UL 94. It can even be run on films and textiles with suitable clamping.

Materials that melt and flow readily will give high LOI values, and in the case of nylon, it is even possible to arrive at two different LOI values depending on whether they are reached from the oxygen side or the high-oxygen side.

6.12.5 UL 94

Underwriters Laboratories UL 94 Test [37] (also known as ASTM D 3801 and ISO 1210) is designed for assessing "flammability of plastic materials for parts in devices and appliances" (see Figure 6.23). It is run on a sample of plastic itself, at specific dimensions. This test measures how long a bar of polymer burns after exposure to a small gas burner flame, and it is a reasonable measure of material's response to a small ignition source, such as a candle or a match. It is accepted for standardization internationally. Five different classifications from the vertical burn configuration are included, but only V-0, V-1, and V-2 classification are introduced because they are most often cited in the literature.

To assign this set of classifications, a plastic bar of 125 mm length x 13 mm width, with smooth edges, is positioned vertically and held from the top. Depending on the application of the plastic, bars may have a thickness of 3.2, 1.6, or 0.8 mm.

Thinner specimens are usually more flammable. A wad of surgical cotton is placed 300 mm below the specimen to detect flaming drips that will ignite the cotton. The Bunsen burner flame (ca. 19 mm high, fed with methane) is applied twice to the test bar (10 s each). After each application, the burning time is recorded. The second application of the flame should follow immediately after the bar ceases to burn from the first application. A V-0 classification is given when the 5 test bars extinguish in less than 10 seconds after any flame application. The average burning time for 5 bars tested (10 flame applications) should not be greater than 5 seconds, and there should be no burning drips that ignite the cotton. The individual specimen afterflame plus afterglow must not exceed 30 s. A V-1 classification is given to samples with maximum total combustion time < 50 s and average burning time for 5 specimens < 25 s. The individual specimen afterflame plus afterglow must not exceed 60 s. There should be no flaming drips. The sample is classified V-2 if it meets the flaming time criteria of V-1 or V-0 but has flaming drips that ignite the cotton.

6.12.6 Steiner Tunnel Test (ASTM E 84)

This Stein tunnel test (ASTM E 84) [38] is used mainly for building materials in the United States and is written in various building codes. The test apparatus is a tunnel measuring 8.7 × 0.45 × 0.31 m (25 × 1.5 × 1 ft) with viewing windows along its length. The sample to be tested is 7.6 m (24 ft) long and 0.51 m (1.67 ft) wide, mounted on the ceiling position. It is exposed from 10 min at one end of the tunnel to an 88-kW (5000 Btu/min) gas burner, with a forced draft through the tunnel, from the burner end, at an average initial air velocity of 1.2 m/s (240 ft/min). The flame spread index (FSI) is calculated on the basis of the area under the curve of the flame tip location as a function of time, compared to assigned values of 0 for an inert board and 100 for a standard red oak flooring. The smoke developed index (SDI) is calculated from the area under the curve of the light obscuration versus time as measured at the tunnel outlet, and is compared to the reference values of 0 for inert board and 100 for the red oak. Some codes require that the test be run on the minimum and maximum thickness and with construction similar to the manner in which the laminate will be used.

6.13 Ablation Properties

The testing of ablative materials involves simulating the extreme conditions to which they are exposed in actual applications, while measuring various metrics. Additional measurements can be made after the screening test is completed. The specific experimental inputs for ablation testing of a sample include extremely high heat flux, high temperature, and typically surface erosion from high-velocity fluid or particulate matter. Some common metrics for ablative performance measurement include mass loss, recession and/or recession rate, front face temperature, back face temperature, in-depth temperature, char thickness, and off-gas composition. The metrics hold different weights depending on the specific application of the ablative material. The material's resistance to heat is determined via measurements of the back face, front face, and heat soak (internal) temperatures during and after the testing procedure. These measurements show how well the material inhibits heat flow and can determine whether it fails to meet specific requirements regarding back face temperature. For example, if the back face temperature is too high, it may result in failure of the adhesive to bond the ablative to the supporting structure, which would result in a failure of the overall material. In addition, post-test measurements of the char thickness and material erosion can provide additional insight into the resistant capabilities of the material relative to exposure time. If the reaction progresses too quickly through the material or the protective char layer erodes too quickly due to surface shear, then the material may not provide thermal protection for a long enough time period.

There are no commercially available ablation testing systems, but a number of independently designed and managed apparatus are available [39]. These can be roughly split into two classes of apparatus: large scale and small scale. Large-scale ablation testing is usually done in the final stages of development for ablative materials and it is intended to actually predict and rate how the material will perform in its real-world application. They are used to validate any tests or models employed in development and to determine whether materials qualify for their intended use. Because of this, large-scale testing setups are typically fine-tuned to very accurately recreate the conditions that thermal protection system (TPS) materials will experience in application and are very adjustable so that a variety of conditions can be simulated. As a result, large-scale ablation testing is very expensive and time consuming to perform, and there are a limited number of facilities that can reliably do it. The VKI Plasmatron developed and operated by the von Karman Institute for Fluid Dynamics in Belgium and the Arc Jet Complex at NASA's Ames Research Center are two examples of large-scale ablation testing facilities [40].

Ablation and thermal performance of the resulting nanostructured materials for propulsion and reentry applications are evaluated using established laboratory devices, such as simulated solid rocket motor (SSRM), subscale solid rocket motor (Char motor), and oxyacetylene test bed (OTB). These three ablation-testing devices are briefly discussed in this section. Besides these three laboratory devices, two other capabilities relating to ablation research are including in this section, namely the in situ ablation recession and thermal sensor and a device to measure the char strength of posttest polymer nanocomposite ablatives.

6.13.1 Simulated Solid Rocket Motor (SSRM)

The SSRM is an established testing approach developed by Koo et al. in 1990s [41] and has been used extensively in the development of ablative materials [42–51]. It is a small-scale, supersonic, liquid-fuel rocket motor burning a mixture of kerosene and oxygen as shown in Figures 6.24 and 6.25. Aluminum oxide particles are injected

Figure 6.24. Ablative material testing using SSRM.

Figure 6.25. Close-up of ablative material after SSRM testing.

into the plume to simulate the particle-laden flow of solid rocket exhaust plumes. The SSRM is a controlled laboratory device capable of producing a particle-laden exhaust environment with measured heat fluxes from 40 to 1,250 Btu/ft^2-s (454 to 14,200 kW/m^2). The flame temperature is approximately 3,992°F (2,200°C), and the velocity of the particle-laden exhaust is approximately 2,000 m/s. Figure 6.25 shows a close-up picture of an SSRM posttest ablative specimen. Calibration of the SSRM was performed using a Medtherm water-cooled heat flux gauge. The total heat flux calibration of the SSRM is shown in Figure 6.26. The highest heat flux is 1,250 Btu/ft^2-s (14,200 kW/m$_2$) at 2 inches (5.08 cm) from the nozzle exit, and the lowest heat flux is about 40 Btu/ft^2-s (454 kW/m$_2$) at 14 inches (35.56 cm) from the nozzle exit. The standard sample size was 4 by 4 by 1/4- or 1/2-inch thick (10.2 by 10.2 by 0.64- or 1.28-cm thick). The composite materials were bonded to 4 by 4 by 1/8-inch (10.2

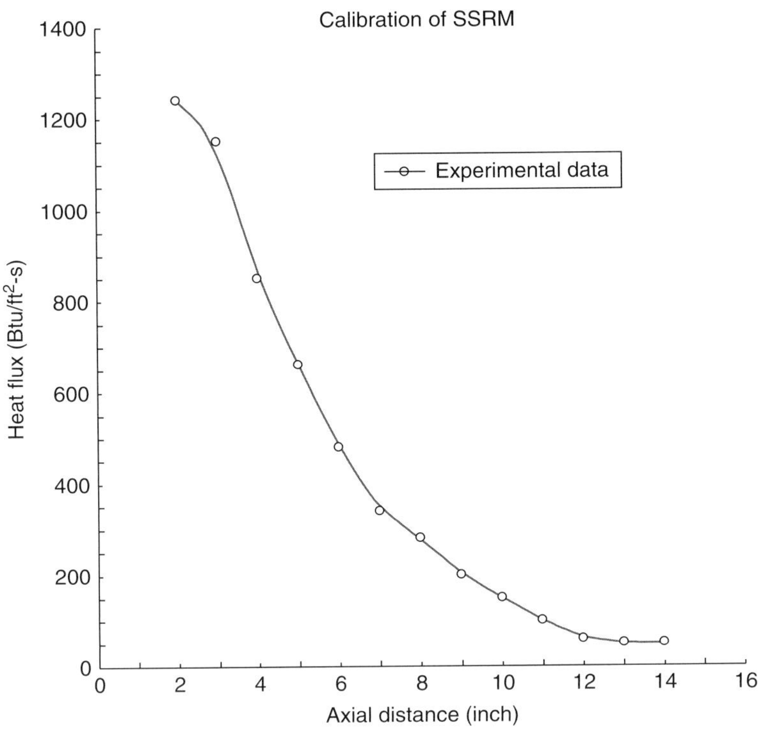

Figure 6.26. Calibration of SSRM in heat flux versus distance from nozzle exit.

by 10.2 by 0.32-cm) steel substrate. A narrow slot was machined into the bondline side of the steel substrate. The thermocouple was embedded into this slot. The bead of the thermocouple was placed at the center point of the plate. The temperature history of the bondline was recorded with this thermocouple. C-clamps were used to clamp the ablatives to the steel substrate during testing. The test samples were placed in a fixture downstream from the SSRM nozzle exit. The fixture had adjustments in axial distance and impingement angle with reference to the plume centerline. A normal 90° impingement was used for this study. Three axial distances were used to correspond with heat fluxes of 250, 625, and 1,000 Btu/ft^2-s (2,838, 7,094, and 11,345 kW/m^2). These distances were selected to simulate low, medium, and high heating conditions. Prior to the motor burn, the bondline temperature was recorded. Peak erosion was determined by pre- and posttest measurements using a pencil-point dial indicator. Mass loss was determined by pre- and posttest weight loss measurements. Consistent data sets were averaged for this study. The following data were acquired during the experiment:

- Peak erosion (mm)
- Residual mass (%)
- Maximum backside heat-soaked temperature (°C)
- Time to reach maximum heat-soaked temperature (min.)
- Surface temperature (°C)
- Microstructures of post-test specimens using SEM

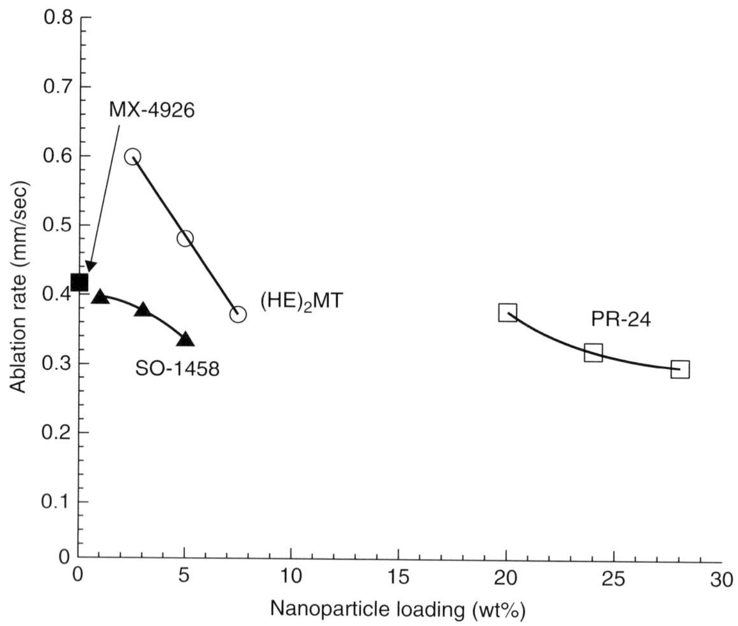

Figure 6.27. Ablation rate of MX-4926 and NRAMs with different types of nanoparticle at various loading levels.

Koo and colleagues used this device to study a family of nanocomposite rocket ablative materials (NRAMs). Three nanomaterials – MMT nanoclays, carbon nanofibers (CNFs), and POSS – were incorporated into a phenolic resole resin (Hexion's SC-1008) and impregnated into carbon fibers to form carbon fiber–reinforced composites [10, 47–49]. The authors' NRAMs were compared with Cytec Engineered Materials' industry standard MX-4926 carbon/phenolic ablative.

The ablation rates (Figure 6.27) of MX-4926 and all three nanoparticles indicate that only 7.5% nanoclay composition shows a lower ablation rate than MX-4926, which is about 0.4 mm/s. All three CNF compositions are lower than MX-4926, with the 28% CNF-NRAM being the lowest. All three POSS® compositions have lower ablation rate, with the 5% POSS®-NRAM being the lowest. Residual mass (Figure 6.28) of MX-4926 is about 92%. The POSS®-NRAMs exhibited the highest of all the nanoparticles in residual mass or about 93%. The clay-NRAMs were comparable to MX-4926 while the CNF-NRAMs were lower. A summary of backside temperature rise is shown in Figure 6.29, with all the nanoparticles compositions exhibiting lower maximum backside heat-soaked temperature rise than MX-4926. The value of MX-4926 is 106°C with CNF-NRAMs being the lowest, from 54° to 72°C. The POSS®-NRAMs were intermediate, from 75° to 86°C, and clay-NRAMs were the highest of the nanoparticles, from 82° to 98°C, but still below the MX-4926 value of 106°C.

An IR pyrometer was used to measure the surface temperatures of all materials during SSRM firings. Figure 6.30 shows the surface temperatures of MX-4926 (1,700°C) and the NRAMs. Surface temperatures of the CNF-NRAMs samples were higher than those of MX-4926, the clay-NRAMs, and the POSS-NRAMs. This

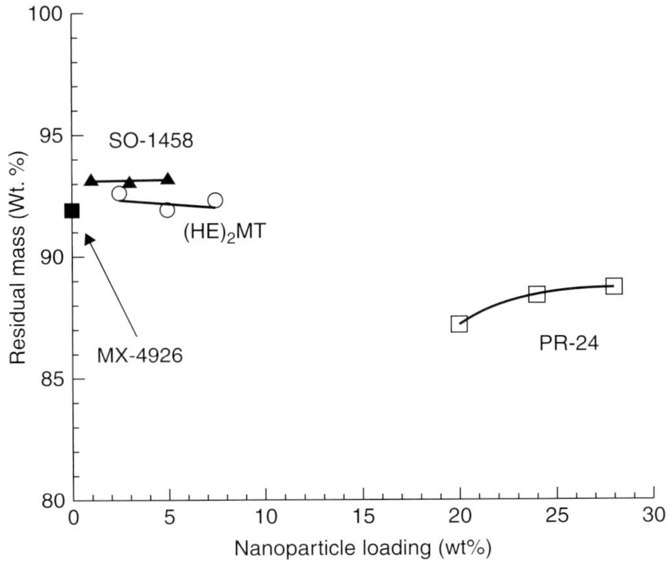

Figure 6.28. Residual mass of MX-4926 and NRAMs with different types of nanoparticle at various loading levels.

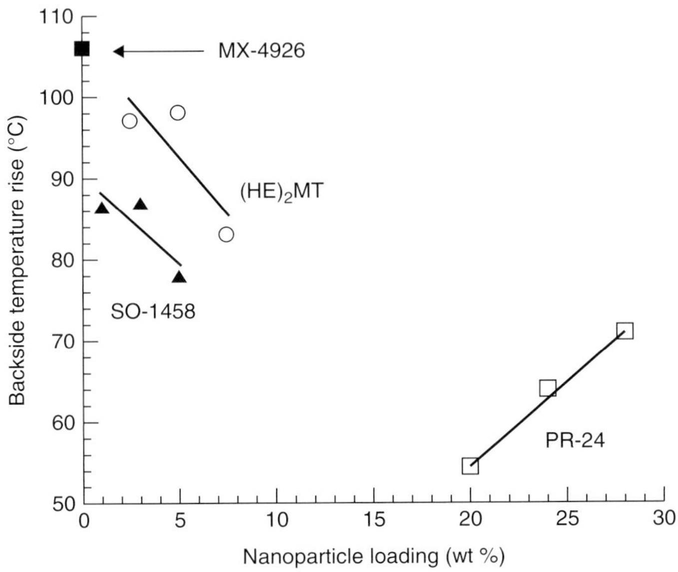

Figure 6.29. Backside temperature rise of MX-4926 and NRAMs with different types of nanoparticle at various loading levels.

finding suggests that better radial heat transfer is occurring rather than axial heat transfer. This is supported by the glowing heat of the material surface observed during material testing. This phenomenon was observed by other researchers [10] and needs further study. The surface temperatures of clay-NRAMs and POSS-NRAMs were lower than that of MX-4926. The amount of nanoclay in the clay-NRAMs had

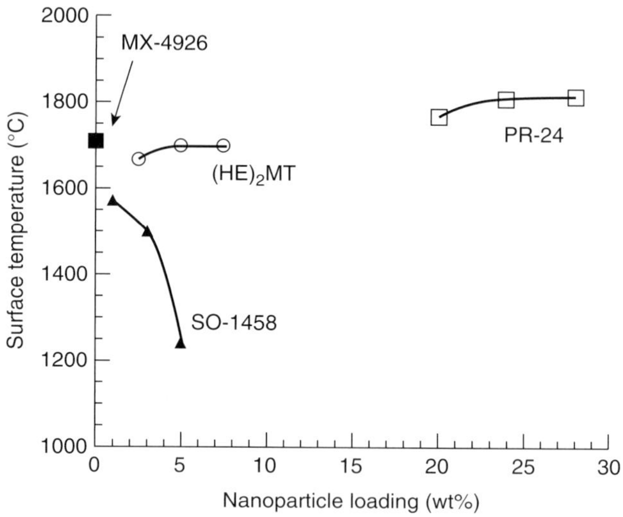

Figure 6.30. Surface temperature of MX-4926 and NRAMs with different types of nanoparticle at various loading levels.

essentially no effect. The amount of POSS in POSS-NRAMs had a significant effect on the surface temperature of the POSS-NRAMs. More detailed discussions can be found in Chapter 10 on ablation properties.

6.13.2 Subscale Solid Rocket Motor (Char Motor)

At Edwards Air Force Base, California, AFRL/PRSM has developed a very cost-effective char motor to screen solid rocket motor (SRM) insulation materials [52–54]. A schematic of the Pi-K SRM experimental setup is illustrated in Figure 6.31. The motor is operated and fired in the upward direction. It usually contains up to 2 lbs (4.4 kg) of solid rocket propellant in an end-grain burning configuration. Both non-aluminized and aluminized propellants have been used for insulation material testing. The average chamber pressure is typically about 800 psia (5.5 MPa), and the test duration is about 8 seconds. For insulation material testing, a cylindrical cone made of glass/phenolic material of different area ratio is used inside the motor. This configuration is used to simulate a range of Mach numbers to which the insulation materials are exposed during combustion of gaseous products. Rapid testing of six samples or more per day can be achieved. For nozzle material testing, the cylindrical cone is removed, and the whole nozzle assembly at the back end of Pi-K SRM is replaced by NRAM candidate materials. The whole assembly is a convergent/ divergent nozzle, and the nozzle throat is a machined graphite nozzle insert, as shown in Figure 6.32. The graphite nozzle insert piece prevents excess erosion in the nozzle throat area. This configuration enables the evaluation of candidate materials in the entrance region and exit cone of the nozzle. Four successful firings were conducted at AFRL/Edwards AFB. Erosion data at different area ratios were collected. Posttest specimens of selective regions of the nozzle assembly were sectioned and analyzed using SEM. More detailed study using this device is discussed in Chapter 10 on ablation properties.

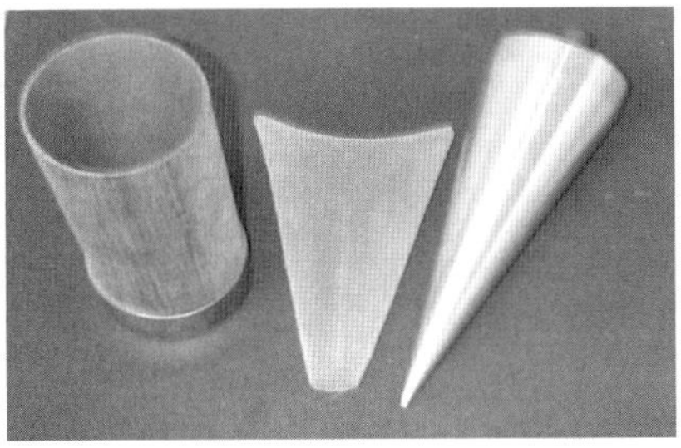

- 4-Inch Char motor
- Only 100 g of insulation material needed (clear POSS® sample shown on the left)
- Very cost-effective tool
- Rapid testing of 6 samples/day (or more)

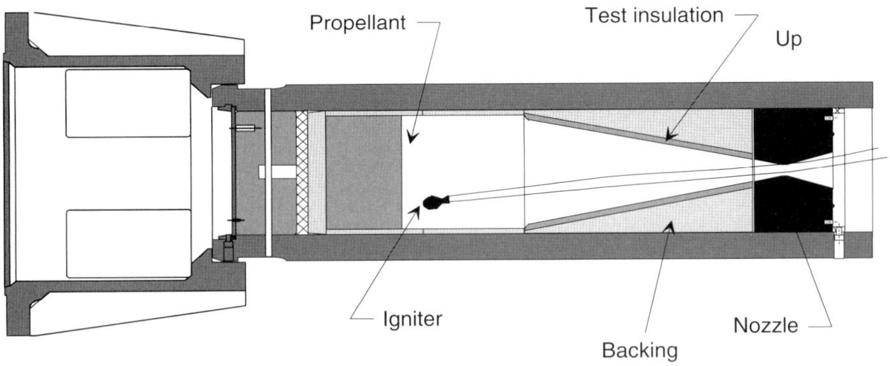

Figure 6.31. Schematic of the AFRL/Pi-K SRM experimental setup for ablation testing. Both non-aluminized and aluminized propellants can be used. Average chamber pressure is about 800 psia and test duration is about 8 seconds when an end-burning propellant grain was used. NRAMs will be tested at the nozzle section.

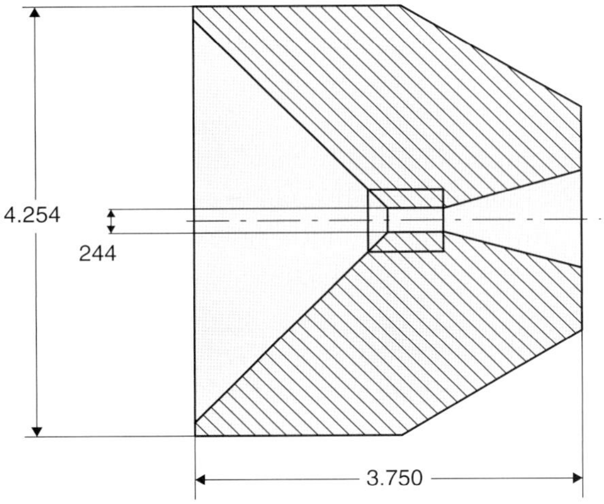

Figure 6.32. Cross-sectional diagram of the motor nozzle assembly showing the graphite insert at the nozzle throat of the NRAM material. This nozzle assembly fits into the back end of the Pi-K motor. The direction of the combustion gaseous products from the rocket chamber is moving from right to left. All dimensions are in inches.

6.13.3 Oxyacetylene Test Bed (OTB)

Small-scale ablation testing is used in the early stages of development to determine the relative worth of a potential ablative through comparison to well-established ablatives. They recreate the ablative environment to a degree that the material will undergo ablation similar to what it would see in the real world, but do not control or account for certain variables in order to allow for lower cost and easier testing. These inaccuracies are why small-scale testing is only used in early stages as a screening tool, and large-scale testing must be used to validate these early test results. However, the primary advantage of small-scale testing is in cost and speed of testing. Two instances of small-scale ablative testing systems are the apparatus developed by M. Natali et al. [55] of the University of Perugia in Terni, Italy and by G. Pulci et al. [56] of the Sapienza University of Rome.

The ablation testing setup used in Koo Research Group is a small-scale setup developed and managed at the University of Texas at Austin [57, 58]. It is known as the University of Texas Oxyacetylene Test Bed (UT-OTB or OTB). M. Natali provided insight and advice aiding in the development of the OTB, and the overall design is based on the ablation test setup developed by Natali et al. The OTB is based on an oxyacetylene torch that is used to create the very high temperatures needed for ablation and that has its flame concentrated on a small area, such that very high heat flux values are also produced. The OTB is a small-scale apparatus, so high temperature and heat flux are the only two experimental conditions of an ablative environment intended to be accurately simulated. Acceptable values for these parameters were determined to be a temperature of at least 2,200°C and a heat flux in excess of 1,000 W/cm^2. Small sample size is also desirable as it reduces the cost of each test and enables a greater number of tests to be run.

Once the OTB was completed, initial testing was done to confirm that the experimental conditions met the required parameters. A literature search of the stoichiometric combustion temperature of acetylene with oxygen confirms that our oxyacetylene torch will produce a temperature in excess of 3,000°C, much greater than the 2,200°C that was required. The researchers were also able to design their setup such that the area of the sample exposed to the ablation conditions was roughly 5–10 mm in diameter. The standard sample holder for the OTB is capable of holding samples with diameters ranging from roughly 1 cm to 2 cm and thicknesses ranging from 1 cm to 5 cm. The produced heat flux had to be measured and calibrated based on distance between the torch tip and the sample surface to allow for a degree of customization of the testing parameters. Figure 6.33 shows the OTB in operation.

The power output of the torch is measured using a slug calorimeter in which a copper slug of known mass is exposed to the oxyacetylene torch [59]. A thermocouple is embedded within the copper slug and provides a temperature vs. time curve for the slug. The slope of this curve, along with the properties of the slug, is then used to calculate the heat flux produced by the oxyacetylene torch. Depending on the distance from the sample to the torch, the OTB is capable of producing heat flux values in excess of 1,000 W/cm^2 (Figure 6.34).

For the materials tested in this chapter, specific settings within the OTB's range of operation were selected and all the material formulations were tested at these conditions. The nanocomposite formulations were cut into 1.27 × 1.27 × 5 cm (0.5 × 0.5 x 2 in.) samples for testing with three samples tested for each

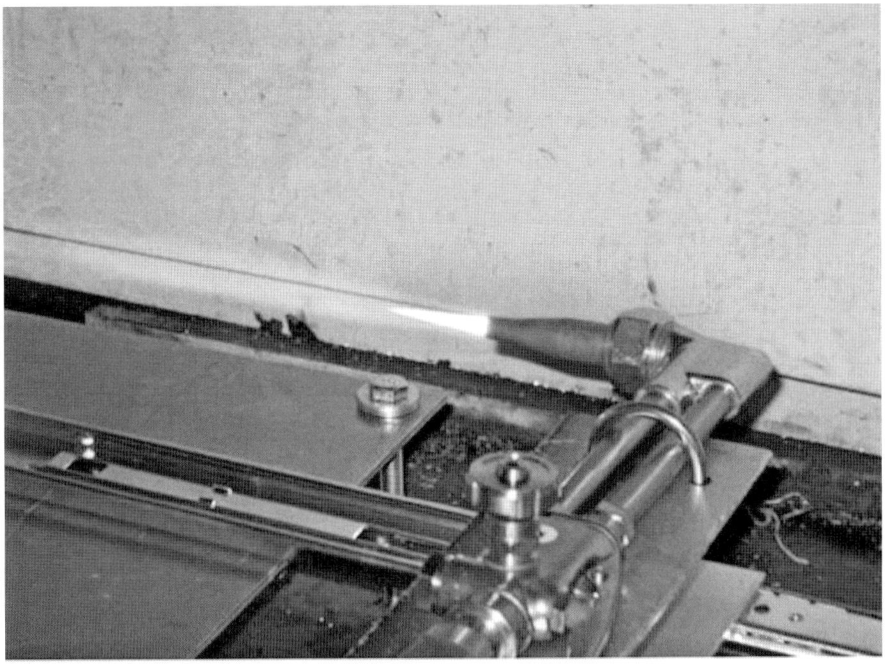

Figure 6.33. Oxyacetylene test bed in operation.

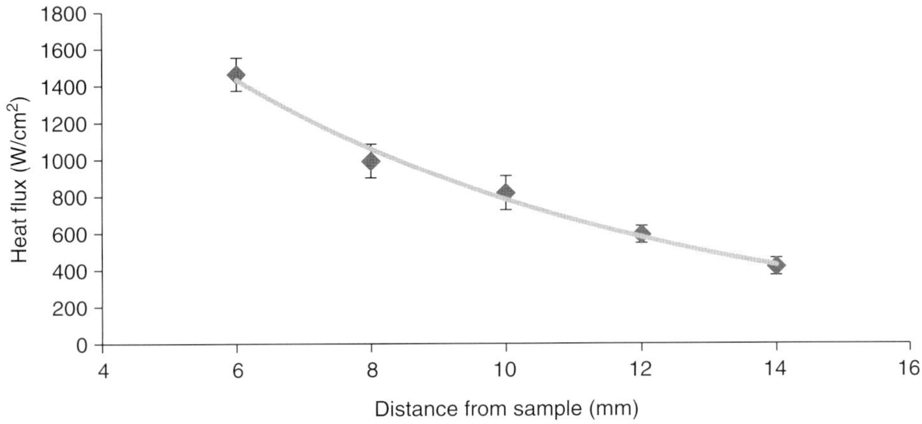

Figure 6.34. Calibration of heat flux for OTB.

formulation to increase the statistical confidence in the results. The square face of the samples was exposed to the oxyacetylene torch at a calibrated heat flux of approximately 1,000 W/cm^2 for a duration of 25 seconds. The OTB facilitates three K-type thermocouple measurements during testing, and for our study these thermocouples were placed within the samples at distances of 5 mm, 10 mm, and 15 mm from the exposed surface with a heat shield in place to protect the protruding wires. In addition to these in situ temperature curves, the mass loss, recession, and char thickness of each sample was determined based on pre- and posttest measurements.

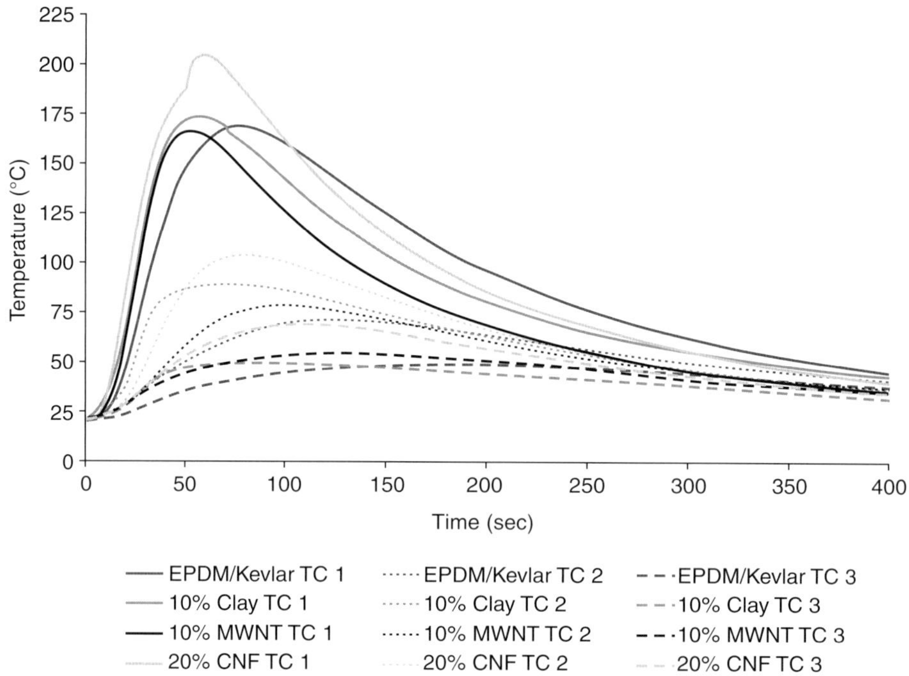

Figure 6.35. Representative oxyacetylene torch EPDM/Kevlar profile and PNCs.

Lee and Koo used the OTB to conduct high-heat flux ablation experiments [60–63]. The thermoplastic elastomer nanocomposites (TPUNs) containing different loading MMT nanoclays, CNFs, and MWNTs were processed via twin-screw extrusion test specimens fabricated by injection molding. These TPUNs is exposed to the hottest portion of the flame, with a 7 mm flame diameter, for 25 seconds. Samples of 12.7 × 12.7 × 50 mm are cut out of a 12.7 mm × 10.16 cm x 10.16 cm (½" × 4" × 4") compression pressed sheet. The torch is directed toward the square face (12.7 × 12.7 mm). In this orientation, the sample may be tested twice. At 5 mm increments, three thermocouples are embedded into the center of the specimen to measure the in-depth heat soak temperature during testing. Thermocouples are drilled into the specimen at both ends for the possibility of testing a single specimen twice depending on the ablation rates.

Thermal Results – A representative temperature profile of each specimen is shown in Figure 6.35 [60]. The specific materials are color coded and the thermocouples (TC) are distinguished as: TC 1 solid line, TC 2 dotted line, and TC 3 dashed line. After 25 seconds, the oxyacetylene flame is removed and heat is transferred through the specimen. In this set of tests, the peak temperature of TC 1 in the CNF sample is 205°C and occurs at 60 seconds. The CNF has the highest heat soak temperature. The MWNT and nanoclay TC 1 peaks at 166°C and 174°C and occur at 50 seconds and 64 seconds, respectively. The EPDM/Kevlar TC 1 peak, 168°C, shows the most delay, 75 seconds. Figure 6.36 shows the peak in-depth temperatures of EPDM/Kevlar and PNCs between thermocouples [60]. The nanoclay, MWNT, and CNF PNCs lose 13.5%, 33.7%, and 46.5% more mass, respectively, compared to the mass loss of Kevlar/EPDM.

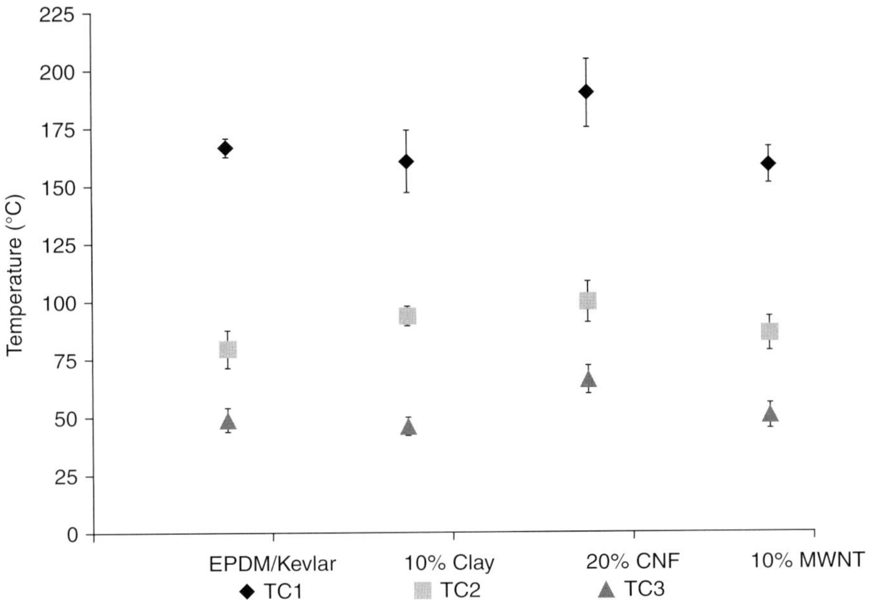

Figure 6.36. Peak in-depth temperatures of temperature.

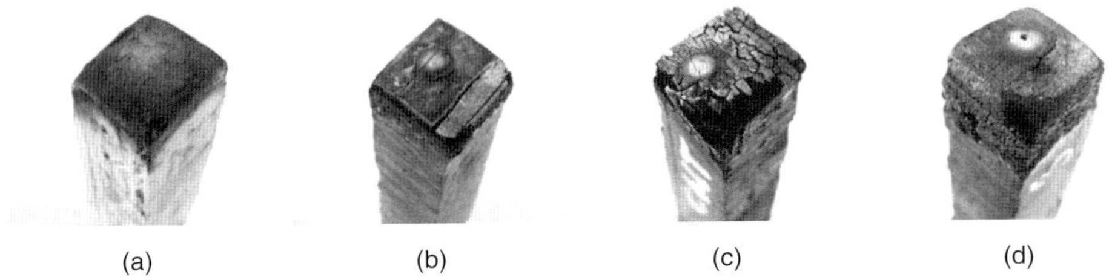

Figure 6.37. (a) Kevlar-EPDM, (b) 10% Clay, (b) 10% MWNT, and (c) 20% CNF PNC post-oxyacetylene torch burn.

Post-firing Images – The samples after firing are shown in Figure 6.37 [60]. A 1 mm compact char layer is left at the end of the firing of Kevlar/EPDM. Although compact, the virgin and char material does not have good adhesion. The nanoclay PNC posttest material has about twice the char thickness of the Kevlar/EPDM. Although the char has a crack, the char was maintained well within the virgin material. Similar to the results found in the vertical UL 94 test, many cracks are observed in the MWNT char surface and result in a weak 2 mm thick char. The CNF PNC char is very thick, greater than 5 mm, as the first thermocouple was in the char layer. The CNF char surface is flat without cracks; however, it is still very weak. The results from this set of experiments are consistent with the flammability and hybrid rocket tests by Lee et al. [61, 62].

SEM images of the posttest samples also were taken. The Kevlar/EPDM char material is compact and shows that the individual Kevlar fibers hold the material together as shown in Figure 6.38 [60]. The nanoclay char is also a compact composition. A granular morphology is also observed (Figure 6.39) [60]. Bubble formations

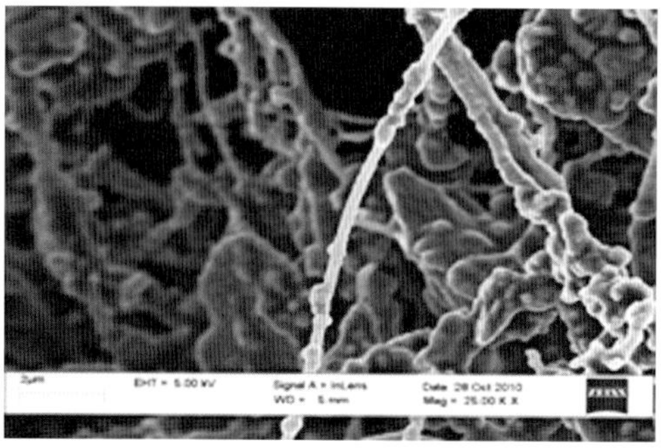

Figure 6.38. SEM of Kevlar/EPDM char (scale bar is 2μm) exposed to post oxyacetylene torch.

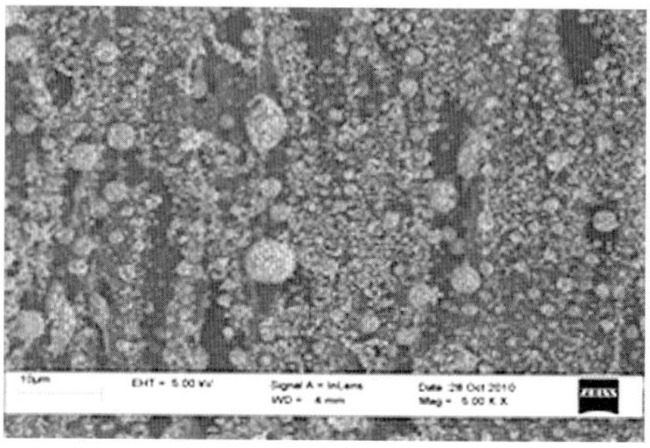

Figure 6.39. SEM of 10% nanoclay PNC char (scale bar is 10μm) exposed to post-oxyacetylene torch.

Figure 6.40. SEM of 10% MWNT PNC char (scale bar is 200nm) exposed to post-oxyacetylene torch.

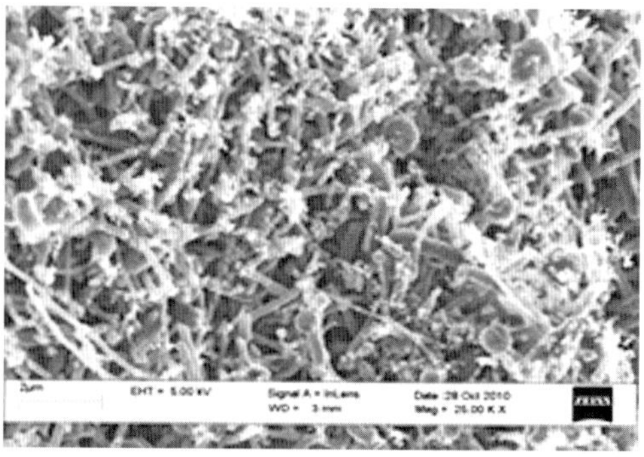

Figure 6.41. 20% CNF PNC char (scale bar is 2μm) exposed to post-oxyacetylene torch.

observed in SEM images found in the UL 94 burn tests are observed on the MWNT char surface as well as crack formations (Figure 6.40) [60]. An in-depth SEM image shows the bundle form of MWNT that is also observed in the UL 94 burn test. The weak CNF char is shown to have a very porous structure (Figure 6.41) [60]. Individual CNFs are also observed.

In summary, the results from the oxyacetylene torch test show that both the nanoclay PNC and MWNT char layers are about twice the thickness of the Kevlar/EPDM char. However, the MWNT char layer is weak compared to the nanoclay PNC char. The CNF char is also very weak and was more than five times the thickness of the Kevlar/EPDM char. However, the mass loss from the CNF sample is almost twice that lost from the Kevlar/EPDM sample. In-depth thermal measurements also show that the thermal protection of the PNCs is comparable to that provided by the Kevlar/EPDM, except for the CNF sample, which had a much higher peak temperature due to the fact that the first embedded thermocouple ended up in the char region. A conventional advanced fiber is therefore needed to boost up the PNC ablation performance. More detailed ablation test results of these TPUNs can be found in Allcorn et al. [64, 65] and Wong et al. [66, 67], as well as in Chapter 10 on ablation properties.

6.13.4 Char Strength Sensor

In order to maximize the payload of solid rocket motors (SRMs), ethylene-propylene-diene-monomer rubber (EPDM) has shown to be an ideal candidate matrix for internal SRM insulation and it practically replaced all other elastomers, such as nitrile and silicone, as insulators in advanced SRMs [68–72, 64–68]. Among fibrous reinforcements, aramid fibers (Kevlar®) in the form of fibers or pulp have been incorporated in the EPDM rubber because of their high thermal capacity, chemical stability, fire resistance, low thermal conductivity, high ablation resistance, and high physical adhesion [71, 72].

The evaluation of the mechanical properties of the charred layer is very important, since it can enable the possibility to better tune the thermoplastic polyurethane nanocomposites (TPUNs) as discussed by Koo et al. (Allcorn et al. [64, 65] and Wong et al. [66, 67]), as well as in Chapter 10 on ablation properties. This char strength property also will produce a more comprehensive screening of the TPUN materials. However, due to the fragile nature of the char, ordinary testing method, such as Rockwell hardness is not feasible, thus requiring the development of a new testing protocol and sensor.

In 2009, a team from The University of Texas at Austin (UT) contributed toward the final success of this new char strength sensor. The team, Redondo et al. [73], provided background and evaluated available testing methods. The preferred testing method selected was to crush the char samples, and thus a prototype was created. However, the data obtained from the prototype proved unreliable due to poor structural tolerances, mechanical vibration, and signal noise. As a result of the prototype not being able to complete the task, it was used as a "proof of concept" allowing for a complete redesign. Lastly, this research utilizes information from the Reshetnikov et al.'s study that details another compression testing method used to test the strength of intumescent char. These Russian scientists cited various ways to test char material, such as centrifugal shear, compression, and fluid flow shear [74].

Air Force Research Laboratory (AFRL)/Edwards Air Force Base in California currently uses a method to test char strength. However, the test is entirely qualitative and subjective. The tester brushes a charred sample with a wire brush and determines the relative strengths of different samples visually by the remaining thickness [75]. The lack of a repeatable, quantitative test prompted the need for a scientific testing method, since the brush test contains significant human error in applied force. Two sensors were developed – a compression sensor and a shear sensor – which are discussed briefly here.

Compression Sensor – The schematic of the compression char strength tester is shown in Figure 6.42 [76–78]. It consists of three main components: the sensor, the power supply, and the stand to hold the sensor. Figure 6.43 shows the compression char strength technical drawing and schematics and Figure 6.44 shows the char compression sensor (left), electronic control box (center), and the char shear sensor (right) [78]. In order to generate a correlation for the LabVIEW program to be used, the potentiometer and force sensor are calibrated. The potentiometer is calibrated to give a correlation

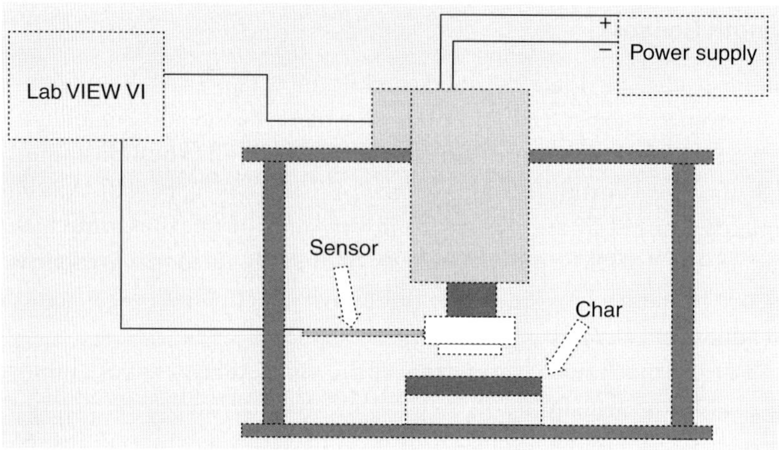

Figure 6.42. Char compressor schematic diagram showing the power supply, LabVIEW VI, sensor, and the char specimen.

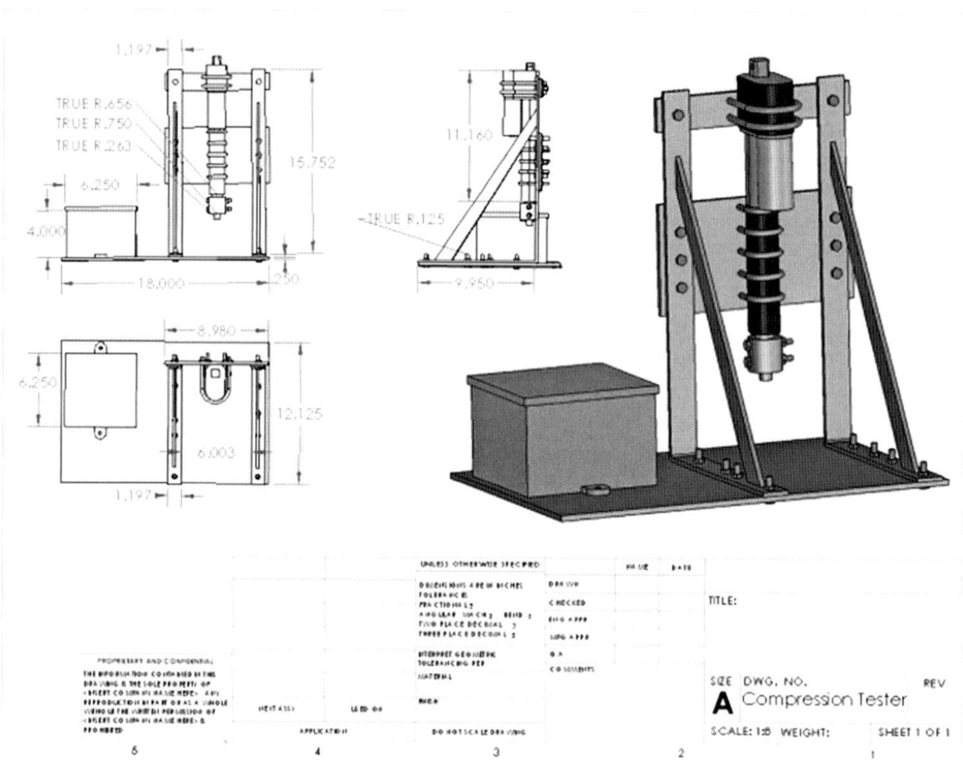

Figure 6.43. Char compressor technical drawing and schematics.

of the actuator's probe tip position as a function of voltage across the potentiometer. The force sensor inside the probe tip is calibrated to determine the force output as a function of voltage. The potentiometer calibration data is obtained using calipers to measure the distance traveled by the actuator and a multimeter to measure the voltage

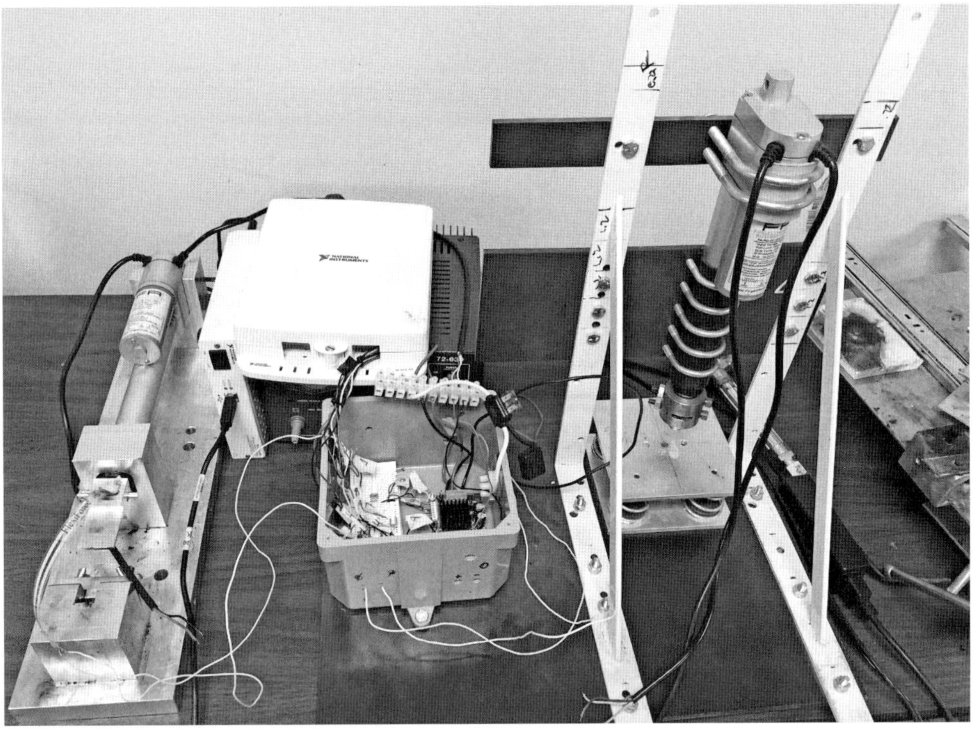

Figure 6.44. Char shear sensor (left), control box (center) and char compression sensor (right).

drop across its potentiometer. The force sensor is calibrated using an apparatus made up of four springs sandwiched by two metal plates. As the probe presses down on the top metal plate, the four springs are compressed, and the force sensor inside the probe tip sends a voltage signal to the control program in LabVIEW. Using the potentiometer in the linear actuator, the distance that the top plate was pressed down is known. Using the distance that the springs are compressed and the stiffness of the springs, the force can then be calculated. The final step is then to generate a correlation of force as a function of voltage. Sample data of the calibration are shown in Figure 6.45 [78].

Shear Sensor – The shear sensor uses a horizontally oriented linear actuator to shear a char sample. The shear sensor uses the same control box, power supply, and DAQ system as the compression sensor does, but a different LabVIEW control program. The shear sensor schematic is shown in Figure 6.46. Figure 6.47 shows the shear sensor without all the control components (control box, DAQ, power supply, etc.) [77, 79]. The shear sensor uses the same type and brand load cell as the compression sensor. It is calibrated using the same method as described for the compression sensor.

Analysis Method – Since the criteria for the strength of the char material could consist of a variety of factors, the sensor gathers force and distance (depth) data as the probe compresses or shears off the charred material. Using these data, two methods were developed to determine the strength of the char.

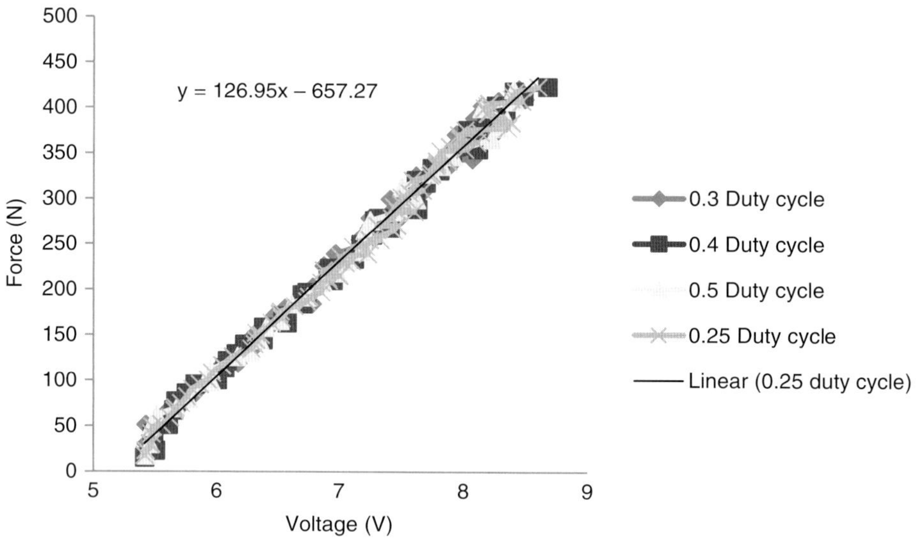

Figure 6.45. Sensor force versus voltage using several duty cycles with correlation equation.

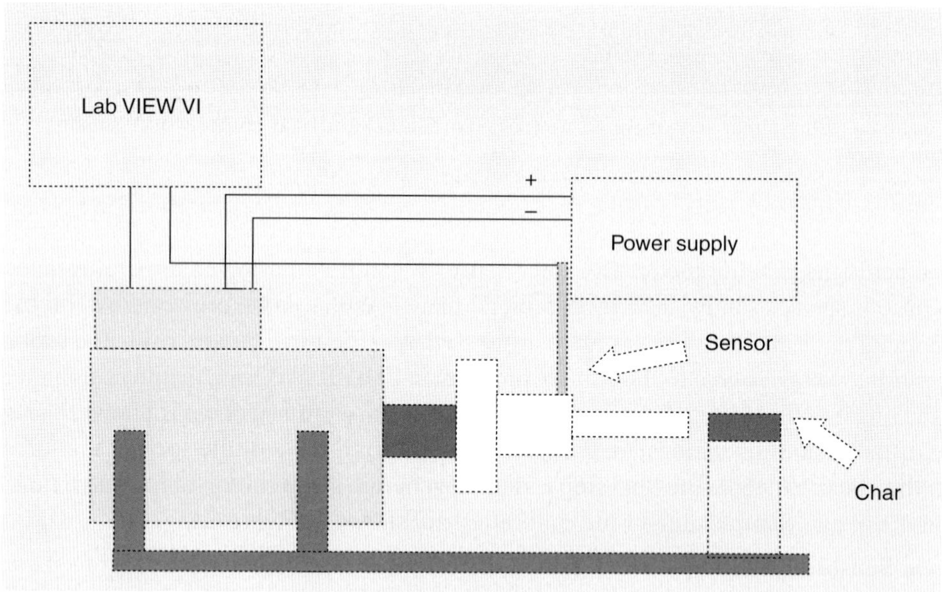

Figure 6.46. Char shear schematic diagram showing the power supply, LabVIEW VI, sensor, and the char specimen.

In the first method, the force data can be used to observe the yield strength of the char. As the piezoresistive load cell implemented in the design changes resistance in response to pressure, the applied force is recorded. A peak in the force followed by a significant decrease in applied force on the char indicates the ultimate strength has been reached, revealing the maximum allowable stress for the material. Using this method, one can rank the strength of several samples, the strongest being the sample that has the largest ultimate stress.

Figure 6.47. Photograph of shear strength sensor.

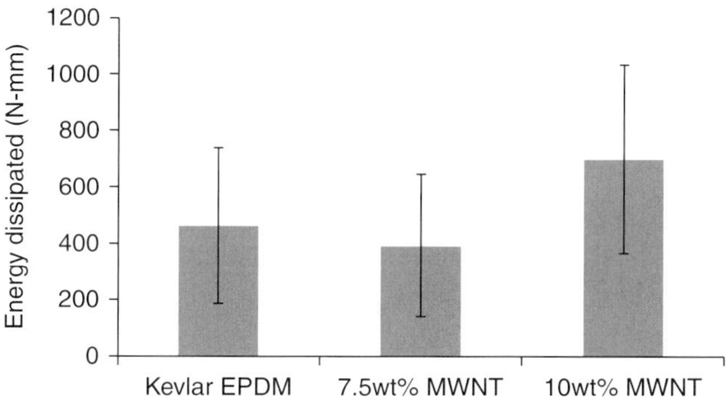

Figure 6.48. Average energy dissipation of Kevlar EPDM, 7.5wt% MWNT, and 10wt% MWNT.

The second method to determine the strength of the char material is to generate a plot of force versus distance (depth). If one integrates the force vs. distance plot to find the area under the curve (F*D = W), one can determine the energy required to destroy the char. The strongest char will produce the highest energy of destruction or energy dissipation.

Compression Char Strength Results – The average energy dissipation results in Figure 6.48 show that TPUN-10wt% MWNT ranks higher with a value of 703 ± 335 N-mm, followed by Kevlar®-filled EPDM with a value of 466 ± 278 N-mm, and lastly TPUN-7.5wt% MWNT ranks lowest with a value of 396 ± 254 N-mm [78].

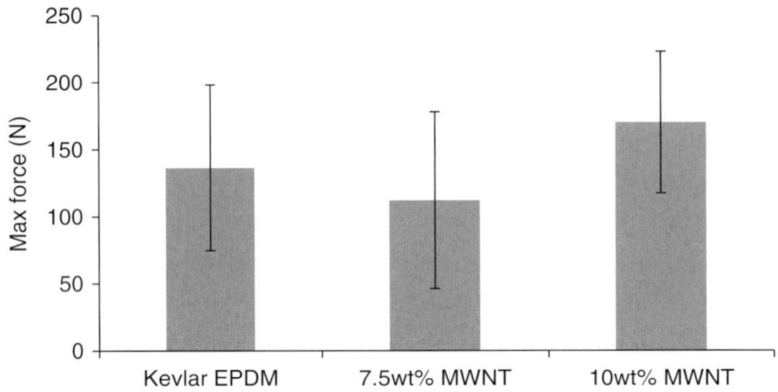

Figure 6.49. Average maximum force of Kevlar EPDM, 7.5wt% MWNT, and 10wt% MWNT.

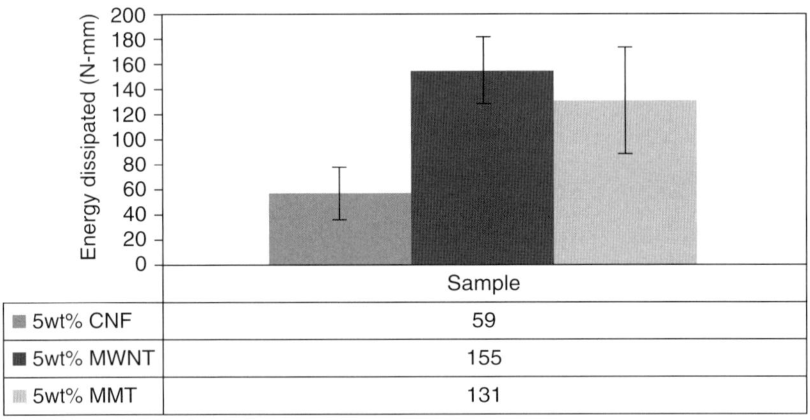

Figure 6.50. Average energy dissipation of each type of thermoplastic elastomer nanocomposites.

The average maximum force results in Figure 6.49 show that TPUN-10wt% MWNT ranks higher with a value of 170 ± 53 N, followed by Kevlar®-filled EPDM with a value of 136 ± 62 N, and lastly TPUN-7.5wt% MWNT again ranks lowest with a value of 112 ± 66 N [78]. As it can be seen from Figures 6.48 and 6.49, TPUN-10wt% MWNT has the potential to replace Kevlar®-filled EPDM, since it is shown to rank higher in average energy dissipation and average maximum force than Kevlar®-filled EPDM if no other material properties are considered [78].

Shear Char Strength Results – Although it was determined from the compressive strength tests conducted in this study that MWNT was the strongest TPUN, Figures 6.50 and 6.51 show that MMT is not that far behind in strength. To determine for sure which of the two materials is stronger, and therefore warrant further testing to determine if it may replace Kevlar®-filled EPDM, shear strength testing was conducted on Kevlar®-filled EPDM, TPUN-7.5wt% MMT, and TPUN-7.5wt% MWNT [77]. The final average energy dissipation results are shown in Figure 6.52 [79]. TPUN-7.5wt% MWNT ranks higher with a value of 292 ± 51 N-mm, followed

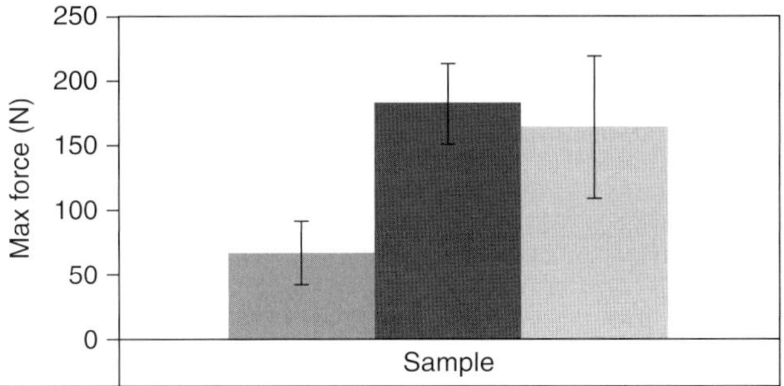

	Sample
5wt% CNF	66
5wt% MWNT	183
5wt% MMT	164

Figure 6.51. Average maximum force of each type of thermoplastic elastomer nanocomposites.

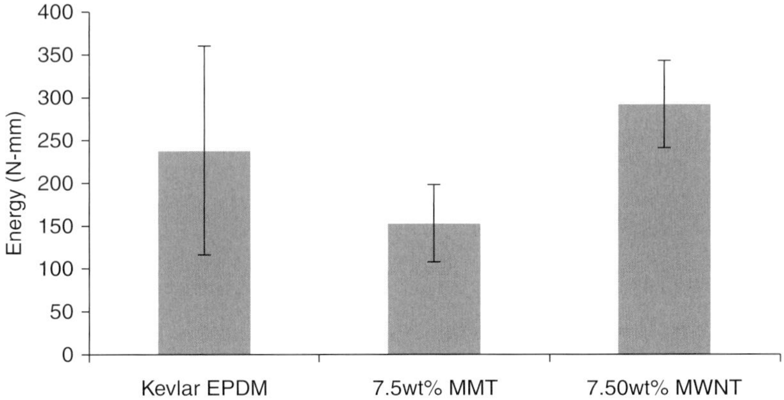

Figure 6.52. Average energy dissipation in shear of Kevlar EPDM, 7.5wt% MMT, and 7.5wt% MWNT.

by Kevlar®-filled EPDM with a value of 239 ± 122 N-mm, and lastly TPUN-7.5wt% MMT ranks lowest with a value of 154 ± 45 N-mm.

Average maximum force analysis was not conducted on the shear tests because the force vs. distance plots on shear tests contain several peaks that could potentially be the yield strength [79]. Additional char strength results can be found in Chapter 10.

6.13.5 In Situ Ablation Recession and Thermal Sensors

Two types of in situ ablation recession and thermal sensor were developed by Koo and colleagues [80–88] based on break-wire concept using ultra-miniature thermocouples (TCs): one sensor was based on 0.25 mm diameter TCs [80, 81] and the other sensor was based on 0.55 mm diameter TCs [80, 81]. These sensors were demonstrated using low-density carbon-carbon (LDCC) and high-density carbon/carbon (HDCC) composites and carbon/phenolic composite. Different configurations: four-levels

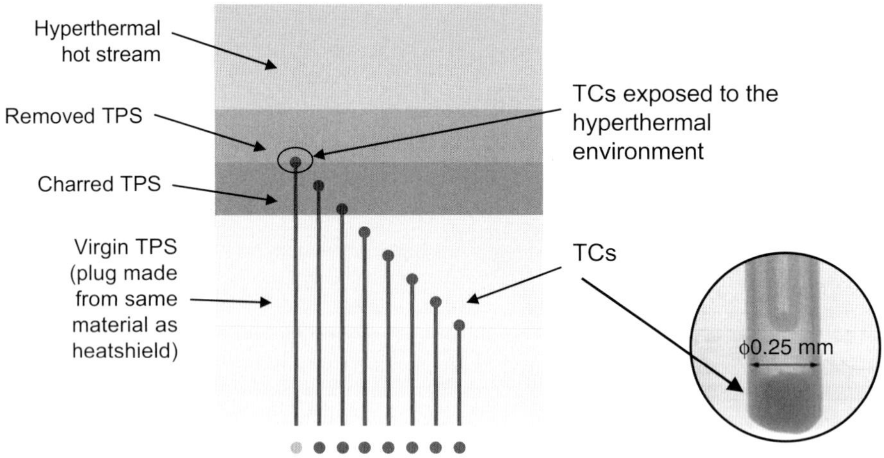

Figure 6.53. Scheme of the proposed thermocouple-based recession sensor.

(four TCs) sensors, eight-levels (eight TCs) sensors, and nine-levels (nine TCs) sensors were designed, fabricated, and tested under high heat flux (1,000W/cm^2) via the oxyacetylene test bed. Representative results of this novel ablation and thermal sensing technique are discussed in the section.

6.13.5.1 The 0.25 mm Diameter TC Ablation Sensor [80, 81]

A break-wire-like recession sensor based on commercial, ultrafine thermocouples (250 μm diameter) was designed and tested. A series of thermocouples (TCs) are embedded in the thermal protection systems (TPS), perpendicularly to the ablator surface (Figure 6.53). Each sensing head of the TC is positioned at a well-established depth from the surface. During the heating of the ablator, the TCs would first work as a temperature sensor providing invaluable information about the state of the TPS. When the temperature rises above the melting point of the metal sheath and of the Seebeck junction, the TC would break. Due to this dual nature of the sensing heads – as a traditional Seebeck junction and as a position marker – it would be possible to obtain a wide range of data on the recession state of the TPS. This ablation recession sensor was tested on carbon/carbon (CC) composites under a severe hyperthermal environment produced by an oxyacetylene test bed (OTB). Two types of CC materials with different densities (LDCC=1.34 g/cc and HDCC=1.70 g/cc) were considered. At the same time, two different configurations were tested (4 levels and 8 levels). The obtained results were quite encouraging showing the TC-based approach can provide accurate data on the recession rate of the TPS material. Due to the uniqueness of the proposed sensing technique, it can be applied both on TPS of spacecraft as well as on rocket nozzles. Moreover, the proposed approach uses low-cost, commercially available TCs, and industry-standard processing techniques.

Materials – The sensing technique was tested on 15 × 15 × 15 mm carbon/carbon cubes (CCCs). Two types of carbon/carbon materials were considered: a low-density material (LDCC) and a high-density material (HDCC). These cubes were machined with a

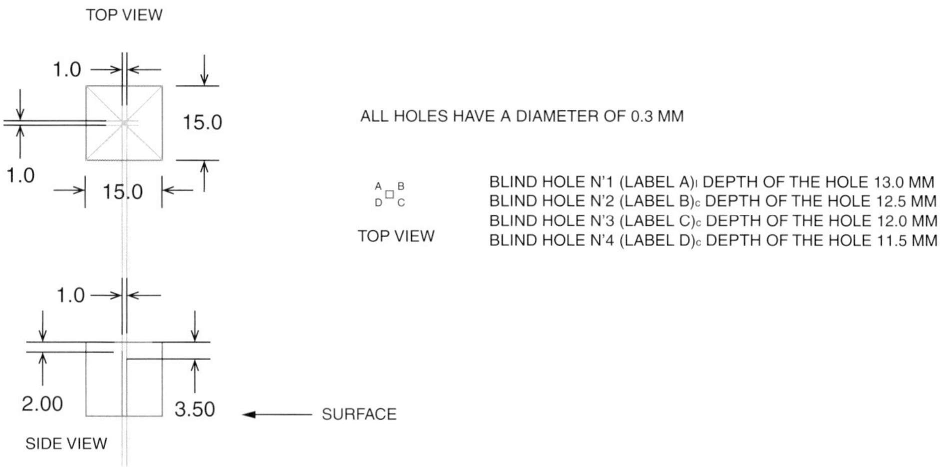

Figure 6.54. Drilling layout of the CCC plug.

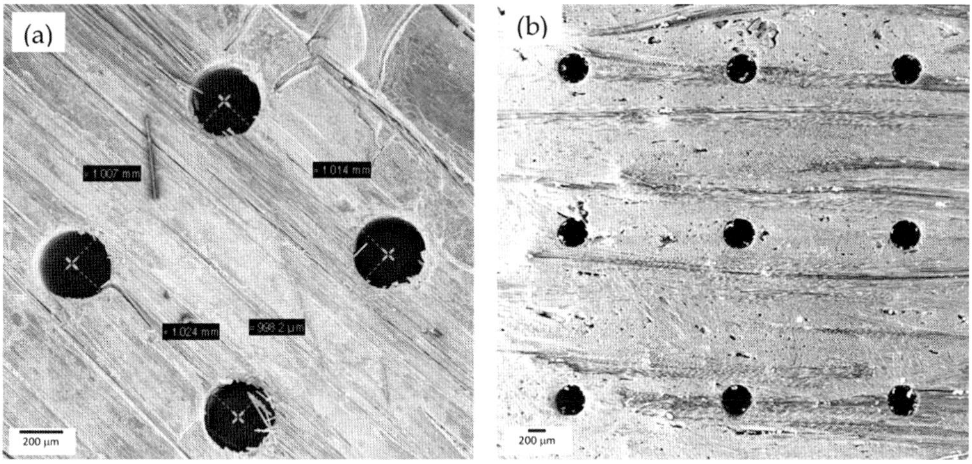

Figure 6.55. SEM of EDM-drilled holes of the four-hole configuration (left) and nine-hole configuration (right) of the carbon/carbon composites.

waterjet cutter starting from a slab of carbon/carbon composite. Miniature K-type TCs having a stainless steel sheath and with an outer diameter of 0.25 mm were chosen.

Production of the Sensor Plug – First of all, it was necessary to identity a suitable technology to drill the CC plug (a cube) to produce a series of blind holes at an increasing depth from one of the faces of the cube. Using blind holes and knowing the depth at which the TC must be inserted, a defined hole arrangement can be produced (Figure 6.54). The drilled samples were produced using two techniques: a laser source and electrical discharge machining (EDM).

In the first part of the research, in order to keep down the costs, only four levels of depth were considered. This means that only four TCs were embedded in the CCC plug. Figure 6.54 shows a detailed technical drawing of the produced plug. Compared to the laser drilling, the EDM produced holes with a superior quality. Figure 6.55

Table 6.2. *Theoretical and Real Depths of the Produced Blind Holes*

Hole tag	Theoretical depth of hole (mm)	Actual depth (mm)
1	13.00	13.15 ± 0.05
2	12.50	12.55 ± 0.05
3	12.00	12.15 ± 0.05
4	11.50	12.20 ± 0.05

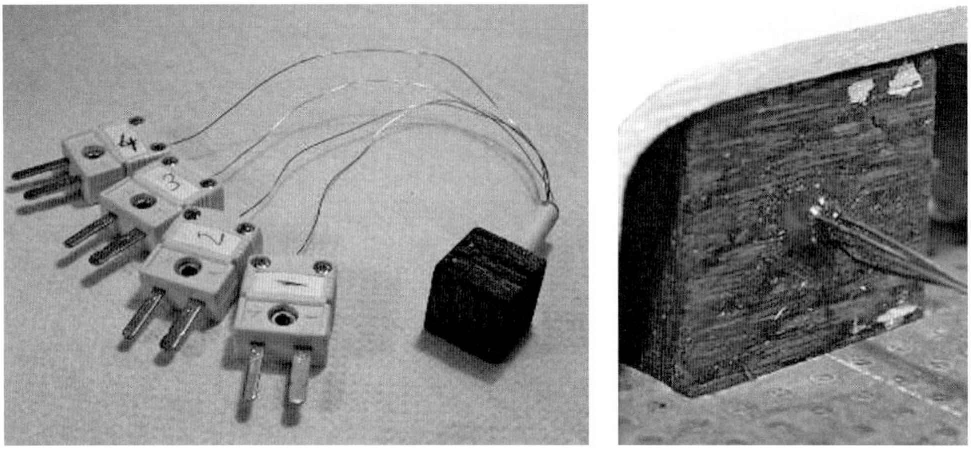

Figure 6.56. Fully assembled CCC sensor plug (left). Details of the cold face of the CCC plug (right).

shows an SEM image of a CCC drilled by EDM. It clearly shows the high quality of the produced holes. After receiving the sample from the drilling company, the actual depth of the produced holes was checked by means of a caliper coupled with gage pins. On the basis of the experimental results it was concluded that the produced holes had the depths reported in Table 6.2.

Table 6.2 shows the theoretical and actual of holes 1 to 3 matched; the actual depth of hole number 4 did not match the required depth. Since most of the holes in the sample were drilled with the right depth, and due to economic considerations, the sample was used: the CCC was considered adequate to test the sensing technique. Once the holes were drilled in the cubic sample, the TCs were glued into CCC plug using a high-temperature adhesive suitable for carbon/carbon composites [80, 81].

A special sample holder designed to ensure a correct positioning of the TCs during the curing of the adhesive was designed and produced. Figure 6.56 (left) shows the fully assembled CC sensor. Also, the details of the cold face of the plug are shown in Figure 6.56 (right).

Sensor Calibration Results – Prior to testing the recession-sensing technique with the OTB, a nondestructive test was carried out in order to obtain some preliminary data on the response of the CCC plug. For this test, the hot face of the sample was heated with a low-power burner. Figure 6.57 represents a typical temperature

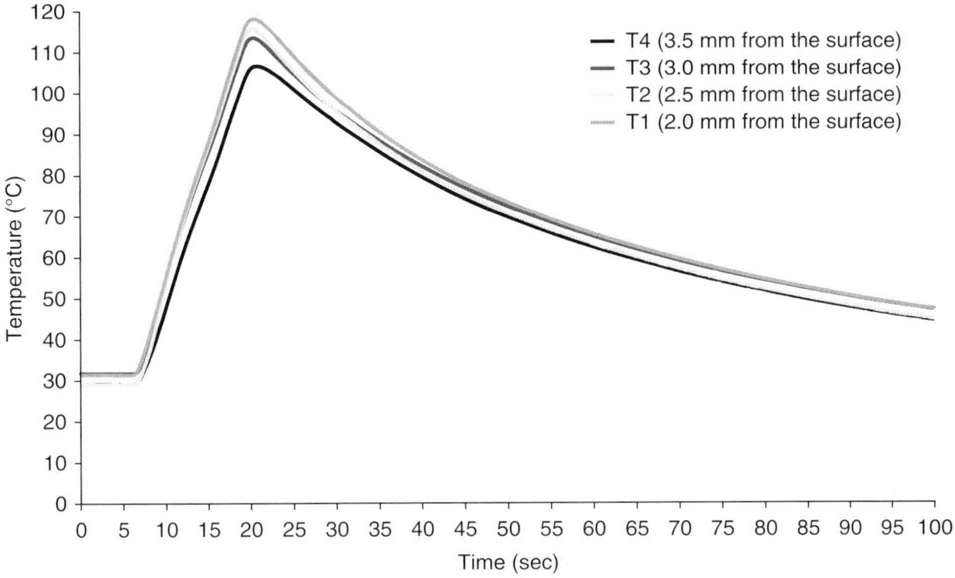

Figure 6.57. Temperature profiles acquired during the heating of the hot face with a low-heat source.

Figure 6.58. CCC tested on the OTB (left). Burnt CCC immediately after the test (right).

profile obtained with this heat source. According to the experimental results, the TC embedded in the deepest hole (closer to the exposed surface) exhibited the higher temperature. In general, during the heating up of the sample, the hierarchy of the displayed temperature profiles exhibited a full dependence on the hole depth: the greater the depth of the hole, the higher the temperature measured.

Ablation Test Results – Figure 6.58 (left) shows the test of the CCC plug and, on the right, the post-burning appearance of the sensor.

The temperature profiles acquired during the experiment are shown in Figure 6.59. All temperature patterns displayed a behavior characterized by three common steps. In this first part (see Figure 6.59), all TCs properly worked as a

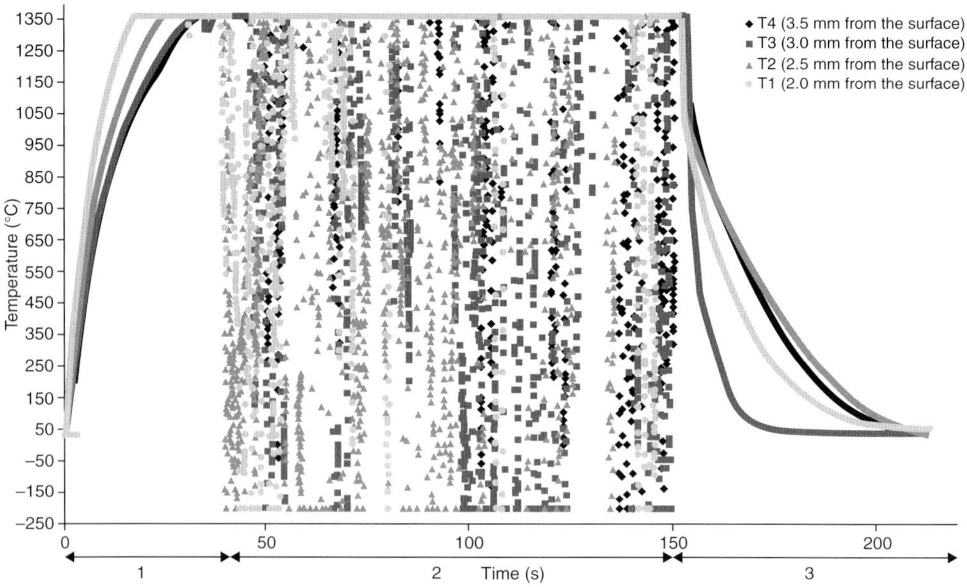

Figure 6.59. Temperature profiles of typical four-level ablation sensor.

temperature sensor. Similar to results obtained during the preliminary study, the smaller the distance of the TC from the surface, the higher the temperature displayed. The data in this first step of the temperature profiles showed that the signals of TC 3 and 4 were very similar: this evidence could confirm that the depth of these two holes was very similar as evidenced by the measure of the hole depths reported in Table 6.2. Unfortunately, once a temperature of 1,360°C was reached by each TC, the DAQ cut the data, which did not allow continued recording of the signal from the TCs. After about 40 seconds, the Seebeck junction (SJ) of the first TC experienced a break: after this period of time, the steel sheath of the TC melted and then the junction directly exposed to the flame lost its integrity, causing a loss of electric contact between the two terminals of the TCs. Consequently, this point marked the exact time at which the TC1 was broken.

The strong discontinuity in the temperature profile clearly marked this event. Figure 6.60 shows the detail of the temperature profiles in the range of time in which all TCs experienced the first break. Starting from the first break of the TC1, the Seebeck junction started to work improperly, providing an intermittent signal. This phenomenon can be explained as follows: due to the presence of the flame, after the TC experienced the first loss of electrical contact, the terminals of the TC continued to melt. These droplets of melt metal were able to close the electrical contacts between terminals, producing a detectable signal. Also, the steel sheath of the TC provided a source of liquid metal able to close the electrical loop. This process was regulated by several uncontrollable factors producing a randomly distributed opening and closing of the junction.

The above-described occurred for all four TCs but, due to their different displacement (depth) in the CC plug, the first break of each TC happened at different times. The working principle of this recession sensor is exactly based on this behavior. The Seebeck junction of the first TC exhibited the loss of integrity after about 39.3 seconds from the beginning of the experiment. The TC2 broke less than one second later than TC1. TC3 and TC4 broke at about 46.5 seconds and 50 seconds, respectively, from the beginning of the test. A possible reason the first two TCs broke at nearly the same time could be that the Seebeck junctions embedded in the metal sheath had a different position with respect to its own outer case. In this case, even if the two TCs were placed at exactly the desired depth into the plug, the actual position of the junctions with respect to the surface of the CCC may not be correct. If this hypothesis is correct, the two TCs would break at nearly the same time. In any case, since commercial TCs were used, there was no way to control the positioning of the SJ in the sheath of the TC: in fact, this parameter strictly depends on the manufacturing process of the thermocouples.

However, this situation does not represent a problem, since the proposed recession-sensing technique is based on the use of many TCs placed at different depths. Moreover, placing different sensing plugs in TPS could be possible to minimize the problem related to the manufacturing tolerances of the TCs. Another possible reason that the first two TCs broke at nearly the same time could be related to local non-uniformity of the erosion. In any case, even considering the response of the first two TCs that broke very close one another, the system was shown to work properly. In fact, the TCs generally broke accordingly to their distance from the surface marking the recession layer of the material. The first break of each TC precisely indicated the recession.

In order to further understand the real behavior of the employed TCs embedded in the CCCs when exposed to a severe hyperthermal environment, the exposure time was extended up to 150 seconds. According to Figure 6.60, it is possible to see that once the TCs experienced the first break, all of them continued to display the previously mentioned intermitting behavior (second step of the graph in Figure 6.60). Finally, in the third step of Figure 6.60, the TCs display the cooling down of the system. However, once broken and open, each original Seebeck junction was altered by the presence of some unknown material between the two terminals of the thermocouple (original composition of the K-TC: chromel [90% nickel and 10% chromium] – alumel [95% nickel, 2% manganese, 2% aluminum and 1% silicon]), thus leading to a practically meaningless reading of the temperature.

In order to get a quantitative evaluation of the real erosion rate provided by the sensor, the actual position of the spatial markers – the TC heads – was plotted as a function of time in which it experienced the first break. This calculation was carried out in two ways. In the first method, the four positions of each TC were considered and correlated with the time of the first break (Figure 6.61). Using this method, the sensor provided a measured recession rate of 0.07128 ± 0.00548 mm/sec. Comparing this data with the value measured on the neat CC samples, it was possible to establish that the sensor provided a very good indication of the actual recession rate of the material. In fact, the erosion rate estimated on the neat CCCs was 0.056 ± 0.007 mm/sec. In the second approach, the recession rate (RR) was calculated considering the

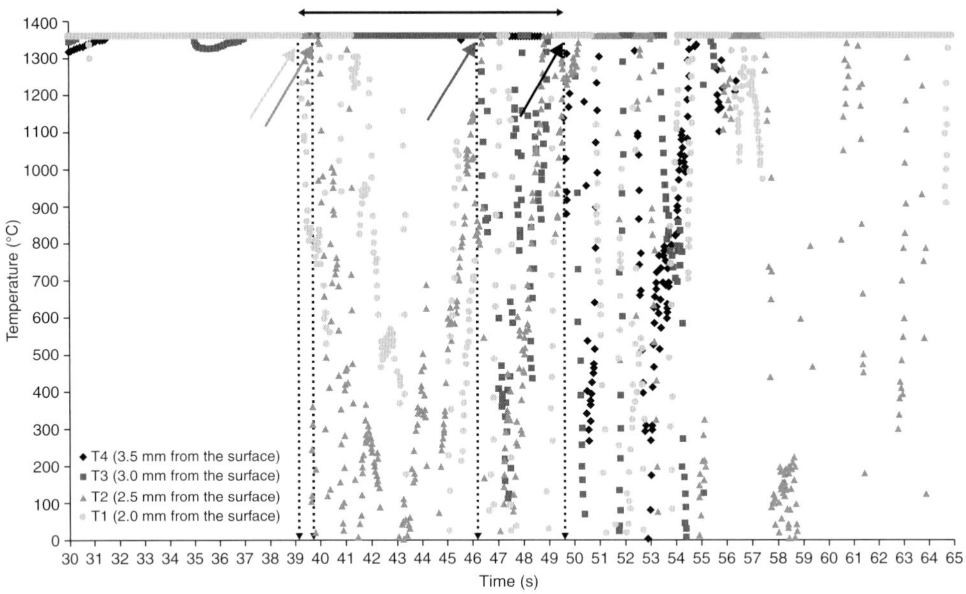

Figure 6.60. Detail of the temperature profiles in the range of time in which all TCs experienced the first break.

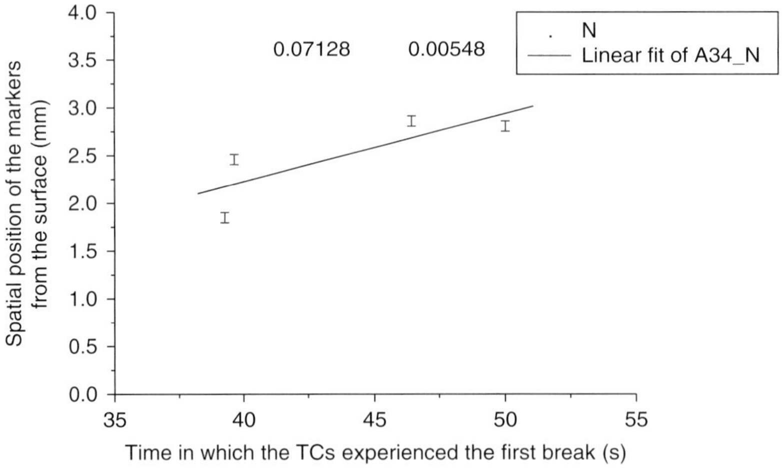

Figure 6.61. Position of the markers versus time of the first break of each marker.

(0; 0) point – in other words, as if there was a virtual TC placed on the surface of the plug experiencing a breaking at the time 0 sec. In this case, the plotting indicated that the recession rate was about 0.05751 ± 0.00124 mm/sec (Figure 6.62), with an impressive agreement with the value found for the neat CCCs. In any case, even considering the limited number of the available experimental points, both approaches showed that the sensor was able to provide a good evaluation of the erosion rate.

After the ablation testing of the sample was conducted, visual inspection of the tested CCC clearly showed the presence of the TC heads on the bottom of the crater produce by the torch (Figure 6.63).

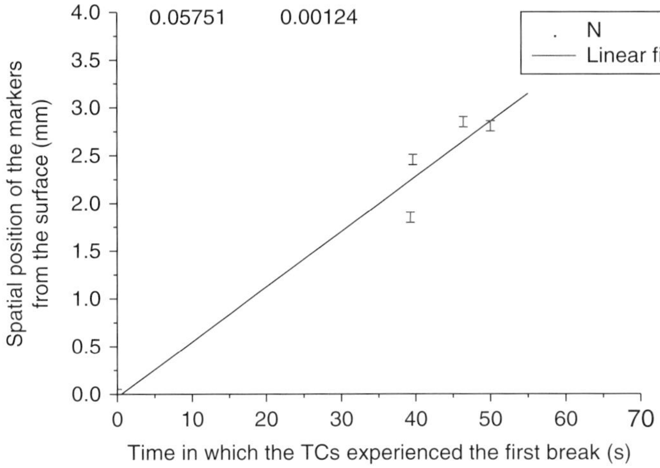

Figure 6.62. Position of the markers versus time of the first break of each marker including (0,0) point.

Figure 6.63. Post-burning image of the CC plug: the arrow indicates the heads of the four TCs.

SEM images were taken of the deepest part of the erosion crater on the CCC sample (Figure 6.64). The first picture (top) clearly shows that each TC head was covered by a layer of material probably made of the steel of the sheath. However, increasing the magnification (bottom), it was possible to see that the cover of each TC was composed of some metal blend in which some chopped carbon fibers were also embedded. Such a layer of electrically conductive material should be responsible of the electrical contact between the two TC terminals.

After all these tests, the adhesion of the burnt TCs on the CCC was also qualitatively evaluated. This test was carried out considering the pullout resistance of

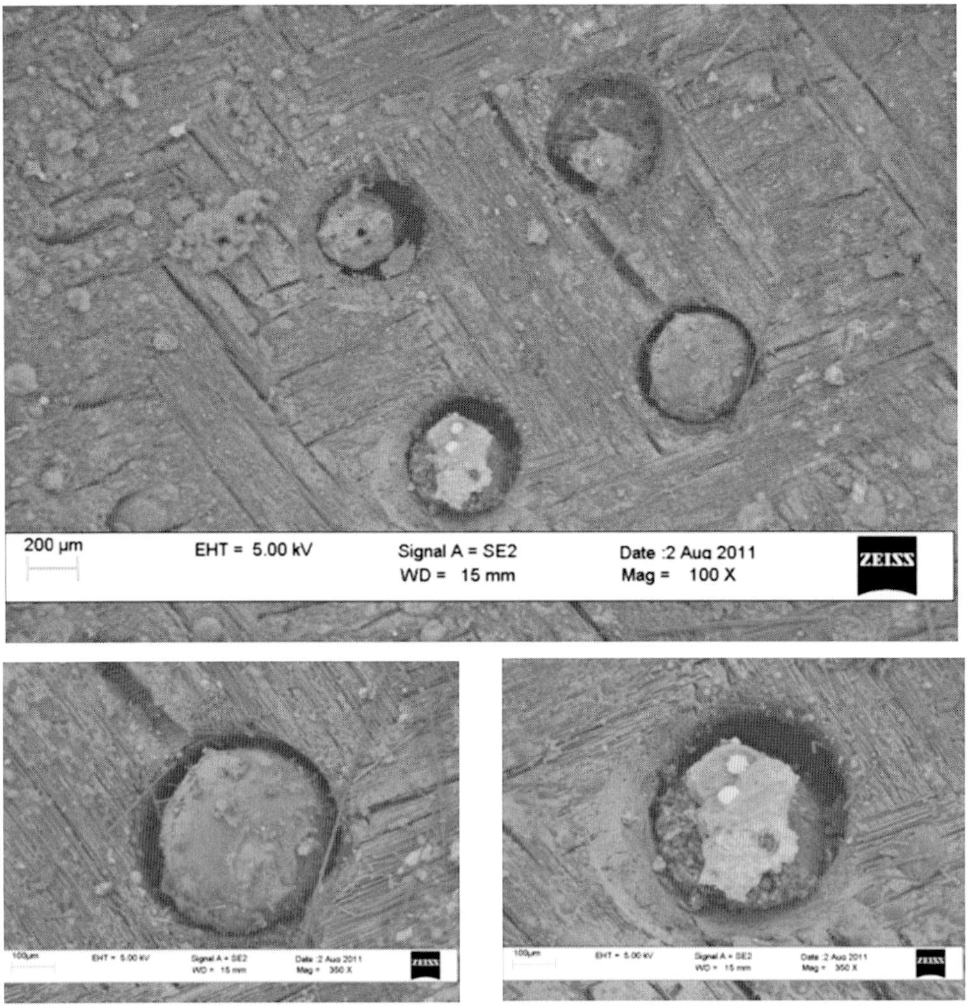

Figure 6.64. SEM images of the bottom of the CC plug.

the TCs. Preliminary results indicated that the TCs exhibited an appreciable pullout resistance. This indication is very important given that, during atmospheric reentry, the role played by the TPS cannot be compromised by the presence of the TCs. In any case, the pulldown resistance of the burnt TCs can be increased with some dedicated designing criteria and a proper holder of the TCs on the cold side of the heat shield. Additional test results for this sensor can be found in [80, 81] and in Chapter 10.

Summary and Conclusions – The proposed system was able to monitor the recession rate of a TPS employing a simple working principle, combined with an inexpensive and commercially available raw materials and processing techniques. It provides real-time monitoring of the ablation rate of an ablative material, and even though the method considered here is an intrusive system, the use of ultrafine thermocouples (250 µm) requires a limited modification of the TPS. To date, due to the selected drilling technique (EDM) used to produce the blind holes in which the thermocouples are hosted, the proposed approach can be applied only to electrically conductive materials, such as carbon/carbon, due to the nature of the EDM drilling technique.

During exposure to the torch, the TCs experienced multiple electrical breaks. The detection of the recession can be strictly related to the first break of each Seebeck junction, which depends on the depth at which the TC is embedded. After this event, the electrical signal provided by the broken thermocouple can be filtered by using some dedicated software able to neglect the following breaks. Table 6.3 summarizes the results of all tests performed on the carbon/carbon plugs. It can be seen that both the LDCC- and HDCC-based plugs using this technique were able to provide a very good real-time evaluation of the ablation rate [80, 81]. These data also indicated that the technology is easily scalable and extremely flexible.

Once combined with a proper hardware/software, this proposed system can provide precious data for ground testing as well as obtaining flight data used as a feedback for reentry TPS application. Moreover, the acquired data could be used to improve the comprehension of ablation mechanism of the ablative material development for ground testing, especially when combined with the data provided by other in situ sensors. Finally, the proposed technique can also be exploited in other fields of the aerospace industry: as an example, it could be employed to monitor the erosion rate of the nozzle throats and nozzle entrance region and exit cone of solid rocket motors. More detailed results can be found in [80–81] and in Chapter 10.

6.13.5.2 The 0.55 mm Diameter TC Ablation Sensor [82–88]

Two types of in situ ablation recession and thermal sensor were developed by Koo and colleagues [80–88] based on break-wire concept using ultra-miniature thermocouples (TCs): one sensor was based on 0.25 mm diameter TC [80, 81] and the other sensor was based on 0.55 mm diameter TC [82, 83]. Earlier in situ ablation sensor designs developed by Natali and Koo involved the use of four 0.25 mm diameter thermocouples and EDM drilling technique [80, 81]. Later designs used four 0.55 mm thermocouples and conventional micro-drilling in order to significantly reduce the cost and lead time of manufacturing each sensor. Further designs involved the use of nine 0.55 mm thermocouples in addition to reduced distances between thermocouples. After manufacturing three samples, testing using the OTB gave results that were compared with data obtained using the 0.25 mm micro-thermocouple design. The results showed that unlike the 0.25 mm micro-thermocouple, the 0.55 mm thermocouple existed well outside of the ablative material and did not disintegrate immediately upon exposing to the flame. Statistical analysis was used to determine the approximate break temperature for each thermocouple. The break temperature of 1,432°C was determined to match well with the sensor-calculated ablation rates and the manually measured average ablation rates. Three more samples were constructed and tested in order to confirm that 1,432°C was a suitable break temperature, in order to achieve repeatable results with the carbon/carbon composite sensor plugs. It was determined that through the 0.25 mm micro-thermocouple, crisp data resolution was provided, since the thermocouples broke upon exposure to the

Table 6.3. *Final Results for the Four- and Eight-Levels Sensors*

Plug tag	# TCs	Material	First approach RR* (mm/s)	First approach mean (mm/s)	First approach *SD* (mm/s)	First approach *mSD* (mm/s)	Second approach RR** (mm/s)	Second approach mean (mm/s)	Second approach *SD* (mm/s)	Second approach *mSD* (mm/s)	MARR***/° (mm/s)	*SD* (mm/s)	Notes
cccld1	4	LDCC	0.0713	0.0645	*0.0055*	*0.0038*	0.0575	0.0576	*0.0012*	*0.0012*	0.056	0.007	°
cccld3	4	LDCC	0.0611		*0.0028*		0.0552		*0.0011*				
cccld4	4	LDCC	0.0610		*0.0030*		0.0602		*0.0012*				
ccnew1	4	HDCC	0.0137	0.0262	*0.0008*	*0.0013*	0.0229	0.0251	*0.0005*	*0.0005*	0.023	0.005	°
ccnew2	4	HDCC	0.0387		*0.0017*		0.0273		*0.0005*				
34	8	HDCC	0.0510	0.0479	*0.0008*	*0.0024*	0.0336	0.0327	*0.0004*	*0.0004*	0.043	0.003	§
42	8	HDCC	0.0421		*0.0057*		0.0308		*0.0003*				
44	8	HDCC	0.0506		*0.0008*		0.0335		*0.0004*				

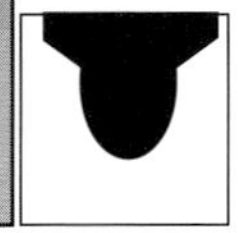

* erosion rate calculated with the first approach
** erosion rate calculated with the second approach
*** MARR : mean erosion rate measured as a function of the final depth of the crater
° measured on a 4mm depth crater
§ measured on a 8mm depth crater

environment. The 0.55 mm thermocouples utilizing break temperature analysis provided the ability for the larger thermocouples to obtain higher-resolution recession rates. Three more samples using nine thermocouples were fabricated and tested in order to determine if a higher thermocouple count would compromise the structural integrity of the plug. Additionally, four samples were fabricated and tested using carbon/phenolic as the base ablative material in place of carbon/carbon composite. In this section only the second round of four levels C/C sensor results are presented; additional results can be found in [82, 83] and in Chapter 10.

Design Specifications – The sensor design was intended to be similar enough to the setup used by Natali and Koo [80, 81], so that the data gained from testing would be easily comparable. The ablation sensor developed by Natali and Koo consisted of four ultra-miniature thermocouples installed at staggered depths from the back of the C/C plug [80, 81]. As the plug is heated, the thermocouples provide data about the temperature distribution within the plug, and finally break when the ablation front reaches the thermocouple and destroys the junction in the tip of the thermocouple. By knowing the depth of the thermocouple and the time at which it broke, the position of the ablation front and rate of ablation can be calculated.

The C/C plugs have dimensions of 15 mm^3, so that they could be mounted and used with the OTB. The thermocouples had to be larger than 0.25 mm but small enough to fit four within the area under the flame generated by the OTB's flame envelope. The deepest thermocouple hole to be drilled was 13 mm deep, 2 mm from the front surface of the plug, and each following hole was to be shallower by 0.5 mm. The holes that house the thermocouples had to be created through conventional drilling. All parts had to be commercially available. The primary purpose of these constraints was to determine whether the structural integrity of the C/C would be significantly affected during testing by the increase in thermocouple diameter from the 0.25 mm version. Additionally, the larger thermocouple size and manufacturing requirements would greatly reduce the cost of producing the sensor plugs.

Production of 0.55 mm TC Sensor – All production had to be accomplished in-house at The University of Texas at Austin (UT) for cost saving and shorter lead time. The actual manufacturing of the sensor systems consisted of two major steps: using subtractive manufacturing processes to prepare the thermal protection material to hold the thermocouples at the proper depth, and installing the thermocouples with the use of a durable high-temperature adhesive.

A. Design – The spacing between each drilled thermocouple hole must be small enough so that during testing the oxyacetylene test bed applies its heat flux on each thermocouple equally. The OTB torch acts on a very fine area that is roughly three millimeters in diameter. As seen in Figure 6.33, the oxyacetylene torch acts on the sample in a cone. By keeping the holes closer together, the same heat flux is applied to all four thermocouples and the thermocouples remain at the deepest point of the ablation area.

However, the holes cannot be spaced too close together for two important reasons. First, the hole spacing must reflect the increasing thermocouple size. If the holes are too close, the structural strength of the ablative material could be weakened enough such that the walls between the thermocouples collapse, resulting in a significant weak point that may be blown through by the environment. Second, the hole spacing must be large enough to prevent thermocouple cross-talk within the conductive C/C sample in order to reduce signal noise during testing. After testing, the shield for each thermocouple is constituted by a metal blend embedded with the carbon fibers of the sample. Since C/C is a conductive material, the effectiveness of the shielding is reduced. By maintaining enough distance between each thermocouple, cross-talking between each thermocouple will not occur.

This design will maintain similar dimensions as the EDM drilled sample from the tests performed by Natali and Koo [80, 81]. With the larger 0.5 mm thermocouples, this resulted in hole spacing of roughly 2 mm. The drill size and hole diameter were determined by the amount of clearance needed between the thermocouple and the hole wall. Figure 6.65 shows the dimensions of the final working print for the sensor plugs utilizing four thermocouples.

Originally the holes in the C/C samples were to be drilled by EDM, but microscale drilling was used instead due to longer lead time and additional funds required in offsite EDM micro-drilling. Having already determined the thermocouple hole size for the samples, a methodology for manufacturing the test samples in a fast and reliable way has to be developed.

B. Fabrication – The C/C acquired from South Korea was extremely tough and had an average density of 1.7 g/cm^3, which is roughly 50% higher than the C/C used in the original micro-thermocouple tests by Natali and Koo [80, 81]. In order to improve manufacturing speeds, improve accuracy, and reduce the risk of drills breaking down-hole, a three-step process was developed to produce the holes.

The carbon/phenolic was significantly more difficult to drill because the phenolic would melt slightly during drilling. Melted phenolic would stick to the drill bit, resulting in excessive torque on the drill bit and breakage when attempting to drill deeper than 13 mm. Spindle speeds and feed speeds were reduced in an attempt to prevent the heating of the carbon/phenolic, though this did not prove successful and drill bits continued to break inside the samples. The thermocouple depths were reduced by 0.5 mm each on the carbon/phenolic samples in order to carry on with the tests [76, 77].

Experimental Setup – The burn tests involved using the UT OTB to generate a heat flux of 940 W/cm^2 to erode the C/C plug. The heat flux of 940 MW/m^2 was used by Natali in his original tests, so the tests were done at the same heat flux in order to make sure that the data would be easily comparable with the Italian team's test results. The four thermocouples were connected to a DAQ, which in turn interfaced with a LabView program on a laptop. The temperatures of the thermocouples were recorded over the length of the burn test, which ranged from 90 seconds to 180 seconds.

The OTB torch as shown in Figure 6.33 was held at 6 mm from the sample in order to reach the proper heat flux as determined by calibration performed by Allcorn et al. [65]. A vice holds the brass sample holder, which in turn holds the C/

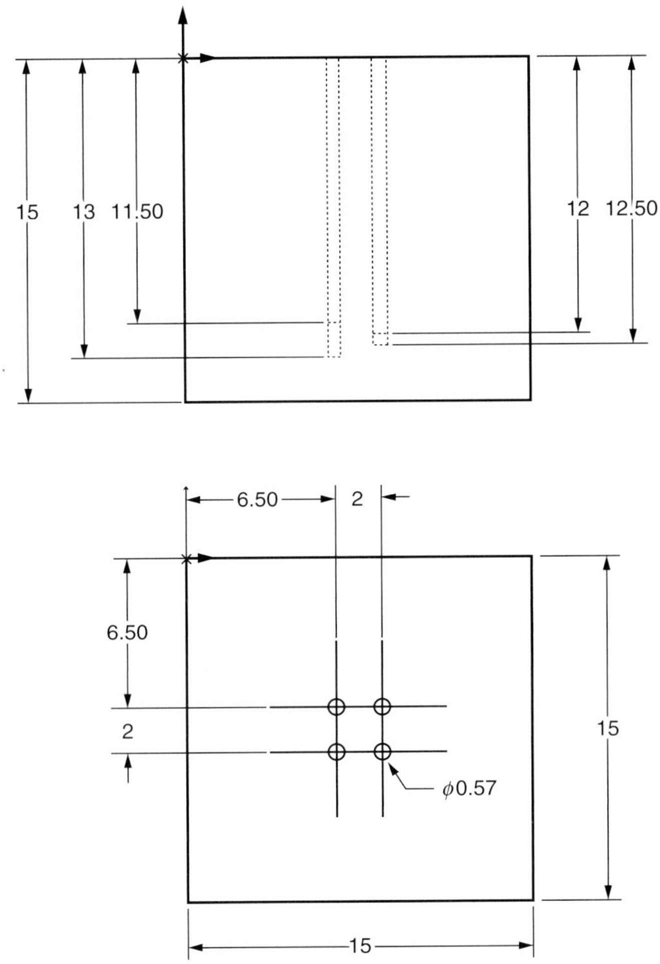

Figure 6.65. Side (top) and top (bottom) views of the sample. All dimensions are in mm for 0.50mm diameter TC C/C ablation recession sensor.

C sample in place during the test. The brass sample holder was originally designed for use with cylindrical samples where the thermocouples were not fixed inside the test sample, such as the copper slugs used to calibrate the heat flux. The holder was modified by cutting a slot into the side so that the thermocouples on the completed C/C samples could be properly inserted behind the sample. Additional holes were drilled and tapped so that cube-shaped samples could be securely held in place, as seen in Figure 6.66.

Ninety-second burn tests were performed on two virgin C/C samples in order to ensure that the OTB was calibrated to have the same heat flux as the OTB that Natali's team used. The first sample was composed of the same C/C that Natali used, while the second sample was made of the high-density C/C from South Korea. After the samples were burned, the crater depths were measured and the average ablation rates were calculated. The ablation rate of the first sample, 0.054 ± .007 mm/s,

Figure 6.66. The modified OTB sample holder to accommodate this C/C sensor.

matched the ablation rate of Natali's test sample, 0.057 ± .007 mm/s, within experimental uncertainty [83, 84]. The ablation rate of the second sample was found to be 0.032 ± .007 mm/s. This recession rate was used to determine a test time of 180 seconds based on the recession rate times the desired burn depth of at least 3.5 mm, which was necessary to burn all four thermocouples and show that the thermocouples could act in a similar capacity to break-wires.

Results and Discussion – Three rounds of three tests each were conducted using the testing procedures described above in order to demonstrate repeatability in using the TC-based ablation sensor. The results of the first round were not as expected and so a second round of testing was performed using the same setup. Between the first six tests, a consistent method for using the thermocouple-based in situ ablation sensor was demonstrated. The last three tests involved the use of plugs with nine thermocouples instead of four, and focused primarily on determining whether or not increasing the number of TCs would have an adverse effect on the structural integrity of the C/C plug. In this study, Phase I testing refers to tests utilizing the 4 TC configuration, while Phase II testing refers to the 9 TC configuration. Only the test results of Tests 4, 5, and 6 are discussed in this section; more detailed results can be found in [83] and Chapter 10.

A. *Results of Tests 4, 5, and 6* – In the second round (Phase I) of testing, the precautions listed above were taken into account. Despite these adjustments, the data were found to be similar to the first round of testing. The data from the second round of testing are shown in Figures 6.67 to 6.69. The average recession rate was calculated both with the thermocouples and manually using the same method as for round one. The recession rates as calculated by the TCs are shown in Figures 6.70 to 6.72. As with the first round of testing, it can be seen that the average recession rates calculated using both methods agree within uncertainty in Figure 6.73.

These results further demonstrate the validity of using a break temperature of 1,432°C. The results presented above demonstrate that using the thermocouple ablation sensor design with the break temperature method should yield consistent and reliable measurement of ablative recession rates. To demonstrate the use of this

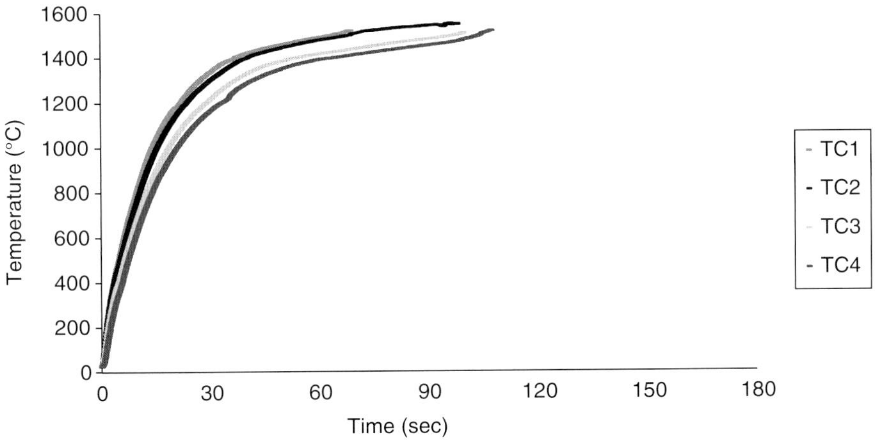

Figure 6.67. Test 4 temperature profile for 0.50 mm TC ablation C/C sensor.

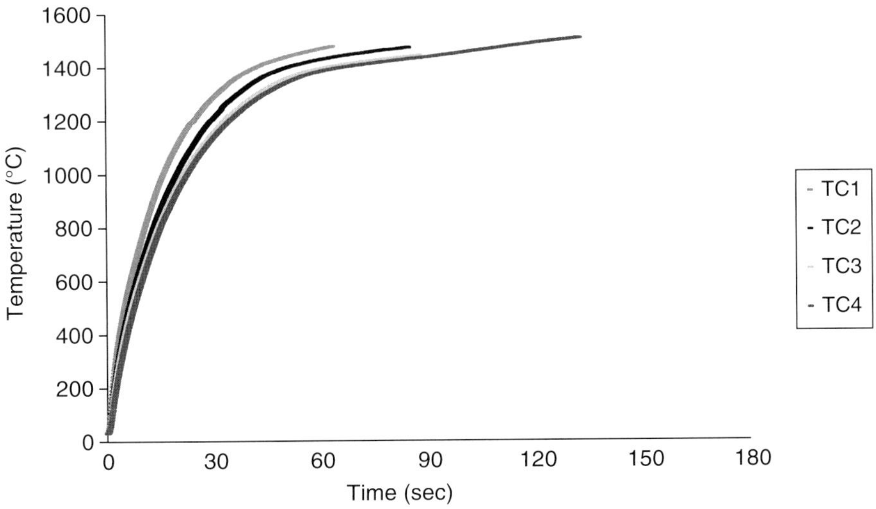

Figure 6.68. Test 5 temperature profile for 0.50 mm TC ablation C/C sensor.

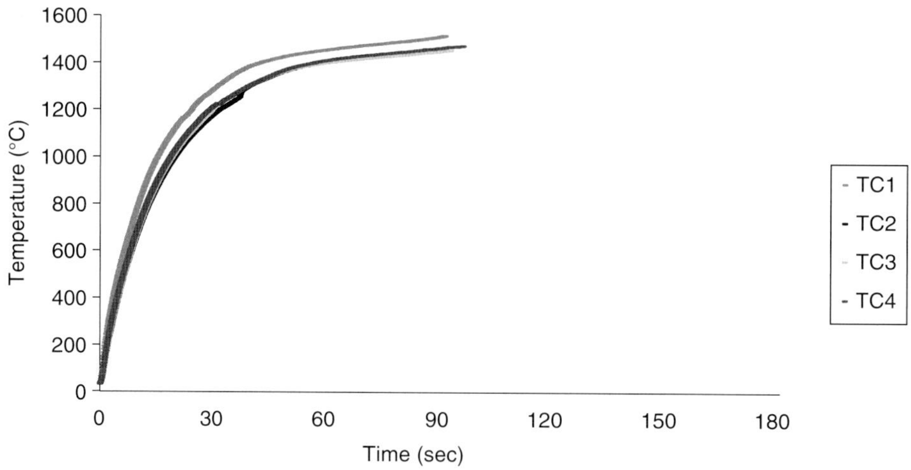

Figure 6.69. Test 6 temperature profile for 0.50 mm TC ablation C/C sensor.

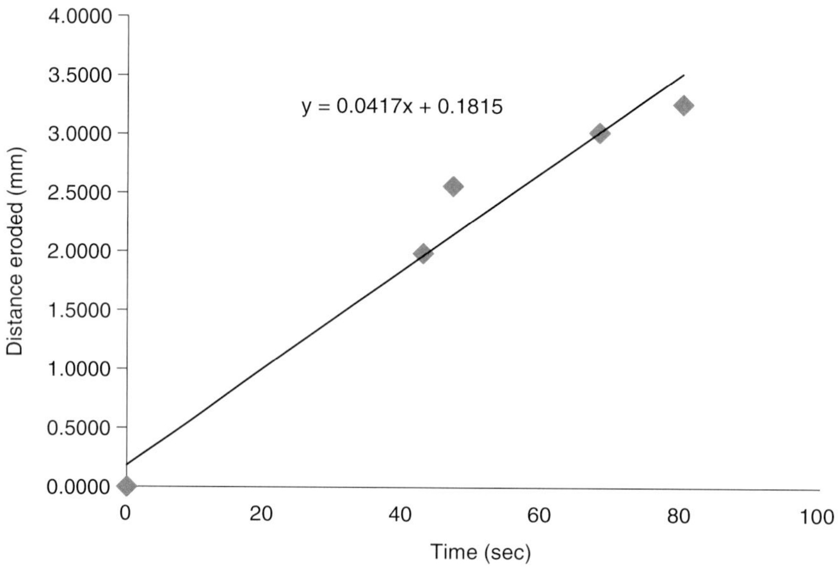

Figure 6.70. Erosion versus time of Phase I, Test 4.

method in practice, a graph of the average recession rate calculated at each thermocouple was prepared as shown in Figure 6.74.

In the real-world application, this ablation sensor and a computer can produce a graph of the average recession rate to make in-flight corrections using current state of ablation of the thermal protection system. Furthermore, it is believed that the thermocouple-based ablation sensor with the break temperature method would work with a large variety of TPS materials as long as the thermocouple remains constant, as the break temperature method only relies on the melting temperature of the thermocouple. Additional ablative materials using this sensing technology are currently being researched at The University of Texas at Austin [80–88].

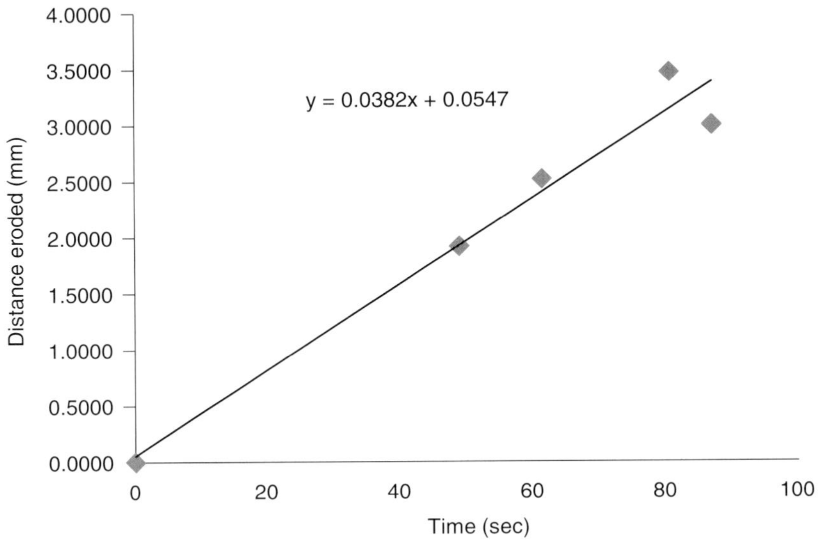

Figure 6.71. Erosion versus time of Phase I, Test 5.

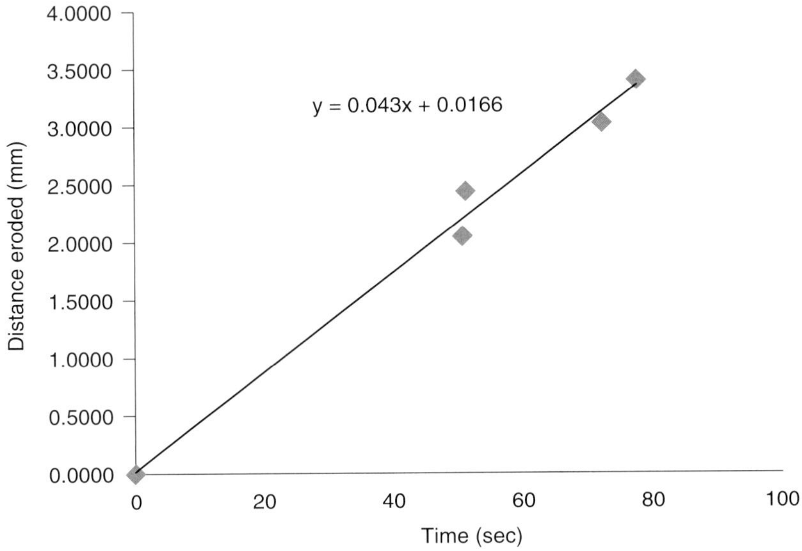

Figure 6.72. Erosion versus time of Phase I, Test 6.

Summary and Conclusion – The four 0.50 mm diameter thermocouple sensor design is an extremely cost-effective way to accurately measure the in situ recession rate of a material exposed in a hyperthermal environment when used with the cutoff temperature methodology developed, and the sensor does not compromise the structural strength of the original ablative material. Crosstalk between thermocouples is not an issue within conductive materials when shielded thermocouples are used. While utilizing more thermocouples may increase the resolution of measurement, it also significantly affects the structural integrity of the material. A different design

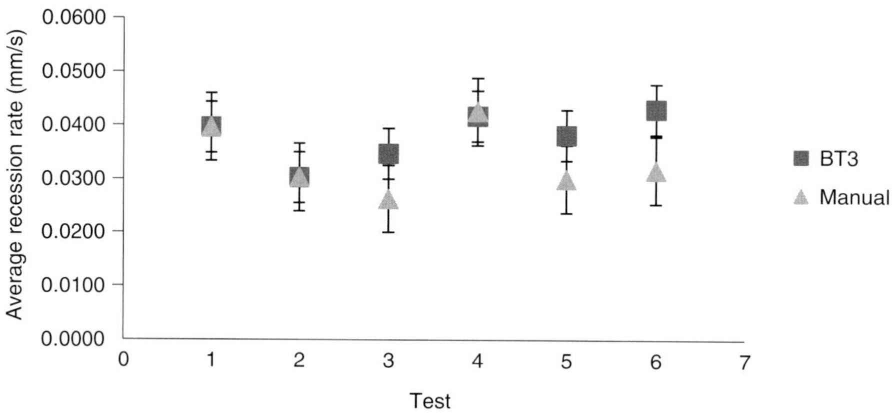

Figure 6.73. Average recession rate of samples at varying break temperatures.

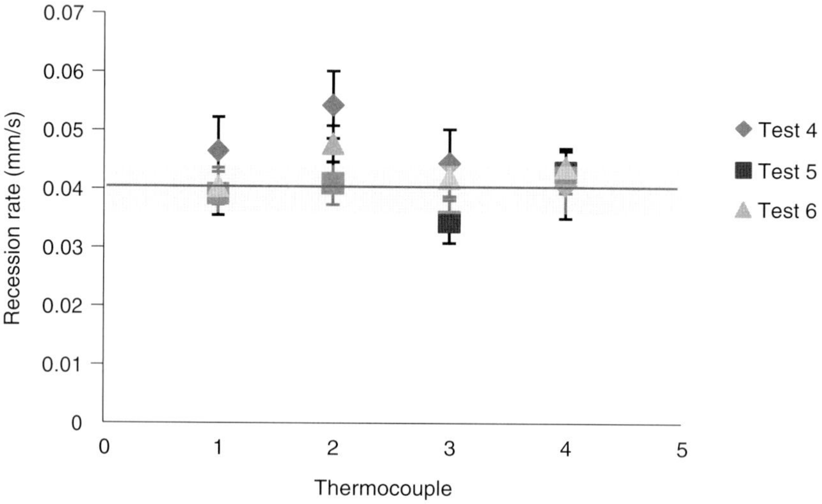

Figure 6.74. Recession rate at TC at 1,432°C for Tests 4 to 6.

will need to be developed in order to properly utilize more thermocouples without creating weak points in the ablative. Additionally, a different DAQ should be used to take measurements from large numbers of thermocouples, so as to prevent major signal noise problems. Additional test results of this sensor technology applying to PICA, AVCOAT, and SiC/glass ceramic matrix composites can be found in [80–88] and later in Chapter 10.

6.14 Electrical Properties

Electrical conductivity of polymer nanocomposite samples is measured typically by the four-point probe method. A commercial electrical conductivity setup, such as Hioki Megaohmmeter or specialized multimeter from Keithley, a U.S.-based company, can be used to measure electrical conductivity. For EMI shielding characterization, ASTM

D4935-10 [89] is used for most cases using network analyzer. However, the limitation for this method is that it requires a large sample size, which is a circular plate having a diameter of 6 inches (15.24 cm). The amplitudes of incident wave, transmitted wave, and reflected wave are used to determine the effectiveness of EMI shielding.

Additional test results of this technology can be found later in Chapter 11. A few conventional electrical properties for polymers are discussed by Shah [90], such as (a) dielectric strength (ASTM D 149, IEC 243-1), (b) dielectric constant and dissipation factor (ASTM D 150, IEC 250), (c) electrical resistance tests – volume resistance and surface resistance (ASTM D 257, IEC 93), (d) ARC resistance (ASTM D 495), and (e) EMI/RFI shield.

6.15 Optical Properties [91]

Optical spectroscopic techniques are widely used in the study of optical properties of nanomaterials and polymer nanocomposites. The different techniques are usually based on measuring absorption, scattering, or emission of light that contains information about properties of the materials. The commonly used techniques include electronic absorption (UV-vis), photoluminsecent (PL), infrared (IR) absorption, Raman scattering, dynamic light scattering, as well as time-resolved techniques, such as transient absorption and time-resolved luminescence [91]. These different techniques can provide different information about the molecular properties of interest. In this section, two common spectroscopic techniques are reviewed, with an emphasis on their principle of operation and instrument required to perform the experiment. The main objective is to explain how one can get useful physical information about the nanomaterials and nanocomposites under study from the optical spectrum measured experimentally.

6.15.1 UV-Visible Electronic Absorption Spectroscopy [91]

6.15.1.1 Operating Principle: Beer's Law

The basic operating principle of electronic absorption spectroscopy is based on the measurement of light absorption due to electronic transitions in a material. Since wavelength of light required for electronic transitions is typically in the UV and visible region of the electromagnetic radiation spectrum, electronic absorption spectroscopy is often known as UV-visible (UV-vis) spectroscopy. It is named electronic absorption spectroscopy because the absorption in the UV-visible regions involves mostly electronic transitions. Based on Beer's law, the absorbance, A, is related to the incident light intensity, I_o, and transmitted light intensity, I, concentration of a solution sample, c, path length of the sample, l, absorption coefficient, α, and molar absorptivity, ε, using the following equation:

$$A = \log I_o/I = \varepsilon l c = \alpha c \tag{6.3}$$

Both I_o and I can be experimentally measured, and thus A can be determined. If l and c are known, the absorption coefficient can be determined by equation (6.3).

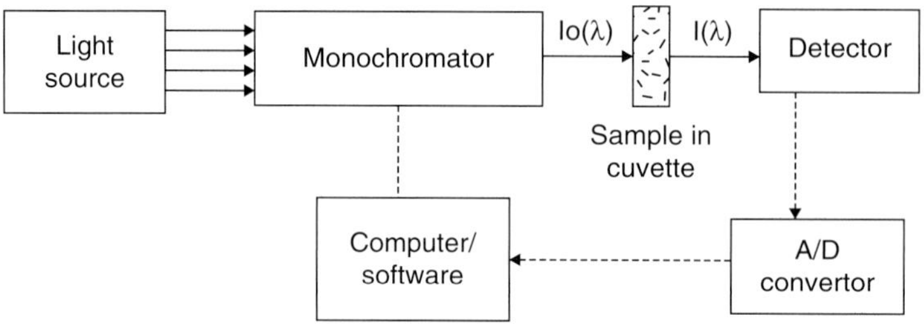

Figure 6.75. Schematic diagram of key components of a typical UV-vis spectrometer that includes a light source, monochromator to disperse the incident light, sample cuvette and holder, detector, analog-to-digital (A/D) converter, and computer with software to control the scan of the monochromator and data acquisition.

The absorption coefficient is wavelength dependent, and a plot of α as a function of wavelength λ is often characteristic of a given sample and reflects the fundamental electronic properties of the sample. For solid samples, concentration, c, has no meaning, and similar expression can be derived in terms of the thickness of the sample. The solid sample has to be thin enough to avoid saturation or absorption.

6.15.1.2 UV-Visible Spectrometer

UV-visible spectroscopy is one of the simplest and most useful techniques for studying optical and electronic properties of nanomaterials. This technique is based on the measurement of light absorption by a material, typically using commercially available spectrometers. Figure 6.75 shows the intensity of light from a light source, such as a lamp, is measured by a light detector, such as a photodiode, photomultiplier tube (PMT), or charge coupled device (CCD) detector, without (blank) and with a sample between the light source and detector [91]. If the sample absorbs light at some wavelength, the transmitted light will be reduced. The intensity of the transmitted light plotted as a function of light wavelength will provide a spectrum of the sample absorption. Most spectrometers cover the wavelength range from about 200 nm to 800 nm. Extending the measurement beyond 800 nm is possible but usually requires different light source, optics, and detector.

UV-vis spectrum is relatively easy to measure. There are still some common mistakes that researchers tend to make. First, the sample concentration cannot be too high. An optical density (OD) or absorbance (A), defined as $-\log_{10}(T)$ with T is the transmittance I/I_o, should usually be around 1 and not more than 3. Too high a concentration leads to saturated absorption and distorted, unreal spectrum. Second, proper blank needs to be taken before the sample spectrum is measured. If there are other species in the sample that could have absorption in the region measured but are of no interest, they should be part of the "blank" measurement. Third, one needs to make sure that the sample is clear – that there are no floaters in a solution sample that are visible to the eye. It will cause significant scattering and distort the measured spectrum. One common cause for floaters in nanomaterial solutions is

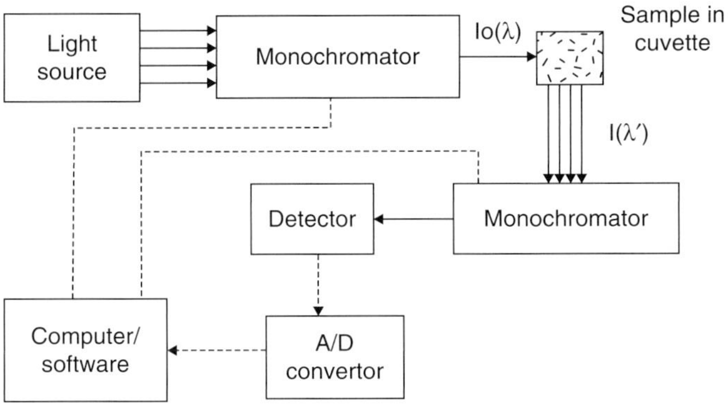

Figure 6.76. Schematic illustration of key components of a typical spectrofluorometer that include a light source, monochromator to disperse the incident light, sample cuvette and holder, a second monochromator to disperse the emitted light, detector, an analog-to-digital (A/D) converter, and computer with software to control the scan of the monochromators and data acquisition.

agglomeration or aggregation of particles to form large structures with dimension on the scale of a few hundreds of nm or larger and thereby visible to eye.

6.15.2 Photoluminescence and Electroluminescence Spectroscopy [91]

6.15.2.1 Operating Principle

The principle underlying photoluminescence (PL) spectroscopy is very similar to that of electronic absorption spectroscopy. They both involve electronic transition of initial and final states coupled by the electrical dipole operator. The main difference is that the transition involved in PL is from a higher energy level to a lower energy level. There is also an important practical difference between the two techniques in that PL is a zero background experiment – that is, no signed detected when there is no PL, which is in contrast to absorption spectroscopy that is a nonzero background experiment. Zero background experiments are intrinsically more sensitive than nonzero background experiments. Therefore, PL is typically more sensitive than an electronic absorption experiment.

A typical PL spectrum is a plot of the PL intensity as a function of wavelength of a fixed excitation wavelength, A, photoluminescence excitation (PLE) spectrum is a measure of PL at a fixed emission wavelength as a function of excitation wavelength. PLE is useful for studying samples for which electronic absorption spectrum is challenging to obtain, for example due to low transmission as a result of thickness or high concentration of the sample.

6.15.2.2 Spectrofluorometer

Figure 6.76 shows a schematic of the key components of a typical spectrofluorometer used for PL measurement. Several components are similar to that in a UV-vis spectrometer, including light source, sample cuvette, and detector, but the detector scheme is different in that emitted light from the sample is detected, rather than transmitted light in UV-vis spectroscopy. Rayleigh scattering, which has the same wavelength as the excitation light, should be avoided in the detection of PL. Raman scattering can show up in PL spectrum, especially when the Raman signal is strong while PL is relatively weak [91].

Basically, in a typical PL measurement, a specific wavelength of light is selected from a light source by a monochromator and directed at the sample of interest. Light emitted from the sample is collected through lenses, dispersed by another monochromator, and detected by a photo detector. The analog electrical signal generated by the photodetector is converted by software on a computer. The spectrum is displayed in terms of intensity of emitted PL light as a function of the wavelength of emitted light.

6.15.2.3 Other Optical Properties

Besides the aforementioned optical properties, Shah [92] suggested the following commonly used tests to measure optical properties for polymers: (a) refractive index (ASTM D 542, ISO 489), (b) luminous transmittance and haze (ASTM D 1003), (c) photoelastic properties, (d) color, and (e) specular gloss (ASTM D 2457, D 523). Readers should also refer to Zhang's recent textbook [91] for a more detailed discussion on the subject. Additional test results of this technology can be found later in Chapter 12 of this volume.

6.16 Other Properties

Besides the properties characterization mentioned throughout this chapter, other properties, such as weathering properties, chemical properties, analytical tests, conditioning properties, and a few other properties characterization can be found in Shah's *Handbook of Plastics Testing and Failure Analysis* [92] and later in Chapter 13.

6.17 Summary, Future Needs, and Assessments

No one "single" technique provides sufficient analytical information related to the structural characteristics of a polymer nanocomposite. More than one technique is recommended to understand the structures and properties of the polymer nanocomposites. It is time consuming and costly to perform all the necessary experiments and to accumulate all of the data to verify the development of each polymer nanocomposite formulation. One needs to be selective in the choice of analytical tools to be used in their studies.

There are other techniques that show promise for polymer nanocomposite analysis but that need to be validated, such as:

- Neutron scattering: it has similar issues to WAXD and also limited access
- Nuclear magnetic resonance (NMR): currently only works for clays with iron in the clay structure. Technique would need to be adapted for synthetic clays with no iron.

Polymer nanocomposites present a complex analytical problem. Properties of PNCs are dependent on many factors. No one analytical technique can analyze all of these factors. Thermal gravimetric analysis, wide-angle X-ray diffusion, and transmission electron microscopy are currently the major tools to characterize these materials. Combining these techniques will give us a better description and analysis of these PNCs. Each technique has its own advantages and shortcomings. To shorten the time to complete analysis on polymer nanocomposites, new analytical techniques are necessary to facilitate analyses more rapidly to reduce the time lag for complete analyses of PNCs. Furthermore, techniques that probe polymer-nanoparticle/polymer-organic treatment interactions are warranted.

6.18 Questions and Problems

1. Besides the characterization methods described in the chapter, identify other common characterization techniques used to evaluate nanomaterials and polymer nanocomposites properties and discuss the advantages and disadvantages of these characterization methods. Present your answers in a format similar to the one used in this chapter.
2. Describe the different methods that are commonly used to characterize the degree of dispersion of nanomaterials in a polymer nanocomposite material. Provide examples. Compare these characterization methods and discuss the advantages and disadvantages of these methods.
3. What are the limitations and benefits of each scanning method: optical microscopy; wide-angle X-ray diffraction (WAXD); scanning electron microscopy (SEM); transmission electron microscopy (TEM); and scanning tunneling microscopy (STM)?
4. What are some advantages and disadvantages of the WAXD method? How can d-spacing be used to determine the structural characteristics of polymer nanostructured materials?
5. Why is transmission electron microscopy (TEM) such a useful technique compared to other characterization methods? What are some methods for preparing TEM samples?
6. How is energy-dispersive X-ray spectroscopy used in conjunction with TEM?
7. What are some advantages that small-angle X-ray scattering (SAXS) provide in comparison to other methods?
8. Briefly describe how cone calorimetry is performed. How are the properties measured by the cone calorimeter to characterize the flammability properties of polymer nanocomposite materials?
9. Are there correlations and general trends for these industry test methods, such as cone calorimeter, microscale combustion calorimeter, limiting oxygen index (LOI), UL 94, and Stein tunnel test, for flammability properties of polymer nanocomposites?

10. Briefly describe the working principle of the in situ ablation recession and thermal sensor. On how many types of ablative materials have this technique been demonstrated?
11. Briefly describe the working principle of the char strength sensor to evaluate polymer nanocomposite chars?

REFERENCES

1. Z. L. Wang (Ed.) (2000). *Characterization of Nanophase Materials*. Weinheim, Germany: Wiley VCH, pp. 37–80.
2. N. Yan and Z. L Wang (Eds.) (2005). *Handbook of Microscopy for Nanotechnology*. Boston, MA: Kluwer Academic Publishers.
3. Z. L. Wang (Ed.) (2000). *Characterization of Nanophase Materials*. Weinheim, Germany: Wiley VCH, pp. 13–36
4. G. Cao (2004). *Nanostructures and Nanomaterials: Synthesis*. London: Properties & Applications, Imperial College Press, pp. 329–390.
5. G. L. Hornyak, H. F. Tibbals, J. Dutta, and J. J. Moore (2009). *Introduction of Nanoscience & Nanotechnology*. Baca Raton, FL: CRC Press, pp. 107–175.
6. A. V. Crewe (1970). The current state of high resolution scanning electron microscopy. *Quarterly Review of Biophysics* 3(1), 137–175.
7. P. Buseck, J. M. Cowley, and L. Eyring (Eds.) (1988). *High Resolution Transmission Electron Microscopy and Associated Techniques*. New York: Oxford University Press.
8. N. D. Browing, M. F. Chisholm, and S. J. Pennycook (1993). Atomic-resolution chemical analysis using a scanning transmission electron microscope. *Nature* 366, 143–146.
9. S. Y. Hobbs and V. H. Watkins (2000). Morphology Characterization by Microscopy Techniques. In *Polymer Blends, vol. 1: Formulation*. D. R. Paul and C. B. Bucknall (Eds.). New York: John Wiley & Sons, pp. 239–289.
10. J. H. Koo, H. Stretz, A. Bray, J. Weispfenning, Z. P. Luo, and W. Wootan (2004). Nanocomposite Rocket Ablative Materials: Processing, Microstructure, and Performance. AIAA-2004-1996 paper, *44th AIAA/ASME/ASCE/AHS Structures*, *Structural Dynamics, and Materials Conference*, Palm Springs, CA, April 19–22.
11. A. Guinier and G. Fournet (1955). *Small-Angle Scattering of X-Rays*. New York: Wiley.
12. O. Glatter and O. Kratky (1982). *Small-Angle X-Ray Scattering*. London: Academic Press.
13. A. Levine (1991). *Quantum Chemistry*, 4th ed. Upper Saddle River, NJ: Prentice Hall.
14. N. Jalili and K. Laxminarayana (2004). A review of atomic microscopy imaging systems: application to molecular metrology and biological sciences. *Mechatronics* 14, 907–945.
15. K. Nakajima, D. Wang, and T. Nishi (2012). AFM Characterization of Polymer Nanocomposites. In *Characterization Techniques for Polymer Nanocomposites*, V. Mittal (Ed.). Weinheim, Germany: Wiley-VCH, pp. 185–228.
16. S. A. Garea, and H. Iovu (2012). Following the Nanocomposites Synthesis by Raman Spectroscopy and X-Ray Photoelectron Spectroscopy (XPS). In *Characterization Techniques for Polymer Nanocomposites*, V. Mittal (Ed.). Weinheim, Germany: Wiley-VCH, pp. 115–142.
17. V. Shah (2007). *Handbook of Plastics Testing and Failure Analysis*. Hoboken, NJ: Wiley & Sons, pp. 17–93.

18. S. C. Lao, J. H. Koo, et al. (2010). Flame-retardant Polyamide 11 and 12 nanocomposites: Processing, morphology, and mechanical properties. *Journal of Composite Materials* 44(25), 2933–2951.
19. R. A. Vaia, R. K. Teukolsky, and E. P. Giannelis (1994). Interlayer structure and molecular environment of alkylammonium layered silicates. *Chemistry of Materials* 6(7), 1017–1033.
20. M. A. Osman, M. Ploetze, and P. Skrabal (2004). Structure and Properties of Alkylammonium Monolayers Self-Assembled on Montmorillonite Platelets. *Journal Physical Chemistry B* 108(8), 2580–2588.
21. H. Eslami, M. Grmela, and M. Bousmina (2009). A mesoscopic tube model of polymer/layered silicate nanocomposites. *Rheological Acta* 48(3), 317–331.
22. H. Eslami, M. Grmela, and M. Bousmina (2009). Structure Build-Up at Rest in Polymer Nanocomposites: Flow Reversal Experiments. *Journal of Polymers Science Part B* 47(17), 1728–1741.
23. S. S. Ray (2006). Rheology of Polymer/Layered Silicate Nanocomposites. *Journal of Industrial and Engineering Chemistry* 12(6), 811–842.
24. M., Song and J. Jin (2012). Characterization of Rheological Properties of Polymer Nanocomposites. In *Characterization Techniques for Polymer Nanocomposites*, V. Mittal (Ed.). Weinheim, Germany: Wiley-VCH, pp. 251–281.
25. R. Krishnamoorti, R. A. Vaia, and E. P. Giannelis (1996). Structure and dynamics of polymer-layered silicate nanocomposites. *Chemistry of Materials* 8(8), 1728–1734.
26. Standard Test Method for Thermal Diffusivity by the Flash Method (ASTM E1461-11). American Society for Testing and Materials. Philadelphia, PA.
27. V. Shah (2007). *Handbook of Plastics Testing and Failure Analysis*. Hoboken, NJ: Wiley & Sons, pp. 94–116.
28. J. Troitzsch (Ed.) (2004). *Plastics Flammability Handbook*, 3rd edition. Cincinnati, OH: Hanser.
29. V. Shah (2007). *Handbook of Plastics Testing and Failure Analysis*. Hoboken, NJ: Wiley & Sons, pp. 218–250.
30. ASTM Fire Standards and Flammability Standards (https://www.astm.org/Standards/fire-and-flammability-standards.html).
31. V. Babrauskas (1996). The Cone Calorimeter. In *Heat Release in Fires*, V. Babrauskas and S. J. Grayson (Eds.). London: E & FN Spon, pp. 61–91.
32. Standard Test Method for Heat and Visible Smoke Release Rates for Materials and Products Using an Oxygen Consumption Calorimeter (ASTM E1354). American Society for Testing and Materials, Philadelphia, PA.
33. Fire Tests – Reaction to Fire – Part 1: Rate of Heat Release from Building Products. ISO DOS 5660. International Organization for Standardization, Geneva, Switzerland.
34. Standard Test Method for Screening Test for Mass Loss and Ignitability of Materials (ASTM E2102). American Society for Testing and Materials, Philadelphia, PA.
35. Standard Test Method for Determining Flammability Characteristics of Plastics and Other Solid Materials Using Microscale Combustion Calorimetry (ASTM D7309-11). American Society for Testing and Materials, Philadelphia, PA.
36. Standard Test Method for Measuring the Minimum Oxygen Concentration to Support Candle-Like Combustion of Plastics (Oxygen Index) (ASTM D2863-12e1). American Society for Testing and Materials, Philadelphia, PA.
37. UL 94, the Standard for Safety of Flammability of Plastic Materials for Parts in Devices and Appliances testing, Underwriters Laboratories, Northbrook, IL.

38. Standard Test Method for Surface Burning Characteristics of Building Materials (ASTM E84-13a). American Society for Testing and Materials, Philadelphia, PA.
39. D. L. Schmidt and H. S. Schwartz (1963). Evaluation methods for ablative plastics. *SPE Transactions* 3, 238–250.
40. B. Botton, O. Chazot, M. Carbonaro, V. Van Der Haegen, and S. Paris (1999). The VKI Plasmatron characteristics and performance. DTIC Compilation Part Notice ADP010745. Rhode-Saint-Genese, Belgium, October.
41. J. H. Koo, M. Kneer, and M. Schneider, (1992). A cost-effective approach to evaluate high-temperature ablatives for military applications. *Naval Engineers Journal* 104, 166–177.
42. M. J. Miller, J. H. Koo, and S. Lin (1993). Evaluation of Different Categories of Composites Ablative for Thermal Protection. AIAA-93–0839, 31st AIAA Aerospace Sciences Meeting, Reno, NV, January.
43. F. Cheung, J. H. Koo, et al. (1995). Prediction of Thermo-Mechanical Erosion of High-Temperature Ablatives in the SSRM Facility. AIAA-95–0254, 33rd Aerospace Sciences Meeting, Reno, NV, January.
44. M. VanMeter, J. H. Koo, et al. (1995). Mechanical Properties and Material Behavior of a Glass Silicone Polymer Composite. *Proceedings of the 40th International SAMPE Symposium*, Covina, CA, SAMPE, pp.1425–1434.
45. J. H. Koo et al. (1998). Effect of Major Constituents on the Performance of Silicone Polymer Composites. *Proceedings of the 30th International SAMPE Technical Conference*, Covina, CA: SAMPE.
46. J. H. Koo et al. (1999). Thermal Protection of a Class of Polymer Composites. *Proceedings of the 44th International SAMPE Symposium*. Covina, CA: SAMPE, pp.1431–1441.
47. J. H. Koo, H. Stretz, J. Weispfenning, Z. Luo, and W. Wootan (2004). Nanocomposite Rocket Ablative Materials: Subscale Ablation Test. *Proceedings International SAMPE 2004 Symposium on Disc* [CD-ROM]. Covina, CA: SAMPE.
48. J. H. Koo, H. Stretz, J. Weispfenning, Z. Luo, and W. Wootan (2004). Nanocomposite Rocket Ablative Materials: Processing, Microstructures, and Performance. AIAA-2004-1996, AIAA, Reston, VA, April.
49. J. H. Koo, L. Pilato, and G. Wissler (2007). Polymer nanostructured materials for propulsion systems. *Journal of Spacecraft and Rockets* 44(6), 1250–1262.
50. J. H. Koo, M. J. Miller, J. Weispfenning, and C. Blackmon (2011). Silicone polymer composite for thermal protection of naval launching system. *Journal of Spacecraft and Rockets* 48(6), 904–919.
51. J. H. Koo, M. J. Miller, J. Weispfenning, and C. Blackmon (2011). Silicone polymer composites for thermal protection system: Fiber reinforcements and microstructures. *Journal of Composite Materials* 45(13), 1363–1380.
52. R. Blanski, J. H. Koo, et al. (2004). Polymer Nanostructured Materials for Solid Rocket Motor Insulation-Ablation Performance. *Proceedings of the 52nd JANNAF Propulsion Meeting*, CPIAC, Columbia, MD, May.
53. J. H. Koo, D. Marchant, et al. (2004). Polymer Nanostructured Materials for Solid Rocket Motor Insulation–Processing, Microstructure, and Mechanical Properties. *Proceedings of the 52nd JANNAF Propulsion Meeting*, CPIAC, Columbia, MD, May.
54. P. Ruth, R. Blanski, and J. H. Koo. (2004). Preparation of Polymer Nanostructured Materials for Solid Rocket Motor Insulation. *Proceedings of the 52nd JANNAF Propulsion Meeting*, CPIAC, Columbia, MD, May.
55. M. Natali, M. Monti, J. M. Kenny, and L. Torre (2011). A nanostructured ablative bulk moulding compound: Development and characterization. *Composites: Part A* 42(9), 1197–1204.

56. G. Pulci, J. Tirillo, F. Marra, F. Fossati, C. Bartuli, and T. Valente (2010). Carbon-phenolic ablative materials for re-entry space vehicles: manufacturing and properties. *Composites: Part A* 41(10), 1483–1490.
57. E. Allcorn, S. Robinson, D. Tschoepe, J. H. Koo, M. Natali (2011). Development of an experimental apparatus for ablative nanocomposites testing. AIAA-2011–6050, *47th AIAA/ASME/SAE Joint Propulsion Conference*, San Diego, CA, August 1–4.
58. L. Gutierrez, J. H. Koo, et al. (2015). Design of Small-scale Ablative Testing Apparatus with Sample Position and Velocity Control. AIAA-2015-1584, *AIAA SciTech 2015*, Kissimmee, FL, January 5–9.
59. Standard Test Method for Measuring Heat-Transfer Rate Using a Thermal Capacitance (Slug) Calorimeter (ASTM E457-08). American Society for Testing and Materials, Philadelphia, PA.
60. J. C. Lee (2010). Characterization of Ablative Properties of Thermoplastic Polyurethane Elastomer Nanocomposites. Ph.D. dissertation, The University of Texas at Austin, Austin, TX, December.
61. J. C. Lee, J. H. Koo, and O. A. Ezekoye (2011). Thermoplastic Polyurethane Elastomer Nanocomposite Ablatives: Characterization and Performance. AIAA-2011–6051, *47th AIAA/ASME/SAE Joint Propulsion Conference*, San Diego, CA, August 1–4.
62. J. C. Lee, J. H. Koo, et al. (2009). Thermoplastic Polyurethane Elastomer Nanocomposites: Density, Hardness, and Flammability Properties Correlations. AIAA-2009–5273, *AIAA Joint Propulsion Conference*, Denver, CO, August 2–5.
63. J. C. Lee, J. H. Koo, et al. (2009). Heating Rate and Nanoparticle Loading Effects on Thermoplastic Polyurethane Elastomer Nanocomposite Kinetics. AIAA-2009–4096, *AIAA Thermophysics Conference*, San Antonio, TX, June 22–25.
64. E. Allcorn, M. Natali, and J. H. Koo (2011). Ablation Performance and Characterization of Thermoplastic Elastomer Nanocomposites. *Proceedings of the SAMPE 2011 ISTC*, Fort Worth, TX, October 17–20.
65. E. K. Allcorn, M. Natali, and J. H. Koo (2013). Ablation performance and characterization of thermoplastic polyurethane elastomer nanocomposites. *Composites: Part A* 45, 109–118.
66. D. Wong, J. H. Koo, et al. (2013). Thermoplastic Polyurethane Elastomer Nanocomposites: Ablation and Charring Characteristics. *Proceedings of the SAMPE 2013 ISSE*, Long Beach, CA, May 6–9.
67. D. Wong, D. Pinero, M. Jaramillo, J. H. Koo, P. Ambuken, and H. Stretz (2013). Ablation and Combustion Characteristics of Thermoplastic Polyurethane Nanocomposites. AIAA-2013–3862, *49th AIAA/ASEM/SAE/ASEE Joint Propulsion Conference*, San Jose, CA, July 14–17.
68. A. Donskoy (1996). Elastomeric heat shielding materials for internal surfaces of missile engines. *International Journal of Polymer Materials* 31(1), 215–236.
69. Solid Rocket Motor Internal Insulation, *NASA Space Vehicle Design Criteria*. NASA-SP-8093, 1976.
70. M. S. Bell and W. Tam (1992). *ASRM Case Insulation Design and Development*. NASA-CR-191947.
71. C. M. Bhuvaneswari, S. D. Kakade, V. D. Deuskar, A. B. Dange, and M. Gupta (2008). Filled ethylene-propylene dieneterpolymer elastomer as thermal insulator for case-bonded solid rocket motors. *Defence Science Journal* 58(1), 94–102.
72. C. M. Bhuvaneswari, M. S. Sureshkumar, S. D. Kakade, and M. Gupta (2006). Ethylene-propylene diene rubber as a futuristic elastomer for insulation of solid rocket motors. *Defence Science Journal* 56(3), 309–320.

73. H. Redondo, M. Atreya, M. Kan, and J. H. Koo. (2010). Evaluation of Char Strength of Polymer Nanocomposites for Propulsion Systems. *Proceedings of the SAMPE 2010 ISSE* [CD-ROM]. Covina, CA, May.
74. S. Reshetnikov, A. N. Garashenko, and V. L. Strakhov (2000). Experimental Investigation into Mechanical Destruction of Intumescent Chars. *Polymers for Advanced Technologies* 11, 392–397.
75. H. Nguyen. (2012). Air Force Research Laboratory, Edwards AFB, CA, private communication.
76. M. Jaramillo, J. H. Koo, A. Edd, and D. Wells (2011). An Experimental Investigation of Char Strength of Polymer Nanocomposites for Propulsion Applications. *Proceedings of the SAMPE 2011 ISTC* [CD-ROM]. Covina, CA, October.
77. M. Jaramillo, D. Forinash, D. Wong, M. Natali, and J. H. Koo (2013). An Investigation of Compressive and Shear Strength of Char from Polymer Nanocomposites for Propulsion Applications. AIAA-2013–3864, *49th AIAA/ASEM/SAE/ASEE Joint Propulsion Conference*, San Jose, CA, July 14–17.
78. M. Jaramillo, J. H. Koo, and M. Natali (2014). Compressive char strength of polyurethane elastomer nanocomposites. *Polymers for Advanced Technology* 25(77), 742–751.
79. D. M. Forinash, R. J. Alter, S. B. Clatanoff, J. E. Newman, M. Jaramillo, and J. H. Koo. (2012). Development of an Apparatus for Measuring the Shear Strength of Charred Ablatives. *Proceedings of the SAMPE TECH 2012* [CD-ROM]. Covina, CA, October.
80. M. Natali, J. H. Koo, E. Allcorn, and O. A. Ezekoye. (2013). In-situ Ablation Recession Sensor Based on Ultra-Miniature Thermocouples – Part A: 0.25mm Diameter Thermocouples. AIAA-2013–3660, *49th AIAA/ASME/SAE/ASEE Joint Propulsion Conference*, San Jose, CA, July 15–17.
81. M. Natali, J. H. Koo, E. Allcorn, and O. A. Ezekoye (2014). An in-situ ablation recession sensor for carbon/carbon ablatives based on commercial ultra-miniature thermocouples. *Sensors and Actuators B: Chemical* 196, 46–56.
82. C. Yee, M. Ray, F. Tang, J. Wan, J. H. Koo, and M. Natali (2013). In-situ Ablation Recession Sensor Based on Ultra-Miniature Thermocouples – Part B: 0.50mm Diameter Thermocouples. AIAA-2013–3659, *49th AIAA/ASME/SAE/ASEE Joint Propulsion Conference*, San Jose, CA, July 15–17.
83. C. Yee, M. Ray, F. Tang, J. Wan, J. H. Koo, and M. Natali (2014). In-situ ablation recession and thermal sensor based on ultra-fine thermocouples. *Journal of Spacecraft and Rockets* 51(6), 1789–1796.
84. B. Lisco, E. Yao, D. Pinero, and J. H. Koo (2014). In-situ Ablation Recession and Thermal Sensors for Low Density Ablators – Revisited. *Proceedings of the CAMX 2014*, Orlando, FL, October 13–16.
85. S. Cameron, A. Astley, S. Leggett, G. Sirgo, and J. H. Koo. (2015). In-situ Ablation Recession and Thermal Sensor Based on Ultra-fine Thermocouples. *Proceedings of the SAMPE 2015 ISTC*, Baltimore, MD, May 18–21.
86. J. H. Koo, M. Natali, et al. (2015). A Versatile In-situ Ablation Recession and Thermal Sensor Adaptable for Different Ablatives. AIAA-2015-1122, *AIAA SciTech 2015*, Kissimmee, FL, January 5–9.
87. T. Grantham, J. H. Koo, et al. (2015). Ablation, Thermal, and Morphological Properties of SiC Fibers Reinforced Glass Ceramic Matrix Composites. AIAA-2015-1581, *AIAA SciTech 2015*, Kissimmee, FL, January 5–9.
88. J. H. Koo et al. (2015). A Versatile In-situ Ablation Recession and Thermal Sensor Based on Ultra-fine Gage Thermocouples for Ablative TPS Materials. *Proceedings of the National Space & Missile Materials Symposium (NSMMS)*, Chantilly, VA, June 22–25.

89. ASTM D4935 – 10 Standard Test Method for Measuring the Electromagnetic Shielding Effectiveness of Planar Materials. American Society for Testing and Materials, Philadelphia, PA.
90. V. Shah (2007). *Handbook of Plastics Testing and Failure Analysis*. Hoboken, NJ: Wiley & Sons, pp. 157–175.
91. J. Z. Zhang (2009). *Optical Properties and Spectroscopy of Nanomaterials*. Singapore: World Scientific Publishing.
92. V. Shah (2007). *Handbook of Plastics Testing and Failure Analysis*. Hoboken, NJ: Wiley & Sons.

PART TWO

MULTIFUNCTIONAL PROPERTIES AND APPLICATIONS

7 Mechanical Properties of Polymer Nanocomposites

7.1 Introduction

In the previous chapters, the three key steps in the development of polymer nanocomposites – (a) nanomaterial selection, (b) polymer nanocomposite processing, and (c) polymer nanocomposite property characterization and performance or device fabrication – were discussed. In the following eight chapters, specific polymer nanocomposite properties' characterization will be investigated, including mechanical, thermal, flammability, ablation, electrical, optical, and other properties.

This chapter is organized into three subchapters to discuss the mechanical properties of (a) thermoplastic-based, (b) thermoset-based, and (c) thermoplastic elastomer-based nanocomposites.

7.2 Thermoplastic-Based Nanocomposites

In this section, examples will be focused on the research conducted to analyze the enhancements of mechanical properties for thermoplastic-based nanocomposites. It is grouped in nanoclays, carbon-based, and other nanomaterials.

7.2.1 Nanoclay-Based Thermoplastic Nanocomposites

Montmorillonite (MMT) clay platelets are probably the most widely researched nanoparticles. These nanoparticles have unique effects on the mechanical properties of the polymer that are influenced by the type of clay, surface treatment of the clay, and how the clay is processed in the polymer matrix.

7.2.1.1 Glass Fiber/MMT Thermoplastic Bio-Nanocomposites

Kord reported a study conducted on glass fiber–reinforced high-density polyethylene (HDPE) [1]. In the study the fibers were to be reinforced by rice husk flour, and Cloisite 30B clay platelets were added to the mixture to increase mechanical and water absorption properties. The purpose of this study was to see if the clay platelets could be added effectively and if their benefits justified the further studying of this process.

Table 7.1. *Compositions Used in the Making of the Glass Fiber/MMT Nanocomposite [1]*

Composition of the Studied Formulations

High-density polyethylene wt%	Rice husk flour wt%	Glass fiber wt%	Nanoclay (phc)	Coupling agent (phc)
50	50	0	0	2
50	45	5	2	2
50	40	10	4	2
50	35	15	6	2

Phc, per hundred compounds.

The raw materials used in the experiment included HDPE at 0.954 g/cm^3, rice husk flour with a particle size of around 100 meshes, glass fibers, and Cloisite 30B clay platelets. During mixing, maleic anhydride was used as a coupling agent. The portions that were mixed in the experiment can be seen in Table 7.1 [1]. Before mixing, the rice husk flour was dried at 65°C for 24 hours. The mixing was conducted using a Haake internal mixer at melting temperature for the HDPE. After the polymer was melted, MMT and the coupling agent were added. After five minutes, the glass fibers were added along with the rice husk flour. The total mixing time was around 13 minutes. Samples for tensile testing were made by injection molding. The flexural and tensile tests were conducted at 2 mm/min under ASTM D790-03 and D638-03 standards, respectively. There were no TEM images presented to characterize the nanocomposite, but the report did mention that the particles did not disperse as well as the researchers would have liked. This uneven dispersion causes degradation in properties such as impact strength.

According to the results that follow, the mechanical properties generally increased with increased loadings of glass fibers and MMT. The data for tensile strength and flexural strength seem to peak at 4 wt% nanoclay loading (Figures 7.1a and 7.1b) [1]. The nanoclay increases both the flexural strength and modulus but decreases the impact strength of the material (Figure 7.1c) [1]. This follows the typical trend that anytime a nanoclay is added, the elongation or elasticity of the nanocomposite is decreased due to a more rigid matrix.

Kord concluded that this method did indeed enhance certain mechanical properties. It also, however, had some negative affects at higher loadings [1]. For 4 wt% MMT loading, the properties seemed to be optimal with reasonable loss in impact strength. The study also indicates that the assimilation of the clay platelets into the polymer matrix may produce better results with the addition of a compatibilizer.

7.2.1.2 Maximizing Exfoliation and Properties for Starch-Based MMT Nanocomposites

The morphology has a direct influence over the performance of nanocomposites under mechanical loading. The enhancement of mechanical properties depends highly on finding the right process to make the nanocomposite. The following two research groups [2, 3] worked on ways to advance the mixing process for starch-MMT

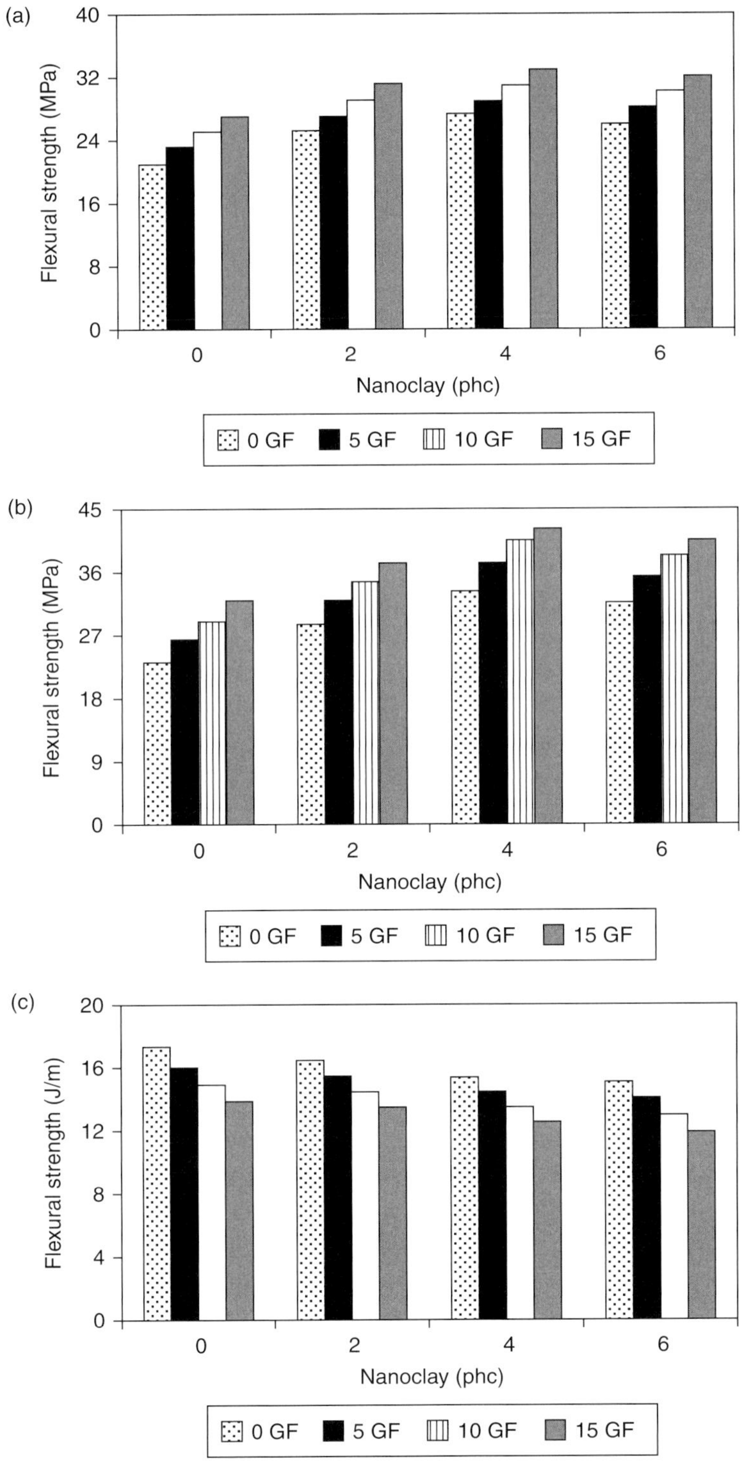

Figure 7.1. Effect of different glass fiber/MMT loadings on (a) tensile strength, (b) flexural strength, and (c) impact strength [1].

Table 7.2. *Mechanical Properties of TPS-MMT Nanocomposites at Different Loadings [2]*
Tensile Properties of TPS and TPS-MMT Nanocomposites

Nanocomposite	Tensile strength (Mpa)	Young's modulus (MPa)	Elongation at break %
TPS	1.5 ± 0.4	8.0 ± 3.5	33.5 ± 5.6
TPS.1% MMT	2.0 ± 0.2	6.7 ± 2.4	31.2 ± 2.3
TPS.2% MMT	2.3 ± 0.5	11.4 ± 1.4	40.1 ± 8.4
TPS.3% MMT	2.4 ± 0.1	15.3 ± 2.5	36.1 ± 3.8
TPS.5% MMT	2.8 ± 0.4	23.8 ± 4.6	37.8 ± 4.4

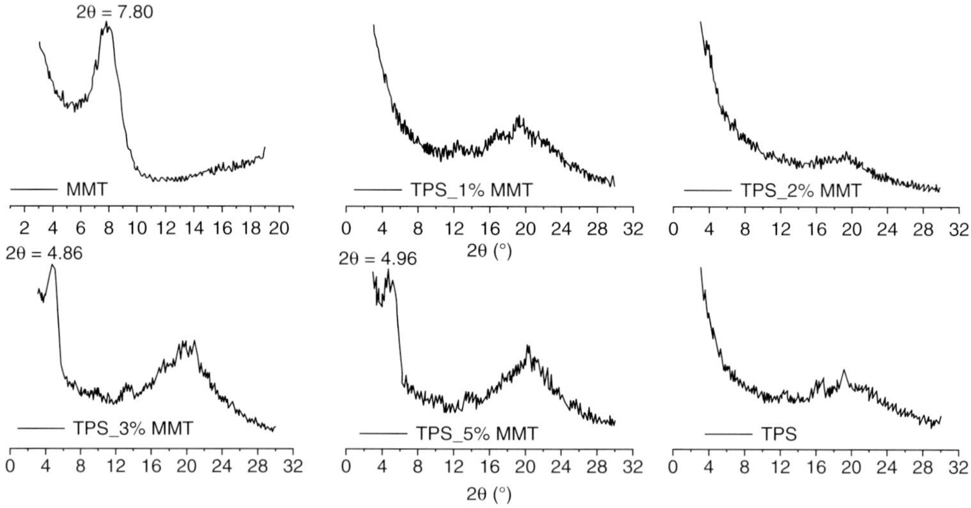

Figure 7.2. XRD of TPS-MMT mixtures at different clay loadings [2].

nanocomposites to get better exfoliation and maximize the effect on mechanical properties.

To prepare the MMT platelets for assimilation, both researching groups used a solution-processing technique that involved adding glycerol as a plasticizer in order to break up the starch (TPS) particles and get a well-exfoliated product. The research group from Brazil [2] focused on providing groundbreaking techniques for the mixing process, while the Iranian research group [3] focused on finding out which mixing steps or ingredients have the most influence on the mechanical properties [3]. Both groups also used a combination of sonication and mechanical mixing as their sole way of blending the MMT platelets into the TPS. Characterization was done through X-ray diffraction in both groups; the Iranian group also included TEM micrographs [3].

The Brazilian group prepared their samples by placing MMT in 200 milliliters of distilled water and placing the resulting solution in a sonic bath for 2 hours. Afterward, the corn starch was added and mixed with a magnetic stirrer for 30 minutes, at which point the glycerol was added and mixed in at high speeds (1,500 rpm). The solution was then placed in an oven at 90°C in order to evaporate the water.

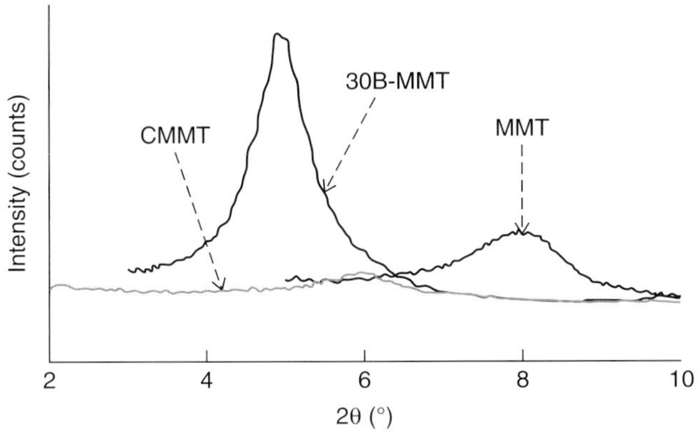

Figure 7.3. XRD comparison between the three different clays tested [3].

The last step was to place the solution in a batch mixer at 150°C at 50 rpm for 20 minutes. [2]. The results of the tensile tests and XRD scans are shown below. The XRD (Figure 7.2) indicates that for 1 and 2 wt% loadings of MMT, the particles were completely exfoliated [2]. However, for loadings greater than 2 wt%, there were peaks in the XRD curve that indicate aggregation of the platelets. Table 7.2 shows that the tensile strength and Young's modulus increased gradually with addition of MMT, but as usual the elongation at break was best at lower loadings (max at 2 wt%) [2]. The group concluded that the process of blending both solution and melt compounding techniques creates complete exfoliation at lower loadings and gives the material excellent tensile properties [2].

On the other hand, the Iranian team tried to maximize the effects to the mechanical properties by investigating the roots of the phenomenon. One type of MMT platelet was purchased from Southern Clay Products (Cloisite 30B) and the other was prepared in the lab. The lab MMT was cleansed in a citric acid solution before mixing. The next variable was the amount of glycerol added (10, 20, and 30 wt%), which was placed in 60 milliliters of distilled water. The second variable was three different types of starch (corn, wheat, and potato). While stirring, the solution was heated to 70°C for an hour. The MMT was mixed in by alternating between mechanical mixing and a sonication probe [3]. The last variable was the different loadings of clay that were made in a different batch with the previous conditions set at their optimum levels. The loadings tested were 0, 3, 4, 5, 6, 7, 10, and 15 wt%. In the XRD (Figure 7.3), it is shown that the citric acid montmorillonite (CMMT) did not have a large diffraction peak indicating that out of the three types of MMT tested, it was the one with the most exfoliation or spacing between the clay layers [3]. The tensile tests performed on all the samples (Table 7.3) indicated also that the CMMT was the best at enhancing the modulus of the material [3].

The researchers concluded that the cation of the clay used was detrimental in determining the outcome of the mechanical properties. While the 30B platelets had little interaction with the polymer, the citric acid from the CMMT reacted highly with it causing the best platelet gap widths. Also, while sonication and mechanical mixing techniques do not affect the quality separately, it does make a large difference

Table 7.3. *Mechanical Properties of All Tested Samples [3]*

XRD data			Mechanical properties		
Trial	2θ of the peak (°)	Δd (nm)	Tensile stress (MPa)	Young's module (MPa)	Elongation at break (%)
1	4.72	0.11	2.0	7.5	73.3
2	5.00	0.66	9.2	131.5	25.2
3	4.52	0.47	5.0	84.1	62.8
4	4.30	0.29	2.5	41.7	76.6
5	6.59	0.23	2.4	19.1	84.8
6	3.80	0.84	28.1	187.5	57.3
7	4.50	0.20	2.9	17.8	68.3
8	6.26	0.30	3.8	54.2	91.9
9	4.20	0.62	3.1	112.4	29.4

if a combination of the two is used. This is due to the fact that both dispersive and distributive techniques will be used. Overall, the combination that seemed to produce the best results were 6 wt% CMMT, potato starch, 20 wt% glycerol, and a mix of both mechanical and sonication mixing techniques [3].

7.2.2 Carbon-Based Thermoplastic Nanocomposites

7.2.2.1 MWNT Added to Polypropylene with Electron Irradiation Treatments

MWNT have been used to reinforce many polymer matrices. The process of assimilating them into the polymer matrix is a delicate one and can depend on the smallest of factor. A small excitation of the particles in the polymer while mixing could cause them to end up in a different place. Ritter and colleagues [4] studied the effect of electron irradiation on the interfaces between MWNT and isotactic polypropylene (iPP).

In the study, 0.1, 0.5, 1, 3, and 5 wt% samples were created by a hot extruder spun at 450°K and 50 revolutions per minute. Plates of the nanocomposites were created by extruding cylindrical pieces from the molten solution and then slicing them into disks. Characterization was done by X-ray diffraction (XRD) on all the different weight distributions of MWNT (Figure 7.4) [4]. The samples were then placed in an electron accelerator where they absorbed electron irradiation. The absorption dose varied up to 4 MGy, and the radiation temperature did not exceed 323°K.

After the material was irradiated, XRD was done on the individual samples. From Figure 7.4 we can see that the more MWNT added, the more impure an exfoliation we achieve. This is common considering that an increase of the MWNT would result in a greater chance of them interacting with one another. Also, microhardness tests were conducted on the samples to get an idea of what the irradiation would do the mechanical properties [4]. From Figure 7.5 we can see that the non-irradiated samples were largely inferior to the irradiated samples in terms of hardness [4]. Although for small content of MWNT the hardness actually dips, overall the results are still higher than the non-irradiated samples.

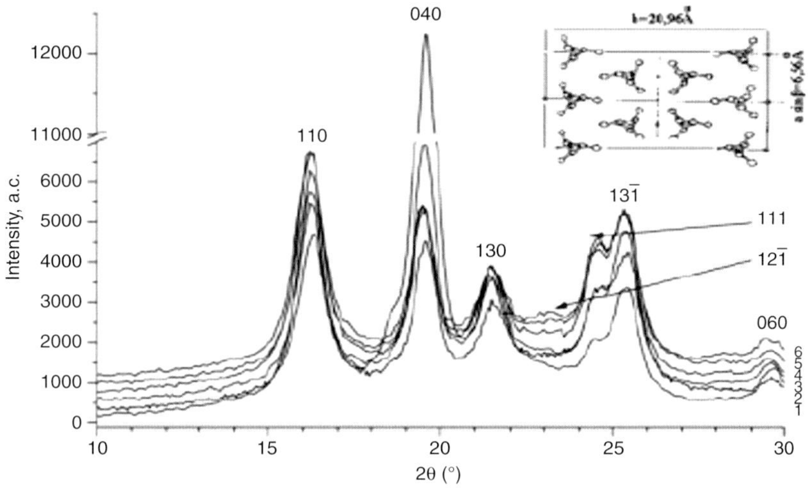

Figure 7.4. XRD taken of irradiated nanocomposite samples (a) 0%, (b) 0.1%, (c) 0.5%, (d) 1%, (e) 3%, and (f) 5% by weight [4].

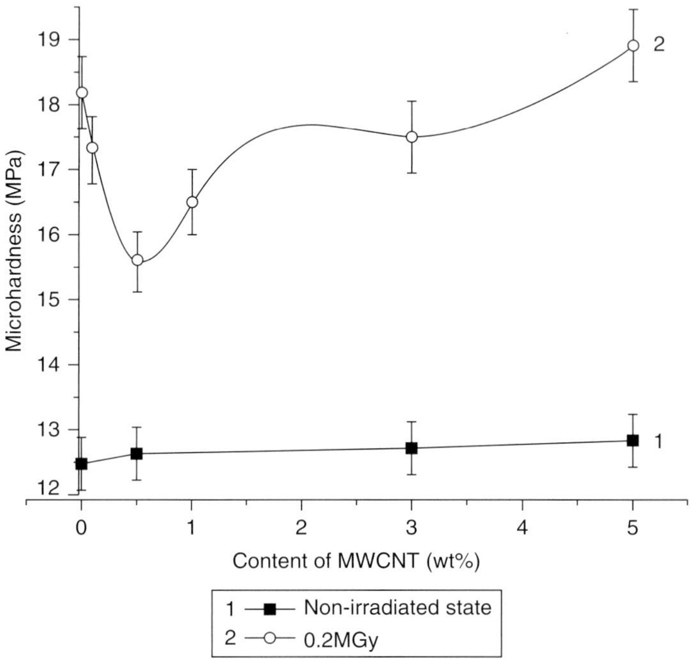

Figure 7.5. Microhardness comparison between (1) non-irradiated sample and (2) irradiated sample [4].

At the end, the study found that the electron irradiation created positive effects toward enhancing the hardness of iPP/MWNT nanocomposites. The results concluded that the irradiation was increasing the degree of crystallinity in the nanocomposite and therefore affecting the hardness [4].

Consumer items that have been made out of graphite or some other type of carbon-based variant are common in the market. These materials are very strong, and the products that include them are known to be durable. In polymers, nano-sized tubes and graphene sheets are created in reactors in order to create a similar strengthening effect in polymer matrices.

7.2.2.2 Forming Multi-Walled Carbon Nanotubes into Buckypaper

Fernandez-d'Arlas [5] includes a report by several scientists working in Spain that focused on two different processing methods for mixing multi-walled carbon nanotubes (MWNT) into a polyurethane matrix. In the experiment, the scientists looked specifically at the mechanical effects of using either a solvent processing method or implementing buckypaper into the matrix. The buckypaper was theorized to help the strengthening process of the MWNT and create a consistent network within the polymer matrix [5].

In the case of the solvent method, the scientists chose to prepare the MWNT with a combination of dimethylformamide and tetrohydrofuran at a 1-to-1 ratio. These solvents were chosen because of their low boiling points, meaning that it would be easy for them to be evaporated out. The MWNTs were placed in a 6 milliliter solution of the solvent and mixed with a sonication tip. Then the polymer was pipetted from the same solvent solution and combined to make a total composite weight of 300 grams [5]. On the other hand, the buckypaper method was started by acid-treating the MWNT in a dimethylformamide solution. The buckypaper is created by forming a network of MWNT that is very porous in nature. That way when the polymer is added, it creates passages for it to seep through. The results for the buckypaper method were as predicted, reducing the porosity of the overall mixture and creating a nanocomposite with enhanced mechanical and electrical properties [5]. In the TEM image shown in Figure 7.6, the buckypaper network was formed in the polymer [5].

Samples of the nanocomposite were put through stress and strain tests in order to see what the enhancements were to the mechanical toughness of the polymer. Figure 7.7 shows the mechanical properties of the polymer nanocomposites [5]. As shown in Figure 7.7, the higher loading of MWNTs causes the yield stress and modulus to increase. However, around 1 wt% loading, the maximum strain of the material begins to degrade. This is due to the carbon nanotube network taking over and becoming the dominant component. The MWNTs that make up the buckypaper are brittle and as more of them are added to the polymer, the buckypaper begins to feel more of the stress [5].

7.2.2.3 Graphene-Reinforced Thermoplastics for Strain Sensors in Structural Health Monitoring

Throughout the years smart materials have been researched in order to provide better response time for sensory equipment in the medical field. Recently, a research group [6] published an article about their study conducted using graphene-reinforced polyvinylidine fluoride (PVDF) for structural strain sensing.

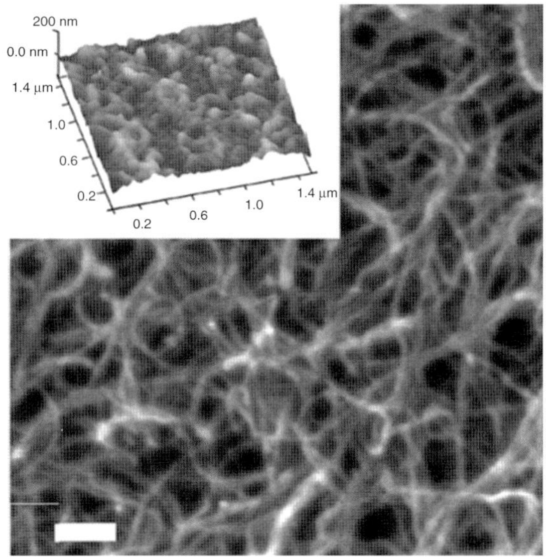

Figure 7.6. TEM image of the buckypaper network in the polyurethane matrix [5].

To synthesize the nanocomposite, a simple solvent casting technique was used. Initially, the graphene was placed in the dimethylformamide (DMF) solvent and stirred for 60 minutes. The same was done to a specific amount of PVDF polymer in a magnetic stirrer for 30 minutes. Afterwards, the two solutions were mixed together by means of an ultrasonicator for a period of 1 hour and then moved to a shear mixer to be mixed at room temperature for 2 hours and at 80°C for 30 minutes at 4,000 rpm. Figure 7.8 shows the dispersion characterized by a TEM image [6]. The image shows the graphene sheets having arranged themselves in a random crumpled order, creating a disordered solid.

Figure 7.8 shows good dispersion of graphene within the polymer matrix. The samples made for testing were cut into thin films (containing 2 wt% graphene) and adhered to the back of a piece of aluminum using epoxy glue. The tensile, compression, and ramping load tests were conducted using a servo hydraulic test machine. A constant electrical current ran through the material as it was tested. The results seen in Figure 7.9 show that the samples containing the graphene sheets reacted very dramatically under strain [6]. For a 0.1 percent increase in strain, the voltage changed by 20 millivolts, which is high.

In conclusion, graphene proved to be a viable additive to increase the mechanical sensitivity of the PVDF polymer [6]. Because the results were positive, for the material's response it is reasonable to use it in a medical application where stress must be measured.

7.2.3 Other Nanomaterial-Based Thermoplastic Nanocomposites

So far only two different types of nano-fillers – MMT clay and carbon nanotubes/graphene sheets – were discussed. Materials like clay and graphite are strong and

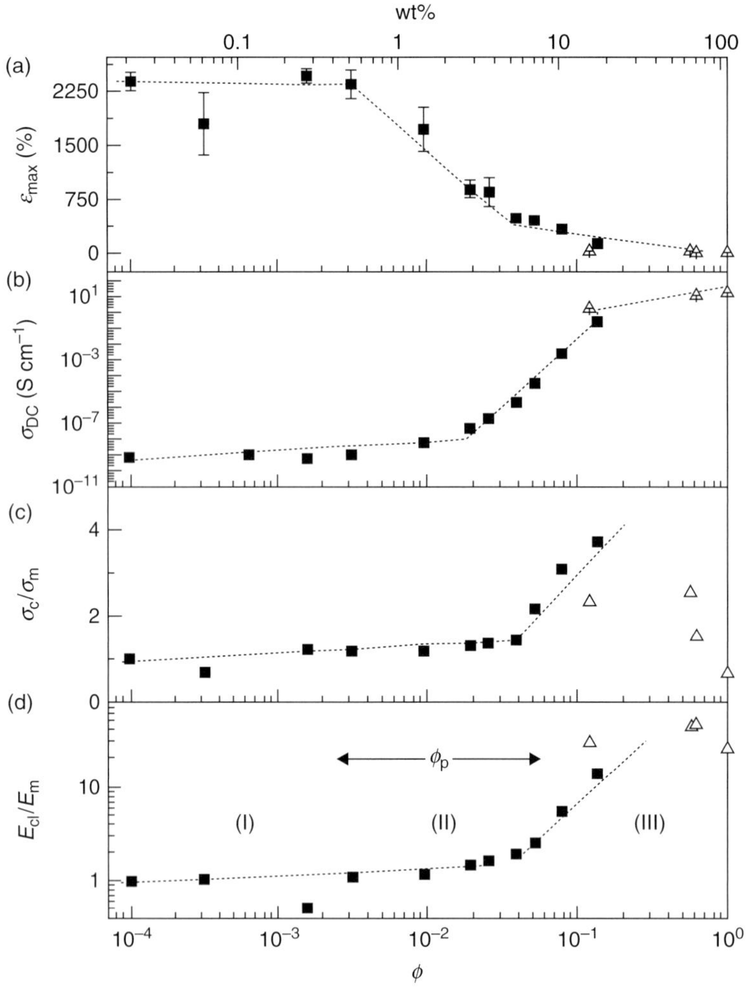

Figure 7.7. Mechanical properties of the new nanocomposite [5].

helpful in boosting mechanical properties in polymers. However, there are other types of nanomaterials that may have a similar effect. When speaking of mechanical strength, exotic materials like diamond are at the peak of the spectrum. Can it be used effectively and properly, though, to make a viable nanocomposite? Also, after the nanocomposite is formed, what shapes can we make while still ensuring good dispersion? These are some of the topics discussed in the following sections.

7.2.3.1 Using Electrospinning to Process the Mixed Nanocomposite into Fibers

There are many different shapes that can be molded using nanocomposites after they are mixed. A group of scientists in Beijing [7] conducted a study on preparing polymer/nanoparticle fibers for insertion into a polyurethane matrix by electrospinning.

The polyurethane was solvent surface treated in a dimethyl formamide solution. The idea was to create small composite fibers that would reinforce the

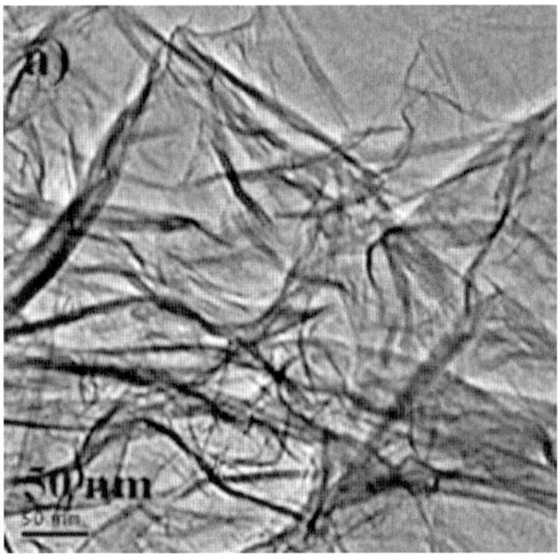

Figure 7.8. Graphene platelets in a polyvinylidine fluoride matrix [6].

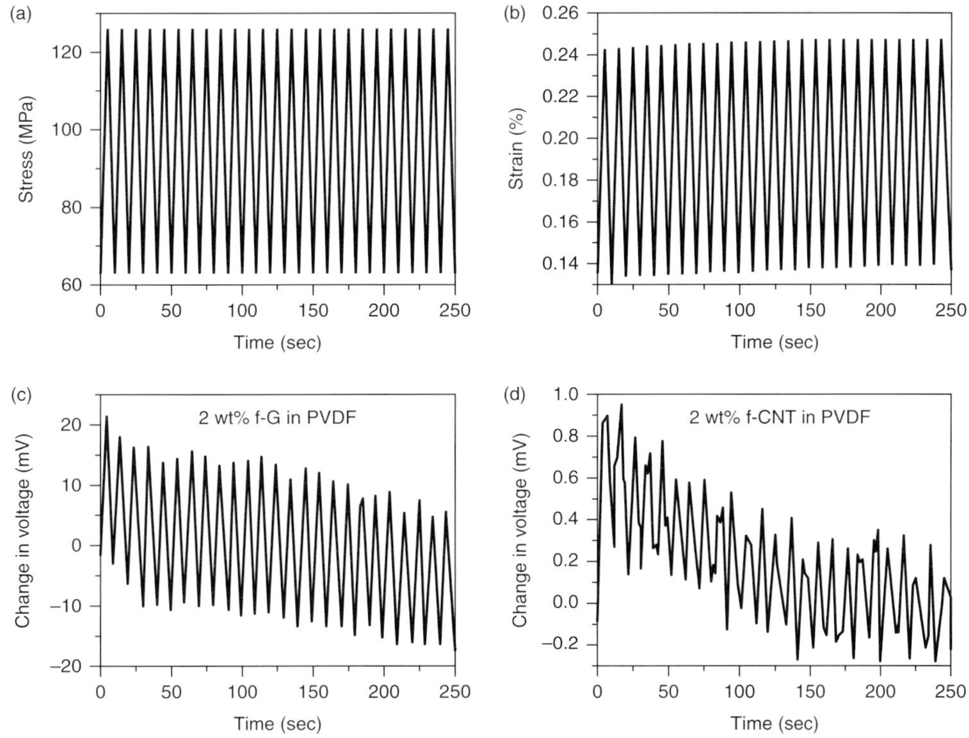

Figure 7.9. (a) stress and (b) strain cycles applied to samples and (c) the effect on voltage running through it compared to (d) the same effect on a sample with carbon nanotubes [6].

nanocomposites that were added, thus enhancing the mechanical properties [7]. The electrospinning process works by pressing the composite through a spindle hooked to a voltage source. As the liquid is released, the charged solution is stretched and a long fiber is formed. In this case the experiment was done at a constant 24 kilovolts and the solution was passed through the spindle at 0.3 milliliters per hour [7]. A set

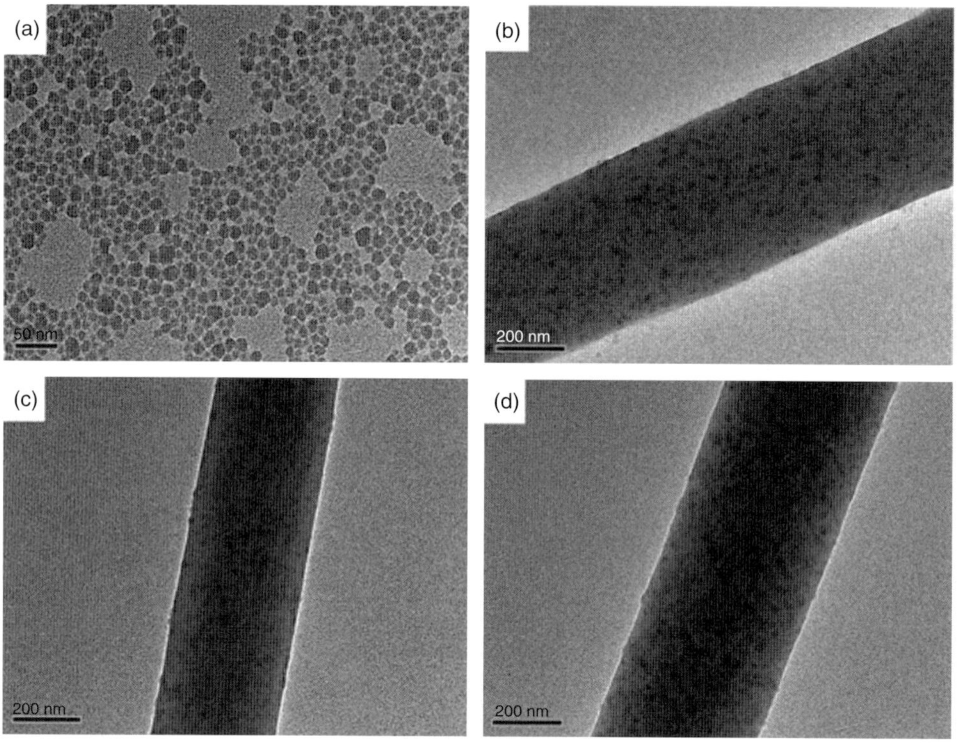

Figure 7.10. TEM images of polyurethane/silica nanocomposite fibers [7].

of TEM images of the resulting mixture is shown in Figure 7.10. It shows that the silica particles are well dispersed throughout the fibers. Figure 7.10 has four TEM images (a, b, c, d) at three different silica loadings [7]. Figure 7.10 (b, c, d) shows fibers that are made up of three different silica loadings containing 1, 5, and 10 wt% silica. The mechanical properties of the electrospun fiber were found using a variety of mechanical testing techniques including tensile tests on some samples [7]. The stress-strain plots for each of the separate loadings (0, 1, 5, 10 wt%) are shown in Figure 7.11 [7].

The stress-strain plot shows that as the silica loading increases, the material is able to take on more stress before yielding. It also experiences a large amount of strain reaching values of up to almost 1,500%. Surprisingly, the strongest material (the 10% load) experiences almost as much strain as the 1% loaded case. We can also see from the TEM images that all the cases studied showed even dispersion within the fibers [7].

7.2.3.2 Thermoplastics Modified with Nanodiamonds

From a mechanical property standpoint, diamonds possess desirable attributes in both strength and hardness. A group of researchers in Russia experimented by incorporating nanodiamonds (NDs) into several thermoplastic matrices. The two thermoplastics used in their experiments were an amorphous styrene-acrylonitrile copolymer and an amorphous polysulfone (PSP). Another goal of the project was to

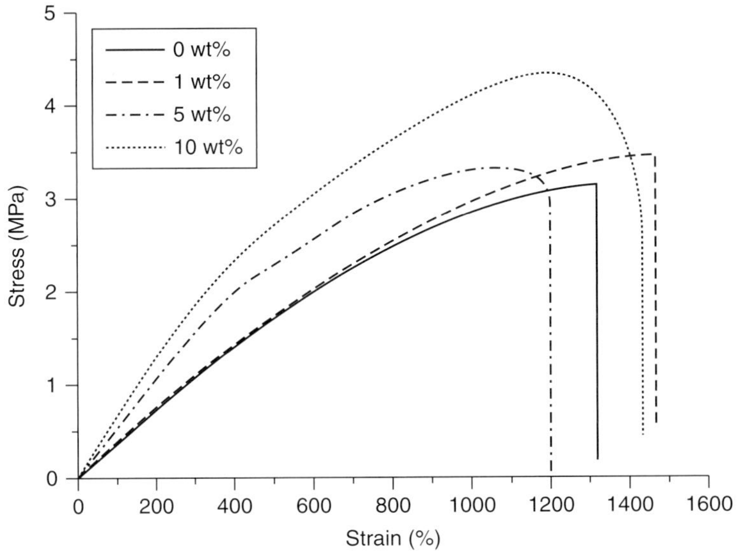

Figure 7.11. Stress-strain results for nanocomposite fibers at different silica loadings [7].

experiment with processing conditions. The differences in microstructure and properties were examined when applying both normal melt mixing and mixing under spurt conditions [8].

The nanodiamonds used were a PUOO-CX DND 96 brand that came in the form of grey powder with a density of 3.3 g/cm^3. The nanocomposite was processed by mechanical blending in the molten state with a worm-piston mixer in 4–5 milliliter samples. The copolymer was mixed at 190–200°C and the PSP was mixed at 270–280°C. Samples were made for tensile testing that contained 0.5, 1, 2.5, and 5 wt% nanodiamonds (NDs) for both thermoplastics [8]. The researchers also experimented with mixing during elastic turbulence. This type of mixing exposes the solution to high degrees of stress, which in theory is supposed to break open nanodiamonds even further in order to obtain smaller particles. In Figure 7.12, the micrographs indicate that the elastic/spurt mixing did help produce finer nanodiamonds particles (Figure 7.12b), compared to the conventional normal melt mixing shown in Figure 7.12a [8].

Figure 7.13a (left) shows the ratio between the elastic moduli of copolymer-ND composites and the modulus of the unfilled polymer vs. the loading of NDs from 0 to 5 wt%. Figure 7.13b (left) shows the ratio between the tensile strengths of copolymer-ND composite and the unfilled polymer vs. the loading of NDs from 0 to 5 wt%. Composites were prepared through (1, 3) blending in the spurt/elastic conditions and (2, 4) blending by conventional normal melt. For the copolymer prepared by the spurt method, as the content of NDs increased, the elastic modulus significantly increased, while the elongation at break for the neat polymer remained unchanged [Figure 7.13a (left)]. For the systems prepared through the conventional normal melt method, the main mechanical parameters tended to decrease as the loading of the filler increased [Figure 7.13b (left)].

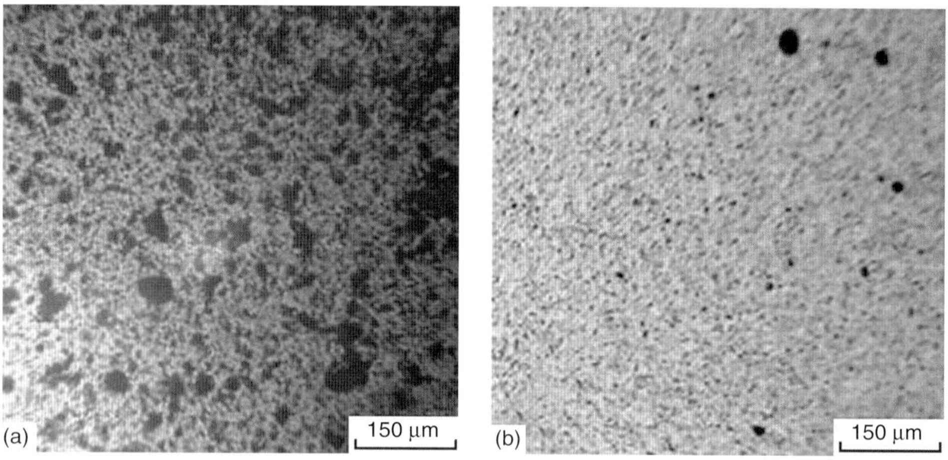

Figure 7.12. Micrographs of PNC made by (a) normal melt blending and (b) spurt/elastic conditions [8].

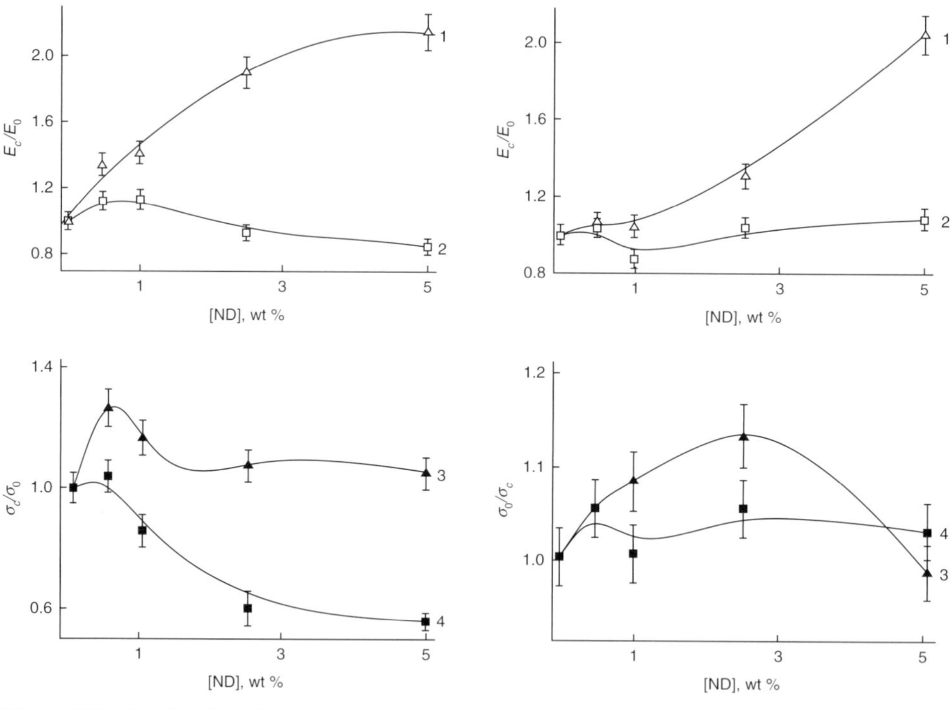

Figure 7.13. (a) the right-hand side is the modulus ratios for PSP while the left-hand side is the same for the copolymer; (b) max stress ratios for both PSP and copolymer (order is as same as modulus). *Note*: Samples 1 and 3 were blending through spurt/elastic conditions and 2 and 4 by conventional normal melt conditions [8].

Figures 7.13a (right) and 7.13b (right) present similar dependences of the PSP-based systems. The mechanical properties of PSP before and after treatment in mixer remain invariable. From these data, blending in the spurt mode provides

Table 7.4. *Nanocomposites Created at Different Polymer Molecular Weights and Nanoparticle Ratios [9]*

Material No.	$\overline{M}_w$ of PEG kg·mol^{-1}	PEG/POSS ratio
1	10	3
2	10	4
3	10	6
4	10	8
5	20	6
6	20	8
7	20	10
8	20	12
9	35	8
10	35	10
11	35	12

higher strength parameter than the conventional melt blending. The elastic modulus increased with the increase of NDs loading [Figure 7.13a (right)]. The modifying effect of NDs for the composites prepared by the conventional melt method is very statistically insignificant and within the scatter of the experimental data [Figure 7.13a (right)]. The modulus and max stress ratios shown in Figure 7.13b (right) are higher for the samples mixed under spurt/elastic conditions [8]. The researchers also found that for the spurt samples the elongation at break remained the same as in the neat polymer. However, all properties degrade under small loadings with conventional normal mixing.

This was explained by the fact that the larger nanodiamond particles in the normally mixed polymers acted as defects in the polymer matrix. Furthermore, the spurt conditions seemed to help the overall assimilation by creating non-uniform shearing rates within the polymer. These conditions allowed the nanodiamonds to assimilate into weak spots within the polymer matrix. Overall, the shearing seemed to create a better dispersion of the nanoparticles [8].

7.2.3.3 Microhardness and Indentation Effects when Adding POSS to Polyurethane

Another important mechanical property that can be enhanced by the addition of nanoparticles is hardness. Lach et al. [9] focused on adding polyhedral oligomeric silsesquioxanes (POSS) to a polyethylene glycol (PEG) matrix to enhance the hardness properties.

The processing techniques were not fully outlined in this research, since the researchers themselves did not make the nanocomposite. Instead they were made by a third party that used a melt compounding technique and created sheets approximately 0.35 mm in thickness. There were different samples created with different molecular weights of PEG and different ratios of POSS added (Table 7.4) [9].

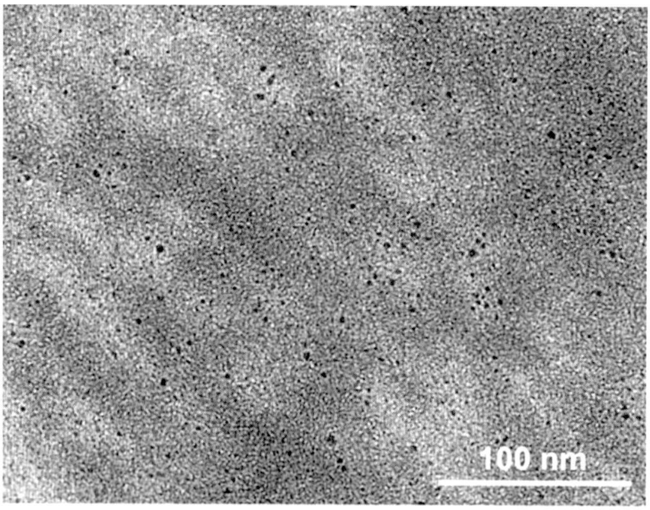

Figure 7.14. TEM of PEG/POSS nanocomposite [9].

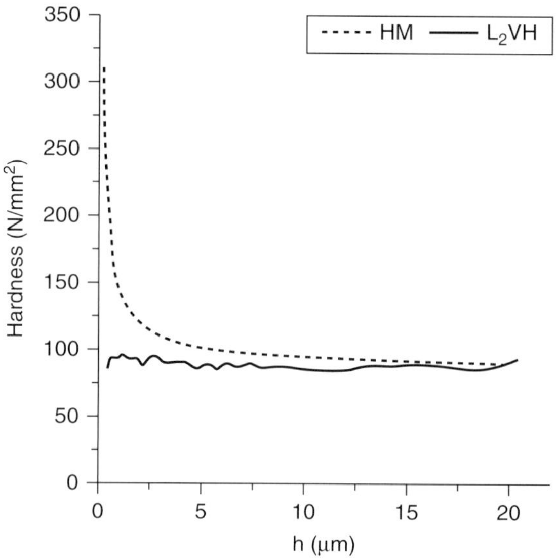

Figure 7.15. Vickers and Martens hardness of nanocomposite under load [9].

Hardness measurements were taken on the Vickers and Martens hardness scales. The force applied did not exceed 1 newton and was applied at a rate of 50 nanonewtons per second. In Figure 7.14, TEM images were taken of the nanocomposite before testing [9].

As shown in the TEM images, the POSS is distributed well (Figure 7.14). The POSS aggregates shown in the TEM are all smaller than 10 nm, proving that there were no significant impurities associated with the processing technique. The microhardness results can be seen in a plot in Figure 7.15 [9]. Aside from the initial loading peaks, the hardness seems to remain uniform as the diamond continues to apply load.

This study results indicated that the softer segments of the PEG were prevented from crystallizing by the POSS particles. Also, it seemed that the variable molecular weights of PEG and the different ratios of POSS added did not affect the results of the hardness test. Each of them came out to be relatively the same value as the other materials [9].

7.2.4 Summary of Thermoplastic-Based Nanocomposites

In summary, the mechanical properties of polymers can be strengthened dramatically by adding nanoparticles if the right process and materials are chosen. Incorrect combinations may produce no changes or, in some cases, even adverse effects. Also, it is not only the process that matters; the fabrication of nanocomposite parts is also important.

7.3 Thermoplastic Elastomer-Based Nanocomposites

This section reviews thermoplastic elastomer-based nanocomposites, subdivided into thermoplastic elastomer-clay nanocomposites, thermoplastic elastomer-carbon-based nanocomposites, and thermoplastic elastomer with other nanomaterials research.

7.3.1 Nanoclay-Based Thermoplastic Elastomer Nanocomposites

7.3.1.1 High-Performance Elastomeric Nanocomposites via Solvent-Exchange Processing

Liff, Kumar, and McKinley examined a thermoplastic elastomer nanocomposite formed from Elasthane 80A, a commercially available thermoplastic polyurethane, and unmodified Lamponite RD, a synthetic smectic clay [10]. Nanocomposites with weight loading of 2, 4, 10, 15, and 20 wt% Lamponite RD were prepared via a "novel solvent exchange process." This study describes this process as using two separate solvents to first dissolve the polymer and nanomaterial separately, and then to combine the resulting solutions. The solvents are then extracted, and the remainder is a well-dispersed nanocomposite.

As shown in Figure 7.16, the solution exchange method employed in this study achieved good dispersion and exfoliation of the Lamponite platelets [10]. The clay had a noticeable effect on the mechanical properties of the polyurethane. Figure 7.17 shows that as clay loading increased, so did the modulus of elasticity and the ultimate tensile strength [10]. The nanocomposite also developed a marked yield point at 20 wt% clay, and the ultimate tensile strength doubled compared to virgin polymer. Elongation at break seemed slightly improved by the addition of nanoclays, although the 15 wt% loading had markedly superior performance compared to the other weight loads. The 15% weight load also exhibited the highest ultimate tensile strength. The study did not explore if this was an artifact of testing or a more optimal loading [10].

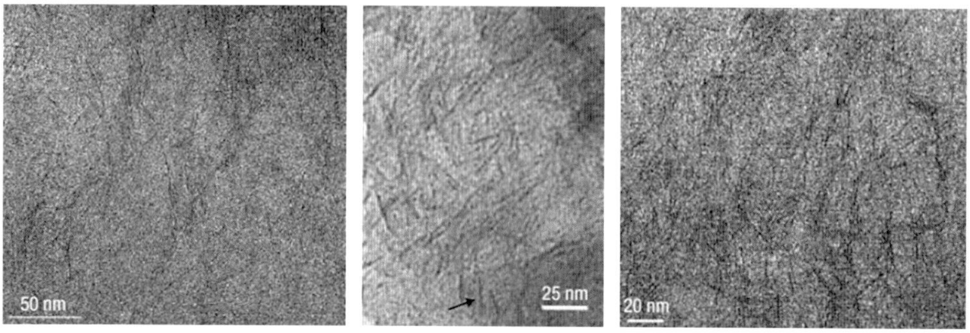

Figure 7.16. TEM images of nanocomposites containing (left to right) 4, 10, and 20 wt% Lamponite RD [10].

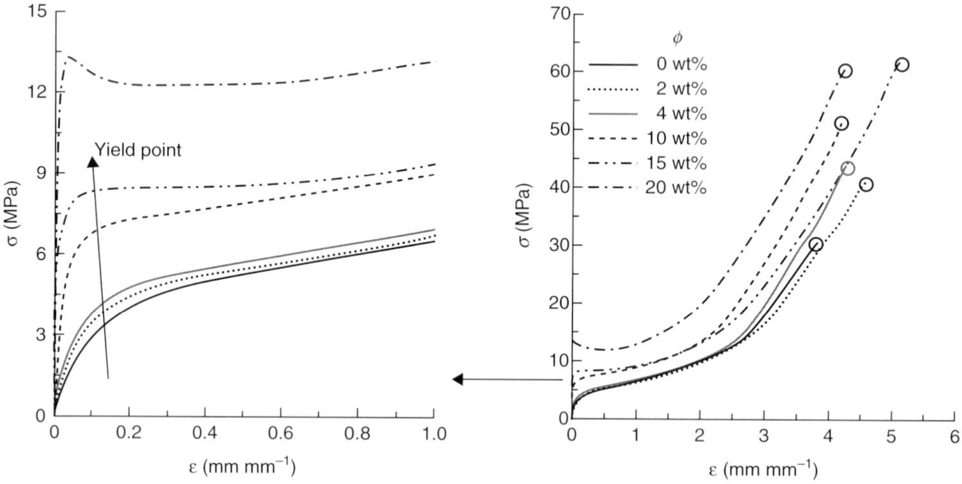

Figure 7.17. Stress vs. strain graphs for the virgin and modified elastomers [10].

7.3.1.2 Morphology and Properties Of Thermoplastic Polyurethane Nanocomposites: Effect of Organoclay Structure

Chivarria and Paul [11] examined how varying the structure of organoclays would affect the properties of the resulting polymer. They produced nanocomposites from five varieties of montmorillonite clays from Southern Clay Products. The clays varied in how they had been surface-treated by the supplier, which affected the surface structures of the clay platelets. Two varieties of polymer were also utilized, one medium hardness and the other high hardness. The nanocomposites were prepared in a twin-screw, co-rotating and intermeshing extrusion machine to a variety of clay loading levels.

As can be seen from the TEM micrographs in Figures 7.18 and 7.19, some of the nanoclays dispersed better than others [11]. In most cases, the addition of nanoclays seemed to have significantly increased the modulus of elasticity as well as the ultimate tensile strength. The stress vs. strain graphs in Figure 7.20 show that more clay did not necessarily yield greater improvements in qualities [11]. Most cases showed

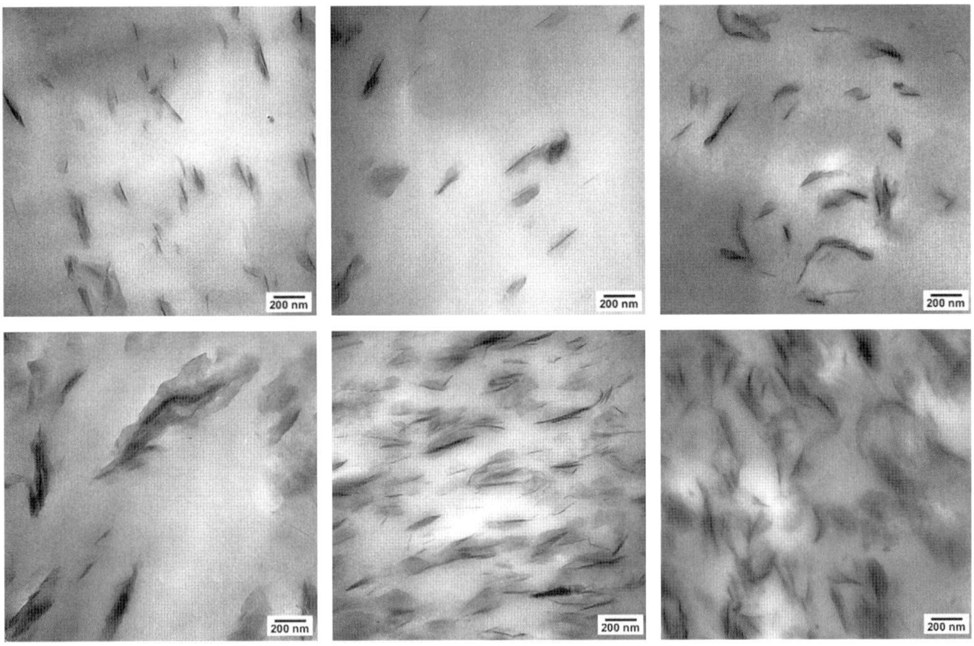

(a) $M_2(HT)_2$ (b) $M_3(HT)_1$ (c) M_3T_1

Figure 7.18. TEM images of 0.48 to 0.94 wt% (above) and 4.61 to 6 wt% (below) for three of the clays [11].

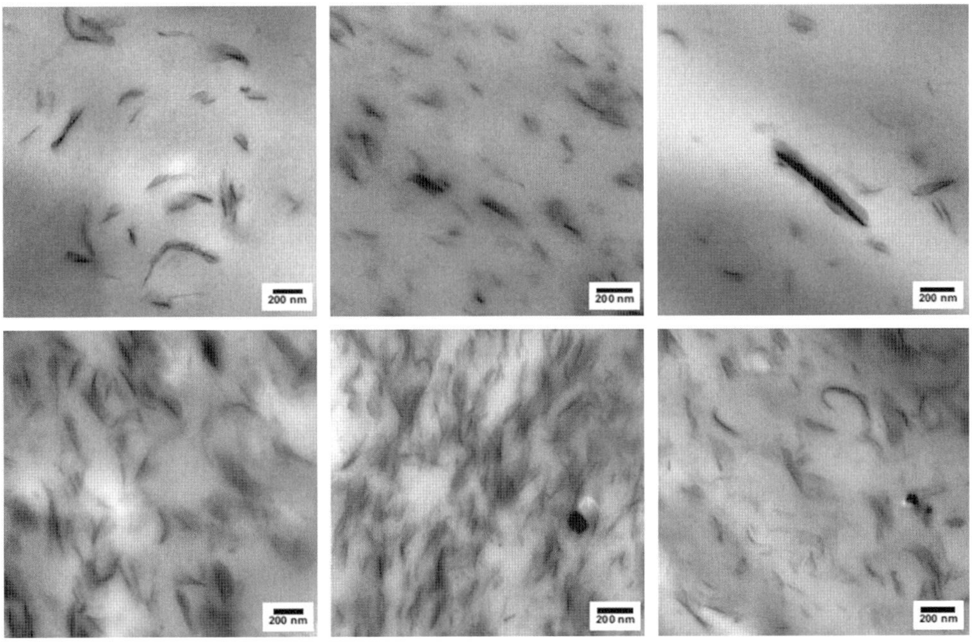

(a) M_3T_1 (b) $(HE)_2M_1T_1$ (c) $(HE)_2M_1C^*_1$

Figure 7.19. TEM images of 0.9 to 0.95 wt% (above) and 5.46 to 6 wt% (below) for three of the clays [11].

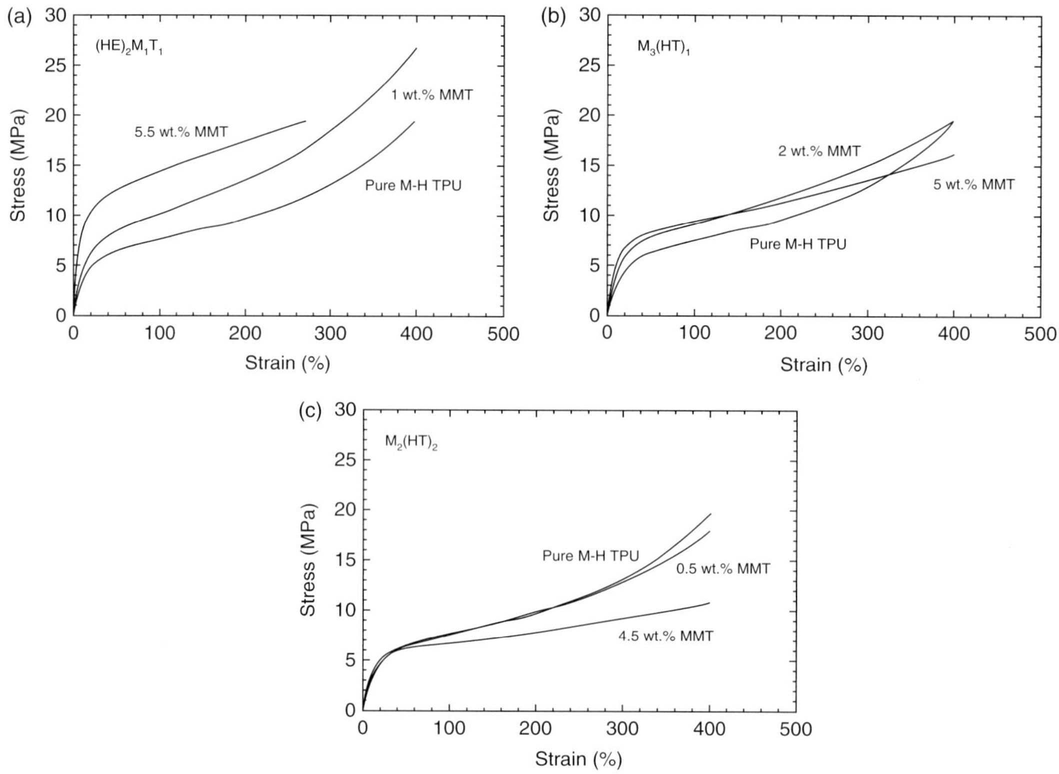

Figure 7.20. Stress vs. strain curves for three different nanoclay composites. M-H denotes medium-hardness polyurethane; H-H denotes high-hardness polyurethane [11].

diminishing returns, or an outright reduction in ultimate tensile strength as loading of clay increased. $M_2(HT)_2$ treated clays did not increase the ultimate tensile strength of the resulting polymer, but instead made it progressively worse as more clay was added. Elongation at break was not explored greatly here, as the tests only ran to 400% elongation, which evidentially was less than what the nanocomposites could withstand. The 5.5 wt% MMT of $(HE)_2M_1T_1$ was noted to have an ultimate shear of less than 400% in medium hardness polyurethane when it broke at approximately 275%. The high-hardness polyurethane became much more brittle as more $(HE)_2M_1T_1$ was added. The 3 wt% MMT loading had an ultimate shear one-third of what was to be expected from virgin high-hardness polyurethane. As it can clearly be seen here, the surface treatments of the nanomaterial have a tremendous influence on the behavior of the resulting nanocomposite. This research showed impressive depth and examined a great deal of material behavior [11].

7.3.1.3 Tunable Properties of Self-Assembled Polyurethane Using Two-Dimensional Nanoparticles: Potential Nano-Biohybrid

Mishra et al. explored selectively manipulating the shape of nanoclays, and how this ultimately changes the properties of the nanocomposites as potential biomaterials

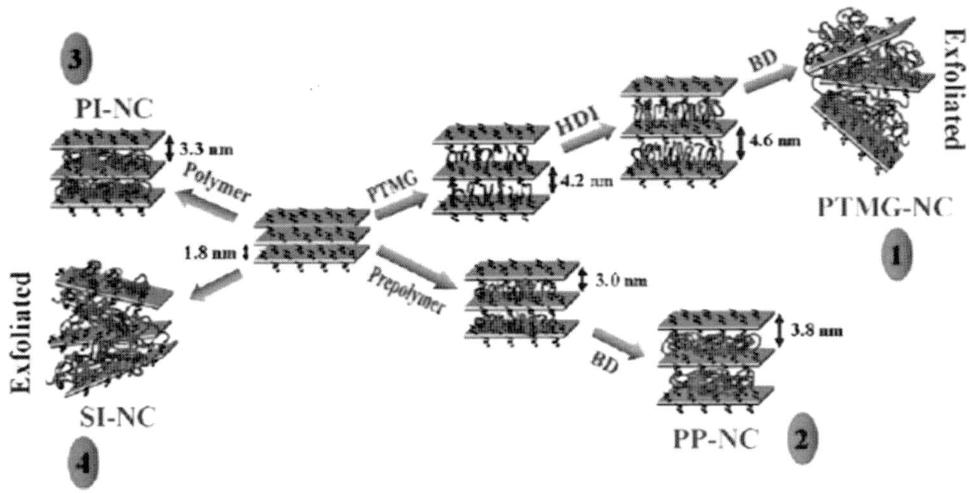

Figure 7.21. Visualization of the various shapes and dispersions taken by the nanoclays depending on when they were introduced to the nanocomposite [12] (Courtesy of ACS).

[12]. Polyurethanes were synthesized in situ from commercially purchased prepolymers, an extender and a catalyst. Rather than employ a mechanical mixing device, nanoclays were directly added during various stages of the polymerization process, leading to different dispersion mechanisms working on the clay platelets, producing different dispersion patterns. The nanoclay utilized was Southern Clay Products Cloisite 30B. Four possible dispersal schemes were discussed, as the research explains: "PTMG-NC (PTMG intercalation; nanoclay added to PTMG), PP-NC (prepolymer intercalation; nanoclay added to prepolymer), PI-NC (polymer intercalation; nanoclay added to polymer), and SI-NC (solution intercalation: adding nanoclay into dilute solution of polymer)" [12]. PTMG-NC, PP-NC, and PI-NC samples were prepared for this study, each with 4 wt% nanoclay.

Figure 7.21 explains well the differences between the nanocomposites [12]. PI-NC and PP-NC were only intercalated, with PP-NC being the more expanded of the two. PTMG-NC exfoliated, which, as can be seen in Figure 7.22, lead to a greater level of nanoclay dispersion when compared to the other two nanocomposites [12]. Figure 7.23 shows that in all three cases, the addition of 4 wt% nanoclay markedly increased the ultimate tensile strength and modulus of elasticity, and slightly to dramatically improved toughness [12]. The overall trend seems to be that the more intercalated or exfoliated the clay, the less improvement in the ultimate tensile strength, but also the less diminishing of the elongation at break. Interestingly, the exfoliated nanoclay composite actually improved the elongation at break by about 35% over the virgin polyurethane. The notion that the properties of a nanocomposite can be tuned by when the nanoparticles are added is a very intriguing idea, and this research clearly shows that it is a promising way to adjust nanocomposite properties [12].

Figure 7.22. TEM images of the three prepared nanocomposites, showing less successful dispersion from left to right [12] (Courtesy of ACS).

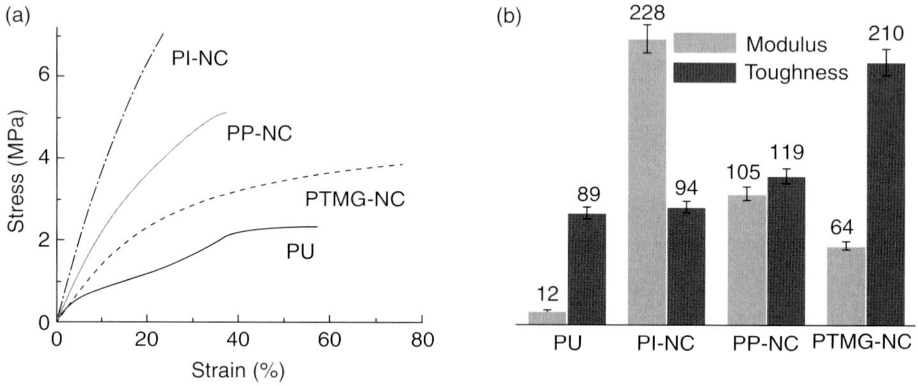

Figure 7.23. (a) Stress vs. strain diagram for virgin polyurethane and the three experimental nanocomposites and (b) a bar graph of the modulus of elasticity (MPa) and the toughness (MJm^{-3}). Note the relatively small error bars [12] (Courtesy of ACS).

7.3.2 Carbon-Based Thermoplastic Elastomer Nanocomposites

7.3.2.1 A Nanostructured Carbon-Reinforced Polyisobutylene-Based Thermoplastic Elastomer

Thermoplastic elastomers have a great track record of being used as biomaterials, so it is only natural that work should be done on exploring thermoplastic elastomer nanocomposites to use inside a living body. Puskas et al. explored the properties of a carbon black nanoparticle reinforced copolymer [13]. The copolymer was formed by polymerizing polyisobutylene and poly(isobutylene-co-para-methlystyrene) together in the lab, rather than purchasing them commercially. Two batches were prepared, one that was stock polymer and the other being loaded with 37.5 wt% carbon black nanoparticles. The nanoparticles conformed to ASTM N234, indicating a particle size of less than 100 nm. Blending was performed in a Brabender mixing, a laboratory-sized internal mixer. The finished polymers were denoted as D_IB(OH)-MS and as D_IB(OH)-MS_CB, the latter of which contained the carbon black. Silicone, a thermosetting elastomer, is commonly used in bio-implantation applications, so it served as the control for this experiment.

Table 7.5. *Experimental Copolymer Tested Ultimate Tensile Strength (UTS) and Ultimate Strain (US) Compared with Silicon and Two Biomaterials [13]*

ID	Shore A	Modulus at 100% strain (MPa)	UTS (MPa)	US (%)
D_IB (OH)-MS	52	0.5	6.8	660
D-IB(OH)-MS_CB	56	2.3	13.8	410
SIBS	–	0.5	7.6	650
Silicone	80	1.5	10.2	850
Hexsyn	60	3.4	13.8	314

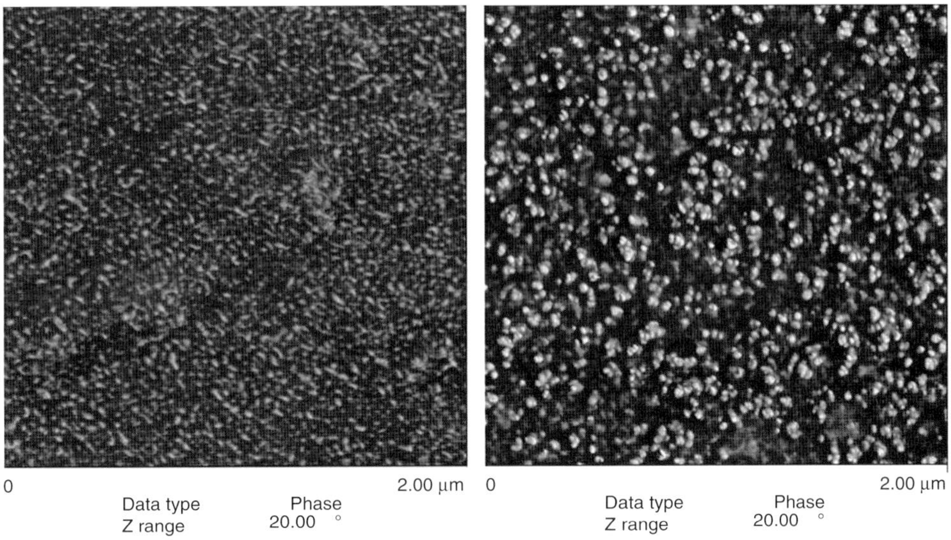

Figure 7.24. AFM of virgin copolymer (left) and copolymer with carbon black added (right). Dark material is elastomeric phase, light material is carbon black or plastic phase [13].

As shown in Figure 7.24, the carbon black seems to have been fairly well dispersed into the copolymer matrix [13]. The resulting nanocomposite seems to be quite homogenous, with little agglomeration of the 100 nm or less carbon black particles. Table 7.5 and Figure 7.25 show that in testing, the carbon black dramatically doubled the ultimate tensile strength of the experimental copolymer, but reduced its ultimate shear (elongation at break) by approximately 40% [13]. Even so, the heavily loaded copolymer (37.5 wt% carbon black) still stretched four times its length before breaking. The modulus of elasticity of the nanocomposite was also observed as being more than four times that of the unmodified copolymer [13].

7.3.2.2 Carbon Nanotube-Reinforced Polyurethane Composite Fibers

Chen, Tao, and Liu were interested in other work done with nylon-6 and carbon nanotubes [14]. Those studies had shown dramatic increases in the mechanical properties of thermoplastic polymers with the addition of nanotubes, and the authors were curious how thermoplastic elastomers performed with similar additions of

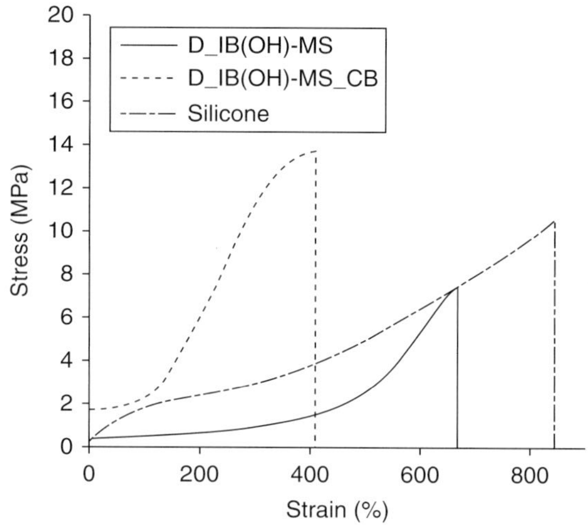

Figure 7.25. Graph of stress versus strain for tests of the experimental copolymer and silicone [13].

carbon nanotubes. Polyurethane was chosen for this study's elastomer matrix, along with multi-wall carbon nanotubes (MWNTs) with diameters between 40 nm and 60 nm. The MWNTs were chemically functionalized by washing them in an acid and then washing the acid away with water. After washing, the MWNTs were dispersed into the polyurethane in a counter-rotating twin-screw extruder, which then extruded 1.75 mm fibers out through a die. These fibers were then melt drawn down to final diameters of about 200 microns. Batches were prepared with 5.6, 9.3, and 17.7 wt% MWNTs.

The combination of acid washing and twin-screw extrusion did a good job of dispersing the carbon nanotubes into the polyurethane matrix. Figure 7.26 shows that the MWNTs were dispersed randomly and seemingly homogenously at the lower two loadings, but at the 17.7 wt% loading, fiber crowding led to some low-level agglomeration [14]. Mechanical testing resulted in promising data, which can be seen in Figure 7.27 [14]. Additional MWNTs loading lead to an increase in Young's modulus, ultimate tensile strength, and elongation at break. The increase in Young's modulus was noted with all loadings, but the increases in tensile strength and elongation at break had seemingly optimal loading levels. Ultimate tensile strength was most greatly improved at the 5.6 wt% level, and elongation at break was highest recorded at the 9.3 wt% level. The 17.7 wt% was noticeably stronger than the virgin polymer, but had a marked decrease in elongation at break. The bar graph in Figure 7.27 indicates than an overall optimal balance of increased tensile strength and elongation at break seems to exist relatively close to 9.3 wt% MWNT [14]. The authors note that, to their knowledge, simultaneous improvement in Young's modulus, tensile strength, and elongation at break in polyurethane had not been addressed before [14].

7.3.2.3 Micromechanics and Macromechanics of Carbon Nanotube-Enhanced Elastomers

Polyurethane is not the only commonly used elastomer on the market, so it should come as no surprise that studies similar to the previously mentioned paper were

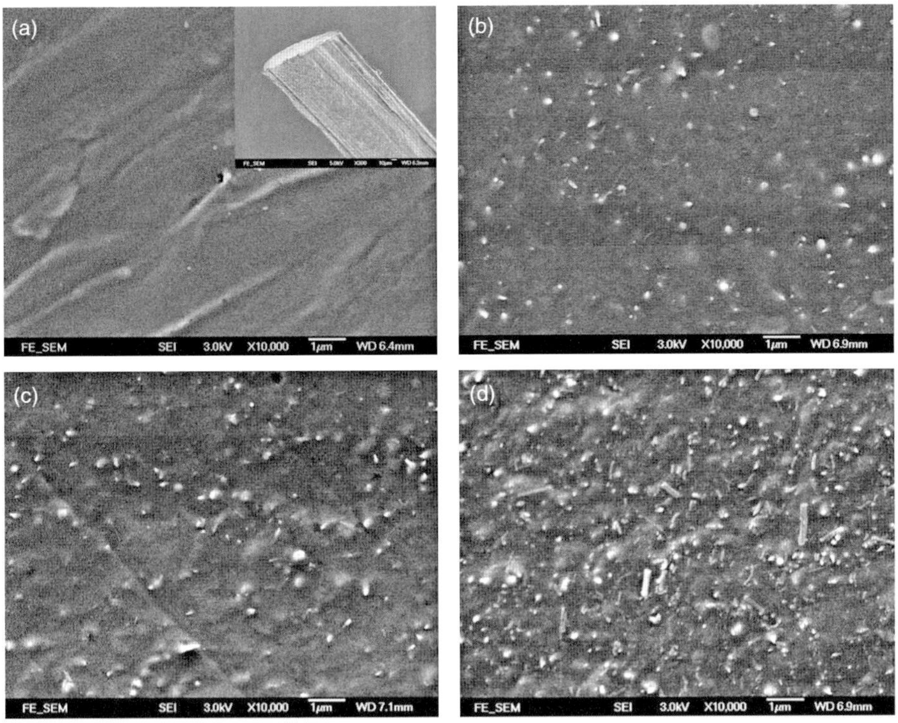

Figure 7.26. SEM images of neat polymer and inset of (a) cross-sectional composite fiber, (b) 5.6 wt% MWNT composite, (c) 9.3 wt% MWNT composite, and (d) 17.7 wt% MWNT composite [14].

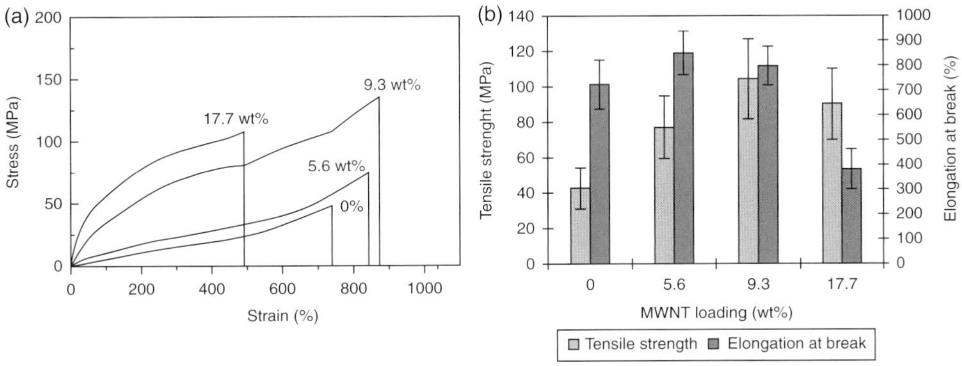

Figure 7.27. Stress vs. strain graph of the nanocomposites (left) and a bar graph summarizing the mechanical test results (right) [14].

carried out on nanocomposites of MWNT and other elastomers. Cantournet, Boyce, and Tsou experimented with a nanocomposite of a copolymer and MWNTs that had average diameters of 70–150 nm [15]. The copolymer contained two phases of isobutylene and paramethylstyrene and was sold under the ExxonMobil Chemical Company's trade name of Exxpro 3745. The nanocomposite was prepared in an internal mixer with mixing ratios of 2.5, 7.5, 10, and 15 parts per hundred of polymer

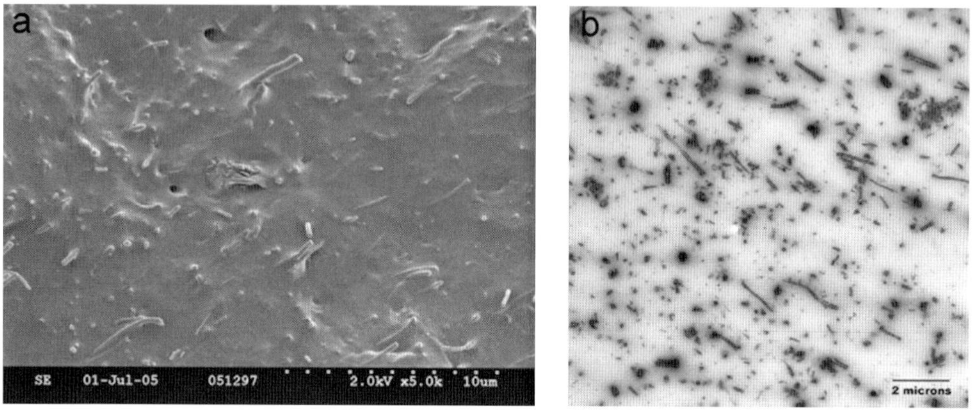

Figure 7.28. SEM (a) and TEM (b) images of freeze-fractured 15 wt% MWNT nanocomposite [15].

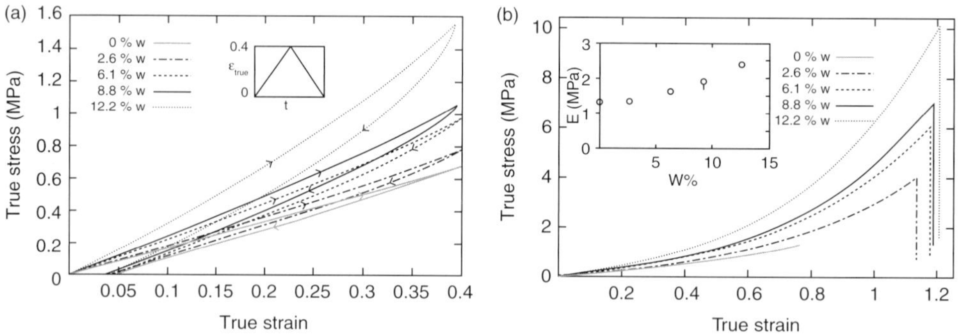

Figure 7.29. Uniaxial tension load-unload true stress vs. true strain (a) and true stress vs. true strain (b) graphs for neat polymer and prepared nanocomposites. The inserted plot charts initial modulus of elasticity vs. wt% MWNT [15].

(phr), which are equivalent to 2.6, 6.1, 8.8, and 12.16 wt% MWNT, respectively. A roll-milling operation was also carried out in order to introduce 1 phr zinc oxide and 1.5 phr stearic acid as curatives. This study also explored the construction of a model to predict the nanocomposite's mechanical behavior. The theory behind the model was based on the Arruda-Boyce eight-chain model of rubber elasticity and Spencer's framework for anisotropic hyperelastic materials.

Figure 7.28 shows the most heavily loaded batch of nanocomposites, and the random, homogenous distribution of MWNTs can be clearly seen. Similarly good distribution was reported for the other three MWNT loadings [15]. When tested, the model constructed by the authors agreed quite well with the experimental data for strains under 0.7. Uniaxial load-unload testing indicated that increasing amounts of MWNT loading increased the stiffness and hysteresis experienced by the nanocomposite (Figure 7.29) [15]. Adding more MWNTs also dramatically increased the ultimate tensile strength, modulus of elasticity, and elongation at break. At 12.2 wt% MWNT, ultimate tensile strength increased by more than a factor of 5 and Young's modulus more than doubled compared to neat polymer. The addition of even small

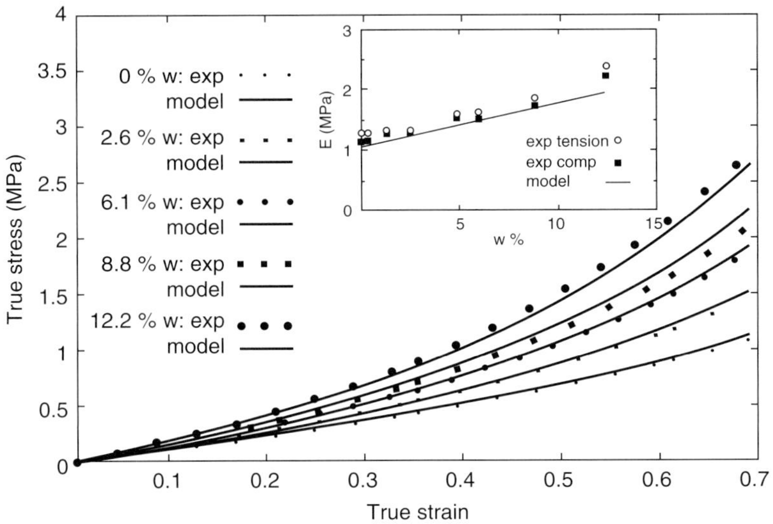

Figure 7.30. True stress vs. true strain graph for experimental and modeled MWNT-elastomer nanocomposite [15].

amounts of MWNT lead to a huge jump in elongation at break, but diminishing returns were seen with increased loading. This data can all be seen in Figure 7.30 [15]. As with the work done on polyurethane by Zhou and Burkhart, MWNTs were found to simultaneously increase the modulus of elasticity, ultimate tensile strength, and elongation at break in this particular polymer [15].

7.3.2.4 High-Shear Processing Induced Homogenous Dispersion of Pristine Multi-Walled Carbon Nanotubes in a Thermoplastic Elastomer

In many things, exactly how someone goes about achieving a goal matters just as much as the goal itself. Li and Shimizu explored a similar notion regarding elastomer nanocomposites [16]. They investigated the relationship between how an unmodified nanomaterial is mechanically mixed into a polymer and the resulting nanocomposites properties. They chose to do this by making batches of polymers in a modified high-shear screw extruder at different screw rotation speeds. The extruder used was specially modified for this study and had only a short screw with an L/D ratio of 1.78. The elastomer used in this experiment was poly(styrene-b-butadiene-co-styrene) (SBBS) and the MWNTs had tube diameters ranging from 10 nm to 40 nm and lengths between 5 μm and 20 μm. Nanocomposites with 3 wt% MWNT loading were prepared at 300, 1000, and 2000 rpm, which were referred to as CP300, CP1000, and CP2000, respectively.

The SEM images in Figure 7.31 clearly show that better dispersion of MWNTs was achieved with faster extrusion rate [16]. The MWNTs agglomerated to a high degree at the lowest screw rotation speed, and still remained agglomerated within the 1,000 rpm screw-rotation nanocomposite. However, when produced with a screw rotation speed of 2,000 rpm, the MWNT became finely dispersed into the polymer matrix. These observations were confirmed by dissolving nanocomposite samples in

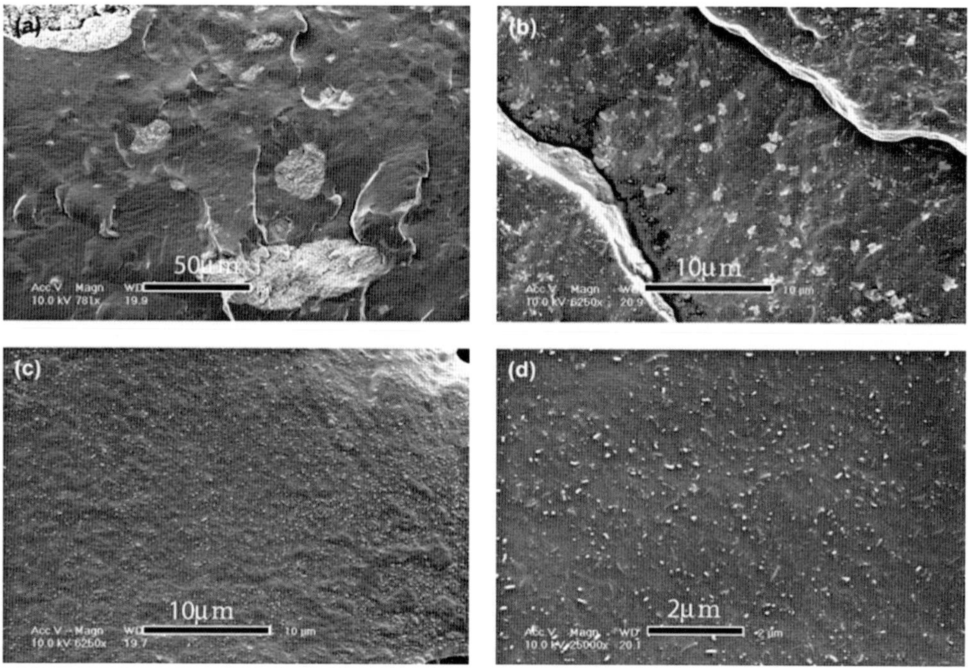

Figure 7.31. SEM images of fractured (a) CP300, (b) CP1000, and (c and d) CP2000. The light material is MWNTs [16].

toluene and examining the resulting film under a high-power optical microscope. In all three cases, the higher the rpm of the screw was used, the smaller MWNT agglomerates were observed in the film. Noticeable differences in mechanical properties were also observed among the nanocomposites, as can be seen in Figure 7.32 and Table 7.6 [16]. In all cases, the addition of MWNT to the SBBS increased the Young's modulus and the residual strain and decreased the elongation at break. Increasing the screw rotation speed was noted to lead to a greater increase in the Young's modulus, a lesser reduction in elongation at break, and a less amount of residual strain. Ultimate tensile strength was highly dependent on screw rotation speed, as it could be less than that of neat polymer at low rotation speed or more than neat polymer at high rotation speed [16]. The takeaway from this study is that superior dispersion of MWNT into SBBS elastomer leads to superior nanocomposites.

7.3.3 Other Nanomaterial-Based Thermoplastic Elastomer Nanocomposites

7.3.3.1 Mechanical and Fracture Behaviors of Elastomer-Rich Thermoplastic Polyolefin/SiC Nanocomposites

Liao and Tjong explored how the addition of 45–55 nm silicon carbide (SiC) nanoparticles to an elastomer-rich thermoplastic polyolefin (ETPO) affected fracture behavior [17]. The thermoplastic polyolefin utilized was a copolymer of 30 wt% commercial polypropylene and 70 wt% commercial poly(styrene-ethylene-butylene-styrene)

Table 7.6. *Summary of Mechanical Properties of Neat Elastomer and Nanocomposites [16]*

	Modulus (MPa)	Elongation at break (%)	Strength (MPa)	Residual strain (%)
Neat SBBS	12.50	1,108	19.59	20.34
CP300	14.31	628	9.45	34.41
CP1000	23.74	870	14.99	30.22
CP2000	25.31	917	24.08	22.63

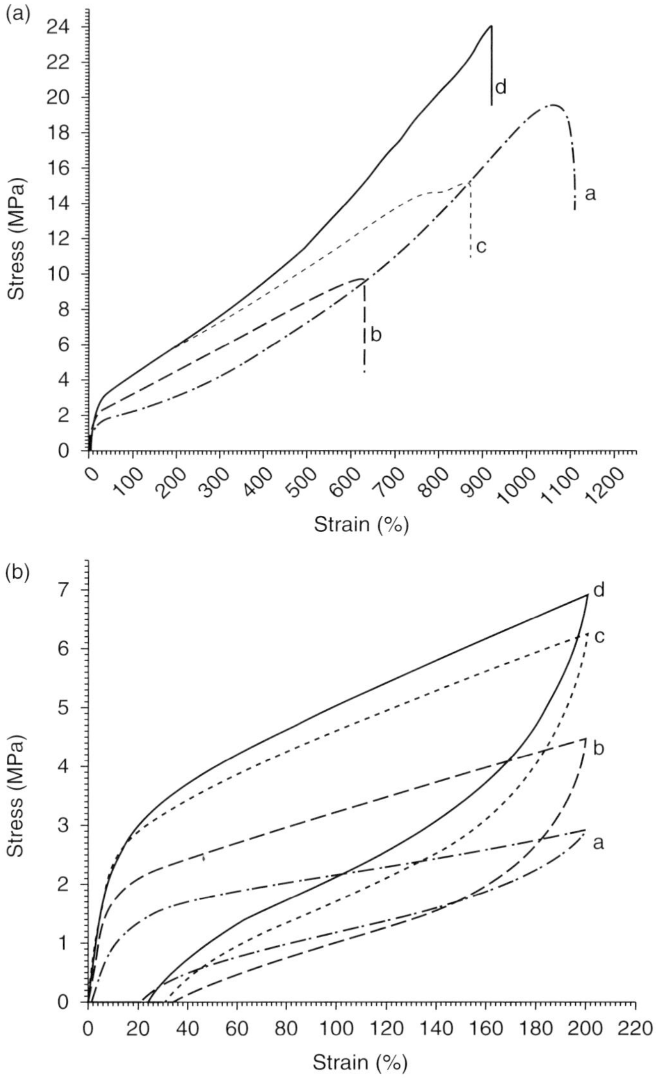

Figure 7.32. Graphs for stress vs. strain (left) and strain relaxation (right) for (a) neat SBBS, (b) CP300, (c) CP1000, and (d) CP2000 [16].

grafted with maleic anhydride. The copolymer was prepared in a twin-screw extruder to which silicon carbide nanoparticle loadings of 1, 3, and 5 wt% were introduced. Izod impact tests were also conducted, in addition to traditional tensile tests at two different crosshead speeds.

Table 7.7. *Mechanical Properties of ETPO and SiC Nanocomposites [17]*

Sample	HDT (°C)	Young's Modulus (MPa)	Tensile strength (MPa)				Impact strength (kJ/m^2)
			10 mm/min		300 mm/min		
		10 mm/min	50%	300%	50%	300%	
ETPO	53.64	269 ± 5	6.92 ± 0.02	8.00 ± 0.11	8.57 ± 0.10	9.30 ± 0.05	2.84 ± 0.02
ETPO-1 wt% SiC_p	54.32	255 ± 7	6.61 ± 0.22	7.80 ± 0.02	8.30 ± 0.12	9.18 ± 0.14	2.90 ± 0.02
ETPO-3 wt% SiC_p	55.28	236 ± 10	6.21 ± 0.13	7.54 ± 0.16	7.81 ± 0.09	8.96 ± 0.13	2.92 ± 0.02
ETPO-5 wt% Sic_p	56.57	238 ± 12	6.22 ± 0.08	7.86 ± 0.09	7.92 ± 0.21	9.10 ± 0.07	3.11 ± 0.25

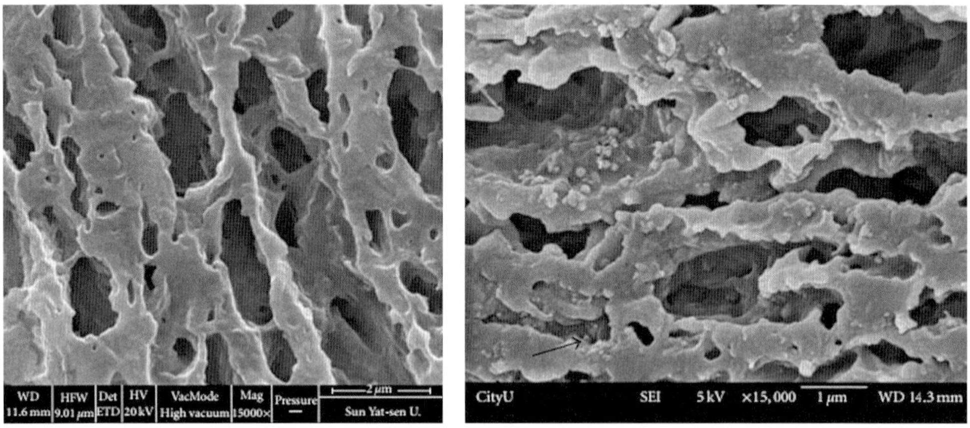

Figure 7.33. SEM images of etched virgin ETPO (left) and etched EPTO with 3wt% SiC (right). The etchant removed the elastomer and left the polypropylene behind. The small light-colored particles in the right image are SiC nanoparticles [17].

Figure 7.33 shows the SiC particles left behind in groups after etching [17]. The study notes that the nanoparticles were not well dispersed and tended to agglomerate within the polymer matrix, and that a sizeable percentage of the SiC particles cannot be seen in the SEM images, as they were removed along with the elastomer during etching. Table 7.7 and Figure 7.34 list and chart the mechanical properties of the ETPO-SiC nanocomposites [17]. As SiC loading increased, the Young's modulus and tensile strength at a given strain or strain rate decreased slightly. Elongation at fracture was not explored in this work, but all nanocomposite compositions demonstrated capacities to withstand strains in excess of 600%. A modest increase in impact strength can be seen with increased SiC nanoparticle content, and the paper notes that the fracture toughness of the material was five times higher in the 5 wt% SiC nanocomposite compared to the virgin copolymer [17]. The authors made no mention of this, but it may be possible that the reduction in tensile strength may have been related to the poor dispersion of the SiC nanoparticles within the polymer matrix.

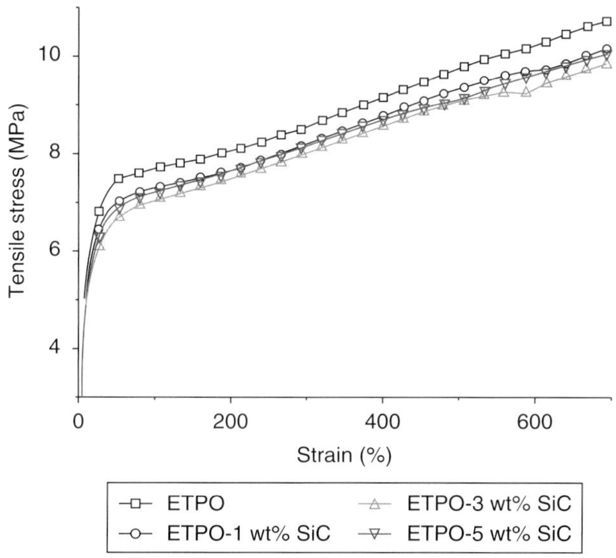

Figure 7.34. Stress vs. strain graph for (ETPO)-SiC nanocomposites [17].

7.3.3.2 The Influence of Surface Modification on the Structure and Properties of a Nanosilica-Filled Thermoplastic Elastomer

Some of the previous studies have shown the importance of surface treatments for organic nanoclays in thermoplastic elastomers, and Aso, Eguiazabal, and Nazabal explored the same question for nanosilica [18]. The study examined nanocomposites formed by dispersing unmodified or surface-modified nanosilica into Hytrel (a DuPont brand name of a thermoplastic copolyetherester elastomer) using a twin-screw extruder. The unmodified nanosilica particles had an average size of 16 nm, while the modified particles averaged 12 nm in size. Composites of 0.5, 1, 2, 3, and 6 wt% nanosilicates were prepared. The XRD patterns of neat Hytrel, 6 wt% modified nanosilicate, and unmodified nanosilicates are shown in [18].

Even without the modifier, the nanosilica was fairly well dispersed, as can be seen in Figure 7.35d, but the structures formed were as large as 1 μm in some cases [18]. The surface modification helped reduce the size of these agglomerates, however, as can be seen by comparing Figures 7.35b to 7.35d. The surface modification reduced most of the nanosilica agglomerates to between 15 nm and 150 nm. This difference in agglomerate size likely contributed to the resulting differences in mechanical properties seen in the nanocomposites. Figure 7.36 shows that the Young's modulus increased as more modified or unmodified nanosilica was blended into the Hytrel matrix, but this was not the case with elongation at break [18]. The addition of modified nanosilica, with its smaller agglomerates, led to a modest increase in elongation at break. Conversely, the addition of unmodified nanosilica reduced the Hytrel's elongation at break to a greater degree than the modified nanosilica increased it. The paper also makes note that the addition of modified nanosilica improved the ability of the polymer to resist creep, and the addition unmodified nanosilica improved creep resistance even more [18]. Clearly, just as with clays, the surface modification

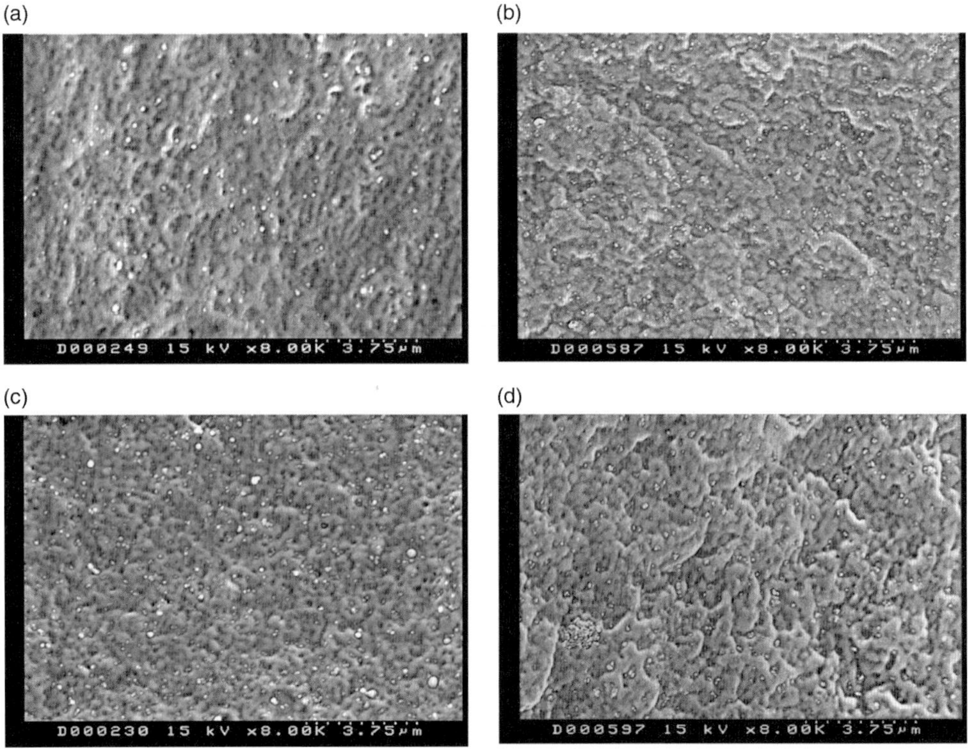

Figure 7.35. SEM of modified nanosilicate composite (a) 1 wt%, (b) 3 wt%, (c) 6 wt%, and (d) 3 wt% unmodified nanosilicate [18].

applied to nanosilica is highly important when determining a nanocomposite's mechanical properties.

7.3.3.4 Thermal and Mechanical Properties of Poly(ether ester)-Based Thermoplastic Elastomer Composites Filled with TiO_2 Nanoparticles

Different nanoparticles affect different materials in different ways, resulting in a great number of possibilities that need to be investigated. Zhou and Burkhart explored one such set of possibilities with their mechanical and thermal investigation of elastomer and titanium dioxide nanocomposites [19]. The elastomer used was a poly(butylene terephthalate)-block-poly(tetramethylene glycol) (PBT-PTMG) and the nanomaterial was Aeroxide 25P, unmodified hydrophilic TiO_2 nanoparticles from Evonik Industries with an average diameter of 21 nm. Batches of nanocomposites were processed in a twin-screw extruder with 1.3, 2.6, and 4.9 vol% TiO_2.

The level of dispersion achieved by the TiO_2 in this study was poor. Figure 7.37 clearly shows the hydrophilic TiO_2 agglomerating into micron-sized structures throughout the elastomer matrix [19]. The study does note, however, that the matrix coated these agglomerates with fiber-like structures, which the authors claim indicates a strong adhesion between the polymer and the nanoparticles near the surface of the agglomerates. Figure 7.38 summarizes well the results of mechanical testing, which was that the nanoparticles changed little mechanically [19]. None of the

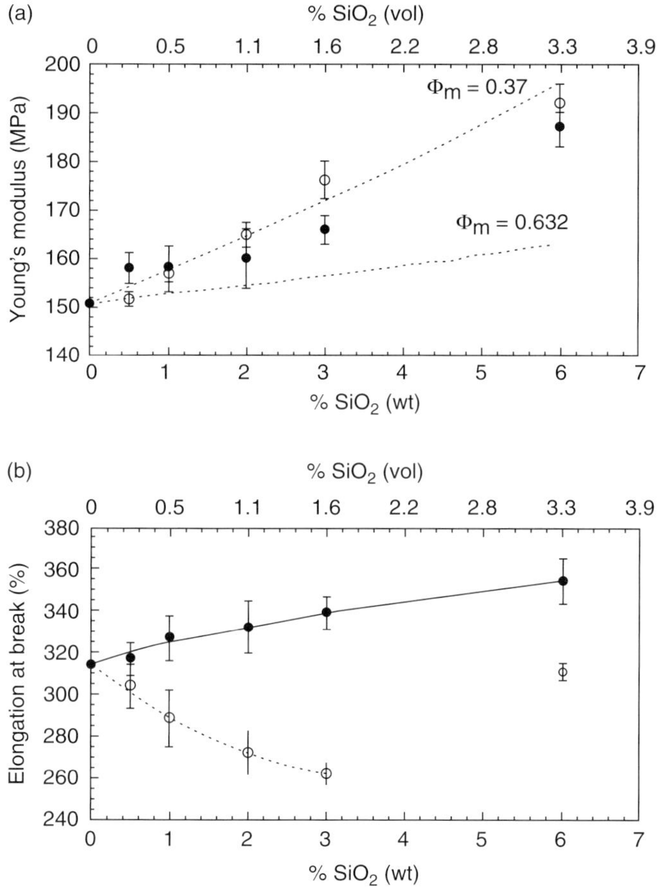

Figure 7.36. Graph of Young's modulus vs. wt% nanosilica (left) and elongation at break vs. wt% nanosilica (right). Solid dots are modified nanosilica composites; hollow circles are unmodified nanosilica [18].

nanocomposites broke prior to the tests limit of 300% shear, so the TiO_2's effect on ultimate tensile strength and ultimate shear is unknown. The one thing that is clear is that the addition of the nanoparticle modestly increased the modulus of elasticity. This corresponded to slightly higher tensile strengths between 0 and 200% strain with increased TiO_2 loading. The lackluster mechanical performance of this nanocomposite is likely due the highly agglomerated nature of the nanoparticles. While not covered here, the paper does note a noticeable improvement in the thermal characteristics of the nanocomposites over the neat elastomer [19].

7.3.4 Summary of Thermoplastic Elastomer-Based Nanocomposites

The first general observation that can be derived from this review is that when adding a nanomaterial to a thermoplastic elastomer, it is crucial that the surface of the nanomaterial be compatible with the polymer. If not, the nanomaterial will just agglomerate into micron-sized structures. The method of processing also must suitably

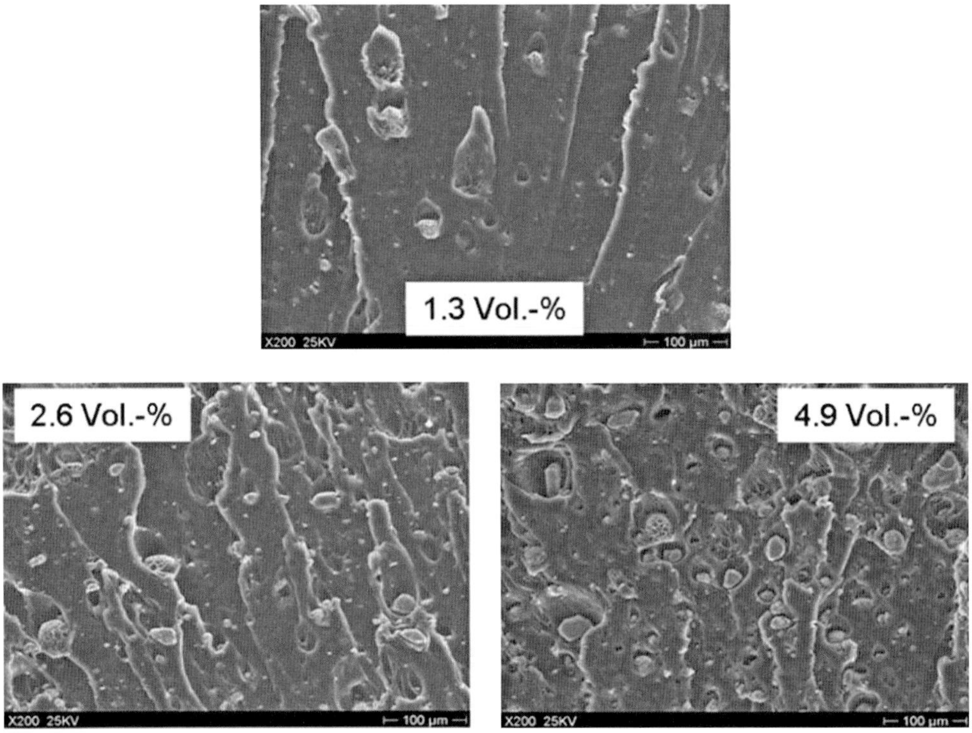

Figure 7.37. SEM images of the elastomer and TiO_2 nanocomposites [19].

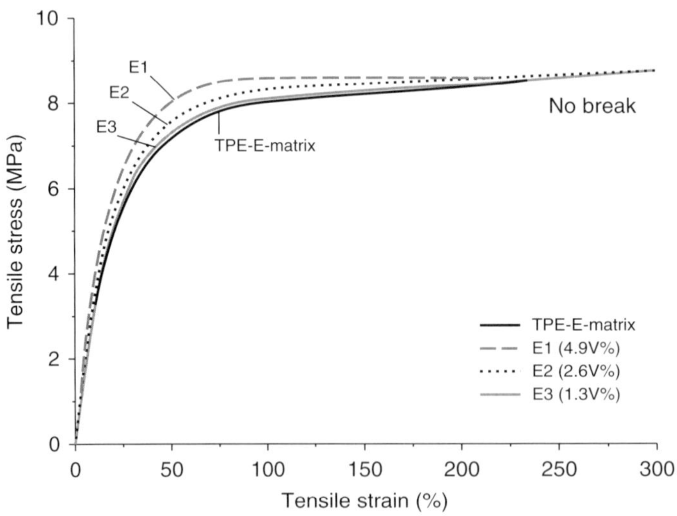

Figure 7.38. Stress vs. strain graph of stock and nanocomposite elastomer [19].

disperse the nanomaterial, or else, again, it will just agglomerate. If sufficiently dispersed, the material will almost always display an increased Young's modulus and will frequently have a higher ultimate tensile strength. The amount of elongation at break is typically negatively impacted by the addition of nanomaterials, but in certain cases it can actually be improved along with Young's modulus and ultimate tensile strength.

As the base polymers are elastomers, the final elongation at break may still be considerable even after being reduced by adding a nanomaterial. Also, the same nanomaterial may preform differently in different thermoplastic elastomer, so it is crucial that perspective formulators of nanocomposites be aware of what is possible from what combinations of nanomaterials and polymers. Hopefully researches will continue to explore thermoplastic elastomer nanocomposites, because there certainly seems to be a great deal of promise there.

7.4 Thermoset-Based Nanocomposites

Thermoset nanocomposites are widely used in many applications, including the aerospace industry. For the past few decades, composite materials have come into use in defense applications, due to the improved efficiencies caused by the weight reduction compared to their metallic counterparts. The higher performance demands of new applications require composites with improved material properties. Composites enhanced with nanomaterials serve to improve properties, such as Young's modulus, toughness, and yield strength while still maintaining the weight advantage these materials offer. In this section, examples focus on the research conducted to analyze the enhancements of mechanical properties for epoxy-based nanocomposites with nanoclay, carbon-based, nanosilica, polyimide-based nanocomposites, cyanate ester-based nanocomposites, and other thermoset-based nanocomposites.

7.4.1 Epoxy Nanocomposites

7.4.1.1 Epoxy-Clay Nanocomposites

Research conducted by Kinloch and Taylor [20] tests the mechanical properties and fracture toughness of DGEBA, a digliycidylether of bisphenol A modified with montmorillonite clay. A comparison to theoretical mechanics models was included in the research. The types of clay used were Cloisite 25A, Cloisite 30B, Cloisite Na^+, and Nanomer I30E. Plates of epoxy composite were produced by stirring the clay modifier into the epoxy and then baking in a vacuum oven at 75°C. The vacuum was removed and the plates were baked at this temperature for another 24 hours. Hardener was added to the mixture, stirred, and cured for 3 hours at 75°C, followed by another round of curing for 12 hours at 110°C.

Characterization of the dispersion of the nanomaterial was performed using wide-angle X-ray diffraction (WAXD) and transmission electron microscopy (TEM) analyses. Mechanical testing was performed at a displacement rate of 1 mm/min and the Young's modulus was determined.

Several analytical models were employed for comparison to experimental data. For the prediction of the tensile modulus, the modified rule of mixtures, the Halpin-Tsai model, and the Mori-Tanaka model were utilized. For the prediction of the fracture energy, the Farber and Evans analysis was utilized.

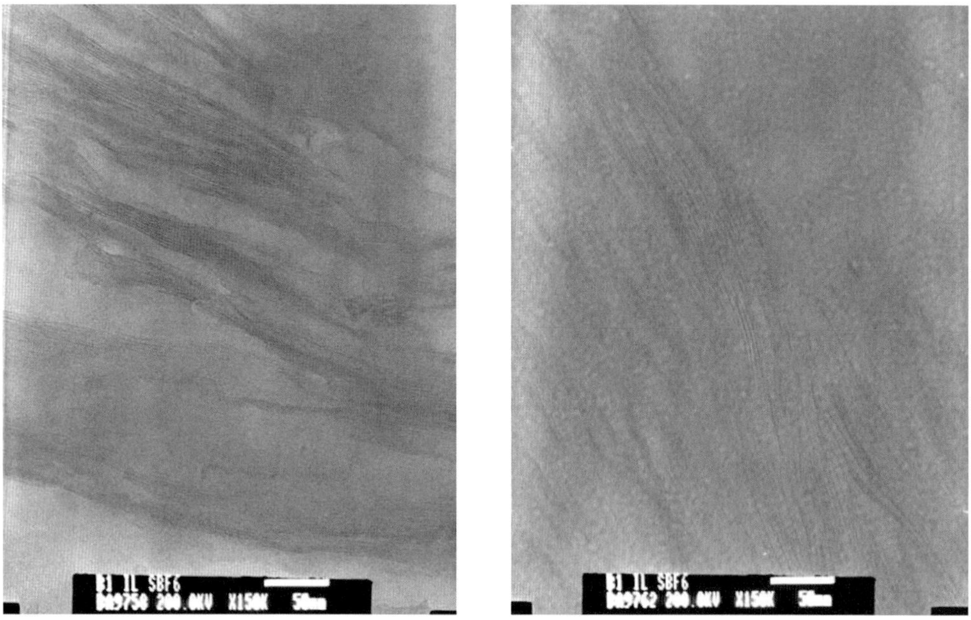

Figure 7.39. Transmission electron micrograph of epoxy nanocomposite with 5 wt% Cloisite 25A (left), and 5 wt% Nanomer 130E (right).

WAXD data showed differing levels of dispersion of the nanoparticles. A loading of 5 wt% Nanomer I30E reveals intensity with no '001' peak, indicating an exfoliated nanostructure. A loading of 5 wt% Cloisite 25A and Cloisite 30B revealed a '001' peak that indicated an intercalated nanostructure, with darker areas representing the agglomerated clay. TEM analysis confirmed these observations, further adding that the exfoliated structure formed in the I30E composite was of an ordered nature, as shown in Figure 7.39 [20].

The tensile modulus experiments show that as the mass of the modifier increases, the Young's modulus increases [20]. The nanoparticle seen as the best modifier of the Young's modulus was Nanomer I30E. However, at 10 wt%, the I30E mixture is too viscous and reaches the maximum amount of modifier that can be added to the filler material. The other modifiers follow a similar modulus trend as nanomaterial is added to the epoxy, but reach a viscosity limit at 15 wt%. The exception to this trend is Cloisite Na^+, which reaches a viscosity limit of 25 wt% and a modulus improvement of 2.5 GPa (unmodified epoxy is 3 GPa).

The fracture toughness experiments show an initial increase of the toughness at the smallest amount of nanomaterial added, but further additions of material eventually reveal a peak in the toughness followed by a decrease that occurs roughly around 2–4 wt% loading, with the exception of Cloisite 25A (~10%) (Figure 7.40) [20].

The analytical models for the tensile modulus agreed fairly well with the tensile data, specifically the Mori-Tanaka model. The model used to predict fracture toughness did not produce acceptable correlation with the experimental data. This is thought be caused by a fracture mechanism dominated by cavities formed by plastic

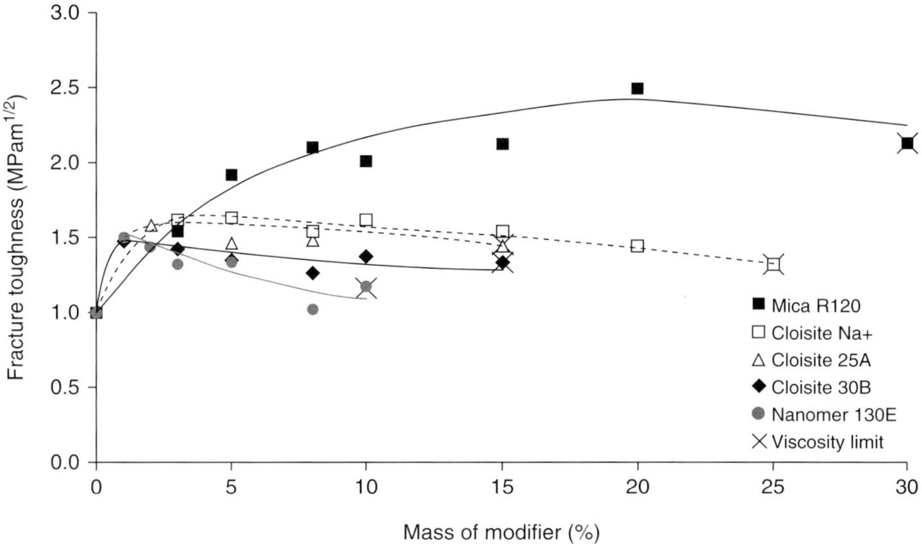

Figure 7.40. Fracture toughness of modified epoxy nanocomposites.

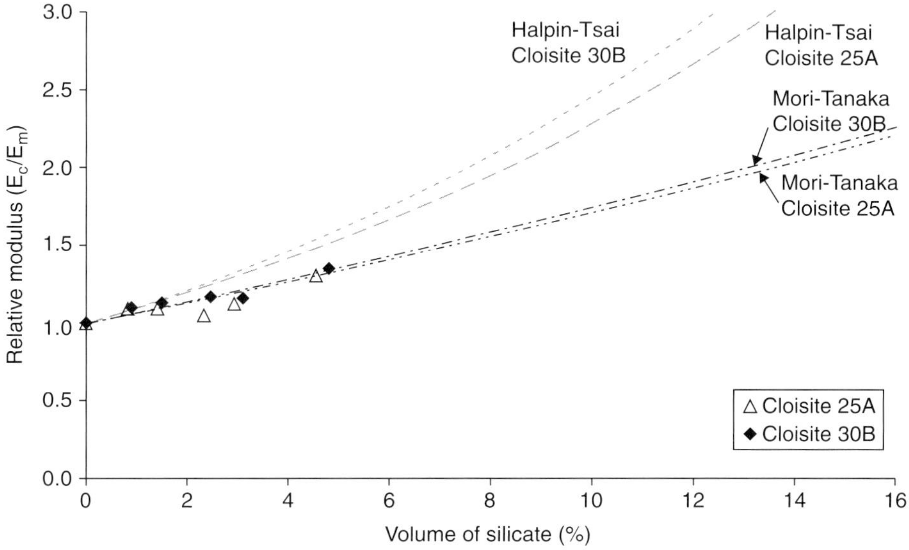

Figure 7.41. Relative modulus (composite modulus divided by unmodified epoxy modulus) versus volume percentage of silicate for Cloisite 25A and 30B.

deformation, as opposed to a crack deflection mechanism described by the Farber and Evans model. Model data are shown in Figure 7.41 [20].

7.4.1.2 Epoxy Carbon-Based Nanocomposites

A study done by Gojny et al. [21] explored the influence of different kinds of carbon nanotubes on the mechanical properties of epoxy matrix composites. The epoxy matrix used in their study was a DGEBA epoxy resin modified with single-walled nanotubes (SWNT) with an average diameter of 2 nm, double-walled nanotubes (DWNT) with an average diameter of 2.8 nm, and multi-walled nanotubes (MWNT) with an average diameter of 15 nm. Additionally, DWNT and MWNT functionalized with ammonia were used as resin nano-modifiers (-NH_2). Resin samples were prepared with 0.1, 0.3, and 0.5 wt% nanomaterial for each of the aforementioned materials.

The epoxy resin was prepared by coarsely mixing the nanotubes into the resin without hardener, followed by high-shear mixing with a gap space of 5 microns and speed ranging from 20 rpm to 180 rpm. Following the mixing, a hardener was added to the mixture and then cured for 24 hours at room temperature, followed by additional curing at 60°C for 24 hours.

Mechanical characterization was performed according to DIN EN 527.1/2 with dog bone specimens using a Zwich universal tensile tester at a crosshead speed of 1 mm/min. Fracture toughness testing was performed according to ASTM standards and at a crosshead speed of 1.3 mm/min.

TEM images revealed adequate dispersion for all samples used. MWNT samples showed the best dispersion of the modifier, with individual nanotubes spaced evenly throughout the TEM sample as shown in Figure 7.42 [21]. Both SWNT and DWNT showed a mixture of small broken-up agglomerates and individual nanotubes. The ammonium-functionalized MWNT and DWNT showed better dispersion of the nanotubes and an overall smaller size agglomerate diameter when compared to their non-functionalized counterparts.

In general, the addition of carbon nanotubes to the epoxy resin increased the Young's modulus, ultimate tensile strength, and fracture toughness for all types of carbon nanotube and weight percentage. The experimental data are shown in Table 7.8 [21].

The most notable property improvement is the functionalized DWNT, resulting in a 14% improvement in Young's modulus, an 8% increase in ultimate tensile strength, and a 90% improvement in fracture toughness. In general, fracture toughness was greatly enhanced for all modifiers and weight percentages when viewed on a percentage basis. Mechanical properties improve as weight percentage increases from 0.1% to 0.3% for all nano-modifiers, but decreases occur at 0.5 wt% for unfunctionalized DWNT and MWNT modifiers and in fracture toughness for SWNT nanomaterial.

In summary, the study revealed that the addition of carbon-based nanomaterials to the DGEBA epoxy matrix improved Young's modulus and ultimate tensile strength while increasing the toughness of the material.

Research performed by Zheng et al. [22] explored the properties of epoxy resin TDE-85 modified with multi-wall carbon nanotubes. Two types of multi-wall carbon nanotubes were utilized for dispersion: untreated multi-wall carbon nanotubes (MWNTs) and amino-functionalized carbon nanotubes (MWNTs-NH_2). This chemical modification is thought to enhance the compatibility of the particle with the matrix. The diameter of the nanotubes ranged from 30 nm to 100 nm. The MWNTs were heated and sonicated for 30 minutes in a trithyleneyreamine ethanol solution. The

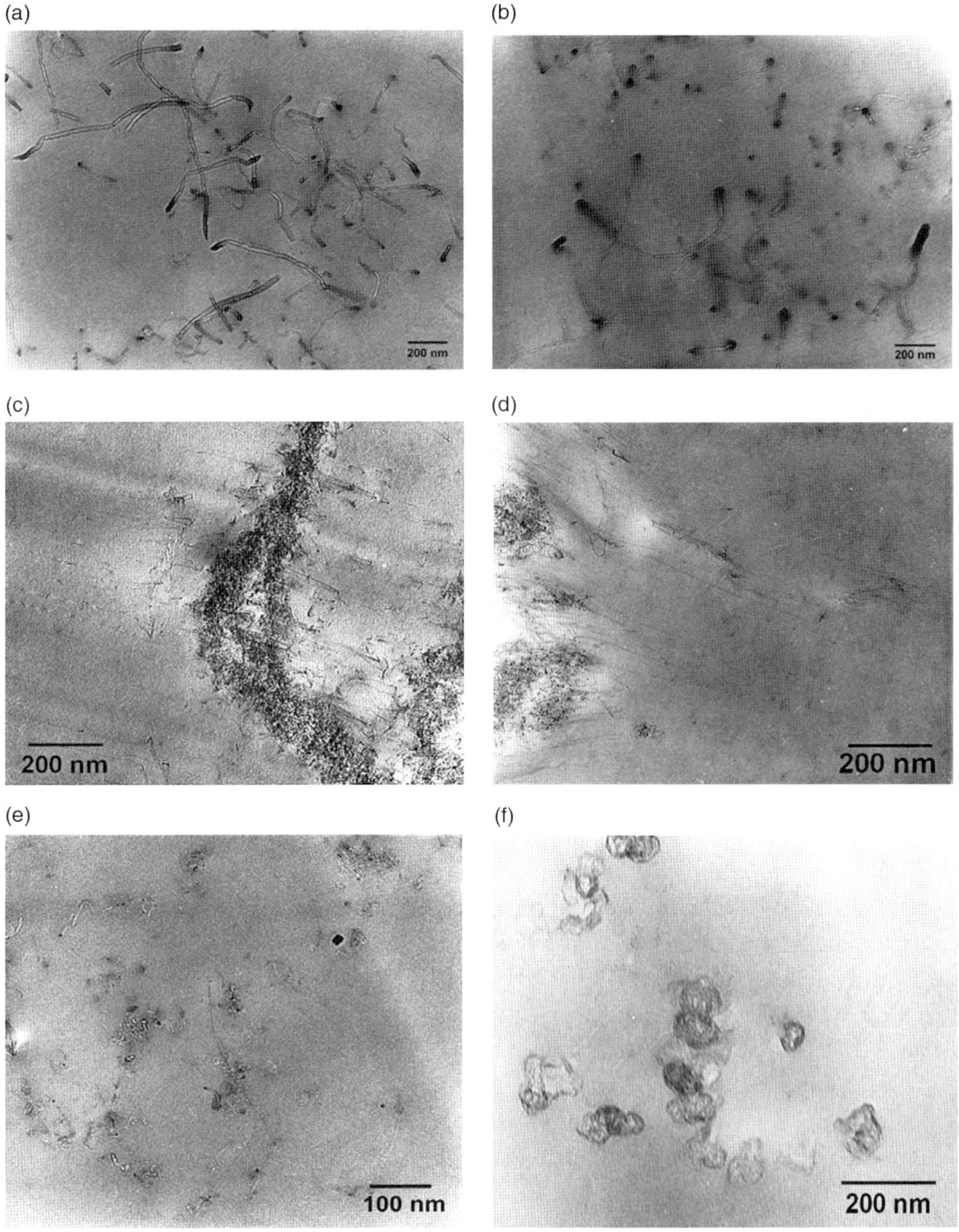

Figure 7.42. TEM images of nanocomposites: (a) epoxy/MWNT, (b) epoxy/MWNT–NH_2, (c) epoxy/DWNT, (d) epoxy/DWNT–NH_2, (e) epoxy/SWNT, and (f) epoxy/CB.

nanotubes were then added to the epoxy resin at 120°C to lower the viscosity of the resin. The nanocomposite slurry was then dispersed in a high-speed homogenizer of 20,000 rpm for 20 minutes and degassed at 90°C for 30 minutes. The epoxy resin was mixed with 0.2%, 0.4%, 0.6%, and 1.0% of MWNTs, and MWNTs-NH_2 separately.

The chemical modification of the carbon nanotubes proved to be an effective way to improve dispersal. TEM images of the epoxy resins showed the improvement

Table 7.8. *Mechanical Properties of Epoxy Nanocomposites [21]*

	Filler type/ content (wt%)	Young's modulus (MPa)	Ultimate tensile strength (MPa)	Fracture toughness KIc (MPa $m^{1/2}$)
Epoxy	0.0	2599 (±81)	63.80 (±1.09)	0.65 (±0.062)
Epoxy/CB	0.1	2752 (±144)	63.28 (±0.85)	0.76 (±0.030)
	0.3	2796 (±34)	63.13 (±0.59)	0.86 (±0.063)
	0.5	2830 (±60)	65.34 (±0.82)	0.85 (±0.034)
Epoxy/SWCNT	0.05	2681 (±80)	65.84 (±0.64)	0.72 (±0.014)
	0.1	2691 (±31)	66.34 (±1.11)	0.80 (±0.041)
	0.3	2812 (±90)	67.28 (±0.63)	0.73 (±0.028)
Epoxy/DWCNT	0.1	2785 (±23)	62.43 (±1.08)	0.76 (±0.043)
	0.3	2885 (±88)	67.77 (±0.40)	0.85 (±0.031)
	0.5	2790 (±29)	67.66 (±0.50)	0.85 (±0.064)
Epoxy/DWCNT-NH2	0.1	2610 (±104)	63.62 (±0.68)	0.77 (±0.024)
	0.3	2944 (±50)	67.02 (±0.19)	0.92 (±0.017)
	0.5	2978 (±24)	69.13 (±0.61)	0.93 (±0.030)
Epoxy/MWCNT	0.1	2780 (±40)	62.97 (±0.25)	0.79 (±0.048)
	0.3	2765 (±53)	63.17 (±0.13)	0.80 (±0.028)
	0.5	2609 (±13)[a]	61.52 (0.19)[a]	[a]
Epoxy/MWCNT-NH2	0.1	2884 (±32)	64.67 (±0.13)	0.81 (±0.029)
	0.3	2819 (±45)	63.64(0.21)	0.85 (±0.013)
	0.5	2820 (±15)	64.27 (±0.32)	0.84 (±0.028)

[a] High-viscosity disabled degassing – composite contained numerous voids.

in the dispersal of the carbon nanotubes, indicated by the appearance of individual nanotubes in the chemically modified resin. The experimental data from impact strength, bending modulus, and bending strength show improved performance for the chemically modified MWNT, further indicating enhanced properties due to improved particle dispersion. The mechanical properties discussed later will address the improvements in performance from the inclusion of the chemically modified MWNTs. TEM image of dispersion is shown in Figure 7.43 [22].

The addition of MWNTs to the neat resin showed an increase of impact strength, bending modulus, and bending strength up to 0.6 wt% carbon nanotube. The peak of property performance occurs at 0.6 wt% for the aforementioned properties. A maximum enhancement of 80%, 50%, and 100% were observed for impact strength, bending modulus, and bending strength, respectively. It was observed that the addition of carbon nanotubes past 6 wt% resulted in a decrease of the aforementioned properties (up to 1 wt%). Additionally, it should be noted that the addition of MWNT to the neat resin increased the density with greater wt%, but this increase in density is minimal. Experimental data for the above observations are shown in Figure 7.44 [22].

Research conducted by M. Rafiee and J. Rafiee [23] considered the previously mentioned MWNT and SWNT, but also included graphene platelets (GPL) as a modifier in an epoxy matrix. A graphene platelet consists of multiple stacked sheets of single-atom thick graphene. Young's modulus, tensile strength, and fracture toughness were measured in a series of experiments. The graphene platelets, SWNTs, and

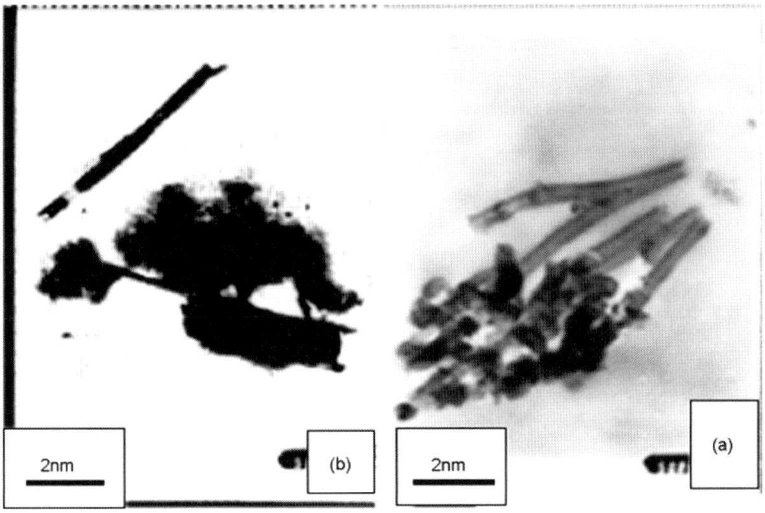

Figure 7.43. Transmission electron microscopy image of (a) unmodified MWNTs and (b) MWNTs-NH_2.

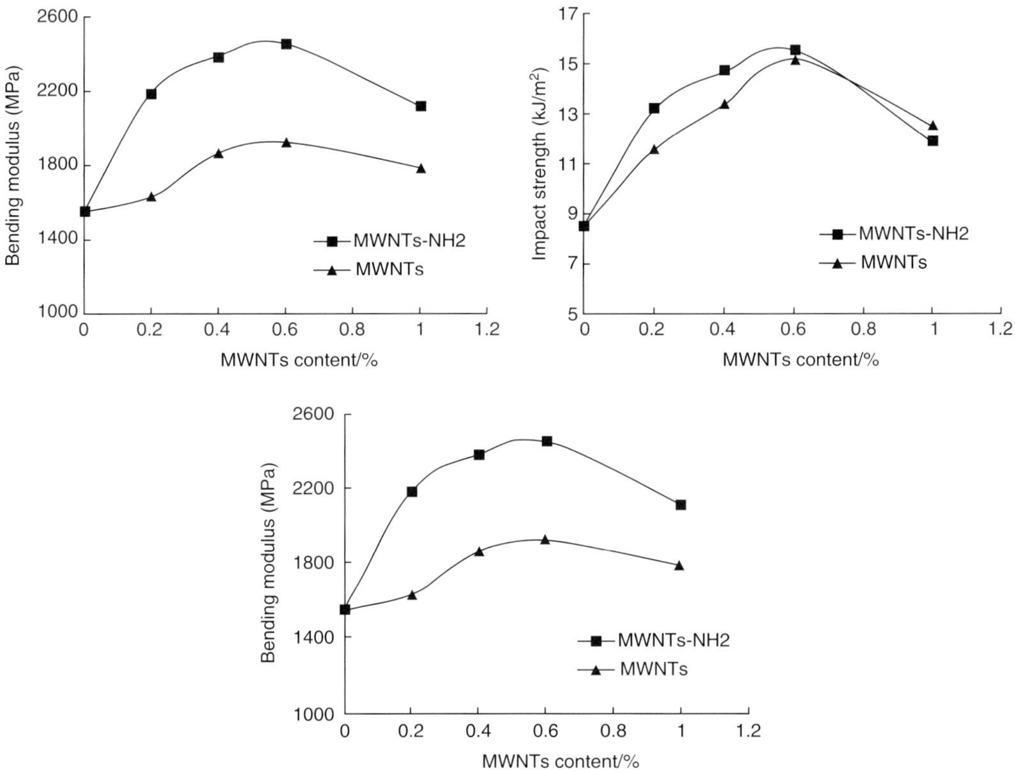

Figure 7.44. Plots of mechanical properties for modified and unmodified MWNT [22].

MWNTs were dispersed in a bisphenol A epoxy. The sample preparation is as follows: nanoparticles are sonicated in a solvent and stirred with a Teflon magnet. The solvent is removed and the mixture is processed in a high-shear mixer. The resulting mixture is then degassed under a vacuum and cured in the final part molds. All

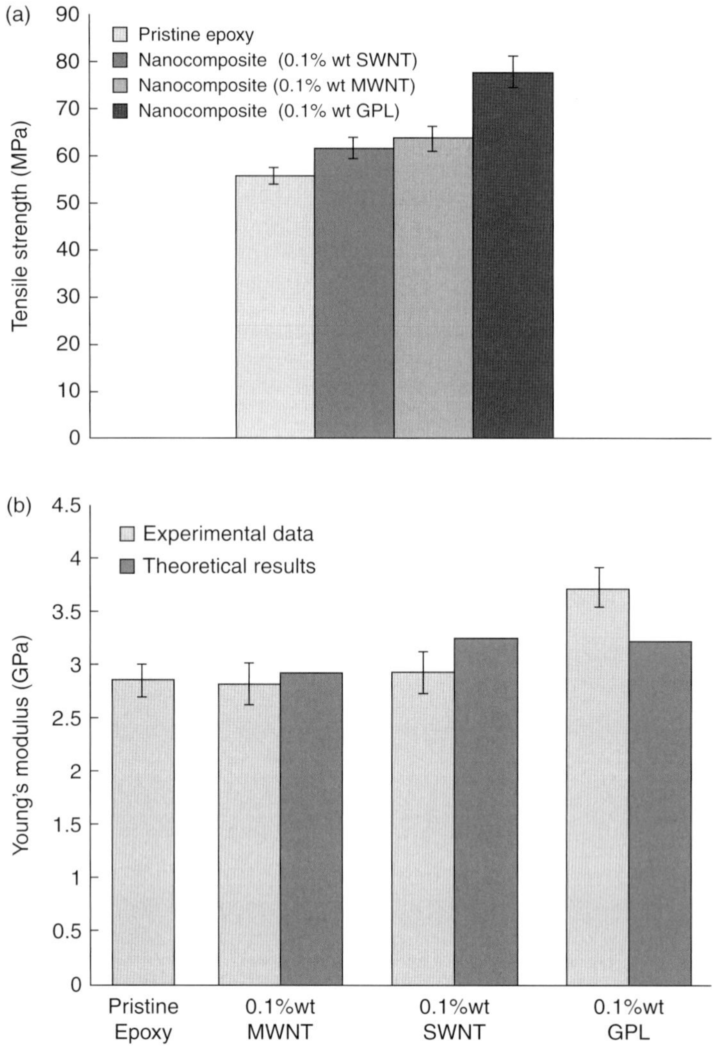

Figure 7.45. Tensile strength and Young's modulus plots for carbon-modified epoxies [23] (Courtesy of ACS).

mixtures prepared contained 0.1 wt% of nanomaterials. TEM images revealed that the preparation described above produced a high level of dispersion of the nanomaterials within the epoxy matrix.

Tensile strength tests revealed that all nanoparticle additives improved the property performance when compared to the pristine epoxy. The graphene platelets were shown to improve the performance the most, followed by MWNT and SWNT. The graphene platelets improved the pristine epoxy tensile strength performance by 45%, as shown in Figure 7.45 [23].

The Young's modulus was also shown to improve the performance of the pristine epoxy, with the graphene platelets leading to the greatest improvement, followed by SWNT and MWNT. The GPL improved the pristine epoxy modulus by ~30%. The results for these experiments are shown in Figure 7.45 [23].

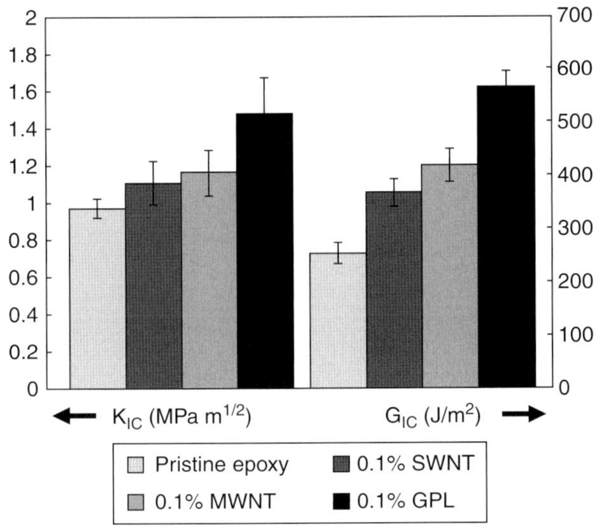

Figure 7.46. Fracture toughness and fracture energy plots for carbon-modified composites [23] (Courtesy of ACS).

Fracture toughness and fracture energy showed similar trends seen in the tensile strength tests. The graphene platelets performed the best, followed by MWNT and SWNT. All nanoparticles improved the pristine epoxy performance, with GPL improving the fracture toughness by 50% and the fracture energy by 125%. The results of these fracture toughness tests are shown in Figure 7.46 [23].

7.4.1.3 Epoxy with Other Nanomaterials

Blackman et al. [24] explored the fracture and fatigue behavior of diglycidyl ether of bisphenol A (DGEBA) epoxy resin modified with silica (SiO_2) nanoparticles. The nanoparticles are dispersed via the sol-gel method and cured for 1 hour at 90°C followed by 2 hours at 160°C. Specimens were prepared for 4.0, 7.8, 14.8, and 20.2 wt% silica. Atomic force microscopy characterized the level of dispersion of the nanoparticles, and revealed good dispersion of the nanomaterial. An AFM image of the dispersion is shown in Figure 7.47 [24].

Glass temperature and tensile and fracture toughness tests were applied to the silica-enhanced specimen. Tests revealed that there was no substantial increase in glass temperature with increasing amounts of weight percent silica. The tensile tests revealed that increasing weight percent silica led to an increase in Young's modulus. Furthermore, the addition of silica nanoparticles was positively correlated with fracture toughness. A 20.2 wt% silica addition to the neat resin led to a 30% increase in modulus and a 70% increase in fracture toughness. An analysis of the crack growth rate per cycle revealed that the value of the range of the applied stress intensity factor at threshold increases steadily as the fracture toughness increases, or as the concentration of silica increases. Table 7.9 shows the experimental data of these epoxy-SiO_2 nanocomposites [24].

Al-Turaif studied the effect of nano titanium dioxide (TiO_2) on the mechanical properties of cured epoxy resins [25]. Particles of titanium oxide with sizes of 220 nm, 50 nm, and 17 nm and at 1, 3, 5, and 10 wt% were dispersed into a DER331

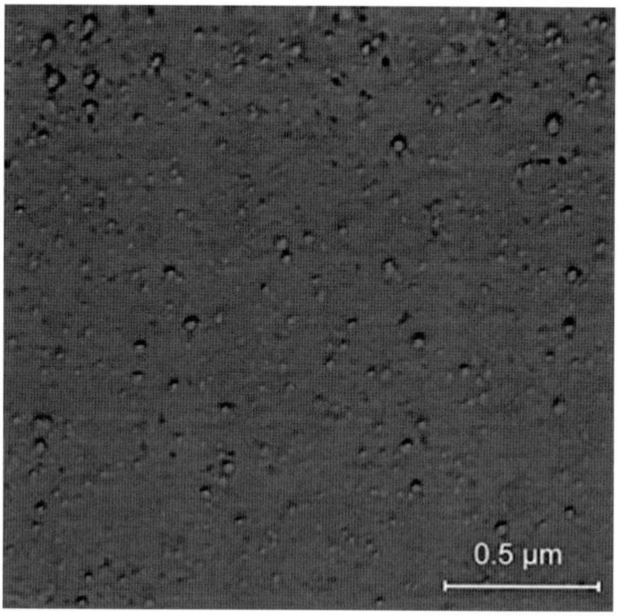

Figure 7.47. Atomic force microscopy image showing the dispersion of SiO_2 nanoparticles in the epoxy matrix [24].

Table 7.9. *Mechanical Properties of Nanomodified Composites [24]*

wt.% nano-SiO_2	T_g (°C)	E (GPa)	K_{Ic} (MPa $m^{1/2}$)
0.0	153	2.96	0.51
4.0	152	3.20	0.65
7.8	154	3.42	0.79
14.8	152	3.60	0.83
20.2	150	3.85	0.88

epoxy, an undiluted DGEBA-based liquid epoxy resin. The composite samples were prepared by using a moderate-speed paddle mixer for 30 minutes followed by a 6-hour degassing at 80°C and 700 mbar.

Tensile tests were conducted on an Instron 5568 using ASTM standard samples and an extension rate of 0.5 mm/min. The tensile tests provided data to determine toughness, ultimate tensile strength, ultimate elongation at break, and Young's modulus. A test was performed to determine the flexural stress of the material according to ASTM D-790 standards and at a crosshead rate of 2.5 mm/min.

The tensile tests revealed that at 1 wt% and 3 wt%, smaller particle sizes result in improved UTS. Conversely, at 5 wt% and 10 wt%, larger particle sizes resulted in improved UTS. The peak in UTS was observed to occur at 3 wt% for the two smaller particle sizes and at 5 wt% for the larger 220 nm particle size. The maximum value of UTS occurred at 17 nm at 3 wt% with a value that was 14% greater than the unfilled epoxy matrix. For Young's modulus, a smaller particle size resulted in the largest value. Like in the UTS data, a peak occurred at 3 wt% for all particle sizes. The maximum value for the modulus was observed to occur at 3 wt% for the smallest particle

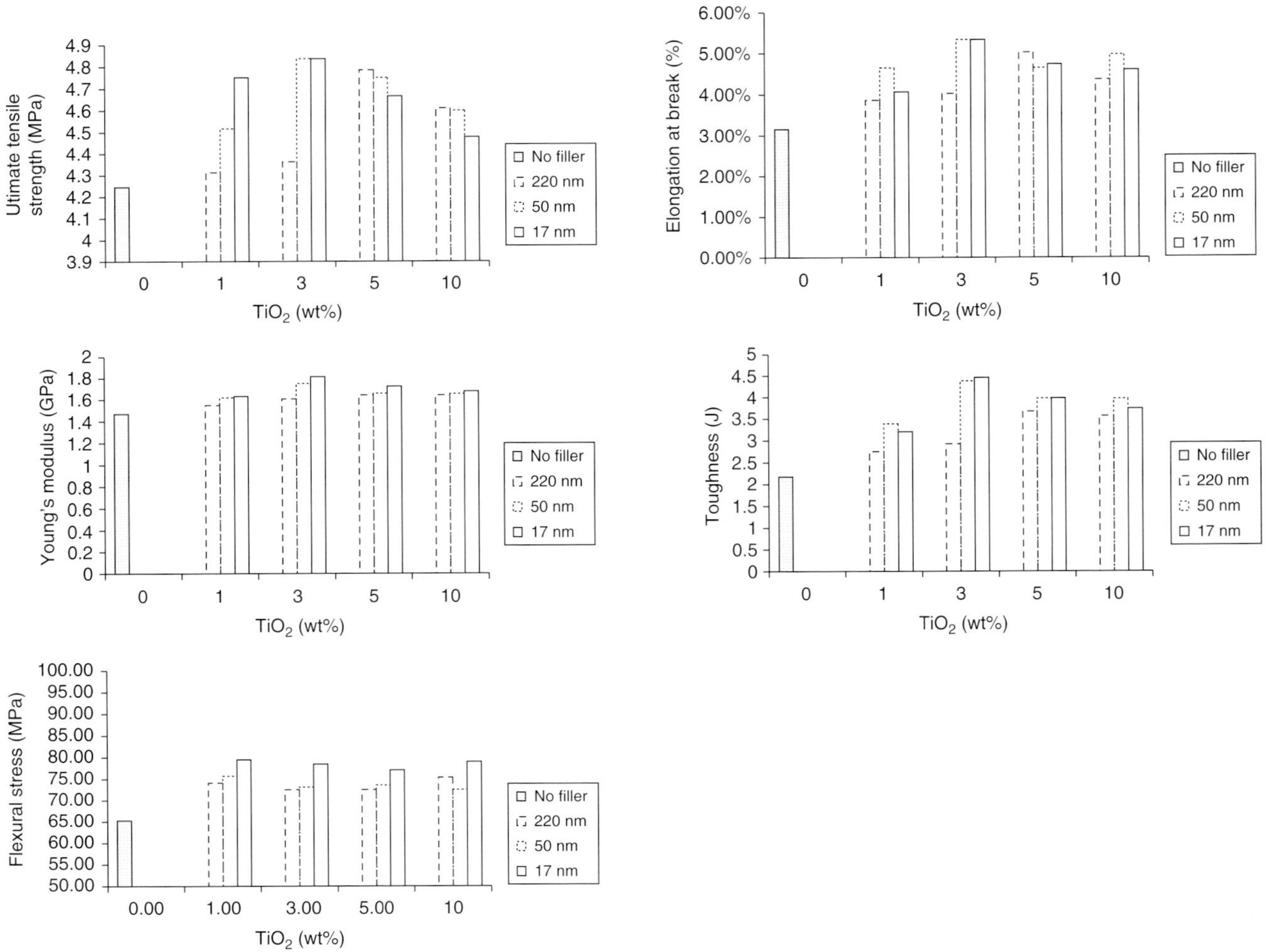

Figure 7.48. Mechanical property plots of titanium oxide modified composites with varying particle size [25].

size with a value that was 24% greater than the unfilled epoxy matrix. There was no noticeable correlation between particle size and elongation at break. However, the largest value of elongation occurred at 3 wt% and smallest overall modified value occurred at 1 wt%, a trend similar to UTS and Young's modulus data. The maximum value of elongation at break occurred for 17 nm particles at 3 wt% with a value that was 70% greater than the unfilled epoxy matrix. Experimental data are shown in Figure 7.48 [25].

Toughness tests continued the general trend of smaller particle size leading to improved material toughness. The common trend of peak values occurring at 3 wt% for all particle sizes was also observed. The maximum value of toughness occurred at 17 nm and 3 wt% with a value that was 100% greater than the unfilled epoxy matrix. Experimental data are shown in Figure 7.48.

Flexural stress also showed a similar trend, but reflected for the desire to minimize the stress values. The lowest values for stress occurred at the larger particle sizes and increased as the particle size decreased. The lowest values for the stresses were observed at 3 wt% and 5 wt% for the 220 nm and 50 nm particle sizes and at 5 wt% for the 17 nm particle size. Overall, the flexural stress for all modified composites was larger than the unfilled epoxy matrix. The smallest flexural stress increase occurred at 3 wt% and 5 wt% for the 220 nm particle size with a value that was 14% larger than the unfilled epoxy. The largest increase occurred at 1 wt% for the 17 nm particle size with a value that was 23% larger than the unfilled epoxy, as shown in Figure 7.48 [25].

It should be noted that even though this study describes the use of X-ray photoelectron spectroscopy (XPS) to confirm the dispersion of the nanoparticles within the epoxy matrix, no data are presented displaying the actual dispersion that was achieved in the samples.

A study done by Cheema and Lichtner explored the mechanical properties of a bisphenol-A-based epoxy resin modified with ZrO_2 nanoparticles [26]. The sizes of the nanoparticles are 10 nm, and the epoxy resin is based on bisphenol-A (RIMR 135i) cured with a mixture of two amine-based curing agents (RIMH 134 and RIMH 137). The nanomaterial chosen for this polymer matrix is treated with sorbic acid (SA) and diphenyl hydrogen phosphate (DPP) ligands because of the different anchoring abilities to the resin matrix, resulting in high dispersibility.

A loading of 2 wt% ZrO_2 was added to the epoxy resin and mixed on a magnetic stirrer for 15 min at ambient conditions. The mixture was then treated using a vacuum dissolver system fitted with a d25 mm dissolver disk and operated at 3,000 rpm for 2 hours at 50°C. The mixture was cured with the aforementioned curing agents and placed in an oven for 3 hours at 65°C, cooled to room temperature, and then tempered for 8 hours at 80°C. Mechanical properties, such as maximum tensile strength and Young's modulus, were determined from tensile testing.

The nanocomposites were characterized by TEM to determine the particle dispersion in the polymer matrix. TEM images, transmittance measurements, as well as SAXS indicated a high level of dispersion of both DPP and SA modified ZrO_2 nanoparticles within the polymer matrix. The mechanical properties of the nanocomposites are summarized in Table 7.10 [26].

Table 7.10. *Mechanical Properties of Zirconium Oxide Modified Epoxy Resin [26]*

Sample	Maximum tensile strength (MPa)	Relative change (%)	Young's modulus (MPa)	Relative change (%)
Pure polymer	69±0.9	–	2630±80	–
NC-ZrO_2-SA	75±0.5	9.3	2640±60	0.4
NC-ZrO_2-DPP	77±3.8	11.9	2960±120	12.5

Both the treated SA and DPP nanoparticles improved the pure polymer maximum tensile strength by 9.3% and 11.9%, respectively. Both treated nanoparticles improved the Young's modulus of the pure polymer as well, with the DPP-treated particle improving the pure polymer by 12.5% and the SA particle improving the polymer by 0.4%. It should be noted that the SA improvement is within in the margin of error for this set of tests. The DPP tests also exhibited larger margins of error, hinting that the trials for the tensile testing resulted in relatively high variance. In summary, the improvement of the mechanical properties is thought to be caused by the enhanced polymer-particle interactions caused by the treatment of the nanoparticles.

Zeng explored the use of cellulose nanocrystals to reinforce an epoxy resin [27]. Cellulose nanocrystals are a natural, abundant, environmentally friendly material. The nanocrystals are produced by the acid hydrolysis of cellulose fibers. The hydrolysis conditions are mild to an extent that a complete breakdown of the fiber does not occur. The average dimensions of the crystallites produced are 4 nm in diameter and 300 nm in length. The epoxy resin used in the study is EPON 863, a low-viscosity, clear resin.

The nanocrystals were dispersed in an organic solvent of EGME and isopropyl alcohol. The nanocrystals and resin were mixed and rotary-evaporated with a water bath set to 90°C. CURE W hardener was added and the mixture was sonicated in order to obtain dispersion of the nanocrystals. Another type of preparation techniques was used involving freeze-dried (FD) cellulose. The FD material was mixed mechanically by hand into the EPON 863 resin and then subjected to ultrasonic irradiation for 1 to 6 hours with mild stirring while being heated to 50°C. Another preparation involved the use of supercritically dried nanocrystal materials. The use of this supercritical drying process is thought to preserve the delicate nanocrystal structure, therefore preserving any inherent property advantages. Supercritically dried material was mixed with resin and hardener, which were kept warm at 60°C. The mixture was then poured under vacuum and completely covered the cellulose. The composite was then cured using the standard cure cycle for this epoxy resin.

The organically dispersed nanocrystals showed a slight decrease in UTS and a larger decrease in the tensile modulus. However, the work to break and elongation at break were improved, hinting that the organic solvent used provided greater elasticity. The results are shown in Figure 7.49 [27].

However, the dispersion of the nanocrystals was revealed to be unsuccessful, with agglomeration of cellulose causing a gel to form within the mixture instead of

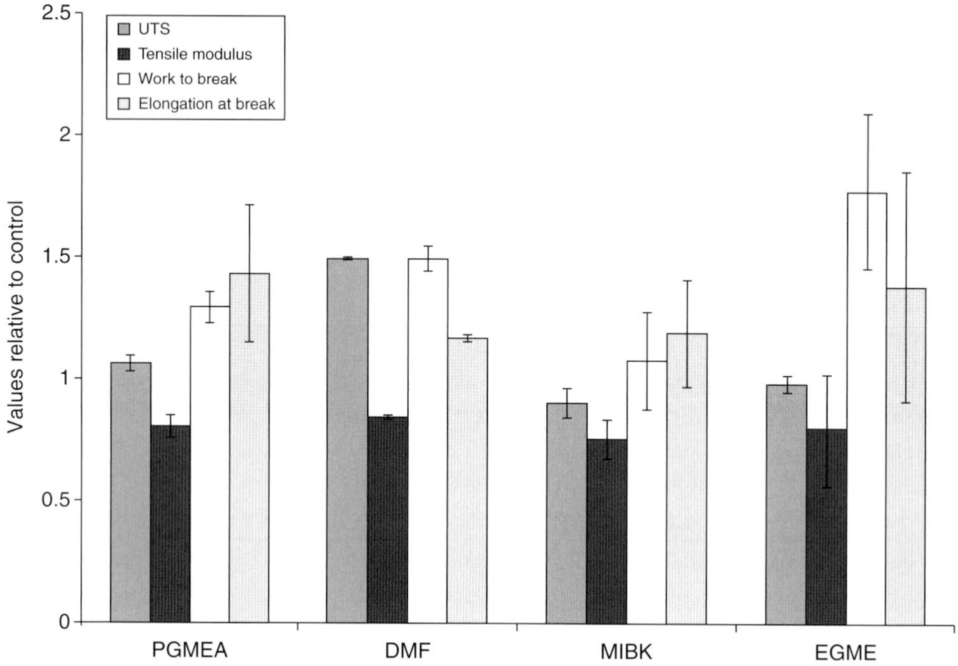

Figure 7.49. Organic solvents used and the effect on mechanical properties [27].

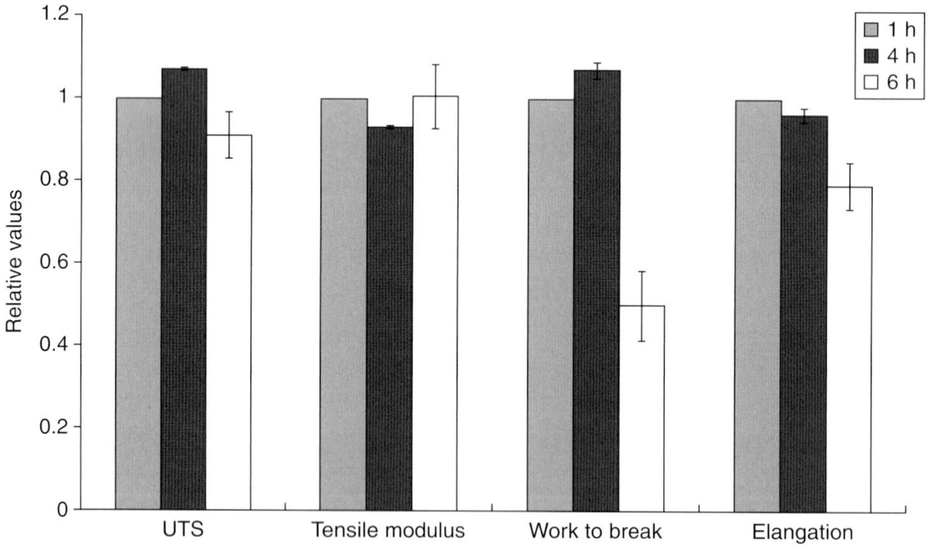

Figure 7.50. Effect of sonication duration on the mechanical properties of nanocrystal-modified resin [27].

being dispersed. Using organic solvents was unsuccessful due to the cellulose's high hydrophilicity and relatively high reactivity. The freeze-dried nanocrystal data show that the duration of sonication had no discernible effect on the tensile properties of UTS, tensile modulus, work to break, and elongation, as shown in Figure 7.50 [27].

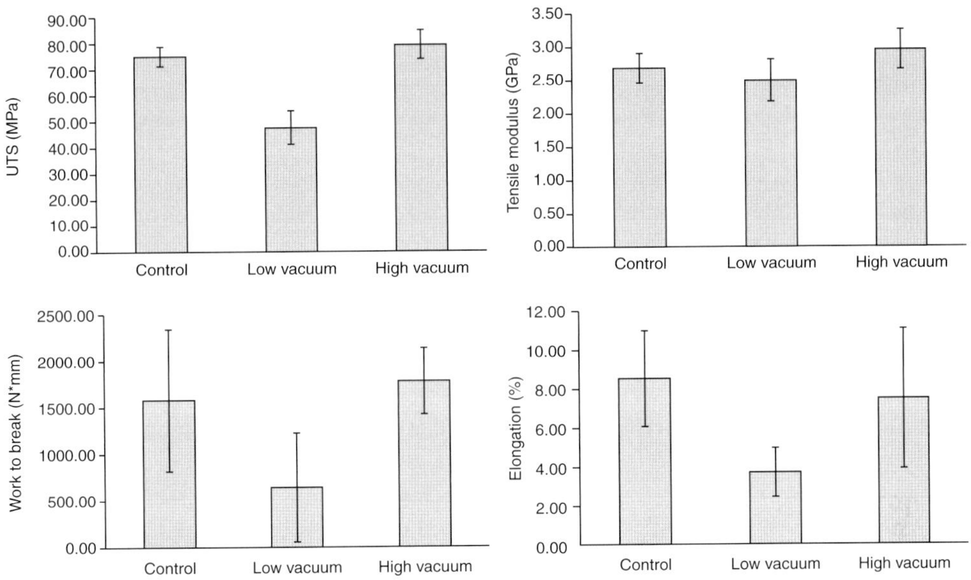

Figure 7.51. Effects of vacuum quality on the mechanical properties of nanocrystal-modified epoxy [27].

Samples were prepared with 1 wt% and 2 wt% of nanocrystals. The results show that for UTS, work to break, and elongation, a higher weight percentage decreases performance. The tensile modulus shows increase in value from 1 wt% to 2 wt% nanomodifier, but both values are still below the neat resin value. This decrease in performance is attributed to the poor formation of freeze-dried cellulose. Observation of the cellulose powder formed suggests an agglomerated nanocrystal structure, which would be difficult to disperse in the epoxy matrix.

Samples of the supercritically dried nanofibers were prepared in both high and low vacuums. Low vacuums lead to large bubble formation in the sample, while high vacuums form virtually bubble-free samples. The data are shown in Figure 7.51 [27].

Figure 7.51 shows that high vacuums improved performance in UTS, tensile modulus, and work to break, while low vacuums produce degradations in performance when compared to the control. Elongation data show a decrease for both low- and high-vacuum specimens, but the high-vacuum specimen shows a lower amount of degradation than the low-vacuum specimen. The bubbles formed in the low vacuum are thought to create a greater susceptibility to crack formation and therefore weaken the mechanical properties of the material.

Overall, the supercritically formed nanocrystals prepared in a high vacuum were the only specimens shown to actually improve the neat resin. Values for UTS, tensile modulus, and work to break improved by 5.6%, 10.1%, and 12.4%, respectively. However, it should be noted that WAXD and TEM analysis was not performed in this study, meaning that the dispersion described in this source cannot be independently verified.

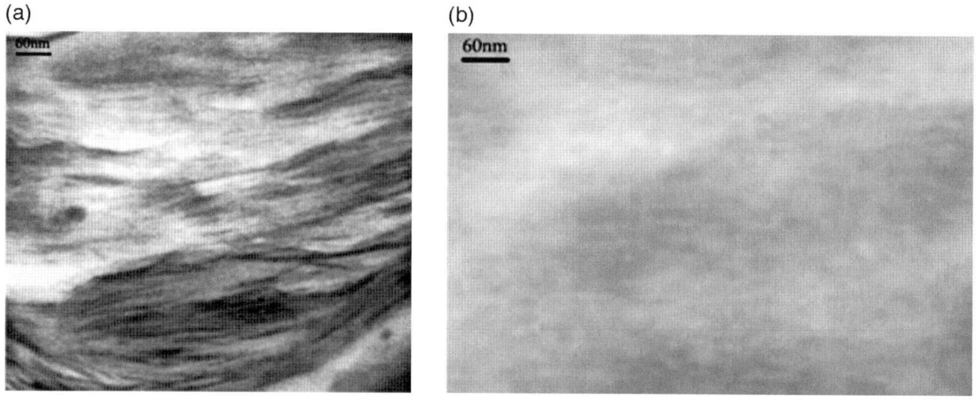

Figure 7.52. TEM image of 3 wt% Cloisite 15A loaded epoxy. Intercalated and exfoliated structure (a) and fully exfoliated structures (b) are shown [28].

7.4.2 Other Thermosetting Polymer Nanocomposites

7.4.2.1 Polyester-Clay Nanocomposite

A study of low-profile unsaturated polyester resin-clay nanocomposites by Beheshty et al. explored the mechanical properties of a clay-modified resin [28]. Cloisite 15A was added to an unsaturated polyester (UP) resin (BUSHPOL 81715) in 1, 3, 5, 7, and 9 wt%. Mechanical properties, such as Young's modulus, Barcol hardness, and Izod impact strength were recorded.

The sample was prepared by adding the UP resin to the clay powder, followed by two sets of mechanical mixing at 85 rpm for 24 hours each. The fourth and fifth steps involved more mechanical stirring at 185 rpm and 400 rpm for 24 hours. The final step involved high-shear mixing at 2,000 rpm for 8 hours at 40°C.

TEM and WAXD data reveal that a mixed level of intercalation and exfoliation for 3 wt% and 9 wt% nanomaterial. TEM images for 7 wt% samples show a lower level of dispersion. These TEM images are shown in Figure 7.52 [28].

Hardness tests show that the 1 wt% samples produce the highest values, and further wt% loadings of modifier decrease the hardness to levels that are lower than the unfilled resin. Izod impact tests show a similar trend, with the maximum impact strength occurring at the smallest weight percentage loading of the nanomaterial. Further additions of the modifier decrease the amount of impact strength achieved at the peak. However, all weight percentages of nano-modifier are above the value recorded for the unfilled resin. Data are shown in Figure 7.53 [28].

The flexural and storage moduli were also tested. Both moduli show a peak occurring at 3 wt% of the nano-modifier. A dip is found at 5 wt%, but the following weight percentages show an increase in the moduli, but not to the same levels as seen at 3 wt% loading. All weight percent loadings create a higher value for the flexural modulus when compared to the unfilled resin. However, the same cannot be said for the storage modulus, as 5 wt% and 7 wt% loadings are of a lower value than the unfilled resin. The data are shown in Figure 7.54 [28].

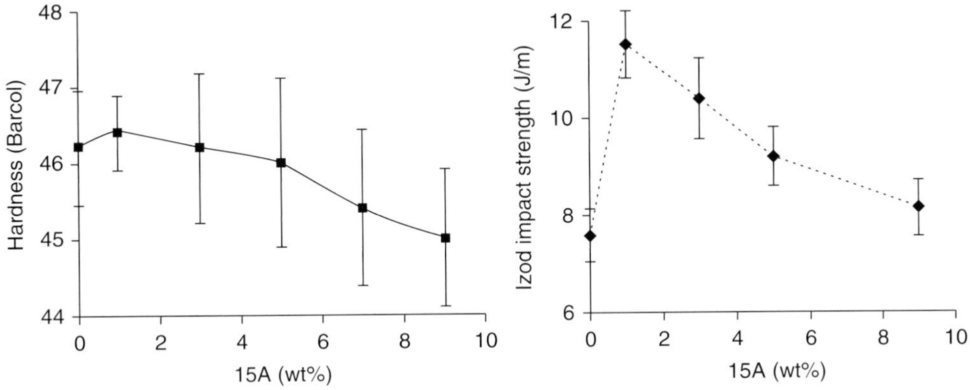

Figure 7.53. Izod impact strength and hardness plots as a function of varying clay loading [28].

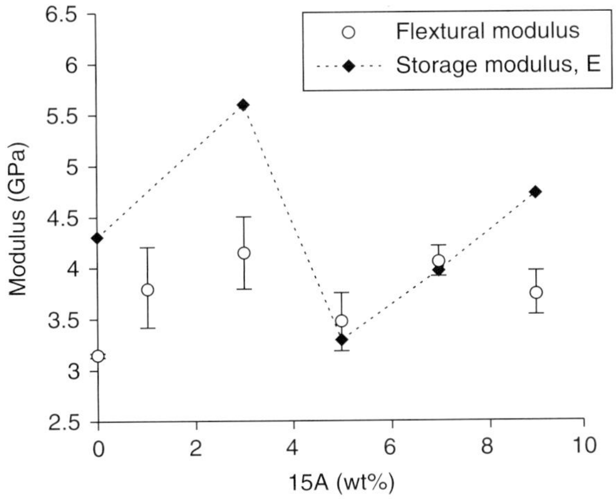

Figure 7.54. Flexural and storage moduli as a function of varying clay loading [28].

7.4.2.2 Polyimide-Ceramic Nanocomposites

A study by Irwin, Cao, and Bansal tested the mechanical properties of polyimide nanocomposites [29]. In this study, tensile and scratch harness measurements were conducted. The material under consideration was a thermosetting polyimide enamel modified with 4 wt% of 40 nm ceramic particles. Tensile measurements were made on 13 mm x 38 mm samples using an Instron tensile tester machine (Model 1331). Scratch hardness testing was done using a scratch tester with a 100 μm diameter stylus resting on the sample with a force of 100 grams.

Mechanical tensile testing revealed an improved elongation and overall strength. The unfilled polyimide sample strength was improved by 260% and elongation by 700%. Since tensile properties are directly related to the hardness of the material, the improvement in ultimate tensile strength was expected to increase material hardness. The ceramic nanoparticles increased hardness when compared to a weight percent filler of micron-sized particles, as shown in [29].

7.4.3 Special Types of CNT-Based Thermoset-Based Nanocomposites

Chou et al. recently published an excellent review article that examines the recent advancements in the science and technology of carbon nanotube (CNT)-based fibers and composites [30]. The assessment is based on the hierarchical structural levels of CNTs used in composites, from 1D to 3D. The authors defined the 1D level as fibers composed of pure CNTs or CNT embedded in a polymer matrix. The 2D level is defined as CNT-modified advanced fibers, CNT-modified interlaminar surfaces, and highly oriented CNTs in planar form. At the 3D level, examples of mechanical and physical properties of the CNT/polymer composites, CNT-based damage sensing, and textile assemblies of CNTs are included. The authors also discussed the opportunities and challenges in basic research at these hierarchical levels. The 1D level of CNT fibers is beyond the scope of this text; the 2D level of CNT-modified interlaminar surfaces and highly oriented CNTs in planar form studies are included in this section. The main focus of this text includes the 3D level describing the mechanical and other material properties of the CNT/polymer composites (CNT-based polymer nanocomposites) in Chapters 7 to 13.

7.4.3.1 CNT-Modified Interlaminar Surfaces

Recently, many attempts have been made by researchers in modifying the interlaminar surface by CNTs for improved mechanical and transport properties. Wardle's group at MIT has made significant advancement and contribution in this research area [31–39]. Their recent approach described by Garcia et al. [31, 32] is particularly interesting in that the group successfully integrated aligned CNTs with existing carbon fiber prepregs and processing. This is accomplished by first growing a vertically aligned CNT forest at high temperature. The CNTs are then transferred using a rolling transfer scheme to prepregs at room temperature, taking advantage of the tack of the prepregs. The aligned CNT forest can readily draw up the polymer through capillary action when laminated and form a bond with the polymer matrix as shown in Figure 7.55.

Garcia et al. [31] claimed that the ideal aligned CNT reinforcement would be comprised of a short forest (<20μm). The CNTs penetrate each ply by about ~ 10 μm. As a result, a ~20 μm forest would introduce no additional thickness to the ply interlayer, which can be detrimental to in-plane properties. Figure 7.56a shows the "transfer-printing" process, where the prepreg is attached to a cylinder that is rolled across the Si substrate containing the CNT forest [31]. Figure 7.56b shows the fully transferred CNT forest on a carbon fiber/epoxy prepreg. The CNT forest maintains vertical alignment and is neither broken nor buckled as shown in Figures 7.56c and 7.56d. The addition of a second ply of prepreg then creates two-ply laminate with an aligned CNT interlayer. Two aerospace-grade unidirectional laminates with CNT interlayer in Mode I and Mode II fracture properties were characterized. In their initial testing, the CNT-modified interface was observed to increase fracture toughness 1.5 to 2.5 times in Mode I and 3 times in Mode II. Evidence of CNT bridging was observed in fracture micrographs. Figure 7.57 shows an illustration of ideal hybrid

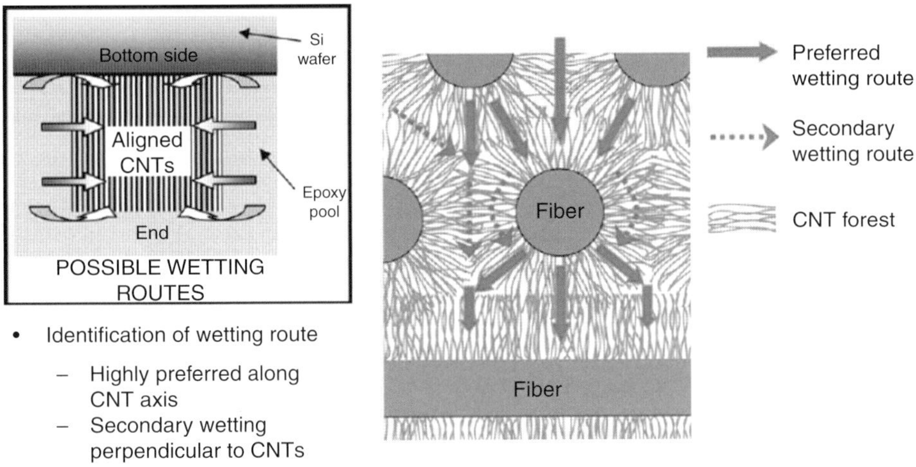

Figure 7.55. Processing of nano-engineered composites enabled via capillary-induced wetting of aligned CNTs. (Courtesy of Prof. Wardle).

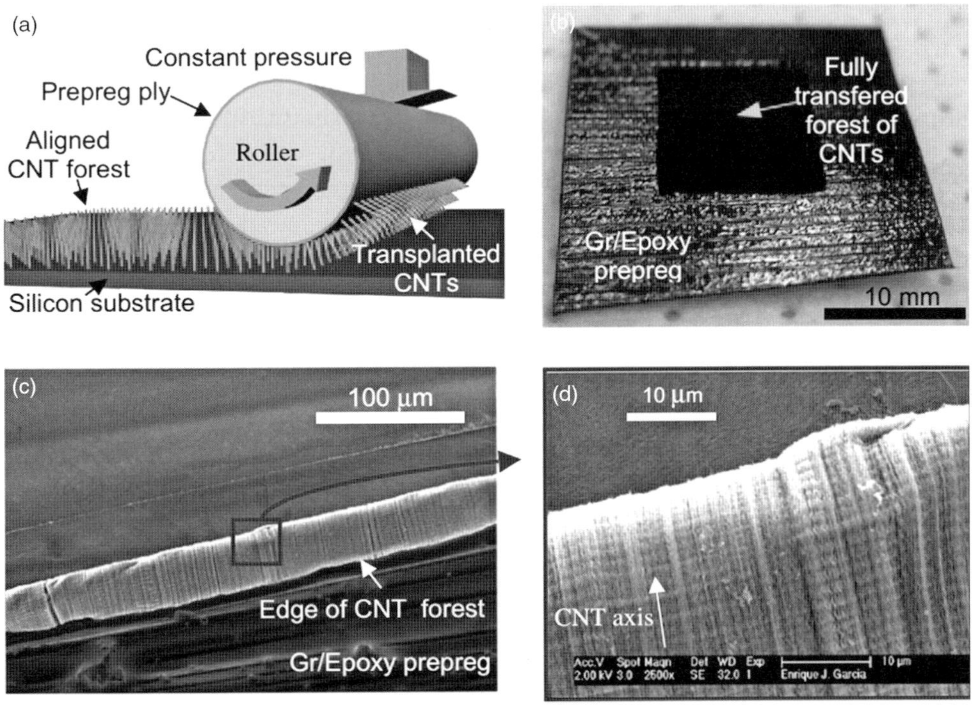

Figure 7.56. Transfer-printing of vertically aligned CNTs to prepreg: (a) Illustration of the "transfer-printing" process; (b) CNT forest transplanted from its original silicon substrate to the surface of a Gr/Ep prepreg ply; (c and d) SEM micrographs of the CNT forest, showing CNT alignment after transplantation [32].

interlaminar architecture showing the aligned CNTs placed in between two plies of a laminated composite [31].

Wicks et al. [37] further explored the 3D reinforcement of woven advanced polymer-matrix composites using CNTs experimentally and theoretically. Aligned

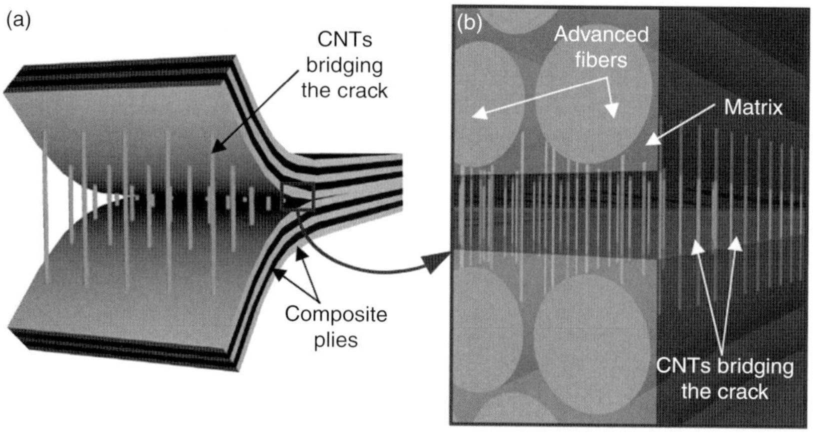

Figure 7.57. Illustration of the ideal hybrid interlaminar architecture: (a) CNTs placed in between two plies of laminated composite and (b) close-up of the crack, showing CNTs bridging the crack between the two pliers. Illustrations are not to scale [32].

CNTs are introduced in both the interlaminar and intralaminar regions of the laminate, providing 3D reinforcement as shown in Figure 7.58 [37]. This architecture has been introduced previously by Garcia et al. in the literature as fuzzy fiber-reinforced plastic (FFRP) laminate [32]. Radially aligned CNTs grown in situ on the surface of fibers in a woven cloth provide significant 3D reinforcement, as measured by Mode I interlaminar fracture testing and tension-bearing experiments. Aligned CNTs bridge the ply interfaces, enhancing both initiation and steady-state toughness, improving the already tough system by 76% in a steady state (more than 1.5 kJ/m^2 increase). CNT pullout on the crack face is the observed toughening mechanism, and an analytical model is correlated to the experimental fracture data. In the plane of the laminate, aligned CNTs enhanced the tension-bearing response with an increase of 19% in bearing stiffness, 9% in critical strength, and 5% in ultimate strength accompanied by a clear change in failure mode from shear-out failure (matrix dominated) without CNTs to tensile fracture (fiber dominated) with CNTs.

As discussed by Wicks et al. [37], areas for future research in aligned CNT-reinforced advanced composites are numerous and can include: (a) additional mechanical characterizations, especially toughness (Mode II, where z-pinning and stitching are particularly ineffective), impact resistance, and damage tolerance; (b) determination of any additional toughening mechanisms at the interface beyond CNT bridging; (c) modeling of the multiscale elastic and damage response for these systems [36]; (d) multifunctional property testing, such as electrical and thermal conductivities [39–41]; (e) development of other fabrication routes beyond hand layup for the FFRP system; (f) replication of the FFRP architecture using carbon fibers, where CNT growth on the carbon fibers must be shown to maintain fiber tensile properties; (g) continuous rather than batch production of aligned CNTs; and (h) particle exposure studies during the fabrication and machining of CNT-containing laminates, such as the FFRP [42–44].

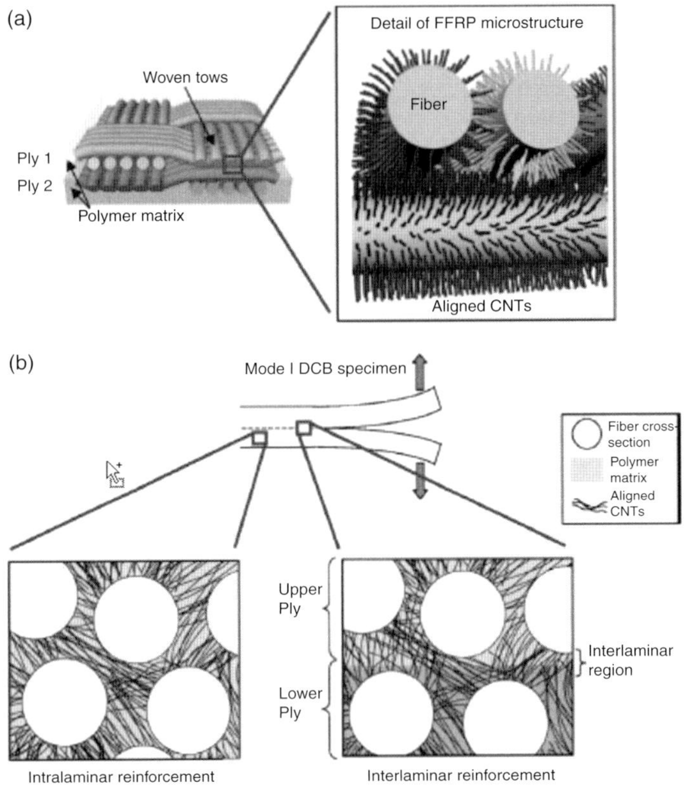

Figure 7.58. Illustration of fuzzy-fiber-reinforced plastic (FFRP): (a) Radially aligned CNTs grown on advanced fiber cloth; (b) CNT intralaminar and interlaminar reinforcement [37].

7.4.4 Summary of Thermoset-Based Nanocomposites

The addition of clay, SWNT, DWNT, MWNT, GPL, SiO_2, TiO_2, and ZrO_2 nanoparticles improved the mechanical properties, such as Young's modulus, ultimate tensile strength, and toughness of the different epoxy resin systems. WAXD, TEM, and AFM techniques were used to determine the degree of nanoparticle dispersion in the resin matrix. The experimental data from impact strength, bending modulus, and bending strength show improved performance for the chemically modified MWNT, further indicating enhanced properties due to improved particle dispersion. The analytical models for the tensile modulus agreed fairly well with the tensile data, specifically the Mori-Tanaka model. The model used to predict fracture toughness did not produce acceptable correlation with the experimental data.

Overall, the supercritically formed nanocrystals prepared in a high vacuum were the only specimens shown to actually improve the epoxy resin. Values for UTS, tensile modulus, and work to break were improved. The polyimide with ceramic nanoparticles sample strength was improved by 260% and elongation by 700%. The ceramic nanoparticles also increased hardness when compared to a weight percent filler of micron-sized particles.

Aligned CNTs are introduced in both the interlaminar and intralaminar regions of the laminate, provide 3D reinforcements as well as enhanced thermal and electrical conductivity values of the prepreg laminates.

7.5 Overall Summary

In summary, the mechanical properties of the thermoplastics, thermoplastic elastomers, and thermosets can be strengthened dramatically by adding a small amount of nanoparticles if the right processing techniques and conditions are selected. Incorrect combinations may produce no changes at all or, in some cases, even adverse effects. Studies that relied on WAXD techniques to determine the dispersion of the nanoparticles within the polymer matrix can be further investigated by obtaining TEM micrographs of the sample to gauge whether the degree of dispersion within the polymer matrix is optimal. It is highly recommended to assess nanoparticle dispersion first, before launching into the expensive and lengthy material properties characterization.

However, all the studies demonstrated that even at low loading percent of nanoparticles, significant improvements in mechanical properties were observed. Also, it is not only the process that matters; the fabrication of the polymer nanocomposite parts is also important to achieve enhanced mechanical properties of polymer nanocomposites. With the possibility of obtaining even more uniformly dispersed nanoparticles, the future application for nanocomposites in terms of cost for the benefit it provides seems promising.

7.6 Questions and Problems

1. How does the nanomaterial dispersion affect the mechanical properties of polymer nanocomposites in general? Provide examples to illustrate your claims.
2. How are the mechanical properties of polymer-clay nanocomposites influenced by the type of clay, surface treatment of the clay, and how clay is processed in the polymer matrix? Provide examples.
3. Enhancement of mechanical properties of thermoplastic-based nanocomposites using MMT clay, carbon-based (MWNT, buckypaper, and graphene), fibers, nanodiamonds, and POSS had been described in this chapter. Describe the mechanisms by which each type of nanomaterials provides the enhanced mechanical properties. What other nanomaterials can be used for thermoplastic-based nanocomposites? Provide examples.
4. Enhancement of mechanical properties of thermoplastic elastomer-based nanocomposites using MMT clay, carbon-based (MWNT), SiC, nanosilica, and TiO_2 had been discussed in this chapter. Describe the mechanisms by which each type of nanomaterials provides the enhanced mechanical properties. What other nanomaterials can be used for thermoplastic elastomer-based nanocomposites? Provide examples.

5. Enhancement of mechanical properties of epoxy-based nanocomposites using MMT clay, carbon-based (MWNT), nanosilica, TiO_2, ZrO_2, and cellulose nanocrystals had been discussed in this chapter. Comment on the mechanisms by which each type of nanomaterials provides the enhanced mechanical properties. What other nanomaterials can be used for epoxy-based nanocomposites? Provide examples.
6. The elongation at break of the polymer nanocomposites tends to degrade significantly with the additional of MMT clay and carbon-based nanomaterials. What will be a good strategy to maintain this property in the polymer nanocomposites?

REFERENCES

1. B. Kord (2012). Studies on mechanical characterization and water resistance of glass fiber/thermoplastic polymer bionanocomposites. *Journal of Applied Polymer Science* 123(4), 2391–2396.
2. F. A. Aouada, H. Luiz, and E. Longo (2011). New strategies in the preparation of exfoliated thermoplastic starch-montmorillonite nanocomposites. *Industrial Crops and Products* 34(3), 1502–1508.
3. K. Majdzadeh-Ardakani, A. H. Navarchian, and F. Sadeghi (2010). Optimization of mechanical properties of thermoplastic starch/clay nanocomposites. *Carbohydrate Polymers* 79(3), 547–554.
4. U. Ritter, P. Scharff, O. Dmytrenk, N. Kulish, Y. Prylutsky, Y. Grabovskiy, et al. (2010). Strength improvement of iPP/MWCNT nanocomposites. *Polymer Composites* 31(1), 179–184.
5. B. Fernandez-d'Arlas, U. Khan, L. Rueda, L. Martin, J. A. Ramos, et al. (2012). Study of the mechanical, electrical and morphological properties of PU/MWCNT composites obtained by two different processing routes. *Composites Science and Technology* 72(2), 235–242.
6. V. Eswaraiah, K. Balasubramania, and S. Ramaprabh (2011). Functionalized graphene reinforced thermoplastic nanocomposites as strain sensors in structural health monitoring. *Journal of Materials Chemistry* 21(34), 12626–12628.
7. X. Zhang, Y. Chen, J. Yu, and Z. Guo (2011). Thermoplastic polyurethane/silica nanocomposite fibers by electrospinning. *Journal of Polymer Science Part B* 49(23), 1683–1689.
8. V. Karbushev, A. Semakov, and V. Kulichikhin (2011). Structure and mechanical properties of thermoplastics modified with nanodiamonds. *Polymer Science Series A* 53(9), 765–774.
9. R. Lach, G. H. Michler, and W. Grellman (2010). Microstructure and indentation behaviour of polyhedral oligomeric silsesquioxanes modified thermoplastic polyurethane nanocomposites. *Macromolecular Materials and Engineering* 295(5), 484–491.
10. S. M. Liff, N. Kumar, and G. H. McKinley (2007). High-performance elastomeric nanocomposites via solvent-exchange processing. *Natural Materials* 6(1), 76–83.
11. F. Chavarria and D. R. Paul (2006). Morphology and properties of thermoplastic polyurethane nanocomposites: Effect of organoclay structure. *Polymer* 47(22), 7760–7773.
12. A. Mishra, B. P. D. Purkayastha, J. K. Roy, V. K. Aswal, and P. Maiti (2010). Tunable properties of self-assembled polyurethane using two-dimensional nanoparticles: Potential nano-biohybrid. *Macromolecules* 43(23), 9928–9936.
13. J. E. Puskas, E. A. Foreman-Orlowski, G. T. Lim, S. E. Porosky, M. M. Evancho-Chapman, et al. (2010). A nanostructured carbon-reinforced polyisobutylene-based thermoplastic elastomer. *Biomaterials* 31(9), 2477–2488.

14. W. Chen, X. Tao, and Y. Liu (2006). Carbon nanotube-reinforced polyurethane composite fibers. *Composites Science and Technology* 66(15), 3029–3034.
15. S. Cantournet, M. C. Boyce, and A. H. Tsou (2007). Micromechanics and macromechanics of carbon nanotube-enhanced elastomers. *Journal of the Mechanics and Physics of Solids* 55(6), 1321–1339.
16. Y. Li, and H. Shimizu (2007). High-shear processing induced homogenous dispersion of pristine multiwalled carbon nanotubes in a thermoplastic elastomer. *Polymer* 48(8), 2203–2207.
17. C. Z. Liao, and S. C. Tjong (2010). Mechanical and fracture behaviors of elastomer-rich thermoplastic polyolefin/SiC nanocomposites. *Journal of Nanomaterials 2010, article ID 327973*, 9 pages.
18. O. Aso, J. I. Eguiazábal, and J. Nazábal (2007). The influence of surface modification on the structure and properties of a nanosilica filled thermoplastic elastomer. *Composites Science and Technology* 67(13), 2854–2863.
19. R.-J. Zhou and T. Burkhart (2011). Thermal and mechanical properties of poly(ether ester)-based thermoplastic elastomer composites filled with TiO nanoparticles. *Journal of Materials Science* 46(7), 2281–2287.
20. A. J. Kinloch and A. C. Taylor (2006). The mechanical properties and fracture behaviour of epoxy-inorganic micro- and nano-composites. *Journal of Materials Science* 41, 3271–3297.
21. F. H. Gojny, et al. (2005). Influence of different carbon nanotubes on the mechanical properties of epoxy matrix composites – A comparative study. *Composites Science and Technology* 65, 2300–2313.
22. Y. Zheng, et al. (2006). Functionalized effect on carbon nanotube/epoxy nanocomposites. *Materials Science and Engineering A* 435–436, 145–149.
23. M. A. Rafiee and J. Rafiee (2009). Enhanced mechanical properties of nanocomposites at low graphene content. *ACS Nano* 3(12), 3884–3890.
24. B. R. K. Blackman, et al. (2007). The fracture and fatigue behavior of nano-modified epoxy polymers. *Journal of Materials Science* 42, 7049–7051.
25. H. A. Al-Turaif (2010). Effect of nano TiO_2 particle size on mechanical properties of cured epoxy resin. *Organic Coatings* 69, 241–246.
26. T. A. Cheema, et al. (2011). Fabrication of transparent polymer-matrix nanocomposites with enhanced mechanical properties from chemically modified ZrO_2 nanoparticles. *Journal of Materials Science* 47, 2665–2674.
27. J. Zeng (2009). An Experimental Study on Tensile Properties of Cellulose Nanocrystal Reinforced Epoxy Nanocomposite Material. MS thesis. Oregon State University, Corvallis, OR.
28. M. H. Beheshty, M. Vafayan, and M. Poorabdollah (2008). Low profile unsaturated polyester resin-clay nanocomposite properties. *Polymer Composites* 30(5), 629–638.
29. P. C. Irwin, Y. Cao, and L. S. Schadler (2003). Thermal and mechanical properties of polyimide nanocomposites. *2003 Annual Report Conference on Electrical Insulation and Dielectric Phenomena*, pp. 120–123.
30. T.-W. Chou, L. Gao, E. T. Thostenson, Z. Zhang, and J.-H. Byun (2010). An assessment of the science and technology of carbon nanotube-based fibers and composites. *Composites Science and Technology* 70, 1–19.
31. E. J. Garcia, B. L. Wardle, and A. J. Hart (2008). Joining prepreg composite interfaces with aligned carbon nanotubes. *Composites: Part A* 39(6), 1065–1070.
32. E. J. Garcia, B. L. Wardle, A. J. Hart, and N. Yamamoto (2008). Fabrication and multifunctional properties of a hybrid laminate with aligned carbon nanotubes grown *in situ*. *Composites Science and Technology* 68, 2034–2041.
33. E. J. Garcia, A. J. Hart, and B. L. Wardle (2008). Long carbon nanotubes grown on the surface of fibers for hybrid composites. *AIAA Journal* 46(6), 1405–1412.

34. J. Blanco, E. J. Garcia, R. Guzman, D. Villoria, and B. L. Wardle (2009). Limiting mechanisms of mode I interlaminar toughening of composites reinforced with aligned carbon nanotubes. *Journal of Composite Materials* 43(8), 825–841.
35. N. Yamamoto, A. J. Hart, E. J. Garcia, S. S. Wicks, H. M. Duong, et al. (2009). High-yield growth and morphology control of aligned carbon nanotubes on ceramic fibers for multifunctional enhancement of structural composites. *Carbon* 47, 551–560.
36. M. C. Ray, R. Guzman de Villoria, and B. L. Wardle (2009). Load transfer analysis in short carbon fibers with radially-aligned carbon nanotubes embedded in a polymer matrix. *Journal of Advanced Materials* 41(4), 82–94.
37. S. S. Wicks, R. Guzman de Villoria, and B. L. Wardle (2010). Interlaminar and intralaminar reinforcement of composite laminates with aligned carbon nanotubes. *Composites Science and Technology* 70, 20–28.
38. N. Lachman, E. Wiesel, R. Guzman de Villoria, B. L. Wardle, and H. D. Wagner (2012). Interfacial load transfer in carbon nanotube/ceramic microfiber hybrid polymer composites. *Composites Science and Technology* 72, 1416–1422.
39. N. Yamamoto, E. J. Garcia, B. L. Wardle, and A. J. Hart (2008). Thermal and electrical properties of hybrid woven composites reinforced with aligned carbon nanotubes. *Proceedings of the 49th AIAA Structures, Dynamics, and Materials Conference*, Schaumburg, IL, April 7–10.
40. S. Vaddiraju, H. Cebeci, K. K. Gleason, and B. L. Wardle (2009). Hierachical multifunction composites by conformally coating aligned carbon nanotube arrays with conducting polymer. *Applied Materials & Interfaces* 1(11), 2565–2572.
41. A. M. Marconnet, H. Yamamoto, M. A. Panzer, B. L. Wardle, and K. E. Goodson (2011). Thermal conduction in aligned carbon nanotube-polymer nanocomposites with high packing density. *ACS Nano* 5(6), 4818–4825.
42. D. B. Bello, et al. (2009). Exposures to nanoscale particles and fibers during handling, processing, and machining of nanocomposites and nano-engineering composites reinforced with aligned carbon nanotubes. In *17th International Conference on Composite Materials (ICCM) proceedings*, Edinburgh, Scotland, July 27–31.
43. D. Bello, A. J. Hart, K. Ahn, M. Hallock, N. Yamamoto, et al. (2008). Particle exposure levels during CVD growth and subsequent handling of vertically-aligned carbon nanotube films. *Carbon* 46(6), 974–977.
44. D. Bello, B. L. Wardle, N. Yamamoto, R. Guzman de Villoria, E. J. Garcia, et al. (2009). Exposure to nanoscale particles and fibers during machining of hybrid advanced composites containing carbon nanotubes. *Journal of Nanoparticles Research* 11(1), 231–249.

8 Thermal Properties of Polymer Nanocomposites

8.1 Introduction

This chapter discusses recent advancements in the improvement and understanding of the thermal behavior of polymer nanocomposites. Many different kinds of polymer nanocomposite systems have been analyzed, focusing on their thermal properties. Most systems analyzed had thermal properties enhanced by the addition of nanoparticles, such as nanoclays or layered double hydroxides. The addition of certain nanoparticles also had a positive impact on mechanical and electrical properties. Some systems, as in the homogeneously nucleated PMMA/pSWNT system, had other interesting thermal behavior. Polymer nanocomposites have found themselves in all kinds of applications. Their light weight offers highly specific properties that make them preferable for a wide range of applications. This chapter focuses on thermal properties of polymer nanocomposites. The nanocomposites were compared to their neat polymer host and sometimes to other macrocomposites and competing nanocomposites. The composites analyzed are organized by the polymers and the type of nanoparticles used.

8.2 Thermoplastic-Based Nanocomposites

8.2.1 Polypropylene-Clay Nanocomposites

Nanoclay has attracted a lot of attention in polymer nanocomposite due to its relative abundance, good performance, and environmentally friendly properties. There are different kinds of commercially available nanoclays. Some are treated to become organophilic in order to be compatible with the polymer matrix. One of the challenges involved in any kind of composite, especially nanocomposites, is dispersing, as well as exfoliating particles in the host material matrix.

Ellis and D'Angelo [1] produced a polypropylene-organophilic montmorillonite clay containing approximately 4 wt% clay. Although the clay was fully dispersed in the polypropylene (PP) matrix, the particles were not fully exfoliated. Figure 8.1 shows that the XRD pattern of the PP nanocomposite had a peak corresponding to

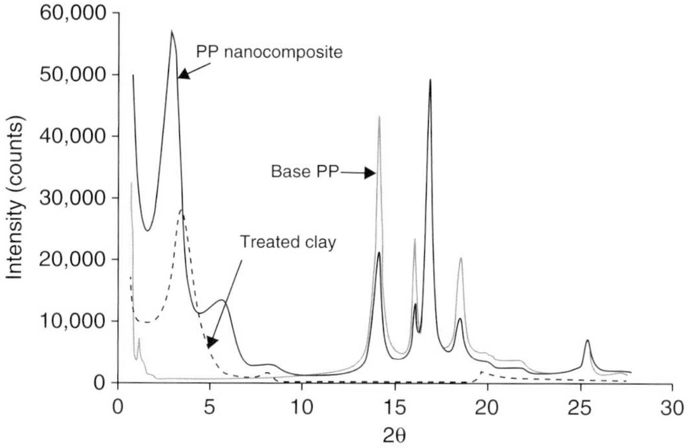

Figure 8.1. XRD pattern of PP, treated clay, and PP nanocomposite.

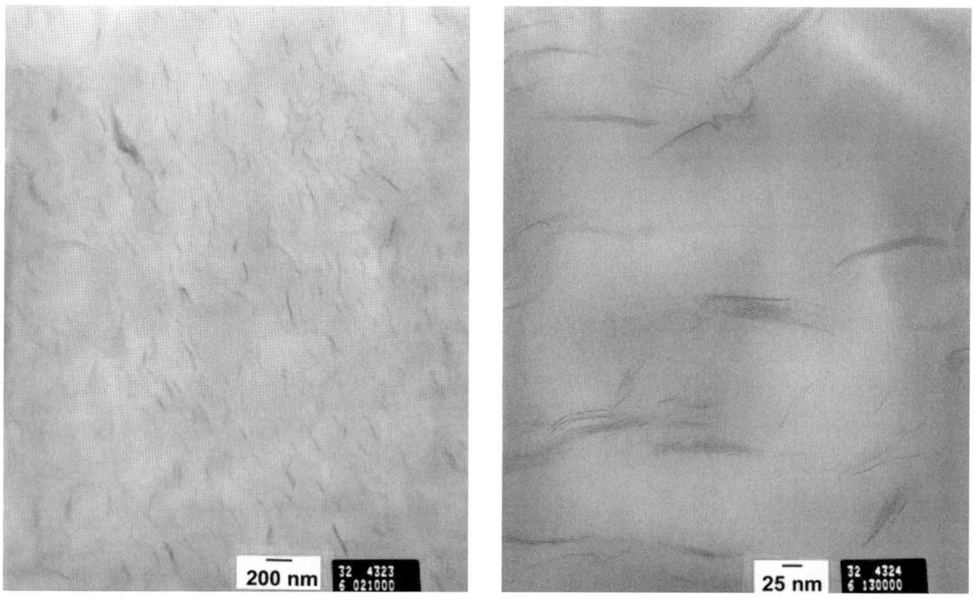

Figure 8.2. TEM images of PP-clay nanocomposite.

the treated (organophilic treatment) clay [1]. This peak indicates that the clay particles were not fully exfoliated in the PP matrix.

Furthermore, Figure 8.2 shows TEM images of the PP nanocomposite at two different magnifications [1]. The image on the left shows that the particles were successfully dispersed in the polymer matrix. However, the image on the right shows that the particles were not fully exfoliated; they are in an intercalated state.

The PP nanocomposite had overall better mechanical properties than did neat PP. The thermal properties of the composite were relatively unaffected. The percent of ash produced was slightly higher, but still lower than a conventional talc macrocomposite. The coefficient of thermal expansion (CTE) was slightly improved. In

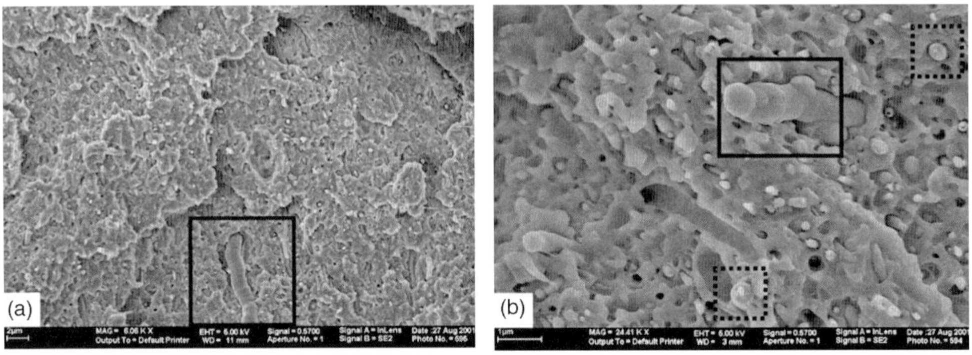

Figure 8.3. SEM micrographs of the fracture surface of (a) 10 wt% and (b) 15 wt% vapor-grown CNFs. The squares represent polymer coated CNFs.

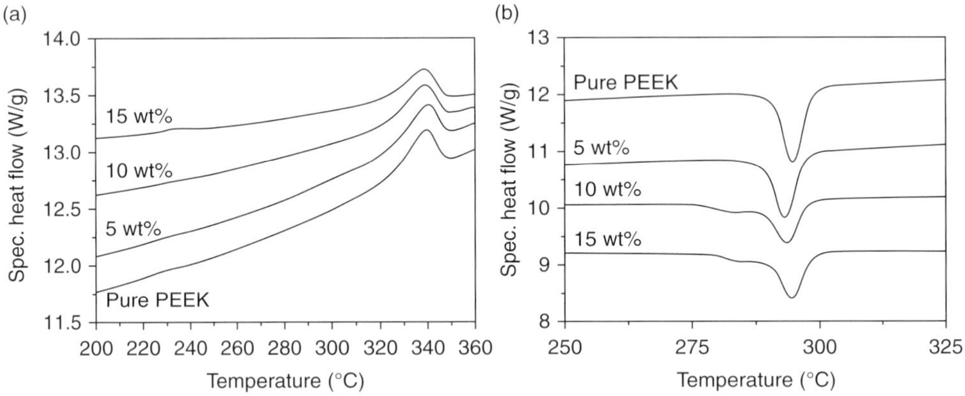

Figure 8.4. Specific heat flow recorded for the PEEK nanocomposites as a function of temperature during an isothermal DSC runs: (a) heating curves and (b) cooling curves. Individual curves are shifted vertically for clarity.

general, the use of organophilic clay in a PP matrix may lead to improved mechanical properties and slightly improved thermal properties. Due to the intercalated nature of the nanocomposite produced, there is reason to believe that even better results can be obtained from a fully exfoliated nanocomposite [1].

8.2.2 PEEK-Carbon Nanofibers Nanocomposites

Sandler et al. [2] produced a carbon nanofiber–reinforced poly(ether ketone) (PEEK) nanocomposite. PEEK is an important material that is already being used in a variety of applications, such as seals in the oil industry. The carbon nanofibers (CNFs) in this composite were successfully dispersed and exfoliated. Figure 8.3 shows the fracture surface of the PEEK-CNF nanocomposite [2]. The images show that the carbon nanofibers did not agglomerate and were coated by the polymer.

The PEEK-CNF nanocomposite had improved mechanical properties while maintaining good thermal properties. The glass transition temperature, among other thermal properties, was not significantly changed. However, Figure 8.4 shows

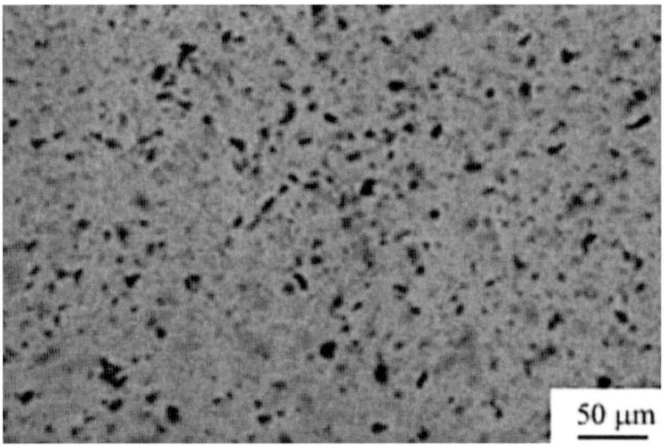

Figure 8.5. Optical micrograph indicating considerable pSWNT agglomeration.

interesting crystallization behavior [2]. Lozano and Barrera previously reported that vapor-grown CNFs in an isotactic polypropylene homopolymer served as nucleation sites [3]. However, the CNFs had no significant impact in the crystallization of the composite. Nucleation started homogeneously in the PEEK matrix.

The PEEK-CNF nanocomposite showed improved mechanical properties while maintaining good thermal properties. Unlike as reported before, the vapor-grown CNFs in this nanocomposite did not function as nucleation sites during crystallization. Homogeneous nucleation of this nanocomposite is an important feat that becomes even more important as the size of the filler material further decreases (if, for example, carbon nanotubes were to be used) [3].

8.2.3 PMMA-Carbon Nanotubes Nanocomposites

Du et al. produced a carbon nanotube-poly(methyl methacrylate) (PMMA) nanocomposite by a new fabrication method known as coagulation [4]. This novel method resulted in better dispersion of CNTs in the polymer matrix, leading to improved mechanical properties, electrical conductivity, and thermal stability.

The coagulation processing method is characterized by immediate precipitation of polymer/composite solution that prevents CNTs from conglomerating due to van der Waals force. Without the newly developed coagulation method, the CNTs will agglomerate into macro particles, as seen in Figure 8.5 [4].

The PMMA-SWNT nanocomposite produced showed improved elastic modulus, electrical conductivity, and thermal stability. Figure 8.4 shows that although both the pure polymer and the polymer nanocomposite start to lose weight at the same temperature, the temperature at maximum mass loss rate is about 60 degrees higher for the composite. Also, the maximum mass loss rate is about twice as high for pure PMMA than for the CNT nanocomposite. This improved thermal stability is consistent with recent reports that pSWNT addition into PMMA results in improved fire retardancy [5].

The novel coagulation processing method used to fabricate a PMMA-pSWNT nanocomposite resulted in improved elastic modulus, electrical conductivity, and thermal stability. The coagulation method can be adapted to other thermoplastic-nanoparticle systems to improve dispersion and exfoliation of composites [4].

8.2.4 PVC-Layered Double Hydroxides Nanocomposites

Liu et al. prepared a poly(vinyl chloride) (PVC)/layered double hydroxide (LDH) using dodecyl sulfate (DS^-) or stearate anion [6]. The polymer nanocomposite showed significantly improved thermal stability when compared to pure PVC. XRD patterns [6] showed no significant difference between the peaks displayed in pure PVC and the different LDH composites, indicating that the ordered layered structure had been successfully destroyed and the nanoparticles dispersed in the PVC matrix.

Furthermore, TEM images (Figure 8.6) indicate that the LDHs were successfully exfoliated in the PVC matrix [6]. The results from XRD patterns and TEM images indicate that organo-LDHs containing DS^- and stearate anions can be exfoliated into nanometer scale and homogeneously dispersed in PVC through the solution intercalation procedure.

The PVC-LDH nanocomposites' thermal stability was evaluated by the conventional Congo Red test. Figure 8.7 compares the thermal stability of pure PVC, a previously developed PVC-LDH composite, and the newly developed LDH composites [6]. The thermal stability of LDH-DS is slightly better than that of the previously developed LDH composite but significantly better than that of pure PVC. Also, LDH-stearate is significantly better than pure PVC and more than twice as good as the previously developed LDH composite.

Furthermore, thermal stability as a function of LDH loading was also analyzed. Figure 8.8 shows the effect of LDH loading on a conventional fire retardant LDH-AP composite and the new LDH-stearate composite [6]. The stearate hydroxide loading has a much bigger effect than the conventional fire retardant composite in thermal stability, and material loading as high as 10 wt% can be used to improve thermal stability.

The PVC-LDH nanocomposite developed by solution intercalation (DS^- and stearate) exhibits significantly improved thermal stability. This composite can be further developed and adapted to promote PVC resins with improved thermal stability, flame retardancy, and smoke suppression properties [6].

8.2.5 Hybrid Systems

Piszczyk et al. prepared rigid polyurethane-polyglycerol-clay nanocomposites using different kinds of polyglycerols and layered silicate nanoclays (montmorillonite) [7]. A polymeric foam-nanoclay composite with improved thermal properties was created. Figure 8.9 is a set of SEM micrographs of the polyurethane-polyglycerol foam with different kinds of clays [7]. The cell structure is maintained by the addition of nanoclay with improved mechanical properties.

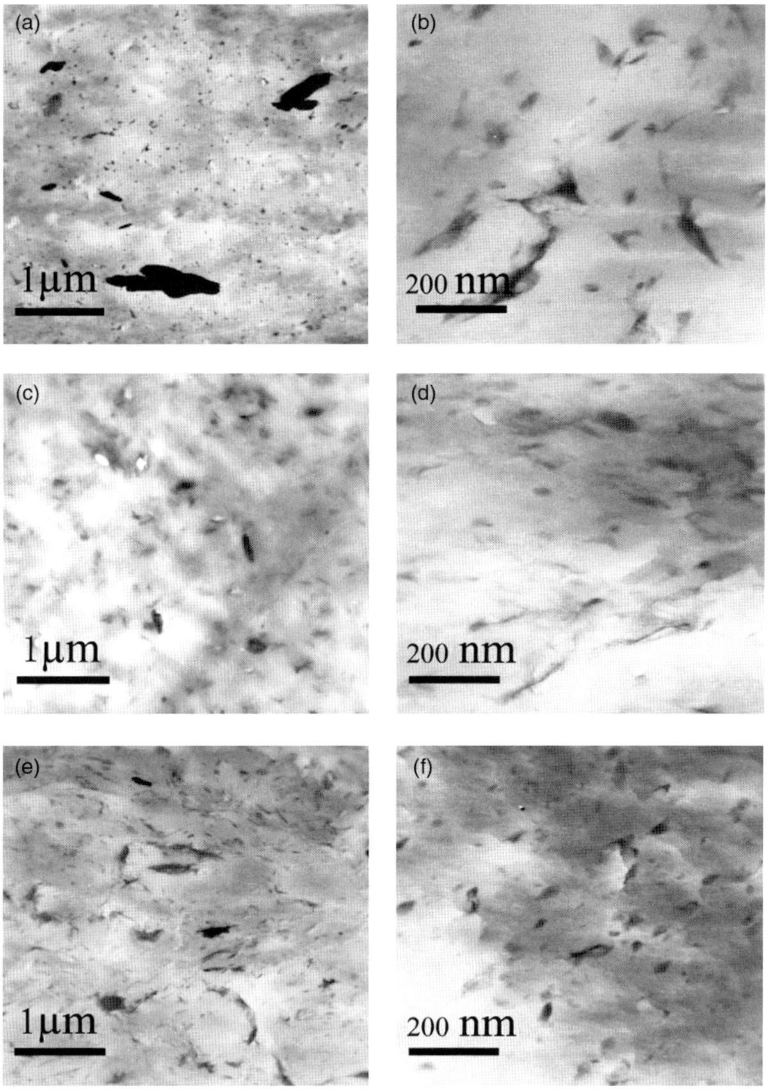

Figure 8.6. TEM micrographs of (a, b) 5 wt% PVC/LDH-NO_3 composite, (c, d) 5 wt% PVC/LDH-DS nanocomposite, and (e, f) 5 wt% PVC/LDH-stearate nanocomposite.

The XRD pattern of pure foams and nanocomposite foams in [7] showed that the nanocomposites did not show diffraction maximums corresponding to clay, indicating a good exfoliation of nanoclays in the foams.

Thermogravimetric analysis data of the nanocomposites are represented in Figure 8.10 and Table 8.1 [7]. Figure 8.10 compares the effect of different clays in conventional polyurethane foams. Cloisite 30B is clearly the best clay in terms of mass loss. Table 8.1 includes the temperature at which the nanocomposites experience 5%, 10%, and 50% weight loss. Three different polyurethane-polyglycerol ratios were analyzed, and three different kinds of clays were used for each kind of

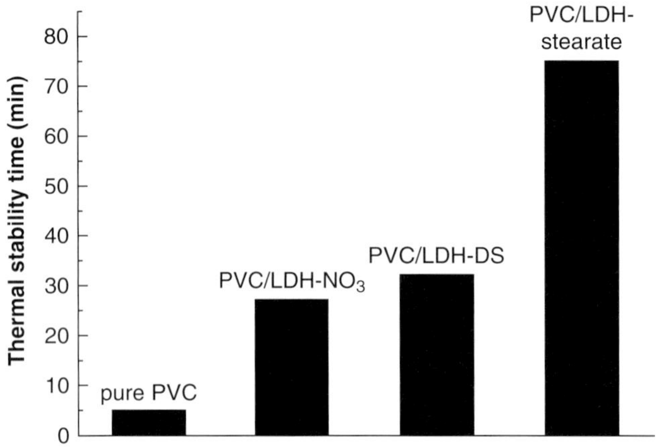

Figure 8.7. Thermal stability time of PVC, PVC/LDH-NO_3 composite, PVC/LDH-DS nanocomposite, and PVC/LDH-stearate nanocomposite measured by the Congo Red test.

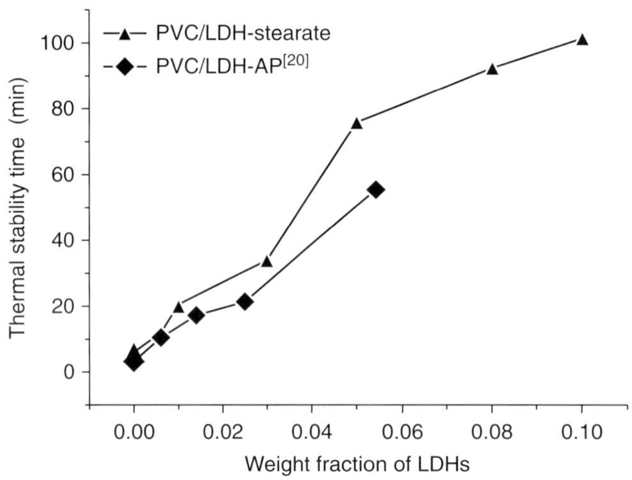

Figure 8.8. Thermal stability time as a function of LDH weight fraction for PVC/LDH stearate and PVC/LDH-AP nanocomposites.

foam. The best thermal stability was obtained for foams containing Cloisite 30B. The bio-based polyglycerol was successfully used to manufacture rigid polyurethane foams. The addition of nanoclay, especially Cloisite 30B, improved not only the mechanical performance (compressive strength) but also the thermal stability of the polyurethane-polyglycerol foams [7].

Shen et al. have tested the effects of nanoclay on a woven glass fiber–reinforced polyamide 6 (PA6) nanocomposite [8]. The XRD pattern in [8] shows that the clay was fully dispersed in the PA6 matrix. Cone calorimetry was also used to evaluate combustion performance of composites. The results can be found in Figure 8.11 [8]. The addition of clay, especially at amounts higher than 4 wt%, had a significant positive effect on time to ignition and time of peak HRR, but not significant effect on peak HRR value.

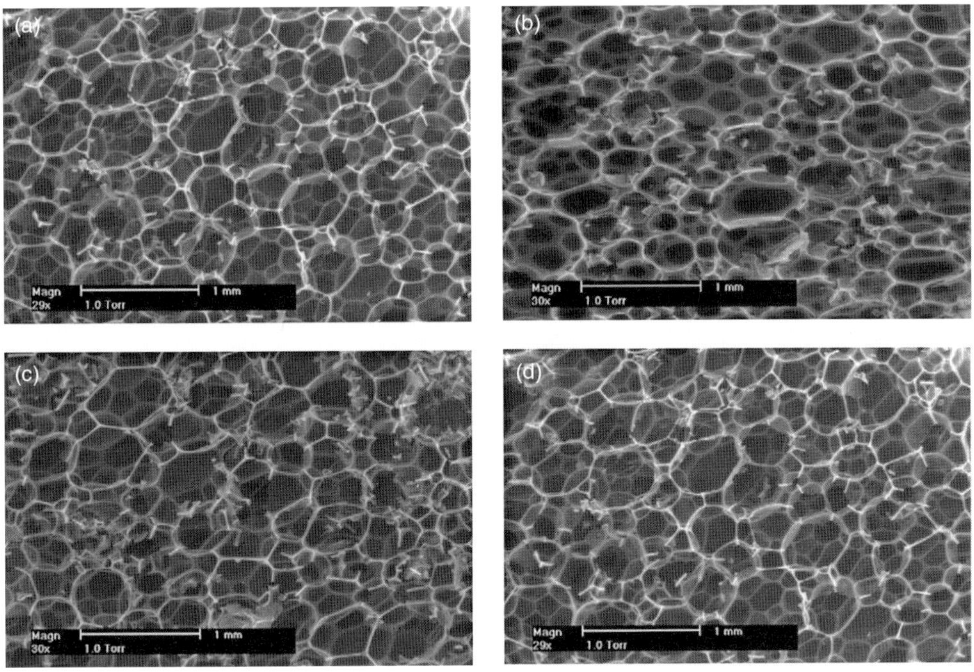

Figure 8.9. SEM micrographs showing cell structure of (a) polyurethane-polyglycerol foams and nanocomposites foams with (b) Cloisite 30B, (c) Laponite RD, and (d) Bentonite.

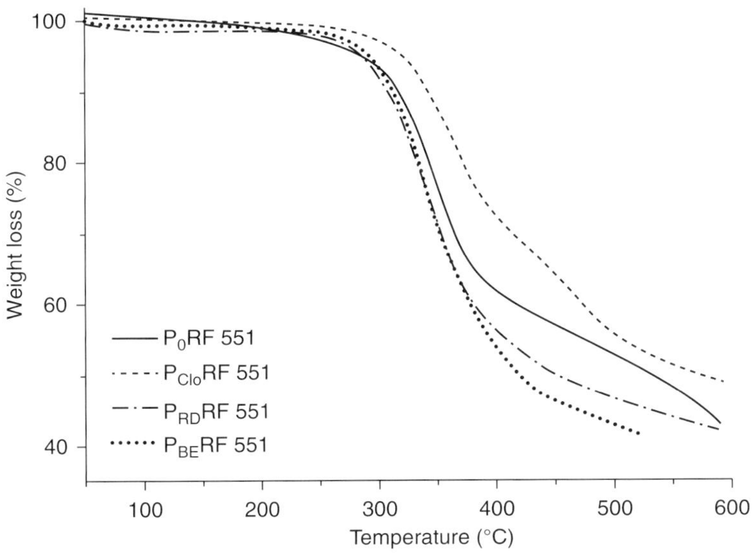

Figure 8.10. TG curves (15°C/min) for conventional rigid polyurethane nanocomposite foams.

The HDT is shown on Figure 8.12 and indicates that the addition of clay to the polymer matrix had no significant impact on laminate composite HDT [8]. The glass transition temperatures of the laminate composites are recorded in Table 8.2 [8]. The

Table 8.1. *Thermogravimetic analysis of rigid polyurethane-polyglycerol nanocomposite*

Clay in foam	RF 551:Polyglycerol	Weight loss (%)		
		Temperature (°C)		
		5	10	15
None	100:0	286	316	480
Cloisite 30B		302	328	533
Laponite RD		300	318	410
Bentonite		280	298	392
None	65:35	271	316	433
Cloisite 30B		289	325	430
Laponite RD		287	306	448
Bentonite		279	312	418
None	30:70	237	285	450
Cloisite 30B		272	326	442
Laponite RD		276	316	439
Bentonite		269	309	441

glass transition temperature of the composites was significantly enhanced by the addition of clay (27% increase at 6 wt% clay).

Woven glass fiber–reinforced laminates using a PA6/Clay polymer nanocomposite matrix were successfully fabricated. The addition of nanoclay delayed the time to ignition and peak heat release rate by 55% and 120%, respectively. The glass transition temperature was also increased by 27% at a clay loading of 6 wt%. It should also be mentioned that mechanical properties were generally slightly enhanced by the addition of clay [8].

8.2.6 Summary of Thermal Properties of Thermoplastic-Based Nanocomposites

The thermal performance of PP. PEEK, PVC, PU, and PA6/glass fiber–reinforced polymer nanocomposite systems with clay, CNF, SWNT, LDH, and clay/glass fiber were analyzed. In general, their thermal performance can be enhanced by the addition of nanoparticles. In particular, the Clay/PA6/glass fiber laminates have improved time to significant positive effect on time to ignition and time of peak HRR, but not significant effect on peak HRR value, higher heat distortion temperature, and improved glass transition temperature. The mechanical and electrical performance of the polymeric matrix can also be improved by the addition of nanoparticles. Sometimes simply maintaining good thermal performance while improving other properties is a significant feat.

It was found that the performance of nanoparticles on the basis of improved properties is highly dependent on the dispersion and exfoliation of the nanoparticles in the polymeric matrix. Certain surface treatments to nanoparticles and specialized processing techniques can help the dispersion and exfoliation of nanofibers in the

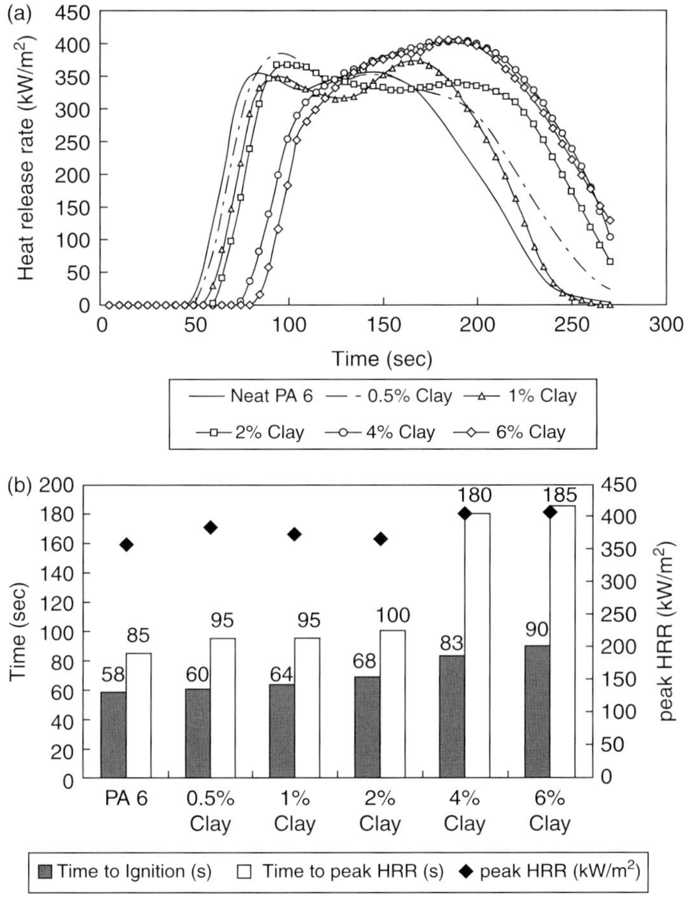

Figure 8.11. Cone test results on Clay/PA6/GF laminates: (a) HRR curves and (b) comparison of ignition time, peak HRR occurring time and peak HRR values (on the right y-axis) (Clay wt% in PA6).

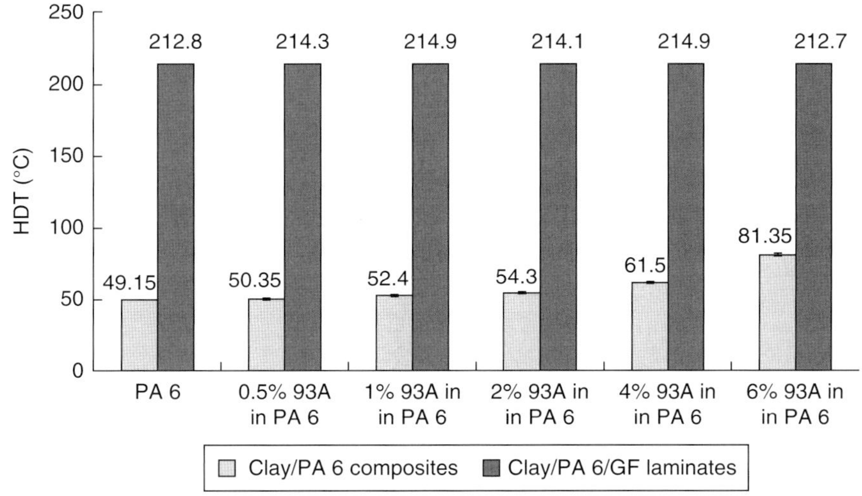

Figure 8.12. HDT values of Clay/PA6 composites and Clay/PA6/GF laminates at various Clay loadings (wt%) in PA6.

Table 8.2. *Glass transition temperatures of Clay/PA6/GF laminates*

Sample	PA6	+0.5 wt.% Clay 93A	+1 wt.% Clay 93A	+2 wt.% Clay 93A	+4 wt.% Clay 93A	+6 wt.% Clay 93A
T_g (°C) of laminates	56	60	63	65	68	71

composite. Polymer nanocomposite is a promising and growing field of study. It has already led to important advancements and continues to be in high demand in the industry.

8.3 Thermoplastic Elastomer-Based Nanocomposites

Over the years, research into the development, testing, and implementation of polymer nanocomposites has gained both interest and attention. The reinforcement of commodity polymers has allowed for the creation of new materials with enhanced properties to fulfill market demands.

Among the many polymers currently being explored, the class of thermoplastic elastomers (TPEs) offers a broad range of attractive properties. TPEs are a physical mixture of polymers, referred to as copolymers, that behave as both thermoplastics and elastomers. This means that they can be used in injection molding, along with other processing methods that exclude elastomeric thermosets. This is because the cross-linking in TPEs occurs via weaker, physical bonds rather than the chemical bonds that form during the curing process of a thermoset. As such, they are prime candidates for reinforcement by nanomaterials. Thermoplastic elastomers are replacing vulcanized rubber due to favorable mechanical properties and recyclability. Properties of TPE are similar to rubber in softness, flexibility, extensibility, and resilience and are more easily processed by the conventional thermoplastic processes of extrusion and injection molding.

Developmental TPE must meet three criteria: clarity, thermal resistance, and softness. These properties must be idealized for thermoplastic elastomers for their use as thermal interface materials for heat dissipation in electronic devices. These properties are important because heat produced during the operation of electronics is generally dissipated by heat conduction. To transfer this heat away from the electronics, heat sinks are generally used; however, heat dissipation capacity of a heat sink decreases due to the interfacial thermal resistance arising from the mismatch of surface roughness of both the device and the heat sink from a lack of a good thermal contact. Therefore, TPEs can act as thermal interface materials to fill the remaining gaps at this interface [9].

This section provides an overview of recent developments in the field of TPE nanocomposites from the perspective of thermal properties including thermal stability, glass transition temperature, crystallization temperature, thermal expansion, and thermal conductivity. New ways to improve the thermal properties of TPEs by using of nanoadditives focusing primarily on increasing thermal conductivity and decreasing the coefficient of thermal expansion (CTE) [10] are discussed.

Sections 8.3.1 to 8.3.5 describe research involved with enhancement of thermal stability, glass transition temperature, and crystallization property, and Sections 8.3.6 to 8.3.10 describe research involved in enhancement of thermal expansion and thermal conductivity properties. Step-by-step review of a set of recent experiments and research and their results as far as thermal properties are concerned are presented along with a summary at the end of this section.

8.3.1 Thermoplastic Polyurethane-MMT Clay

The effects of organically modified montmorillonite (OMMNT) on thermoplastic polyurethane (TPU) based nanocomposites were studied by the Rubber Technology Centre of the Indian Institute of Technology [11]. The composite was prepared by melt intercalation using the Thermo Scientific HAAKE PolyLab OS Rheomox at 185°C and 100 rpm for 6 minutes. Cloisite 15A was the nanoclay used in the following amounts: 1, 3, 5, 7, and 9 wt% (TPU1A, TPU3A, TPU5A, TPU7A, and TPU9A). Wide angle X-ray diffraction showed no evidence of d-spacing, which suggests that the nanoparticles may be well dispersed; the transmission electron microscopy analysis showed that there was a mixed morphology of partially exfoliated and intercalated structures. Although the group was unable to thoroughly exfoliate the clay, there was still significant interaction between the TPU and OMMNT [11].

TPU is a copolymer, which consists of two different segments. The hard segment is comprised of alternating diisocynate and diols or diamine units while the soft segment is comprised of linear long-chain diols. Due to the organic modifier on the nanoclay, there is a strong interaction between the highly polar soft segments. There is hydrogen bonding that occurs between the OMMNT and the TPU, and the clay particles appear to preferentially attract to the soft segments.

TGA was performed to compare the residual weight of the pure TPU and the composites (Figure 8.13) [11]. Note that there are two distinct peaks in the derivative of the TGA curve, which shows that there are two separate stages of degradation for the hard and soft segments, respectively (Figure 8.13). The temperature at the onset of each stage is denoted as Td1 and Td2 in Table 8.3 [11].

Increasing the amounts of Cloisite 15A shifts the peak degradation temperatures (DTG1max, DTG2max) higher, which shows that the TPU-OMMNT composite has better thermal stability than the pure TPU polymer (Table 8.3). These two temperatures correspond to the peak rate of degradation of the hard segment and the soft segment, respectively. At DTG2max, the degradation of the base polymer is dominant [11]. Although the peak degradation temperatures are a maximum at 5 wt%, the residual mass at 500°C (500 wt%) still increases almost linearly. While the clay itself does not burn, the organic modifier is combustible and accounts for some of the mass loss, which explains why the residual mass is less than the amount of Cloisite 15A added.

The authors explain the effect of the nanoclay by the "Labyrinth barrier effect," which is caused by the anisotropic silicate platelet mesh that impede the diffusion of volatile degradation products [11]. Since the structure of TPU has two distinct hard

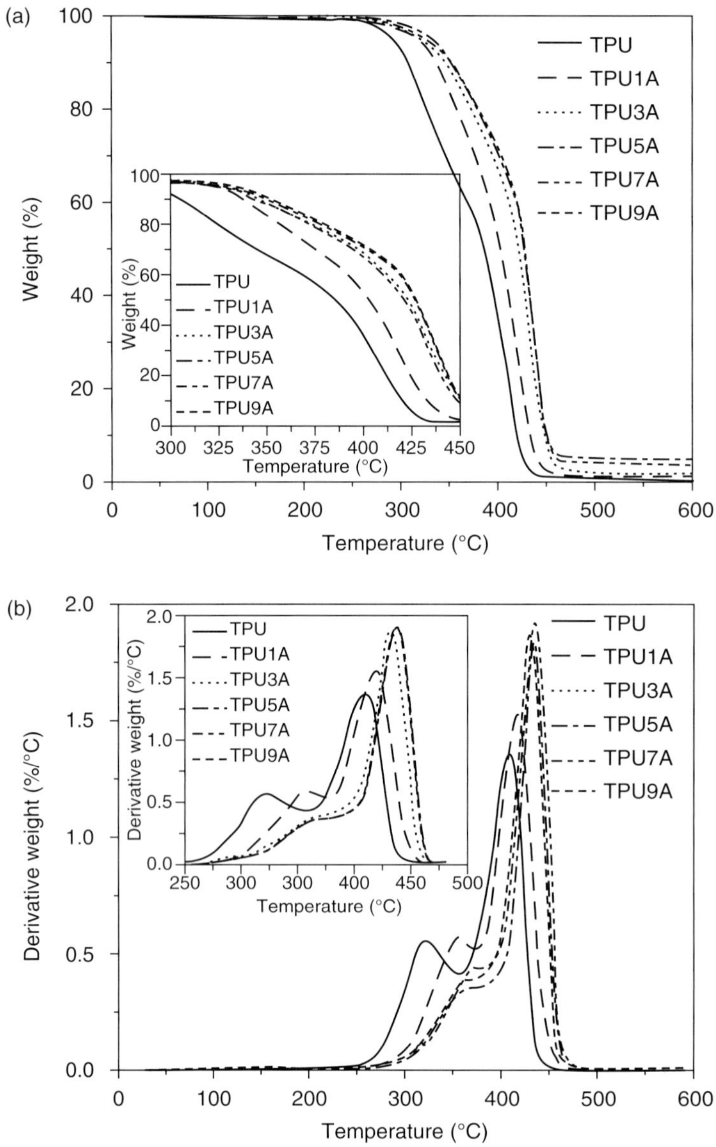

Figure 8.13. (a) Residual mass TGA curve and (b) its derivative of TPU and TPU composites.

and soft segments, two melting temperatures are expected. However, in the DSC curve, only one endothermic phase is detected which is where the differential heat flow is at a local minimum [11]. It is speculated that the reason for detecting only one melting point is that there is limited movement of the hard segment. Additionally, the hard segment has only a slight change in specific heat and is uniformly dispersed throughout the composite matrix [11]. As a result, all of the data acquired by means of DSC represents the soft portion of the composite [11].

It was observed that increasing amounts of Cloisite 15A slightly increased the glass transition temperature for the soft segment. The authors concluded that the small amount of dispersed nanoclay is insignificant to produce significant changes

Table 8.3. *TGA results of neat TPU and their composites*

Sample codes	DTG_{1max} (°C)	DTG_{2max} (°C)	$T_{50\,wt\%}$ (°C)	T_{d1} (°C)	T_{d2} (°C)	$T_{95\,wt\%}$ (°C)	$r_{500\,wt\%}$ (%)
TPU	321.37	408.65	386.73	297.31	356.85	425.58	0.801
TPU1A	352.21	418.03	407.52	326.21	380.32	433.24	1.451
TPU3A	365.19	432.26	425.33	336.99	386.38	455.54	2.119
TPU5A	373.29	436.65	427.33	332.63	390.41	462.07	4.132
TPU7A	362.16	436.05	427.63	329.03	378.38	461.34	4.209
TPU9A	362.14	432.39	421.81	320.64	376.29	563.67	5.346
Cloisite 15A	–	316.65	–	253.67	–	–	–

in T_g of the composite due to the TPU comprising most of the bulk-free volume of the composite and the interactions between the organic and inorganic parts. Consequently, the soft segments become more rigid and have limited movement, which results in a slight increase in T_g [11].

The change in specific heat of the composite across the glass transition was observed for increasing amounts of nanoclay. It is speculated that the difference in specific heat is the result of the nanoclay layers hindering the motion of polymer chains [11]. As the amount of OMMNT is increased, the difference in specific heat decreases. However, at a certain amount of OMMNT, the difference in specific heat begins to increase but still remains significantly below that of pure TPU. This phenomenon is probably due to the specific heat of the OMMNT being significantly higher than that of the TPU.

The addition of OMMNT increases the melting temperature of the composite, which suggests that the nanoclay alters the crystal structure of the soft segments. The data show that the melting temperature plateaus at about 3 wt%, which is a 2.33% improvement over neat TPU [11]. However, better results may be obtained for higher loadings of OMMNT if the nanoclays can be more uniformly dispersed and fully exfoliated.

It was observed that the heat of fusion increased with OMMNT loading. This value plateaued around 3 wt%, which is 211% greater than neat TPU. Increasing amounts of OMMNT increases the degree of crystallinity of the soft segments, and since the enthalpy change during the phase change is small, it is speculated that only some of the polymer linkages formed phase separation domains. The authors attribute the plateau in the heat of fusion to be caused by filler agglomeration at high concentrations, which reduces the effective interfacial surface area of the filler to the polymer [11].

The authors suggest that the peak shown in the DMA curve is associated with the main glass transition temperature of the system due to the mobilization of the soft segments in the TPU matrix (Figure 8.14) [11]. As Cloisite 15A was added, the tan(δ) peak shifted to higher temperatures due to resistance in motion of the polymer chains caused by the nanoclay [11]. The value of the glass transition temperature appears to plateau at a certain loading of OMMNT, resulting in a maximum of 3.9%

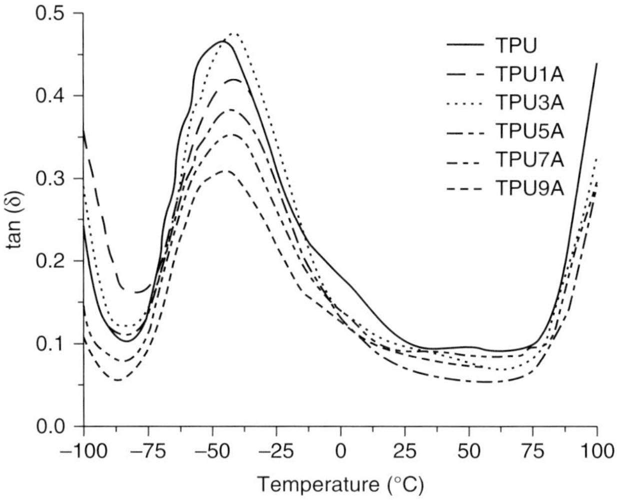

Figure 8.14. DMA curve and T_g for TPU and TPU composites.

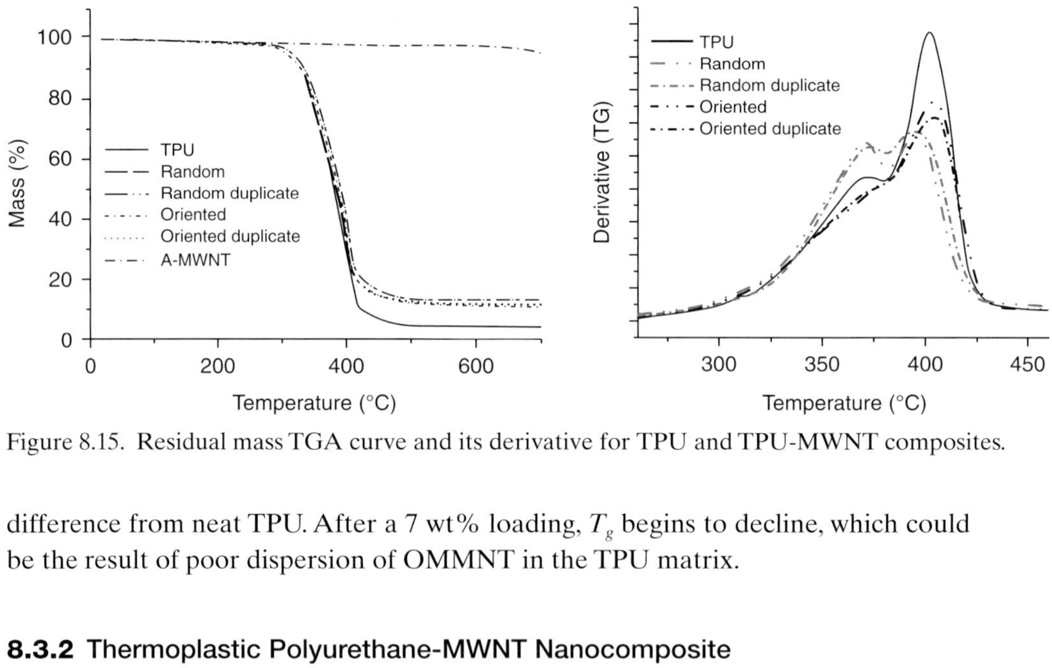

Figure 8.15. Residual mass TGA curve and its derivative for TPU and TPU-MWNT composites.

difference from neat TPU. After a 7 wt% loading, T_g begins to decline, which could be the result of poor dispersion of OMMNT in the TPU matrix.

8.3.2 Thermoplastic Polyurethane-MWNT Nanocomposite

A study done by Rice University researchers showed the effects of nanotube orientation on the properties of multi-wall carbon nanotubes (MWNT). The carbon nanotube arrays, created using a chemical vapor deposition process, grew to lengths of 3.5 mm with an average diameter of 75 nm. Different samples of oriented composites and randomly dispersed composites loaded with 6 wt% MWNT were produced for testing. The base polymer used was TPU [12].

The TGA was performed in a nitrogen environment, so that the MWNTs do not combust (Figure 8.15) [12]. Overall, there was virtually no change in the onset of thermal degradation from the derivative TG regardless of MWNT orientation

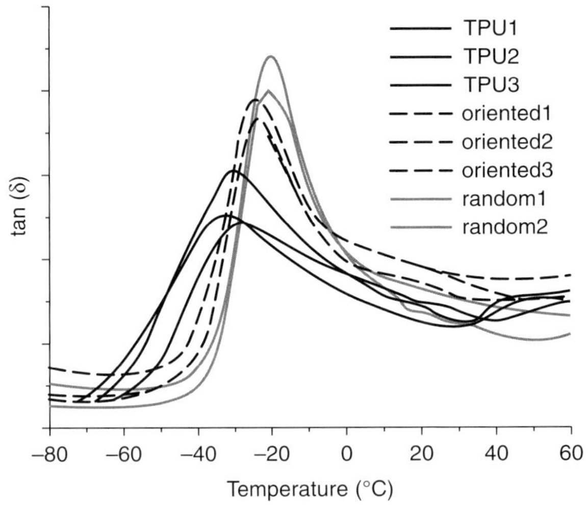

Figure 8.16. DMA curve for TPU-MWNT composites showing tan(δ).

(Figure 8.15). In general, it appears that the carbon nanotubes do not influence the degradation of TPU in an inert environment. The nanotube content was determined from the difference in residual mass at 600°C and there was virtually no change [12]. Unlike OMMNT, the MWNT do not have the proper aspect ratio to replicate the "Labyrinth effect" discussed earlier (Section 8.3.1). Additionally, since they are made of carbon, the MWNT will probably combust in an oxygen environment and will be detrimental to the degradation of TPU.

Interestingly, the results of the DMA showed that the randomly oriented MWNTs consistently had a higher glass transition temperature than the oriented samples (Figure 8.16) [12]. This suggests that the random dispersion induced an amorphous phase that is more closely packed than the oriented samples [12]. The oriented samples had a 2.89% increase in T_g while the randomly dispersed samples had a 4.54% increase.

8.3.3 Thermoplastic Polyurethane Mixed with Organo-Modified Layered Silicates (OMLSs)

Aurilia et al. tested various properties of a nanocomposite made by mixing organoclay with thermoplastic polyurethane [13]. The neat resin was Desmopan DP 9370AU produced by Bayer. Its polyol is poly tetramethylene ether glycol, its chain extender is butane-diol, and its isocyanate is methylene-diphenyl diisocyanate. The nanoadditive was Cloisite 30B from Southern Clay Products Inc. that was dried in a vacuum oven at 90°C for 12 hours prior to use. The two components were mixed in a Haake Rheocord internal mixer at 160°C and 60 rpm for 6 minutes at clay weight percentages of 0.1, 0.5, 2.0, 6.0, and 10.0.

The authors used SAXS and FT-IR spectroscopy to characterize the dispersion in the resultant materials. SAXS results comparing the neat polymer and 10 wt% Cloisite 30B are shown with the diffraction peaks marked [13]. The rest of the weight percentages' SAXS results along with the Lorentz-corrected plots of all the samples

Table 8.4. *Summary of thermal property data for TPU mixed with Cloisite 30B [13]*

Clay wt%	Glass transition temprature SDs [°C] peak loss tangent DMA	Melting enthalpy of soft domains [J/g] DSC	T1 DSC [°C]	Enthalpy of T2 endotherm [J/g] DSC	Heat of crys tallization soft domains isothermal scan [J/g] DSC	Crystallization temprature of SDs [°C]
0	−55.61	13.11	59.31	9.02	61.31	−19.49
0.1	−54.47	15.69	77.05	1.95	67.88	−20.66
0.5	−52.99	13.76	66.67	11.68	44.64	−19.00
2.0	−54.83	14.35	64.73	12.24	27.95	−19.68
6.0	−54.22	12.44	65.74	7.54	14.44	−19.20
10.0	−56.52	12.57	67.64	7.63	22.13	−21.16
10.0 (quenched)	−56.78	8.24	48.10	6.73	//	//

are shown in reference [13]. As can be observed, diffraction peaks for the OMLS appearing in 2.0, 6.0, and 10.0 wt% around 1.6 nm^{-1} indicate intercalation. No peaks were observed for 0.1 and 0.5 wt%, indicating partial or complete exfoliation.

Furthermore, FT-IR was used to examine the degree of phase separation (DPS). The DPS of 0.1 and 0.5 wt% went down compared to the neat resin and increased in the higher wt% samples with respect to the neat resin. The authors also used DMA to obtain T_g results for the various samples and DSC for the other properties (e.g., melting enthalpy of soft domains/SDs, crystallization temperatures, etc.). Table 8.4 presents a summary of their findings [13].

Glass transition temperature initially decreases with increasing wt% of clay, to the lowest with 0.5 wt% at -52.99°C. It then increases with increasing wt% all the way to 10 wt% and -56.52°C. Crystallization temperatures show no discernible pattern with respect to wt% of clay. Figure 8.17 shows the DSC thermogram used to obtain some of the data in Table 8.4 [13].

These curves have been shifted for clarity, but changes based on the different clay compositions can be observed. The authors also performed an isothermal DSC experiment (-10°C) on the various samples and observed their heat flows over time. The results are presented in Figure 8.18 [13]. Samples with 6.0 wt% and 10.0 wt% show the lowest heat flow consistently over time. Samples with 2.0 wt% and 0.5 wt% slightly reduce the heat flow versus the neat resin, while the 0.1 wt% sample actually had increased heat flow over the pure polymer.

8.3.4 Thermoplastic Polyurethane Mixed with Laponite and Cloisite

Mishra et al. [14] compared two varieties of clay in preparing TPU-based nanocomposites. The neat resin was Desmopan KU2 8600E, which is a polyether-based TPU prepared from MDI, polytetramethylene glycol, and 1,4-butanediol. The nanoadditives were Laponite RD and Cloisite 20A obtained from Rockwood additives Ltd. (United Kingdom). Laponite RD was modified with dodecylamine hydrochloride in order to assist in dispersal via a solvent (named L in the figures). Cloisite 20A

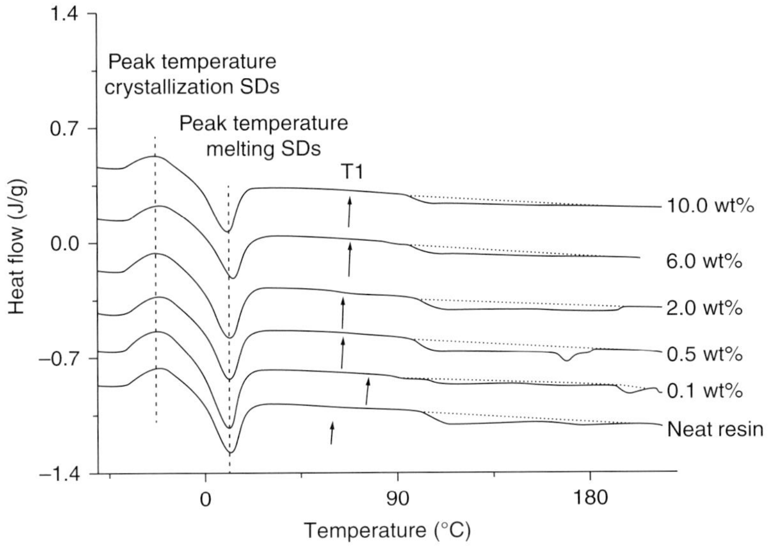

Figure 8.17. DSC thermogram of various TPU/clay nanocomposite samples [13].

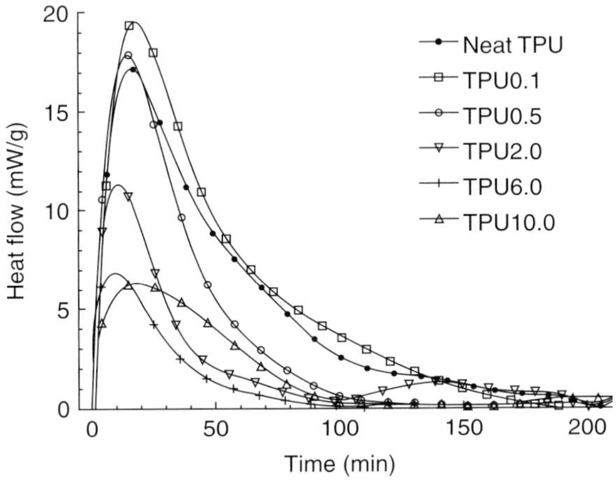

Figure 8.18. Heat flow (mW/g) versus time (minutes) for isothermal DSC experiment at −10°C [13].

was modified by dimethyl dehydrogenated tallow quaternary ammonium ion (C in figures). Samples were thus prepared using 20% solution of the TPU in tetrahydrofuran (THF) and mixing the calculated amounts of Laponite/Cloisite while stirring and sonicating. For final preparation, samples were then held in a vacuum oven at 70°C.

The authors used WAXD, SEM, TEM, and AFM to characterize the material morphology. FTIR was used to differentiate between the modified and unmodified nanoadditives. WAXD results for neat resin (T0), 1 wt% Cloisite 20A (T1C), 3 wt% Cloisite 20A (T3C), 5 wt% Cloisite 20A (T5C), 10 wt% Cloisite 20A (T10C), and pure Cloisite 20A (C) for a lower angular range are shown in [14].

The authors claim that the peak at roughly 7.1° corresponds to a fraction of unmodified clay remaining in the composite, while the peak at roughly 3.7°

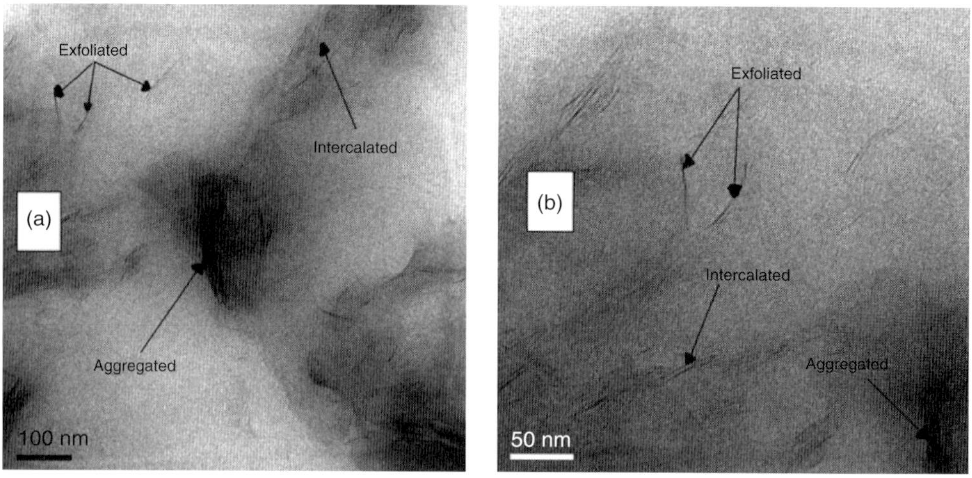

Figure 8.19. TEM of 5 wt% Cloisite TPU at (a) 100nm and (b) 50nm [14].

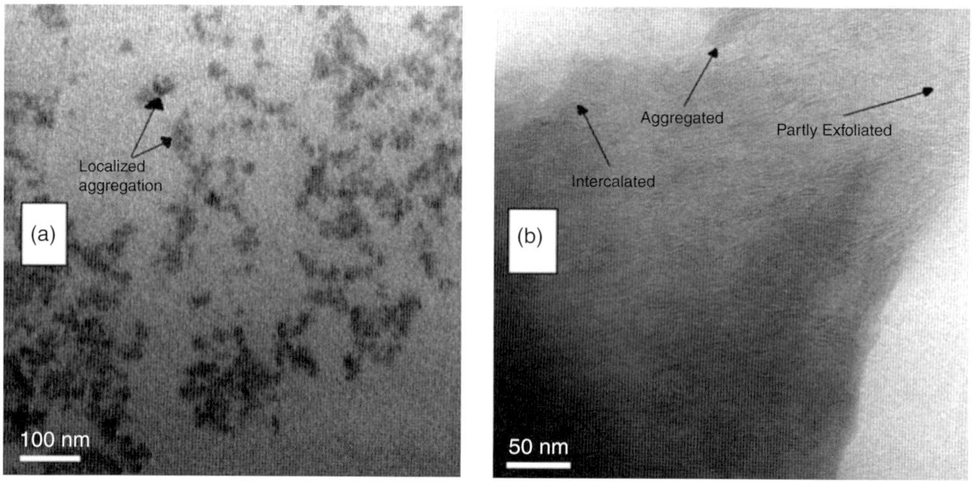

Figure 8.20. TEM of 3 wt% Laponite at (a) 100nm and (b) 50nm [14].

corresponds to the modified clay. The WAXD intensity spectra for the full angular range for Cloisite are shown in [14].

Compare the WAXD results for Laponite at low and full angular ranges in the neat resin (T0), 1 wt% Laponite (T1L), 3 wt% Laponite (T3L), 5 wt% Laponite (T5L), 10 wt% Laponite (T10L), and pure Laponite (L) as shown in [14]. No prominent peaks are observed because of the lack of a well-defined stacking of platelets in Laponite.

The mixed results reported by WAXD are confirmed by the TEM images. Although the authors only report images for T5C and T3L, regions of aggregation, intercalation, and exfoliation are all revealed. Figure 8.19 shows two different scales for T5C (5 wt% Cloisite) [14]. Results are even more mixed for the 3 wt% Laponite (Figure 8.20) [14].

From the TEM images it seems that better dispersion was achieved with the Cloisite. Thermogravimetric analyses of the Cloisite- and Laponite-reinforced TPU

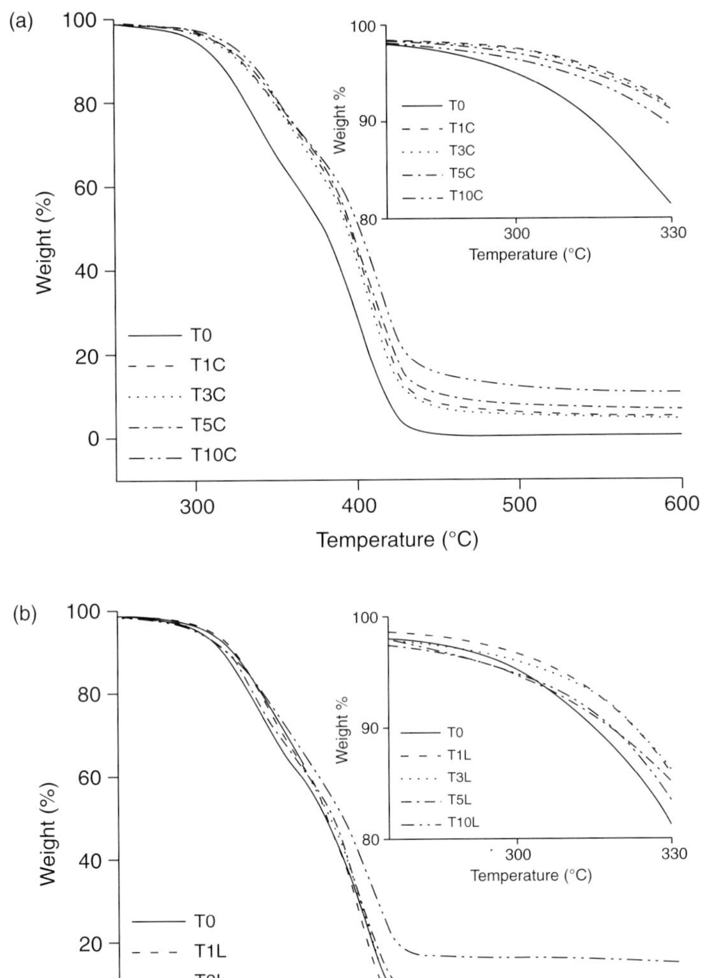

Figure 8.21. TGA results for various samples of (a) Cloisite (C, top) and (b) Laponite (L, bottom) reinforced TPU [14].

show essentially delayed thermal degradation as the amount of Cloisite or Laponite was increased. This applies to the onset of degradation, which was delayed 17.5°C by the 1 wt% Cloisite loading and 8.3°C by the 1 wt% Laponite loading. As the loading was increased, further increases in the temperature of degradation onset were observed in both types of clay. TGA results are shown in Figure 8.21 [14].

DSC thermograms reveal almost no change in T_g as Cloisite/Laponite compositions varied, remaining at roughly ~40°C (trends confirmed by DMA, although the value of T_g measured there is ~22°C). Increasing Cloisite loading generally made heat flow more negative, as shown in the thermogram in [14]. This was also the case for increasing Laponite loading, except for the 10wt% case, as observed in [14].

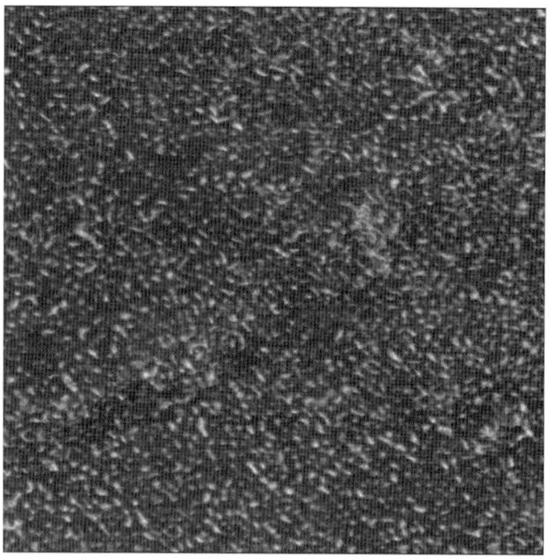

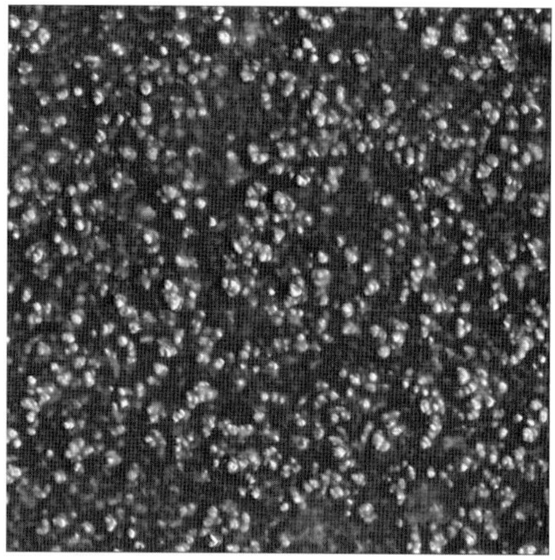

Figure 8.22. AFM of neat resin (left) and composite (right) [15].

8.3.5 Polyisobutylene-Based TPE Mixed with Carbon Black

Puskas et al. [15] tested various properties of a nanostructured composite made using TPE and carbon black. The neat polymer consisted of a branched polyisobutylene core with hydroxyl functional groups at each branch point flanked by blocks of poly(isobutylene-*co-para*-methylstyrene). This was mixed with 37.5 wt% N234 carbon black as received from Cabot Corp. in a Brabender mixer. Dispersion characteristics of the materials are demonstrated using the AFM technique as shown in Figure 8.22 [15]. The dark regions are the elastomeric continuous phase, while the light regions are the plastic phase and/or the carbon black. The good dispersion is confirmed by WAXS and SAXS results as shown in Figures 8.23 and 8.24 [15].

The authors attribute the second, broad peak in the composite to an order in the dispersion, and support that claim with their SAXS results, shown in Figure 8.24 [15]. The marked hump in the composite plot corresponds to the ordered dispersion. Additionally, the authors cite the high volume resistance, which is abnormal in a carbon-reinforced TPE, as an indicator of good carbon dispersion (quoted as 3×10^{11} Ohm cm).

The study reports an improvement in the rate of thermal degradation as well as an increase in the high T_g as a result of the addition of carbon black. The neat resin had 1% mass loss at 265.6°C, while onset of degradation was delayed until 327.7°C in the composite. The neat resin reached 100% mass loss at 375°C, while the polymer part of the composite fully degraded at 400°C (leaving the carbon, which degraded at 600°C) [15]. The authors also performed DSC analysis, which revealed an increase in high T_g from 105°C in the neat resin to 126°C in the composite. The low T_g was unchanged (at –73°C).

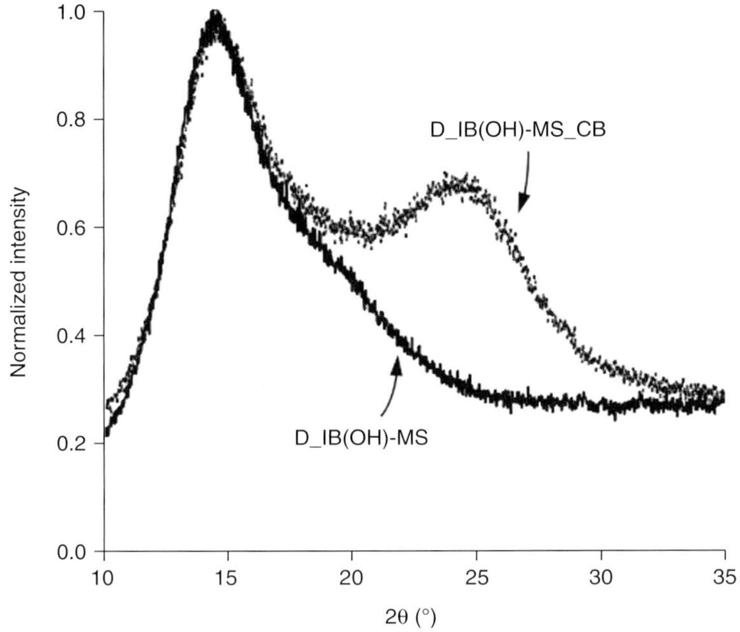

Figure 8.23. WAXS of neat resin (D_IB(OH)-MS) and composite (D_IB(OH)-MS_CB) [15].

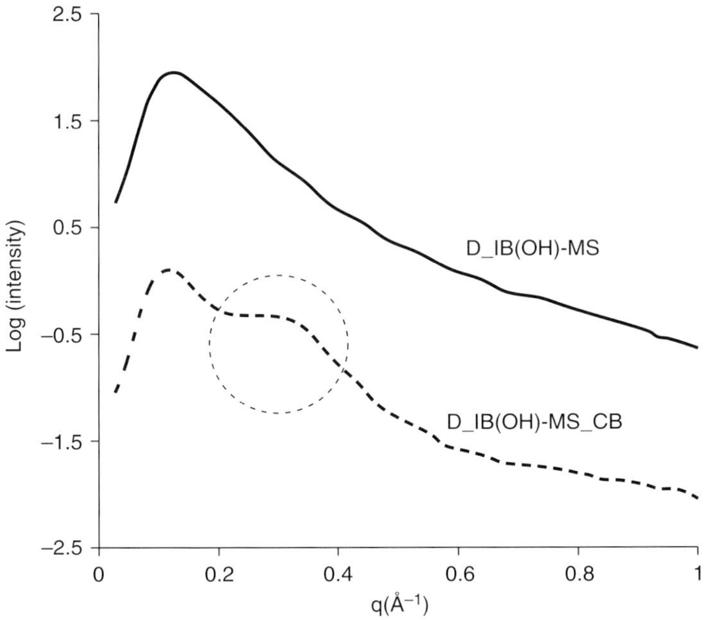

Figure 8.24. SAXS of neat resin (D_IB(OH)-MS) and composite (D_IB(OH)-MS_CB) [15].

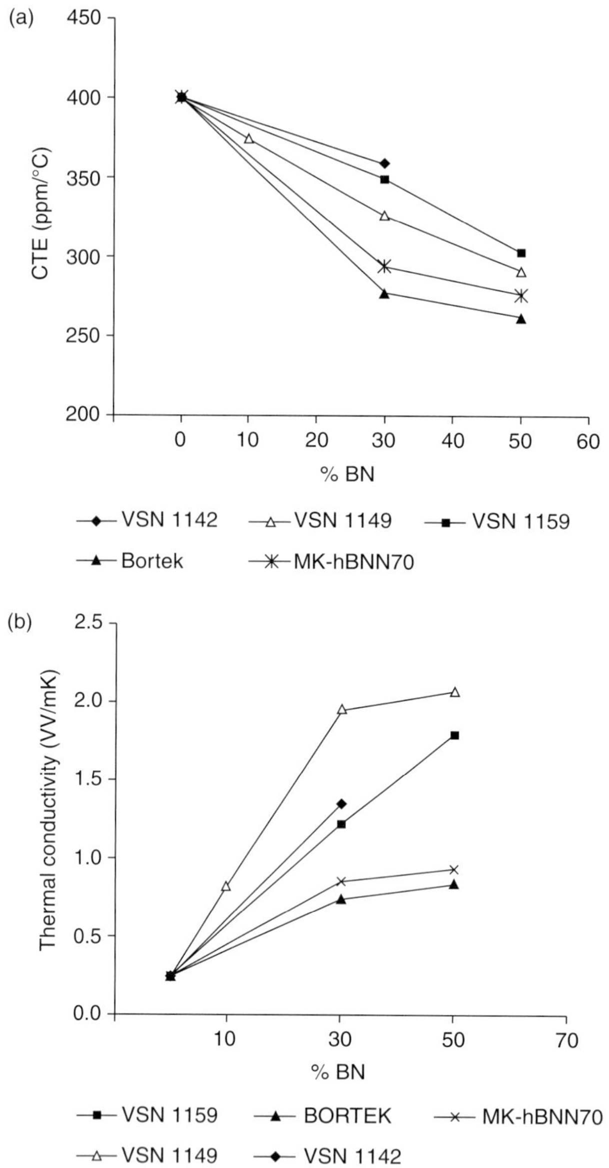

Figure 8.25. Effect of BN type filler concentration on (a) coefficient of thermal expansion and (b) thermal conductivity.

8.3.6 Poly(dimethyl Siloxane)/Boron Nitride

The silicone rubber used with boron nitride was vinyl terminated poly(dimethyl siloxane) PDMS. The filler loading of different types of BN was varied at 0, 10, 30 and 50 wt%. Figure 8.25 shows the results for coefficient of thermal expansion (CTE) as well as thermal conductivity for the PDMS/BN composite with varying filler loading [10]. The incorporation of boron nitride particles into silicone elastomer results in a reduction of the CTE. This is due to the restriction of mobility of the silicon molecules from the adsorption of the BN surfaces. As the particle size of boron nitride decreased, the CTE also decreased for any given loading level. Therefore, the lowest values for CTE occurred when using nano-sized fillers. This is likely due to the high surface/volume ratio of nano-sized fillers, which promotes a higher level of interaction at the interface.

For thermal conductivity, however, we see the opposite trend. For this measurement it is seen that as particle size decreases, we obtain the less desirable result of a lower thermal conductivity as opposed to the micron-sized fillers. This is because the ability to form a conductive network of the filler in the matrix depends on the aspect ratio and concentration of the filler. When concentration and aspect ratio of the filler increase, after a critical concentration, particles percolate and touch one another to form a conductive network. Due to the lower aspect ratio of the nano-sized fillers, we see lower conductivities with these fillers. Regardless of BN type, the addition of filler material enhances the thermal conductivity of the composite [10].

8.3.7 Polyethylene/Single-Wall Carbon Nanotube

For this study, multiple polymer matrices were used to host the SWNTs. These polymers were low-density and high-density polyethylene (PE) with PE crystallinities of 33% and 78%, respectively. The composites were prepared with SWNT volume fractions of 1 and 30 vol% for LDPE and 1, 5, 10, and 30 vol% in HDPE [16].

The thermal conductivity results for both LDPE and HDPE are presented in Figure 8.26 [16]. The thermal conductivity of neat LDPE is 0.26 W/mK while the thermal conductivity of neat HDPE is 0.5 W/mK. For both HDPE/SWNT as well as LDPE/SWNT we see only moderate increases in thermal conductivity at low SWNT volume fractions. This absence of a significant increase in thermal conductivity at low loadings is attributed to the modest difference of the constituent thermal conductivities and the large interfacial thermal resistance between the polymer and SWNTs, with this large interfacial thermal resistance likely due to weak thermal coupling. At the highest SWNT loading we see a significant increase in the HDPE/SWNT samples. This is likely due to changes at the interface between PE and SWNT, which reduces the interfacial thermal resistance [16].

8.3.8 Ethylene Propylene Diene Monomer/ZnO

The polymer used to host the nano-sized ZnO was ethylene propylene diene monomer (EPDM). The filler amount of ZnO was calculated on a volume basis with values ranging from 0 to 225 loading of filler/phr. The thermal conductivity results for both untreated and in situ modified composites are shown in Figure 8.27 [17].

Figure 8.27 indicates that thermal conductivity of both untreated and in situ modified nano-ZnO filled composites increase linearly with filler loading. Maximum

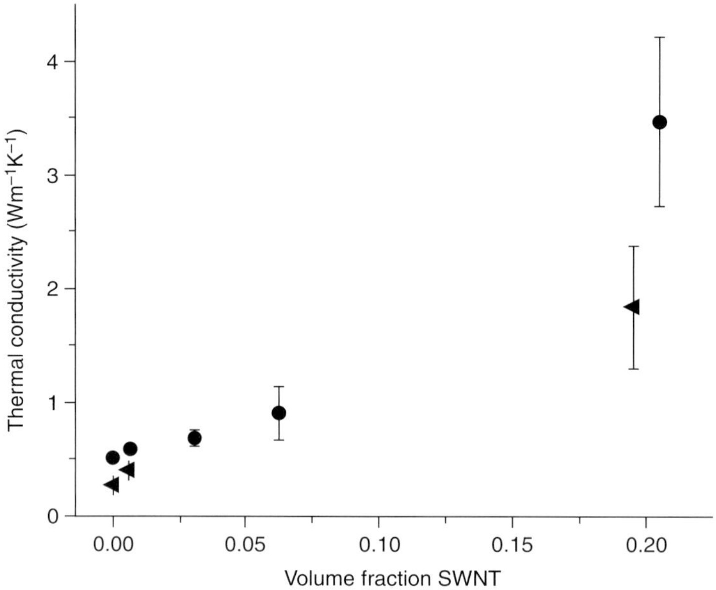

Figure 8.26. Effect of SWNT volume percent on the thermal conductivity of PE. HDPE is represented by circles and LDPE is represented by triangles (Courtesy of ACS).

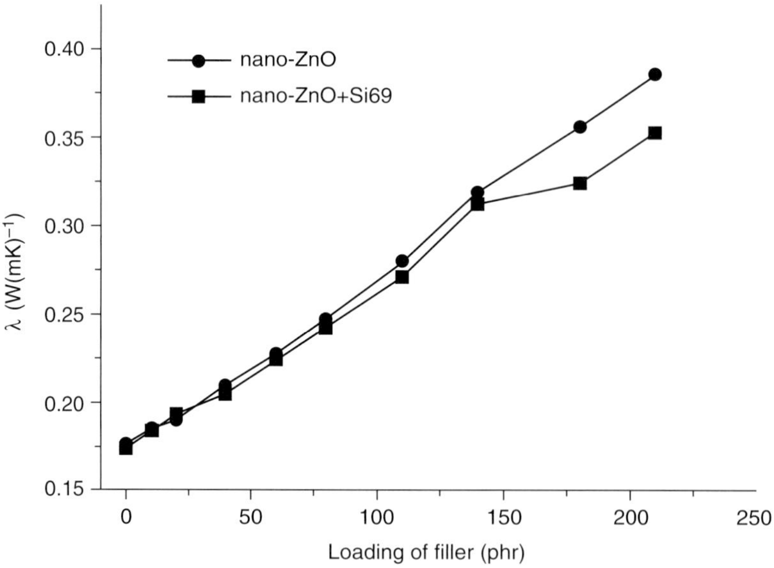

Figure 8.27. Thermal conductivity results for the EPDM/ZnO composite for both untreated and in situ modification as a function of filler loading.

thermal conductivity obtained is 0.386W/mK at a loading of 210 phr. This is more than two times greater than that of a 10 phr nano-ZnO filled composite. It is also observed that, at the same filler content, thermal conductivity of untreated nano-ZnO is a little higher than that of the in situ modified nano-ZnO system. One possible reason for this is that in situ modification improves the dispersion, which harms

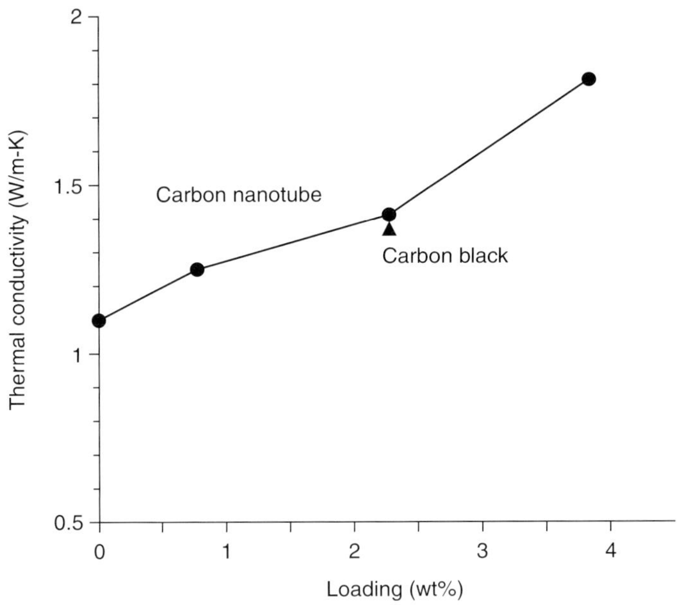

Figure 8.28. Thermal conductivity results for the silicone elastomer/CNT composite with varying weight loadings of CNTs.

the formation of thermal conductive chains composed of filler particles and increases the total quantity of filler-rubber interface thermal resistance. Also, it is possible that parts of Si69 molecules grafted on the particle surface can enhance the original thermal resistance and decrease the thermal conductivity for the untreated nano-ZnO composite. One thing to note from these results is that excessive pursuit of higher thermal conductivity will result in a deterioration of the mechanical properties [17].

8.3.9 Silicone Elastomer/CNT

The polymer host used with carbon nanotubes is a commercially purchased silicone elastomer from Dow Corning. Carbon nanotubes were loaded from 1 wt% to 4 wt%. For comparison purposes, a silicone elastomer with carbon black was also prepared. The thermal conductivity results can be seen in Figure 8.28 [18]: there is a linear relationship between weight loading of carbon nanotubes and thermal conductivity for the composite. These results show a great improvement for only small amounts of loading. With a 3.8 wt%, a 65% enhancement in thermal conductivity with a final value of approximately 1.8 W/mK was observed. Also when comparing equal loadings of CNTs and carbon black at 2.3 wt%, a slightly greater thermal conductivity provided by the silicone elastomer/CNT composite was shown [18].

8.3.10 Silicone Elastomer/SiO_2 or Al_2O_3

The host polymer used for silicon dioxide and aluminum oxide fillers was HTV silicone rubber. Filler contents of 2, 4, 10, and 40 wt% were investigated as well as samples with no filler. The results for thermal conductivity can be found in [19]. All samples display a linear increase in thermal conductivity with increasing filler concentration. It is interesting to note that the nano-sized fillers actually performed worse than the micron-sized fillers at increasing filler concentration for their respective filler type. One possible reason is that nano-sized fillers can increase the thermal degradation of a polymer enough to actually hurt its thermal conductivity in comparison with micron-sized fillers. With that being said, all types of fillers provided very similar results, and an excellent improvement in thermal conductivity was observed at high wt% loadings [19].

8.3.11 Summary of Thermal Properties of Thermoplastic Elastomer-Based Nanocomposites

This section provided an overview of the current technology in enhancing the thermal properties of thermoplastic elastomers by using nanoadditives. These studies analyzed the thermal properties of different thermoplastic elastomers reinforced by carbon black, synthetic, and organic clays. Consistently, nanoadditives in increasing weight percentages resulted in delayed and/or reduced thermal degradation (as measured by TGA) and reduced heat flow (as measured by DSC). Sometimes, transition temperatures, such as T_g were affected by the additives, but often they were unchanged. Different levels of dispersion were achieved in the different studies, and some performance differences can be attributed to that.

Organic clays have become a mainstay as a nanoadditive because of their low cost and consistent performance. The comparison with synthetic clay (Laponite) is important to review where synthetics stand with respect to organics. Additionally, of the three-dimensional nanoadditives, carbon black is one of the least expensive, and so is a more accessible material. The economies of these reinforcements make it important to study them, as other, more expensive nanomaterials will not be practical for industrial implementation for some time.

In the case of TPU, the organic modifier made the nanoclays preferentially attracted to the softer segments, and as a result, affected the properties of these softer segments much more than the harder segments. In the case of polymer foams, the porosity of the foam was also a large contributor to the thermal conductivity.

On the basis of reviews in this section, the effects of nanomaterials can range from extremely useful to negligible, and are highly dependent on the type of filler used as well as its relative size. For all cases, there were enhanced properties with increasing loading of filler, leading us to believe that the more filler, the better properties your composite will display. Interestingly enough, nano-sized fillers are not always ideal when compared to micron-sized fillers. For all studies in which thermal conductivity data was taken for both nano- and micron-sized fillers, the micron fillers performed better. Based on these results, we see that many factors can affect the thermal properties of polymer nanocomposites.

8.4 Thermoset-Based Nanocomposites

Thermosets are a class of polymers known for the curing behavior. They begin in a malleable or liquid form, and after being cured the material becomes hard and remains in that form. This means that once the material is cured, it cannot be reheated and reshaped or recycled. The curing behavior also means that this material holds up well under elevated temperatures, expanding its range of applications. Thermosets are highly cross-linked, which gives them good mechanical properties, such as a high modulus and strength and good resistance to creep, but they have poor ductility due to their rigid behavior [20]. As a polymer, they have low thermal stability, high electrical resistivity, and low thermal conductivity [21]. Many thermoset polymers have been known for their excellent thermal performance, which makes them prime candidates for the primary base in which to develop further advancements with nanocomposites.

8.4.1 Epoxy Nanocomposites

Epoxy is known for its excellent chemical and corrosion resistance, good adhesive properties, low shrinkage, and low price [20]. As mentioned earlier, curing gives them a highly cross-linked microstructure, which results in a high modulus and strength, good resistance to creep, good performance at elevated temperature, but poor ductility [20]. To address different shortcomings, nanofillers have been investigated. With the addition of ceramic, the stiffness of the material is improved and the cost is reduced, but the material becomes even more brittle [20]. Soft particle addition results in a tougher material but at the sacrifice of the stiffness [20]. A popular epoxy system is based on di-glycidyl ether of bis-phenol A (DGEBA) or -phenol F (DGEBF) [22].

8.4.1.1 POSS in Modified Epoxy Nanocomposites

In a study performed by Nagendiran et al. [23], DGEBA was mixed with tetraglycidyl diamino diphenyl methane (TGDDM), and was then formed with 4,4'-diaminodiphenylsulfone (DDS) in the presence of the octa-aminophenyl silsesquioxane (OAPS) form of POSS. The addition of TGDDM to the DGEBA contributes to increased thermal stability of the composite through the former's higher epoxy functionality. The POSS and DGEBA in this case are formed through in situ polymerization. Even though POSS composites can be formed through physical blending, in other studies the authors found that this preparation method has a tendency to show immiscibility toward the epoxy resin [23].

Characterization of the thermal properties of this epoxy-POSS composite was performed with differential scanning calorimetry and thermogravimetric analyses. Calorimetry was performed on 10 mg samples with a continuous flow of nitrogen and a heating range from ambient to 400°C at 10°C/min [23]. Glass transition temperature was taken as the midpoint of this change in capacity. TGA was performed

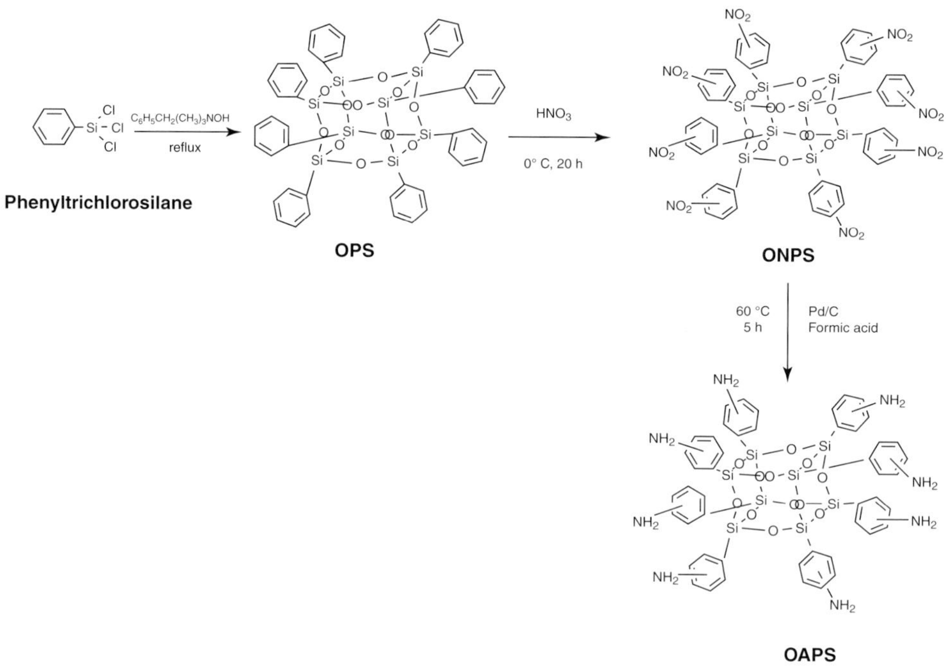

Figure 8.29. Three-step process denoting the reactions to form OAPS [23].

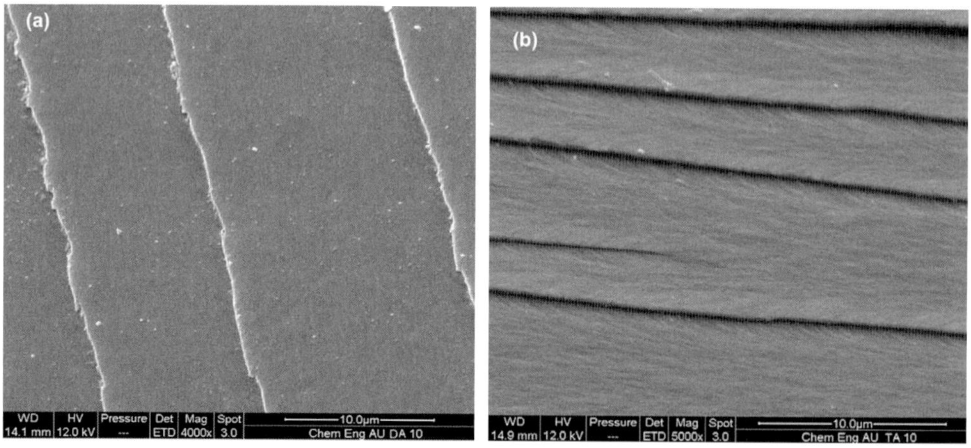

Figure 8.30. SEM micrographs of (a) 10 wt% OAPS in DGEBA and (b) 10 wt% OAPS in TGDDM [23].

on 50 mg samples that were heated from ambient to 900°C under continuous nitrogen flow at a rate of 10°C /min. The structure of the composite was also observed using XRD and SEM micrographs [23]. The OAPS form of POSS was synthesized in a three-step reaction, characterized in the scheme shown in Figure 8.29 [23].

The SEM micrographs taken of the composite do not show the usual dispersion among thermoset nanocomposites (Figure 8.30) [23]. The OAPS appears to have blended into a striated unidirectional form in the polymer matrix, which is more

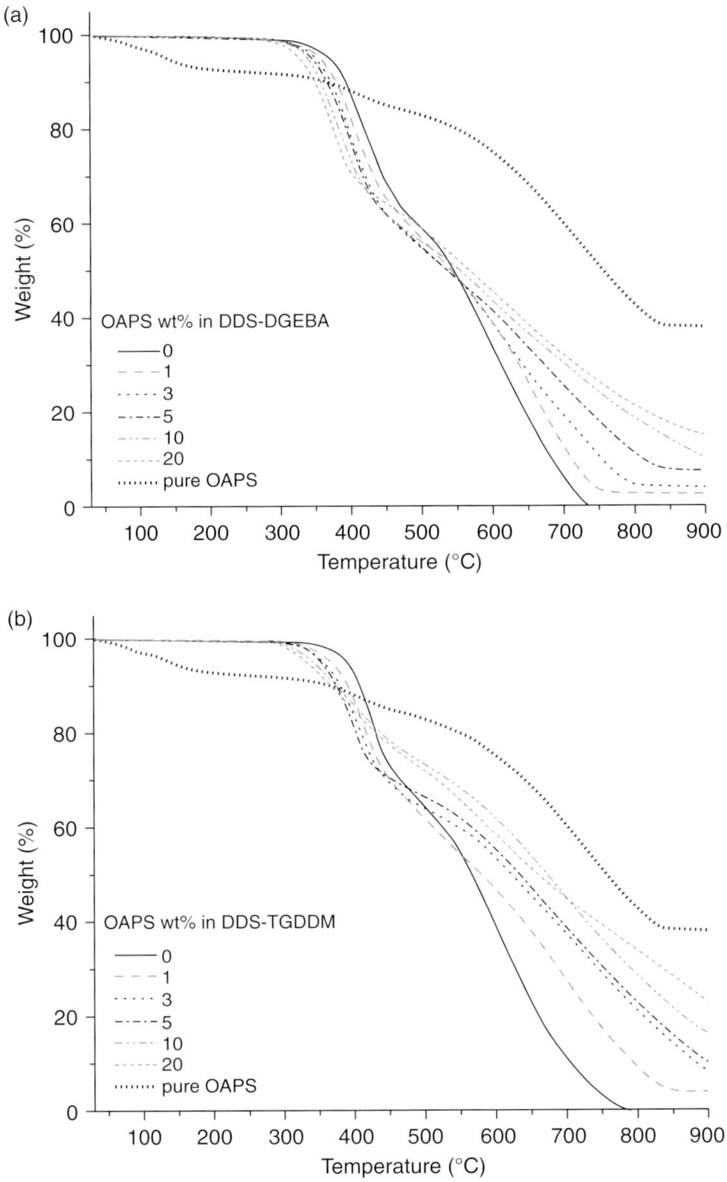

Figure 8.31. TGA curves for OAPS-epoxy nanocomposites.

indicative of intercalated structure than it is of the more common exfoliated structure than thermosets form.

The TGA curves show a fairly steady improvement to the percent mass loss at final decomposition for all weight percentages of OAPS-epoxy combinations tested (Figure 8.31) [23]. This is concurrent with – and supportive of – the findings of the other previously reviewed research on this subject. Though it shows similar trends, the overall improvement of thermal decomposition from that of neat epoxy resin does not appear to be significant, or even different, from the other thermal stability studies seen here. Furthermore, the overall improvement is someone subjective.

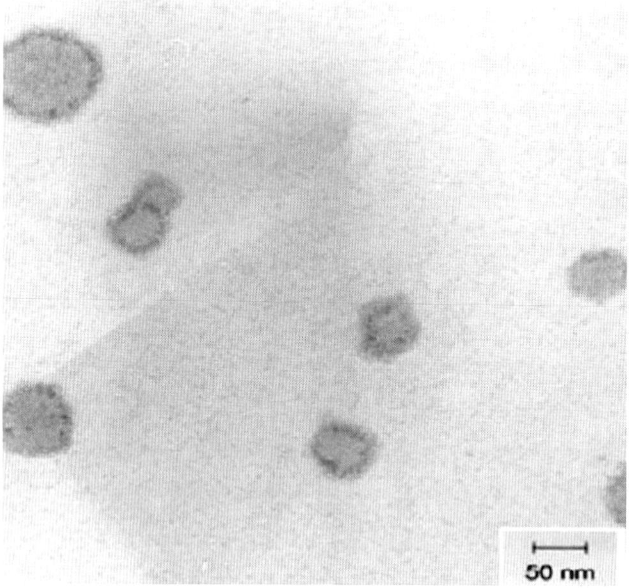

Figure 8.32. TEM image of the hybrid TiO_2-SiO_2 nanocomposite [24].

The best improvement, aside for pure POSS material, is seen at 20 wt% OAPS. The issue in this case is that this is a high weight percentage for a nanocomposite, since most general nanocomposites, are limited to a weight percentage of 5 or less. This could create cost-effectiveness issues or take its toll on other mechanical or electrical properties.

8.4.1.2 TiO_2-SiO_2 in Epoxy (DGEBA) Nanocomposites

Omrani et al. [24] investigated the effect of TiO_2-SiO_2 in an epoxy resin (DGEBA) to form an organic-inorganic hybrid system. These composites have become of increasing interest recently because of the balanced property improvement the inorganic particles provide to an organic polymer matrix. Commonly explored inorganic particles include TiO_2, SiO_2, Al_2O_3, clay, SiC, and many others. TiO_2 has been of interest in the research of thermal properties where it has been seen to greatly improve a polymer's characteristics. Though they have an improving effect on host polymers, TiO_2 nanoparticles have demonstrated instances of poor mechanical properties and some less thermally stable structure behaviors. To smooth out some of the variant characteristics, SiO_2 has been added to the TiO_2 filler particles. It is worth noting that forming a hybrid with both particles cannot create a material that embodies all the benefits of both materials; however, the mix of resulting properties that stem from each can further widen their current areas of application to broader horizons. The polymer matrix used was DGEBA that had been cured using diaminodiphenyl sulfone. Mixing this solution with diamine hardener produces a series of cross-linking reactions between the ends of epoxide groups.

TEM images for the microstructure show a very low level of particulate saturation within the polymer matrix (Figure 8.32) [24]. Given that the scale is 50 nm, the principle cause does not seem to be the result of agglomeration or otherwise poor

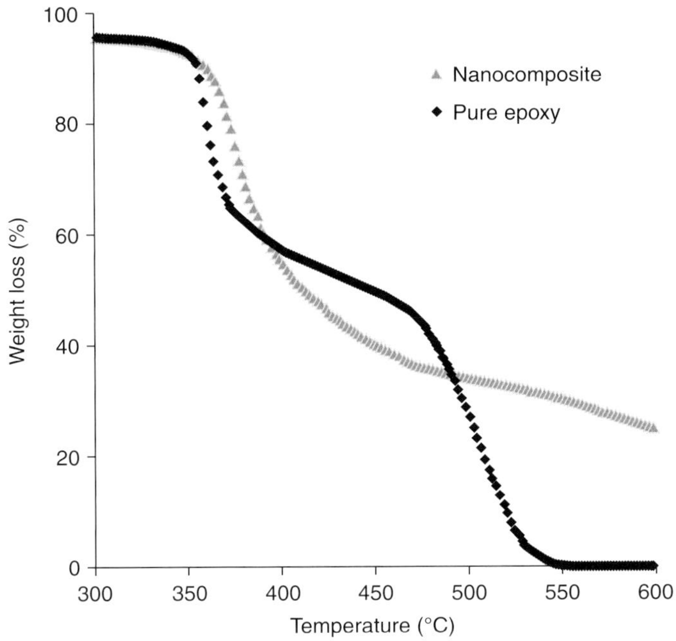

Figure 8.33. TGA curves for the two cured samples [24].

dispersion. If each dark circle is a hybrid TiO_2-SiO_2 nanomolecule, then the dispersion, at least from the limited scope of the TEM, appears to be fairly good. Thermal stability was measured using TGA analysis up to 1,000°C at a rate of 10°C/min. SEM micrographs were taken to document the resulting structures and dispersion of the nanoparticles in the polymer matrix. Tests were performed on only one nanocomposite sample and on a sample of the neat epoxy.

The TGA data show that the TiO_2-SiO_2 hybrid composite had an increase in the onset thermal decomposition temperature of 16°C from the neat epoxy as shown in Figure 8.33 [24]. This can possibly be explained by its contribution to catalyze during curing reactions and its behavior as a thermal stabilizer [24]. The char yield of the nanocomposite was 25% and 0% for neat epoxy at 600°C. These high yield differences can be attributed to the interactions of polymer chains inorganic particles. Overall, the desired results from the experiment were to some level achieved, though no real designed magnitude was addressed or stated so we can only attribute it to a positive result of overall principle.

8.4.1.3 Polythiophene in Diglycidyl Ether Bisphenol (Epoxy) Nanocomposites

Polythiophene is a type of conducting polymer that exhibits low density, environmental stability, and good electrical conductivity [25]. Polythiophene has gained interest for development of polymer nanocomposites, because they show signs of good thermal stability while maintaining good electrical conductivity, which makes it incredibly promising as filler for electrically conducting polymers. Prior to research on polythiophene, electrically conductive polymers were primarily using metallic particles or carbon black as conducting filler. The problem with these fillers

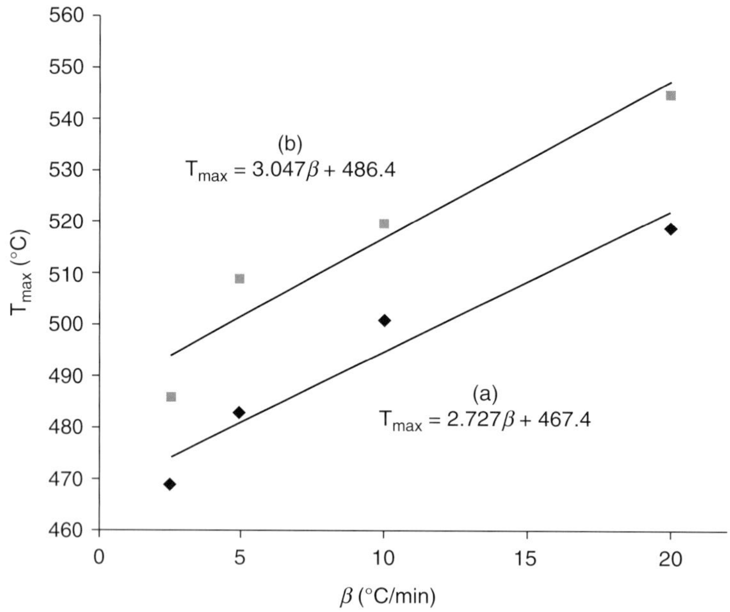

Figure 8.34. T_{max} versus heating rates of (a) pure epoxy and (b) epoxy PT nanocomposite.

is that metallic particles have an unfortunately common tendency to form an oxide layer on its edges. Carbon black does not suffer this problem, but it is far more expensive [25].

Zabihi et al. [25] conducted thermal testing on samples of polythiophene in a diglycidyl ether bisphenol matrix that was cured with phthalic anhydride. Thermal tests were conducted using thermogravimetric analysis and differential scanning calorimetry. Additionally, Kissinger plots were used to map heating rate against maximum temperature (Figure 8.34) [25]. TGA tests were performed on samples of 1, 5, 10, and 20 wt% PT, then another TGA test was performed on samples of 1 wt% at heating rates of 2.5, 5, 10, and 20°C (Figure 8.35) [25].

The thermal tests used a composite sample with 1 wt % polythiophene. The Kissinger plot in Figure 8.34 shows that the addition of 1 wt% PT was consistently able to reach a higher maximum temperature at each heating rate than in the pure epoxy sample [25]. The difference in maximum temperature between the epoxy and epoxy nanocomposite remained relatively stable for each heating rate tested (Tables 8.5 and 8.6) [25]. This confirms that polythiophene nanoparticles additives can enhance the conducting performance of a material; however, since no data is currently available on the range of temperatures that conductive materials need to reach in application, the true applicable value of these results is indeterminate. Further, since data are only being published for 1 wt% PT, there is only a very narrow view of the possible property enhancements made by the incorporation of PT nanocomposites. This limits the value of the research by limiting the extent to which it can be applied to real applications. The data show that the addition of PT nanoparticles to the polymer matrix increases the final decomposition temperature by 5%.

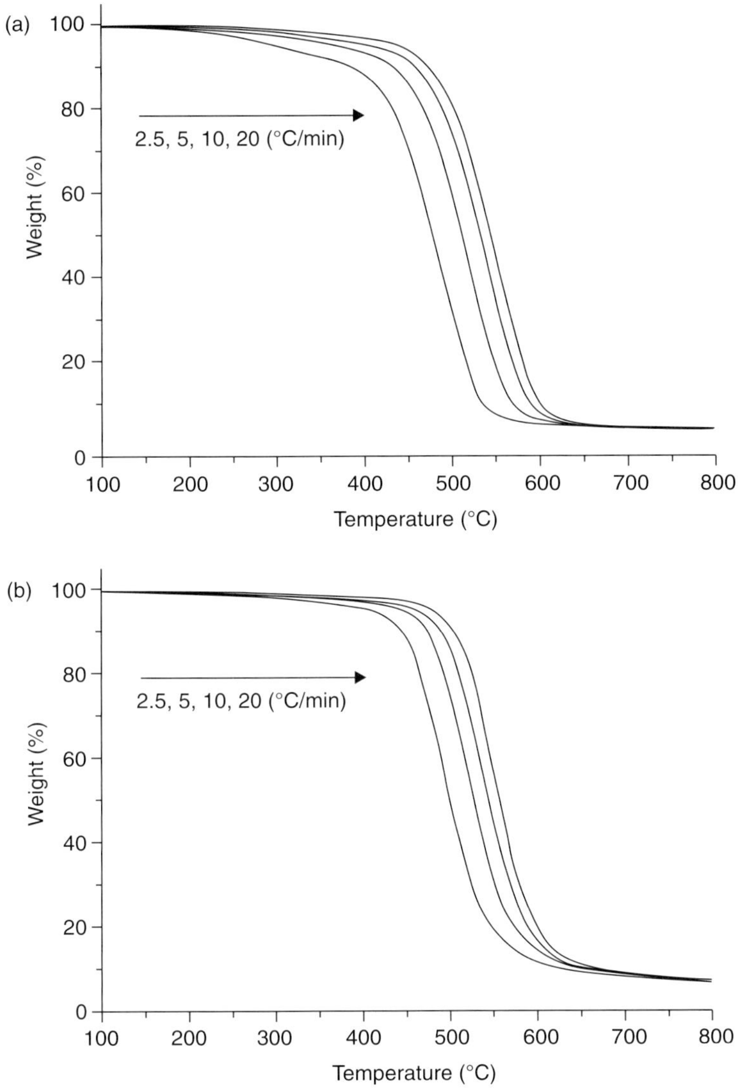

Figure 8.35. TGA curves for (a) pure epoxy and (b) polythiophene/epoxy.

The test results incorporated into the TGA on the effect of heating rate seems to suggest that thermal stability in the material is higher with a higher heating rate. From the TGA data (Figure 8.35) one can observe that the difference between heating rates remains largely the same for both pure epoxy and the epoxy nanocomposite, and that the two effects are mutually exclusive of.

8.4.1.4 Thermal Conductivity of Epoxy-Based Nanocomposites

Thermally conductive thermoset polymers are an important material when dealing with electronics. As new technology is created to miniaturize electronics, the material used to package the technology, thermal interface materials (TIMs), must be able to dissipate heat quickly and effectively in order to maintain the operating

Table 8.5. *TGA data for neat epoxy and epoxy/PT at a heating rate of 10°C/min*

Type	$T_{initial}$ (°C)	T_{max} (°C)	IPDT (°C)
Neat epoxy	346	501	464
Epoxy/1% nano-PT	384	520	489
Epoxy/5% nano-PT	380	509	473
Epoxy/10% nano-PT	366	487	459
Epoxy/20% nano-PT	320	439	436

Table 8.6. *TGA data for thermal degradation at different heating rates*

β (°C/min)	$T_{initial}$ (°C)	T_{max} (°C)
Neat epoxy		
2.5	248	469
5	296	483
10	346	501
20	386	519
Epoxy nanocomposite		
2.5	302	486
5	345	509
10	383	520
20	410	545

temperature of devices [26]. As the components become smaller, the material must also be able to be made smaller without the loss of function.

Common thermally conductive fillers are AlN, BN, Al_2O_3, and SiC, which are cost effective but require very high loading (about 60 vol% or higher) in order to satisfy the percolation thresholds while still achieving the thermal conductivity and heat conducting network that is desired [26]. As a result of the high loading, these nanocomposites have a high bulk density and poor mechanical properties, which are undesirable [26].

Carbon-based nanofillers have been very popular due to their excellent mechanical, electrical, and thermal properties [20–22, 26–29]. In general, carbon nanofillers in epoxy resins improve the thermal conductivity significantly without destroying the density or mechanical properties [26]. This section discusses the effects of carbon-based nanofillers on the thermal conductivity of epoxy based nanocomposites.

Han and Fina published an excellent review on "Thermal conductivity of carbon nanotubes and their polymer nanocomposites" [30]. The authors summarized the status of research of CNTs and their polymer nanocomposites through 2011. The dependence of thermal conductivity of CNTs on the atomic structure, tube size, morphology, defect, and purification is reviewed. The roles of particle/polymer and particle/particle interfaces on the thermal conductivity of polymer/CNT nanocomposites are discussed in detail, as well as the relationship between the thermal conductivity and the micro- and nano-structure of the composites.

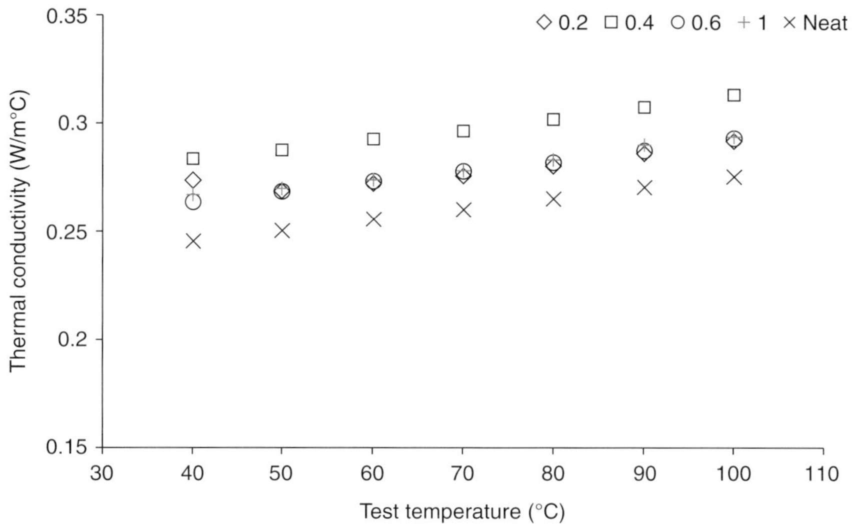

Figure 8.36. Results of the 3ω method measuring thermal conductivity of MWNT/epoxy composite system [22].

8.4.1.4.1 CARBON NANOTUBES–BASED DGEBA EPOXY. Carbon-based nanofillers have excellent mechanical, electrical, and thermal properties. Specifically, carbon nanotubes (CNTs) have been reported to have a stiffness of 1 TPa, strength up to 100 GPa, thermal conductivity greater than 3,000 W/mK [22], as well as a high aspect ratio [28, 29]. Due to these extraordinary characteristics, the addition of CNTs has the potential to improve the mechanical, electrical, and thermal properties of certain materials. A high aspect ratio and good thermal conductivity of the nanofiller is important in improving the overall thermal conductivity of a material, because these characteristics determine the potential for an effective heat flow network, which is the essence of thermal conduction [29].

Unfortunately, due to the van der Waals interactions of the CNTs, they have a tendency to bundle or aggregate when mixed into a polymer system. This aggregation results in poor dispersion, which hinders the property enhancements by creating defects in the material. Certain methods have been explored in order to promote dispersion, including mechanical methods (high-shear mixing and shortened CNTs), surfactants, acid oxidation, in situ polymerization, and other chemical methods [29]. The advantages and disadvantages are described in more detailed in the following sections, along with the resulting thermal conductivity evaluations.

Balakrishnan and Saha [22] researched MWNTs at low weight percentages in toughened epoxy resins and the resulting tensile behavior and thermal conductivity. They used unmodified MWNTs with a diameter of 5–10 nm and length of 10–20 μm. DGEBA was used as the polymer material. In order to toughen the epoxy, rubber particles with a diameter of about 250 nm were added. MWNTs were dispersed into the epoxy resin by sonication at loadings of 0.2, 0.4, 0.6, and 1 wt%. In order to measure thermal conductivity, the 3ω method was used (as described in Chapter 5), and results are shown in Figure 8.36 [22]. Compared to the neat system, the addition of

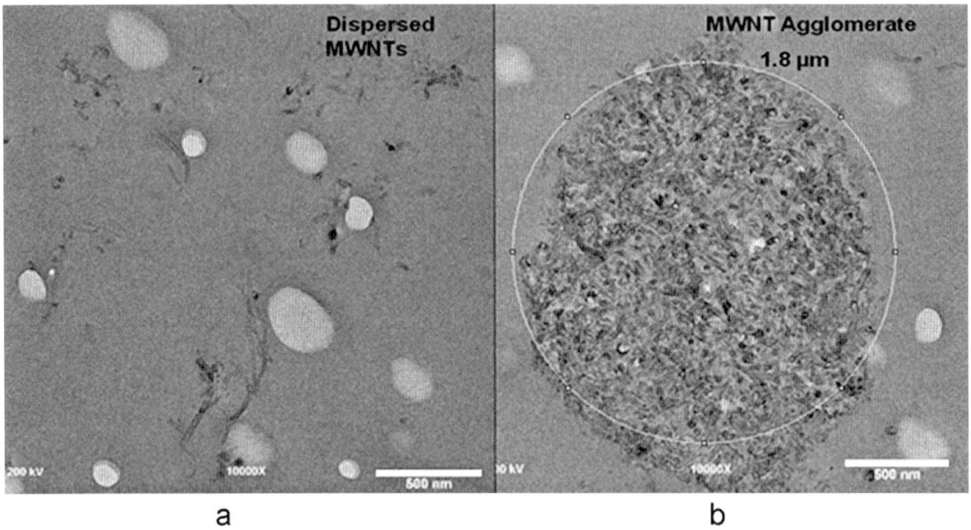

Figure 8.37. TEM image of 0.6 wt% MWNT loading in the toughened epoxy resin [23].

MWNTs improved the thermal conductivity at all temperatures. The most improved thermal conductivity for all temperatures was the addition of 0.4 wt% MWNT, which resulted in about a 16% increase in thermal conductivity. It is also noted that the thermal conductivity variance between loading content for all temperatures was about the same.

The loading of 0.2, 0.6, and 1 wt% of MWNT had about the same increase in thermal conductivity compared to the baseline. TEM images were taken in order to visualize the quality of dispersion of MWNTs achieved in their study. Figure 8.37 shows that several states of dispersion can be found in the material [22]. Also, as the content of MWNTs increased in the epoxy system, larger aggregates were found. From these observations and the results of the thermal conductivity analysis it can be assumed that something is interfering with heat transport; perhaps the rubber particles get in the way or the MWNT-epoxy interface scattered the phonons. The Effective Medium Approach was used to measure the thermal interfacial resistance with and without the addition of MWNTs, and the authors found that the interfacial resistance was not decreased by the addition of MWNTs as expected. It was concluded that this reason was the cause for the low thermal conductivity values.

Yang et al. conducted a study on the effect of functionalized MWNTs in an epoxy composite on thermal conductivity [29]. They used the Friedel-Craft method of functionalization in order to avoid damaging the MWNTs, thereby enhancing the thermal conductivity more than previous functionalized MWNTs [29]. The MWNTs were functionalized by mechanically stirring a mixture of (benzenetricarboxylic acid) BTC, MWNTs, PPA, and P_2O_5 and purging with nitrogen. The weight ratio of BTC to MWNT was 1:1 while the overall solid content in PPA was controlled below 6 wt%. The mixture was heated and mechanically stirred for 2 hours. Then PA was added and slowly stirred for 24 hours. After being cooled in an ice bath, PPA, P_2O_5, and PA were dissolved in distilled water and the solution was filtered through a

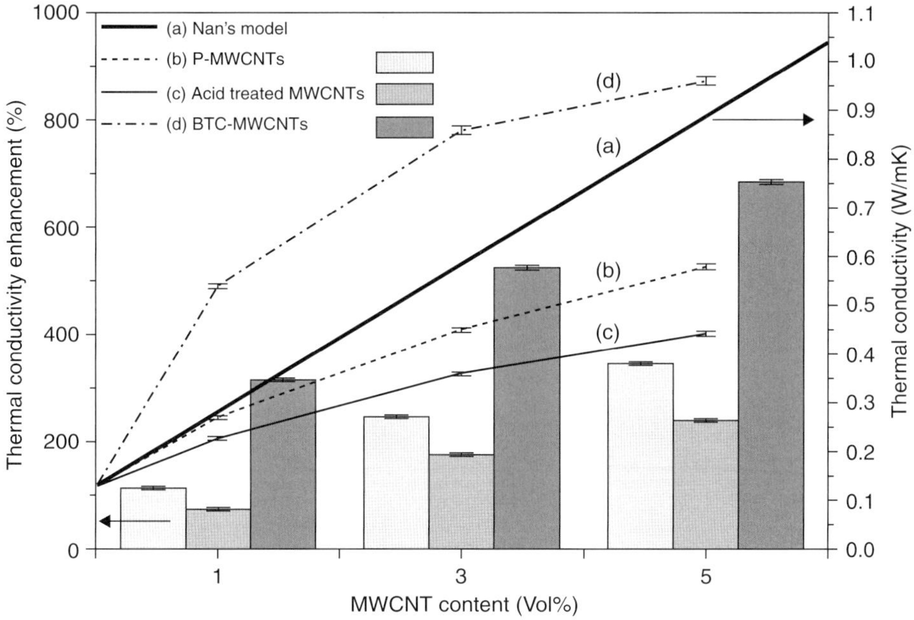

Figure 8.38. Thermal conductivity of MWNT/epoxy composite with varying content of pristine MWNTs, acid-treated MWNTs, and BTC-MWNTs as well as Nan's model [29].

polytetrafluoroethylene (PTFE) membrane. The collected powder was mixed with acetone and filtered from the PTFE membrane two or three more times in order to filter out all the excess uncreated BTC [29]. The BTC-MWNT/epoxy composite was characterized by the hot disk method using a TPS2500 at room temperature, and the results were compared to Nan's model:

$$\frac{K_e}{K_m} = 1 + \frac{f \times P}{3} \frac{\frac{K_c}{K_m}}{P + \frac{2a_k K_c}{d \times K_m}}$$

where K_e is the effective thermal conductivity of the composites, K_m is the thermal conductivity of the epoxy resin (0.13 W/mK), K_c is the thermal conductivity of the carbon nanotube (2500 W/mK), f is the content of CNTs filled, a_k is the Kapitza radius, d is the diameter of the carbon nanotube, and P is the aspect ratio of CNTs (~350). This model considers the interface effect [29]. The results of the experiment are shown in Figure 8.38 [29].

For all MWNT types tested, the addition of the nanotube increased the thermal conductivity of the material. While the thermal conductivity with pristine MWNTs (P-MWNTs) increased, the values reported are lower than Nan's model, which can be explained by the formation of agglomerates, and confirmed by SEM and TEM images shown in Figures 8.39a, 8.39b, and 8.40a [29]. The formation of MWNT bundles in the composite restricts the transport of phonons and can cause a

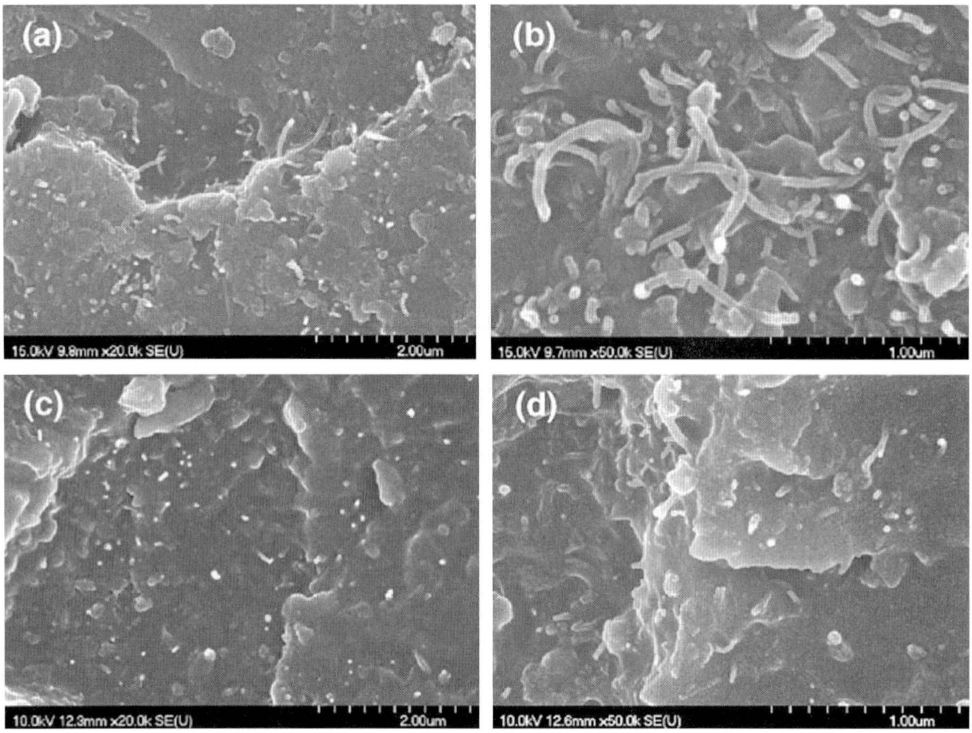

Figure 8.39. SEM images taken of the fracture surface of (a and b) P-MWNT/epoxy composite and (c and d) BTC-MWNT/epoxy composites [29].

phenomenon called reciprocal phonon vector, which describes the creation of a heat reservoir that restricts the diffusion of heat. Aggregates can cause this because they reduce the aspect ratio, decrease contact area, and trap or scatter phonons. BTC-MWNT/epoxy composite resulted in the highest thermal conductivity enhancement (a 684% increase with the addition of 5 vol% of BTC-MWNT), and the experimental values were higher compared to the theoretical values predicted by Nan's model. With only 1 vol% BTC-MWNT, thermal conductivity was increased by 315%. This can be attributed to the good dispersion, which can be seen in Figures 8.39c, 8.39d, 8.40c, and 8.40e [29].

In the case of BTC-MWNT/epoxy composites, the interface has been improved by covalent bonding between the epoxide group of the epoxy and the carbonyl group of the BTC-MWNTs, thereby reducing the thermal interfacial resistance and increasing thermal conductivity [30]. From these results, the Friedel-Crafts functionalization method was shown to prevent damage on the surface of MWNTs and result in an enhancement of thermal conduction in the material.

8.4.1.4.2 GRAPHITE-BASED: EXPANDED GRAPHITE (EG) FLAKES. Graphite is a naturally occurring and synthetically producible form of carbon that is crystalline and highly conductive [21]. Expanded or exfoliated graphite (EG) is a form of graphite that has been treated by acid and rapidly heated in order to create an expanded

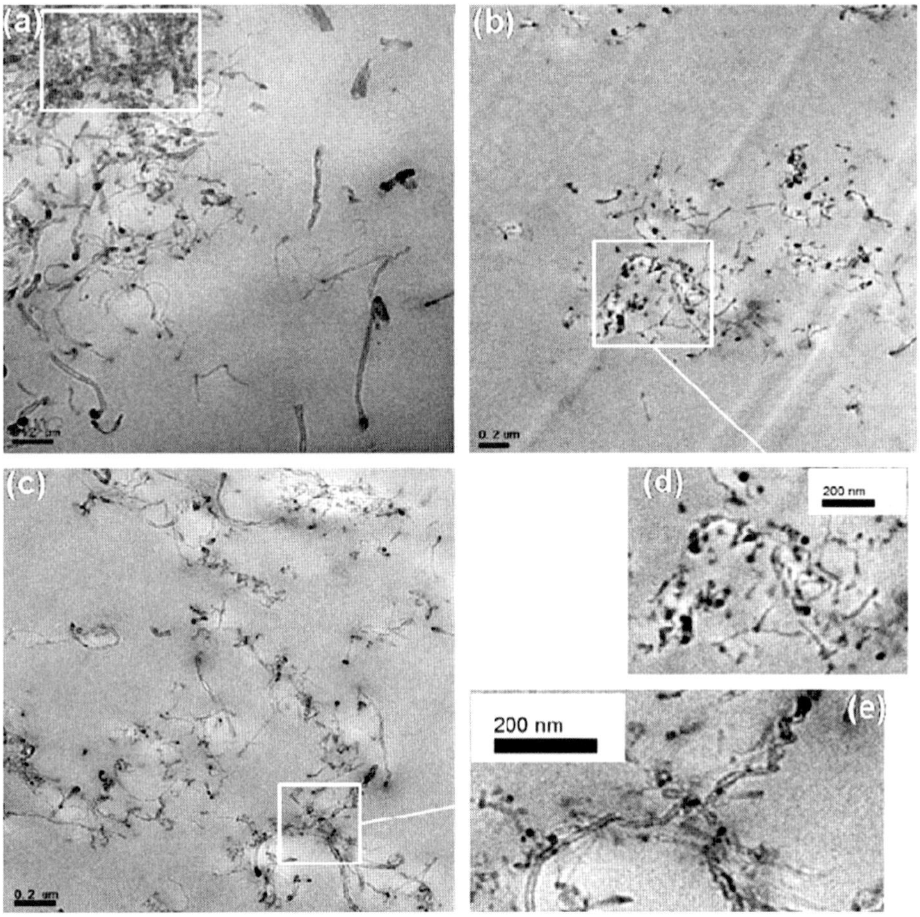

Figure 8.40. TEM images of (a) P-MWNT/epoxy composite, (b) acid-treated MWNT/epoxy composite, and (c) BTC-MWNT/epoxy composites, with (d) showing an enlarged region from (b) and (e) showing an enlarged region from (c) [29].

material. In order to make exfoliated graphite, natural graphite flakes are mixed with nitric acid and sulfuric acid in order to form intercalations [21]. The result is an intercalated graphite compound. After being washed and dried, the compound is quickly heat-treated at a high temperature so that the intercalations decompose and expansion of the compound occurs, resulting in exfoliated graphite [21]. The idea behind exfoliating graphite is that a lower concentration of the material will be needed in order to achieve the desired thermal conductivity [21].

Debelak and Lafdi investigated the effect of the size of exfoliated graphite nanofillers on the physical properties of EG/epoxy composites [21]. In this experiment, exfoliated graphite flakes were separated into large, medium, and small flakes and 0.1, 0.5, 1, 2, 4, 8, 12, 16, and 20 wt% EG were prepared. The polymer used was EPON 862 resin, which is an epoxy bisphenol F resin. To prepare the composite, the EG flakes were placed in a solvent and highly sheared under a homogenizer in order to disperse the flakes. This process was followed by ultrasonication to further

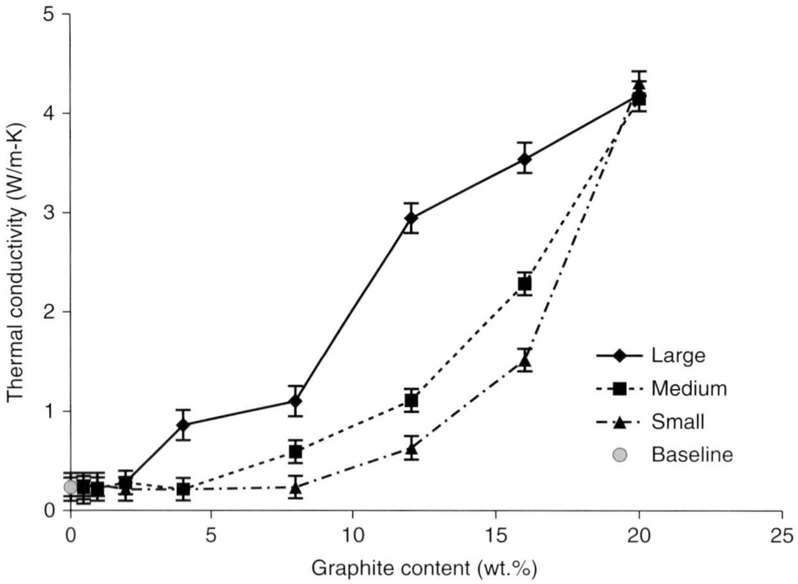

Figure 8.41. Thermal conductivity results for EG/epoxy composite with large, medium, and small EG flakes [21].

improve dispersion. Then the composite was molded and cured [21]. The EG/epoxy composite was characterized using a LFA 447 Nanoflash apparatus, which measured the thermal diffusivity of the specimen. Based on the Nan's model comparison equation mentioned earlier, thermal conductivity was determined. The results are shown in Figure 8.41 [21]. For all particle sizes, the increase in EG flake content resulted in an increase in thermal conductivity.

At 4 wt% of large EG flakes, thermal conductivity increased 300% from the baseline epoxy resin. The largest increase in thermal conductivity occurred from 8 wt% to 12 wt% of the large flake content. For medium EG flakes, a higher content of EG flakes was required at lower weight percentages to achieve the same increase in thermal conductivity as for the large flakes. The same was true for small flakes – to achieve the same increase in thermal conductivity, a higher loading of small EG flakes was required compared to medium and large flakes. At 20 wt% loading, there was a 2,000% increase in thermal conductivity for all particle sizes, which is about a thermal conductivity of 4.3 W/mK. These results show that an adequate conductive network has been formed by the nanofillers for heat transport, but perhaps at 20 wt% loading there is enough filler that the particle size no longer matters for thermal conductivity. Each particle size had a different percolation threshold of 3, 6, and 10 wt% for large, medium, and small EG flakes, respectively [21]. The reason for the drastic increase in thermal conductivity can be attributed to the high aspect ratio of EG flakes [21].

Another study was conducted on the effect of EG flakes in an epoxy resin on the thermal conductivity of the material. This study focused on the difference between treated and untreated EG flakes and the relation to the thermal conductivity. Ganguli,

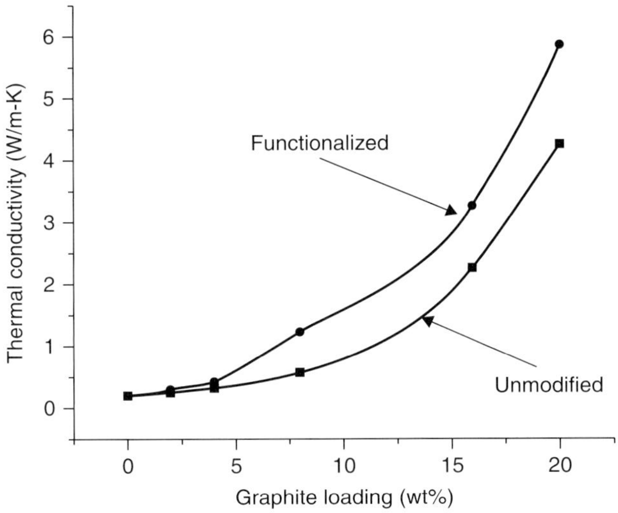

Figure 8.42. Thermal conductivity of unmodified and functionalized EG flakes in EG/epoxy composite [27].

Roy, and Anderson believed that by chemically functionalizing the EG flakes, thermal conductivity of the composite would improve due to enhanced interaction between the nanofiller and matrix [27]. For the polymer, they also used Epon 862 as the polymer, plus Epicure W (an aromatic amine curing agent) and Epicure 537 (an organic accelerating agent used to quicken the rate of curing) in order capture the dispersed morphology [27]. The chemical modification was carried out in a mixture of 25 vol% water and 75 vol% ethanol. First, 3 g of 3-aminopropoxyltriethoxy silane was added into 1,000 ml of the water/ethanol mixture while the temperature was maintained at 80°C. Followed by the addition of 10 g of EG whereby the grafting reaction took place under shearing. Afterwards, the reaction product was filtered, washed, and freeze-dried before being ground and stored. In order to create the nanocomposite, a Flacktek Speedmixer with very high shear was used to disperse the EG flakes in the epoxy resin, and then the curing and accelerating agents were mixed into the nanocomposite. Samples were loaded with 2, 4, 8, 16, and 20 wt% of treated or untreated EG flakes for testing. The thermal conductivity results are shown in Figure 8.42 [27].

From the graph it is apparent that the functionalized EG flakes increased the thermal conductivity at lower loadings, but the addition of modified or unmodified EG flakes lead to an increase in thermal conductivity. At loadings before 4 wt% for both types of EG flakes, the thermal conductivity was barely affected and the chemically modified and unmodified flakes behaved similarly. At 8 wt% loading of untreated EG flakes, the thermal conductivity increased from 0.2 to 0.5 W/mK, but the treated EG flakes at the same loading resulted in a much better improvement in thermal conductivity. Overall, for untreated EG flake nanofillers, there was a 19-fold increase in thermal conductivity at 20 wt% loading, which was a 4.0 W/mK measurement of thermal conductivity. This result shows agreement with the previous study mentioned by Debelak and Lafdi [21] who reported a high thermal conductivity of 4.3 W/mK for their untreated EG flakes. The chemically modified EG flakes saw

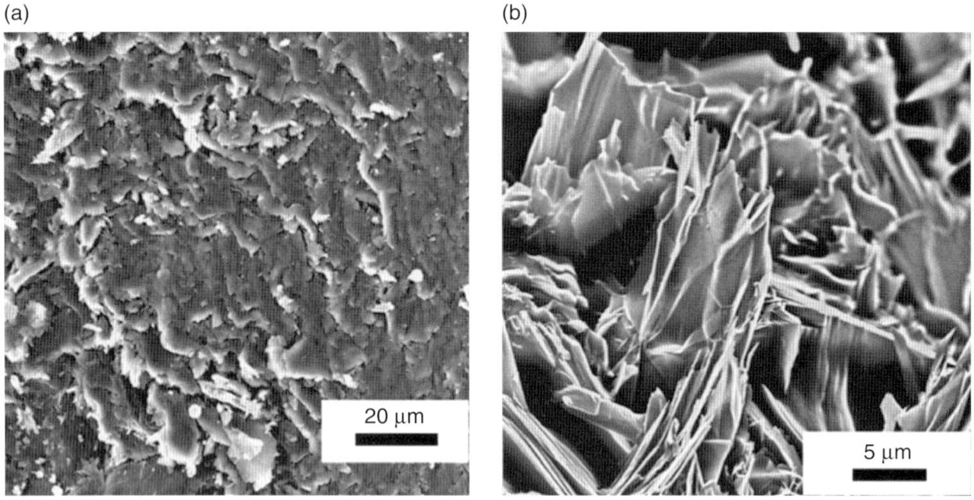

Figure 8.43. SEM micrograph of untreated EG flakes at 4 wt% loading in the EG/epoxy composite [27].

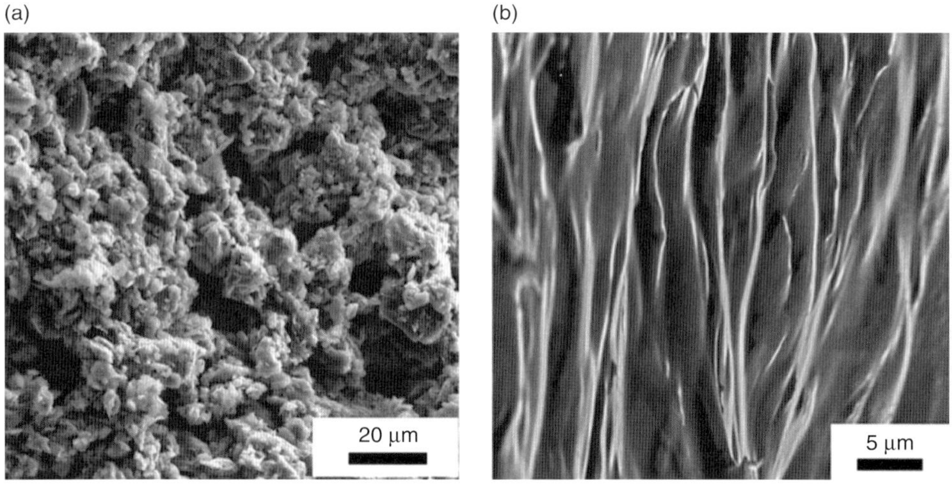

Figure 8.44. SEM micrograph of chemically treated EG flakes at 4 wt% loading in the EG/epoxy composite [27].

a 28-fold enhancement of thermal conductivity at 20 wt%, which measured to be 5.8 W/mK.

SEM micrographs (Figures 8.43 and 8.44) show the microstructure of the unmodified and chemically modified EG flakes at 4 wt% loading used in this experiment [21]. Both SEM micrographs show that agglomerations formed in both composites, but better EG-epoxy interactions were made in the chemically modified sample compared to the unmodified composite. These results corroborate the findings that the composite with functionalized EG flakes resulted in higher thermal conductivity and also shows the importance of achieving good interaction between nanofiller and matrix. The fact that agglomerations were present at a loading of 4 wt% and thermal conductivity improved at higher loadings challenges the idea that a uniform dispersion is absolutely required for enhanced thermal conductivity. While it is known that the dispersion is important in determining a material's properties, perhaps efforts should be focused more on improving the interaction between nanofiller and epoxy matrix. The interface between the nanofiller and the polymer is where the heat transfer occurs; therefore, to improve thermal conduction in a material, the interface must be optimal. This will reduce the acoustic impedance mismatch between the two materials, which minimizes the interfacial phonon scattering resulting in an improved thermal conductivity.

8.4.1.4.3 EXPANDED GRAPHENE NANOPLATELETS (EGNPS). Graphene has been determined to have a high aspect ratio, two-dimensional geometry, good stiffness, and low interface thermal resistance [20], which have resulted in the ability to improve thermal conductivity of materials with poor heat dissipation qualities. For this reason, Chatterjee et al. decided to "investigate the influence of expanded graphene nanoplatelets (EGNPs) on the tensile, nano-mechanical and thermal properties in epoxy" [20]. Natural graphite flakes were used to make the expanded graphene nanoplatelets. First, the natural graphite flakes were immersed in sulfuric acid and nitric acid to cause intercalation of the sulfuric acid into graphene layers of the flakes. Then the flakes were washed, neutralized, and dried before being thermally shocked in to expand the graphene layers and create EGNPs. The EGNPs were further functionalized with dodecyl amine (DDA) to create amine-EGNPs. To create the nanocomposite, the amine-EGNPs were dispersed in acetone under ultrasonication in order to break up the aggregation of the material, then the EGNPs were continued to be dispersed in a high-pressure homogenizer and placed in an ultrasonic bath. Afterwards, the dispersed EGNPs were incorporated into the epoxy resin via three-roll mill calendaring. Samples of 0.1, 0.5, 1, 1.5, and 2 wt% EGNPs were measured prior to hardening. The composite was hardened, molded, and then cured at 80°C for 12 hours, and then post-cured at 120°C for 4 hours. In order to characterize thermal conductivity, thermal diffusivity was measured using the LFA 447 Nanoflash apparatus [20].

Chatterjee et al.'s study determined that a conductive network was formed in the epoxy matrix, thereby increasing thermal conductivity of the composite [20], as shown in Figure 8.45. As the content of EGNP increased in the composite, the thermal conductivity increased as well, showing an almost linear relationship between EGNP content and thermal conductivity. While other researchers added much more than 2 wt% EGNP, the trend shows that the addition of more than 2 wt% EGNP will continue to increase thermal conductivity. Compared to pristine epoxy, the addition of 2 wt% of EGNP leads to a 36% increase in thermal conductivity value.

This study also investigated the dispersion of the material, because it has been shown to be correlated to the evaluation of a material's properties. From TEM images the researchers concluded that uniform dispersion was attained at up to 0.5 wt% EGNP loading, but as the content of EGNP increased, the realization of uniform dispersion became more challenging. This explains the decrease in thermal conductivity improvement from 0.5 to 1.0 wt% of EGNP. Without uniform dispersion, thermal interface resistance may increase and the acoustic phonons may not

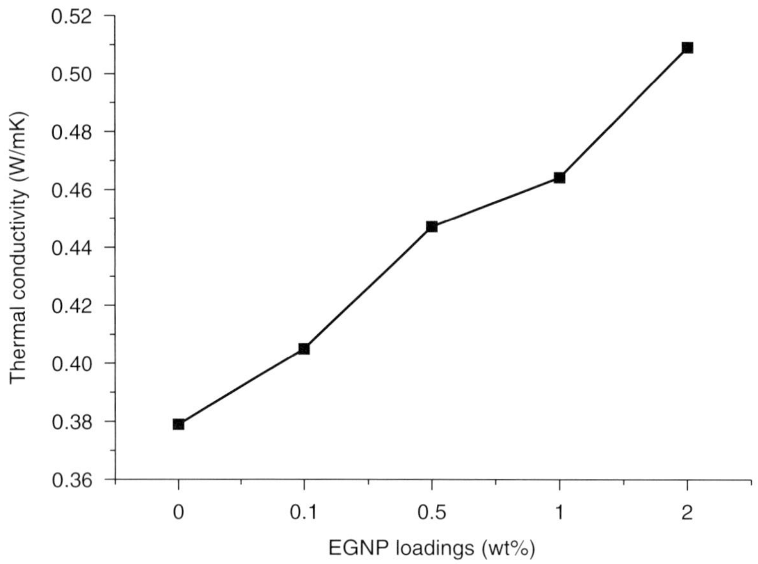

Figure 8.45. Resulting measurements for thermal conductivity of EGNP/epoxy composite [20].

have a network path to travel or can be scattered. For these reasons, dispersion must be perfected.

8.4.1.4.4 GRAPHENE OXIDE/MULTIWALL NANOTUBE HYBRID. Im and Kim conducted a study on thermal conductivity of graphene oxide/MWNT hybrid in an epoxy composite [26]. The polymer matrix for this experiment was made out of DGEBF. The MWNTs were synthesized by chemical vapor decomposition (CVD) and were then carboxyl acid functionalized in order to improve the interface between nanotubes and the matrix. The graphene oxide (GO) was made using the Hummers method. In order to make the GO/MWNT/epoxy composite, a wetting process was used. Both GO and MWNT were suspended separately and then mixed in order to create hybrid fillers, and the MWNT content varied among fillers. The hybrid fillers were then poured into a glass mold and filtered, and then annealed to remove the solvent. Epoxy was then dropped onto the GO/MWNT mold and the "dropped epoxy penetrated into the fabricated cake easily because of sufficient wettability between GO and the MWNTs" [26]. Then the composite was cured at 120°C for 30 minutes. A schematic of the fabrication process is shown in Figure 8.46 [26].

In order to characterize thermal conductivity, LFA and DSC were used. Im and Kim found that at 50 wt% of filler, the thermal transport properties were optimal due to the fixed pore volume [26]. At less than 50 wt%, an insulating layer of epoxy could form on the upper surface because the amount of filler used to make the cake was insufficient. Above 50 wt% filler, there is not enough epoxy to permeate the whole GO/MWNT network. With the hybrid/epoxy ratio fixed at 50 wt% each, the results of thermal conductivity were measured with varying GO/MWNT content

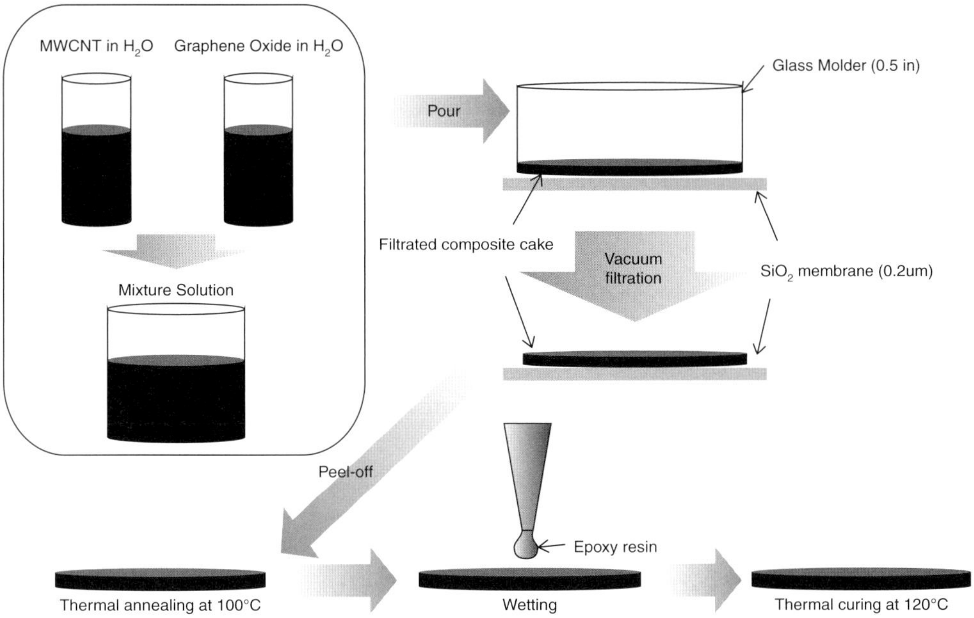

Figure 8.46. The fabrication process for GO/MWNT/epoxy composite [26].

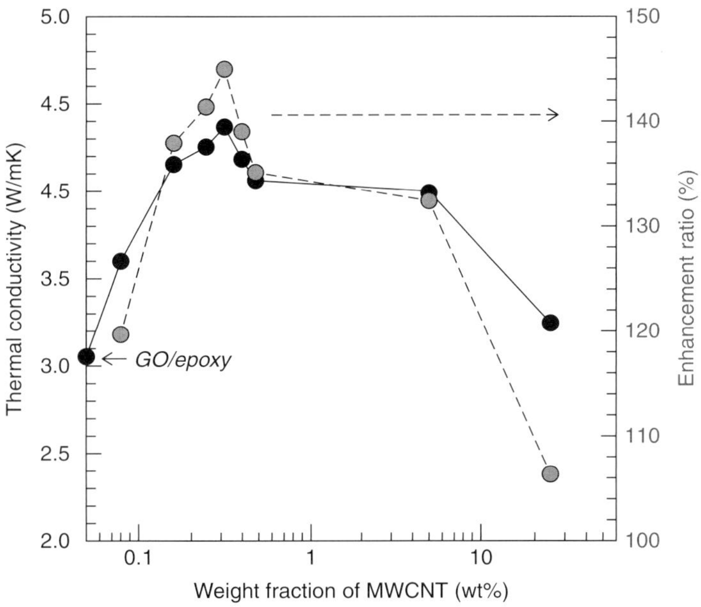

Figure 8.47. Thermal conductivity of GO/MWNT/epoxy composite with varying ratio of GO to MWNT, shown by black dots [26].

and are shown in Figure 8.47 [26]. At 0.36 wt% of MWNT, the filler achieved maximum thermal conductivity. Below 0.4 wt% MWNT, thermal conductivity was higher than expected, while above 0.4 wt% MWNT, thermal conductivity was lower than predicted.

Possible explanation of these results is that the hybrid fillers may have formed a three-dimensional heat conduction pathway, which resulted in the higher thermal conductivity. Because MWNTs are one-dimensional and GO is two-dimensional, the MWNTs cold have acted like bridges between the GO particles due to the MWNT's high aspect ratio, and this created an efficient heat transport network and good thermal conduction. But increasing the MWNT content resulted in an excess of MWNT disrupting the network and scattering the phonons in order to decrease thermal conductivity.

8.4.1.4.5 AL_2O_3 IN HTTE/DGEBA EPOXY. Research by Chen et al. [31] into the thermal behaviors of Al_2O_3 in DGEBA was aimed at obtaining a nanocomposite that enhanced thermal conductivity, thermal stability, and mechanical toughness. Particles. such as AIN, Al_2O_3, ZnO, and SiO_2 are often used as filler material in composite blends over other commonly used particles, such as organoclays, because of their greater thermally conductive performance. The negative drawback of these filler particles is their high tendency to agglomerate. The inability to disperse into the polymer matrix is a limiting factor on possible property enhancements. In order to overcome this hurdle, the DGEBA epoxy resin was mixed with epoxy-terminated low-viscosity liquid thermosetting aromatic polyester hyper-branched epoxy resin (HTTE) [31]. HTTE is a hyper-branched polymer that has often been used for its ability to increase polymer toughness. Because of the unique structures of HTTE, such as lower melt and solution viscosities, small molecular dimensions, and high functional end group density, it is able to increase the performance of desired properties in the DGEBA resin without sacrificing other key properties or overall processability. Additionally, high end group density, cavity structure, and low liquid viscosity can greatly improve the dispersion of nanoparticles in the epoxy resin matrix [31].

For experimentation, the polymer matrix was composed of 85 wt% DGEBA and 15 wt% HTTE. The matrix and subsequent composite was formed through a series of low- and high-temperature mixing and magnetic stirring, before being degassed, cured, and then post-cured to ensure the complete reaction of the thermoset. Nanoparticles compositions of 5 wt% and 40 wt% were tested in the HTTE/DGEBA matrix.

Measurements of the thermal conductivity utilized a thermal conductivity analyzer. Thermal stability data was collected through TGA. SEM was used to investigate the structure and dispersion of the nanocomposite formation as shown in Figure 8.48 [31]. For the TGA, 10 mg samples were heated to 1,000°C at a rate of 10°C/min under atmospheric conditions. The initial decomposition temperature was taken as the temperature at which 5% mass loss occurred. Samples of unfilled neat epoxy, 15 wt% HTTE/85 wt% DGEBA, 5 wt% Al_2O_3, and 40 wt% Al_2O_3, were tested.

From the SEM micrographs of the DGEBA/Al_2O_3 composite we can see large regions of well-dispersed particles in image (b), while in image (a) there appears to be a certain level of agglomeration present. In image (b), however, the dispersion of other particles in the background is still noticeable, indicative of an overall good level of dispersion of particles in the polymer matrix. The results of the thermal conductivity experiments are shown Table 8.7 [31].

Table 8.7. *Thermal conductivity results of the Al_2O_3-DGEBA composite [31]*

Filler	Filler content	Thermal conductivity (W/m-K)				
		EP	HTTE-1/EP	HTTE-2/EP	HTTE-3/EP	HTTE-4/EP
Unfilled	0	0.20	0.20	0.20	0.20	0.20
Al_2O_3-A151	5 wt%	0.3	0.31	0.26	0.27	0.29
	40 wt%	0.36	0.45	0.43	0.44	0.45
Al_2O_3-KH570	5 wt%	0.26	0.26	0.25	0.24	0.29

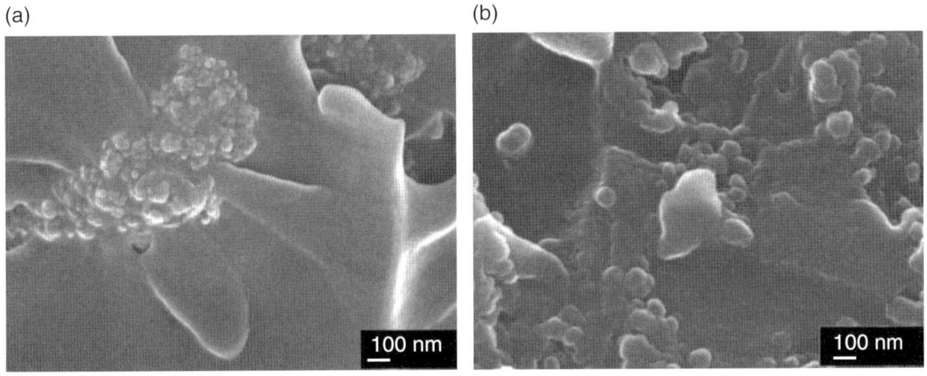

Figure 8.48. SEM images of Al_2O_3 particles dispersed in the DGEBA matrix [31].

From Table 8.7, the thermal conductivity of the HTTE/DGEBA IPNs is 0.20 W/m-K. Al_2O_3/HTTE/DGEBA had an optimum conductivity value of 0.31 W/m-K at 5 wt%. At 40 wt% they achieved a maximum value of 0.45 W/m-K. This corresponds to a 55% maximum increase in conductivity for the 5 wt% and a 125% increase in conductivity for the 40 wt% composite. Since conductivity depends only on level of dispersion, and not on the dispersed structure formation, the increased conductivity could be seen as a product of the hyper-branched epoxy creating better particle dispersion among the polymer matrix.

8.4.1.4.6 SUMMARY OF THERMAL CONDUCTIVITY PROPERTY OF THERMOSET-BASED NANOCOMPOSITES. In general, good thermal conductivity in an epoxy nanocomposite is achieved by adding highly conductive nanofiller and creating a uniformly dispersed, conductive network and a good nanofiller-epoxy interface. Epoxy is not a conductive material, and as such it requires fillers if it is to be used in applications requiring heat dissipation.

In amorphous polymers like epoxy, heat flows through lattice vibrations known as phonons [27]. Each material has its own acoustic impedance, and at the interface of two materials there is an acoustic impedance mismatch. This mismatch causes the phonon to scatter depending on the degree of mismatch. The interfacial thermal resistance is a measure of compatibility between the two contacting surfaces and is another way to characterize the interface and determine thermal conductivity.

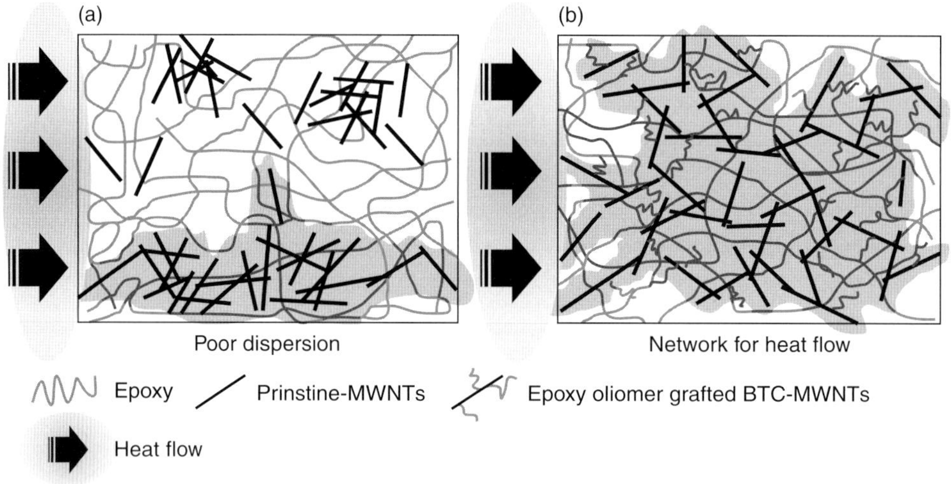

Figure 8.49. A model of heat flow in (a) unmodified MWNT/epoxy composites and (b) modified MWNT/epoxy composites [29].

The lower interfacial thermal resistance, the better thermal conductivity of the material is.

One way to improve the interface is by functionalizing the nanomaterial. This will create a bridge between the nanofillers and the matrix. For MWNTs, Friedel-Craft method resulted in the BTC-MWNTs that could covalently bond with the epoxy matrix [29]. A schematic of this interaction is shown in Figure 8.49 [29]. For exfoliated graphite, functionalization improved the thermal conductivity, but the viscosity of the resulting material at higher loadings was determined to be outside the processing window for conventional composite processing [27]. Another way to enhance thermal conduction of a material is to use the wetting process described by Im and Kim [26]. Using their method, they created a heat conduction path composed of MWNT bridges to GO fillers, but phonon scattering may occur if the balance between MWNT and GO is not optimal.

Based on Han and Fina's extensive review [30], the authors concluded that CNTs remain one of the most promising thermally conductive filler types for polymers. However, significant advances are still needed to obtain thermally conductive composites sufficiently efficient to meet the requirements of most market applications.

8.4.2 Phenylethynyl Polyimide-Graphene Oxide Nanocomposites

A high volume of research has gone into novel applications of graphene ever since it was first discovered due its potential to improve electrical, mechanical, and thermal properties [32]. The problem is that pure graphene has very poor dispersibility, effectively eliminating its practical use in the realm of composites. However, changing the processing of graphite by adding oxidative treatment helps establish large functional surface moieties in the graphene oxide structure that greatly improves dispersibility [32] and reopens an approach to use graphene oxide as composite nanofiller.

Table 8.8. *TGA data for 5% decomposition, 10% decomposition, final temperatures, and residual weight [33]*

Sample	T(5%) °C	T(10%) °C	T(final) °C	Residual Weight
Pure PI	480	542	750	0%
1 wt % GO	505	550	800	15%
3 wt % GO	505	551	800	29%
5 wt % GO	530	571	800	9%

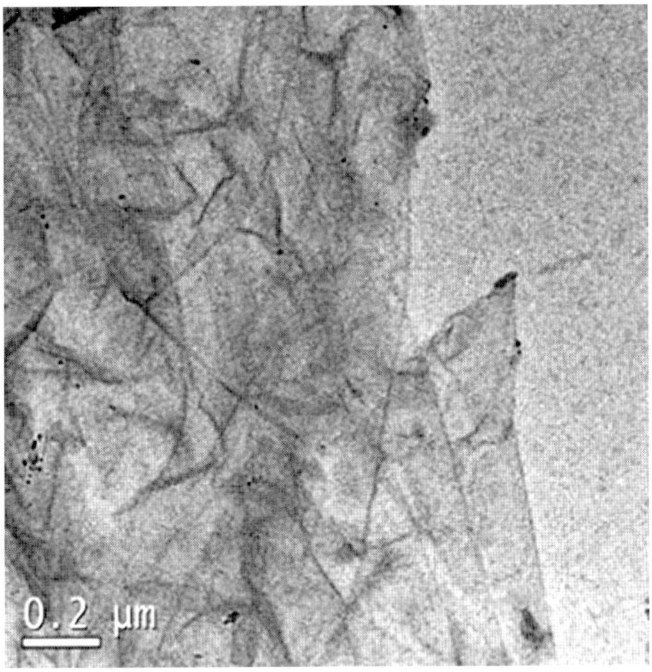

Figure 8.50. TEM of layered graphene oxide platelets in a polyimide matrix [32].

Hong et al. [32] performed a series of experiments exploring the thermal stability of fully exfoliated graphene oxide a phenylethynyl terminated polyimide thermosetting polymer matrix. Testing was performed using TGA from 20–800°C and TEM imaging. The TEM image of the graphene oxide particles in polyimide matrix shows that this particular micro-scale test area exhibits fairly decent dispersion (Figure 8.50) [32]. The dark thin lines signify the sides of the grapheme platelets, in the light grey matrix. However, there are regions, such as the top left corner, where the overall darker grey appearance suggests higher concentrations of graphene platelets stacked against each in comparison to the much lighter background color of the neat polymer. The data in Figure 8.51 are further summarized in Table 8.8 [32].

The TGA shows that thermal stability increases as graphene oxide nanoparticles were increasingly added into the composite at higher wt%. This trend reached a plateau at 3 wt% graphene oxide, after which point thermal stability began to decrease.

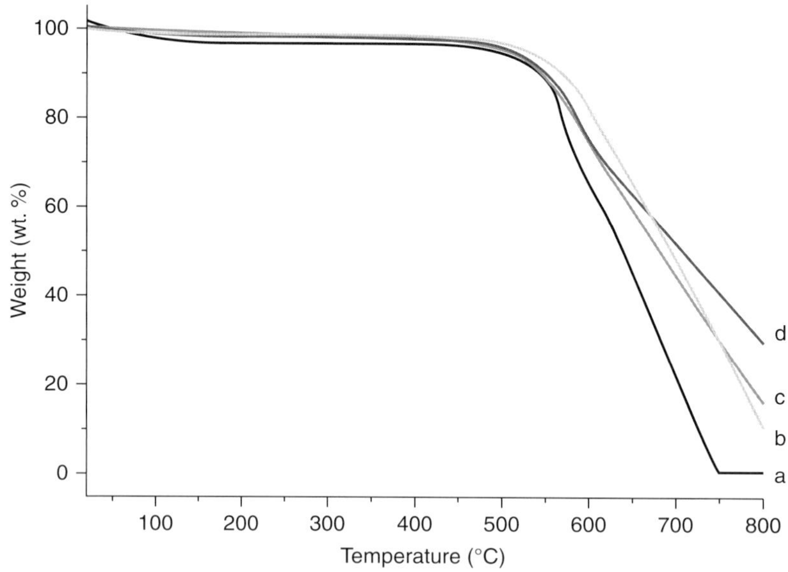

Figure 8.51. TGA curves of (a) pure PI, (b) 5 wt%, (c) 1 wt%, and (d) 3 wt% GO/PI composites at a heating rate of 10°C/min in air [32].

The increased thermal stability is hypothesized to be a result of the graphene oxide form trapping polyimide moieties [32], which induces chemical reactions in the graphene network that increase the overall material's stability. Once the optimal 3 wt% of graphene oxide is surpassed, the excess nanoparticles block and decreases the amount of network cross-linking because of its reduced ability to thoroughly disperse in the polymer matrix. This reduction in cross-linking makes the material more susceptible to decomposition as thermal energy is imported into the system.

8.4.3 Polyester with Bamboo-Reinforced Clay Nanocomposite

Bamboo is used in this research because of its good mechanical performance, anatomical properties, and fracture mechanisms conducted by Kushwaha and Kumar [33]. Bamboo has a much higher tensile strength than other natural fibers due to its high composition of cellulose, lignin, and relatively small microfibrillar angle [33]. Thermal stability of the bamboo/clay/polyester composite was determined using TGA run under nitrogen atmosphere with a 200mL/min flow rate, 20°C/min heating rate, and a temperature range up to 800°C. Samples of 0.25, 0.5, and 1 wt% nanoclay were tested. The tested samples are abbreviated in tables and further discussion in the following manner: BE5N is bamboo epoxy, E1 is bamboo epoxy with 1 wt% nanoclay, BP5N is bamboo polyester, and P1 is the bamboo polyester with 1 wt% nanoclay. Figure 8.52 shows the SEM micrographs of several bamboo epoxy-clay samples [33].

The following table (Table 8.9) summarizes the data collected for the thermal degradation of the various composites tested [33].

The Table 8.9 parameters are as follows: T_d is the decomposition onset temperature, WL_d is the weight loss percentage corresponding to T_d, T_p is the maximum

Table 8.9. *Thermal decomposition of bamboo epoxy, bamboo polyester nanocomposites [33]*

	T_d	WL_d	T_p	WL_p	T_s	WT_s	Percentage residue at 650°C
BE5N	331.3	12.67	386.93	53.52	411.30	70.42	9.28
E1	324.35	19.01	380	50	418.26	69.72	17.14
BP5N	331.3	14.08	428.69	64.20	460	79.58	2.86
P1	300	13.38	380	42.96	460	80.28	5

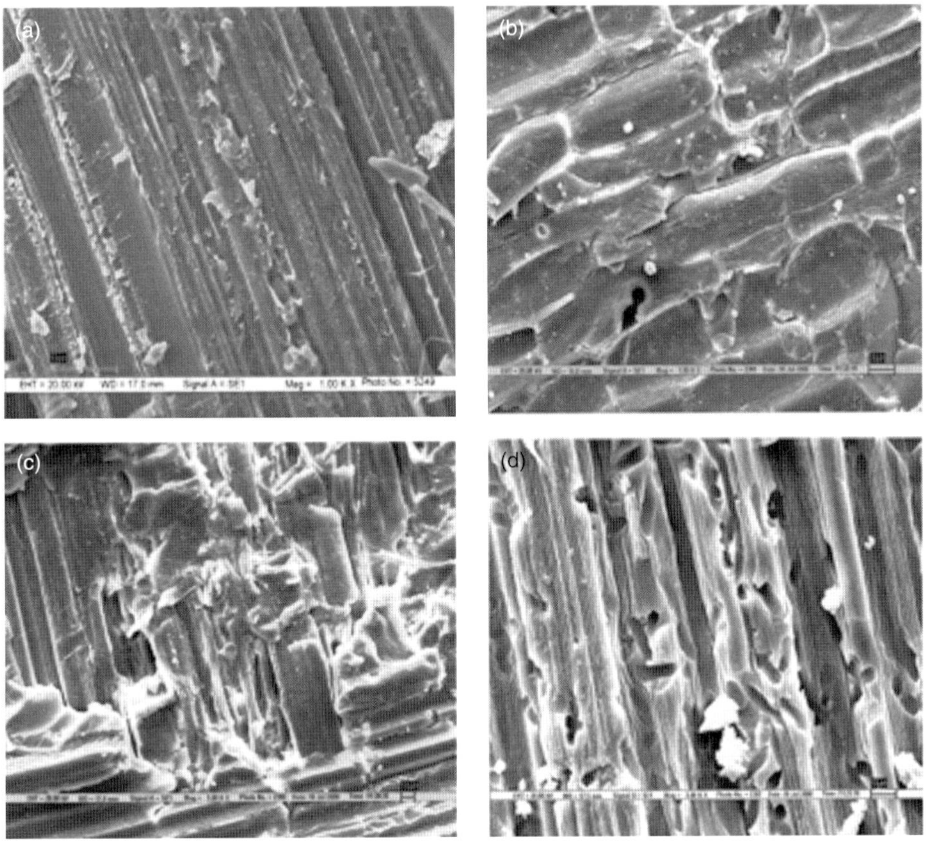

Figure 8.52. SEM micrographs of bamboo epoxy: (a) 0%, (b) 0.25%, (c) 0.5%, and (d) 1% nanoclay [33].

decomposition rate, WL_p is the weight loss at T_p, T_s is the shift temperature, which defines the end of primary decomposition, and WL_s is the weight loss corresponding to T_s. The clay-incorporated epoxy composite experienced an increase in initial weight loss, as well as an increase in initial mass loss temperature of 14 °C. At the end temperature of 650 °C, there was a final retained mass of 9.28% for BE5N and 17.14% for E1. The data suggest that the increase in onset temperature is more significant than the higher pace of the onset degradation. Final percentage residual mass at 650°C was 2.86% for BP5N, 5% for P1, 10.71% for untreated bamboo-polyester, and 5% for bamboo-polyester w/ 1 wt% nanoclay. Maximum decomposition temperature decreased for both epoxy and polyester composites, so its significance is not considered any further. From the data we can observe that the thermal stability of the clay included hybrid composite is slightly improved from that of the base epoxy.

8.4.4 Summary of Thermal Properties of Thermoset-Based Nanocomposites

Thermal properties of thermoset-based nanocomposites are very important features that hold significant implications for the future performance of systems. There exist an astounding number of different thermoset nanocomposite combinations, all capable of an array of different properties. It cannot be universally stated what polymers and nanoparticles composite is the greatest for any specific thermal property. The data seen also suggest that unlike other polymer forms, thermoset-based nanocomposites do not always abide by the golden upper limit of 5 wt% nanoparticles content being a compositional maximum [34]. Another key finding is that oftentimes, the greatest difference in property enhancements are not due to the addition of nanofiller material, but to the treatment or chemical additives applied to the nanofiller material [35]. This was especially true in the case of fire retardancy, where sizable differences could be seen between the same nanocomposites that only differed in processing and chemical additives applied to the nanofiller [35, 36]. Comparatively, thermal stability can be improved greatly, but at the risk of overweighing the composite with nanofiller material and increasing deleterious performance of mechanical properties. In terms of future development, it seems DGEBA and different forms of clay, such as hectorite or montmorillonite, hold the greatest potential for widespread application, due in part to more thorough research on the materials, ease of processing, and a quickly growing repertoire of results leading to increasingly predictable, and clearly enhanced, thermal characteristics.

In general, the addition of nanoclay improves the overall thermal properties of the resulting composite. Interestingly, the organic modifier not only helps in dispersion of the nanoclays but also induces hydrogen bonding with the polymer chains and results in greater interaction between the filler and the polymer. However, at higher loadings, the amount of the organic modifier is large enough to detrimentally affect the thermal properties due to its combustibility. The best thermal performance was observed for loading between 3 wt% and 5 wt%. However, the addition of the nanoclay affects the processabilty of the composites as the viscosity increases as well as the melting temperature and heat of fusion, yet these properties are generally desired for the finished application.

The use of MWNT for improving the thermal properties did not seem effective from the Rice University study. The group performed scanning electron microscopy and showed that the MWNT seemed reasonably dispersed. Since the TGA was performed in an inert environment, the residual mass in an oxygen environment was not obtained. It was interesting to note that the random dispersion of the MWNT showed a higher change in T_g, which is not surprising because of greater interfacial coverage. Further work must be conducted in order to determine what loadings are required in order to obtain the best trade-off between properties. Additionally, there may be some synergistic effects from combining certain nanoparticles, which may improve overall properties that also need to be explored.

The polymer structure is a large factor in the thermal performance of these composites. Both the nature of the material and the compatibility of the filler will determine the composite effects of combining the nanomaterial and the polymer. It was seen that sometimes the addition of a nanocomposite will exhibit multifunctionality behavior of improving multiple properties, while other times the increase in one property results in the decrease of another. Interestingly, properties like thermal stability and fire retardancy are sometimes do not have a positive correlation despite intuition. It was also observed that at certain loadings, interactions between the filler and the polymer form certain complexes that serve as an inflection point for the properties, such as the crystallinity that in turn affects other properties. It is probably good practice to require a TEM micrograph of the composites produced, as it is difficult to compare results whenever it is uncertain whether the nanoparticle is well dispersed.

8.5 Overall Summary

An overview of the current technology in enhancing the thermal properties of thermoplastic-, thermoplastic elastomer-, and thermoset-based polymer nanocomposite systems was analyzed. In general, their thermal performance can be enhanced by the addition of nanoparticles. The mechanical and electrical performance of the polymeric matrix can also be improved by the addition of nanoparticles. Sometimes simply maintaining good thermal performance while improving other properties is a significant feat.

Organoclays have become a mainstay as a nanoadditive because of their low cost and consistent performance. The comparison with synthetic clay (Laponite) is important to review where synthetics stand with respect to organics. In general, the addition of nanoclay improves the overall thermal properties of the resulting composite. Interestingly, the organic modifier not only helps in dispersion of the nanoclays but also induces hydrogen bonding with the polymer chains and results in greater interaction between the filler and the polymer. However, at higher loadings, the amount of the organic modifier is large enough to detrimentally affect the thermal properties due to its combustibility. The best thermal performance was observed for loading between 3 wt% and 5 wt%. However, the addition of the nanoclay affects the processabilty of the composites as the viscosity increases as well as the melting temperature and heat of fusion, yet these properties are generally desired for the finished application.

The effects of nanomaterials can range from extremely useful to negligible and are highly dependent on the type of filler used as well as its relative size. For all cases, there were enhanced properties with increasing loading of filler, leading us to believe that the more filler, the better properties your composite will observe. Interestingly enough, nano-sized fillers are not always ideal when compared to micron-sized fillers. For all studies were thermal conductivity data was taken for both nano- and micron-sized fillers, the micron fillers performed better. Based on these results, we see that many factors can affect the thermal properties of polymer nanocomposites.

It was found that the performance of nanoparticles on the basis of improved properties is highly dependent on the dispersion and exfoliation of the nanoparticles in the polymeric matrix. Certain surface treatments to nanoparticles and specialized processing techniques can help the dispersion and exfoliation of nanofibers in the composite. Polymer nanocomposite is a promising and growing field of study. It has already led to important advancements and continues to be in high demand in the industry.

8.6 Questions and Problems

1. How does the nanomaterial dispersion affect the thermal properties of polymer nanocomposites in general? Provide examples to illustrate your claims.
2. How are the thermal properties of polymer-clay nanocomposites influenced by the type of clay, surface treatment of the clay, and how clay is processed in the polymer matrix? Provide examples.
3. Enhancement of thermal properties of thermoplastic-based nanocomposites using MMT clay, carbon-based (CNF, SWNT, LHDH), and clay/woven glass fibers was included in this chapter. What other nanomaterials can be used for thermoplastic-based nanocomposites? Provide examples and comment on the mechanical properties of these nanocomposite systems.
4. Enhancement of thermal properties of thermoplastic elastomer-based nanocomposites using MMT clay, synthetic clay (Laponite), carbon-based (carbon black, SWNT), boron nitride, zinc oxide (ZrO), nanosilica, and aluminum oxide was included in this chapter. What other nanomaterials can be used for thermoplastic elastomer-based nanocomposites? Provide examples and comment on their mechanical properties of these nanocomposite systems.
5. Enhancement of thermal properties (thermal stability and thermal conductivity) of epoxy-based nanocomposites using POSS, TiO_2-SiO_2, conducting polymer (polythiophene), carbon-based (MWNT, expanded graphite flakes, expanded graphene nanoplatelets, graphene oxide/MWNT hybrid), and Al_2O_3 was included in this chapter. What other nanomaterials can be used for epoxy-based nanocomposites? Provide examples and comment on their mechanical properties of these nanocomposite systems.
6. What will be the effects of hybrid systems, such as MMT clay-CNT, CNT-nanographene platelet, and others? What will be a good strategy to optimize the thermal properties in the polymer nanocomposites? Comment on their mechanical properties of these nanocomposite systems.

REFERENCES

1. T. S. Ellis and J. S. D'Angelo (2003). Thermal and mechanical properties of a polypropylene nanocomposite. *Journal of Applied Polymer Science* 90, 1639–1647.

2. J. Sandlera, P. Wernerb, M. S. P. Shaffera, V. Demchukc, V. Altstatd, and A. H. Windlea (2002). Carbon-nanofibre-reinforced poly(ether ketone) composites. *Composites: Part A* 33, 1033–1039.
3. K. Lozano and E. V. Barrera (2000). Nanofiber-reinforced thermoplastic composites: Thermoanalytical and mechanical analyses. *Journal of Applied Polymer Science* 79, 125–133.
4. F. Du, J. E. Fischer, and K. I. Winey (2003). Coagulation method for preparing single-walled carbon nanotube/poly(methyl methacrylate) composites and their modulus, electrical conductivity, and thermal stability. *Journal of Polymer Science: Part B: Polymer Physics* 41, 3333–3338.
5. T. Kashiwagi, F. Du, K. I. Winey, R. H. Harris, J. R Shields, and J. F. Douglas (2005). Nanoparticle networks reduce the flammability of polymer nanocomposites. *Nature Materials* 12(4), 928–933.
6. J. Liu, G. Chen, and J. Yang (2008). Preparation and characterization of poly(vinyl chloride)/layered double hydroxide nanocomposites with enhanced thermal stability. *Polymer* 49, 3923–3927.
7. Ł. Piszczyk, M. Strankowski, M. Danowska, J. T. Haponiuk, and M. Gazda (2012). Preparation and characterization of rigid polyurethane–polyglycerol nanocomposite foams. *European Polymer Journal* 48, 1726–1733.
8. S. Z. Shen, S. Bateman, P. McMahon, M. Dell'Olio, J. Gotama, et al. (2010). The effects of clay on fire performance and thermal mechanical properties of woven glass fiber reinforced polyamide 6 nanocomposites. *Composites Science and Technology* 70, 2063–2067.
9. J. H. Wu, C. H. Li, Y. T. Wu, M. T. Leu, and Y. Tsai (2010). Thermal resistance and dynamic damping properties of poly(styrene-butadiene-styrene)/thermoplastic polyurethane composites elastomer material. *Composite Science and Technology* 70, 1258–1264.
10. S. Kemaloglu, G. Ozkoc, and A. Aytac (2010). Properties of thermally conductive micro and nano size boron nitride reinforced silicon rubber composites. *Thermochimica Acta* 499, 40–47.
11. A. K. Barick and D. K. Tripathy (2009). Effect of organoclay on the morphology, mechanical, thermal, and rheological properties of organophillic montmorillonite nanoclay based thermoplastic polyurethane nanocomposites prepared by melt blending. *Polymer Engineering & Science* 50, 484–498.
12. G. G. Silva, M. F. Rodrigues, C. Fantini, R. S. Borges, M. A. Pimenta, B. J. Carey, et al. (2010). Thermoplastic polyurethane nanocomposites produced via impregnation of long carbon nanotube forests. *Macromolecular Materials and Engineering* 296, 53–58.
13. M. Aurilia, F. Piscitelli, L. Sorrentino, M. Lavorgna, and S. Iannace (2011). Detailed analysis of dynamic mechanical properties of TPU nanocomposite: The role of the interfaces. *European Polymer Journal* 47, 925–936.
14. A. K. Mishra, G. B. Nando, and S. Chattopadhyay (2008). Exploring preferential association of laponite and Cloisite with soft and hard segments in TPU-clay nanocomposite prepared by solution mixing technique. *Journal of Polymer Science: Part B: Polymer Physics* 46, 2341–2354.
15. J. E. Puskas, E. A. Foreman-Orlowski, G. T. Lim, S. E. Porosky, M. M. Evancho-Chapman, et al. (2009). A nanostructured carbon-reinforced polyisobutylene-based thermoplastic elastomer. *Biomaterials* 31, 2477–2488.
16. R. Haggenmueller, C. Guthy, J. Lukes, J. Fischer, and K. Winey (2007). Single wall carbon nanotube/polyethylene nanocomposites: thermal and electrical conductivity. *Macromolecules* 40, 2417–2421.
17. Z. Wang, Y. Lu, J. Liu, Z. Dang, L. Zhang, and W. Wang (2011). Preparation of nano-zinc oxide/EPDM composites with both good thermal conductivity and mechanical properties. *Journal of Applied Polymer Science* 119, 1144–1155.

18. C. Liu, H. Huang, Y. Wu, and S. Fan (2004). Thermal conductivity improvement of silicone elastomer with carbon nanotube loading. *Applied Physics Letters* 84, 84–86.
19. S. Ratzke and J. Kindersberger (2005). Erosion behaviour of nano filled silicone elastomers. *Proceedings of the 14th International Symposium for High Voltage Engineering*, n. p.
20. S. Chatterjee, J. W. Wangb, W. S. Kuo, N. H. Tai, C. Salzmann, et al. (2012). Mechanical reinforcement and thermal conductivity in expanded graphene nanoplatelets reinforced epoxy composites. *Chemical Physics Letters* 531, 6–10.
21. B. Debelak and K. Lafdi (2007). Use of exfoliated graphite filler to enhance polymer physical properties. *Carbon* 45, 1727–1734.
22. A. Balakrishnan and M. C. Saha (2011). Tensile fracture and thermal conductivity characterization of toughened epoxy/CNT nanocomposites. *Materials Science and Engineering A* 528, 906–913.
23. S. Nagendiran, M. Alagar, and I. Hamerton (2010). Octasilsesquioxane-reinforced DGEBA and TGDDM epoxy nanocomposites: Characterization of thermal, dielectric, and morphological properties. *Acta Materialia* 58, 3345–3356.
24. A. Omrani, S. Afsar, and A. Safarpour (2010). Thermoset nanocomposites using hybrid nano TiO_2SiO_2. *Materials Chemistry and Physics* 122, 343–349.
25. O. Zabihi, A. Khodabandeh, and S. M. Mostafavi (2012). Preparation, optimization and thermal characterization of a novel conductive thermoset nanocomposites containing polythiophene nanoparticles using dynamic thermal analysis. *Polymer Degradation and Stability* 97(1), 3–13.
26. H. Im and J. Kim (2012). Thermal conductivity of a graphene oxide-carbon nanotube hybrid/epoxy composite. *Carbon* 50, 5429–5440.
27. S. Ganguli, A. K. Roy, and D. P. Anderson (2008). Improved thermal conductivity for chemically functionalized exfoliated graphite/epoxy composites. *Carbon* 46, 806–817.
28. M. Abdalla, D. Dean, P. Robinson, and E. Nyairo (2008). Cure behavior of epoxy/MWCNT nanocomposites: The effect of nanotube surface modification. *Polymer* 49, 3310–3317.
29. S. Yang, C. M. Ma, C. Teng, Y. Huang, S. Liao, et al. (2010). Effect of functionalized carbon nanotubes on the thermal conductivity of epoxy composites. *Carbon* 48, 592–603.
30. Z. Han, and A. Fina (2011). Thermal conductivity of carbon nanotubes and their polymer nanocomposites: A review. *Progress in Polymer Science* 26, 914–944.
31. L.-Y. Chen, Y. Chen, J.-F. Fu, L.-Y. Shi, and Q.-D. Zhong (2010). Thermally conductive nanocomposites based on hyperbranched epoxy and nano-Al_2O_3 particles modified epoxy resin. *Polymer Advanced Technologies* 22, 1032–1041.
32. L. Hong, Y. Li, T. Wang, and Q. Wang (2012). In-situ synthesis and thermal, tribological properties of thermosetting polyimide/graphene oxide nanocomposites. *Journal of Materials Science* 47(4), 1867–1874.
33. P. Kushwaha and R. Kumar (2011). Reinforcing effect of nanoclay in bamboo-reinforced thermosetting resin composites. *Polymer-Plastics Technology and Engineering* 50, 127–135.
34. A. Shojaei and M. Faghihi (2010). Physico-mechanical properties and thermal stability of thermoset nanocomposites based on styrene-butadiene rubber/phenolic resin blend. *Materials Science & Engineering, A: Structural Materials: Properties, Microstructure and Processing* 4–5, 917–926.
35. X. Zheng and C. Wilkie (2003). Flame retardancy of polystyrene nanocomposites based on an oligomeric organically-modified clay containing phosphate. *Polymer Degradation and Stability* 81, 539–550.
36. G. Chigwada, P. Jash, D. Jiang, and C. Wilkie (2004). Synergy between nanocomposite formation and low levels of bromine on fire retardancy in polystyrenes. *Polymer Degradation and Stability* 88, 382–393.

9 Flammability Properties of Polymer Nanocomposites

9.1 Introduction

This chapter provides an overview on the fundamentals, properties, and applications of flammability of polymer nanocomposites (PNCs). The structures, properties, and surface treatment of the different types of commonly used flame-retardant (FR) nanoadditives were introduced in Chapter 2. These FR nanomaterials are subdivided based on their nanoscale dimensions: one-dimensional nanoscale, such as nanoclay and nanographene platelets; two-dimensional nanoscale, such as carbon nanofibers (CNF), carbon nanotubes (CNT), and Halloysite nanotubes (HNT®); and three-dimensional nanoscale, such as nano-silica, nano-alumina, nano-magnesium hydroxide, and polyhedral oligomeric silsequioxanes (POSS®). The test methods that are commonly used to evaluate thermal and flammability of polymer nanocomposites were briefly reviewed in Chapter 6. The effects of one-dimensional, two-dimensional, three-dimensional, combined nanoadditives, as well as combined nanoadditives with conventional FR additives on thermal and flammability properties of the polymer nanocomposites are discussed. The mechanisms of the effect of nanoadditives on flammability of the PNCs are proposed. The section on concluding remarks and trends for the study of polymer nanocomposites completes this chapter. The readers who need to have a more detailed description of the nanomaterials and their detailed influence on a specific matrix polymer may pursue some reviews, books, and papers in this emerging research area, which is listed in the reference section of this chapter and other chapter in this volume.

9.2 Thermal and Flame Retardancy of Polymer Nanocomposites

Although polymer nanocomposites show improved thermal and flame retardancy, it has been shown that different nanoadditives have different effects on improvement in flame retardancy. The thermal and flammability properties of polymer nanocomposites depend on the types of nanofillers, types of polymers, and the nanostructures obtained. In this section, the thermal and flammability properties of these polymer nanocomposites will be subdivided into (a) one-dimensional nanomaterial polymer nanocomposites that include clay-based and graphene-based nanocomposites;

(b) two-dimensional nanomaterial polymer nanocomposites that include carbon nanofibers-based, carbon nanotube-based, halloysite nanotube-based polymer nanocomposites; and (c) three-dimensional nanomaterial polymer nanocomposites that include nano-silica-based, nano-alumina-based, nano-magnesium hydroxide-based, and POSS-based polymer nanocomposites. Different types of polymers, such as thermosets, thermoplastics, and elastomers, are included.

9.2.1 One Nanoscale Dimension-Based Nanocomposites

9.2.1.1 Polymer-Clay Nanocomposites

Clay-based nanomaterial is the most commonly used for nanocomposites because it has been around the longest and can be obtained in high purity at low cost. In the following discussions, the clay used to create the polymer nanocomposites is montmorillonite (MMT) unless otherwise identifying differently. More information can be found in several recent books on polymer-clay nanocomposites [1–3].

The polyamide 6-clay nanocomposite was the first discovered polymer nanocomposite. Flame retardancy (FR) of polymer-clay nanocomposites has received much attention, because FR was one of the first desirable properties noted by researchers. Initially all the polymers exhibit improved FR, as evaluated by cone calorimeter, upon the incorporation of clay. The following sections summarize the research progress on clay nanocomposites using different polymers.

9.2.1.1.1 FLAMMABILITY OF POLYAMIDE-CLAY NANOCOMPOSITES. The polyamide 6 (PA6)-clay nanocomposite was the first discovered polymer nanocomposite and was studied by Gilman et al. [4] at NIST (United States). Cone calorimeter was used as the characterization tool to evaluate the PA6-clay nanocomposites. The peak heat release rate (PHRR) is reduced by 32% and 63% in a PA6-clay nanocomposite containing, respectively, 2% and 5% of the clay. The PA6-clay nanocomposite not only showed improved FR, but the mechanical properties also improved compared to neat PA6. Foster Corp. (United States; www.fostercomp.com) demonstrated that high levels (13.9%) of clay can be added to PA12 elastomers to achieve UL 94 V-0 ratings at 1/8-in. thickness. The clay can be used as a char former that allows the typical 50% loading of halogen/antimony oxide flame-retardant system to be cut in half, which reduces the detrimental effects on physical properties of the polymer. Foster Corp. introduced PA12-clay nanocomposites for tubing and film in 2001.

Patel et al. [5] used organically modified montmorillonite (OMT) to disperse catalytically active platinum (Pt) nanoparticles that influence the char formation processes of polyamide 6 (PA6). The Pt nanoparticles of varying particle size are generated in organically modified montmorillonite (Pt-OMTs), which are subsequently melt-blended with PA6. The average particle size of Pt nanoparticles ranges from 9.2 nm to 21.6 nm as found by X-ray diffraction and field emission scanning electron microscopy. The char formation mechanism was studied by thermogravimetric analysis coupled with Fourier transform infrared spectroscopy (TG–IR). Infrared spectroscopy on the residue reveals the formation of unsaturated hydrocarbons

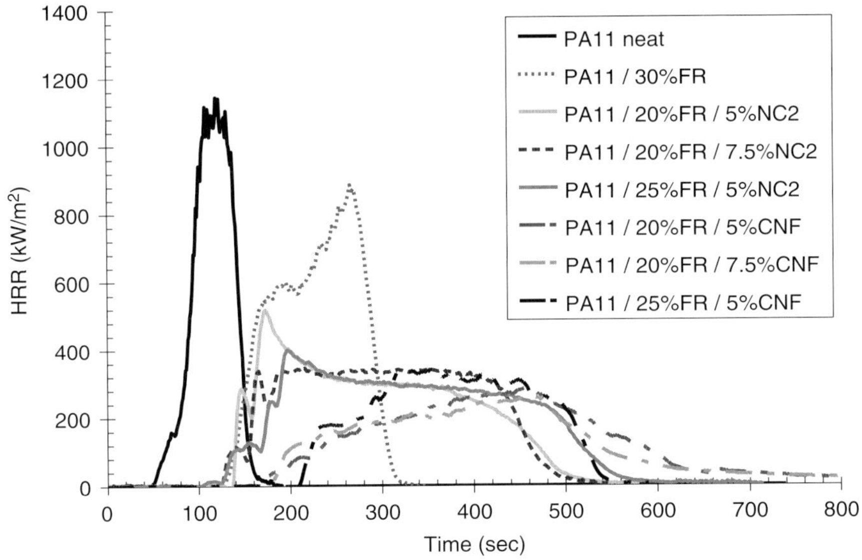

Figure 9.1. Heat release rate of neat and nanoparticle-filled PA11 with and without FR additive as a function of time (cone calorimeter, 35 kW/m^2). FR is flame-retardant additives, NC2 is nanoclay, and CNF is carbon nanofiber [6].

during decomposition of PA6 nanocomposite, which are cross-linked into a highly conjugated structure. The Pt nanoparticle intercalated OMTs enhanced the formation of char by the formation of unsaturated compounds that are cross-linked during pyrolysis in nitrogen atmosphere forming larger amounts of char compared to nanocomposites without Pt nanoparticles. The degradation of PA6 nanocomposites in the presence of Pt nanoparticles in oxidative conditions resulted in larger amounts of CO_2 evolution in two steps. MCC analyses demonstrate the decrease of evolved volatiles also in heating conditions, similar to what would be expected in a fire scenario.

Lao and Koo recently have done extensive research on flame-retardant PA11-clay and PA12-clay composites for selective laser sintering processing in additive manufacturing [6–8]. His research group demonstrated that with PA11 at 7.5% and 10% of nanoclay (NC), the PHRR was reduced by 68% and 73%, respectively [6]. Figure 9.1 shows the PHRR of the neat PA11 and several PA11 polymer blends at a radiant heat flux of 35 kW/m^2 [6]. When 20% of conventional intumescent FR additives and 7.5% NC were added to PA11, PHRR was reduced to 70% and achieving a UL 94 V-0 rating [6]. The heat deflection temperatures (HDT) of neat PA11 and PA11/20% FR/7.5% NC were 48° and 97°C, respectively. A synergistic effect exists between conventional FR intumescent additives with nanoclay, allowing the authors to lower the FR intumescent additive from 30% to 20% by incorporating 5% to 7.5% NC in PA11.

9.2.1.1.2 FLAMMABILITY OF EPOXY-CLAY NANOCOMPOSITES. Epoxy resins are used in a variety of applications, such as coatings, adhesives, and electronics or in composites in the aerospace and transportation industries. Although the polyfunctional reactivity of most epoxy systems leads to a high cross-link density, brittleness of epoxy systems are always a concern. In most applications, the polymer matrix is combined with one or more phases, such as short or long fibers or a rubbery phase for toughening to form epoxy fiber-reinforced composites. Epoxy nanocomposites have attracted a lot of attention within nanocomposites research to enhanced modulus, strength, fracture toughness, impact resistance, gas and liquid barrier, and improve flame retardancy. However, like all organic materials, epoxy is flammable, and their use for replacing traditional nonflammable materials has increased the fire hazard in the past years. Evolved smoke and toxic gases from these fires create hazards for both people and the environment. At the present, nanoclays are by far the most investigated nanofillers in flame retardancy. In this section, we focus on epoxy nanocomposites based on nanoclays.

The additive effect of resorcinol bis(diphenyl phosphate) (RDP) and nanoclay on epoxy was studied by Katsoulis et al. [9]. The epoxy resin, tetraglycidy1-4,4'-diaminodiphenylmethane (TGDDM), and the curing agent, 4,4'diaminodipheny sulfone (DDS), RDP flame retardant, and nanoclay of different loadings were synthesized. TEM images of epoxy-RDP-I.30E (80/15/5) at low magnification show intercalated nanoclay with gallery spacing at 5–10 nm [9]. Formulations with RDP alone at 5% and 10% gave the best flame resistance in LOI and UL 94 experiments, while going beyond 10% RDP had an adverse effect. In cone calorimeter, the addition of RDP at 10% results in remarkable reductions in PHRR and enhanced char formation. Addition of nanoclay alone did not improve the thermal stability of the epoxy resin, and neither did it improve its flame resistance using cone calorimeter. The addition of clay with RDP did not show an advantage. It also did not produce any significant improvement in the fire properties as measured by cone calorimeter and LOI. This may be attributed to possible antagonistic interactions between the constituent components. The addition of a known flame retardant, RDP, and nanoclay yielded mixed results with respect to thermal stability and improved fire performance.

In nitrogen, residue yields from epoxy-clay nanocomposite reveals little improvement in the carbonaceous char yield once the presence of silicate in the residue is accounted for [9–12]. In air, the char yield increase is more relevant [13]. Hussain et al. [14] studied the effect of clay and an organophosphorus epoxy modifier, DOPO on DGEBA and TGDDM epoxies cured with amine curing agent. They prepared 3% P-containing epoxies by reacting DGEBA or TGDDM with DOPO. Standard and modified epoxies were used to prepare clay nanocomposites that show a mixed intercalated-exfoliated structure. The clay enhanced char formation significantly. The results are shown in Figure 9.2 [14]. The char yield at 600°C increase from 14% in the near DGEBA to 38 and 42% for 5% clay-DGEBA nanocomposite and 3% P-modified epoxy, respectively. Similar results were observed for TGDDM-based formulations.

Polymer nanocomposites showed that clays can reduce the amount of tradition flame retardants required to comply with the standards, the approach to nanocomposites itself alone is not sufficient to facilitate passing commercial tests. In terms

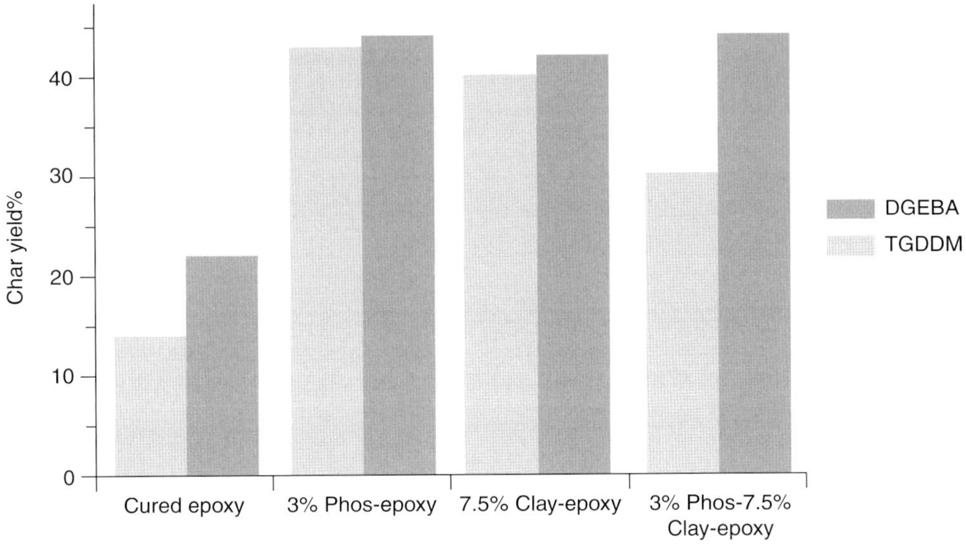

Figure 9.2. Char yields at 600°C for DGEBA and TGDDM control samples, 3% P-epoxies, and 7.5% clay nanocomposites [14].

of limiting oxygen index (LOI), the improvement achieved by using clays is generally not relevant. A typical increase between 1% and 2% in LOI values is seen with 5% nanoclay [15]. Hussain et al. [14] reported an increase in LOI values from 25% for neat DGEBA to 32.7% and 34.5% for 5.0% and 7.5% nanocomposites, respectively. For a tetrafunctional TGDDM epoxy, the LOI values increase from 26.3% in neat polymer to 35.1% and 36.7% in nanocomposites containing 5% and 7.5% clay, respectively.

Cone calorimeter results of nanoclay in DGEBA cured with either methylenedianiline (MDA) or benzyldimenthylamine (BDMA) [16] were impressive. The PHRR and the average heat release rate (HRR) are significantly improved with 6% nanoclay. The HRR plots for DGEBA-MDA and the DGEBA-MDA clay nanocomposites are compared in [16]. The PHRR, HRR, and average mass loss rate (MLR) decrease by about 40%. The heat of combustion, smoke obscuration, and carbon monoxide yields are unchanged. The results suggest that the clay nanocomposite is operating primarily in the condensed phase. A shorter ignition time is observed in the nanocomposite may be due to the low stability of the organic modifier.

9.2.1.1.3 FLAMMABILITY OF ELASTOMER-CLAY NANOCOMPOSITES. The combustion behavior of thermoplastic elastomer has been studied by Koo et al. [17, 18]. Two types of commercially available thermoplastic polyurethane elastomer (TPU) (Pellethane® and Desmopan®) were melt blended with various loadings (2.5–10%) of MMT nanoclay (Cloisite® 30B) using twin-screw extrusion. The morphological, physical, thermal, flammability, and thermophysical properties and kinetic parameters of these two families of polymer-clay nanocomposites were characterized. The *processing-structure-property* relationships of this class of novel thermoplastic

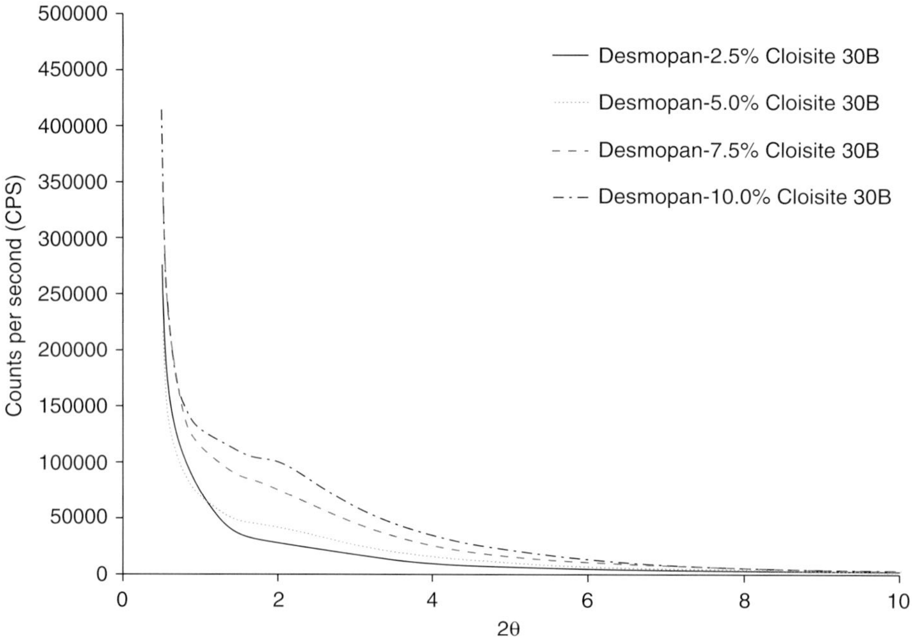

Figure 9.3. Wide-angle X-ray diffraction of Desmopan-clay TPUNs.

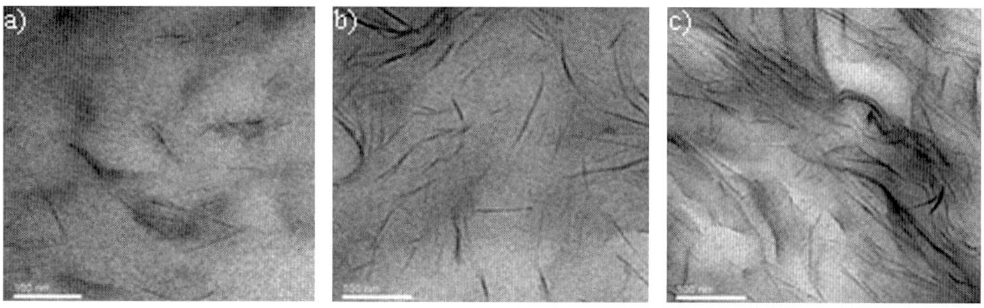

Figure 9.4. TEM micrographs at 40 kX of Desmopan with (a) 2.5%, (b) 5%, and (c) 10% Cloisite 30B nanoclay TPUNs with a unit bar of 100 nm.

polyurethane elastomer-clay nanocomposites (TPUNs) were established. The WAXD plots of the four different nanoclay loading of Desmopan TPUNs are shown in Figure 9.3 [17, 18]. No peaks were observed in WAXD, which indicated the all loadings of nanoclays were exfoliated in the TPU polymer. No stacks are observed in any of the three 40 kX TEM images in Figure 9.4; the scale bar is 100 nm [17]. The lack of platelet stacks indicates that the nanoclay is exfoliated and well dispersed in the Desmopan TPU material.

Cone calorimeter experiments at 50kW/m^2 are tested on the neat Desmopan TPU, neat Pellethane TPU, Desmopan-5% Cloisite 30B TPUN, and Pellethane-5% Cloisite 30B TPUN. Two or three experiments are performed to test for repeatability. The neat material experiments are the least repeatable due to the fact that the material melted and dripped off the test fixture; for this reason, the test with the

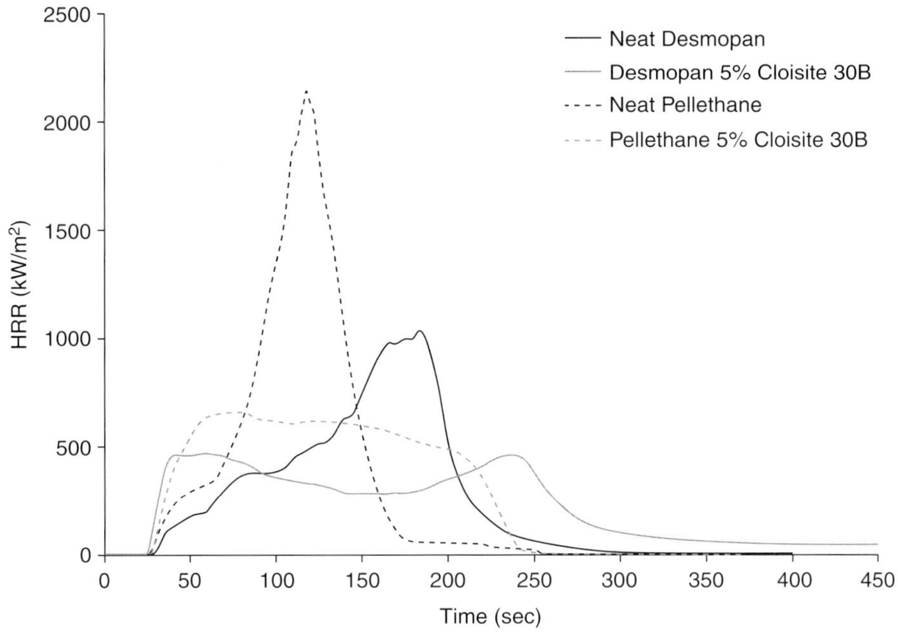

Figure 9.5. Heat release rate of Desmopan and Pellethane TPU and Cloisite 30B TPUNs.

higher peak heat release rate (PHRR) is used in the comparisons. The heat release rate (HRR) graph of the materials is shown in Figure 9.5 [17]. The flammability properties are shown in Table 9.1 [17]. The material's time to sustained ignition is not changed drastically with the addition of Cloisite 30B, it is +2 sec for Pellethane TPUN, and –1 sec for the Desmopan TPUN. However, a dramatic decrease in peak heat release rates (PHRR) is observed. Pellethane TPU PHRR decreases by 73% and Desmopan TPU PHRR decreases by 50% with the addition of 5wt% Cloisite 30B. In the first 60 seconds, the HRR is higher in the TPUN. This is due to a higher rate of decomposition at this stage observed in the mass loss graph.

9.2.1.2 Polymer-Graphene Nanocomposites

Researchers started using nanographene (NG) and graphene oxide (GO) in polymer in recent years aiming to enhance the electrical and thermal conductivities of polymer resins. Both NG and GO also have the potential to work as heat barrier and char promoter when the polymer nanocomposites are subjected to fire burning, thus improving the flammability properties. However, this research area is not yet well developed. The majority of exploratory work on the use NG and GO as a flame retardant has been done when the two are added to epoxy resin.

9.2.1.2.1 FLAMMABILITY OF EPOXY-GRAPHENE NANOCOMPOSITES. Avila et al. [19] incorporated 3wt% of NG into epoxy resin and showed that fire resistance of the resin was enhanced. Specimens were burned under a large heat flux using an oxy-acetylene torch for two minutes. The unburned thickness of the neat resin and nanocomposites at 60 seconds was ~1 mm and ~3.2 mm, respectively, while at 120 seconds

Table 9.1. *Summary of Cone calorimetry data at irradiance heat flux of 50 kW/m²*

Material	t_{ig} (s)	PHRR (kW/m²)	Avg. HRR, 60s (kW/m²)	Avg. HRR, 180s (kW/m²)	Avg. Eff. H_c (MJ/kg)	Avg. SEA (m²/kg)
Neat Pellethane	32	2,290	406	653	30	237
Pellethane-5% Cloisite 30B	34	664 (71% reduction)	560	562	25	303
Neat Desmopan	28	1,031	228	515	27	311
Desmopan-5% Cloisite 30B	27	518 (50% reduction)	442	376	28	256

t_{ig} = Time to sustained ignition; PHHR = Peak heat release rate; Avg. HRR = Average heat release rate after ignition; Avg. Eff, H_c = Effective heat of combustion; Avg. SEA = Average specific extinction area

it was 0.16 mm and 2.74 mm. Also, the peak mass loss in TGA of the neat resin and the NG nanocomposite was 1.04 mg/min and 0.63 mg/min, respectively. These results indicate that the nanoparticle can act effectively as a thermal barrier protecting the resin underneath the char layer. Koo et al. [20] tested Avila's epoxy-graphene materials and showed similar results using cone calorimeter. Guo et al. [21] also found that by incorporating 0.3–5 wt% of NG, GO, and FGO (functionalized graphene oxide) into epoxy resin increased the char yield under TGA of 14–16%. The peak heat release rate (PHRR) and total heat release (THR) under microscale combustion calorimeter (MCC) was reduced up to 45% and 24%, respectively. Although all three nanoparticles could reduce the heat release, NG provided the greatest reduction. Guo concluded that there was a good flame retardant effect of graphite sample in epoxy composites and the nanoparticles were good char promoters. Wang et al. [22] showed that an epoxy nanocomposite with 1 wt% NGO showed 28.5% and 11% reduction in THR and PHRR, respectively, under a heat flux of 50 kW/m² in a cone calorimeter.

9.2.1.2.2 FLAMMABILITY OF HIPS-GRAPHENE OXIDE NANOCOMPOSITES. Higginbotham et al. [23] incorporated 1, 5, and 10 wt% of GO into high-impact polystyrene (HIPS). As the weight loading of GO increased in the HIPS resin (0–10 wt%), the THR reduced from 37.5 kJ/g to 34 kJ/g (10% reduction), PHRR reduced from 840 W/g to 610 W/g (28% reduction), and the char yield increased from 1 wt% to 6 wt% (500% improvement) as observed from the MCC results. However, since HIPS is a very flammable material, dripping was observed when the HIPS-GO nanocomposites were burned in a vertical flame test, and fire did not self-extinguish. Therefore, no correlation between flame retardancy and flammability was observed in this nanocomposite system.

9.2.1.2.3 FLAMMABILITY OF ABS-GRAPHENE OXIDE NANOCOMPOSITES. Higginbotham et al. [23] incorporated 1, 5, and 10 wt% of the same GO as in the HIPS nanocomposites into the acrylonitrile butadiene styrene (ABS) resin. The GO had similar effects on the THR, PHRR, and char yield on the ABS polymer. As the weight loading of GO increased, the THR decreased (up to 11%), PHRR decreased (up to 17%), and the char yield decreased (up to 600 wt%). ABS-GO nanocomposites also exhibit dripping under the vertical flame test. Although self-extinguishment of fire was also observed, there was still no correlation between flame retardancy and flammability in this system.

9.2.1.2.4 FLAMMABILITY OF PC-GRAPHENE OXIDE NANOCOMPOSITES. Higginbotham et al. [23] also incorporated the same GO into polycarbonate (PC) with the same weight loadings. A general trend of decreases in THR (up to 13%) and PHRR (up to 13%) with increasing GO weight loading was observed, with a few exceptions that were probably random errors. The char yield increased 21 wt% when 10 wt% of GO was added. Since the PC itself was not very flammable (fire self-extinguished in 14 seconds in a vertical flame test), the addition of 1 wt% GO decreased the time to self-extinguish to 4 seconds, and thus no dripping was observed. The incorporation of 5 wt% and 10 wt% of GO even made the PC nonflammable.

9.2.2 Two Nanoscale Dimensions-Based Nanocomposites

9.2.2.1 Polymer Carbon Nanofiber Nanocomposites

9.2.2.1.1 FLAMMABILITY OF THERMOPLASTIC-CNF NANOCOMPOSITES. Polyamide 11 (PA11) was melt-blended by dispersing low concentrations of carbon nanofibers (CNFs) by Lao and Koo [6] via twin-screw extrusion. These PA11-CNF nanocomposites will be used for selective laser sintering in additive manufacture. To enhance their thermal and flame retardant (FR) properties, an intumescent FR additive was added to the mechanically superior PA11-CNF formulations. Good dispersion of the CNFs and FR additives in the PA11 polymer was achieved. For neat and CNF-reinforced PA11 as well as for PA11 reinforced by both intumescent FR and select CNF, decomposition temperatures by TGA, flammability properties by UL 94, and cone calorimeter values were measured. All PA11 polymer systems infused with CNFs and FR additive had higher decomposition temperatures than those infused with solely FR additive. For the PA11/FR/CNF formulations, all Exolit® OP 1311 (FR1), OP 1312 (FR2) and OP 1230 (FR3) FR additives passed the UL 94 V-0 requirement with 20%.

Based on the decomposition temperatures at 50% mass loss, thermal stability of the PA11 was enhanced by CNF significantly. All FR/CNF-reinforced PA11 blends had higher decomposition temperatures than those reinforced with CNF only. The FR3 intumescent additive provided the best synergistic effect to CNF since PA11/20%FR3/7.5%CNF sample had a significant increase of $T_{50\%}$ ($\Delta \cong 66$ °C).

It was shown that 30% FR3 was needed for PA11 to achieve a UL 94 V-0 rating. We also observed that a minimum of 20% of either FR1, FR2, or FR3 is needed with either 5 wt% or 7.5 wt% of CNF to achieve the V-0 rating. All three FR additives worked well with CNF. Since 15 wt% of FR intumescent additive will only achieve a V-1 rating, the optimal percentage of intumescent FR additive that is needed with CNF lies between 15% and 20% to achieve a V-0 rating. The CNF blend had no combustion at the first flame application and a lower total combustion time than that

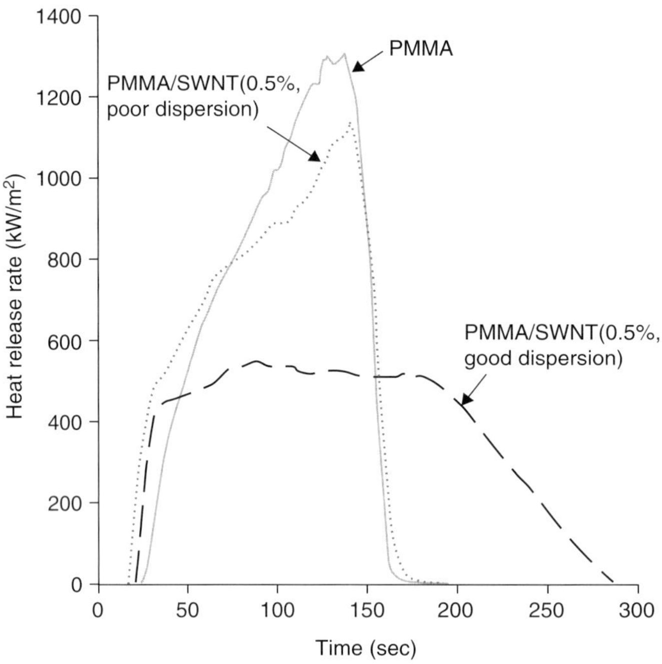

Figure 9.6. Effect of SWNT dispersion on heat release rate of PMMA-SWNT (0.5%) nanocomposites at an external radiant flux of 30 kW/m^2 [24].

of the NC blend. Therefore, the FR/CNF blend outperformed its NC counterpart. Cone calorimeter peak heat release rates of FR/CNF blends were lower than that of the FR/NC blends. This may suggest that CNF has a good synergistic effect with the FR2 additive.

9.2.2.2 Polymer Carbon Nanotube Nanocomposites

Carbon nanotubes (single-walled and multi-walled) based polymer nanocomposites have been studied intensively by Kashiwagi and others. Heat release rates of three different samples – poly(methyl methacrylate) (PMMA), PMMA-SWNT (0.5%, good dispersion), and PMMA-SWNT (0.5%, poor dispersion) – were measured in a cone calorimeter at an external radiant flux of 50 kW/m^2 as shown in Figure 9.6 [24, 25]. The 0.5% SWNT (good dispersion) showed more than 50% reduction in the PHRR compared with neat PMMA. It was proposed that the SWNTs formed a continuous network that acts as a heat shield to slow the thermal degradation of PMMA. The effects of SWNT concentration on flammability properties of PMMA-SWNT nanocomposites by cone calorimeter that have good dispersion of SWNT at levels from 0.1% to 1% prepared by the coagulation method are shown in Figure 9.7 [24]. The additional of 0.1% of SWNT did not significantly reduce the heat release rate of PMMA. The most reduction in heat release rate was achieved by 0.5% SWNT. A 60% PHHR reduction was achieved by 0.5% SWNT.

Fina et al. [57] studied the chemical activity of carbon nanotubes and polyhedral oligomeric silsesquioxanes during thermal degradation and combustion of polymer

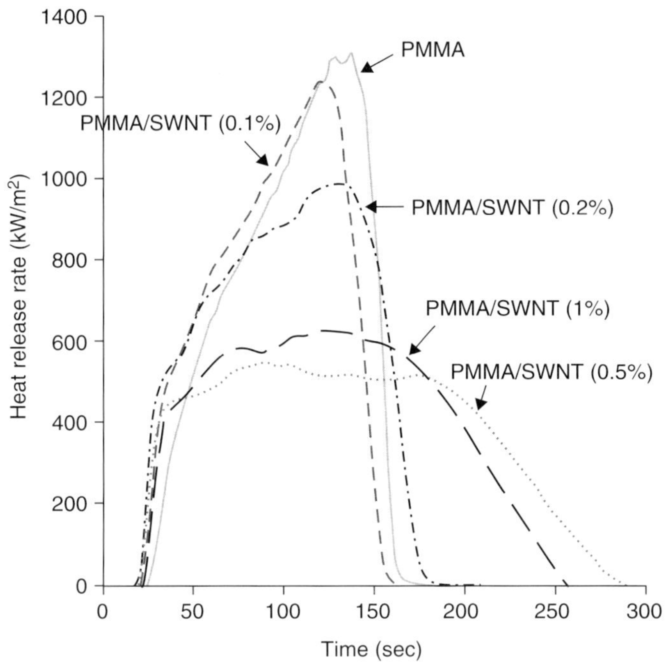

Figure 9.7. Effect of SWNT concentration on the heat release rate of PMMA-SWNT at 50 kW/m^2 [24].

nanocomposites. Polymer-nanofiller systems may exhibit chemical effects capable of thermal stabilization of polymers as well as reduction of combustion rate and heat released, owing to catalytic effects induced by the nanofillers at high temperature. Carbon nanotubes in the presence of oxygen are shown to promote oxidative dehydrogenation in polyethylene with production of a stable surface layer of carbon char that provides an effective oxygen barrier effect. A similar action is performed by metal-containing polysilsesquioxanes dispersed in polypropylene. With either carbon nanotubes or metal POSS, partial carbonization of the polymer matrix occurs during combustion, subtracting part of the organic polymer from combustion, targeting one of the major fire retardancy aim. Heat release rate plots for PP/MWNT, PB/MWNT, and PE/MWNT, compared with their reference polymers, are shown in Figure 9.8 [26]. All three referenced polymers showed the typical thermally thin behavior, with an almost symmetric HRR curve and a single heat release rate peak. The plots for the corresponding MWNT nanocomposites show significant reductions of the peak HRR compared with reference polymers, amounting to about 42% for both PP and PB and about 46% for PE. Such a behavior is typical of partially charring system producing physical effects during combustion with data reported in the literature for polyolefin/CNT nanocomposites.

Bocchini et al. [26] studied the multi-wall carbon nanotubes (MWNT)/linear low-density polyethylene (LLDPE) nanocomposites to understand the stabilization mechanism for their thermal and oxidative degradation. Thermogravimetry coupled with infrared evolved gas analysis and pyrolysis gas chromatography–mass

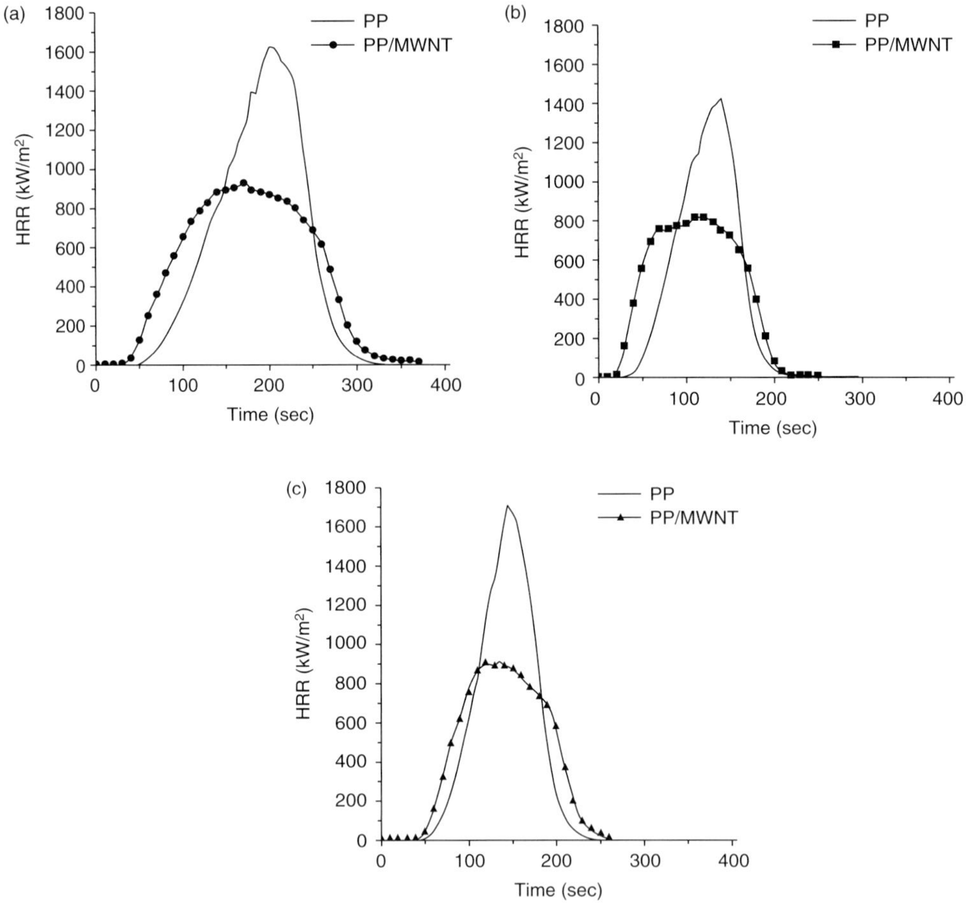

Figure 9.8. HHR plots for (a) PP/MWNT, (b) PB/MWNT, and (c) PE/MWNT compared with their reference polymers [26].

spectrometry demonstrate that MWNT presence slightly delays thermal volatilization (15–20°C) without modification of thermal degradation mechanism, whereas thermal oxidative degradation in air is delayed by about 100°C independently from MWNT concentration in the range used here (0.5–3.0 wt%). In thermal degradation, the stabilization of LLDPE is due to a thin film of MWNT net formed by migration at the surface and accumulation due to LLDPE volatilization. In oxidative conditions the MWNT surface layer favors formation of a polyaromatic carbon char to produce a MWNT/char composite layer protecting the underlying polymer from oxygen shown by SEM and ATR FTIR of degradation residues. Thus, the thermal stabilization of MWNT/LLDPE nanocomposites can be reached with low concentration of MWNT (0.5%).

9.2.2.3 Polymer Halloysite Nanotube Nanocomposites

Butler et al. [27] developed an enhanced polyamide 11 (PA11) composite with improved mechanical, thermal, and flammability properties through the addition of Halloysite nanotubes (HNT®) and a conventional intumescent flame-retardant (FR) additive. Polyamide 11 (nylon 11) is of particular interest, as it is one of the major polymers used in the selective laser sintering (SLS) for additive manufacturing. Unfortunately, PA11 has poor flammability properties that impose limitations on the applications for this particular material via the additive manufacturing process. HNTs are an attractive additive as they occur naturally, are relatively easy to process, and are less detrimental to the elongation and toughness properties of the polymeric material compared to other additives. HNTs and FR additives were melt-compounded via twin-screw extrusion processing in PA11. Test specimens in this study were injection-molded. The principal objective of this work is to study the combined effects of the addition of HNTs and a conventional intumescent FR additive to PA11. A matrix of nine formulations of PA11-HNT-FR was studied; properties are characterized using tensile and Izod impact testing, UL 94 flammability testing, thermogravimetric analysis (TGA), and scanning electron microscopy (SEM).

None of the PA11-HNT nanocomposites (2.5, 5, 7.5, and 10 wt%) were able to achieve a UL 94 V-0 rating. When a Clariant FR intumescent component (Exolit® OP 1312) was added to the PA11-FR-HNT nanocomposites, they appear to become valuable high-performance materials for the SLS process. With only 2.5 wt% HNT in PA11, the tensile modulus is increased from 424 MPa to 1,867 MPa, and an elongation at break of 40% is obtained. In traditional composites, a substantial increase in modulus almost always results in a brittle material. A 40% elongation at break may be considerably lower than the neat PA11, which had an elongation at break of 227%; however, 40% elongation at break is still a considerable toughness for a material and is much higher than many other polymer nanocomposites that incorporate nanoparticles, such as MMT nanoclays [3, 4], carbon nanofibers [3, 4], and carbon nanotubes [28]. Traditional polymer nanocomposites often see elongations at break around 20% and lower for a similar increase in modulus. The combination of stiffness and toughness in the PA11 nanocomposites is therefore very intriguing. In the PA11 nanocomposite formulation comprising 20 wt% FR and 5 wt% HNT, a non-dripping formulation that achieved the UL 94 V-0 rating, 940 MPa tensile modulus, and 19% elongation at break was achieved. This balance of flame retardancy, stiffness, and toughness is also very promising. A Design of Experiment approach was adapted in this study to identify additional PA11-FR-HNT formulations. Future work will involve the development of surface treatments for HNT to improve the interfacial adhesion and thus the tensile strength and elongation at break of FR-HNT nanocomposites.

9.2.3 Three Nanoscale Dimensions-Based Nanocomposites

9.2.3.1 Polymer Nano-Silica Nanocomposites

Nano-silica can have a huge interfacial area as long as the diameter of the particles is in the range of nanometers. Although they do not have the narrow gallery structure of the layered nanoclay, improvement in physical properties [29–32] and in thermal stability [33, 34] by the additional of nano-silica to polymer have been

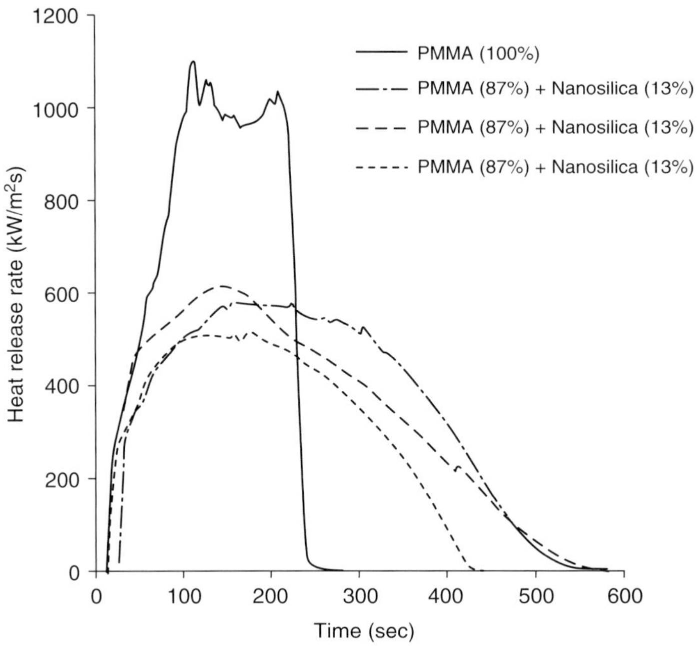

Figure 9.9. Effects of nano-silica addition on heat release rate of PMMA at 40 kW/m^2. The dashed lines are the results of three replicas of nanocomposites made at three different times [38].

reported. It was also reported that the addition of mesoscale silica to different polymers significantly reduced the HRR of polymers [35, 36]. Flammability properties of PMMA-nano-silica [37–39] and polyimide-nano-silica nanocomposites [39] have been reported. Dispersion of the particles in a polymer is critical for obtaining better flame retardant performance. Roughly 50% reduction in PHRR was reported with the addition of 13 wt% of silica particles [38] as shown in Figure 9.9. Little to no improvement was reported in limiting oxygen index (LOI) measurement with up to 10 wt% of nano-silica of diameter of 7 nm. Although the LOI values increased from 36 to 44, the addition of 20 wt% of silica particles (diameter 50 to 300 nm) was required. The addition of nano-silica hardly reduced the HRR at the early stage of burning, and it was demonstrated that the addition of nano-silica did not significant modify the UL 94 rating [39]. The overall FR effectiveness of nano-silica appears to be less than that of the nanoclays as described in earlier section in this chapter.

9.2.3.2 Polymer Nano-Alumina Nanocomposites

Three types of nano-alumina (X-0 needle, X-25SR, and X-0SR) with different organic treatments manufactured by SASAL NA were incorporated into polyamide 11 via twin-screw extrusion by Lao et al. [40]. A total of ten formulations of polyamide 11/n-alumina nanocomposites, including the neat PA11, were melt-compounded and injection-molded to test specimens. TEM demonstrated that n-alumina surface-treated particles are well dispersed in the polyamide 11 matrix. TGA showed that n-alumina improved the thermal stability of the nylon 11 substantially. The UL 94 test showed that all formulations did not pass the ideal V-0 rating. All of them exhibited flammable dripping that ignited the cotton under the specimens during the test. The X-0 needle formulations performed the best, followed by X-25 SR, and the X-0 SR performed the worst. The addition of selective flame-retardant additives or anti-dripping agents should be incorporated into the system together with the *n*-alumina for enhanced flammability properties in the future.

Rallini et al. reported using nano-alumina (average diameter of 13 nm) to improve flame resistance of carbon fiber-reinforced composites [41]. Nano-alumina was dispersed in an epoxy resin as a matrix for carbon fiber-reinforced composites by simple mechanical stirring. The 5 wt% nano-alumina allowed the authors to obtain a fiber-reinforced composite with improved fire resistance. After exposure to high temperature, the nano-alumina acts as a high-temperature adhesive coating and has sintered onto the carbon fibers. This coating binds together the carbon fibers, promoting residual structural integrity of the burnt composite. TGA data also demonstrated that the sintered alumina coating of the fibers acts as a barrier for the oxidation of carbon fibers. The morphological investigation of the samples burnt in a muffle has shown further evidence on the role of alumina particles. The 1 wt% content is insufficient to improve the residue strength of the burnt sample, whereas the 5 wt% content is sufficient to produce a ceramic skeleton on the entire volume of the composite.

9.2.3.3 Polymer Nano-Magnesium Hydroxide Nanocomposites

Magnesium hydroxide [$Mg(OH)_2$ / MH] has been used in polymer as a flame-retardant additive. However, it commonly requires 60–70 wt% of the micron-sized (2–5 μm) MH to be added into polymer to make it truly fire retardant [42]. Since flame-retardant efficiency of MH is related to inter-particle distance, which decreases with decreasing particle size at a given volume fraction of filler [43, 44], it is believed that by using MH of smaller particle size – that is nano-MH – effective flame retardancy can be achieved at a lower loading of filler.

9.2.3.3.1 FLAMMABILITY OF EVA-MG(OH)$_2$ NANOCOMPOSITES. Kalfus [42] incorporated 26 wt% of nano-MH (comparing to conventional loading of 60–70 wt% of micro-MH) into ethylene vinyl acetate (EVA) resin, and found that the rate of weight loss was about the same as the micro-MH of the same weight loading. However, the thermal stability of the EVA-MH nanocomposite was slightly better. Qiao [45] added 33 wt% nano-MH into EVA and observed a 43% reduction in PHRR, 43% reduction in mean HRR, and 182% increase in residue. Ly et al. [46] also compared the effect of particle sizes on the flammability of EVA. Nano-MH and micro-MH of 80, 100, 120, and 150 phr (parts per hundred polymer) were incorporated into EVA, respectively. It was found that the limited oxygen index (LOI) of the composite was increased from 18% to 39% by micro-MHs and to 46% by nano-MHs, respectively. However, all of the EVA nanocomposite, beside the 150phr, failed the UL-94 burning test (while EVA-150phr nano-MH achieved the V-2 rating). The addition of 100 phr nano-MH increased the time to ignition (TTI) from 87 seconds to 148 seconds

(70%), decomposition temperature from 438°C to 470°C (7%), and char yield from 0 to 6 wt%; it also reduced the PHRR from 1,655 kW/m^2 to 423 kW/m^2 (74%) and average HRR from 439 kW/m^2 to 251 kW/m^2 (43%). Ly then incorporated one more component (microcapsulated red phosphorus [MRP]) into the nanocomposites and found a synergistic effect. While EVA with 100 phr nano-MH failed the UL-94 test, EVA with 100 phr of nano-MH and MRP (e.g., 95 phr nano-MH and 5 phr MRP) achieved a V-0 rating. The synergism also increased the LOI to 60%, TTI to 176 seconds (102% increase comparing to neat EVA), decomposition temperature to 490°C (12%), char yield to 16 wt%, and reduced the PHRR to 148 kW/m^2 (91%) and HRR to 72 kW/m^2 (83.5%).

9.2.3.3.2 FLAMMABILITY OF PP-MG(OH)$_2$ NANOCOMPOSITES. Mishra [47] incorporated 2, 4, 8, and 12 wt% of nano-MH into polypropylene (PP). It was found that the rate of burning per second was reduced with the increasing nanofiller composition. The reduction in the flammable property was as high as 35% comparing to the neat PP. Mishra claimed that the reduction in flammability was due to the endothermic nature of the evenly dispersed nano-MH in the PP matrix.

9.2.3.3.3 FLAMMABILITY OF EPDM-MG(OH)$_2$ NANOCOMPOSITES. Zhang [48] mixed 20–100 phr MH into ethylene-propylene-diene monomer rubber (EPDM) with particle sizes of 800 mesh, 1,250 mesh, 2,500 mesh, and nano. A decrease in PHRR with decreasing particle sizes was observed. Tang [49] incorporated 100 MP (mass portion) of nano-MH into an EPDM(100 MP)/Paraffin(100 MP) system and found an increase of LOI from 17% to 28%. Tang claimed that greater LOI value was obtained with higher content of nano-MH added, and it was obvious that the addition of the nano-MH promote flame resistance.

9.2.3.3.4 FLAMMABILITY OF ABS-MG(OH)$_2$ NANOCOMPOSITES. Cao [50] investigated the acrylonitrile-butadiene-styrene (ABS) system with nano-MH. The ABS/1 wt% nano-MH nanocomposite had similar flammability properties to those of the neat ABS polymer. However, when incorporating 5 wt% of the nano-MH, the PHRR reduced from 663 kW/m^2 to 430 kW/m^2 (35%), and THR from 107 MJ/m^2 to 96.8 MJ/m^2 (9.5%); TTI increased from 30 seconds to 67 seconds (123%), and time to reach PHRR from 169 seconds to 245 seconds (45%). However, the combustion time was extended from 400 seconds to 580 seconds (45%).

9.2.3.3.5 FLAMMABILITY OF SBR-MG(OH)$_2$ / PBR-MG(OH)$_2$ NANOCOMPOSITES. Patil et al. [51] incorporated 2–10 wt% of both conventional MH and nano-MH into styrene-butadiene rubber (SBR) and polybutadiene rubber (PBR), respectively. It was observed that the flame retardancy (improvement) in a burning test increased with increasing MH from 1.5 mm/sec to 2.1 mm/sec while conventional MH was used and from 1.6 mm/sec to 2.3 mm/sec when nano-MH was used. Similar trend was observed in the PBR/MH systems. The flame retardancy increased from 1.38 mm/sec to 1.54 mm/sec and 1.41 mm/sec to 1.58 mm/sec when conventional MH and nano-MH were used in the PBR resin, respectively. According to the data, flame retardancy was noticeably enhanced by the MHs; however, the difference between the conventional MH and the nano-MH in both polymer systems was insignificant, although Patil et al. [51] claimed that the nano-MH provided a significant increase in flame retardancy in the SBR resin comparing to the conventional filler.

9.2.3.3.6 FLAMMABILITY OF EPOXY-MG(OH)$_2$ NANOCOMPOSITES. Suihkonen et al. [52] compared the flammability of epoxy resins filled with micro-MH and nano-MH. Weight loadings of 1%, 5%, and 10% of each of both fillers were incorporated into the epoxy. TTI of all samples were about the same. However, PHRR was reduced significantly from 933 kW/m^2 (neat) to 572 kW/m^2 (10% micro-MH) and 539 kW/m^2 (10% nano-MH). THR and TSP (total smoke produced) were also decreased, but the effects of micro-MH and nano-MH were about the same. THR was reduced from 124 MJ/m^2 to 114 MJ/m^2 while TSP from 40 m^2 to 35–36 m^2. There was a significant decrease (71%) in the burning rate of the sample in the UL94HB burning test when 10 wt% of nano-MH was added, comparing to the neat epoxy. Suihkonen et al. [52] also commented that the poor dispersion of the micro-particles did not obstruct the functionality of the flame-retardant filler, since the efficiency of MH in flammability was not significantly affected by the particle size of the filler or the surface treatment in this study.

9.2.3.4 Polymer POSS Nanocomposites

Polyhedral oligomeric silsesquioxanes (POSS)-based nanocomposites have received increasing attention because of the unique three-dimensional structure of the POSS macromonomer [53], as shown in Chapter 2. Early examples were presented with siloxane [54, 55], followed by numerous applications showing enhancement of thermal stability and improving the flammability properties of polymers. Thermal analysis study demonstrated the enhancement of thermal stability of polymer-POSS nanocomposites and suggested that there is a potential to improve the flammability properties of matrix polymers. Studies clearly demonstrating such improvement by means of the use of POSS-based nanocomposites are limited. Kashiwagi et al. [56] studied polytetramethylene ether-glyco-b-polyamide-12, 1% polyimide-12 (PTME-PA), polystyrene-polybutadine-polystyrene (SBS), and polypropylene (PP) with POSS ranging from 10% to 20% via solution blending in tetrahydrofuran (THF). For comparison purposes, composites based on other silicone compounds, such as polycarbosilane (PCS) and polysilastyrene (PSS), were also prepared by solution blending. The flammability properties of these polymer blends were characterized using a cone calorimeter. The results shown in Table 9.2 [56] reveal that both PCS and POSS are reasonably effective for reducing the heat release rate measured at 35 kW/m^2. However, the total heat release of the nanocomposites was not significantly reduced from that of the polymer matrix. Furthermore, the residual yields are about the same as the theoretical yields as shown in Table 9.2. This means that the addition of POSS to the nanocomposites does not significantly increase the yield of carbonaceous char. The residue is mainly the inorganic component of the POSS.

Table 9.2. *Summary of Cone calorimeter data of PP, PTME-PA, and SBS with siloxanes and POSS at 35 kW/m²*

Sample	Residue Yield[a] (%)	Mean Mass Loss Rate (g/m²-s)	Peak HRR [kW/m² (Δ%)]	Mean HRR [kW/m² (Δ%)]	H_c (MJ/kg)	SEA (m²/kg)	Mean CO Yield (kg/kg)
PP	0	25.4	1,466	741	34.7	650	0.03
PP/POSS 80/20	17 (16)	19.1	892 (40%)	432 (42%)	29.8	820	0.03
PTME-PA	0	34.2	2,020	780	29.0	190	0.02
PTME-PA/PCS 80/10	15 (15)	14.8	699 (65%)	419 (46%)	28.5	260	0.02
PTME-PA/ POSS 90/10	6 (8)	19.8	578 (72%)	437 (44%)	25.5	370	0.02
SBS	1	36.2	1,405	976	29.3	1,750	0.08
SBS/PCS 80/20	20 (15)	18.5	825 (42%)	362 (63%)	26.4	1,550	0.07
SBS/POSS 90/10	6 (8)	31.2	1,027 (27%)	755 (23%)	26.9	1,490	0.07

H_c = mean heat of combustion; SEA = specific extinction area (smoke measurement). Uncertainties: ±5% of reported value of residue yields for heat release rate (HRR) and H_c data; ±10% for carbon monoxide and SEA data.
[a] Theoretical residue yields in parentheses.

Fina and colleagues focused on the preparation and characterization of POSS-based polymer hybrids by melt-blending and reactive blending techniques [57–62]. The chemical activity of POSS and CNT during thermal degradation and combustion of polymer nanocomposites was studied by Fina et al. [62]. Polymer-nanofiller systems may exhibit chemical effects capable of thermal stabilization of polymers and reduction of combustion rate and heat released, owing to catalytic effects induced by the nanofillers at high temperature. POSS in the presence of oxygen is shown to promote oxidative dehydrogenation in polyethylene with the production of a stable surface layer of char that provides an effective oxygen barrier effect. The combustion of polypropylene (PP) is strongly affected by metal-containing POSS, with a mechanism that cannot be related to the simple accumulation of a ceramic phase on the surface as reported for metal-free polysilsesquioxanes [59]. In particular, Al-isobutyl POSS was reported to be an effective fire retardant for PP, resulting in lower heat release rate (HRR) and lower effective heat of combustion (EHC), owing to a catalytic effect of the Al moieties, promoting secondary reactions during polymer degradation and leading to partial charring instead of complete volatilization. To investigate the material surface evolution upon heating PP, PP/oib-POSS, and PP/Al-POSS were heated to 350°C in air with a 15°C/min heating ramp and the residues were studied by means of SEM/EDS and ATR-FTIR analyses of the surface [59].

The PP and PP/oib-POSS undergo a large weight loss of about 44% and 37% of the initial weights, respectively, and appear to be strongly degraded as shown in [59]. PP/Al-POSS evidenced a higher thermal stability and a limited weight loss (5% of the initial weight). SEM micrographs of the surface of residue for PP/oib-POSS and PP/Al-POSS as shown in [59] proved that a thin physical barrier may be obtained when treating PP/oib-POSS, this being responsible for a limited delay in the weight

Table 9.3. *Fire performance of FRNH coaxial cables with EVA-ATH and EVA-ATH-organoclay outer sheaths*

UL-1666 Requirement	EVA-ATH Compound	EVA-ATH-Organoclay Compound
Maximal temperature at 12 ft: <850°F	1,930°F	620°F
Maximal flame height <12 ft	>12 ft	6 ft

loss in air [57]. On the other hand, SEM/EDS results evidence that thermal stabilization of PP/Al-POSS cannot be ascribed to a pure physical effect by inorganic accumulation on the surface, confirming the occurrence of chemical effects induced by the metal-POSS embedded in the PP matrix [59].

9.2.4 Multicomponent FR Systems: Polymer Nanocomposites Combined with Additional Materials

9.2.4.1 Polymer-Clay with Conventional FR Additive Nanocomposites

A combination of 65 wt% alumina trihydrate (ATH) and 35 wt% of a high-level accepting polymer matrix, such as poly(ethylene-co-vinyl acetate)(EVA), must often be used for cable outer sheaths [63]. The performances of two polymer blends were compared by Beyer [64, 65]. One blend was made from 65 wt% ATH and 35 wt% EVA (28 wt% vinyl acetate content), and a second blend was made from 60 wt% ATH, 5 wt% organoclay, and 35 wt% EVA. Both blends were investigated with TGA in air and by cone calorimeter at 50 kW/m^2. TGA in air clearly showed a delay in degradation by a small amount of organoclay [64, 65]. The char of the EVA-ATH-organoclay compound generated by the cone calorimeter was very rigid and showed only a few small cracks. On the other hand, the EVA-ATH compound was much less rigid and with many big cracks. This could be the reason the PHRR of the nanocomposite was reduced to 100 kW/m^2 as compared to 200 kW/m^2 for the EVA-ATH compound. To obtain the same decrease for PHRR by using just ATH in EVA, the content of ATH has to be increased to 78 wt% within the EVA-ATH system. To maintain 200 kW/m^2 at a sufficient peak heat release rate level, ATH could be decreased from 65 wt% to 45 wt% by the presence of 5 wt% organoclay within the EVA-ATH-clay system. The reduction of ATH resulted in improved mechanical and rheological properties of the EVA-based nanocomposite.

This research resulted in the commercial product FRNH cables passing UL-1666 [65]. The outer sheath was based on a nanocomposite with an industrial EVA-ATH-organoclay composition. The analogous coaxial cable was tested with an outer sheath based on EVA-ATH. Table 9.3 shows the results of the fire performance of these two coaxial cables [65]. The insulating and no burning char reduced the emission of volatile produce from polymer degradation into the flame area, and thus minimized the maximal temperature and height of the flames.

Table 9.4. *Peak of heat release rates at 35 kW/m² for various compounds with organoclays and multi-walled carbon nanotubes*

Sample	EVA[a] (parts resin)	MWCNT/ Purified (phr)	MWNCT/ Crude (phr)	Organoclay[b] (phr)	PHRR (kW/m²)
EVA[a]	100.0	–	–	–	580
1[c]	100.0	2.5	–	–	520
2	100.0	5.0	–	–	405
3	100.0	–	–	2.5	530
4	100.0	–	–	5.0	470
5[c,d]	100.0	2.5	–	2.5	370
6a[c]	100.0	–	5.0	–	403
6b[e]	100.0	–	5.0	–	405

[a] Eaxorene UL-00328 with 28 wt% vinyl acetate content.
[b] Nanofil 15.
[c] The screw velocity was 45 rpm and the mass temperature was 136°C.
[d] The nanotubes and the organoclay were premixed before addition.
[e] The screw velocity was 120 rpm and the mass temperature was 142°C.

9.2.4.2 Polymer-Carbon Nanotubes with Conventional FR Additive Nanocomposites

EVA with purified MWNT, crude MWNT, and organoclay were investigated by Beyer [64–69] by cone calorimeter at 35 kW/m². All compounds were melt-blended in a Brabender mixing chamber. It is evident from the results in Table 9.4 that all the filled polymers had improved flammability properties. For the EVA and EVA-based nanocomposites containing 2.5 phr of filler, the PHRR decreased as follows: EVA > organoclays ~ purified MWNTs. For EVA and EVA-based composites containing 5.0 phr of filler, the PHRR decreased as follows: EVA > organoclays > purified MWNTs = crude MWNTs. Crude MWNTs were as effective as purified MWNTs in the reduction of PHRR! Increasing the filler content from 2.5 phr to 5.0 phr caused an additional flame-retardant effect that was most significant when purified or crude MWNTs were used.

A synergistic effect for flame retardancy between MWNTs and organoclays was observed for a nanocomposite containing 2.5 phr of purified MWNTs and 2.5 phr of organoclays [59]. The latter sample was found to be the best flame-retardant compound. The variation of screw velocity from 45 rpm (sample A) to 120 rpm (sample B) did not change the flame-retardant properties for composites containing 5.0 phr of crude MWNTs. There was also no reduction in time to ignition for the EVA MWNT-based composite, in contrast to the EVA organoclay-based composite (Figure 9.10).

9.2.4.3 Polymer-Clay and Polymer-Carbon Nanotubes with Conventional FR Additive Nanocomposites

Johnson and Koo et al. [70] studied the combined effects of MMT clay, CNF, and FR additives on the mechanical and flammability properties of these PA11 nanocomposites for application in selective laser sintering (SLS) for additive manufacturing. Test specimens of PA11 containing various percentages of intumescent FR additive,

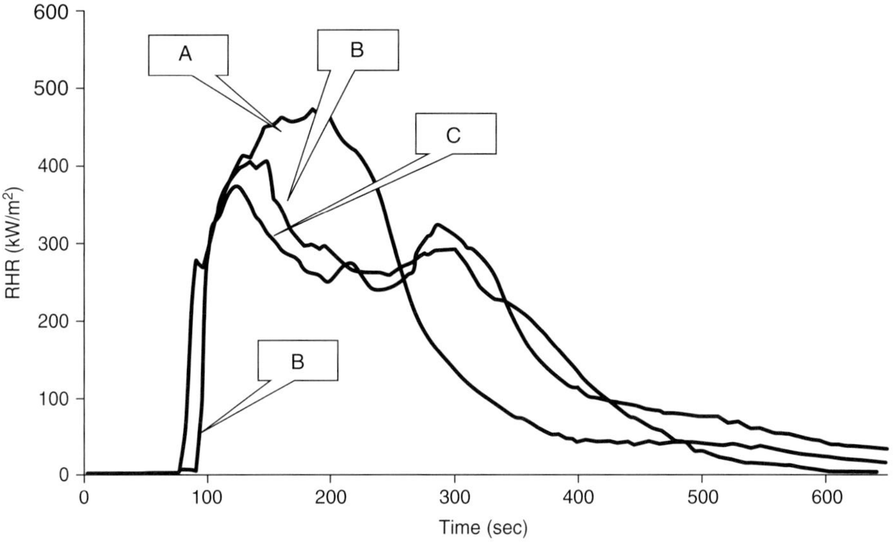

Figure 9.10. Heat release rates at 35 kW/m² for various EVA-based materials: (a) EVA+5.0 phr organoclay; (b) EVA+5.0 phr pure MWNT; and (c) EVA+2.5 phr organoclay+2.5 phr pure MWNT. EVA is Escorene UL-00328 with 28 wt% vinyl acetate content; organoclay is Nanofil 15 [66].

montmorillonite (MMT) clay, and carbon nanofiber (CNF) were prepared via the twin-screw extrusion technique. Izod impact testing, tensile testing, and SEM analysis were used to characterize mechanical properties. UL-94 and SEM analysis of char surfaces were used to characterize the flammability properties of these materials. Results were analyzed to determine any synergistic effects among the additives to the material properties of PA11.

The results for the UL-94 vertical burn test of their materials are shown in Table 9.5. The baseline PA11 material received a V-2 rating because it produced burning drips but was still self-extinguishing in less than 30 seconds. The two formulations with MMT clay and CNF but no FR additive failed the test. In the PA11/2.5% MMT/2.5% CNF formulation, the material burned quickly once ignited and produced numerous viscous drips. The PA11/3.5% MMT/3.5% CNF also burned readily once it was ignited, but it only produced drips in 6 out of 10 tests and the drips were very viscous. All of the formulations with FR additive passed the test with V-0 ratings. They all immediately self-extinguished and produced no drips during testing. Photos of the post-test specimens are show in Figure 9.11 [70].

The results for the formulations without FR additive are as expected. It is not unusual for nanomaterial additives alone to fail to improve a material's flammability rating and they are typically most beneficial when used to improve the effectiveness of conventional FR additives [8]. The results for the remaining formulations are quite promising. In previous tests, where MMT clay and CNF were combined with intumescent FR additives separately, such significant improvements in flammability properties were not observed with such low percentages of FR. In those experiments there were no formulations that were able to reach V-0 rating with only 15% FR,

Table 9.5. *UL-94 vertical burn ratings for PA11 composites*

Formulation	MMT	CNF	FR	UL-94 Rating
Neat PA11	0%	0%	0%	V-2
#1	2.5%	2.5%	0%	Fail
#3	2.5%	2.5%	15%	V-0
#4	2.5%	2.5%	20%	V-0
#2	3.5%	3.5%	0%	Fail
#5	3.5%	3.5%	15%	V-0
#6	3.5%	3.5%	20%	V-0

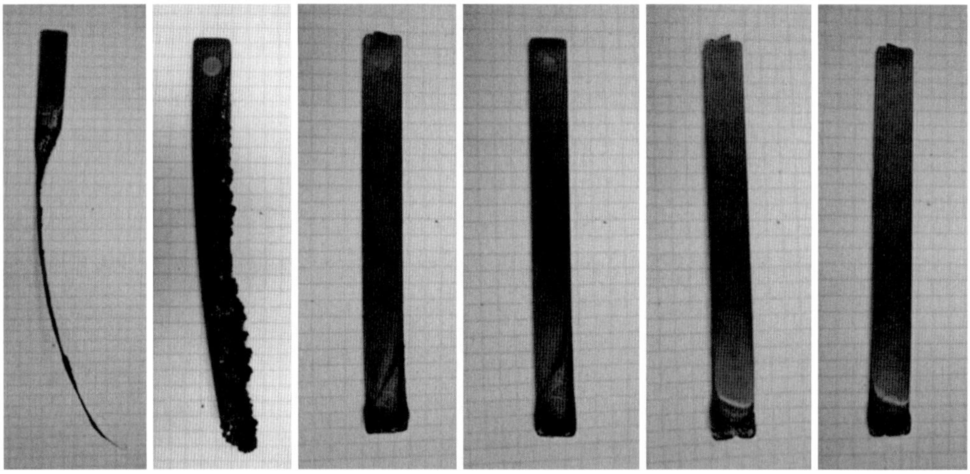

Figure 9.11. Post-test photos of UL-94 test specimens. Formulations 1 and 2 (left-most two photos) can be seen to have burned significantly, while formulations 3–6 (in ascending order from left to right) show no significant burn damage.

and there were several formulations that failed to reach it even at 20% FR [6, 8]. These results suggest that there is a synergistic effect between MMT clay and CNF in improving the flammability of PA11 composites when combined with conventional FR additives.

Gao et al. reported on the clay- and CNT-enhanced EVA composites. Ternary and binary PNCs were investigated through the cone calorimeter test [71]. Figure 9.12 depicts the heat release rate of three types of EVA composites containing CNT, organoclay, and mixtures of clay and MWNT [71]. The results indicate that CNT/clay nanofillers were more effective in reducing heat release rate than the binary PNCs were. Figure 9.13 shows the morphology of the chars of the PNCs from the cone calorimeter test, the natural burning experiment, and the burning test at 600°C for 20 minutes, respectively [71]. The ternary composites show the well-formed char and the smooth surface (Figure 9.13a). In the case of natural burning (Figure 9.13b), it presents reduced crack extent compared to the binary composite. Figure 9.13c exhibits the similar morphology. Hence, the adding of clay into the CNT-EVA composites enhances the flammability properties. This could be explained by the fact

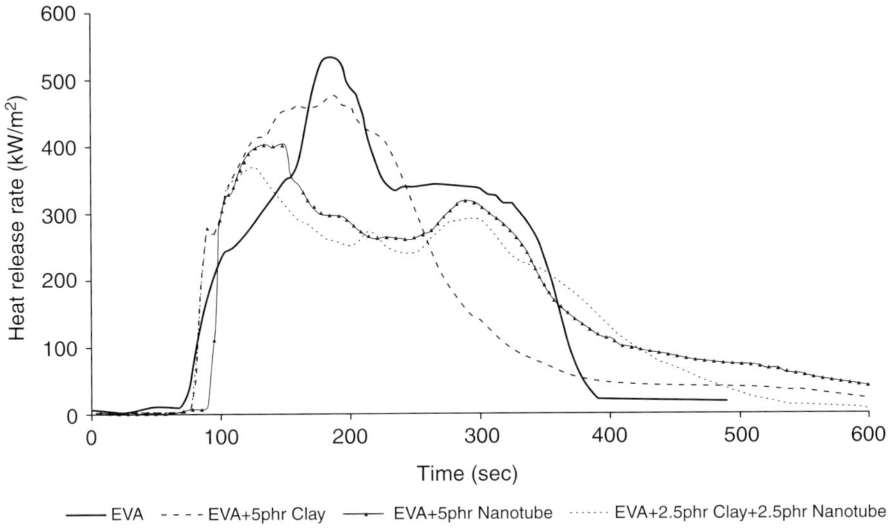

Figure 9.12. The relationship between the rate of heat release rate and burning time for EVA and EVA composites containing 5% multi-walled carbon nanotubes, 5% organoclay, and the combination of 2.5% nanotubes and 2.5% clay, respectively. The data were obtained using cone calorimetry under 35 kW/m^2 heat flux [71].

that the clay promotes the formation of graphitic structure and increases the resistance of char oxidation.

Peeterbroeck et al. studied and announced the effect of the CNT/organo-modified clay on the thermal and mechanical properties of EVA nanocomposites [72]. Table 9.6 presents the result of the cone calorimeter test. The largest reduction of the PHRR (36%) was obtained for the ternary PNCs. Figure 9.14 shows the thermal degradation property using the remaining weight and its derivative [72]. The positive effect on EVA thermal stabilization resulting from the organo-modified clay/CNT is illustrated in Figure 9.14c. Thus, the synergetic influence of the ternary system was observed for the flame retardant properties.

Mechanical properties of the binary and ternary system were evaluated by tensile tests. Binary PNCs present the better strength compared to the ternary ones [72]. However, the ternary PNCs still exhibit the more enhanced mechanical properties than the pure EVA does.

9.3 Flame-Retardant Mechanisms of Polymer Nanocomposites

The effect of nanomaterials on the thermal and flammability properties of polymer nanocomposites has been widely investigated, and different mechanisms have been proposed [73], but a comprehensive understanding of this phenomenon is not yet available. The enhancement of the thermal and flammability properties for the PNCs is very dependent on the matrix polymer as well as how the different types of nanomaterials are used to fabricate the PNCs. For polymer-clay nanocomposites, it is hypothesized that migration of the nanoclays to the surface of the matrix

Table 9.6. *Comparison between binary and ternary PNC*

Sample	Ignition Time (s)	PHRR (kW/m^2)	Crack density
EVA	84	580	+++
EVA /OM clay (4.8%)	67	470	+
EVA /CNT (4.8%)	83	405	++
EVA /OM clay (2.4%)/ CNT (2.4%)	71	370	0

+++, almost complete elimination/volatilization of EVA matrix; ++, formation of holes; +, presence of cracks; 0, almost no visible cracks on the surface.

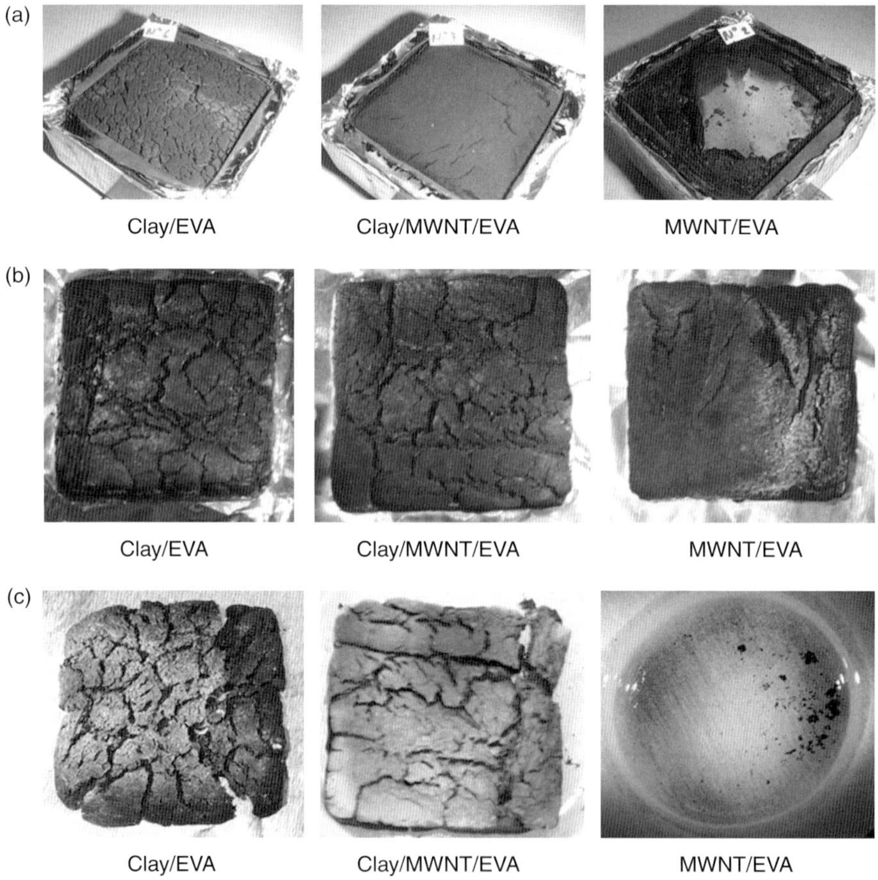

Figure 9.13. The morphology of the chars produced from clay/EVA, clay/MWNT/EVA, and MWNT/EVA nanocomposites following (a) the cone calorimeter test, (b) the natural burning, and (c) the burning at 600°C for 20 minutes [71].

polymer occurs during combustion [73, 74]. One mechanism is based on the force of the volatile products produced in the bulk of the polymeric material. The numerous rising bubbles during combustion push the nanoclays upward from the burning area. Another mechanism is based on the recession of polymer from the surface during

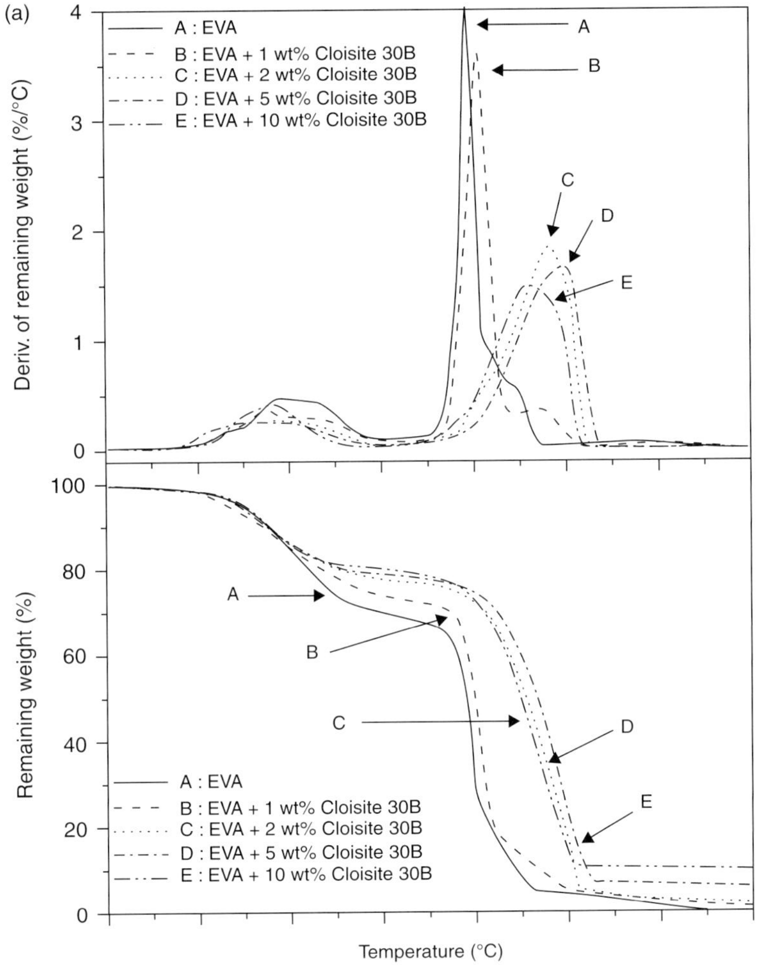

Figure 9.14. Thermogravimetric analysis of (a) binary EVA nanocomposites based on various amounts of Cloisite 30B, (b) binary EVA nanocomposites based on various amounts of crude and purified MWNTs, and (c) EVA nanocomposites filled with 3 wt% Cloisite 30B and either 0, 0.5, or 1.0 wt% purified MWNTs under air flow at 20K/min [72].

pyrolysis, leaving the de-wetted nanoclay platelets behind. The decomposition of the organic modified in the clay will lead to aggregation of the clay. The migration of clay onto the surface was also confirmed by XRD and attenuated total reflectance Fourier transform infrared (ATR-FTIR) measurements on the isothermally heated samples [75]. As observed, a minimum amount of clay platelets is needed to be present in order to have an effective char formation on the surface of the polymer nanocomposite.

It is proposed that the accumulation of clay at the surface of the degraded polymer acts as a barrier (ceramic shield) to both mass transport of degradation productions and thermal transfer of energy from the heat source to the polymer underneath the clay platelets. Figure 9.15 shows a schematic of the mechanism on how the clay barrier forms [73]. This clay platelet barrier slows the rate of mass loss

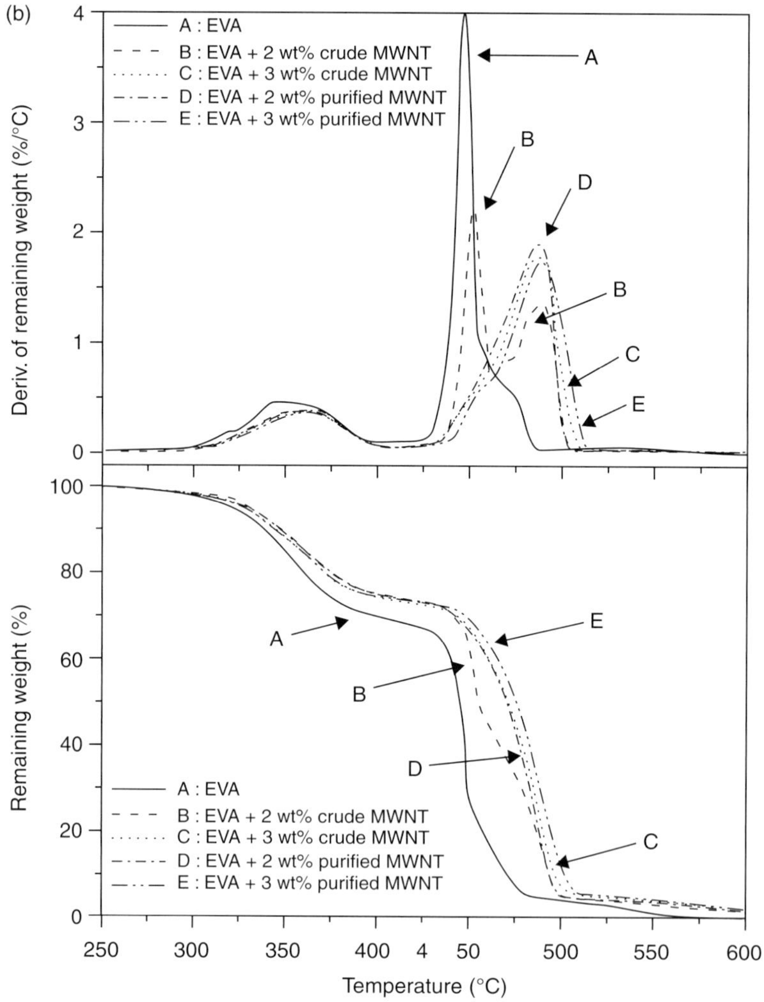

Figure 9.14 (*continued*)

and will in turn lower the HRR in a cone calorimeter experiment. The barrier function of the clay platelets can provide thermal insulation for the condensed phase and thus increased the thermal stability of the matrix polymer. No char formation were observed by burning the polyamide 6, but a significant amount of char was formed for polyamide 6 with 2 wt% and 5 wt% clay. If the char is continuous, it will be more effective in enhancing flame retardancy [76].

The protective mechanism for carbon nanofibers and carbon nanotubes is similar, since they are both nanoscale-diameter carbon fibers, but they are different from MMT nanoclay platelets. The carbonaceous residues from CNFs and CNTs form a continuous carbon network; as a result, the CNFs and CNTs need to be uniformly distributed in the polymer nanocomposite. As the polymer degrades, a carbon network is formed. As observed, a minimum amount of carbon network needs to be present in order to have an effective char formation on the surface of the polymer nanocomposite. Formation of tougher char layer during the combustion process can

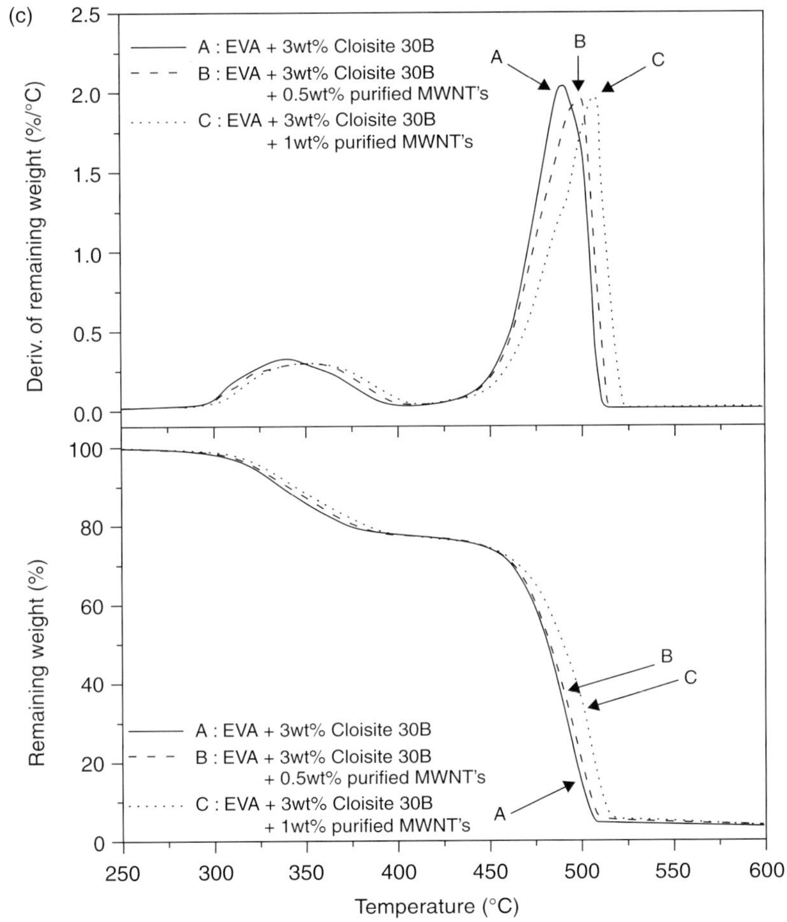

Figure 9.14 (*continued*)

also be observed [76] for PP/MWNT and is essential to enhance the flammability properties of the polymer nanocomposites. As the surface of the matrix polymer is the primary pathway by which the nanomaterials can escape from the polymer, crack-free and carbonaceous network barriers can significantly shield the heat and oxygen from the matrix polymer. This mechanism will reduce the heat release rate by slowing down the combustion process [76].

Fina, Bocchini, and Camino prepared an excellent article describing the Thermal behavior of nanocomposites and their fire test performance [77]. The authors addressed the mechanisms of thermal evolution of nanocomposites occurring during combustion, reviewed the present state of the art, and provided a critical discussion of experimental results from different widely used fire tests. The mechanisms involved in PNCs fire retardancy are based on (a) *physical effects* – formation of a ceramic protective floccules on the sample surface acting as a thermal insulation layer; and (b) *chemical effects* – formation of a carbonaceous char catalyzed by the nanofiller during combustion.

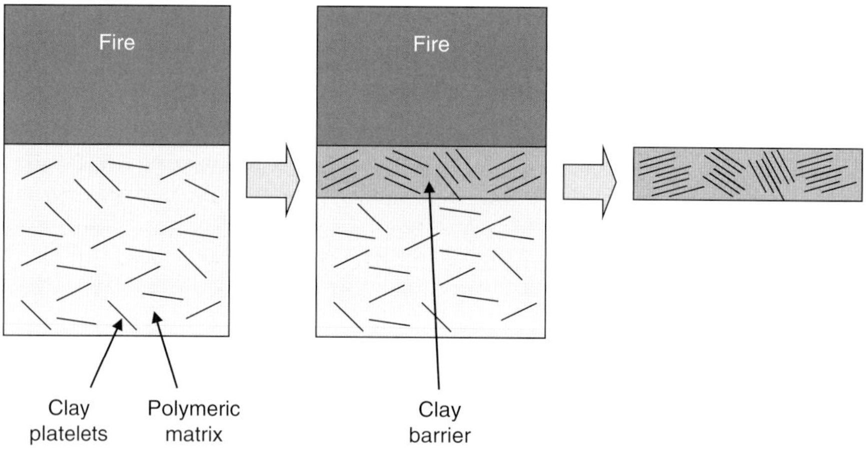

Figure 9.15. Schematic showing the mechanism of how the clay barrier forms.

Combustion of PNCs in different testing scenarios was examined by Fina et al. [77, 78]. At the present, the three different fire tests used extensively are: vertical UL 94, limiting oxygen index (LOI), and cone calorimeter. UL 94 and LOI are generally referred to as flammability tests, in terms of the materials' capability to ignite and to self-sustain a flame, thus representing scenarios in which the material is at the fire's origin. The cone calorimeter test is representing a forced combustion, in which the material is forced to burn under controlled heat flux. This test addresses the ignition time, the rate of combustion, and total heat released. These flammability and combustion tests also differ in terms of specimen positioning. With PNCs, the differences in performance obtained in the flammability and forced combustion test are usually very significant, having caused ongoing discussions about the actual effectiveness of nanoparticles as fire retardants.

Typical nanocomposites' cone calorimeter results reported in the literature show a decrease in peak HRR of about 50–70% compared to reference unfilled polymers, with either nanoclays, carbon nanotubes, POSS, or other inorganic nanoparticles [77]. The general observation is that the presence of dispersed nanoparticles switches the typical non-charring behavior of most thermoplastic polymers to that of charring materials to develop a surface protective layer when exposed to heat. This often leads to the formation of a solid residue at the end of the test, with variable degree of compactness. The compactness of the residue during burning is generally related to the efficiency in HRR reduction: the higher the compactness, the lower HRR. The simple comparison of the peak HRR values to quantitatively assess relative fire risk of virgin polymer and corresponding PNC may be misleading. In most cases, the reduction of combustion rate of polymer matrix in PNCs is explained with the barrier effect on nanoparticle and with polymer char accumulation on the surface of the burning sample by mechanisms described previously in this section.

Among flammability tests, the most widely used are the UL 94 and LOI tests. UL 94 was originally developed for devices and appliances. From the scientific point of view, there is little knowledge about the phenomenological behavior of material in the simple pass/fail result. To better describe the flaming behavior during the test, such as the height of the flames, the burning rate, and the end-of-test residue, complementary information can be obtained via LOI test, which represents the opposite configuration compared to the vertical UL 94, the specimen being ignited from the top. Relatively few papers reported an LOI test for PNCs, generally showing either a decrease or no significant increase in the LOI values compared to their reference polymers. The LOI setup appears to be critical for the production of an effective ceramic protective layer on the material surface, as the specimen below the flame front is not licked by the flame and there is no external irradiation. This prevents the formation of the protective char-ceramic surface layer on the material prior to being reached by the flame front.

Nanocomposites thus take a step toward the reduction of fire risk and hazard for polymers because they avoid the spread of the flame by way of dripping and reduce the rate of combustion. For the first time in polymer fire retardant records we have polymeric materials available in which fire risk and hazard are reduced with simultaneous improvement of other polymer material properties and using an environmentally friendly technology. Most of the fire retardants used now have to be loaded into polymers in relatively large amount (10–70 wt%), with negative effects on the polymers' physical and mechanical properties and on the environment.

9.4 Concluding Remarks and Trends in Polymer Nanocomposites

Nanotechnology has a very bright future in improving flame retardancy of polymeric materials. Although the nanomaterial by itself cannot achieve the fire properties required in industrial applications, the presence of the nanomaterial will enhance other desirable properties, such as mechanical, barrier, or electrical properties, and in combination with other conventional flame retardants can provide better flame retardant solutions. Finding new ways of applying nanomaterials cost-effectively in polymers with desired properties, such as thermal and flammability, have been of increasing interest and is highly challenging. Adaption of nanomaterials in polymers to promote thermal and flame retardancy as well as other desired properties is warranted [79].

To promote thermal and flame retardancy and multifunctional properties of the polymer, one trend is to combine two or more nanomaterials into the polymer matrix to achieve a PNC that will consist of multi-components. If synergism exists between these nanomaterials, the resulting PNC will exhibit better performance. Currently, MMT clay is considered as one component of the combined nanomaterials due to its low cost, availability, and its comprehensive performance. Of increasing interest is the use of halloysite nanotubes and synthetic clays, such as fluorinated synthetic mica, magdiite, and layered double hydroxides (LDHs) to fabricate PNCs. However, the high cost and limited sources of the synthetic clays has been an obstacle for the wide usage of these nanomaterials.

Another trend is the use of other conventional FR additives, which are micron-sized, to enhance the flame retardancy of the polymer matrix, such as using an additive that can form a glassy barrier to seal the crack in the char, resulting in a self-extinguishing PNCs. One common and practical way is to use nanomaterials in the presence of conventional flame retardants, such as halogen-, or phosphorus-based, or as a nano-sized component in a conventional FR system to achieve a flame retardant polymeric composite. This includes nanoparticles, such as nano-$Mg(OH)_2$ and nano-$Al(OH)_3$, obtained from conventional inorganic FR additives that have original particle size in microns. Again, clay will be of great interest because of its excellent performance, such as the anti-dripping and charring properties, and/or the synergism with many conventional FR additives.

Other nanomaterials, such as CNFs, CNTs, POSS, and NGPs, although expensive when comparing with MMT clay when used alone in the polymer matrix, can also be combined with conventional FR system to reduce the raw material cost and to improve the flame retardancy of the resulting PNCs.

Due to the thermal instability of the organic modifier of the nanoclays, it is most likely that advances in the future will occur with advanced polymers that can be processed below about 200°C. As new surfactants are developed and commercialized, flame-retardant polymer nanocomposites that can be used for polymers that must be processed at higher temperatures may be developed. As the prices of CNFs, CNTs, and POSS are coming down, they may be used as the next generation of flame-retardant nanomaterials. The PNC does offer many advantages – enhanced barrier, flame, mechanical, and electrical properties – so there will be used in multifunctional materials and the enhanced flame retardancy will be an added "bonus."

Alongi et al. [80] conducted a comprehensive review on current emerging techniques to impart flame retardancy to fabrics. The authors review the state of the art on the novel and emerging techniques recently developed in the textile field for conferring flame-retardant properties to natural and synthetic fibers. The current emerging techniques reviewed by the authors are showing very significant potentialities for effectively improving the FR properties of different natural and synthetic fibers/fabrics. They may represent a possible efficient alternative to the conventional RF systems, some of which (e.g., halogenated products) are encountering limitation in use due to their suspected/assessed toxicity and poor eco-sustainability. The in situ assembly of nanocoatings on the fabric surface through nanoparticle adsorption, layer by layer deposition, sol-gel, and dual-cure processes, or plasma treatments has clearly demonstrated that nanotechnology represents a valued tool for the textile field. The unusual use of proteins and nucleic acids may overcome the current limitations of flame retardants toward novel, sustainable, and green products.

Industrial development of PNCs seems now under way because since the appearance of the first clay-based nanocomposites two decades ago, a wide selection of nanofillers has become available on a commercial basis. A parallel extensive worldwide research effort has strongly improved our knowledge about preparation and characterization of PNCs, now available for industrial exploitation.

9.5 Questions and Problems

1. How does the nanomaterial dispersion affect the flammability properties of polymer nanocomposites in general? Provide examples to illustrate your claims.
2. How is the flammability properties of polymer-clay nanocomposites influenced by the type of clay, surface treatment of clay, and how clay is processed in the polymer matrix? Provide examples.
3. Enhancement of flammability properties of thermoplastic-based nanocomposites using one-, two-, and three-dimensional scaled nanomaterials were discussed in this chapter. Which types of nanomaterials are more effective for thermoplastic-based nanocomposites? Provide examples and summarize the mechanical properties of these thermoplastic-based nanocomposites.
4. Enhancement of flammability properties of thermoplastic elastomer-based nanocomposites using one-, two-, and three-dimensional scaled nanomaterials were discussed in this chapter. Which types of nanomaterials are more effective for thermoplastic elastomer-based nanocomposites? Provide examples and summarize the mechanical properties of these thermoplastic elastomer-based nanocomposites.
5. Enhancement of flammability properties of thermoset-based nanocomposites using one-, two-, and three-dimensional scaled nanomaterials were discussed in this chapter. Which types of nanomaterials are more effective for thermoset-based nanocomposites? Provide examples and summarize the mechanical properties of these thermoset-based nanocomposites.
6. What will be the effects of these multicomponent FR systems with two or more nanomaterials and conventional FR additive systems? What will be a good strategy to optimize the flammability properties in these polymer nanocomposites? Summarize the mechanical properties of these multicomponent FR nanocomposite systems.
7. What are the flame-retardant protective mechanism for these one-, two-, and three-dimensional scaled nanomaterials, as well as the hybrid FR systems? Summarize and compare them.
8. Are these any correlations or observed trends when these polymer nanocomposites were evaluated with different flammability test methods?

REFERENCES

1. T. J. Pinnavaia and G. W. Beall (Eds.) (2000). *Polymer-Clay Nanocomposites*. West Sussex, England: John Wiley & Sons.
2. L. A. Utracki (2004). *Clay-Containing Polymeric Nanocomposites*. Shropshire, England: Rapra Technology Limited.
3. S. N. Bhattacharya, R. K. Gupta, and M. R. Kamal (2007). *Polymeric Nanocomposites-Theory and Practice*. Düsseldorf: Carl Hanser Verlag.
4. J. W. Gilman, T. Kashiwagi, and J. D. Lichtenhan (1997). Nanocomposites: a revolutionary new flame retardant approach. *SAMPE Journal* 33(4), 40–46.
5. H. A. Patel, S. Bocchini, A. Frache, and G. Camino (2010). Platinum nanoparticle intercalated montmorillonite to enhance the char formation of polyamide 6 nanocomposites. *Journal of Materials Chemistry* 20, 9550–9558.

6. S. C. Lao, C. Wu, T. J. Moon, J. H. Koo, A. Morgan, et al. (2009). Flame-retardant Polyamide 11 and 12 nanocomposites: Thermal and flammability properties. *Journal of Composite Materials* 43(17), 1803–1813.
7. S. C. Lao, W. Yong, K. Nguyen, T. J. Moon, J. H. Koo, et al. (2010). Flame-retardant Polyamide 11 and 12 nanocomposites: Processing, morphology and mechanical properties. *Journal of Composite Materials* 44(25), 2933–2951.
8. S. C. Lao, J. H. Koo, et al. (2011). Flame-retardant Polyamide 11 nanocomposites: Further thermal and flammability studies. *Journal of Fire Sciences* 29, 1-20. doi: 10.1177/0734904111404658.
9. C. K. Katsoulis, E. Kandare, and B. K. Kandola (2009). Thermal and Fire Performance of Flame-Retarded Epoxy Resin: Investigating Interaction between Resorcinol Bis (Dipenyl Phosphate) and Epoxy Nanocomposites. In *Fire Retardancy of Polymers-New Strategies and Mechanisms*, T. R. Hull and B. K. Kandola (Eds.). Cambridge: The Royal Society of Chemistry, pp. 184–205.
10. M. Zammarano (2007). Thermoset Fire Retardant Nanocomposites. In *Flame Retardant Polymer Nanocomposites*, A. Morgan and C. Wilkie (Eds.). Hoboken, NJ: John Wiley & Sons, pp. 235–285.
11. Z. Wang and T. J. Pinnavaia (1998). Hybrid organic-inorganic nanocomposites: exfoliation of magadiite nanoclays in an elastomeric epoxy. *Polymer* 10, 1820–1826.
12. C. S. Triantafillidis, P. C. LeBaron, and T. J. Pinnavaia (2002). Thermoset epoxy-clay nanocomposites: The dual role of alpha, omega-diamines as clay surface modifiers and polymer curing agents. *Journal of Solid State Chemistry* 167, 354–362.
13. G. Camino, G. Tartaglion, A. Frache, C. Manferti, and G. Costa (2005). Thermal and combustion behavior of layered silicate-epoxy nanocomposites. *Polymer Degradation & Stability* 90, 354–362.
14. M. Hussain, R. J. Varley, Z. Mathys, Y. B. Cheng, and G. P. Simon (2004). Effect of organophosphorus and nano-clay materials on the thermal and fire performance of epoxy resins. *Journal of Applied Polymer Science* 91, 1233–1253.
15. A. Hartwig, D. Purtz, B. Schartel, M. Bartholmai, and M. Wenschuh-Josties (2003). Combustion behavior of epoxide based nanocomposites with ammonium and phosphonium bentonites. *Macromolecular Chemistry & Physics* 204, 2247–2257.
16. J. W. Gilman, T. Kashiwagi, M. Nyden, J. E. T. Brown, C. L. Jackson, et al. (1990). Flammability Studies of Polymer-Layered Silicate Nanocomposites: Polyolefin, Epoxy, and Vinyl Ester Resins. In *Chemistry and Technology of Polymer Additives*, S. Ak-Malaika, A. Colovoy, and C. A. Wilkie (Eds.). Malden, MA: Blackwell Science, pp. 249–265.
17. J. H. Koo, K. Nguyen, J. C. Lee, W. K. Ho, M. C. Bruns, and O. A. Ezekoye (2010). Flammability studies of a novel class of thermoplastic elastomer nanocomposites. *Journal of Fire Sciences* 28(1), 49–85.
18. J. H. Koo, O. A. Ezekoye, J. C. Lee, W. K. Ho, and M. C. Bruns (2011). Rubber-Clay Nanocomposites Based on Thermoplastic Elastomers. In *Rubber Clay Nanocomposites*, M. Galimberti (Ed.). New York: John Wiley & Sons, pp. 489–521.
19. A. F. Avila, E. C. Dias, D. T. Lopes da Cruz, M. I. Yoshida, A. Q. Bracarense, et al. (2010). An investigation on graphene and nanoclay effects on hybrid nanocomposites post fire dynamic behavior. *Materials Research* 13, 143–150.
20. J. H. Koo, H. Leo, W. Clay, and J. Conaway (2011). Methodology to Evaluate Epoxy Nanocomposites for Fire Protection Application. *AIAA Paper No. 2011-1799*, Reston, VA.
21. Y. Guo, C. Bao, L. Song, B. Yuan, and Y. Hu (2011). In situ polymerization of graphene, graphite oxide, and functionalized graphite oxide into epoxy resin

and comparison study of on-the-flame behavior. *Industrial and Engineering Chemistry Research* 50, 7772–7783.

22. Z. Wang, X. Z. Tang, Z. Z. Yu, P. Guo, H. H. Song, and X. S. Du (2011). Dispersion of graphene oxide and its flame retardancy effect on epoxy nanocomposites. *Chinese Journal of Polymer Science* 3, 368–376.
23. A. L. Higginbotham, J. R. Lomdea, A. B. Morgan, and J. M. Tour (2009). Graphite oxide flame-retardant polymer nanocomposites. *Applied Materials Interfaces* 1, 2256–2261.
24. T. Kashiwagi, F. Du, K. I. Winey, K. M. Groth, J. R. Shield, et al. (2005). Flammability properties of polymer nanocomposites with single-walled carbon nanotubes: Effects of nanotube dispersion and concentration. *Polymer* 46, 471–481.
25. Kashiwagi, T. (2007). Progress in Flammability Studies of Nanocomposites with New Types of Nanoparticles. In *Flame Retardant Polymer Nanocomposites*, A. B. Morgan and C. A. Wilkie (Eds.). Hoboken, NJ: Wiley & Sons, pp. 285–324.
26. S. Bocchini, A. Frache, G. Camino, and M. Claes (2007). Polyethylene thermal oxidative stabilization in carbon nanotubes based nanocomposites. *European Polymer Journal* 43, 3222–3235.
27. S. Butler, G. Kim, J. H. Koo, et al. (2011). Polyamide 11-Halloysite Nanotube Nanocomposites: Mechanical, Thermal, and Flammability Characterization. Proceedings of the *2011 SAMPE ISTC*, SAMPE, Covina, CA.
28. S. C. Lao, M. F. Kan, C. K. Lam, J. H. Koo, T. Moon, et al. (2010). Polyamide 11-Carbon Nanotubes Nanocomposites: Processing, Morphological, and Property Characterization. Proceedings of the *2010 Solid Freeform Fabrication Symposium*, The University of Texas at Austin, Austin, TX.
29. C. J. T. Landry, B. K. Coltrain, M. R. Landry, J. J. Fitzgerald, and V. K. Long (1993). Poly(vinyl acetate) silica filled materials: Material properties of in-situ vs. fumed silica particles. *Macromolecules* 26, 3702–3712.
30. P. Hajji, L. David, J. F. Gerard, J. P. Pascault, and G. Vigier (1999). Synthesis, structure, and morphology of polymer-silica hybrid nanocomposites based on hydroxyethl methacrylate. *Journal of Polymer Science B* 37, 3172–3187.
31. Y. Ou, F. Yang, and Z. Z. Yu (1998). New conception on the toughness of nylon 6/silica nanocomposite prepared via in situ polymerization. *Journal of Polymer Science B* 36, 789–795.
32. E. Reynaud, T. Jouen, C. Gauthier, G. Vigier, and J. Varlet (2001). Nanofiller in polymeric matrix: A study on silica reinforced PA6. *Polymer* 42, 8759–8768.
33. G. H. Hsiue, W. J. Kuo, Y. P. Huang, and R. J. Jeng (2000). Microstructural and morphological characteristics of PS-SiO_2 nanocomposites. *Polymer* 41, 2813–2825.
34. Y. L. Liu, C. Y. Hsu, W. L. Wei, and R. J. Jeng (2003). Preparation and thermal properties of epoxy-silica nanocomposites from nanoscale colloidal silica. *Polymer* 44, 5159–5167.
35. T. Kashiwagi, J. W. Gilman, K. M. Butler, R. H. Harris, and J. R. Shields (2000). Flame retardant mechanism of silica gel/silica. *Fire Materials* 24(6), 277–289.
36. T. Kashiwagi, J. R. Shields, R. H. Harris, and R. D. Davis (2003). Flame-retardant mechanism of silica: Effect of resin molecular weight. *Journal of Applied Polymer Science* 87, 1541–1553.
37. F. Yang and G. L. Nelson (2004). PMMA/silica nanocomposite studies: Synthesis and properties. *Journal of Applied Polymer Science* 91, 3844–3850.
38. T. Kashiwagi, A. B. Morgan, J. M. Antonucci, M. R. Van Landingham, R. H. Harris, et al. (2003). Thermal and flammability properties of a silica-poly(methylmethacrylate) nanocomposite. *Journal of Applied Polymer Science* 89, 2072–2078.
39. F. Yang, R. Yngard, and G. L. Nelson (2005). Flammability of polymer-clay and polymer-silica nanocomposites. *Journal of Fire Science* 23, 209–226.

40. S. C. Lao, J. H. Koo, T. J. Moon, B. Hadisujoto, W. Yong, et al. (2009). Flammability and Thermal Properties of Polyamide 11-alumina Nanocomposites. Proceedings of the *2009 Solid Freeform Fabrication Symposium*, University of Texas at Austin, Austin, TX.
41. M. Rallini, M. Monti, M. Natali, J. M. Kenny, and L. Torre (2011). Alumina Nanoparticles as a Filler of Carbon Fibre/Epoxy Composites for Improved Fire Resistance. *Proceedings of the 2011 SAMPE ISTC*. Covina, CA: SAMPE.
42. J. Kalfus and J. Jancar (2010). Effect of particle size on the thermal stability and flammability of $Mg(OH)_2$/EVA nanocomposites. *Computer Interfaces* 17, 689–703.
43. K. Vesely, J. Rychly, M. Kummer, and J. Jancar (1990). Flammability of highly filled polyolefins. *Polymer Degradation and Stability* 30, 101–105.
44. J. Rychly, K. Vesely, E. Gal, M. Kummer, J. Jancar, and L. Rychla (1990). Use of thermal methods in the characterization of the high-temperature decomposition and ignition of polyolefins and EVA copolymers filled with $Mg(OH)_2$, $Al(OH)_3$, and $CaCO_3$. *Polymer Degradation and Stability* 30, 57–62.
45. H. Gui, X. H. Zhang, J. M. Gao, W. F. Dong, Z. H. Song, et al. (2007). An EVA/unmodified nano-magnesium hydroxide/silicone rubber nanocomposite with synergistic flame retardancy. *Chinese Journal of Polymer Science* 25, 437–440.
46. J. P. Ly and W. H. Liu (2007). Flame retardancy and mechanical properties of EVA nanocomposites based on magnesium hydroxide nanoparticles/microcapsulated red phosphorus. *Journal of Applied Polymer Science* 105, 333–340.
47. S. Mishra, S. H. Sonawane, R. P. Singh, A. Bendale, and K. Patil (2004). Effect of nano-$Mg(OH)_2$ on the mechanical and flame-retarding properties of polypropylene composites. *Journal of Applied Polymer Science* 94, 116–122.
48. Q. Zhang, M. Tian, Y. Wu, G. Lin, and L. Zhang (2004). Effect of particle size on the properties of $Mg(OH)_2$-filled rubber composites. *Journal of Applied Polymer Science* 94, 2341–2346.
49. G. Song, S. Ma, G. Tang, Z. Yin, and X. Wang (2010). Preparation and characterization of flame retardant form-stable phase change materials composed by EPDM, paraffin and nano-magnesium hydroxide. *Energy* 35, 2179–2183.
50. H. Cao, H. Zheng, J. Yin, Y. Lu, S. Wu, et al. (2010). $Mg(OH)_2$ complex nanostructures with superhydrophobicity and flame retardant effects. *Journal of Physics and Chemistry C* 114, 17362–17368.
51. C. B. Patil, U. R. Kapadi, D. G. Hundiwale, and P. P. Mahulikar (2008). Effect of nano-magnesium hydroxide on mechanical and flame-retarding properties of SBR and PBR: A comparative study. *Polymer-Plastics Technology and Engineering* 47, 1174–1178.
52. R. Suihkonen, K. Nevalainen, O. Orell, M. Honkanen, L. Tang, et al. (2012). Performance of epoxy filled with nano- and micro-sized magnesium hydroxide. *Journal of Materials Science* 47(3), 1480–1488.
53. Hybrid Plastics, Inc. "Home page." Last modified 2010. http://www.hybridplastics.com.
54. R. A. Mantz, P. F. Jones, K. P. Chaffee, J. D. Lichtenhan, J. W. Gilman, et al. (1996). Thermolysis of polyhedral oligomeric silsequioxane (POSS) macromers and POSS-siloxane copolymers. *Chemistry of Materials* 8, 1250–1259.
55. J. J. Schwab, and J. D. Lichtenhan (1998). Polyhedral oligomeric silsesquioxane (POSS)-based polymers. *Applied Orgaometallic Chemistry* 12, 707–713.
56. T. Kashiwagi and J. W. Gilman (2000). Silicon-Based Flame Retardants. In *Fire Retardancy of Polymeric Materials*, A. F. Grand and C. A. Wilkie (Eds.). New York: Marcel Dekker, pp. 353–389.
57. A. Fina, D. Tabuani, A. Frache, and G. Camino (2005). Polypropylene-polyhedral oligermeric silsesquioxanes (POSS) nanocomposites. *Polymer* 46, 7855–7866.

58. F. Baldi, F. Bignotti, A. Fina, D. Tabuani, and T. Ricco (2007). Mechanical characterization of polyhedral oligomeric silsesquioxane/polypropylene blends. *Journal of Applied Polymer Science* 105, 935–943.
59. A. Fina, S. Bocchini, and G. Camino (2008). Catalytic fire retardant nanocomposites. *Polymer Degradation and Stability* 93, 1647–1655.
60. O. Monticelli, A. Fina, A. Ullah, and P. Waghmare (2009). Preparation, characterization, and properties of novel PSMA-POSS systems by reactive blending. *Macromolecules* 42, 6614–6623.
61. A. Fina, D. Tabuani, T. Peijs, and G. Camino (2009). POSS grafting on PPgMA by on-step reactive blending. *Polymer* 50, 218–226.
62. A. Fina, O. Monticelli, and G. Camino (2010). POSS-based hybrids by melt/reactive blending. *Journal of Materials Chemistry* 20, 9297–9305.
63. M. J. Herbert and S. C. Brown (1992). New Developments in ATH Technology and Applications. Paper presented at *Flame Retardants '92 Conference*, London, CT, January 12–13, pp. 100–119.
64. G. Beyer (2007). Flame Retardant Properties of Organoclays and Carbon Nanotubes and Their Combinations with Alumina Tri Hydrate. In *Flame Retardant Polymer Nanocomposites*, A. B. Morgan and C. A. Wilkie (Eds.). Hoboken, NJ: Wiley & Sons, pp. 163–190.
65. G. Beyer (2005). Flame retardancy of nanocomposites: from research to technical products. *Journal of Fire Science* 23, 75–87.
66. G. Beyer (2002). Carbon nanotubes as flame retardants for polymers. *Fire Materials* 26, 291–293.
67. G. Beyer (2002). Improvements of the Fire Performance of Nanocomposites. Paper presented at the *13th Annual BCC Conference on Flame Retardancy for Polymers*, Stamford, CT, June 3–6.
68. G. Beyer (2005). Filler blend of carbon nanotubes and organoclays with improved char as a few flame retardant system for polymers and cable application. *Fire Materials* 29, 61–69.
69. G. Beyer, F. Gao, and Q. Yuan (2005). A mechanistic study of fire retardancy of carbon nanotube/ethylene vinyl acetate copolymers and their clay composites. *Polymer Degradation and Stability* 89, 559–564.
70. B. Johnson, E. Allcorn, M. G. Baek, and J. H. Koo (2011). Combined Effects of Montmorillonite Clay, Carbon Nanofiber, and Fire Retardant on Mechanical and Flammability Properties of Polyamide 11 Nanocomposites. Proceedings of the *2011 SAMPE ISTC*, SAMPE, Covina, CA.
71. F. Gao, G. Beyer, and Q. Yuan (2005). A mechanistic study of fire retardancy of carbon nanotube/ethylene vinyl acetate copolymers and their clay composites. *Polymer Degradation and Stability* 89(3), 559–564.
72. M. Peeterbroeck, J. Alexandre, J. B. Nagy, et al. (2004). Polymer-layered silicate–carbon nanotube nanocomposites: Unique nanofiller synergistic effect. *Composites Science and Technology* 64(15), 2317–2323.
73. J. W. Gilman (2007). Flame Retardant Mechanism of Polymer-Clay Nanocomposites. In *Flame Retardant Polymer Nanocomposites*, A. B. Morgan and C. A. Wilkie (Eds.). Hoboken, NJ: Wiley & Sons, pp. 67–87.
74. T. Kashiwagi, R. H. Harris, X. Zhang, R. M. Briber, B. H. Cipriano, et al. (2004). Flame retardant mechanism of polyamide 6-clay nanocomposites. *Polymer* 45, 881–891.
75. M. Lewin, E. M. Pearce, K. Levon, et al. (2006). Nanocomposites at elevated temperatures: migration and structural changes. *Polymer for Advanced Technology* 17, 226–234. doi:10.1002/pat.684.
76. T. Kashiwagi, F. Du, K. I. Winey, K. M. Groth, J. R. Shields, et al. (2005). Flammability properties of polymer nanocomposites with single-walled carbon

nanotubes: Effects of nanotube dispersion and concentration. *Polymer* 46, 471–481.
77. A. Fina, S. Bocchini, and G. Camino (2009). Thermal Behavior of Nanocomposites and Fire Testing Performance. In *Fire and Polymers*, C.A. Wilkie, A. B. Morgan, and G. L. Nelson (Eds.). Washington, DC: American Chemical Society, pp. 10–24.
78. A. Fina, F. Canta, A. Castrovinci, and G. Camino (2009). Significant Assessment of Nanocomposites' Combustion Behaviour by the Appropriate Use of the Cone Calorimeter. In *Fire Retardancy of Polymer – New Strategies and Mechanism*, T. R. Hull, and B. K. Kandola (Eds.). Cambridge: RSC Publishing, pp. 147–159.
79. A. B. Morgan and C. A. Wilkie (2007). Practical Issues and Future Trends in Polymer Nanocomposite Flammability Research. In *Flame Retardant Polymer Nanocomposites*, A. B. Morgan and C. A. Wilkie (Eds.). Hoboken, NJ: Wiley & Sons, pp. 355–399.
80. J. Alongi, F. Carosio, and G. Malucelli (2013). Current emerging techniques to impart flame retardancy to fabrics: An overview. *Polymer Degradation and Stability* 106, 138-149. doi: 10.1016/j.polymdegradstab.2013.07.012.

10 Ablation Properties of Polymer Nanocomposites

10.1 Introduction

A comprehensive review was conducted on the research and development (R&D) of polymer nanocomposites used as thermal protection systems (TPS) for reentry vehicles as well as rocket nozzle and internal insulation materials for propulsion systems. This chapter summarizes the significant R&D efforts on the studies of nanostructured ablative materials by scientists from China, India, Iran, Italy, Pakistan, South Korea, and the United States for the past decade. Thermoset, thermoplastics, elastomers, and thermoplastic elastomers were used as polymer matrices. Nanomaterials, such as montmorillonite (MMT) organoclays, carbon nanofibers (CNFs), Polyhedral Oligomeric Silsesquioxanes (POSS®), nano-silicas (*n*-silicas), nano-alumina (*n*-alumina), multiwall carbon nanotubes (MWNTs), and others, as well as classical fillers, such as carbon black and ceramics were incorporated into the different resin matrices as nano- or micro-sized fillers. Conventional fibers, such as asbestos, carbon, glass, Kevlar, phenolic, and silica were used as reinforcements. Ablation mechanisms were studied and proposed by the researchers on their nanostructured ablatives.

Thermal protection materials are required to shield structural components of space vehicles during the reentry stage, solid rocket motors (SRMs), liquid rocket engines, and missile launching systems. A thorough literature survey was conducted by Koo et al. [1–4] to review the development of these thermal protection materials for different military and aerospace applications. The series of literature surveys were grouped into: (a) numerical modeling [1], (b) materials thermophysical properties characterization [2], (c) experimental testing [3], and (d) polymer nanocomposites [4].

The introduction of inorganic nanomaterials as additives into polymer matrix systems has resulted in polymer nanocomposites (PNCs) exhibiting multifunctional, high-performance polymer characteristics beyond what traditional polymer composites possess [5–13]. Multifunctional features attributable to PNCs consist of improved thermal resistance, flame resistance, ablation resistance, moisture resistance, decreased permeability, chemical resistance, charge dissipation, and electrical/thermal conductivity [5–13]. Through control/adjustment of the additive at the nanoscale level one is able to maximize property enhancement of selected polymer systems to meet or exceed the requirements of current military, aerospace, and commercial applications.

One of the most important and pioneering research studies on the use of the montmorillonite (MMT) for poly (caprolactam) (nylon 6) was performed by Vaia et al. [14] in 1999 using 2 wt% and 5 wt% loaded layered silicate/poly (caprolactam) nanocomposite. These materials were synthesized via in situ polymerization of ε-caprolactam in the presence of dispersed organically modified MMT nanoclay. This concept was further developed by many researchers worldwide using different types of polymers, such as nylon, phenolic, elastomer, and thermoplastic elastomer, with different types of nanomaterials, such as nanoclay (*n*-clay), POSS, carbon nanofiber, multi-walled carbon nanotube, nano-alumina, nano-silica, nano-graphene, nano-kaolinite, nano-ceramic, and others. Traditional reinforcements, such as asbestos, carbon, glass, Kevlar, phenolic, and silica were also used. This chapter reviews the literature on how researchers processed, characterized, and tested their polymer nanocomposites as thermal protection materials for low- and high-energy hyperthermal environments.

10.2 Behavior of Thermal Protection Materials

Polymeric composites have been used as ablative thermal protection systems (TPS) for a variety of military and aerospace applications. Thermal protection materials, such as carbon phenolic and carbon-carbon composites are widely used for spacecraft heat shields during the reentry stage, thermal protection of missile launching system, and in SRMs as nozzle materials. Thermosetting polymer matrix composites are constructed of layers of woven carbon, glass, quartz, or silica fibers impregnated with a resin matrix. Phenolic resin is the most commonly used polymer matrix for ablatives. Phenolic resins undergo facile carbonization into carbon/carbon composites at elevated temperatures. These ablative materials are exposed to a thermochemical flow and subjected to high temperatures in excess of 3,000°C with very high heating rates.

Thermochemical ablation refers to the phenomenon of surface recession of an ablative due to severe thermal attack by an external heat flux. The initial heat transfer into the ablative occurs by pure conduction, and the resulting temperature rise causes material expansion (swelling) which may be attributed to pure thermal expansion as well as vaporization of any traces of moisture and other volatiles generated by resin decomposition. When the material reaches a sufficiently high temperature, thermochemical degradation or pyrolysis of the polymer matrix begins. The pyrolysis reactions result in the production of decomposition gases and solid carbonaceous char residue. The thermal expansion and the disappearance of solid material due to the decomposition result in an increase in porosity and permeability of the polymeric material. Thus pyrolysis gases begin to escape through the polymeric material. For the thermal protection to be effective, all these reactions must be strongly endothermic. The gases that flow through the char structure remove energy

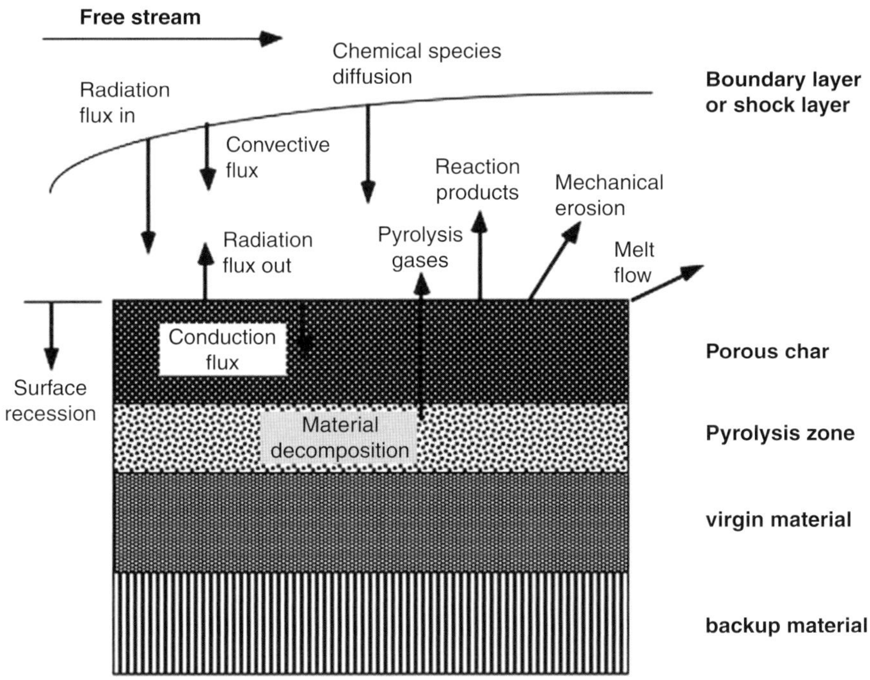

Figure 10.1. Energy accommodation mechanisms of ablative TPS materials [15].

by convention, thus attenuating the conduction of heat to the reaction zone. This phenomenon is depicted in Figure 10.1 by Laub et al. [15].

10.3 Polymer Nanocomposites Review

There are only a small number of research groups globally who are investigating polymer nanocomposites as advanced thermal protection materials:

- In China: Beijing University of Chemical Technology, Harbin Institute of Technology, National University of Defense Technology, and Xi'an Jiaotong University
- In India: Defence Research and Development Organisation (DRDO) and Indian Institute of Technology
- In Iran: Islamic Azad University and Tarbiat Modares University
- In Italy: University of Perugia, Politecnico di Torino, and University of Rome
- In South Korea: Gyeongsan National University
- In Pakistan: National University of Sciences and Technology
- In the United States: Air Force Research Laboratory (AFRL), Mississippi State University, Texas State University–San Marcos, and The University of Texas at Austin

It is more meaningful to subdivide these studies in terms of the types of polymer matrix used in each group's studies. Table 10.1 shows a summary of the different

Table 10.1. *Summary of the different polymer matrix used by the various research groups studying ablation using polymer nanocomposites*

	Polymer Matrix/Institution		Nanoadditive/Reinforcement	Remarks
10.3.1.1	Nylon 6	AFRL, United States	MMT organoclay	Improvement in ablative performance of nanocomposites is associated with the char-forming characteristics of the polymer nanocomposites.
10.3.1.2	Poly(phenylene oxide)/polystyrene (PPO/PS)	Politecnico di Torino, Italy	Modified clays (MMT, hydrotalcite, sepiolite) and equiaxial, oxide and non-oxide powders	Sepiolite-added and alumina-based PPO composites showed the most promising behaviors. Only TGA, DTGA, cone calorimeter, and microstructural characterization were conducted in this study.
10.3.2.1	Different Polymers	AFRL, United States	MMT organoclay	Focused on polymer/layered silicate materials for use in space applications.
10.3.3.1	EPDM	AFRL, United States	POSS	Detailed the strategy employment of using sacrificial nanocomposites along with applied research in the areas of SRM insulation and space-survival materials.
10.3.3.2	EPDM	The University of Texas at Austin, United States	Carbon nanofibers (CNFs)	Detailed TGA data and SEM micrographs were reported on different heating rates.
10.3.3.3	EPDM	Harbin Institute, China	Organo-MMT (OMMT) with Kevlar fibers	Up to 11 phr (parts per hundred parts of EPDM) OMMT in EPDM erosion rate decreased, beyond it ablation resistance decreased.
10.3.3.4	EPDM	Indian Institute of Technology, India	Polyimide micro-particles with nano-silica in maleic anhydride grafted EPDM	Erosion rate at 3 MW/m^2 is 0.14–0.17 mm/s and heat of ablation for the corresponding sample was 4,700 cal/g. Not much details on ablation data were reported.
10.3.3.5	EPDM	University of Perugia, Italy	Nano-silica, aramid pulp, phenolic fibers, and silica fibers	Silica fibers yielded the greatest residual mass during TGA and the greatest mass loss during ablation testing. The aramid pulp had the lowest in-depth temperatures. The amount of fibers used should be increased for better char and more conclusive results.
10.3.3.6	EPDM	National University of Sciences and Technology, Pakistan	MWCNT and carbon black	Backface temperature decreased significantly, ablation rate decreased sharply at low loadings, thermal conductivity increased, ultimate tensile strength and elastic modulus increased, and hardness increased with increased loadings of MWCNTs

10.3.4.1	NR	Tarbiat Modares University, Iran	MMT organoclay	Improvement in aging hardness, curing time, mechanical, flammability properties by the addition of 3 wt% of MMT clay.
10.3.4.2	Hydrogenated nitrile butadiene rubber (HNBR)	Beijing University of Chemical Technology, China	Fumed silica, organically modified montmorillonite (OMMT), and expanded graphite (EG)	OMMT was found to be a better additive for ablation resistance than fumed silica and EG.
10.3.5.1	Thermoplastic polyurethane (TPU)	The University of Texas at Austin, United States	MMT organoclay, CNF, and MWNT	MMT organoclay-based TPU nanocomposite (TPUN) has the best ablation performance, followed by MWNT-based TPUN.
10.3.5.2	Silicone rubber	National University of Defense Technology, China	Zirconium carbide (ZrC) and zirconia (ZrO_2) were incorporated into silicone rubber with SiO_2 and carbon fibers	40 phr (parts per hundred parts of rubber) of ZrO_2, ablation rate was reduced by 72% and the mass loss dropped by 38%. 40 phr of ZrC reduced the ablation rate by 40% and the mass loss by 32%.
10.3.6.1	Phenolic	Mississippi State University, United States	CNF with carbon fibers	The nanoscale dimensions of the VGCF causes major changes in the heat transfer rates and affects the resulting combustion/decomposition chemistry.
10.3.6.2	Phenolic	The University of Texas at Austin, United States	MMT organoclay, POSS, and CNF with and without carbon fibers	High loading of MMT organoclay in phenolic, low loading of POSS in phenolic impregnated with carbon fibers, and high loading of CNF in phenolic without carbon fibers all gave better ablation performance than did the carbon phenolic baseline.
10.3.6.3	Phenolic	Harbin Institute and National University of Defense Technology, China	POSS with carbon fibers; flake graphite reinforced barium-phenolic resin composite	Phenolic-POSS-CF produced the best charred surface, hence has a best ablation performance. Composite with flake graphite diameter of about 0.5 mm has the lowest ablation rate.
10.3.6.4	Phenolic	Xi'an Jiaotong University, China	Graphene oxide (GO) dispersed in phenolic with no fiber reinforcement	The GO particles were found to disperse well into the phenolic and produced a moderate increase in char yield and thermal stability even at low loadings, but no ablation testing was conducted.
10.3.6.5	Phenolic	DRDO, India	Nano-silica with carbon fibers	Ablation resistance increased with the nano-silica content up to 2.0 wt%; beyond it, ablation resistance decreased.
10.3.6.6	Phenolic	DRDO, India	Zirconia (ZrO_2) and MWCNTs with carbon fabric	No ablation resistance decrease resulted from using ZrO_2 and MWCNT coated on carbon fabric to produce composites.

(*continued*)

Table 10.1 (*continued*)

	Polymer Matrix/Institution		Nanoadditive/Reinforcement	Remarks
10.3.6.7	Phenolic	Tarbiat Modares University, Iran	MMT organoclay with asbestos fibers	The 6 wt% MMT with asbestos fibers has the best ablation performance.
10.3.6.8	Phenolic	Tarbiat Modares University, Iran	Kaolinite and graphite nanopowder without carbon fibers. Micron-sized kaolinite, nano-sized kaolinite, and nano-sized graphite powders with chopped carbon fibers	The phenolic nanocomposites without carbon fiber reinforcements have ablation rates of 0.0184 mm/s for neat phenolic resin, 0.0164 mm/s for 3 wt% graphite, and 0.0153 mm/s for 5 wt% graphite. The 9 wt% graphite with carbon fibers decreased ablation rate (measured in g/s and not in mm/s) and thermal diffusivity coefficient by 10%, and 62%, respectively.
10.3.6.9	Phenolic	Islamic Azad University, Iran	Nano-silica with chopped carbon fibers	Mass loss and ablation rates showed greater improvement at higher loadings of nano-silica.
10.3.6.10	Phenolic	University of Perugia, Italy	Nano-silica, carbon black, and MWNT with E-glass fibers	The synergy between traditional and nano-sized fillers was investigated.
10.3.6.11	Phenolic	Gyeongsang National University, South Korea	MWNTs with chopped carbon fibers and woven carbon fibers	Ablation resistance was observed with the Ph/carbon nanotubes, chopped carbon fibers, and woven carbon fibers composites.
10.3.6.12	Phenolic	Texas State University–San Marcos, United States	MWNT with carbon fibers	Mass loss, recession rate, and in-depth temperature all decreased with the increase of MWNT wt% (up to 2 wt%).
10.3.6.13	Phenolic	University of Rome, Italy	Graphitic felt and graphitic foam	All felt-based samples passed undamped the longest oxyacetylene exposure tests.

polymer systems, types of nanomaterials, and conventional fiber reinforcements used in their particular applications [4].

10.3.1 Thermoplastic Nanocomposite Studies

10.3.1.1 Nylon-Clay Nanocomposite Studies by AFRL, United States

The ablative performance of poly(caprolactam) (nylon 6) nanocomposites was examined by Vaia et al. of Air Force Research Laboratory (ARFL) [14]. A relatively tough, inorganic char forms during the ablation of these nylon-clay nanocomposites, resulting in at least an order-of-magnitude decrease in mass loss (erosion) rate relative to the neat polymer. This occurs for a small amount of 2 wt% (~0.8 vol%) exfoliated mica-type layered silicate. The nanoscopic distribution of silicate layers leads to a uniform char layer that enhances the ablative performance. The effect of concentration and size of the silicate layers as well as specific bonds between the polymer and the silicate on degradation kinetics and on the microstructure of the char layer has been well characterized.

Poly(caprolactam) (nylon 6, N, Ube) was compared to 2 wt% (NCH2) and 5 wt% (NCH5) layered silicate/poly(caprolactam) nanocomposites. These materials were synthesized by in situ polymerization of ε-caprolactam in the presence of dispersed organically modified montmorillonite. Additionally, a nominally 5 wt% layered silicate/poly(caprolactam) nanocomposite (NLS5) formed during melt processing of poly(caprolactam) (Capron B125WP, Allied Signal) and Cloisite 30A (organically modified MMT, Southern Clay Products) was examined. This exfoliated nanocomposite was fabricated using a co-rotating twin-screw extruder (MPC/V-30, Haake). In contrast to in situ polymerization of NCH2 and NCH5, melt processing of NLS5 is not expected to result in a large fraction of poly(caprolactam) chains tethered through covalent bonds to the organic modifiers on the silicate surface.

The ablation rate (erosion rate) of the materials was determined from exposure to combustion gases in a mock SRM firing apparatus. Fifty grams of pelletized resin were compression-molded into 7 in. (180 mm) by 3 in. (75 mm) triangular sections approximately 0.2 in. (5 mm) thick in an autoclave. To enhance retention of material during the test, a second series of plaques were produced by consolidating the resins into a 1/8 in. (0.32 mm) polyimide mat. In addition to the poly (caprolactam) nanocomposites, ethylene-propylene diene monomer (EPDM) with 30% silica filler, EPDM with 15% chopped Kevlar fiber (state-of-the-art (SOTA) internal insulation for SRMs), and phenolic impregnated canvas were also examined. Table 10.2 summarizes the char motor ablation rate results [14]. Ablation tests were performed in a 4 in. (10.2 cm) diameter char motor. Two triangular plaques were arranged along the inside of a tapered, rectangular chamber. The propellant consisted of 88% ammonium perchlorate with hydroxylated polybutadiene binder. Temperature of the combustion gases was approximately 3,000°C and the test duration was 8 seconds. The thickness of the remaining virgin material (after removal of the degraded char layer) was determined from the average of at least four measurements.

Table 10.2. *Summary of char motor ablation rate results*

	Erosion rate (mil/s)[a], ± 15%	
	Low mass flux (Mach 0.013, 12.8 m/s)	High mass flux (Mach 0.041, 39.6 m/s)
	(0.10 lb/in²s) (0.69 kPa/s)	(0.40 lb/in²s) (2.75 kPa/s)
N(poly(caprolactam))	nvmr	nvmr
NCH2	5	nvmr
NCH2, mat	4	9
NCH5	4	9[b]
NCH5, mat	3	10
NLS5	5	14
NLS5, mat	5	11
Phenolic canvas	7	8
EPDM/Kevlar	6	10
EPDM/30% silica	6	11

[a] mil/s = 0.0254 mm/s
[b] At 0.35 lb/in²s
nvmr: no virgin material remaining

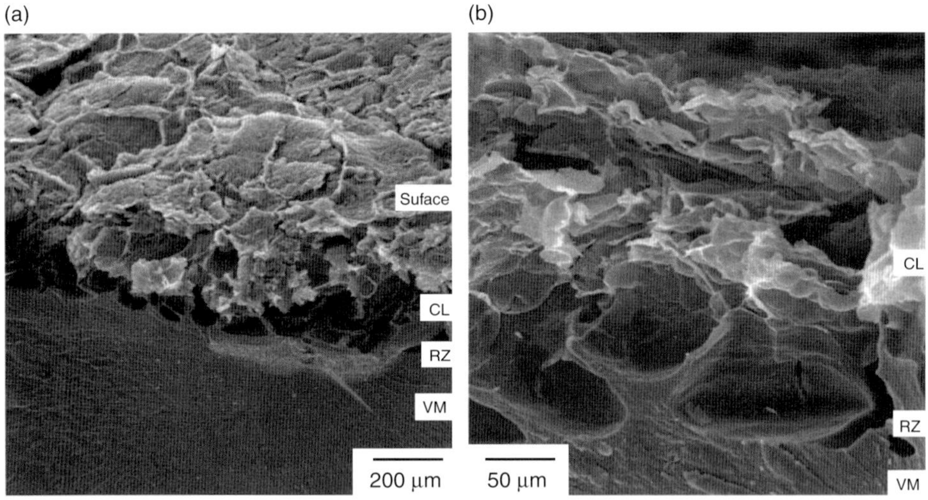

Figure 10.2. Cross-section SEM image of NCH5 ablative material, the charred layered silicate/poly (caprolactam) nanocomposite. In the (a) image, the char layer (CL), the reaction zone (RZ), and the virgin material (VM) are clearly visible. In the (b) image, the nanostructured nature of the char is displayed [14].

Since the presence of the layered silicate does not alter primary polymer decomposition in the reaction zone, the initial nanostructure of the dispersed silicate must influence the microstructure and composition of the char layer, thus enhancing the ablation performance. A series of scanning electron micrographs of the surface char region of several ablative samples were shown. Figure 10.2 shows the scanning electron micrographs of the surface char region of NCH5 ablative sample [14].

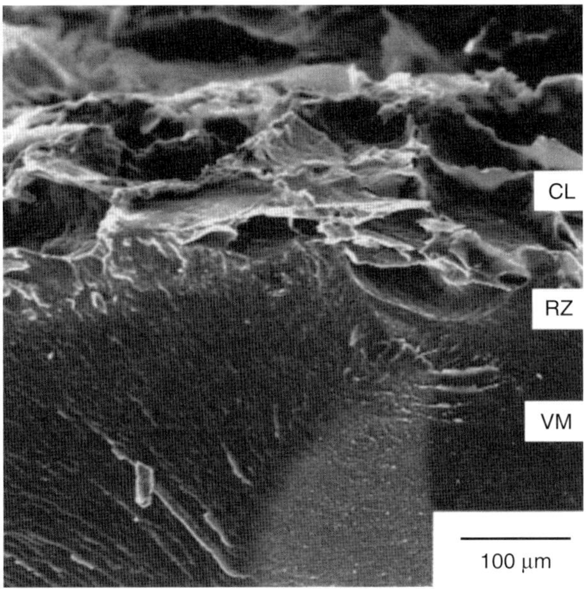

Figure 10.3. Scanning electron micrographs of surface region (tilted to observe cross-section and surface features) of NLS5 ablative sample (VM: virgin material, RX: reaction zone, CL: char layer) [14].

Figure 10.3 shows the tilted angle to observe the cross-section and surface features of the NLS5 ablative sample [14]. In these chars, a nanoscopic (10–20 nm thick) densely packed platelets morphology is observed (Figure 10.4) [14]. The characteristic ablation regions (virgin material, porous reaction zone, and dense char layer) are apparent in all samples. The microstructural features of these chars imply that increasing silicate content results in a tougher char. In these chars, nanoscopic (10–20 nm thick), densely packed plate-like morphology is observed. Increased silicate content results qualitatively in spatially more uniform and thicker char. Specific characteristics of the silicate, such as layer size, uniformity of layer distribution, and strength of the interphase, result in a finer surface texture for NCH5 char than NLS5 char. In general, the spatially uniform arrangement of the silicate layers on an ultra-fine (nanometer) level facilitates the formation of a spatially uniform inorganic char. At 5 wt% the layers are on average only 20–25 nm apart. Thus, the efficient distribution of the inorganic precursors necessary to form a uniform char layer is the key structural characteristic of the nanocomposites leading to the enhanced ablative performance at remarkable low additions to the reinforcing phase.

10.3.1.2 Poly(phenylene oxide)/Polystrene (PPO/PS) Studies by Politecnico de Torino, Italy

Lombardi and colleagues [16–18] investigated PPO-based composites for potential ablative TPS application. The first part of their studies focused on PPO composites containing layered modified clay, such as MMT, hydrotalcite, and sepiolite [16]. A commercial PPO/PS blend, Noryl® 914 resin (GE Plastics) based on polyphenylene oxide, PPO, blended with polystyrene, PS was used as the polymer matrix. It has a density of 1.09 g/cm^3 and a melting range of 280–310°C. The three layered silicates used were: Cloisite® 15A (C, Southern Clay Products), an MMT clay; Perkalite®

Figure 10.4. Bright field transmission electron micrograph of NCH% char [14].

A100 (H, AkzoNobel Polymer Chemicals), an organically modified hydrotalcite; and Sepiolite (S, Tolsa), a hydrated magnesium silicate.

Brabender internal mixer was used to melt-compound these nanomaterials into the PPO/PS matrix blends. Degradation of these polymer composites were studied using thermal analyses (TGA) and cone calorimeter tests. Thermal analyses were conducted in Ar heating up to 1,500°C with heating rate of 20°C/min, and DTA analyses were performed at up to 550°C with a heating rate of 10°C/min. Cone calorimeter tests were conducted with a heat flux of 35kW/m^2. SEM analyses revealed that all the composites still presented micron-sized aggregates platelets or needles. On the basis of the differential thermal analysis (DTA) results, PPO+S and PPO+C presented the best behavior for ablative materials, since they showed a more endothermic decomposition in comparison with the same temperature range of the neat polymer. Based on cone calorimeter results, PPO with C and S added presented a longer combustion time and a higher charring ability. Overall investigation showed the Sepiolite-added PPO composites can be selected as promising materials for innovative ablative shields.

In a second study [17, 18], equiaxial, oxide and non-oxide powders, such as magnesium aluminate spinel (SP), aluminum hydroxide (boehmite) (B), σ- and α-alumina, silicon carbide (SiC), silicon nitride (N), and molybdenum disilicide (M) were used as fillers incorporated into the PPO/PS matrix polymer. Nanometric and micrometric α-alumina and silicon carbide powders were utilized to investigate the effect of particle size. SEM analyses revealed all the composites still presented micron-sized aggregates platelets or needles. From DTA data it is evident that PPO decomposition was unaffected by oxide fillers, except for PPO+SP material. Among the non-oxide fillers, SiC decreased the material ability to absorb heat during decomposition. The alumina-based composites presented the most promising behavior. DTA results indicated that PPO+S and PPO+C presented the best behavior for ablative materials. They exhibited a more endothermic decomposition in comparison to the same temperature range of the neat polymer. Cone calorimeter results showed PPO combined with C and S presented a longer combustion time and a higher charring ability. Thermal behaviors of all fillers were summarized in this study [17, 18]. The overall investigation showed that the Sepiolite-added PPO composites can be selected as promising materials for innovative ablative shields. No experimental data to simulate typical high heat flux for ablative shields were reported.

10.3.2 Polymer-Clay Nanocomposite Studies

10.3.2.1 Polymer-Clay Nanocomposite Studies by AFRL, United States

Lincoln et al.'s study [19] focuses on polymer/layered silicate materials to highlight the potential of nanocomposites for use in space applications. They also focus on fabrication methods for nanocomposites by controlled dispersion of preformed constituents, such as passivated nanoparticles or organically modified mica-type silicates. Filler particles on the nanoscale can alter the fundamental structure of the matrix material, especially in the case of semi-crystalline polymers, and have a profound effect on the physical properties. The filler can also affect the materials' interaction with electromagnetic radiation. The improvement of nanocomposites over the neat polymer has the potential to offer space system designers a choice of a new, relatively less expensive material for use in their systems. These nanocomposites are polymer-containing materials. Lincoln et al. [19] discussed the benefits of using polymer nanocomposites for space applications, such as environmental stability, launch systems (solid rocket motors and liquid fuel tanks), and satellite bus (platform, deployable membranes, photovoltaic arrays, and solar concentrators). Nanocomposites are novel materials that potentially offer drastic cost savings for the space industry. They are lightweight, which is most desirable in the age where it cost thousands of dollars to place spacecraft into orbit. They are also easy to process and manufacture into useable parts, making them inherently less expensive and/or more reliable. The most attractive nanocomposite property is their potential to be space survivable. Nanocomposites can offer protection to vital systems and maintain spacecraft operational possibly for the duration of their lifetimes.

10.3.3 EPDM Nanocomposite Studies

10.3.3.1 EPDM-POSS Studies by AFRL, United States

Air Force Research Laboratory researchers have studied how the incorporation of POSS can be used for protective coatings in oxygen-rich environments. Philips et al. [20] detailed the strategic employment of using sacrificial nanocomposites along with applied research in the areas of SRM insulation and space survival materials. Proprietary POSS monomer (5–50%) was blended into the EPDM pre-polymer using a standard brabender, followed by extended curing in a standard oven. The flexible POSS-EPDM (ethylene-propylene-diene monomer rubber) polymer was then cut and fitted into conical phenolic sections for testing. POSS-dianiline and POSS-polyimide were synthesized and characterized. A film was cast from the resulting POSS-poly(amic acid) on a clean glass plate and placed in a clean oven under nitrogen to cure under specific conditions. The resulting films were dried at room temperature for 24 hours.

The testing of POSS-EPDM samples for ablation performance was achieved using standard tactical and boost SRM propellants. Both the insulation standard and the POSS-EPDM sample were present in the same SRM firing. Post-firing analyses of the samples involved the use of industrial standard removal of the formed char layer followed by measurements of the remaining polymer at various stages in the sample, which represent different mass fluxes that the sample was exposed. At higher mass flux, the POSS-EPDM sample performs significantly better than the standard, with ablation rates of up to 50% less than the industry standard. These results are very significant, since a 50% reduction in the ablation rate of current SOTA materials results in up to a 7.4% increase in payload to orbit. For SRM tests, analysis of post-fired samples showed the formation of a glass SiO_2 layer. Significant reduction in ablation rates was reproducibly demonstrated to the SOTA SRM insulation.

10.3.3.2 EPDM-CNF Studies by The University of Texas at Austin, United States

Polymer nanostructured ablatives were prepared by blending an EPDM elastomer polymer system with carbon nanofibers using standard brabender mixer. Thermogravimetric analyses (TGA) and SEM analyses were used to characterize these materials in a laboratory and a scaled solid rocket motor heating environment. Kevlar®-filled EPDM is used as motor insulation in solid rocket propulsion systems [21]. EPDM blends consisting of Kevlar fibers and carbon nanofibers were also included in this study.

Materials. Ethylene-propylene-diene monomer rubber (EPDM) filled with silica or Kevlar has been used in propulsion systems for both thermal and ablation protection. Carbon nanofibers were blended with Kevlar-filled EPDM.

Thermogravimetric Analysis (TGA). The TGA data compiled in these studies were collected using a TA Instruments Model TGA 2050. The TGA tests were conducted in an air environment, as compared to nitrogen, in order to best simulate the conditions inside solid rocket motors. The TGA has a heating rate range of 0–200°C/min. The maximum temperature that the TGA furnace can achieve is 1,000°C. The TGA was equilibrated at 30°C before ramping to 1,000°C at the specified heating rate.

TGA Tests on EPDM/Nanoparticles Systems. Mathias and Johnson [22] reported an extraordinary high heating rate (100,000°C/min) material property data of carbon cloth phenolic used in space shuttle reusable solid rocket motor nozzles. This

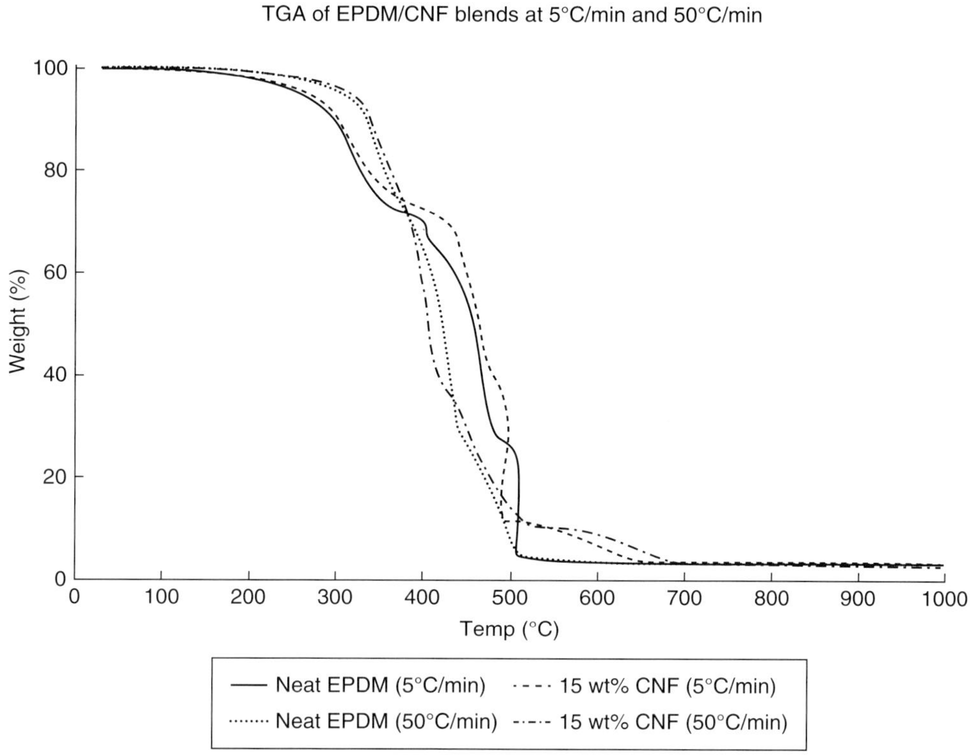

Figure 10.5. Mass loss of EPDM and EPDM/CNF composites at 5 °C/min and 50°C/min.

study prompted the examination of solid rocket motor insulation samples using TGA and SEM [23–25]. Two types of materials were included in this study, Kevlar-filled EPDM (11% Kevlar) and Kevlar-filled EPDM blended with 15 wt% CNF. A low heating rate of 5°C/min and a high heating rate of 50°C/min from TGA tests, as well as a calculated heating rate of 150,000°C/min from AFRL's four-inch char motor, were included in the study. Figure 10.5 shows the TGA of the two samples at 5°C/min and at 50°C/min [23]. The EPDM-Kevlar-15 wt% CNF samples had less mass loss than the EPDM-Kevlar sample at both heating rates. Behavior of EPDM-Kevlar and EPDM-Kevlar-CNF exhibited a strong dependence on heating rate.

SEM Analyses. Scanning electron microscopy (SEM) images were produced to study the surface as well as the cross-sectional behavior of the nanostructured materials under thermal stress. Figures 10.6 to 10.8 show the samples at heating rates of 5°C/min (TGA), 50°C/min (TGA), and 150,000°C/min (four-inch char motor), respectively [23]. At 5°C/min there was no EPDM charring observed in Figure 10.6, and the Kevlar fibers were visible. Slight charring of the EPDM and Kevlar fibers was again quite visible in Figure 10.7 [23]. In Figure 10.8, severe EPDM charring was quite visible in the rocket motor-fired samples [23]. The SEMs show substantial difference in material behavior under these three heating conditions. Figures 10.9 to 10.11 show the material behavior of the three samples at different heating rates of the EPDM-Kevlar-CNF samples [23].

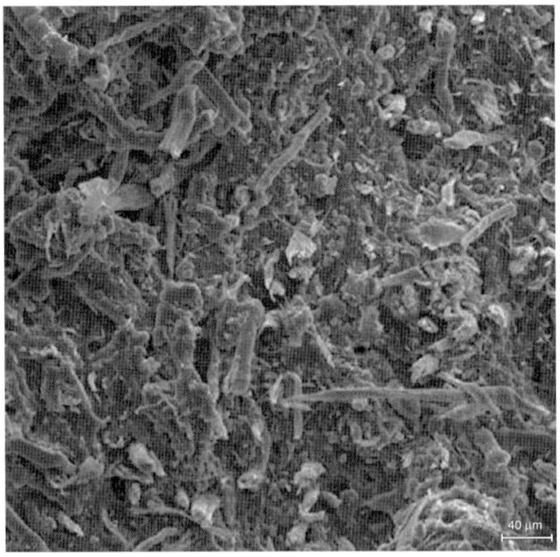

Figure 10.6. Scanning electron microscopy image of EPDM-Kevlar exposed to a heating rate of 5°C/min where the unit bar is 40 μm.

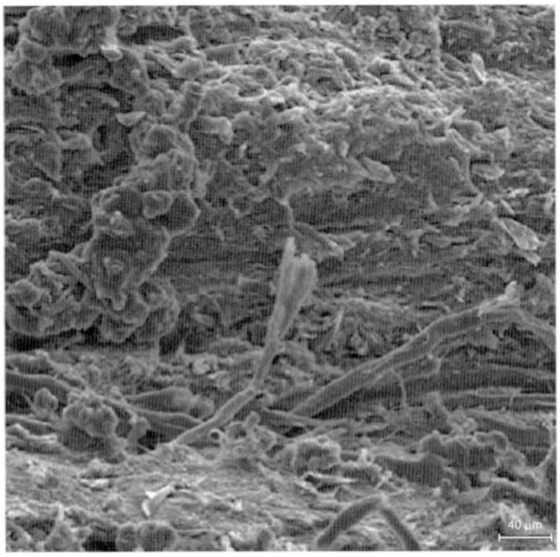

Figure 10.7. Scanning electron microscopy image of EPDM-Kevlar exposed to a heating rate of 50°C/min where the unit bar is 40 μm.

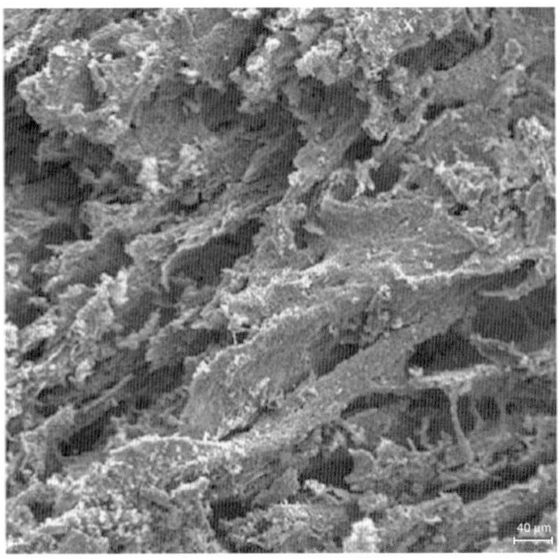

Figure 10.8. Scanning electron microscopy image of EPDM-Kevlar exposed to a heating rate of 150,000°C/min where the unit bar is 40 μm.

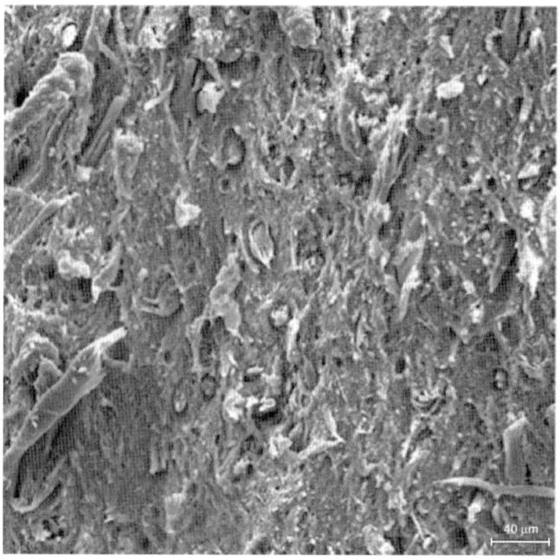

Figure 10.9. Scanning electron microscopy image of EPDM-Kevlar with carbon nanofiber exposed to a heating rate of 5°C/min where the unit bar is 40 μm.

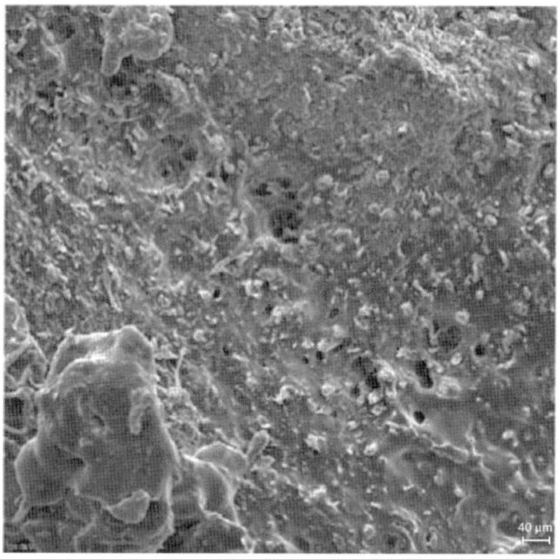

Figure 10.10. Scanning electron microscopy image of EPDM-Kevlar with carbon nanofiber exposed to a heating rate of 50°C/min where the unit bar is 40 μm.

Figure 10.11. Scanning electron microscopy image of EPDM-Kevlar with carbon nanofiber exposed to a heating rate of 150,000°C/min where the unit bar is 40 μm.

Conclusions. Thermogravimetric analysis may not be a suitable method to study polymer nanostructured ablatives because of the extremely high heating rate that occurs inside the solid rocket motor combustion chamber. The SEM images also confirmed that the polymer nanostructured ablatives behave very differently for TGA and scaled rocket motor samples.

10.3.3.3 EPDM-Clay Studies by Harbin Institute of Technology, China

Gao et al. [26] of China focused their attention on nanomodifying the SOTA SRM insulation material, EPDM/Kevlar, by adding an organo-montmorillonite (OMMT) to the EPDM matrix and examining the synergy of these nanofillers with the traditional components of the formulation. The content of OMMT was increased from 1 phr to 15 phr. After all the additives – namely the nano-silica, the short fibers, the flame retardants, the high char retention resin, and the peroxide – were well dispersed in the matrix with a two-roll mill for about 20 minutes, the materials were vulcanized (160°C and 15 MPa for 40 minutes). XRD analysis of the EPDM/OMMT was performed on the cured materials and TEM was used to further evaluate the morphology of nanocomposites. The mechanical properties of the produced materials, mainly the elongation at break, were measured and discussed. This property is very important, since when the ablative is placed in the SRM, it must absorb the mechanical stresses the SRM experiences during propellant casting, storage, transportation, and flight. It also assists in anchoring the propellant grain to the inner wall of the SRM case. As a result, the insulator should have high elongation at break (in excess of 200%). Moreover, the ablative properties were evaluated using an oxy-acetylene torch according to ASTM E285-80 [27].

The results can be summarized as follows. The XRD patterns revealed that the interlayer spacing between the nanoclay layers of the OMMT increased due to the intercalation of EPDM molecules in the filler tactoids. The TEM images also showed that the platelets were homogeneously dispersed in the matrix by the strong shear forces of rollers. In terms of mechanical properties, the elongation at break of the EPDM/Kevlar composites was analyzed as a function of the fiber content and these results were compared with the corresponding data of the EPDM/Kevlar/OMMT system with respect to the OMMT percentage. For the baseline EPDM/Kevlar, it appeared that above a fiber content equal to 6 phr, the elongation at break strongly decreased with the increasing content of the fibers: over 8 phr of fibers, the elongation at break of the EPDM/Kevlar (about 150%) became less effective in meeting the requirements of SRM liners. The 12 phr EPDM/Kevlar formulation was chosen for the next step of the research in which the OMMT were introduced in the heat-shielding material. Even at this high fiber content, due to the presence of the nanoclay, the material exhibited an impressive increase in the maximum strain. This result was confirmed also at a high nanoclay content: at 13 phr OMMT, the elongation at break displayed by the EPDM/Kevlar/OMMT system showed a value equal to approximately 560%.

However, with OMMT content of 7 phr, the elongation at break appeared to decrease. From 1 phr to 7 phr OMMT, the enhancement in the elongation at break

Table 10.3. *Designation of EPDM-based samples*

SI. No.	Sample designation	Polyimide powder (phr)	Nano-silica (phr)
1	EP_0S_0	0.0	0.0
2	EP_5S_0	5.0	0.0
3	EP_1S_5	1.0	5.0
4	EP_3S_5	3.0	5.0
5	EP_5S_5	5.0	5.0
6	EP_7S_5	7.0	5.0
7	$EP_{10}S_5$	10.0	5.0
8	$EP_{10}S_{10}$	10.0	10.0
9	EP_5S_1	5.0	1.0
10	EP_5S_3	5.0	3.0
11	EP_5S_7	5.0	7.0
12	EP_5S_{10}	5.0	10.0

E = EPDM, P = polyimide, S = nano-silica

can be directly attributed to the excellent dispersion of OMMT in EPDM matrix. The SEM images of the EPDM/Kevlar/OMMT fracture surfaces indicated that the nanoclay platelets reduced the interfacial adhesion of the aramid fibers within the matrix, and thus the OMMT acted as a lubricating agent. At the higher filler content, the reduction of the elongation at break can be attributed to the aggregation of the silicate layers. In terms of ablative properties with the increase of the nanoclay content up to 11 phr OMMT, the loss of mass as well as the erosion rate decreased. Quantitatively, the erosion rate decreased from 0.18 mm/s at 0% OMMT phr to about 0.07 mm/s at an OMMT content of 11 phr. Above this threshold, the erosion rate and the loss of mass slightly increased. These results are particularly interesting given that the OMMT-based formulations are potentially attractive candidates to replace the traditional EPDM/Kevlar shielding elastomers.

10.3.3.4 EPDM-Polyimide-Nano-silica Nanocomposites Studies by Indian Institute of Technology

Singh et al. [28] examined the potential synergistic effect of polyimide micro-sized particles with nano-silica in maleic anhydride grafted EPDM. Several formulations were considered, in which the loadings of each filler varied from 0 to 10 wt%, as shown in Table 10.3 [28]. They were mixed in a Brabender Plasti-Corder at 70 rpm and 120°C and vulcanized with vulcanizing agents at 150°C for 30 minutes at 6 MPa.

The authors characterized mechanical properties (ASTM D412), hardness (ASTM2240), density (ASTM 792), morphology via SEM, dispersion via TEM, thermal stability via TGA in N_2 at 10°C/min to 600°C, thermal conductivity (ASTM D5334-08), erosion rate via arc-jet ablation testing (ASTM E285-65T) on a selected formulation, coefficient of thermal expansion (CTE) on a selected formulation via TMA (ASTM E831-12), and specific heat via DSC.

The authors observed that the density of all formulations remained nearly constant, varying within a range of only 0.066 g/cm, and that their mixing techniques had resulted in near-uniform dispersion. They found that all formulations had improved mechanical properties, and the formulation with the most improvement was the one with 10 wt% of each filler. For this reason they chose this loading to be the one used in the arc-jet facility, producing an erosion rate of 0.155 mm/s under 3 MW/m^2 of heat flux. This formulation showed a small improvement in CTE over pure EPDM, as well. TGA showed small improvements in thermal stability for all formulations. The authors concluded that the material showed good promise for multiple thermal protection and insulation applications.

10.3.3.5 EPDM-Nano-Silica Nanocomposite with Aramid Pulp, Chopped Phenolic Fibers, and Chopped Silica Fibers Studies by University of Perugia, Italy

Natali et al. [29] compared three microfiber reinforcement candidates for EPDM-silica nanocomposites. The three candidates considered were aramid pulp, chopped phenolic fibers, and chopped silica fibers. These were added to EPDM at 10 wt% in addition to 20 wt% nano-silica, 15 wt% paraffin oil, and 4 wt% dicumyl peroxide, a vulcanizing agent. The formulations were mixed in a Bausano SD3025 single-screw extruder and vulcanized at 180°C for 2 hours in a heated press at a pressure of 50 bar.

They were studied for thermomechanical and thermal stability as well as ablative characteristics. They were examined via SEM before and after ablation testing. Their mechanical properties were evaluated using a dynamometer and a thermomechanical analyzer, and ablation tests were carried out using the group's oxyacetylene test bed according to ASTM E-285-80 at 500 W/cm^2. Thermogravimetric analysis (TGA) and differential scanning calorimetry (DSC) were conducted on each, as well.

First, the formulations were examined with SEM for dispersion of fibers. This was found to be satisfactory in each, but the dispersion of nano-silica was not verified. There appear to be some micron-sized particles visible in Figure 10.12 [29]. The researchers observed that the material was stiffer with the phenolic fibers and the aramid pulp, but tougher with silica fibers. The unreinforced material was not investigated. They found that while the silica fibers yielded the greatest residual mass during TGA, they showed the greatest mass loss during ablation testing. Posttest SEM inspection showed that the silica fiber–reinforced material produced a char that had high porosity, while the phenolic fiber material had the least. It was the phenolic fiber material that showed the highest char/virgin material adhesion and the highest in-depth temperatures during ablation testing. The aramid pulp–reinforced material had the lowest in-depth temperatures. The researchers concluded that the amount of fibers included in the material should be increased for better char and more conclusive results. They also suggested mixing the fiber reinforcements into a single formulation.

10.3.3.6 EPDM-MWNT-Carbon Black Nanocomposite Studies by University of Sciences and Technology, Pakistan

Iqbal et al. [30] investigated nanocomposites of EPDM with MWNTs and carbon black. The carbon black was held at 40 wt% loading, while the MWNT loading varied

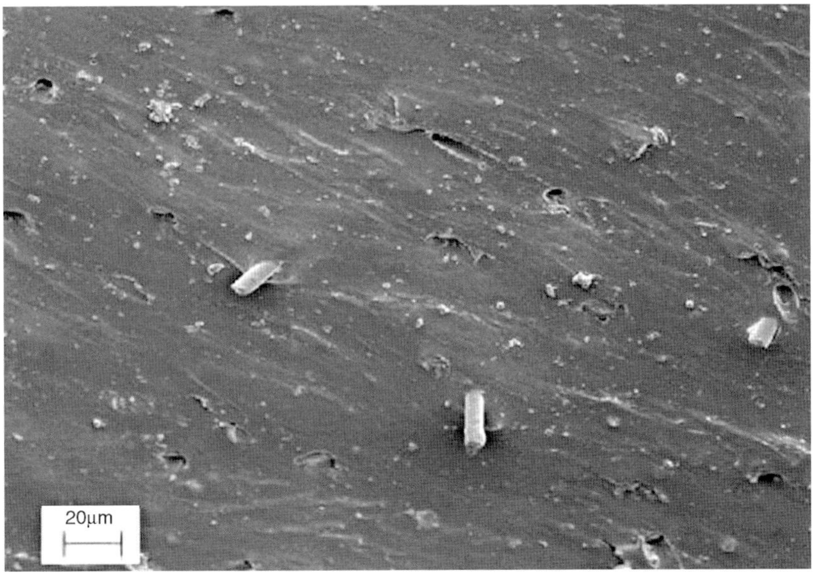

Figures 10.12. Scanning electron microscopy image of the pretest material with silica fibers and nano-silica [29].

between 0.1 wt% and 1 wt%. In addition, a silane coupling agent was added to the mixtures at a 4 wt% loading. The formulations were mixed using dispersion kneader at 110°C for 30 minutes and then using a two-roll mill at 70°C and 40 rpm speed for 20 minutes. They were cured using a hot press at 130°C and 1,500 psi for 40 minutes.

The formulations were characterized for their ablation resistance according to ASTM E285-08 using a 1:1 oxygen/acetylene ratio with a flow rate of 0.35 m^3/h and torch distance of 10 mm for 200 seconds, dispersion via SEM and EDS, thermal stability via TGA to 830°C at 10 °C/min, thermal conductivity according to ASTM E1225-99, mechanical properties according to ASTM D412-98A, and Shore A hardness.

The researchers observed a steady decrease in elongation at break and an improvement in all other properties with increasing loadings of MWNTs. Backface temperature dropped significantly, ablation rate dropped sharply at low loadings, thermal conductivity rose, ultimate tensile strength and elastic modulus rose, and hardness rose with increased loadings of MWCNTs. The authors found that this composite is a good candidate for use as an internal insulation SRM material.

10.3.4 Natural Rubber (NR) and Hydrogenated Nitrile Butadiene Rubber (HNBR) Nanocompostie Studies

10.3.4.1 NR-Clay Studies by Tarbiat Modares University, Iran

The effect of organically modified MMT clay in natural rubber on the thermal, flammability, hardness, and mechanical properties using melt-compounding were investigated by Khanlari and Kokabi [31]. The authors have shown that the addition of 3 wt% of organically modified MMT (Cloisite 15A) to NR decreased curing time by 33%, its aging hardness rise was decreased more than 55%, and elastic modulus increased by 63%. The ignition time was delayed by 150% and a reduction in peak heat release rate of 54% was achieved, compared to pristine NR. The thermal stability for the NR-clay nancomposite was also enhanced.

10.3.4.2 HNBR-Fumed Silica, Nanoclay, and Expanded Graphite Nanocomposite Studies by Beiging University of Chemical Technology, China

Guan et al. [32] conducted an investigation into the effects of organically modified montmorillonite (OMMT), expanded graphite (EG), and nano-silica on the thermal and ablative properties of hydrogenated nitrile butadiene rubber (HNBR) based composites. The composites were reinforced with 40 wt% aramid pulp and mixed with some other agents for vulcanization. The nanofillers were added at a loading of 35 wt%. The formulations were mixed in a 6-inch two-roll mill at a speed ratio of 1:1.1 and cured in a molding compound at 170°C and 10 MPa.

For characterization, the researchers used an oxyacetylene torch with an oxygen/acetylene ratio of 1:1 and total flow rate of 0.7 m^3/h, a torch distance of 10 mm, and an exposure time of 20 seconds to observe ablation resistance. They also used SEM and WAXD to discover morphology and dispersion of pre- and posttest samples. TGA in N_2 to 800°C were conducted at heating rates of 5, 10, 15, and 20 °C/min to determine thermal stability properties.

The researchers found that the processing technique they had used to prepare the EG from its as-received state was insufficient to expand the layers. This prevented the filler from producing an optimal performance. However, the other nanofillers dispersed satisfactorily, and the authors were able to observe that the nano-silica did not form a gas barrier during ablation, resulting in a tighter, more solid char that conducted heat in-depth much more, as shown in Figure 10.13 [32]. The OMMT was found to be a better additive for ablation resistance, since it forces the pyrolysis gases to escape through the char as it forms, making its pores much larger and its structure looser, as well as impeding the incoming torch gases, as shown in Figure 10.14 [32]. The EG char is shown in Figure 10.15 [32]. As per the authors' conclusions, the char is compact. The fillers were not found to have significant effect on the TGA results.

10.3.5 Thermplastic Polyurethane Nanocomposties (TPUN) Studies

10.3.5.1 TPUN Studies by The University of Texas at Austin, United States

Polymer chopped-fiber-filled systems are used as insulative materials in rocket propulsion systems. This research program was aimed at developing new classes of polymer nanostructured materials that are lighter and exhibit better erosion and insulation characteristics than current insulative materials (e.g., Kevlar-filled EPDM rubber).

Materials. The TPSiV™ X1180 thermoplastic elastomer (TPE) is a polyamide–based, vulcanized silicone thermoplastic resin manufactured by Dow Corning that was selected as a potential replacement material. Its typical uses include profiles for automotive, fuel, and vapor line covers, brake hose covers, and industrial applications

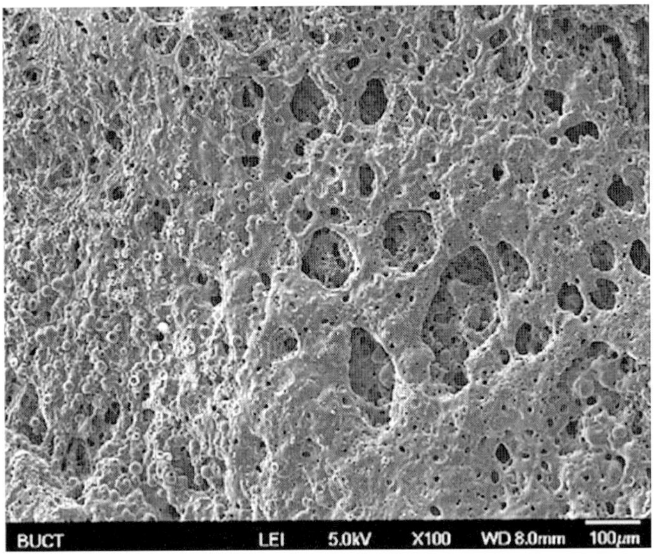

Figure 10.13. Scanning electron microscopy image of the char layer for the nano-silica formulation [32].

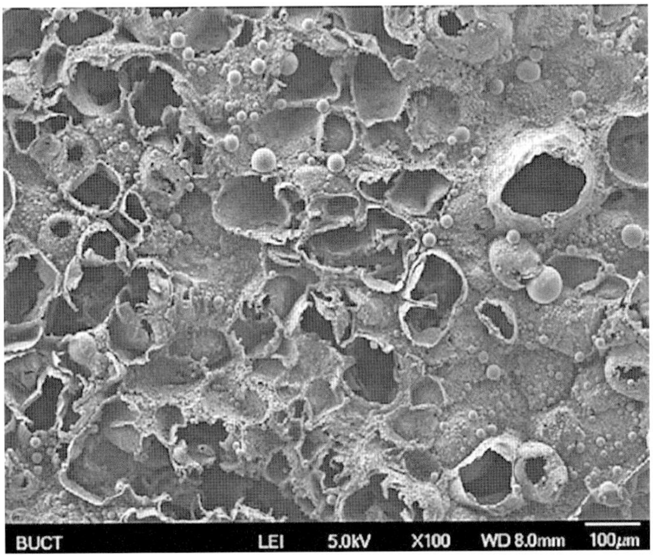

Figure 10.14. Scanning electron microscopy image of the char layer for the OMMT formulation [32].

involving extruded profiles exposed to harsh environments. Table 10.4 shows the chemical compositions of three thermoplastic elastomer/nanoparticle blends that were produced by twin-screw extrusion [33].

Characterization. Transmission electron microscopy analyses were conducted on the aforementioned three blends. Figure 10.16 shows the TEM images of the Dow Corning polyamide silicone TPSiV™ X1180 in two separate phases: the dark color

Table 10.4. *Thermoplastic elastomer/nanoparticle blends*

Material	TPE (wt%)	Nanoparticle (wt%)
1	TPSiV™ X1180 (90%)	$Ph_{12}T_{12}$-POSS (10%)
2	TPSiV™ X1180 (92.5%)	Cloisite 30B (7.5%)
3	TPSiV™ X1180 (85%)	PR-24-PS CNF (15%)

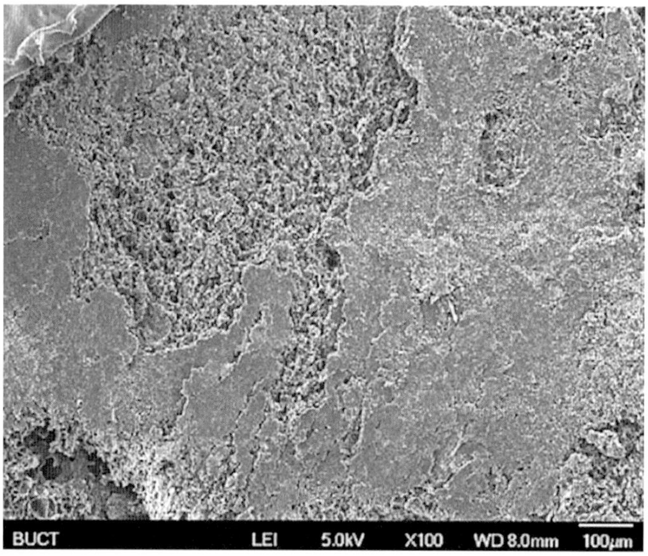

Figure 10.15. Scanning electron microscopy image of the char layer for the EG formulation [32].

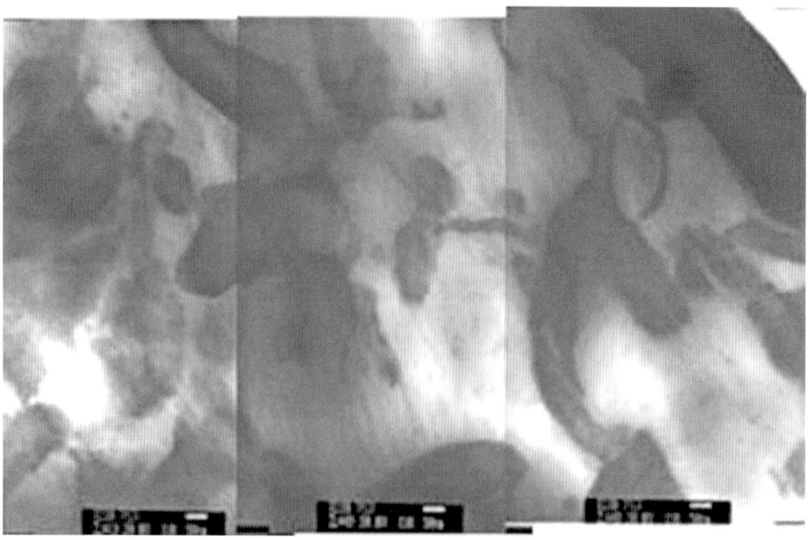

Figure 10.16. Transmission electron micrographs of polyamide silicone TPSiV™ X1180 (scale bar is 500 nm).

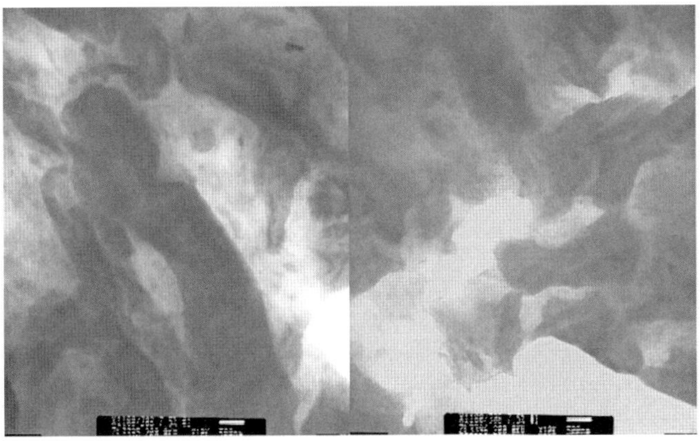

Figure 10.17. Transmission electron micrographs of 92.5% polyamide silicone TPSiV™ X1180 with 7.5% Cloisite 30B (scale bar is 500 nm).

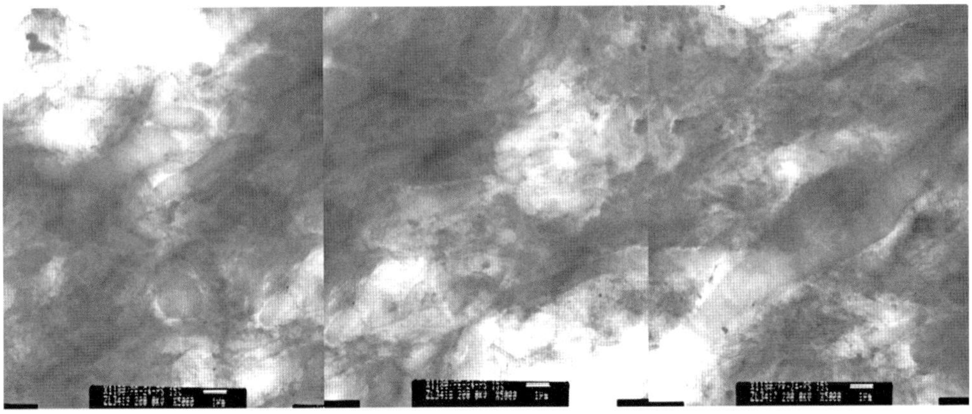

Figure 10.18. Transmission electron micrographs of 85% polyamide silicone TPSiV™ X1180 with 15% PR-19-PS CNF (scale bar is 1 μm).

is the silicone phase, and the light color is the polyamide phase [33]. Figure 10.17 shows the TEM images of the 7.5 wt% Cloisite 30B/92.5 wt% TPSiV™ X1180 blend [33]. We speculate that the Cloisite 30B nanoclays are dispersed only in the polyamide phase, because the silicone phase is already cross-linked. Transmission electron microscopy images of the 15 wt% PR-24-PS CNF/85 wt% TPSiV™ X1180 blend (Figure 10.18) and 10% $Ph_{12}T_{12}$-POSS/90 wt% TPSiV™ X1180 blend (Figure 10.19) showed similar results [33]. It was determined that the TPSiV™ X1180 TPE is difficult to process with any of the three selected nanoparticles. None was compatible with the cross-linked silicone.

Performance. The aforementioned three blends with the baseline Kevlar-filled EPDM rubber were tested for ablation resistance using a subscale solid rocket motor. Figure 10.20 shows the ablation rate of the materials at low, medium, and high Mach number regions inside the rocket motor [33]. The baseline material outperformed all three TPEs. These observations clearly demonstrate that if the nanoparticles are

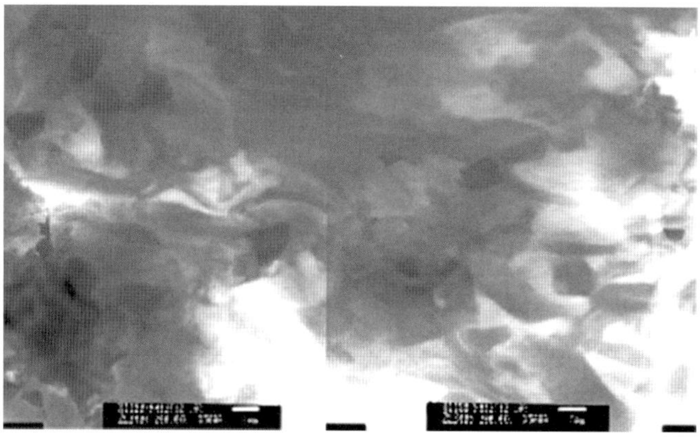

Figure 10.19. Transmission electron micrographs of 90% polyamide silicone TPSiV™ X1180 with 10% Ph12-T12-POSS [scale bars are 1 μm (left) and 500 nm (right)].

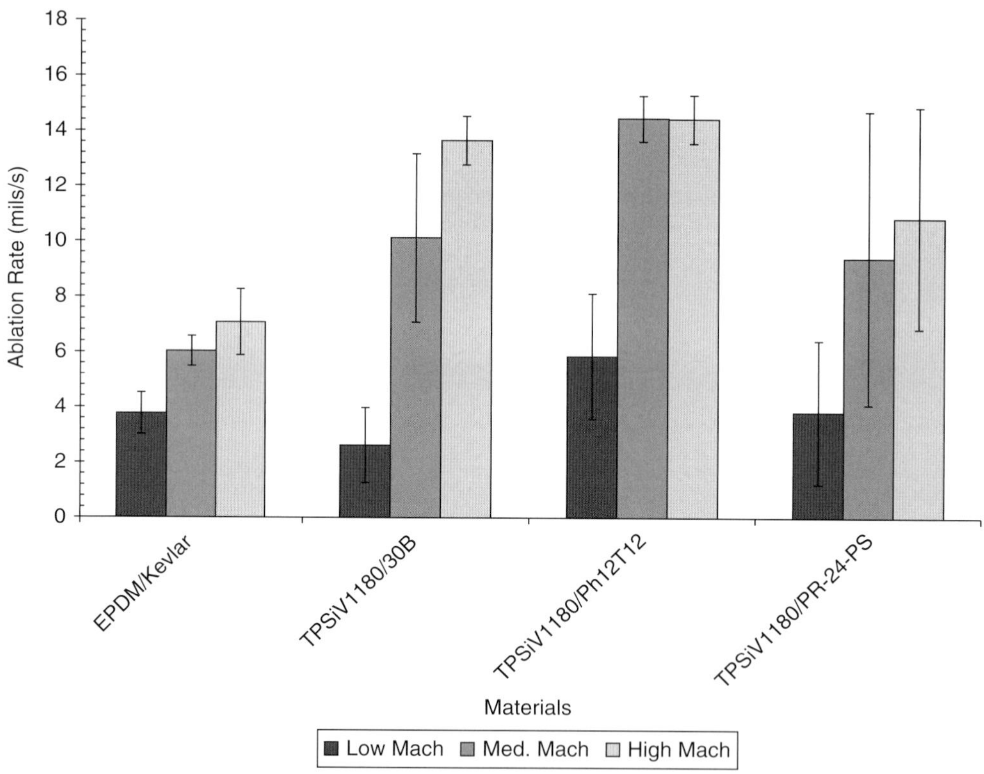

Figure 10.20. Ablation rate of polyamide silicone TPSiV™ X1180 with Cloisite 30B, Ph12-T12- POSS, and PR-24-PS CNF as compared to EPDM/Kevlar in low, medium, and high Mach number regions inside the rocket motor.

poorly dispersed in the polymer matrix, no thermal performance improvement can be expected. The limited dispersion of nanoparticles into the polyamide phase was insufficient for improved thermal performance.

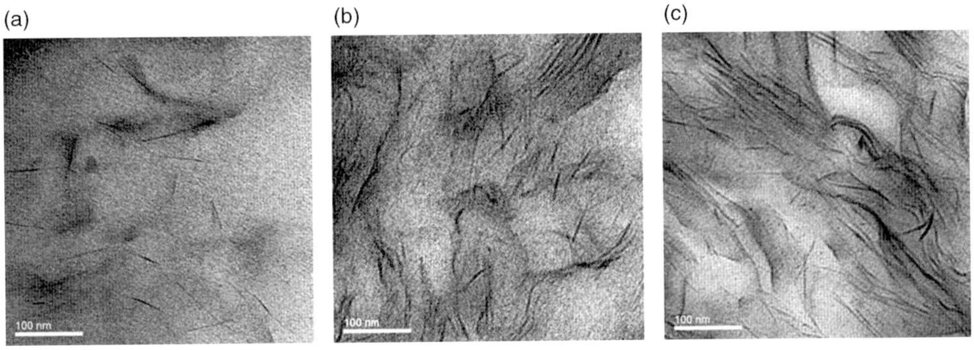

Figure 10.21. TEM images of (a) 2.5%, (b) 5%, and (c) 10% Cloisite 30B in TPUN (scale bars are 100 nm).

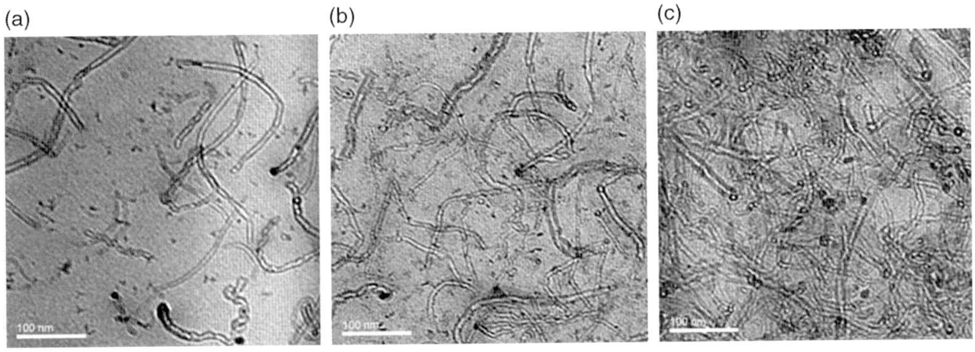

Figure 10.22. TEM images of (a) 2.5%, (b) 5%, and (c) 10% MWNT in TPUN (scale bars are 100 nm).

For advanced SRM internal insulation, Koo and colleagues [33–47] (which include Koo et al. [33–37], Ho et al., Bruns et al., Lee et al., and Allcorn et al.) used two thermoplastic polyurethane elastomers (TPUs) as the base polymer and incorporated them with different amounts of MMT clay, CNF, POSS®, and multi-walled carbon nanotubes (MWNT). Selective properties improvement was observed with these TPU nanocomposites (TPUNs) as compared to industry standard Kevlar®-filled EPDM insulation material. The effects of weight loadings of layered clay, CNF, and MWNT on the thermal and flammability performance of this novel class of materials have been explored using a variety of test protocols and methods, such as thermogravimetric analysis (TGA), radiant panel experimental apparatus, UL 94, and cone calorimeter. Selective TPUNs were then tested using scaled SRM, oxyacetylene torch, as well as scaled hybrid rocket motors with the intent to develop them for SRM insulation application. Two commercially available TPUs were melt-blended with various loadings of these nanofillers using twin-screw extrusion. The morphological, physical, thermal, flammability, thermophysical properties, and kinetic parameters of these three families of TPUNs were characterized. The *processing-structure-property* relationships of this class of novel TPUNs were established and are reported elsewhere.

Morphological Results. TEM analyses of both the MWNT and nanoclay TPUs were carried out to determine the visual degree of dispersion (Figures 10.21 and

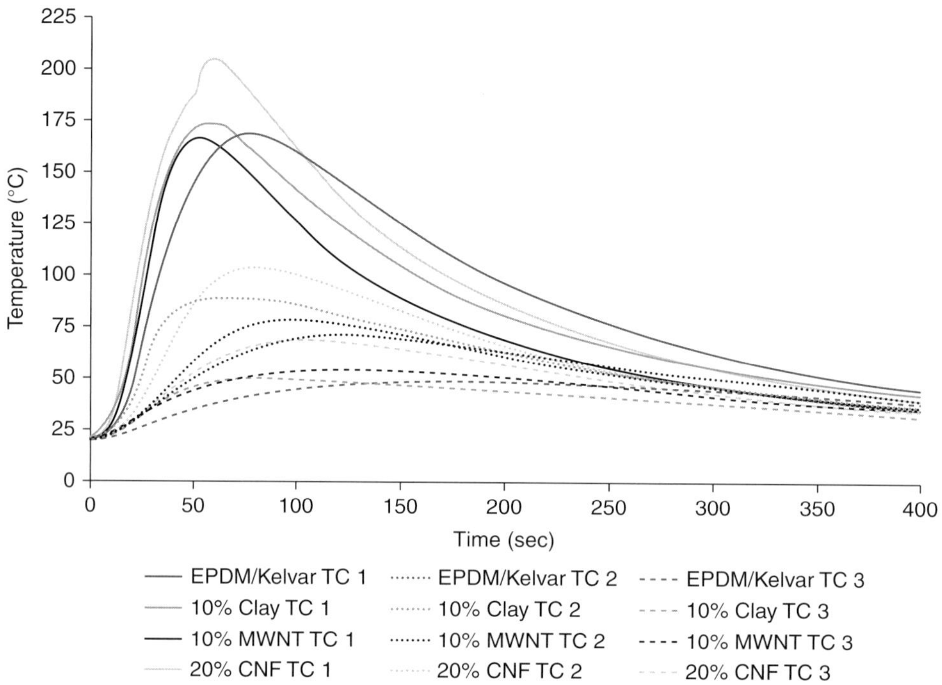

Figure 10.23. Representative temperature profile of each specimen of oxyacetylene ablation tests.

10.22) [33]. In the 2.5 wt% nanoclay sample, the individual platelets are easily identifiable and good dispersion is observed. The 5 wt% and 10 wt% images show some remaining stacks, but the majority of the images show individual nanoclay platelets. It should also be noted that it does not appear to have any directional orientation from the injection-molding process. The MWNT agglomerates are clearly debundled as shown in the two lower weight percentages. In the 10 wt% MWNT TEM image, there is a high density of MWNTs. These are not in bundle form. The bulk material dispersion is assessed by observing TEM images at different locations and view similar images. Again, there does not appear to be any preferred orientation.

Ablation Test Results. High heat flux experiments have been conducted. The material is exposed to the hottest portion of the flame, with a 7 mm flame diameter, for 25 seconds. Samples of 12.7 mm × 12.7 mm × 50 mm are cut out of a 12.7 mm × 10.16 cm × 10.16 cm (½ in. × 4 in. x 4 in.) compression-molded sheet. The torch is directed toward the square face (12.7 mm × 12.7mm). In this orientation, the sample may be tested twice. At 5 mm increments, three thermocouples are embedded into the center of the specimen to measure the in-depth heat soak temperature during testing. Thermocouples are drilled into the specimen at both ends for the possibility of testing a single specimen twice depending on the ablation rates.

Thermal Results. A representative temperature profile of each specimen is shown in Figure 10.23 [44, 45]. The specific materials are color coded and the thermocouples (TC) are distinguished as TC 1 (solid line), TC 2 (dotted line), and TC 3 (dashed

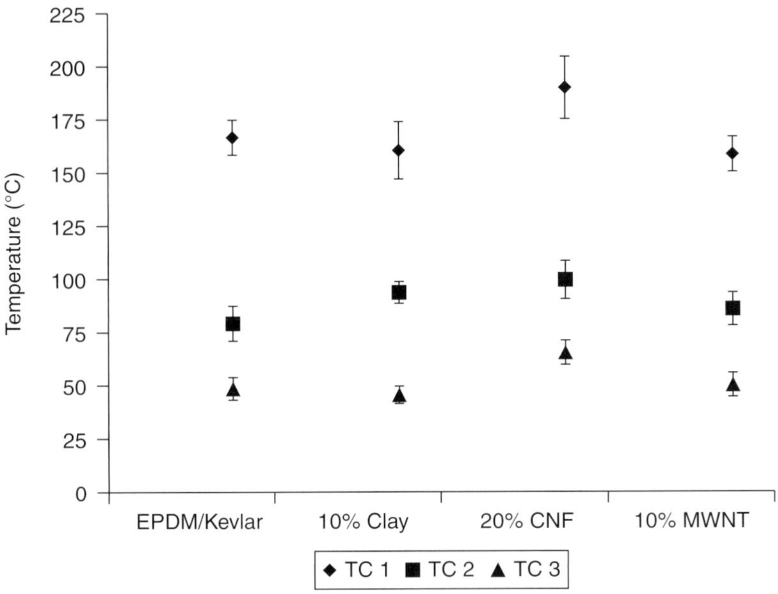

Figure 10.24. Peak in-depth temperatures of temperature profile of EPDM/Kevlar and PNCs.

line). After the 25-second flame exposure is removed, the heat is transferred conductively through the specimen. In this set of tests, the peak temperature of TC 1 in the CNF sample is 205°C and occurs at 60 seconds. The CNF has the highest heat soak temperature. The MWNT and nanoclay TC 1 peak at 166°C and 174°C and occur at 50 seconds and 64 seconds, respectively. The EPDM/Kevlar TC 1 peaks at 168°C and shows the most delay, at 75 seconds. Figure 10.24 shows the peak in-depth temperatures of EPDM/Kevlar and PNCs between thermocouples [44, 45]. The nanoclay, MWNT, and CNF PNCs lose 13.5%, 33.7%, and 46.5% more mass, respectively, compared to the mass loss of Kevlar/EPDM.

Post-Firing Images. The samples after firing are shown in Figure 10.25 [44, 45]. A 1 mm compact char layer is left at the end of the firing of Kevlar/EPDM. Although compact, the virgin and char material does not have good adhesion. The nanoclay PNC posttest material has about twice the char thickness of the Kevlar/EPDM. Although the char has a crack, the char was maintained well within the virgin material. Similar to the results found in the vertical UL 94 test, many cracks are observed in the MWNT char surface and result in a weak 2 mm thick char. The CNF PNC char is very thick, greater than 5 mm, as the first thermocouple was in the char layer. The CNF char surface is flat, without cracks; however, it is still very weak. The results from this set of experiments are consistent with the flammability and hybrid rocket tests [44, 45].

SEM images of the posttest samples also were taken. The Kevlar/EPDM char material is compact and shows that the individual Kevlar fibers maintain the material together as shown in Figure 10.26 [44, 45]. The nanoclay char is also a compact composition. A granular morphology is also observed (Figure 10.27) [47]. Bubble formations observed in SEM images found in the UL 94 burn tests are observed on

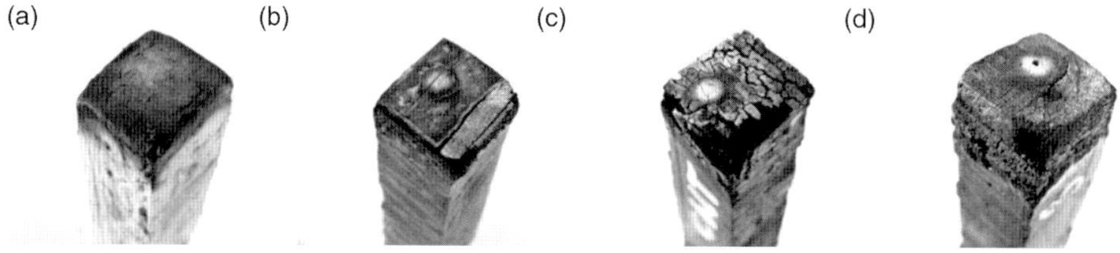

Figure 10.25. (a) Kevlar-EPDM, (b) 10% Clay, (c) 10% MWNT, and (d) 20% CNF PNC post oxyacetylene torch burn.

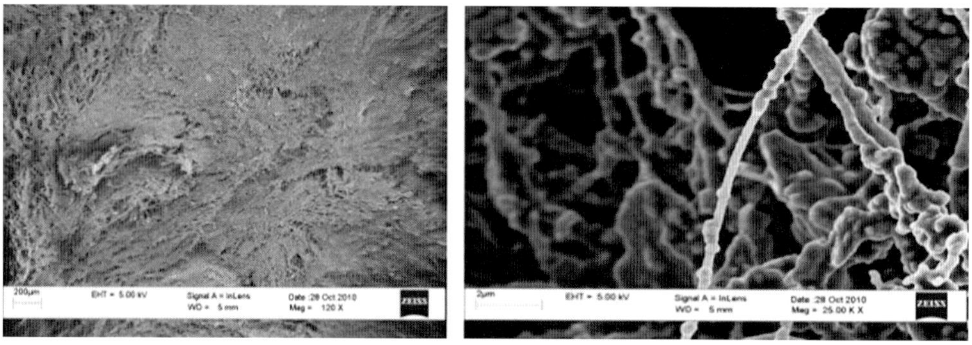

Figure 10.26. SEM of Kevlar/EPDM char post oxyacetylene torch burn (unit bar of left is 200 μm and right is 2 μm).

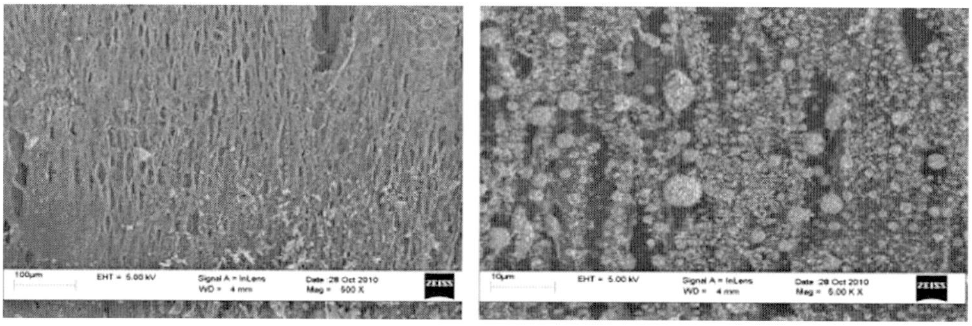

Figure 10.27. SEM of 10% nanoclay PNC char post oxyacetylene torch burn in progressive magnification (unit bar of left is 100 μm and right is 10 μm).

the MWNT char surface as well as crack formations (Figure 10.28) [44, 45]. An in-depth SEM image shows the bundle form of MWNT that is also observed in the UL 94 burn test. The weak CNF char is shown to have a very porous structure (Figure 10.29) [44, 45]. Individual CNFs are also observed.

The initial results from the oxyacetylene torch test show that both the nanoclay PNC and MWNT char layers are about twice the thickness of the Kevlar/EPDM char. However, the MWNT char layer is weak compared to the nanoclay PNC char. The CNF char is also very weak and was more than five times the thickness of the Kevlar/EPDM char. However, the mass loss from the CNF sample is almost twice that lost from the Kevlar/EPDM sample. In-depth thermal measurements also

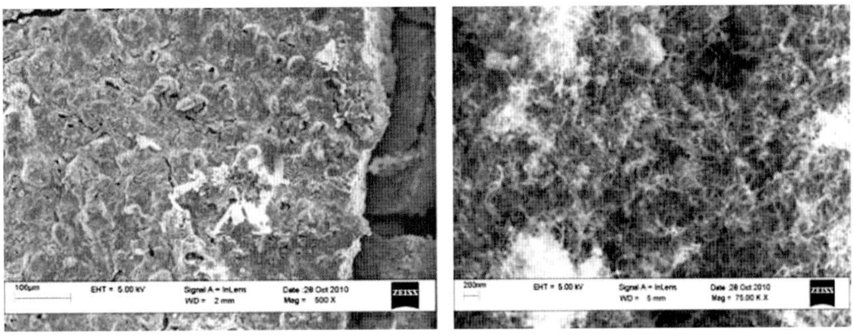

Figure 10.28. SEM of 10% MWNT PNC char post oxyacetylene torch burn in progressive magnification (unit bar of left is 100 μm and right is 200 nm).

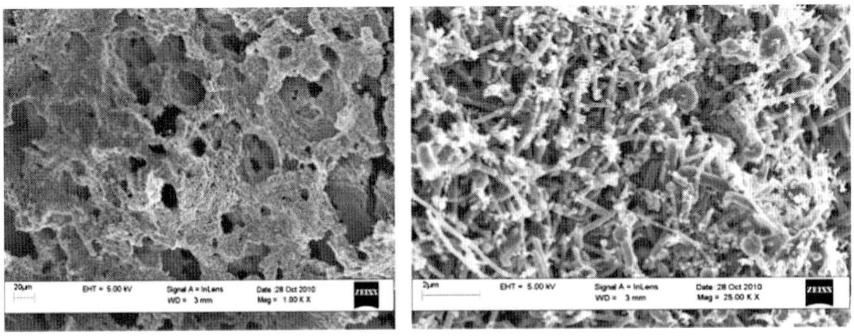

Figure 10.29. SEM of 20% CNF PNC char post oxyacetylene torch burn in progressive magnification (unit bar of left is 20 μm and right is 2 μm).

show that the thermal protection of the PNCs is comparable to that provided by the Kevlar/EPDM, except for the CNF sample, which had a much higher peak temperature due to the fact that the first embedded thermocouple ended up in the char region. A conventional advanced fiber is therefore needed to boost up the PNC ablation performance.

Allcorn et al. [46, 47] continued to investigate the potential of thermoplastic polyurethane elastomer nanocomposites (TPUNs) as alternative elastomeric heat-shielding materials (EHSMs) in ablative applications. Three different nanofillers were tested – montmorillonite clay (MMT Clay), multi-walled carbon nanotubes (MWNTs), and carbon nanofibers (CNFs) – and formulations were made for each material with a range of weight loadings as show in Table 10.5 [47]. Small-scale ablation testing was performed using an oxyacetylene test bed with Kevlar-filled EPDM as the baseline EHSM. Mass loss, recession, and char thickness values were measured in addition to in situ internal temperatures. The TPUN formulations with 5 wt% and 7.5 wt% MMT clay showed the best results, outperforming the baseline in recession and peak temperature values. SEM images of the char from MMT clay containing TPUNs showed ablative reassembly, which likely served to enhance the material performance. Overall, results showed that the TPUN ablatives performed well relative to the baseline EHSM.

Table 10.5. *Composition of nanocomposite formulations*

Nanocomposite Formulation	Weight Percent TPU	Weight Percent MMT Clay	Weight Percent MWNT	Weight Percent CNF
MMT – 2.5%	97.5%	2.5%	0.0%	0.0%
MMT – 5%	95.0%	5.0%	0.0%	0.0%
MMT – 7.5%	92.5%	7.5%	0.0%	0.0%
MMT – 10%	90.0%	10.0%	0.0%	0.0%
MWNT – 2.5%	97.5%	0.0%	2.5%	0.0%
MWNT – 5%	95.0%	0.0%	5.0%	0.0%
MWNT – 7.5%	92.5%	0.0%	7.5%	0.0%
MWNT – 10%	90.0%	0.0%	10.0%	0.0%
CNF – 5%	95.0%	0.0%	0.0%	5.0%
CNF – 10%	90.0%	0.0%	0.0%	10.0%
CNF – 15%	85.0%	0.0%	0.0%	15.0%
CNF – 20%	80.0%	0.0%	0.0%	20.0%

The testing of ablative materials involves simulating the extreme conditions to which they are exposed in actual applications, while measuring various metrics. Some common metrics for ablative performance measurement include mass loss, recession and/or recession rate, front face temperature, back face temperature, in-depth temperature, char thickness, and off-gas composition. The metrics identify different criteria depending on the specific application of the ablative material. The material's resistance to heat is determined via measurements of the back face, front face, and heat soak (internal) temperatures during and after the testing procedure. These measurements show how well the material inhibits heat flow and can determine whether it fails to meet specific requirements regarding back face temperature. For example, if the back face temperature is too high, it may result in failure of the adhesive bonding the ablative to the supporting structure, and ultimately in the failure of the overall material. In addition, posttest measurements of the char thickness and material erosion can provide additional insight into the resistant capabilities of the material relative to exposure time. If the reaction progresses too quickly through the material or the protective char layer erodes too quickly due to surface shear, then the material may not provide thermal protection for a long enough time period.

For the materials tested in this study, specific settings within the OTB's range of operation were selected and all the material formulations were tested under these conditions. The nanocomposite formulations were cut into 1.27 cm × 1.27 cm × 5 cm (0.5 in. × 0.5 in. × 2 in.) samples for testing, with three samples tested for each formulation to increase the statistical confidence in the results. The square face of the samples was exposed to the oxyacetylene torch at a calibrated heat flux of approximately 1,000 W/cm^2 for duration of 25 seconds. The OTB facilitates three K-type thermocouple measurements during testing, and for our study these thermocouples were placed within the samples at distances of 5 mm, 10 mm, and 15 mm from the exposed surface with a heat shield in place to protect the protruding wires.

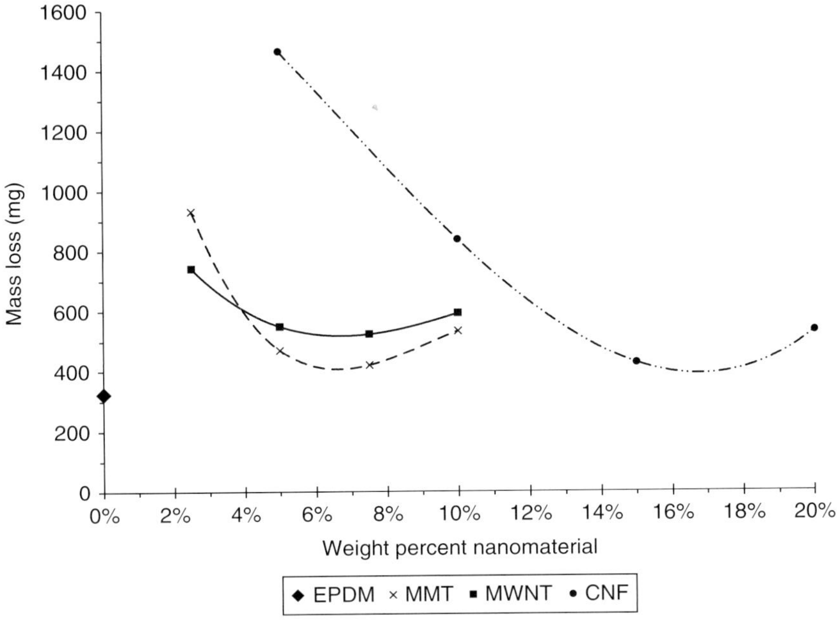

Figure 10.30. Mass loss results for OTB ablation testing of TPUN formulations.

In addition to these in situ temperature curves, the mass loss, recession, and char thickness of each sample was determined based on pre- and posttest measurements.

Mass Loss. The results of the mass loss measurements for the TPUN formulations are shown in Figure 10.30 [47]. Mass loss values can provide an indication of the time duration that ablatives can offer effective thermal protection; low values are desired. For all three nanomaterials a similar trend is observed with a decrease in mass loss up to a certain weight percent and then a slight increase beyond that value. This trend is typical among nanocomposites as greater amounts of the additive material provide benefits to the base matrix until a certain threshold amount where excessive additive content compromises the overall matrix structure of the base material [6, 47]. For both MMT clay and MWNT, this ideal value of minimum mass loss is somewhere between 5% and 10%, while for CNF the ideal value is somewhere between 10% and 20%. When comparing the different nanomaterials, they all appear to have minimum mass loss values that are approximately the same. The minimum mass loss for both MMT clay and CNF is near 400 mg while the minimum for MWNT is slightly worse at approximately 500 mg.

Overall, none of the TPUN formulations outperforms the EPDM baseline, which shows an average mass loss of slightly more than 300 mg. However, it must be pointed out that for SOTA EPDM/Kevlar® materials the amount of the fillers – the fibers, the micron, and nanosized additives like the mineral oxides and flame retardants – is higher than in the case of the produced nanocomposites [49–51]. Consequently, since the filler content of the EPDM/Kevlar® is typically higher than the nanocomposites produced in this research, these results can be considered quite promising. The low weight percentage formulation for each nanomaterial has high

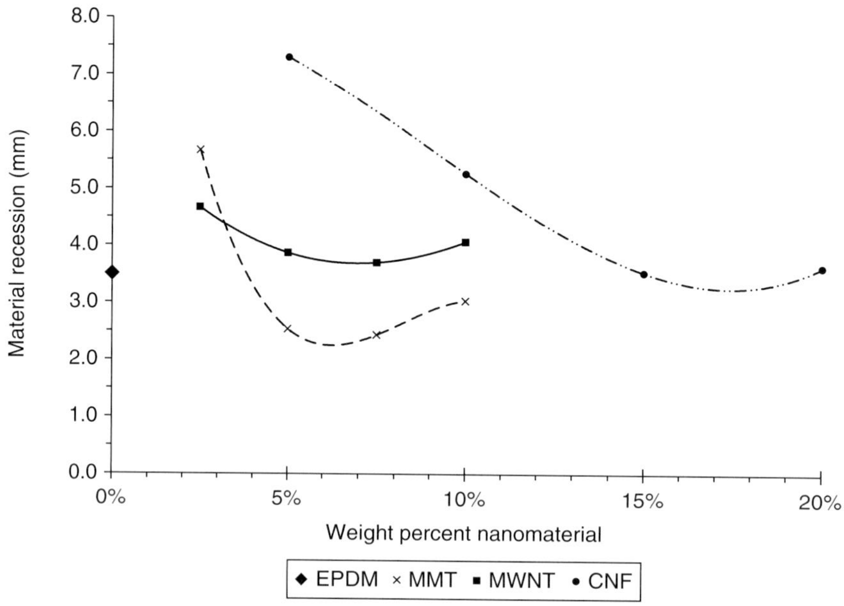

Figure 10.31. Material recession results for OTB ablation testing of TPUN formulations.

mass loss because they are not only losing mass through the pyrolysis of the material, but also through physical erosion at the char surface. The char material for the MMT clay 2.5%, MWNT 2.5%, and CNF 5% TPUNs was too weak to withstand the pressure from the OTB flame and was observed to physically erode off of the sample. The CNF 10% formulation also experienced a small amount of erosion, though more limited.

Recession and Char Thickness. The thickness of each sample was measured before and after testing and was used to determine the overall recession of the sample. For this measurement, the thickness after testing was considered to be the thickness of virgin material only, so the recession value is a measure of the progression of the reaction layer into the virgin material and does not factor in the thickness of the char on the surface. Recession is also an indicator of the time duration of thermal protection, so low values are once again desirable. The results are shown in Figure 10.31 and follow roughly the same trends as the results for mass loss, and the best performing weight percent values match up to those for minimal mass loss: 5% to 10% for MMT clay and MWNT, and 10% to 20% for CNF [47]. When the nanomaterials are compared to one another, both MWNT and CNF TPUNs have similar minimum recession values of roughly 3.5 mm. However, MMT clay performs better than the other materials with minimum recession value of around 2.5 mm. This is superior even to the baseline EPDM-Kevlar material which showed recession values near 3.5 mm.

The char thickness of each sample was also measured after testing (Figure 10.32) [47]. For our measurement, we considered the char thickness to be the distance from the front surface of the posttest sample to the reaction layer (Figure 10.33) [47]. The

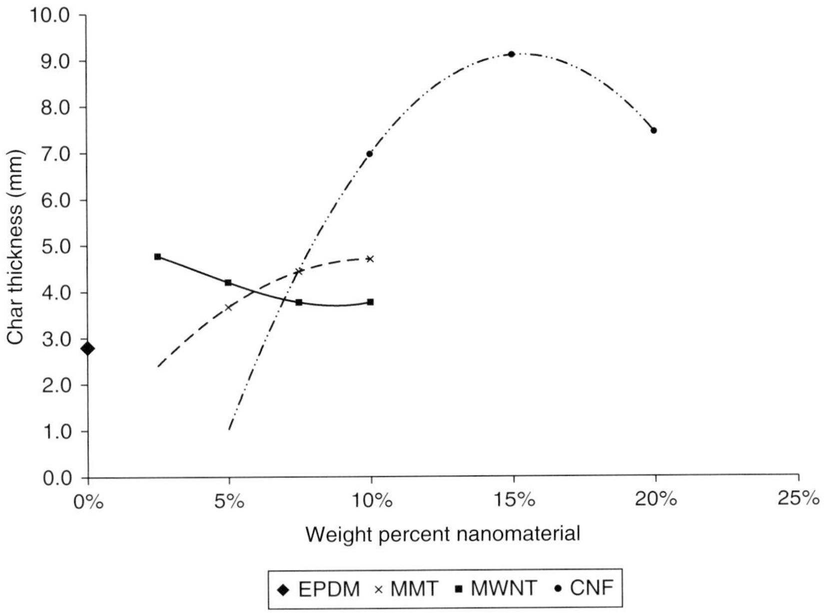

Figure 10.32. Char thickness results for OTB ablation testing of TPUN formulations.

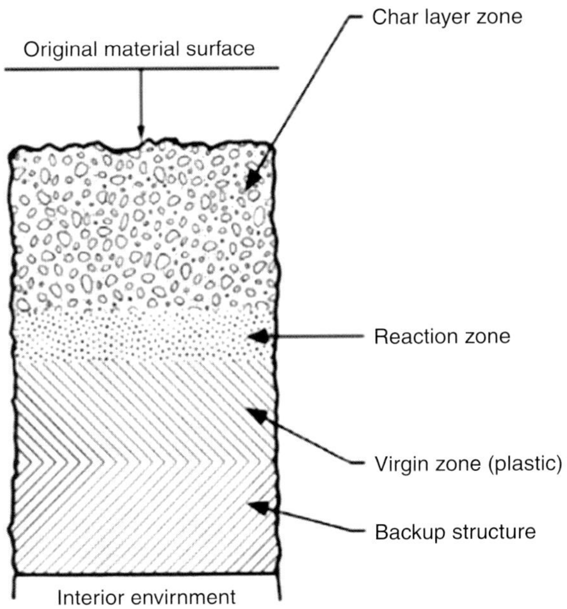

Figure 10.33. Cross-sectional schematic of the ablation process.

trends in this instance are quite different than those seen in other measurements. Both MMT clay and CNF TPUNs have char thickness values that generally increase with weight percent. This is likely due to the intumescent nature of the char for these materials, which causes them to swell under heat exposure. For these materials, the total sample length after testing was actually several millimeters longer than before testing. This trend suggests that the amount of swelling from the char increases with weight percent up to at least 10% for MMT clay and roughly 15% for CNF. The CNF 20% shows less char thickness than the CNF 15%, suggesting that beyond that weight percent the inclusion of more CNF inhibits the intumescent effect, possibly due to a decrease in the amount of polymer matrix. For MMT clay, no decrease is observed in the studied range, suggesting the intumescent effect is likely not inhibited until the weight percent reaches some unspecified value above 10%. The char of MWNT TPUNs behaves differently and show a trend of decreasing thickness with greater weight percent nanomaterial. This is because MWNT char does not swell so the sample thickness before and after testing is roughly equivalent. This trend suggests that MWNT char becomes more effective at protecting the material beneath it at greater weight percentages, which slows the growth of the char layer. This assertion agrees with the observed trends in recession in Figure 10.31 [47]. EPDM showed smaller char thickness than any of the TPUNs tested, though this is not necessarily an indicator of superior or inferior performance, as there appears to be little correlation between char thickness and the other performance parameters measured in this study. It should be noted that for both MMT clay and CNF, no value was recorded for the lowest weight percent loading because the char was either completely eroded or eroded to a degree that no measurement could be taken. Finally, it should be noted that the char of the MMT-based systems was compact and qualitatively it seemed to have the highest structural integrity while the chars of both the MWNT- and CNF-based formulations were quite porous and fragile. In addition, the MMT char did not easily detach from the remaining virgin material, while both the MWNT and CNF chars would detach relatively easy. Thorough analysis of the relative quantitative strengths of these chars may be useful, but it was not considered in this study. However, a char strength sensor is being developed by our research group [52].

Peak Temperature and Time. The three embedded thermocouples in each sample produced a temperature vs. time curve for each sample from which data could be gathered concerning the peak temperature at each depth within the sample as well as the time element to reach peak value. A representative image of these curves can be seen in Figure 10.34 [47]. Peak temperature and time to peak are both indicators of the actual effectiveness of the ablative materials at providing thermal protection and are therefore some of the most important performance parameters. A material that provides effective thermal protection should have both low peak temperatures as well as long time to peak values.

The peak temperature values for each TPUN specimen are shown in Figure 10.35 and compared to the performance of the baseline EPDM-Kevlar composite at the left-most position in the figure [47]. The MMT clay formulations show the lowest peak temperature values, while the MWNT and CNF have mixed results. Some notable formulations are the 5% and 7.5% weight loadings of MMT clay that appear to outperform the baseline EPDM-Kevlar material. The 2.5% and 5% weight loadings of MWNT also perform well, almost duplicating the performance of the baseline. The 15% weight CNF-based formulation exhibited reasonably similar results to the baseline. For both the MMT clay- and CNF-based TPUNs, the

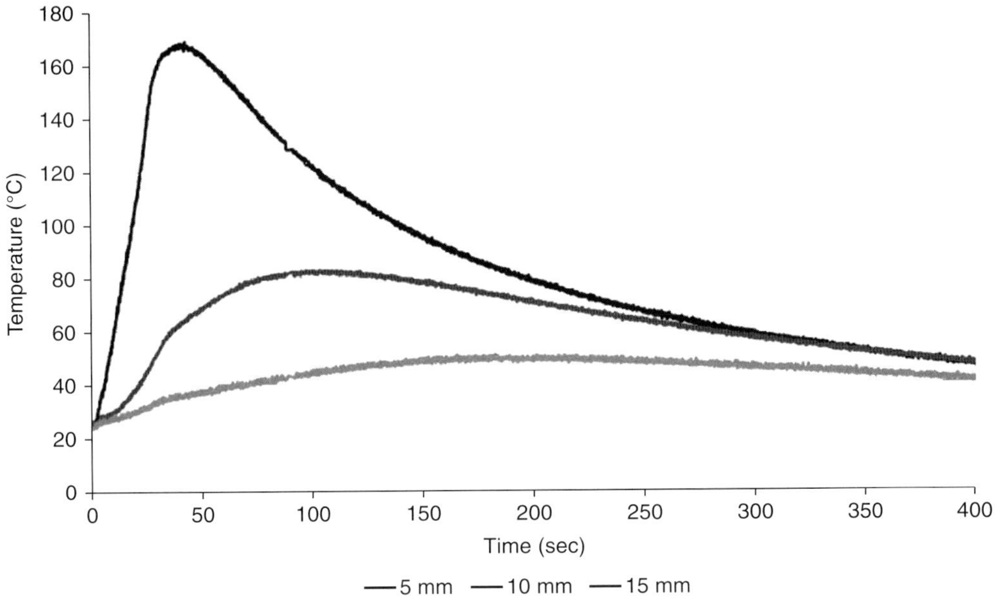

Figure 10.34. Representative temperature vs. time curves for TPUN samples (curves shown are from MMT-10% sample).

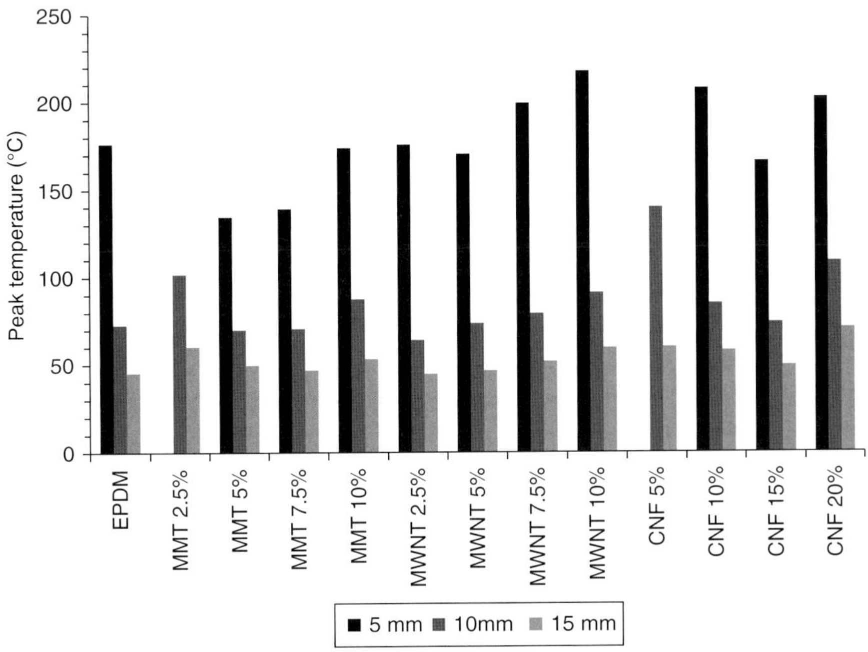

Figure 10.35. Peak temperature values during ablation testing for TPUN formulations.

peak temperature trends seem similar to the mass loss and recession trends, with mid-range weight loadings performing the best. For MWNT-based TPUNs, however, the trends appear to vary, as lower weight loadings seem to perform better in terms of peak temperature values. This may be due to the enhanced thermal

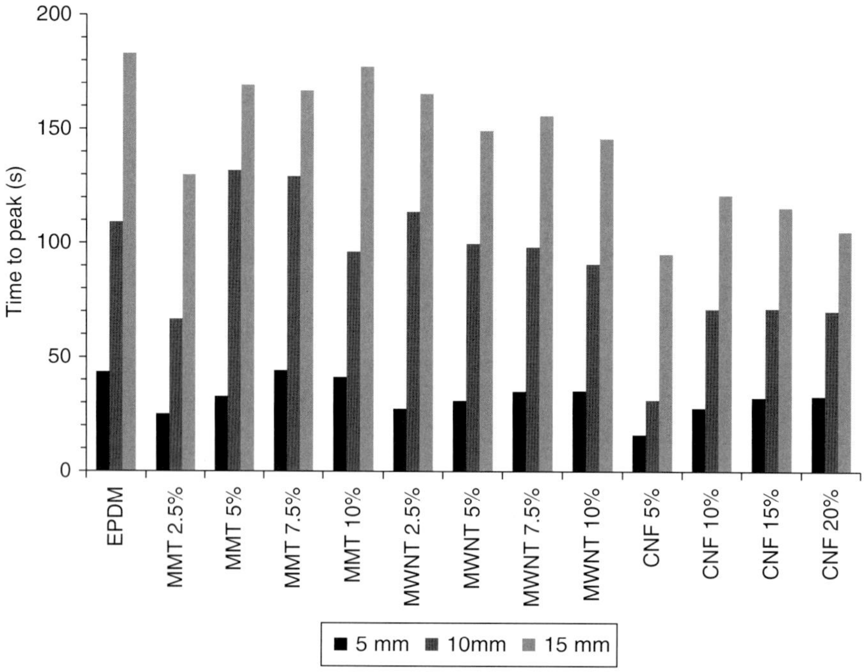

Figure 10.36. Time to reach peak temperature during ablation testing for TPUN formulations.

protection from the more porous char of lower weight loading formulations or the heat energy that is removed as these materials physically erode. However, while these factors may enhance the peak temperature performance, they sacrifice the length of protection that is desired.

The time to peak values for each TPUN specimen can be seen in Figure 10.36 and compared to the performance of the baseline material at the left-most position in the figure [47]. The MMT clay formulations show the highest overall time to peak values, the MWNT have a slightly lower overall values, and CNF formulations overall have very low values. Some notable formulations are 5% and 7.5% weight loadings of MMT clay, as they appear to have the highest time to peak values among the TPUNs and roughly match the baseline material. The 10% weight loading of MMT clay as well as all the MWNT formulations also perform relatively well, though still not as well as the baseline. All the CNF formulations have very poor time to peak values. The MWNT and CNF formulations likely performed slightly worse than MMT clay because CNF increased thermal conductivity of polymeric materials while MMT clay decreased it [53–55]. The effect is probably more prominent in the CNF materials due to their larger size, which allowed for a more connected network through which heat could flow.

Microstructural Analyses. SEM images taken after testing show the microstructure of the char that forms at the surface of these TPS materials during ablation. Figure 10.37 shows the char structures for the 5 wt%, 7.5 wt%, and 10 wt% MMT clay formulations [47]. The 2.5 wt% formulation did not produce stable enough

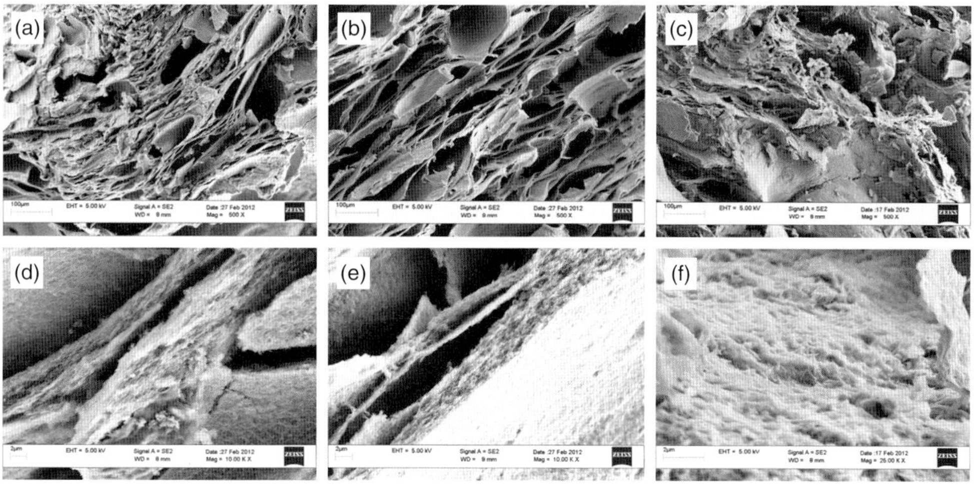

Figure 10.37. SEM images of MMT clay TPUN char microstructures at high magnification (unit bars are 100 um), where u is micron for (a) 5 wt%, (b) 7.5 wt%, and (c) 10 wt% formulations and at low magnification (unit bars are 2 um) where u is micron for (d) 5 wt%, (e) 7.5 wt%, and (f) 10 wt% formulations.

char for imaging. The char of the MMT clay TPUNs is composed of a microstructure of layered platelets, as is seen in the low-magnification images of the MMT clay formulations (Figure 10.37d–f) [47]. Although the MMT clay is not aligned during dispersion or synthesis, the layers formed in the char structure are strongly aligned such that they lie parallel to the sample surface (ablative reassembly) [56]. This enhances the thermal protection effectiveness of the material because the layers are also aligned perpendicular to the direction of heat flow, providing no direct path for heat flow across the spacing between layers. There also appears to be evidence of pockets of gas forming bubbles within the char during ablation, such as at the top of Figure 10.37b [47]. This may be the cause of the intumescent behavior observed in this char. Under higher magnification (Figures 10.37a–c), the platelets appear to be composed of nanoclay that has either been compacted as the polymer matrix is eroded away or sintered together to form the larger platelets [47]. It is likely that there is at least some degree of sintering within these platelets, because SEM images of the ablative surface show spherical particles evident of sintering on all three formulations of MMT clay TPUN (Figure 10.38) [47].

Figure 10.39 shows the char structures of the MWNT TPUNs [47]. The 2.5 wt% formulation has again been excluded because it did not produce structurally stable char. From the low-magnification images (Figures 10.39a–c), the MWNT char appears have a relatively isotropic structure, contrasting the obvious directionality of the MMT clay char [47]. Under higher magnification (Figures 10.39d–f), the MWNT char is clearly composed of a matrix of entangled MWNTs [47]. This structure of entangled conductive nanotubes likely results in a much more thermally conductive char, which could contribute to the poorer performance of the MWNT formulations overall. In addition, the high-magnification images show some bright spots that are evident of eroded MWNTs and related to the metallic catalytic precursors used during their preparation [47].

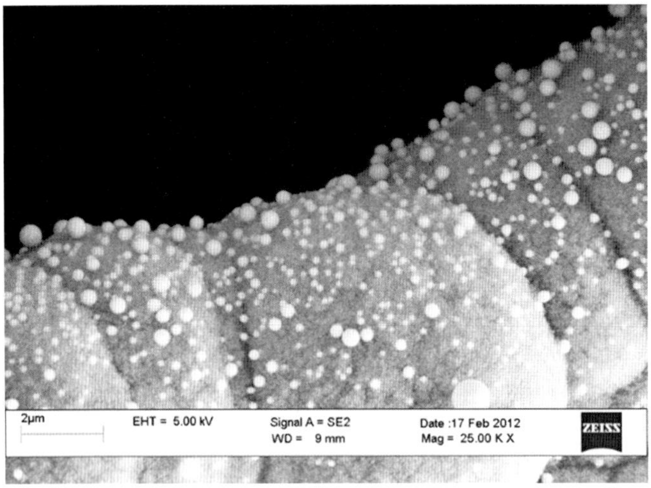

Figure 10.38. SEM images of the 7.5 wt% MMT clay TPUN ablation surface showing evidence of nanoclay sintering.

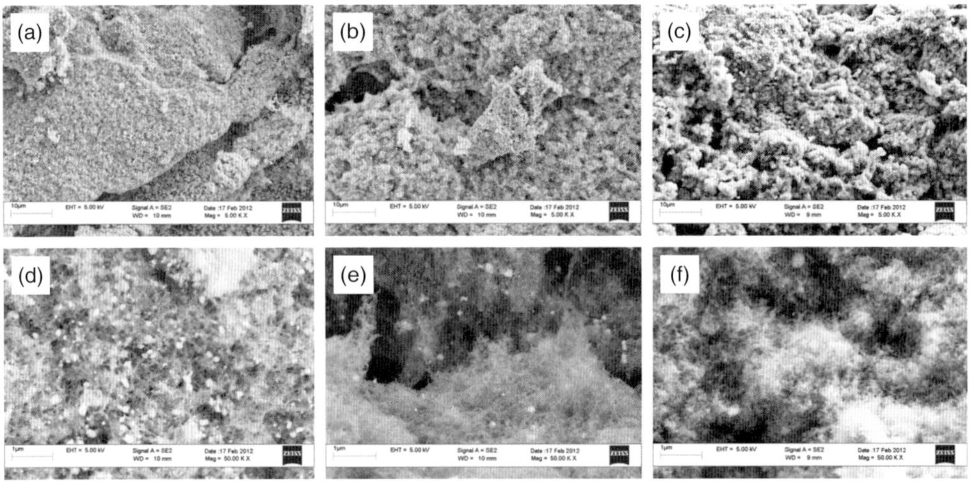

Figure 10.39. SEM images of MWNT TPUN char microstructures at low magnification (unit bars are 10 um) where u is micro for (a) 5 wt%, (b) 7.5 wt%, and (c) 10 wt% formulations and at high magnification (unit bars are 1 um), where u is micro for (d) 5 wt%, (e) 7.5 wt%, and (f) 10 wt% formulations.

Figure 10.40 shows the char structures of the CNF TPUNs. Only the 15% and 20% weight loading formulations are shown, as both the 5 wt% and 10 wt% formulations produced unstable char [47]. From the low-magnification images (Figures 10.40a–b), the CNF char shows a relatively isotropic structure, similar to that of the MWNT char [47]. In Figure 10.40b there is also evidence of gas pockets within the char, which could account for the highly intumescent behavior of the CNF ablatives [47]. Higher magnification images of the char (Figures 10.40c–d) show the char to be composed of a matrix of entangled CNFs [47]. This structure

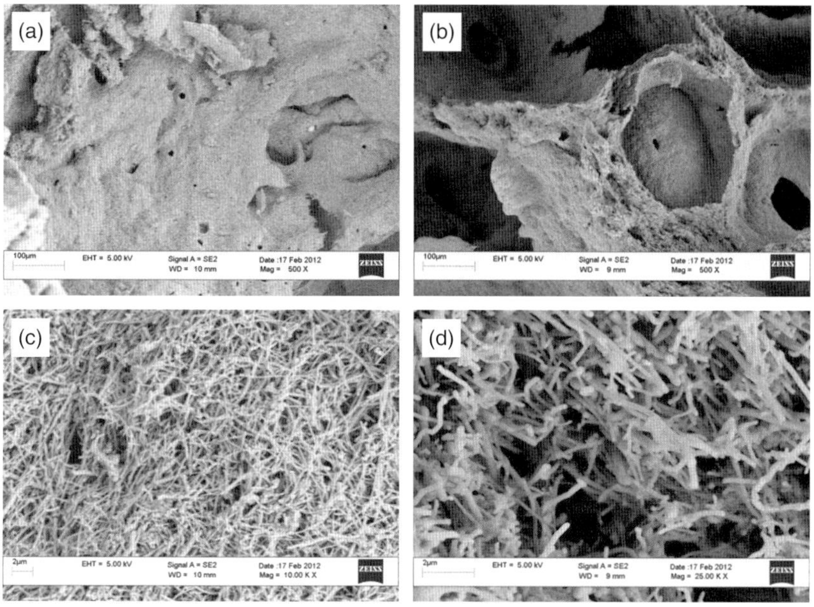

Figure 10.40. SEM images of CNF TPUN char microstructures at low magnification (unit bars are 100 um), where u is micro for (a) 15 wt% and (b) 20 wt% formulations and at high mag (unit bars are 2 um), where u is micro for (c) 15 wt% and (d) 20 wt% formulations.

produces a thermally conductive char, resulting in poorer overall performance of the CNF TPUNs.

Summary and Conclusions. Based on the results of this study, TPUNs appear to offer a potentially attractive alternative to conventional composites as a choice for SRM internal insulation materials. All three of the different nanomaterial additives enabled the polymer system to perform at a level comparable to that of SOTA insulation composite EPDM-Kevlar. Of particular importance were the TPUN formulations of 5% and 7.5% weight loading of MMT clay. These materials performed exceptionally well in every metric measured and actually outperformed the baseline, EPMM/Kevlar® material, in recession and peak temperatures. Some other noteworthy formulations are the 5 wt% MWNT and 15 wt% CNF samples. The 5 wt% MWNT material performed well by roughly matching the baseline in all metrics, but was outperformed by the MMT clay formulations. The 15 wt% CNF material also performed well in all metrics, but was again outperformed by the MMT clay formulations and produced very poor time to peak results, which reflects negatively on the overall performance of the material.

Through this study, polymer nanocomposites are shown to be viable candidates for SRM insulation materials, and the relative effectiveness of various nanomaterial additives has been well documented. The next step in advancing these materials and enhancing their performance is to either focus on the fine-tuning of nanomaterial loading, incorporation of multiple nanomaterial additives, or incorporation of a combination of conventional and nanomaterial additives. In addition, alternative processing techniques or nanomaterial types and preparations can be considered.

Ambuken et al. [58] investigated the high-temperature flammability and mechanical properties of thermoplastic polyurethane (TPU) nanocomposites. From the UL 94 studies it was found that high loading of nanoparticles is needed for all the formulations to have a stable char structure. From the cone calorimeter studies on TPU nanocomposites, the lowest PHRR was observed for TPU-nanoclay formulation, while the highest PHRR was shown by TPU-CNF nanocomposites. The char obtained after firing TPU-nanoclay nanocomposites was swollen and retained its structural integrity while TPU-MWNT and TPU-CNF exhibited fractures.

DMA testing was performed on the same materials to study complex modulus as a function of temperature up to 300°C when the char is expected to form. TPU-MWNT displayed better char modulus in comparison to TPU-nanoclay. For TPU-CNF, tests could not be completed above 220°C as the sample was transformed into a viscous melt. All samples lost modulus (soften) in the range of 120–160°C. For TPU-MWNT and TPU-nanoclay, nanocomposite recovery in modulus (reinforced char formation) was observed at about 230°C with loadings above 7.5 wt%. The temperature range at which modulus had decreased but char formation was not yet significant enough to exemplify modulus recovery was termed the reinforcement gap. A correlation can be developed between crossover of dissipative vs. elastic behavior (dominance of loss modulus vs. storage modulus) and dripping in UL 94 tests.

10.3.5.2 Silcone Rubber-Ceramic Nanocomposite Studies by National University of Defense Technology, China

For their investigation, Yang et al. [59] created nanocomposites of silicone rubber with 30 wt% nano-silica reinforced with 8 wt% chopped carbon fibers. Then they investigated the effects of various loadings, from 0 to 40 phr (parts per hundred parts of rubber), of ZrC and ZrO_2 microparticles. The formulations were mixed for 30 minutes in a two-roll mill, using toluene to reduce the viscosity during mixing so that the carbon fibers would not be broken. Vulcanization lasted for 15 minutes at 160°C and 10 MPa in a compression mold, and for 2 hours at 180°C at 1 atmosphere.

The researchers characterized the formulations for their mechanical properties using a universal testing machine (UTM) according to GB/T 528-1998 at a strain rate of 200 mm/min, and their ablative properties using an oxyacetylene torch with a heat flux of 4152.9 kW/m^2. The calorimetry method used to find this heat flux was not given. They performed SEM, TGA at 20°C/min to 800°C in argon and 1100°C in air, FTIR, and XRD on the pre- and posttest samples as well.

The carbon fibers were found to disperse well without breaking, but to adhere poorly to the rubber matrix, as shown in Figure 10.41 [59]. For this reason, they actually had a negative impact on the composites' mechanical properties. The authors speculated that it was the toluene that prevented strong bonding before and during its evaporation during processing. When the ZrO_2 content increased to 40 phr loading, the ablation rate was reduced by 72% and the mass loss dropped by 38%. The loading of 40 phr ZrC was found to reduce the ablation rate by 40% and the mass loss by 32%. The FTIR analysis revealed a few chemical reactions affecting the ceramics present in the char of each formulation. The TGA results showed a

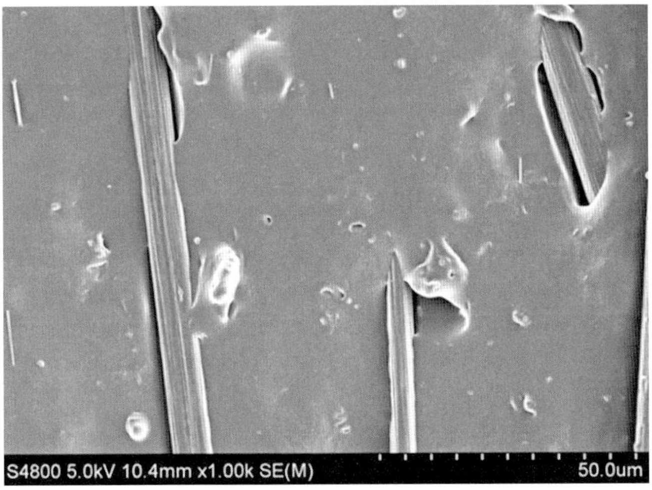

Figure 10.41. SEM of carbon fibers adhering poorly to the silicone matrix [59].

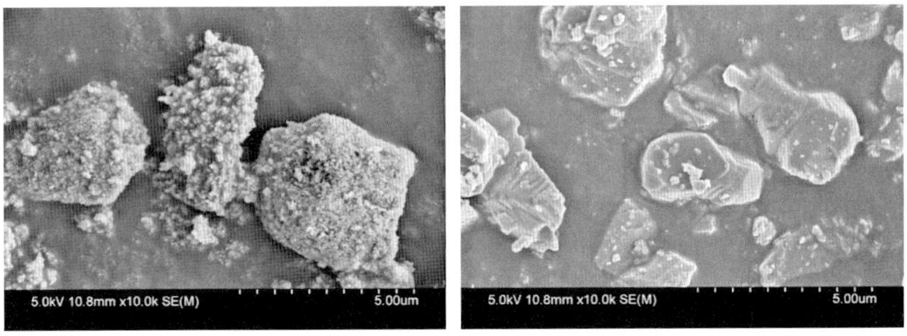

Figure 10.42. SEM images of SEM images of ZrO_2 (left) and ZrC (right) [59].

slight improvement in thermal stability with increased loadings of microparticles. The authors found that the ZrO_2 particles were actually agglomerated nanoparticles, whereas the ZrC microparticles were each single crystals as shown in Figure 10.42, and concluded that this was the reason for the better ablation performance of the ZrO_2 formulations [59].

10.3.6 Phenolic Nanocomposite Studies

10.3.6.1 Phenolic-CNF Nanocomposite Studies by Mississippi State University, United States

The ablation, mechanical, and thermal properties of vapor-grown carbon fiber, VGCF (Pyrograf III™), manufactured by Applied Sciences, Inc. (Cedarville, Ohio), and phenolic resin composites were evaluated by Patton et al. [60, 61]. The objective of Patton et al.'s study was to determine the potential of using this material in SRM nozzles. Composites with varying VGCF loadings (30–50 wt%), including one sample with ex-rayon carbon fiber plies (this is the carbon fiber used in MX-4926, industry-standard ablative material), were prepared and exposed to a plasma torch with a heat flux of 16.5 MW/m^2 at approximate 1,650°C for 20 seconds. Low erosion rates and little char formation were observed, confirming that these materials were promising for SRM nozzle materials. When fiber loadings increased, mechanical properties and ablative properties improved. The VGCF composites had low thermal conductivities (approximately 0.56 W/m-K), indicating they were good insulating materials.

VGCF and phenolic resin, SC-1008, manufactured by Borden Chemical, Inc. (now owned by Hexion), were used in this study. VGCF/phenolic resin composite specimens were prepared by high-shear mixing followed by thermal curing. Nineteen composite specimens were made by different mixing methods. Four types of samples were prepared and identified in wt./wt. as follows: 30/70 VGCF/Phenolic, 40/60 VGCF/Phenolic, and 40/60 VGCF/Phenolic (where ball milled as-received VGCF was used). Two types of fibers were used: as-received and compacted fibers. All specimens used in ablation testing were made as-received VGCF.

The flexural strengths and moduli were lower for these VGCF/phenolic samples compared to the authors' previous data of VGCF/epoxy using identical mixing techniques [61]. The flexural strength of 95.9 MPa and a modulus of 6.24 GPa for VGCF/epoxy composite [60, 61] are reported earlier. When the epoxy resin was replaced by phenolic resin, a flexural strength of 25.6 MPa and a modulus of 1.07 GPa was obtained [60]. This reduction in mechanical properties may be due to poor fiber-to-matrix adhesion for the VGCF/phenolic composites as well as lower mechanical properties of phenolic resin as compared to epoxy resin. Samples made with compacted fiber had high flexural moduli and strengths. The compacted fiber was a different batch of fiber, and this may have affected also the properties of the composite samples made with it.

All the composite samples exhibited similar thermal conductivities (0.54–0.62 W/m-K) as compared to 0.28 W/m-K of cured phenolic resin without fiber. The data did not indicate that increased carbon loading or different manufacturing techniques significantly affect the thermal conductivity. The thermal conductivities in three dimensions were obtained on a 40/60 VGCF/phenolic ablation test composite sample. These values indicated that the material is slightly anisotropic with a difference of less than ± 9% from the average in all three directions. The specimen has a thermal conductivity, which is nearly the average of all three specimens.

Ablation was characterized by erosion rates, specimen weight loss, and the increase in load. NASA's standard nozzle material, MX-4926, was used as the baseline. The MX-4926 composite is composed of 50 wt% woven ex-rayon carbon fiber, 15 wt% carbon black filler, and 35 wt% SC-1008 phenolic resin. The total carbon loading is 65 wt% in MX-4926. The VGCF/phenolic tested in this study contained only 30–45 wt% VGCF (only 46–69% as much as carbon as MX-4926 composite). As compared to the MX-4926 baseline, the VGCF composites experienced (1) higher (58–114%) erosion rates, (2) lower weight loss (10–26%), and (3) lower load changes (60.5–68.5%). Therefore, more weight was being lost or near the surface of the specimen due to higher erosion rates vs. weight loss from subsurface thermal decomposition. The load changes for the VGCF composites are significantly lower than that of the MX-4926 composite. Load change is an indirect measure of neat penetration and char depth; increases in loading change indicate an increase in heat penetration and char depth. Thus, the VGCF composite appears to be a far better insulator than the MX-4926. The higher erosion rates of the VGCF samples probably reflect their lower carbon content. The erosion rates/wt% of carbon present in the composites was very similar between MX-4926 and the VGCF composites.

The series of VGCF/phenolic composites exhibited extremely good erosion resistance (considering their low carbon loadings) while exhibiting less weight loss and load change than current SOTA MX-4926. Higher carbon contents are needed to match the erosion resistance of MX-4926. The carbon loadings used in this study have been much lower (30–45 wt%) that the MX-4926 composite (50 wt%). The mechanical properties of the VGCF composites need improvement. Lower void volume, better surface treatments for fiber/matrix adhesion, and mixed VGCF/carbon fiber weave/carbon filler combinations should be investigated. VGCF/phenolic composites should be prepared with higher carbon loadings (50–65 wt%) and subjected to ablation testing. The nanoscale dimensions of the VGCF causes major changes in the heat transfer rates and affects the resulting combustion/decomposition chemistry. The nano-sized fiber reduces the rate of heat transfer perpendicular to the surface into the composite compared with the rate of heat transfer in the MX-4926 material, which uses continuous carbon fibers.

10.3.6.2 Phenolic Nanocomposites Studies by The University of Texas at Austin, United States

The Univeristy of Texas at Autin group has been involved in the traditional R&D of ablative materials since early 1990 by Koo and colleagues [63–69] (Koo et al. [62–64]; Cheung et al. [65]; Wilson et al. [66]; Shih et al. [67]; Koo et al. [68, 69]) and began the development of polymer nanocomposite (PNC) ablatives in 2002. Koo and colleagues reported two new classes of PNCs (SRM insulation and nozzle materials) that are lighter and exhibit better ablation performance and insulation characteristics than current SOTA ablative materials. The SRM insulation PNCs were reported in previous sections (10.3.3.2 and 10.3.5). This section relates solely to SRM nozzle PNCs.

Ablative materials are required to protect aerospace launching systems against rocket exhaust plumes. The flow environment of solid rocket exhaust is particularly hostile. System components must be protected from extreme heat flow temperatures of 1,000°C to 4,000°C and from highly abrasive particles ejected at velocities greater than 1,000 m/s. Phenolic matrix composites have been used extensively for ablative materials. Current rocket nozzle assemblies are made of carbon phenolic composites, such as MX-4926 from Cytec Engineered Materials (CEM). Three versions of carbon phenolic ablative MX-4926 laminates were fabricated by CEM as controls for this study. The three versions of MX-4926 laminates were fabricated based on fiber orientation: (a) 60° shingle, (b) molding compound, and (c) two-dimensional fabric.

The objective of this nanocomposite for rocket ablative materials (NRAMs) program is to develop a new class of nanostructured material that is lighter and has

Figure 10.43. Ablative material testing using SSRM.

better erosion and insulation characteristics than current ablatives [70–78]. Our previous studies [70–72] have allowed us to explore other nanoparticles besides MMT nanoclays with phenolic for ablative application. The NRAMs exploit the ablation resistance of both phenolic and nanoparticles. Hexion's (formerly Borden Chemical) SC-1008, a resole phenolic, was selected as the resin for this investigation. Several nanoparticles, such as Southern Clay Products (SCP) nanoclay, Applied Sciences CNFs, and Hybrid Plastics POSS, have been evaluated with the SC-1008 phenolic resin. Based on WAXD and TEM, several nanoparticles were selected for further study. Cytec Engineered Materials fabricated several MX-4926 alternates by replacing the carbon black filler with selected nanoparticles for ablation testing.

A small-scale, supersonic, liquid-fueled rocket motor burning kerosene and oxygen was used to study the ablation and insulation characteristics of the ablatives. Testing ablative materials in full-scale firings is not only expensive but also requires lengthy planning and limited exposure time, thus reducing the number of samples and the variety of test conditions available. The simulated solid rocket motor (SSRM) (Figure 10.43) has been demonstrated as a cost-effective laboratory device to evaluate different ablatives under identical conditions for initial material screening and development [62].

Experimental Approach: Simulated Solid Rocket Motor (SSRM). The SSRM [62] was used to evaluate the candidate materials. The SSRM is an established testing apparatus developed earlier that has been used extensively in our previous studies [63–69, 79–83]. It is a small-scale, liquid-fueled rocket motor burning a mixture of kerosene and oxygen as shown in Figures 10.43 and 10.44 [62]. Aluminum oxide (Al_2O_3) particles are injected into the plume to simulate the particle-laden flow of solid rocket exhaust plumes. The SSRM is a controlled laboratory device capable

Figure 10.44. Close-up of ablative material after SSRM testing.

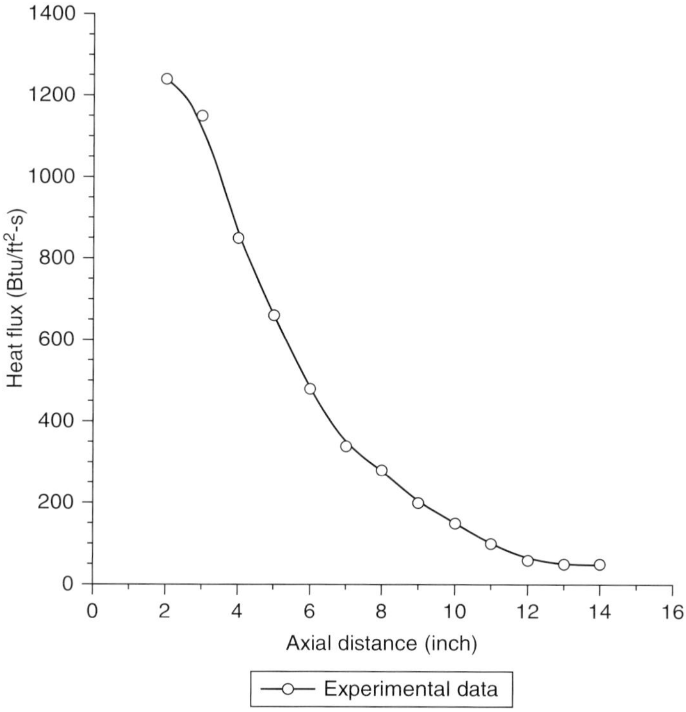

Figure 10.45. Calibration of SSRM in heat flux vs. distance from the nozzle exit.

of producing a particle-laden exhaust environment with measured heat fluxes from 40 to 1250 Btu/ft^2-s (454 to 14,200 kW/m^2). The flame temperature is approximately 3992°F (2200°C), and the velocity of the particle-laden exhaust is approximately 2000m/s. Figure 10.44 shows a close-up photo of a SSRM posttest ablative specimen. Calibration of the SSRM was performed using a Medtherm water-cooled heat flux gauge. The total heat flux calibration of the SSRM is shown in Figure 10.45 [62].

The highest heat flux is 1250 Btu/ft^2-s (14,200 kW/m^2) at 2 inches (5.08 cm) from the nozzle exit, and the lowest heat flux is about 40 Btu/ft^2-s (454 kW/m^2) at 14 inches (35.56 cm) from the nozzle exit.

The standard sample size is 4 by 4 by 1/4- or 1/2-inch thick (10.2 by 10.2 by 0.64- or 1.27-cm thick). The composite materials were bonded to 4 by 4 by 1/8-inch (10.2 by 10.2 by 0.32-cm) steel substrate. A narrow slot was machined into the bondline side of the steel substrate. The thermocouple was embedded into this slot. The bead of the thermocouple was placed at the center point of the plate. The temperature history of the bondline was recorded with this thermocouple. C-clamps were used to clamp the ablatives to the steel substrate during testing. The test samples were placed in a fixture downstream from the SSRM nozzle exit. The fixture had adjustments in axial distance and impingement angle with reference to the plume centerline. A normal 90° impingement was used for this study.

Three axial distances were used to correspond to heat fluxes of 250, 625, and 1,000 Btu/ft^2-s (2838, 7094, and 11,345 kW/m^2). These distances were selected to simulate low, medium, and high heating conditions. Prior to the motor burn, the bondline temperature was recorded. Peak erosion was determined by pre- and posttest measurements using a pencil-point dial indicator. Mass loss was determined by pre- and posttest weight loss measurements. Consistent data sets were averaged for this study. The following data were acquired during the experiment:

- Peak erosion (mm)
- Residual mass (%)
- Maximum backside heat-soaked temperature (°C)
- Time to reach maximum heat-soaked temperature (min.)
- Surface temperature (°C)

Selection of Materials: Polymer Matrix Composites. MX-4926 is a rayon-based, eight-harness weave carbon fabric impregnated with a Mil R-9299 phenolic resin that contains carbon black fillers [84]. The approximate composition of materials for MX-4926 is 50 wt% carbon reinforcements, 35 wt% phenolic resin, and 15 wt% carbon black. Figure 10.46 shows how MX-4926 is currently inserted in a typical solid rocket motor nozzle assembly [85]. As noted in the figure, MX-4926 (shaded areas) is mainly used in the entrance region and nozzle exit cone of the rocket nozzle assembly. The phenolic char formed at high temperature is usually not very strong. The carbon black filler in the phenolic resin was replaced by different nanoparticles, so that the nanoparticles would reinforce the phenolic in the nanometer scale and strengthen the char.

Phenolic Resin. Hexion SC-1008 [86] is the phenolic resin used in the MX-4926. This was our phenolic resin of choice for this investigation. SC-1008 is a solvent (isopropanol)-based resole. Table 10.6 shows some properties of the phenolic resin [86].

MMT Nanoclays. Cloisite 30B, with loadings of 5, 10, and 15 wt%, was dispersed in the phenolic resole (SC-1008). The TEM analyses indicated the degree of dispersion and exfoliation of the nanoclay before committing to a 20 lb run at CEM of these nanoparticle-resin mixtures to make prepregs. This proved to be a very cost-effective

Table 10.6. *Phenolic resin properties*

Resin	Supplier	Formulation	Final Cure Temp (°C)	T_g (°C) by DMTA	Specific Gravity
SC-1008[12]	Hexion	60–64% solids isopropanol solvent	140	110	1.28

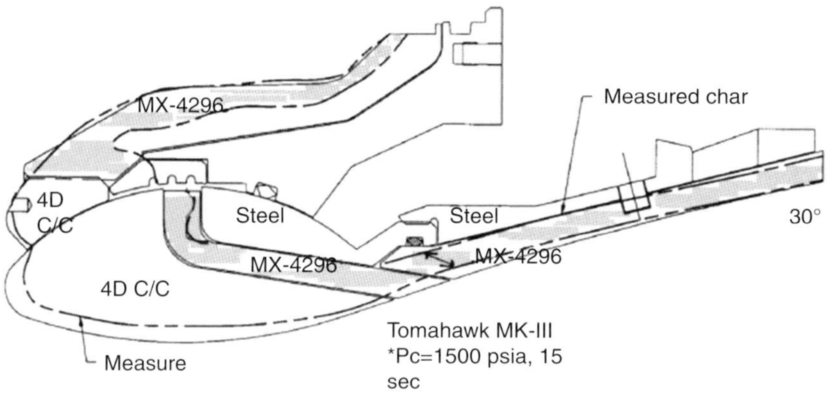

Figure 10.46. Schematic of a typical solid rocket motor assembly showing the usage of MX-4926 (shaded areas).

and efficient technique for screening different formulations. This nanoclay has been used with other resin systems, such as cyanate ester [87–89], thermoplastic elastomer [90–92], epoxy [93–95], and polyamide [96–104] in our research. For more details on MMT nanoclay, see Chapter 2.

Carbon Nanofibers (CNFs). CNFs are a form of vapor-grown carbon fiber that is a discontinuous graphitic filament produced in the gas phase from the pyrolysis of hydrocarbons. The CNFs PR-19-PS and PR-24-PS were used in this study. More details on CNFs can be found in Chapter 2.

Polyhedral Oligomeric Silsesquioxane (POSS). Representing a merger between chemical and filler technologies, POSS nanostructured materials can be used as multifunctional polymer additives, acting simultaneously as molecular-level reinforcements, processing aids, and flame retardants. Trisilanolphenol POSS was selected for this study. More details on POSS can be found in Chapter 2.

Discussion of Results. The candidate materials were evaluated for degree of dispersion using WAXD and TEM prior to full ablation testing using the SSRM. Our first attempts were to modify the phenolic resin by the incorporation of MMT organoclays. Blends of phenolic and nanoparticles were dispersed using high-shear, non-sparking paint-mixing equipment. Neat resin castings (no fiber reinforcement) were prepared for WAXD and TEM analyses.

Transmission Electron Microscopy Analyses and Blending Experiments. Each type of nanoparticles was dispersed in 0.5% IPA solvent overnight to create specimens for TEM analyses. The TEM image in Figure 10.47a shows the carbon black particles

Figure 10.47. Transmission electron microscopy images of (a) carbon black, (b) Cloisite 30B, and (c) PR-24-PS CNF dispersed in the IPA solvent (scale bar is 500 nm).

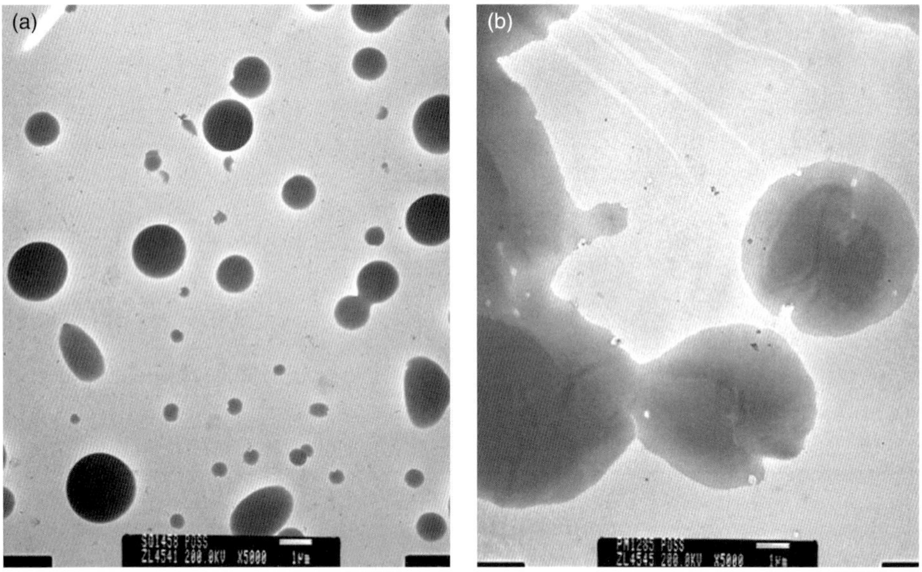

Figure 10.48. Transmission electron microscopy images of (a) Trisilanolphenyl-POSS-SO1458 ($C_{48}H_{38}O_{12}Si_7$) and (b) (PVSQ)-PM1285 $(C_2H_3O_{1.5}Si)_n$ dissolved in the IPA solvent (scale bar is 1 μm).

agglomerated in the IPA [33]. The TEM image in Figure 10.47b shows that the Cloisite 30B nanoclay layers are in an intercalated state [33]. Figure 10.47c shows the PR-24-PS CNFs are entangled [33]. Figures 10.48a and 10.48b show the TEM images of Trisilanolphenyl-POSS-SO1458 ($C_{48}H_{38}O_{12}Si_7$) and polyvinyl silsesquioxane uncured (PVSQ)-PM1285 $(C_2H_3O_{1.5}Si)_n$ dissolved in the IPA solvent [33]. The scale bar is 1 μm in Figures 10.48a-b.

Based on TEM analyses, it was concluded that carbon black (CB) particles caused interference in the dispersion of all nanoparticles in the SC-1008 resin. As

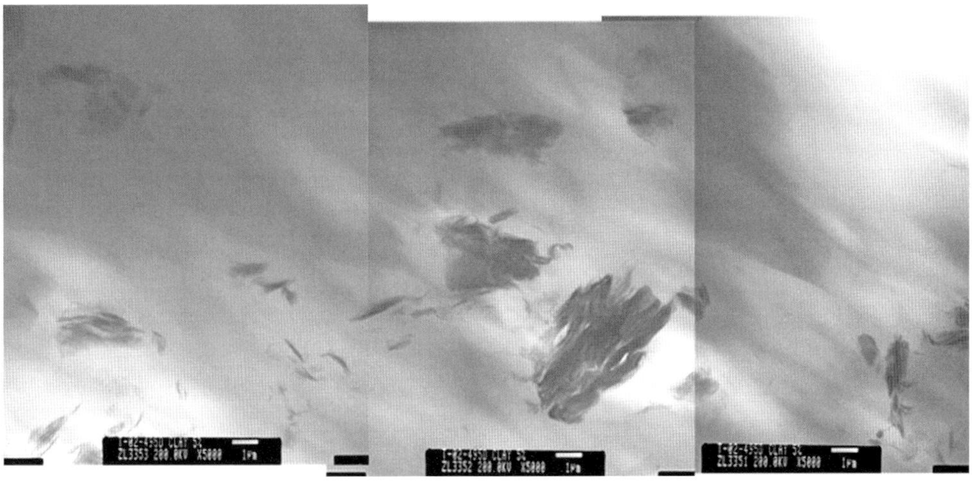

Figure 10.49. Transmission electron microscopy images of 5 wt% Cloisite 30B in 95 wt% SC-1008 (scale bar is 1 μm).

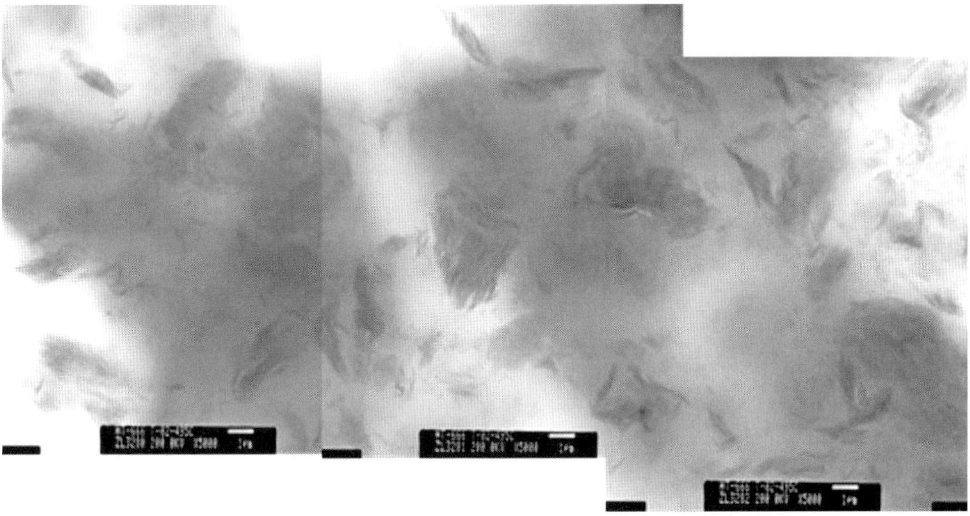

Figure 10.50. Transmission electron microscopy images of 10 wt% Cloisite 30B in 90 wt% SC-1008 (scale bar is 1 μm).

a result, CB was eliminated in subsequent blending. Cloisite 30B, in loadings of 5, 10, and 15 wt%, was dispersed in the SC-1008. Figures 10.49–10.51 show the TEM images of the 5 wt% Cloisite 30B: 95 wt% SC-1008 (Figure 10.49), 10 wt% Cloisite 30B: 90 wt% SC-1008 (Figure 10.50), and 15 wt% Cloisite 30B: 85 wt% SC-1008 (Figure 10.51) [33]. The TEM images identified intercalation, and not exfoliation, of the nanoclay in the resin system. The TEM analyses provided guidance to determine the degree of dispersion and exfoliation of the nanoclay before committing to a 20 lb run at CEM of these nanoparticle-resin mixtures for the preparation of prepregs. This proved to be a very cost-effective and efficient technique for screening different formulations on a small scale.

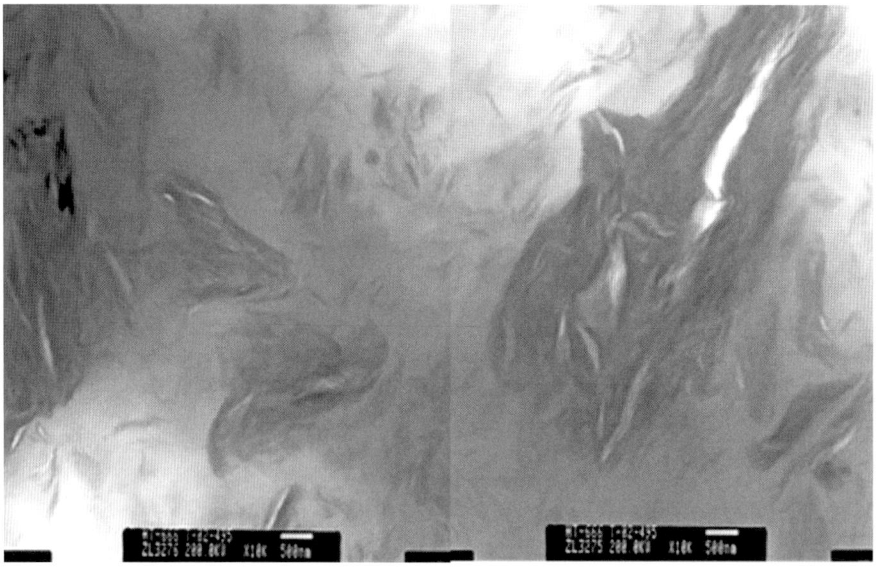

Figure 10.51. Transmission electron microscopy images of 15 wt% Cloisite 30B in 85 wt% SC-1008 (scale bar is 500 nm).

SSRM Test Conditions for Ablation Test. Three test conditions to simulate thermal effect of the exhaust plumes using low, medium, and high heat flux were selected. These different conditions represent different regions in a rocket nozzle assembly. Alumina (Al_2O_3) particles were added to the exhaust stream to simulate the particle impingement effect of the solid rocket exhaust plumes from the aluminized solid propellants. The following six test conditions used are listed in the order of their severity:

- 1000 Btu/ft^2-s (11,345 kW/m^2) with Al_2O_3 particles
- 625 Btu/ft^2-s (7094 kW/m^2) with Al_2O_3 particles
- 250 Btu/ft^2-s (2838 kW/m^2) with Al_2O_3 particles
- 1000 Btu/ft^2-s without Al_2O_3 particles
- 625 Btu/ft^2-s without Al_2O_3 particles
- 250 Btu/ft^2-s without Al_2O_3 particles

MX-4926 as Controls. Cytec Engineered Materials fabricated three sets of carbon phenolic laminates (MX-4926) based on the SC-1008 resin system. Three versions of MX-4926 were used as controls for this investigation. They were made based on fiber orientation:

- Molding compound (1/2 in. squares of chopped fabric, designated by MX-4926 MC)
- Two-dimensional fabric (designated by MX-4926 2D)
- 60° shingle (the most effective orientation for rocket nozzle application designated by MX-4926 SH)

Samples of MX-4926 MC were tested at six conditions, as listed above, with an Al_2O_3 particle flow rate of 15 g/min for a test duration of 15 seconds. Table 10.7 shows a

Table 10.7. *Summary of ablation data for MX-4926 MC*

Ablation Data/ Heat Flux	1,000 Btu/ ft²-s with Particles	625 Btu/ ft²-s with Particles	250 Btu/ ft²-s with Particles	1,000 Btu/ ft²-s without Particles	625 Btu/ ft²-s without Particles	250 Btu/ ft²-s without Particles
Peak Erosion (inch)	**0.235**[a]	0.234	0.073	0.03	0.029	0.
Residual Mass (%)	85.0	86.3	90.2	88.2	88.9	91.6
MBHS Temp Rise (°C)	315	159	133	175	164	133
Time to MBHS Temperature Rise (min)	0.3	0.5	1	1	0.9	0.9

[a] Bold entry stands for total erosion (burned through)

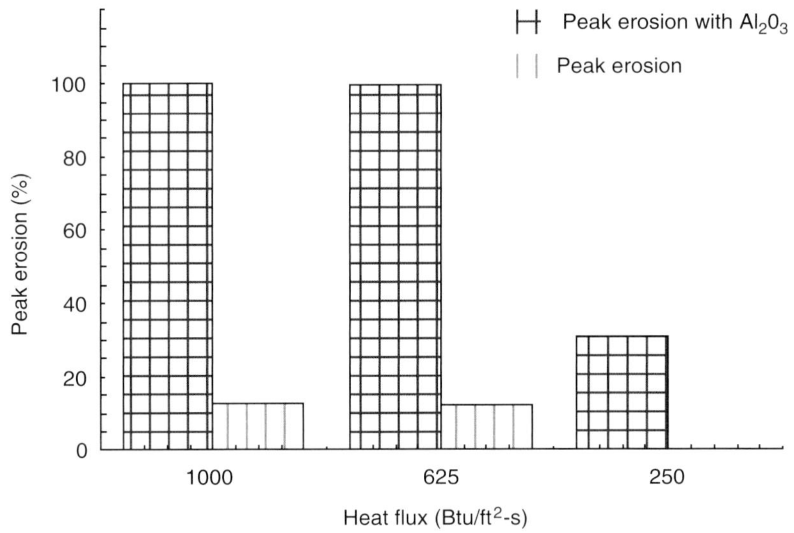

Figure 10.52. Peak erosion of MX-4926 MC exposed to six test conditions.

summary of the ablation data for MX-4926 MC specimens. Figure 10.52 shows the peak erosion of MX-4926 MC exposed to the six test conditions [33]. As shown in Figure 10.52, the peak erosion with Al_2O_3 was significantly higher than those specimens without particles. The residual mass increased with the decrease of heat flux for test cases with and without particles. The maximum backside heat-soaked (MBHS) temperature rise also decreased with the decrease of heat flux.

Figure 10.53 shows that MX-4926 SH has the lowest peak erosion, followed by MX-4926 2D, and MX-4926 MC has the highest peak erosion [33]. The residual mass of the three versions of MX-4926 specimens is similar for the three different heat fluxes. This is an insensitive criterion for comparing material candidates. The data indicated that MX-4926 MC has the best insulative property, followed by MX-4926 2D, and MX-4926 SH has the worst insulative property.

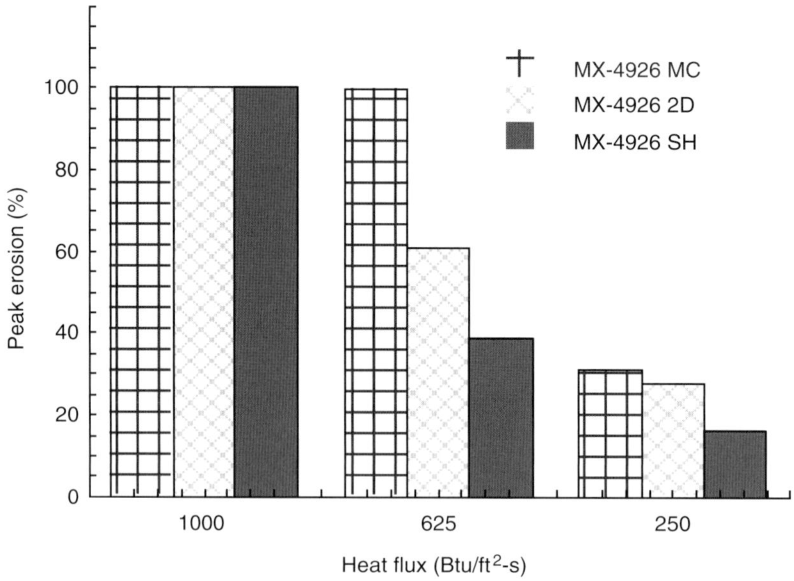

Figure 10.53. Peak erosion of MX-4926 MC, 2D, and SH with Al_2O_3 particles.

In summary, MX-4926 SH has the most erosion-resistant (lowest peak erosion) characteristic and the worst insulative property (highest MBHS temperature rise); MX-4926 MC has the worst erosion-resistant characteristic and the best insulative property; and MX-4926 2D has moderate peak erosion and insulative properties. In terms of manufacturing, MX-4926 SH is the most difficult and MX-4926 2D is the easiest to manufacture. MX-4926 MC is a more representative fabrication method to be used in solid rocket motors than MX-4926 2D. The molding compound (MC) specimens were selected for this study.

Scale Ablation Test. SC-1008 with 5, 10, and 15 wt% Cloisite 30B were selected to replace 15 wt% of carbon black in the original MX-4926 formulation. Cytec Engineered Materials prepared three versions of MX-4926 alternates and designated them MX-4926 ALT 5%, MX-4926 ALT 10%, and MX-4926 ALT 15%. Three loadings of PR-24-PS in 20, 24, and 28 wt% were dispersed in phenolic without the rayon-carbon fiber reinforcement. Three loadings of Trisilanolphenyl-POSS in 2, 6, and 10 wt% were also dispersed in SC-1008 phenolic resin. The POSS/SC-1008 mixture was used with the rayon-carbon fabric to produce prepregs in a manner similar to the nanoclay mixture. Table 10.8 shows the chemical compositions of the laminates used for ablation testing. Figure 10.54 compares the densities of three types of NRAMs with nanoclay, CNF, and POSS at various loading levels of nanoparticles [33]. All clay-NRAMs, CNF-NRAMs, and POSS-NRAMs have densities lower than MX-4926.

At the suggestion of W. Luehmann [85] of Pratt & Whitney Space Propulsion/Chemical System Division (P&WSP/CSD), a lower flow rate of Al_2O_3 particles in the SSRM to simulate a total erosion of 200 mils (0.20 inch) of MX-4926 SH at 1,000 Btu/ft^2-s was adopted. Several MX-4926 SH and MX-4926 2D were tested under

Table 10.8. *Specimen configuration for SSRM laminate fabrication*

Material ID	Density (g/cc)	Rayon carbon fiber reinforcement (wt%)	Resin SC-1008 phenolic (wt%)	Filler (wt%)
MX-4926 (Control)	1.44	50	35	15 carbon black (CB)
MX-4926 ALT Clay 5%	1.42	50	47.5	2.5 Cloisite® 30B [$(HE)_2MT$]
MX-4926 ALT Clay 10%	1.43	50	45	5 Cloisite® 30B [$(HE)_2MT$]
MX-4926 ALT Clay 15%	1.43	50	42.5	7.5 Cloisite® 30B [$(HE)_2MT$]
PR-24-PS 20%/ SC-1008	1.35	None	80	20 PR-24-PS CNF
PR-24-PS 24%/ SC-1008	1.38	None	76	24 PR-24-PS CNF
PR-24-PS 28%/ SC-1008	1.41	None	72	28 PR-24-PS CNF
MX4926 ALT SO-1458 2%	1.41	50	49	1 Trisilanolphenyl-POSS® [SO-1458]
MX4926 ALT SO-1458 6%	1.38	50	47	3 Trisilanolphenyl-POSS® [SO-1458]
MX4926 ALT SO-1458 10%	1.40	50	45	5 Trisilanolphenyl-POSS® [SO-1458]

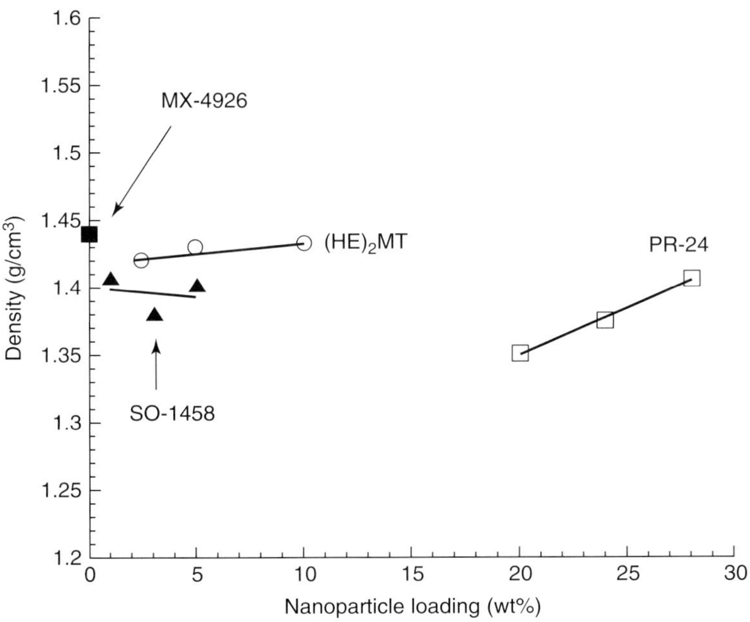

Figure 10.54. Density of MX-4926 and NRAMs with different nanoparticle loadings.

these new test conditions, using a flow rate of 4 g/min Al_2O_3 particles [39]. For this study, test conditions of 1,000 Btu/ft^2-s heat flux with a flow rate of 4 g/min Al_2O_3 particles for a test duration of 15 seconds were adopted.

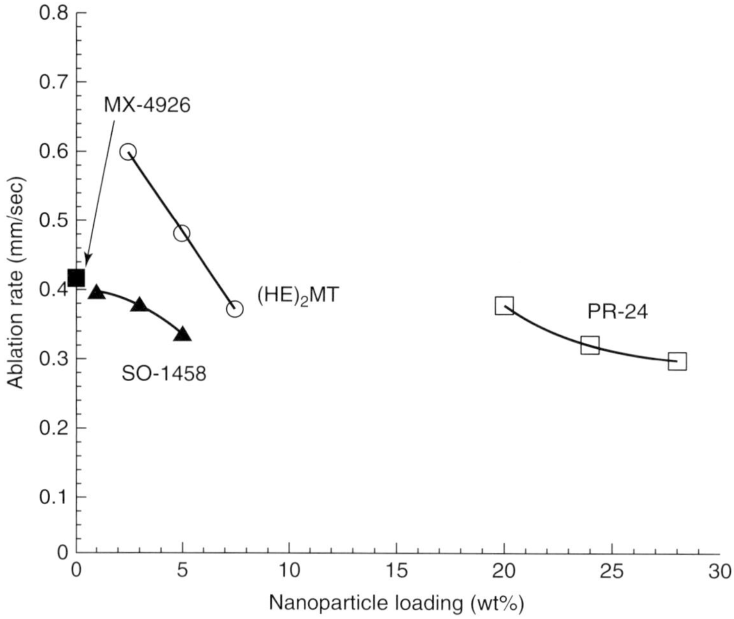

Figure 10.55. Ablation rate of MX-4926 and NRAMs with different types of nanoparticles at various loading levels.

Figure 10.55 shows the ablation rates of MX-4926 and all three groups of NRAMs: Clay-NRAM ([HE]$_2$MT nanoclay), CNF-NRAM (PR-24), and POSS-NRAM (SO-1458) [34]. The ablation rate of MX-4926 was about 0.4 mm/s. For the clay-NRAM group, only the 7.5 wt% clay-NRAM had a lower ablation rate than MX-4926. The 15 wt% clay and 85 wt% phenolic becomes 7.5 wt% clay-NRAM when it is transformed into laminate, since all clay-NRAMs contain 50 wt% carbon fiber reinforcements (Table 10.8) [34]. For the CNF-NRAM group, all three loadings had lower ablation rates than MX-4926, with 28% CNF-NRAM being the lowest. For the POSS-NRAM group, all three loadings had lower ablation rates than MX-4926, with 5% POSS-NRAM being the lowest. The loadings of POSS were 1, 3, and 5 wt%, with the 5% exhibiting the lowest value of the three NRAM groups. The loadings of clay were 2.5, 5, and 7.5 wt%, the medium values of the three NRAM groups. The loadings of CNF were 20, 24, and 28 wt% without the rayon-carbon reinforcements, the highest of the three NRAM groups.

Figure 10.56 shows the residual masses of MX-4926 and all the NRAMs [34]. The residual mass of MX-4926 was about 92%. The POSS-NRAM group had the most residual mass, about 93% for all three loadings. The clay-NRAM group had about the same residual mass as the MX-4926. The CNF-NRAM group had about 86–88 wt% residual mass, the lowest of all materials, control and NRAMs.

Figure 10.57 shows the maximum backside heat-soaked temperature rise of MX-4926 and the NRAMs [34]. All NRAMs had lower maximum backside heat-soaked temperature rises than MX-4926. The backside heat-soaked temperature rise of MX-4926 was about 106°C. The CNF NRAM group had substantially lower

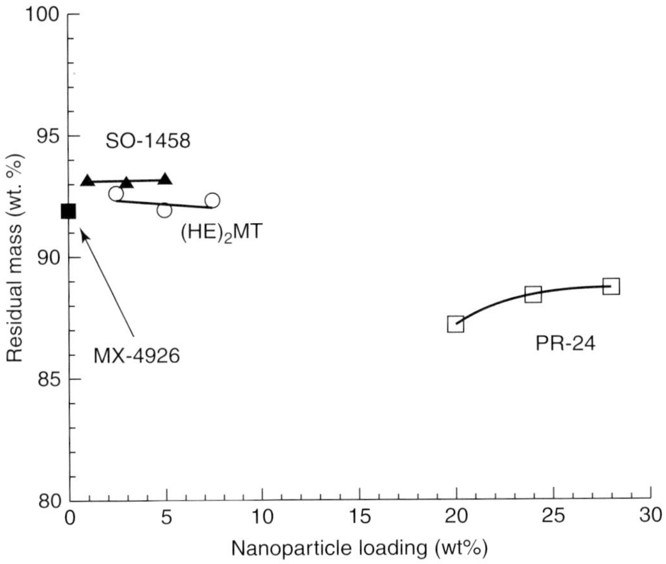

Figure 10.56. Residual mass of MX-4926 and NRAMs with different types of nanoparticles at various loading levels.

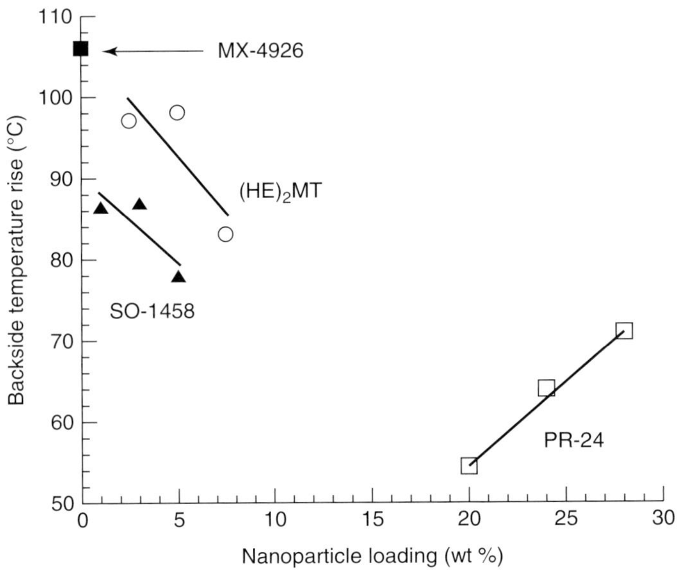

Figure 10.57. Backside heat-soaked temperature rise of MX-4926 and NRAMs with different types of nanoparticle at various loading levels.

maximum backside heat-soaked temperature rise than MX-4926, from 54°C to 72°C. The result for the POSS-NRAM group was from 75°C to 86°C, the second lowest. The clay-NRAM group had the third lowest, from 82°C to 98°C.

An IR pyrometer was used to measure the surface temperatures of all materials during SSRM firings. Figure 10.58 shows the surface temperatures of MX-4926 and

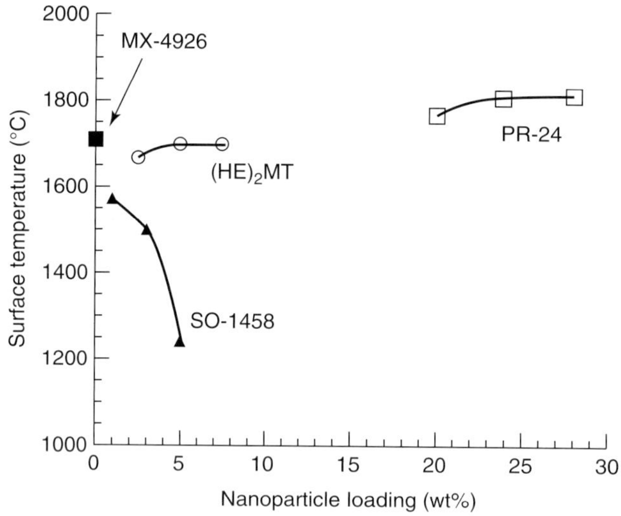

Figure 10.58. Surface temperature of MX-4926 and NRAMs with different types of nanoparticles at various loading levels.

the NRAMs [34]. The surface temperature of MX-4926 was about 1,700°C. Surface temperatures of the CNF-NRAM samples were higher than those of MX-4926, the clay-NRAMs, and the POSS-NRAMs. This finding suggests that we may have better radial heat transfer than axial heat transfer, supported by the glowing heat of the surface observed during material testing. This phenomenon was observed by other researchers [60] and needs further study. The surface temperatures of clay-NRAMs and POSS-NRAMs were lower than that of MX-4926. The amount of nanoclay in the clay-NRAMs had essentially no effect. The amount of POSS in the POSS-NRAMs had a significant effect on the surface temperature of the POSS-NRAMs.

Air Force Research Laboratory Pi-K Solid Rocket Motor Ablation Test. Based on the SSRM test data, selective NRAMs were scaled to 9.1 kg (20 lb) quantities at CEM. These nanoparticle-resin mixtures were transformed into prepregs using T300 carbon fabric in a semi-production R&D facility in Winona, MN. These prepregs were fabricated into molding compound and compression-molded into cylindrical billets [about 4.25 inches (10.8 cm) in diameter and 4 inches (10.16 cm) in length]. The billet was machined into a nozzle assembly similar to the ones shown in Figure 10.59 [33]. Figure 10.59 shows an MX-4926 with SC-1008 phenolic resin, clay-NRAM with SC-1008 phenolic resin, a CNF-NRAM with SC-1008 phenolic resin, and an SM-8029 (quartz fabric-silicone resin composite) machined nozzle assembly of the AFRL/Edwards Pi-K SRM, respectively. The nozzle throat area is replaced by a graphite insert to avoid excessive throat erosion during the motor firing. Using this experimental setup, reasonably accurate testing of materials designed to be used for the entrance region and the exit cone of the solid rocket nozzle assembly was anticipated.

At Edwards AFB, CA, AFRL/PRSM has developed a very cost-effective char motor to screen SRM insulation materials [90–92]. A schematic of the Pi-K SRM

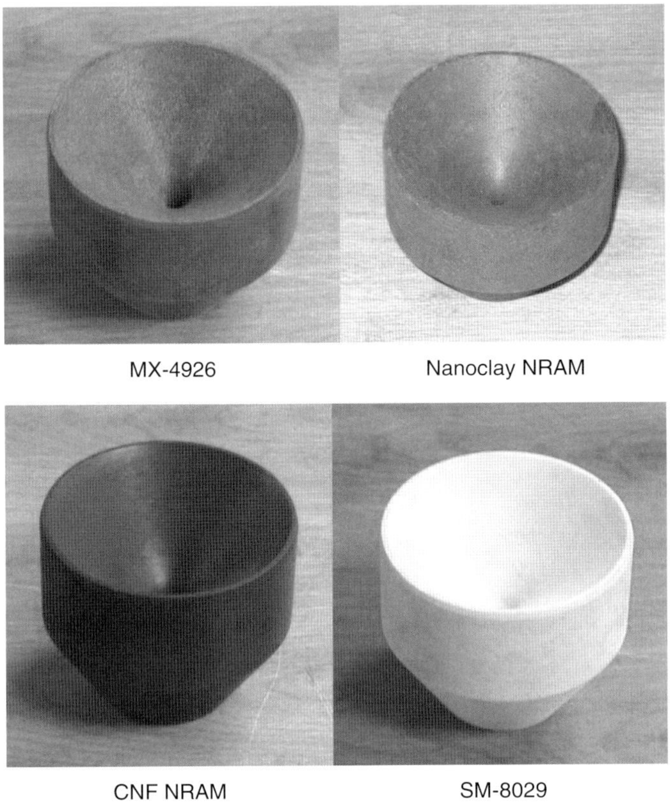

Figure 10.59. Prefired Pi-K rocket motor nozzle assemblies fabricated from MX-4926, clay-NRAM, CNF-NRAM, and SM-8029 materials. This nozzle assembly is shown as the nozzle part in Figure 10.60 slides into the back end of the heavy-walled Pi-K rocket motor.

experimental setup is illustrated in Figure 10.60 [33]. The motor is operated and fired in the upward direction. It usually holds up to 2 lb. (0.91 kg) of solid propellant in an end-grain burning configuration. Both non-aluminized and aluminized solid propellants have been used for insulation material testing. The average chamber pressure is typically about 800 psia and the test duration is about 8 seconds. For insulation material testing, a cylindrical cone made of glass phenolic material of different area ratio is used inside the motor. This configuration is used to simulate a range of Mach numbers that the insulation materials are exposed to as the combustion of gaseous products occurs during motor firings. Rapid testing of six or more samples per day can be achieved.

For nozzle material testing, the cylindrical cone is removed, and the whole nozzle assembly at the back end of Pi-K SRM is replaced by NRAM candidate materials. The whole assembly is a convergent/divergent nozzle, and the nozzle throat is a machined graphite nozzle insert, as shown in Figure 10.61 [33]. The graphite nozzle insert piece prevents excessive erosion in the nozzle throat area. This configuration enables the evaluation of candidate materials in the entrance region and exit cone of the nozzle.

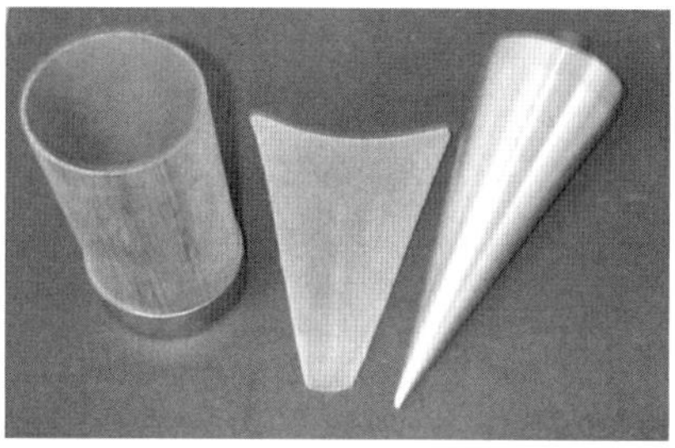

- 4-Inch Char motor
- Only 100 g of insulation material needed (clear POSS® sample shown on the left)
- Very cost-effective tool
- Rapid testing of 6 samples/day (or more)

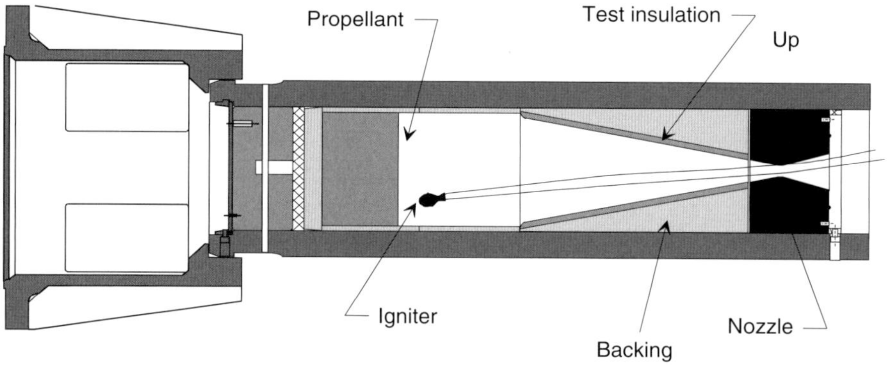

Figure 10.60. Schematic of the AFRL/Pi-K SRM experimental setup for ablation testing. Both non-aluminized and aluminized propellants can be used. Average chamber pressure is about 800 psia, and test duration is about 8 seconds when an end-burning propellant is used. Nanocomposite rocket ablative materials will be tested at the nozzle section.

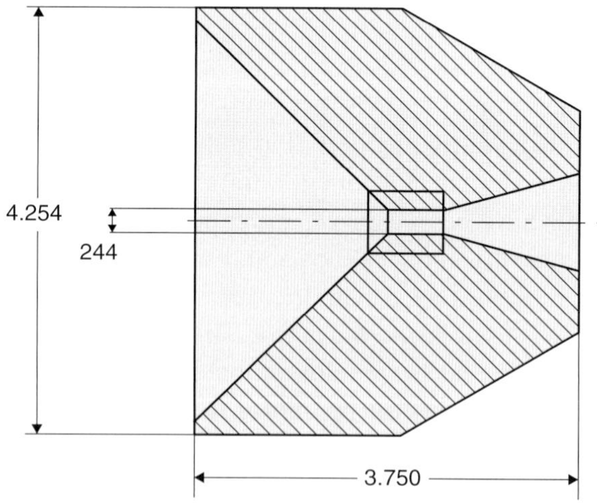

Figure 10.61. Cross-sectional diagram of the motor nozzle assembly showing the graphite insert at the nozzle throat of the NRAM material. This nozzle assembly fits into the back end of the Pi-K motor. The direction of the combustion of gaseous products from the rocket chamber is from right to left. All dimensions are in inches.

Ablation Results. Four successful Pi-K motor firings were conducted. The clay-NRAM specimen with 7.5 wt% Cloisite 30B survived the Pi-K motor firing with no obvious damage and exhibited a better appearance than MX-4926 in visual examination, as shown in Figure 10.62 [105]. In Figure 10.63, it was noted that a deposit of aluminum oxide was formed near the center of the nozzle that resulted from the motor

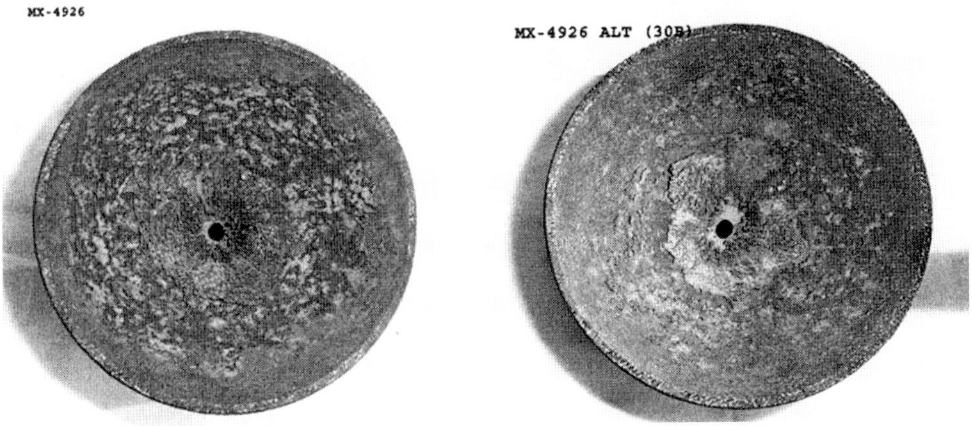

Figure 10.62. Pictures of the MX-4926 (left) and clay-NRAM (MX-4926 ALT/7.5 wt% Cloisite 30) (right) Pi-K rocket nozzles after firing at AFRL/Edwards AFB.

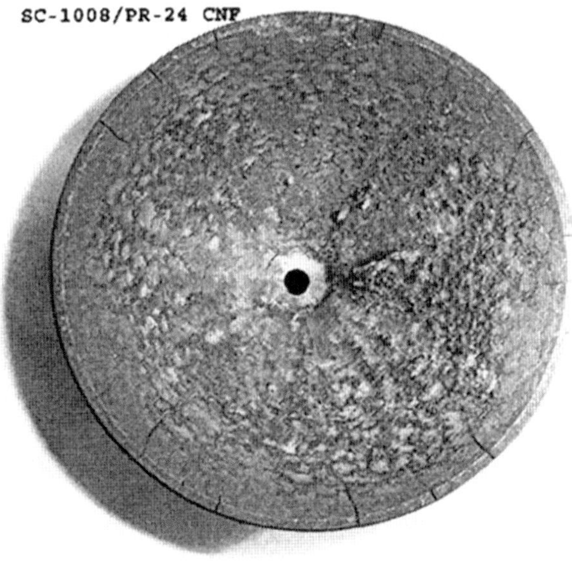

Figure 10.63. Deposit of aluminum oxide in CNF-NRAM (SC-1008/PR-24 CNF).

firing [105]. This deposit was removed for later analyses, resulting in some apparent material loss in the shape of the nozzle. A series of dimensional measurements were recorded on the clay-NRAM nozzles, as shown in Figure 10.64, and which are expressed relative to the MX-4926 control [105]. In all regions but right at the transition from the graphite insert to the NRAM materials (where the aluminum oxide deposit was removed, as noted earlier), there was an advantage to the NRAM in eroded shape. Figure 10.65 is a microscope image of the clay-NRAM nozzle at the edge of the cone (the edge is at the right of the black area in the photo) [105]. Note the darkened area highlighted with the first bar – this is the primary heat-affected zone. Behind this region is a secondary heat-affected zone highlighted with the second bar.

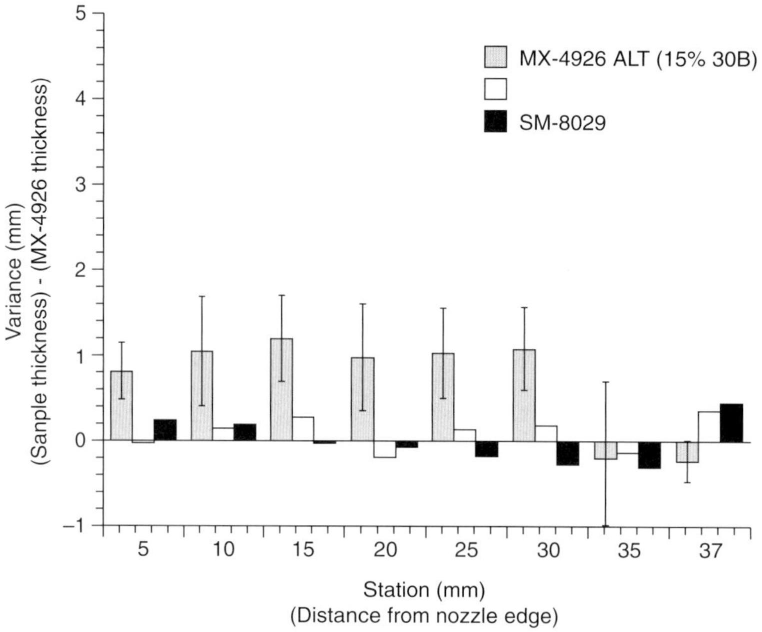

Figure 10.64. Dimension measurement of clay-NRAM (MX-4926 ALT [15% 30B]).

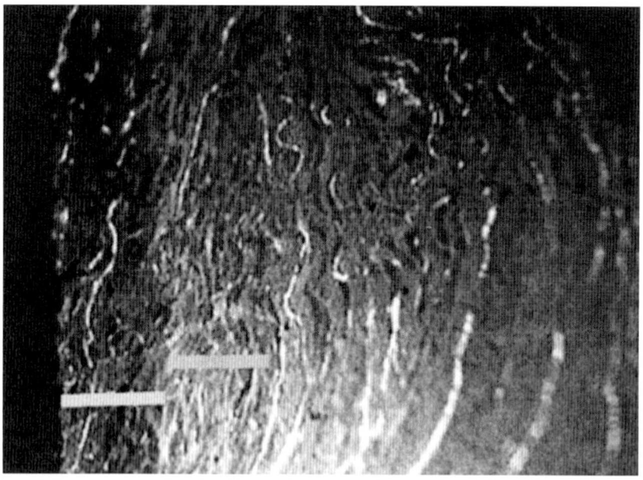

Figure 10.65. Microscope image of clay-NRAM nozzle after firing.

Figure 10.66 is a photo of a posttest CNF-NRAM (without T300 carbon fiber reinforcement) after a Pi-K motor firing [105]. This sample fractured during test or during motor cooldown, as can be seen on the edges of the nozzle in the Figure 10.66. This CNF-NRAM specimen consists of 24 wt% PR-24-PS CNF and 76 wt% SC-1008 phenolic resin and has no carbon fiber reinforcement. It is obvious that the conventional carbon fibers are needed as reinforcement for nozzle ablation.

Mechanical Properties Results. Peak stress and tensile modulus of MX-4926 control, 7.5 wt% Cloisite-NRAM, 24 wt% PR-19-PS CNF-NRAM, and 24 wt% PR-24-PS

Table 10.9. *Clay-, CNF-, and POSS-NRAM and MX-4916 mechanical properties*

Nanoparticles	Peak Stress (MPa)	Elongation @ Peak Load (mm)	% Strain @ Peak Load (%)	Energy to Peak Load (N-mm)	Elongation at Break (mm)	Modulus (GPa)
MX-4926	24.55	0.25	0.41	82.02	0.27	6.85
MX-4926 ALT 7.5% 30B	25.74	0.52	0.87	427.65	0.53	5.05
24% PR-19-PS/ SC-1008	4.66	0.15	0.25	13.88	0.15	2.00
24% PR-24-PS/ SC-1008	2.19	0.15	0.24	4.60	0.15	1.35
POSS-NRAM	–	–	–	–	–	–

Figure 10.66. Photo of posttest CNF-NRAM (without T300 carbon fiber reinforcement) after Pi-K firing.

CNF-NRAM specimens were measured [105]. Figure 10.67 shows that only the tensile modulus of 7.5 Cloisite 30B-NRAM exceeded the MX-4926 control [105]. Table 10.9 shows the mechanical properties of MX-4926 and MX-4916 ALT 7.5% 30B [105]. The increased elongation and strain of the 7.5% Cloisite 30B-NRAM resulted in a much higher energy level needed to reach the peak load than the MX-4926 control, but this was accompanied by a small decrease in modulus as compared to the control. Certainly nothing in the tensile data for the clay-NRAM would preclude it from service based on a comparison to MX-4926.

As expected, the CNF-NRAMs have very inferior mechanical properties than the MX-4926 since there was no carbon fiber reinforcement. Table 10.9 shows the tensile test results for CNF-NRAMs with PR-19-PS or PR-24-PS [105]. The lack of reinforcement in the CNF-NRAMs made them significantly weaker than MX-4926. No POSS-NRAM mechanical properties were measured.

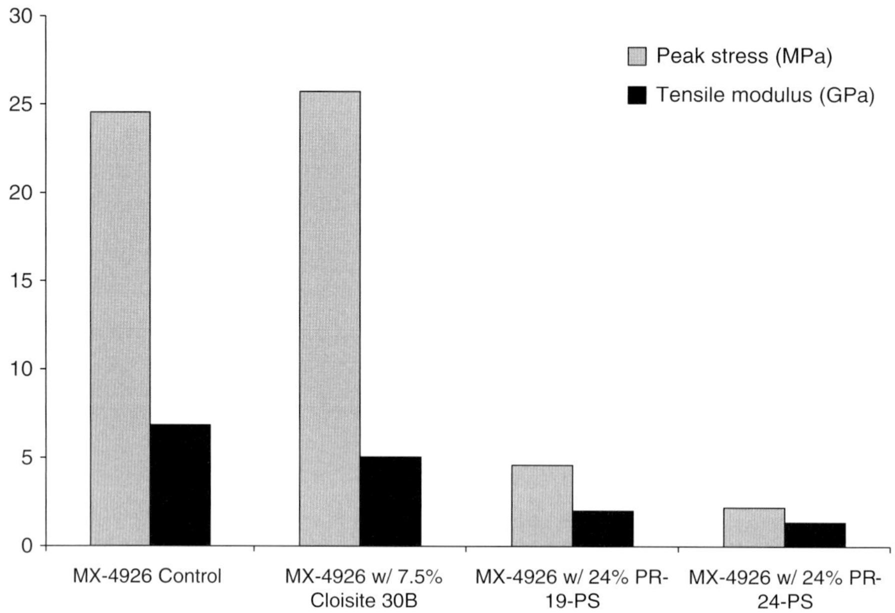

Figure 10.67. Peak stress and tensile modulus of MX-4926 control, MX-4926/7.5% Cloisite 30B, MX-4926/24% PR-19 CNF, and MX-4926/24% PR-24 CNF.

Summary and Conclusions. For this study, we concluded the following:

- The feasibility of using nanocomposite rocket ablative materials (NRAMs) in rocket nozzle assemblies was clearly demonstrated using SSRM subscale and solid rocket motor Pi-K Char motor ablation testing.
- MMT organoclays, CNF, and POSS can be implemented into the existing semi-production line at Cytec Engineered Materials to manufacture fiber-reinforced prepregs.
- A higher loading of MMT, POSS, and CNF improves erosion resistance, and 28% CNF has the lowest erosion rate in the absence of carbon fiber–reinforcing agent using SSRM.
- Backside heat-soaked temperatures of all NRAMs were lower than that of baseline MX-4926, and CNF-NRAM as a group had lower backside heat-soaked temperatures than the MMT-NRAM and POSS-NRAM using SSRM.
- Backside heat-soaked temperature trends were reversed for MMT and CNF, whereas higher loadings of CNF appeared to conduct heat better using SSRM.
- Peak erosion of POSS-NRAM is 20% lower than MX-4926 at very low loadings using SSRM.
- Peak erosion of CNF-NRAM (28% CNF) without rayon fabric is 42% lower than MX-4926 using SSRM.
- Backside heat-soaked temperatures of CNF-NRAM are 68% lower using SSRM.
- Backside heat-soaked temperatures of clay-NRAM are 28% lower using SSRM.
- Backside heat-soaked temperatures of POSS-NRAM are 26% lower using SSRM.

- Clay-NRAM specimen with 7.5 wt% Cloisite 30B survived the Pi-K motor firing with no obvious damage and exhibited a better appearance than MX-4926; mechanical properties of clay-NRAM are comparable to MX-4926, which made this clay-NRAM a suitable replacement of MX-4926.
- Physical, mechanical, and thermophysical properties of selective NRAMs need to be measured.
- Optimal nanoclay dispersion in SC-1008 phenolic resin needs further study.
- Synergism between CNF and other nanoparticles in SC-1008 phenolic resin needs further study.
- Selective high-performance thermosetting resin system, such as cyanate ester resin, can be a potential replacement of phenolic resin due to its high T_g and relative ease of manufacturing compared to phenolic resin used for advanced NRAM development.

10.3.6.3 Phenolic Nanocomposite Studies in China

Based on the current developments, it clearly appears that the Chinese scientific community is diligently conducting the study of nanostructured ablative materials. Liu et al. [106] have focused their attention on the study of carbon/phenolic composites (PR/CF) in the presence of POSS. Two types of functionalized POSS were employed as nanomodifiers of the phenolic resin: a commercial octamethyl-POSS (Me-POSS) and a phenyl-POSS (Ph-POSS). The Ph-POSS was produced with commercial OctaTrimethylsiloxy-POSS (TMS-POSS) as starting material and using styrene according to a procedure developed by Laine [107]. For comparison purposes, a silica precursor tetraethoxysilicate (TEOS) was also selected to produce another nanomodified resin [108] used to impregnate the carbon/phenolic composite (T-SiO_2/PR/CF). The exact compositions of the different formulations in terms of filler percentages were not reported. The ablation performances of the following composites were explored: T-SiO_2/PR/CF, Me-POSS/PR/CF, and Ph-POSS/PR/CF. These materials were studied using an oxyacetylene test bed. The following conditions were selected: the distance from torch tip to sample was 20 mm and the exposure time was 20 seconds.

In terms of linear ablation rate as well as mass loss, the hierarchy displayed by the tested materials was the following: T-SiO_2/PR/CF > Me-POSS/PR/CF > Ph-POSS/PR/CF. With respect to the type of functionalization attached to the POSS, the Ph-POSS displayed a significantly lower erosion and mass loss than did the Me-POSS. This result can be directly related to the higher carbon content of the phenyl groups over methyl structures. The better thermal stability of Ph-POSS over Me-POSS embedded in a polymeric matrix was also confirmed by other studies [109, 110]. The SEM analysis of the eroded samples also confirmed that the Ph-POSS/PR/CFs produced the best charred surface: a homogeneous hybrid carbonaceous/silica skin on the top region of the Ph-POSS/PR/CF samples protected the inner layers of the composites, whereas the other two formulations displayed a porous and irregularly eroded char, leaving the carbon fibers exposed to the flame.

A novel flake graphite–reinforced barium-phenolic resin composite was made by roller coating technology and its thermal ablation was studied by Yu et al. [111] under long pulse laser irradiation. Results show that the ablation rate of the flake graphite–reinforced barium-phenolic composite is 32.8 μg/J at 1,700 W/cm^2 irradiation power density, which is lower than that of the barium resin and the barium-phenolic carbon fiber–reinforced composites. When the composites were exposed to the laser radiation, the flake graphite acted as a mirror and reflected part of the laser energy. As a result, the laser radiation deposited on the composites was reduced. The size of the flake graphite also affects the ablation rate. The composite with flake graphite diameter of about 0.5 mm has the lowest ablation rate.

10.3.6.4 Phenolic-Graphene Oxide Studies by Xi'an Jiaoton University, China

For their experiment, Si et al. [112] prepared nanocomposites of graphene oxide (GO) in phenolic resin for thermal characterization. They prepared GO from graphite at three different temperatures and mixed it into the phenolic resin via ultrasonic bath at loadings ranging from 0.1 wt% to 2 wt%. The mixtures were degassed at room temperature and pressure overnight and then at 50°C and 0.095 MPa until no more weight loss was observable. Finally, the composites were cured at 120°C for 2 hours, 165°C for 4 hours, and 220°C for 2 hours.

The researchers conducted FTIR to verify the success of the transformation to GO, and particle size analyses on the unmixed GO as well. For thermal characterization, TGA in N_2 to 800°C at 10°C/min and DSC in N_2 to 300°C at 10°C/min were conducted on each formulation. Finally, SEM and XRD were conducted on the composites for dispersion, composition, and morphology verification.

The team found that the transformation technique was most effective at higher temperatures, and that the particles were comparable to nanoclay in size. The GO particles were found to disperse well into the resin and produced a moderate increase in char yield and thermal stability even at low loadings. The researchers recommended that the GO-phenolic nanocomposite be investigated for ablation resistance in the future.

10.3.6.5 Phenolic-Nano-Silica Nanocomposite Studies by the Defense Research and Development Organisation, India

The most significant research of the Indian scientific community on nanostructured polymeric ablatives was carried out at the Defence Research and Development Organisation (DRDO). Srikanth et al. [113] studied the effect of nano-silica on the ablation resistance and on the interlaminar shear strength (ILSS) of carbon/phenolic (C–Ph) composites. The through-thickness thermal conductivity was also measured at 300°C under steady-state conditions (ASTM E–1225-09) [114]. Moreover, the microstructure of the ablated materials was investigated by SEM. Rayon-based carbon-fabric was used as a reinforcing material and a resole was used as the matrix resin. The nano-silica had an average diameter of 40–50 nm. Four different formulations were prepared by adding 0.0 (control), 0.5, 2.0, and 4.0 wt% of nano-silica to the phenolic resin. The pristine resin and the produced nanocomposites were used to impregnate the carbon fabric. Three samples from each of the laminates were machined to a size of 10 mm diameter by 20 mm (the length of the sample was perpendicular to the direction of the fabric). The specimens were exposed to a plasma arc jet at a flame velocity of about Mach 1 and at a stagnant flux of 2.5 MW/m^2 for 20 seconds. The ablation performance was determined by dividing the weight loss during the test by the exposure time.

Due to the presence of nano-silica, the ILSS increased with the nano-silica content up to a filler percentage equal to 2.0 wt%: above this threshold, the ILSS decreased. This result indicated that nano-silica particles embedded in the matrix acted as an anti-slippage agent. It was also speculated that above 2.0 wt%, the nano-silica decreased the adhesion between the fabric layers. An increased ILSS allows the composites to be more resistant to the high aerodynamic shear forces, and is expected to result in an enhanced ablation resistance. Plasma arc jet test results were in good agreement with the ILSS trend: in fact, the ablation resistance increased with the nano-silica content up to 2.0 wt%, beyond which it was lower than the control sample. The addition of nano-silica also reduced the thermal conductivity of the C–Ph composites: the thermal conductivity depends on the lattice vibrations and the microstructure (micro-cracks, grain boundaries, porosity, etc.). The introduction of nano-silica increased the number of grain boundaries, resulting in lower thermal conductivity: similar observations for the C–SiC composites have been reported by Kumar et al. [115]. As a result of a reduced thermal conductivity, ablation resistance was improved.

Since the formation of the char is an endothermic process, the amount of heat dissipated during the ablation depends on the efficiency of this process. Moreover, the char dissipates a large fraction of incident heat through surface re-radiation. Accordingly, a complete charring of the resin combined with high char retention provides better heat dissipation during ablation. The SEM microstructure studies showed that the damage to the carbon fibers in the blank C–Ph is considerably higher compared to the damage occurring in the presence of nano-silica. The ablated surfaces of nanostructured C–Ph composites showed that SiC was formed on carbon fibers during ablation. This result was also confirmed by XRD analysis. The authors proposed two main mechanisms for the ablation results and the related microstructural and compositional observations. First, under plasma arc jet testing, the temperature can reach 4,000°C. In the case of C–Ph composites, the matrix cannot effectively maintain the carbon layers together, leading to severe erosion losses. In the case of nano-silica-loaded C–Ph composites, two unique processes occur when the temperature increases beyond 1,600°C: the silica becomes molten [116] and forms a viscous layer that acts like a high-temperature binder able to hold the underlying phenolic matrix and the char. Thus it ensures a higher radiant emission and leads to a complete charring of the resin for absorbing more heat (first process). Moreover, the molten silica reacts with the carbon, forming a hard SiC phase [117] (second process). The resulting carbon–silicon carbide (C–SiC) can withstand the aerodynamic shear forces under high-temperature erosive environments, thereby enhancing the ablation resistance [118].

10.3.6.6 Phenolic-Zirconia and MWNT Nanocomposites Modified Carbon Fabrics Composites Studies by the Defense Research and Development Organisation, India

Srikanth et al. [119] conducted a novel investigation into the effects of ZrO_2 as a coating on carbon fabric for use in carbon-phenolic (C-Ph) composites. MWNTs were also used in nanocomposites of carbon-phenolic without ZrO_2. Then, the best performer of the ZrO_2-coated composites was combined with MWNTs for a nanocomposite with ZrO_2. A solution of $ZrOCl_2$, ethanol, and water was sprayed on rayon carbon fabric and allowed to sit for 96 hours, while the $ZrOCl_2$ and H_2O reacted to produce ZrO_2 and HCl in a coating on each fiber. By modulating the amount sprayed, the researchers produced formulations of up to 9.5 wt% of ZrO_2. Resole-type phenolic resin was applied to the fabric to produce prepregs, and these were cut into 60 squares of 15 cm side length for hand layup and autoclaving at 120°C and 5 bar for 2 hours, and then at 180°C and 5 bar for 4 hours. MWNTs were mixed into the resin at 0.5 wt% and ball-milled for 3 hours at 250 rpm. This resin was applied to uncoated fabric and processed the same way. When the best performer of the ZrO_2 formulations was found, 40 layers of ZrO_2-coated carbon fabric prepregs were laid on 9 layers of uncoated carbon fabric prepregs with MWCNTs.

The authors characterized the formulations for fiber volume fraction (ASTM D 3171), flexural strength (ASTM D 790), ILSS (ASTM D2344), thermal conductivity (ASTM E1225-99), and ablation resistance via exposure to a heat flux of 4.0 MW/m^2 produced by plasma arc-jet for 30 seconds at a flow velocity of Mach 1. Backside temperature was recorded during and for 40 seconds after the test, which was not enough time to reach peak temperature. The posttest ablation samples were examined via SEM and XRD.

The researchers found that the ZrO_2 coating significantly reduced the density, fiber volume fraction, flexural strength, and ILSS of the composite, so much so that the highest loading made the composite so easily delaminated that it could not be machined into usable samples. SEM showed that the ZrO_2 coating often cracked, flaked, or grew in nodules. Higher loadings of ZrO_2 were found to decrease thermal conductivity and increase the ablation rate of the composites (Table 10.10) [119]. While the MWNTs were dispersed well and did improve mechanical properties and increase thermal conductivity, they did not have a significant effect on the ablation rate. Table 10.10 shows the ablation rate and backface temperature of composites tested in plasma arc-jet test. The blank C-Ph composite (control sample) has the best ablation performance.

10.3.6.7 Phenolic-Clay Nanocomposite Studies by Tarbiat Modares University, Iran

The main contribution of the Iranian scientific community on the topic of nanostructured ablatives has been the efforts of researchers at Tarbiat Modares University. They focused mostly on the improvement of asbestos/phenolic ablatives using numerical modeling and experimental approach. Particularly, Bahramian et al. [120–125] studied the numerical and experimental aspects of the ablative performance,

Table 10.10. *Ablation rate and backface temperature of composites in plasma arc-jet test*

Composite Description	Ablation Rate (mm/s)	Backface Temperature (°C)[a]
Blank C-Ph	0.054	169
3.5 wt% Zr-C-Ph (zirconia coated C-fabric)	0.078	114
6.5 wt% Zr-C-Ph	0.105	104
FG-C-Ph (functionally graded C-Ph)	0.103	135

[a] Backface temperature at the end of test (30s)

thermal decomposition, and temperature distribution through the thickness of phenolic/layered silicate nanocomposites impregnated with asbestos cloth and compared it with traditional asbestos/phenolic ablators.

Bahramian et al. [120] modeled the ablating, charring, and thermal degradation behavior of a resole-type phenolic resin/asbestos cloth composite using an oxyacetylene flame test. Their model requires solving heat transfer equations with moving boundaries. Explicit forward finite difference method (FDM) is used for the heat transfer calculation. Moving boundaries are fixed by the Landau transformation. The ablation equation is solved numerically. This model demonstrates the variation of thermophysical properties, such as heat conductivity, density, and specific heat at different temperatures. The authors showed that their model is adequately confirmed by the experimental data of the thermophysical and ablation properties of an asbestos/phenolic composite and can be used as a simple tool in the design of the thickness of a heat shield. In their test case study using conditions of 800 W/cm^2 heat flux and 3,000K temperature of hot gas, and a test duration of 20 seconds, the backface temperature of char surface of a 4 mm thick asbestos/phenolic composite are 468K, while three oxyacetylene flame tests show temperatures of 431, 438, and 423K.

In their work [122], the combination of solution and in situ intercalation methods was used to fabricate the nanocomposite. Ethyl alcohol was used to disperse the layered silicates (Cloisite 15A) and dissolve the phenolic resin. The phenolic resin in an excess of ethyl alcohol could be adsorbed into the nanoclay tactoids. However, once the ethyl alcohol was removed, the phenolic chains remained trapped in the layered silicate sandwiched structure. The asbestos cloth was impregnated with the nanocomposite system. Three composites having different nanoclay percentages were studied: 3 wt% (NMA1), 4 wt% (NMA2), and 6 wt% (NMA3). Flat-panel samples were prepared for the oxyacetylene flame test (heat flux 900 W/cm^2). Type-K thermocouples were placed through the thickness of the sample at 2, 4, and 6 mm from the surface: the surface erosion measurement was carried out on cylindrical shaped samples (10 mm in diameter and 25 mm in height). XRD analysis was used to evaluate the dispersion of the nanoclay in the composites. The specific heat capacity measurements were performed according to ASTM E-1269 [126]. Thermo-gravimetric and DSC analyses were also performed on the resulting materials.

The XRD patterns of nanocomposites exhibited a behavior typical of an exfoliated morphology, confirming the penetration of the resol molecules between the clay sheets. According to the TGA data, the presence of clay improved the thermal stability of the composites. Many factors are responsible of this improvement. Among them, the dispersed large aspect ratio platelets hindered the diffusion of the decomposed volatile products, as a direct result of the decreased permeability. For the thermal oxidation process, the mass transfer of oxygen into the gallery may also be reduced, decreasing the degradation of the matrix. The ablative reassembly of nanoclay platelets and the sintering of them at elevated heat fluxes led to the formation of a homogeneous silicate-carbonaceous char layer that functions as a heat shield during erosion. This is widely accepted as the main flame retardancy mechanism common in this class of nanocomposites [127–130]. It follows that in the range of the quantity of nanoclay content, the higher the nanoclay content, the better the thermal stability. TGA results were also used to acquire the parameters used to produce a model of the ablation process. The heat capacity measurements were carried out to a temperature of 500 K: the values of the specific heat of the asbestos/phenolic composite and of the NMA3 nanocomposite are in the range of 1,000–1,500 and 3,300–5,800 J/kg-K, respectively. The higher value of the NMA3 nanocomposite is one of the key factors to provide the better insulation performance of the NMA3 over the asbestos/phenolic composite. The dynamic DSC scans also indicated that the heat of ablation of the NMA3 composite was considerably higher (about 950 kJ/kg) than that of the control material (45 kJ/kg).

The experimental temperatures of the top surface and through the thickness of asbestos/phenolic and NMA3 samples were also determined. During the same experimental conditions, the steady-state surface temperature of the asbestos/phenolic composite and of the NMA3 were about 1,500°C and 2,100°C, respectively. This difference clearly shows that the nature of top surface of the traditional composite and of the nanocomposite-impregnated asbestos fabric are completely different. The higher value of surface temperature of the NMA3 is another key factor to anticipate a higher insulation performance than the traditional asbestos/phenolic composite. Since the re-radiated energy follows a T^4 law, a decreased heat flux is transferred into the material, thus increasing the lifetime of the NMA3 material. The authors also evaluated the thermal diffusivity of the material as a function of the depth. The effective thermal diffusivity of the asbestos/phenolic composite is twice than the NMA3 composite. This phenomenon suggests a clear difference in the char layer formation and in the ablation mechanism of the composite and of the nanocomposite-based system.

The above study can be summarized as follows: (a) at 900 W/cm^2 and 3,400K of hot gas, the effective thermal diffusivity of the asbestos/phenolic and asbestos/phenolic/6 wt% MMT clay composites were estimated. The ablation performance of NMA3 nanocomposite heat shield is 100% more than the composite counterpart, and (b) in cone calorimeter test at 80 kW/m^2 radiation heat flux, the NMA3 nanocomposites shows 51% HRR and 40% mass loss lower than the asbestos/phenolic composite.

In Paydayesh et al.'s work [125], a comparative study of ablation performance of highly filled phenolic/asbestos/montmorillonite nanocomposite was conducted and evaluated with comparison material. The heat diffusion through the thickness and erosion rate was measured for both systems. Ablation mechanism, thermal degradation kinetics, and thermophysical properties of highly filled ablative nanocomposites were also investigated. The mathematical model was evaluated by the oxyacetylene test data (heat flux 800 W/cm^2). MMT clay (Cloisite 15A) content of 0, 40, 50, 60, and 70 phr (0, 14, 16, 17, and 20 wt% of the total composite weight) were studied with varying amounts of phenolic resin and asbestos cloth. To prepare the highly filled nanocomposites, a combination of solution and in situ interaction methods was used in their study.

Figures 10.68 and 10.69 show the SEM micrographs of the asbestos-phenolic composite and nanocomposite (containing 50 phr (16 wt%) MMT clay) specimens after the oxyacetylene flame test, respectively [125]. These figures show top surface (surface of char zone), lateral surfaces, and heat flux direction. These figures were intended to show the characteristics of various regions: virgin zone, porous pyrolysis zone, and char zone. In the char zone, nanoscopic porous morphologies are observed. The molten combustion products formed particles with special shapes. Comparison of Figures 10.68 (a, b) and 10.69 (a, b) clearly shows that the char formed on the surface of composite sample has higher porosity than those formed on the surface of nanocomposite sample [125].

Several conclusions can be made from the above study [125]:

1. Ablation performance of highly filled nanocomposites is better than their composite counterpart. Nanocomposite sample containing 50 phr of MMT clay (asbestos/phenolic/clay ratio is 52/32/16 wt%) has the best erosion rate (mm/s) of 135% lower than that of the composite sample.
2. Better performance of highly filled nanocomposites is due to the formation of ceramic char layer on the ablator surface. The formation of char layer is easier in the highly filled than the low filled nanocomposites.
3. No significant difference is observed between thermal stability of highly filled nanocomposite and its composite comparison sample.
4. Mathematical model is adequately confirmed by the experimental data of the thermophysical and ablation properties of nanocomposites. It can be used as a design tool to estimate optimal thickness of a nanocomposite heat shield.
5. Asbestos fibers are banned in many countries throughout the world because they are carcinogenic.

10.3.6.8 Phenolic-Kaolinite and -Graphite Nanocomposite Studies by Tarbiat Modares University, Iran

For this investigation, Bahramian [131] developed a mathematical model for the heat release rate and thermal conductivity curves of an ablative and prepared nanocomposites of novolac-type phenolic resin with graphite to compare results. The graphite was dispersed by solution and in situ intercalation in methanol with

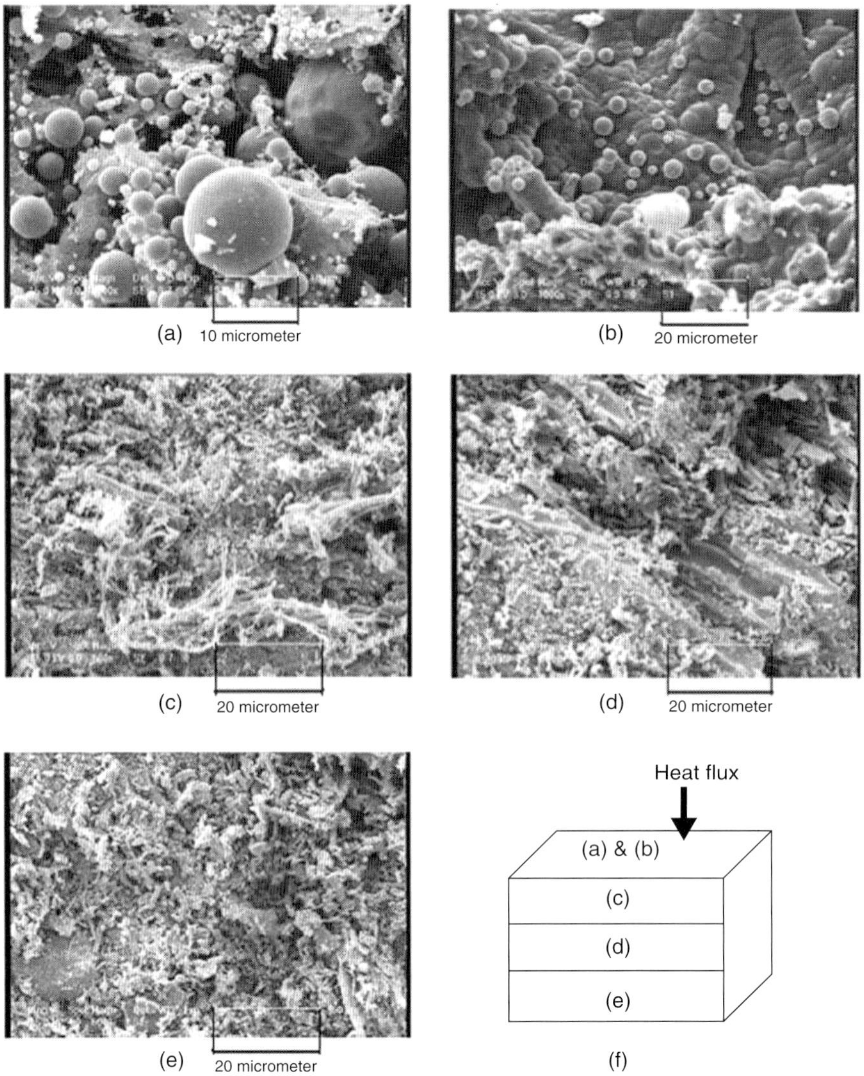

Figure 10.68. Scanning electron micrographs of asbestos-phenolic ablative composite after ablation test: (a) and (b) top surface; (c), (d), and (e) lateral surfaces; and (f) illustrating of heat flux direction [125].

phenolic resin and dried, then held at 80°C for 180 minutes, 120°C for 15 minutes, 150°C for 60 minutes, and 160°C for 30 minutes. Formulations with 0, 3, and 5 wt% graphite were prepared.

The author inspected the samples for morphology via SEM, composition via XRD, dispersion via TEM, specific heat via DSC at 10°C/min according to ASTMD-1269, thermal stability via TGA, thermal conductivity according to ASTM E1225-87, density and porosity according to ASTMD-4018, flammability via cone calorimetry according to ASTM E1354 at a heat flux of about $8x10^4$ W/m^2, and ablation resistance via oxyacetylene flame test according to ASTM-E-285-80 at $2.5x10^6$ W/m^2.

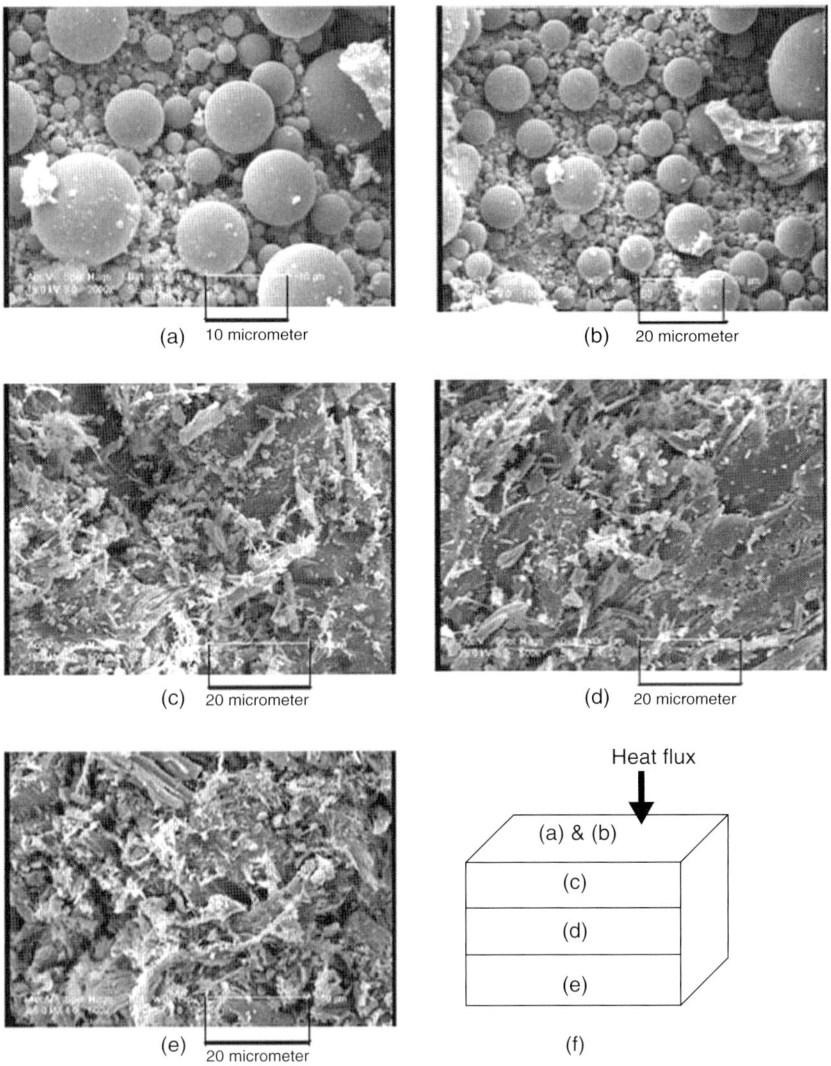

Figure 10.69. Scanning electron micrographs of ablative composite containing 50 phr MMT clay after ablation test: (a) and (b) top surface; (c), (d), and (e) lateral surfaces; and (f) illustrating of heat flux direction [125].

The author found that the graphite was dispersed well in the polymer matrix. He plotted HRR and conductivity curves from the models that agreed well with the experimental results for all three formulations [131]. The peak HRRs were reduced by 41% and 58% for 3 wt% and 5 wt% graphite, respectively. The compositions were found to have ablation rates of 0.0184 mm/s for neat phenolic resin, 0.0164 mm/s for 3 wt% graphite, and 0.0153 mm/s for 5 wt% graphite. The graphite was found to have a positive effect on the thermal stability.

For their most recent investigation, Bahramian et al. [132] prepared composites of 20 wt% kaolinite microparticles, 6–9 wt% kaolinite nanoparticles, and 6–9 wt% graphite nanoparticles in novolac-type phenolic resin for ablation, mechanical, and tribological testing. In addition, chopped carbon fibers were used for reinforcement. The formulations were prepared by dissolving phenolic powder in isopropyl alcohol and stirring in the kaolinite or graphite particles at 1,500 rpm for 1 hour and sonicating for 20 minutes. Then, 40 wt% carbon fiber was added and mixed by hand. Finally, the mixtures were degassed at 350 K for 30 minutes and hot press–molded at 370 K for 1 hour and then at 410 K for 2 hours.

The formulations were inspected for their morphology via XRD, ablation resistance according to ASTM-E-285-80, thermal stability via TGA at 10 °C/min to 950 °C, flexural strength according to ASTM D 790, dispersion via TEM, and thermal insulation/diffusivity and abrasion resistance via the team's own methods. The ablation specimens were 10 mm in diameter and 20 mm tall, and were exposed to 8.5 x 10^6 W/m^2 for 20 seconds. Abrasion testing was conducted by measuring mass loss after a certain amount of exposure to an abrasive surface, and thermal testing was performed by measuring backside temperature response to a heating jacket placed on the sample.

The authors found that for all loadings, the graphite nanoparticles produced a greater ablation resistance than the kaolinite nanoparticles. No nanoparticles were found to have a severely negative effect on the mechanical properties, but the kaolinite microparticles were found to reduce the flexural strength by more than 75%. At the same time, the kaolinite nanoparticles produced the same ablation resistance at their highest loading as the microparticles. The thermal insulation and diffusivity of the 9 wt% graphite formulation were anomalously low, lower even than the kaolinite microparticles. This formulation also showed the least decrease in abrasion resistance, so the researchers concluded that it was the formulation with the best dispersion. One TEM picture was taken of this formulation [132]. The dispersion shown is not excellent. Finally, the TGA results showed no significant change in thermal stability properties.

10.3.6.9 Phenolic-Nano-Silica with Carbon Fibers Studies by Islamic Azad University, Iran

Mirzapour et al. [133] investigated nanocomposites of resole-type phenolic resin with nano-silica and chopped carbon fibers for ablative properties. A loading of 50 wt% carbon fiber was used for all formulations, while the nano-silica loading varied from 0 to 5 wt%. The nano-silica was added to the resin with methanol and simultaneously sonicated and stirred at 25°C for 1 hour. The chopped carbon fiber was then added and mixed by hand. The formulations were degassed in a compression press for 30 minutes at 120°C and 1 bar, before the press was opened to allow the gases to escape. Then it was replaced and the formulations were cured at 160 °C and 10 bar for 5 hours. After sitting for one week in atmosphere, they were post-cured at 200°C for 2 hours.

The authors inspected the formulations' mechanical properties via three-point bending testing at a strain rate of 1 mm/min, thermal stabilities via TGA in air at 10°C/min to 1,000°C, and ablation resistances via oxyacetylene flame testing according to ASTM E285-80 at a heat flux of 8 MW/m^2. Pre- and posttest specimens were examined via SEM, XRD, and EDS.

The researchers found that the nano-silica was dispersed well in the matrix, with particles averaging 50 nm across. They found that the thermal stability and mechanical properties improved somewhat at the lowest loading of nano-silica, but increases in loading beyond that produced marginal improvements at best. However, the mass loss and ablation rates showed greater and greater improvement at higher loadings of nano-silica. XRD patterns revealed the existence of a small amount of SiC in the char, and SEM revealed protective coatings of resolidified silica on the ablated carbon fibers [133].

10.3.6.10 Phenolic Nanocomposites Studies by University of Perugia, Italy

For more than 15 years, the University of Perugia – STM group (Terni) – have extensively worked in the field of ablative materials. At the end of 1990, in collaboration with ALENIA Spazio S.p.A, the group developed and tested a medium-density silicone-based ablative (ALS-051) for the heat shield of the Italian Space Agency recoverable capsule (CARINA). An accurate ablation model of the studied material was also developed by Torre et al. [134–136]. Since the beginning of 2000, the researchers were also focused on the study of nanostructured ablatives. Nano-silica [137, 138] and MWNTs [139] were the main filler candidates. The possibility to exploit the synergy between traditional and nano-sized fillers was also investigated.

Glass and silica fiber phenolic composite ablators are a very important class of ablatives. As an example, they can be effectively used in the production of passively cooled rocket combustion chambers. To reduce the erosion phenomena of these ablators, the studies were focused on the selection of the best types of fiber reinforcement in terms of short or long fibers. Research has shown that in many hyperthermal conditions, liners produced with chopped reinforcements – fibers or fabrics cut in small squares (e.g., 1.27 cm by 1.27 cm [1/2 in. by 1/2 in.]) – could match the performance of configurations based on the use of more expensive long fibers [138]: accordingly, small parts as well as significant portions of a rocket combustion chamber can be effectively produced using ablative Bulk Molding Compounds (BMCs) [140, 141]. A wide range of commercial ablators are sold as molding compounds [142, 143]. Research on glass or silica/phenolic liners also showed that silica powder performs a crucial role in the improvement of the ablative performance of composite liners [144–146]. In fact, when exposed to high temperature, micron-sized silica particles form a highly viscous melt layer that covers the charred surface, working as an antioxidizing protective barrier and without inhibiting radiative cooling. The melt layer also continues to adsorb a significant amount of heat by virtue of endothermic processes and due to the phase transitions of silica.

Natali et al. [137] investigated the possibility of producing glass phenolic composites with nanosized silica as an alternative to micron-sized silicon dioxide. Typically, in glass/phenolic or silica/phenolic liners composed with micron-sized silica, the amount of SiO_2 particles is about 8–10 wt% and the fiber content is about 60 wt% [137, 138]. For example, Cytec MX2600 (an ablative used in SRM nozzle entrance region) consists of 31 wt% resin, 61 wt% chopped strands of silica, and 8 wt% of micron-sized silica particles. The authors first studied two nano-silica-composed nanocomposites [137]: the resulting filler percentages were 5 wt% and 20 wt%. Then the pristine resin and the nanocomposites were used as matrices for the impregnation of E-glass chopped fibers. In all the cases, the fiber percentage remained constant at a value equal to 60 wt%. The amount of nano-silica on the total weight was 8 wt% (BMC3) when the 20 wt% loaded nanocomposite was used to impregnate the short fibers: this composition strongly resembles the formulation of the MX2600 with the difference that the micron-sized silica was replaced by the nanosized material. To further investigate the effects of nano-silica, another BMC composition containing a nanofiller percentage equal to 2 wt% was studied (BMC2): this viscous mixture was produced using the 5 wt% loaded nanocomposite. A BMC composition prepared with no nano-silica was used as a control material (BMC1). Oxyacetylene torch testing (ASTM-E-285-80) [27] was used to study the resulting materials (heat flux equal to 850 W/cm^2 and an exposure time of 60 seconds). The flame was applied to one of the bases of the cylindrically shaped samples. In-depth temperatures were acquired at 5 mm (T1) and 10 mm (T2) from the surface. The morphological structure of charred materials was investigated by SEM analysis.

The charred surface of the control composition (BMC1) summarizes the ablation mechanism for this material [137]. In the absence of nano-silica, the glass fibers melt into low-viscosity material under exposure to the flame, leading to a high erosion rate. Unbounded spheres of glass easily drifted away from the zone impacted by the torch plume, leaving the charred surface unprotected: the glass droplets moved to a well-defined corona on the external edge of the flame. In terms of the contact angle within the charred surface, the marked spherical shape of the glass droplets clearly identified the poor compatibility of the melt with the substrate.

The results of the BMC2 composition are shown in [137]. Under exposure to the flame, the glass fibers and nano-silica melt together to form a blend. Even this low nano-silica percentage was able to increase the viscosity of the blend. The droplets of molten glass fibers exhibited better adhesion on the charred surface. However, this percentage of nanofiller did not completely immobilize the glass droplets under the zone impacted by the flame as summarized in [137]: the shear forces produced by the flame gases were sufficiently vibrant to displace the molten material from the zone impacted by the flame [137].

However, in comparison to the BMC1 composition, glass droplets moved to a corona with a reduced diameter, confirming the enhanced viscosity of the charred surface. The contact angle of the molten droplets within the charred surface indicated that the charred substrate increased its adhesive capability and compatibility within the molten system. Accordingly, the erosion rate experienced by the BMC2 material was decreased. As a matter of fact, the most interesting result was obtained from the BMC3 composition (Figure 10.70a) [138]. In this case, the higher nano-silica content ablative was able to effectively maintain high viscosity, silica rich char, along with droplets of molten glass precisely in the zone impinged by the flame (Figure 10.70b) [138]. The protection of the charred substrate and the preservation of the surface noticeably increased, thus leading to significantly lower erosion. Observing the contact angle of the molten droplets, it was possible to maintain the high compatibility

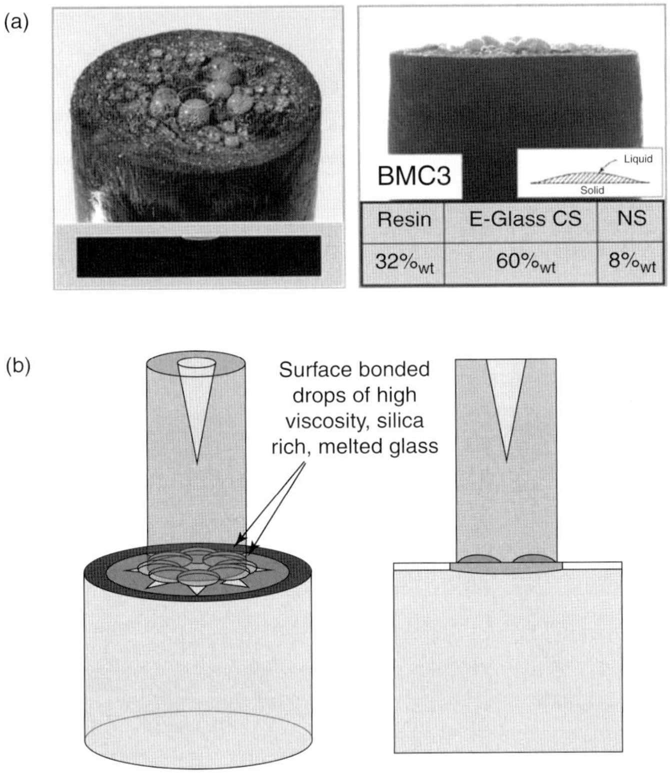

Figure 10.70. Charred surfaced of (a) control composition (BMC3) and (b) ablation mechanism [138].

between the charred surface and the droplets of melt. The adhesion capability of the charred surface was very high, confirming the effectiveness of the use of silica nanoparticles dispersed in an organic matrix. Since the charred substrate was held above the virgin material for a long period, the re-radiation efficiency was increased. Figure 10.71 shows the results of the SEM characterization carried out on the charred samples, confirming the observations with traditional microscopy [138].

To further demonstrate the role of the nano-silica, Figure 10.72 shows the detail of the charred surface of the BMC3 composition: at high magnification, it is possible to identify sintered nano-silica particles. Moreover, the inset photo also demonstrates how nano-silica particles adhere satisfactorily to the glass filaments, acting as a binder. The sintering mechanism of silica nanoparticles was more intently investigated by the same authors in a previous work on silica/phenolic nanocomposites [138]. The in-depth temperatures showed that as the nano-silica content increased, the BMC viscous mixture exhibited improved insulation. The higher the amount of silica nanoparticles, the higher the amount of heat absorbed by the filler. Moreover, increasing the nano-silica content increased the thickness of charred region above the thermocouples, reducing the transferred heat.

For silica-rich chars exposed to a nonoxidizing hyperthermal environment, a reaction producing silicon carbide is proposed: the presence of SiC can enhance the

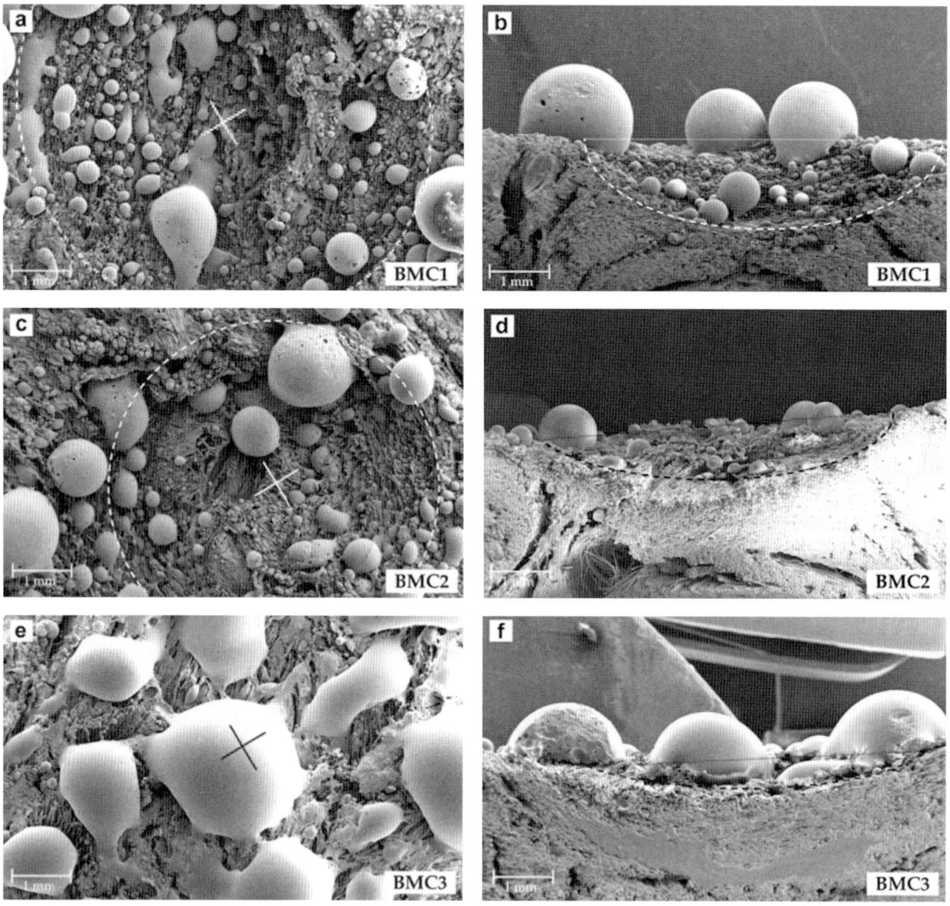

Figure 10.71. SEM micrographs of the charred samples: (a) and (b) refer to BMC1; (c) and (d) refer to BMC2; (e) and (f) refer to BMC3. The crosses on the images (a), (c), and (e) mark the point where the core of the torch impinged the sample. In all the pictures, the dashed lines represent the eroded areas [138].

ablation performance of the char [148–150]. The production of SiC is possible only when a carbothermal reduction between silica and carbon takes place [137]. During oxyacetylene tests, at the core of the plume, in the presence of a neutral flame, the charred surface is relatively protected against the oxidation and SiC can be produced. However, since the temperature of the core exceeds about 2,700°C, the SiC phase undergoes decomposition.

Natali et al. [139] also investigated the ablative properties of two carbon nanofillers. In particular, carbon black (CB) and multi-walled carbon nanotubes (MWNTs) were examined to produce highly loaded (50 wt%) phenolic composites. This study is important because it allowed a distinction of the main differences in terms of ablative performance among traditional carbonaceous filler, such as CB and relatively new filler such as MWNT. In fact, prior to the concept of "nanocomposites" and their uses, carbon black was extensively used in the production of polymeric ablators. For example, Cytec Engineered Materials MX-4926, a standard carbon/phenolic nozzle

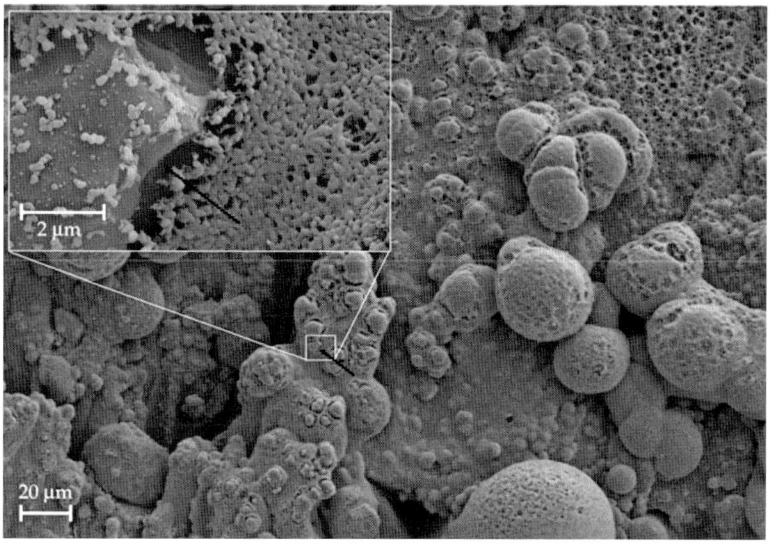

Figure 10.72. Morphology of charred surface of BMC3.

material, is composed by 50 wt% of carbon fibers, 15 wt% of carbon black, and 35 wt% of phenolic resin. Recently, in the class of nano-sized carbon fillers, nano-filaments, such as nanofibers (CNFs) have been studied as filler for polymeric TPSs (see Section 10.3.6.2 of The University of Texas at Austin study). The aim and novelty of the Italian research was to evaluate carbon nanotubes (MWNTs) as potential fillers for ablatives. In fact, MWNTs have been shown to be very effective flame-retardant additives for polymers [151]: the improved flame resistance was attributed to the formation of a continuous protective nanotube felt layer functioning as a heat shield [142–144]. In analogy with the mechanism of layered silicates, such a protective barrier reduces the mass loss rate and the flammability of the material. In agreement with the aforementioned mechanism, flame retardancy was improved at higher MWNT loading. Moreover, the MWNTs embedded in the charred surface re-emit a large amount of the incident radiation into the gas phase, decreasing the heat transferred into the inner layers and thus reducing the polymer pyrolysis rate [155, 156]. As a drawback, carbon-based reinforcements are easily decomposed in an oxidized hyperthermal environment. Nonetheless, to date, after considering the flame-retardant properties of MWNTs, there appeared to be a lack of research on MWNTs used as nanofillers for ablatives. The research of Natali et al. is important not only because the ablative response of MWNTs embedded in a polymeric matrix was investigated, but also because a direct comparison with CB certainly represents a valuable contribution for the scientific community and the industry. In this study two different mixtures were produced, both consisting of 50 wt% of the phenolic matrix and 50 wt% of the nanofiller: one with CB (Vulcan 7H), referred to as PR-CB, constituted by 50%-PR and 50%-CB; the other, with MWNTs, constituted by 50%-PR and 50%-MWNTs (Arkema Graphistrength C100), referred to as PR-MWNT. The results concerning the thermogravimetric analysis, the evaluation of the heat capacity, the oxyacetylene test and the post-burning morphology are summarized below.

In terms of thermal stability, at a heating rate of 20°C/min, the TGA results in air and nitrogen showed that PR-CB and PR-MWNT exhibited a parallel behavior. The study of the heat capacity as a function of the temperature can be very important for the understanding of the ablation mechanism and it can also be used for the modeling of the process. It appeared that in the whole range of temperatures, both composites exhibited a higher heat capacity than the pristine matrix, with the PR-MWNT system having the highest values. The post-charred images of the tested materials showed that in terms of the erosion rate, the PR-MWNT composite exhibited higher erosion than the PR-CB system. The PR-MWNT samples suffered a marked removal of material, and the flame produced a crater exactly in a zone confined under the plume of the torch. During the tests, small flakes of the material eroded off the surface of the PR-MWNT samples: this behavior was also found by other authors for CNF-based NRAMs. The PR-MWNT samples also showed a very thin charred layer above the virgin material. The PR-MWNT samples showed a very limited number of in-depth cracks propagating from the mantel surface to the center of the sample. For the PR-CB composites, the flame produced a broad and deep charred region having relatively good integrity: however, the flame did not produce any crater. Also, during the test, this material did not show any macroscopic release of burning material. The PR-CB charred samples exhibited a remarkable number of macro-cracks emanating from the mantel of the samples up to the lower limit of the charred region. It can be concluded that the PR-CB system suffered char penetration extended on a higher and deeper region of the sample.

The morphology of the charred surfaces was also investigated by SEM. At low magnification the PR-CB revealed the cracking phenomena also identified with the optical analysis. The top surface of the PR-CB showed a morphology characterized by micron- and sub-micron-sized open and closed bubble-like structures: the carbon black particles were not distinguishable from the charred layer. The morphology of the zones beneath the surface was very close to that of the surface, but in some cases, a uniform and compact medium with no voids or bubble was found. If compared to the PR-CB system, the surface of the PR-MWNT samples showed a homogeneous morphology free of any macro-cracks: the ablation appeared to be a surface phenomenon with a very thin transition zone – the char – between the virgin and charred material. The char was constituted by many irregular aggregates mainly composed of entangled MWNTs. Beyond the undamaged and uncharred MWNT filaments, SEM also demonstrated many bright spots clearly related to the metal catalyst precursors used to produce the MWNTs. The temperature profiles showed that the PR-MWNT system exhibited higher thermal diffusivity than the PR-CB composition. The authors also concluded that, if compared to the PR-CB system, the PR-MWNT composite possessed a higher thermal conductivity in the direction of the flame plume. Loss of mass data indicated that both materials experienced a similar removal of mass. As a consequence, since the surfaces of PR-MWNT samples experienced a more severe erosion than the PR-CB specimens, and considering that the amount of loss of mass is of the same order of magnitude for both materials, it can be concluded that the ablation mechanisms of the two materials is considerably different. For the PR-MWNT, the loss of mass was due to the physical erosion of material and the loss of mass of the PR-CB can be attributed to the higher-degree char of the organic matrix since the PR-CB did not erode appreciably. In conclusion, it is clear that CB and MWNTs exhibit a very different charring behavior.

10.3.6.11 Phenolic-MWNT Nanocomposites Studies by Gyeongsang National University, South Korea

In the investigation by Park et al. [157], carbon nanotubes and carbon fibers were used in conjunction with phenolic resole-type resin to produce articles for thermal characterization. Three composites were produced: one with 30 wt% chopped CFs, one with 30 wt% woven CF fabric, and one with 30 wt% woven CF fabric as well as 0.5 wt% CNTs. The chopped CFs were stirred into the phenolic resin with acetone as a diluting agent for 3 hours at 100 rpm. The CNTs were dispersed into the resin according to a method developed previously by the team, and the mixture was applied to the carbon fabric by a hand lay-up method. All neat and composite samples were cured by autoclave at 690 kPa at 70 °C for 1.5 hours, 100 °C for 1 hour, and then 140 °C for 4 hours.

To characterize the formulations, flame retardancy testing, ablation testing, and thermogravimetric analysis (TGA) was conducted on all candidate materials. For the flame retardancy investigation, a butane flame was applied to the specimen according to ASTM E84-11, while a thermal conductivity analyzer using a modified transient plane source technique recorded conductivity data. Thermal distortion and surface damage were both inspected after testing. For the ablation testing, an oxygen-kerosene flame with a 1:1 fuel ratio was applied to the sample. Distance and heat flux values were not reported. For the TGA, samples were taken from the front and back sides of the posttest specimens for data on the amount of carbonization of the material during testing.

It was observed that each of the composites had increased thermal conductivity as compared to the neat resin, with the chopped CF composite having the least increase and the CF fabric/CNT composite having the greatest. This trend persisted in the ablation testing results, where the same ranking occurred for increase in ablation resistance. The researchers conclude that the two properties are related. In addition, it was found that the carbonization of the material during testing may also be indicative of ablation resistance, since it does not occur without sustained elevated temperatures in the material. The neat resin had no carbonization on the back side, but the chopped CF composite had some and the two CF fabric composites had nearly identical carbonization on their front and back sides.

10.3.6.12 Phenolic-MWNT Nanocomposites Studies by Texas State University–San Marcos, United States

The research by Tate et al. was a collaborative effort with Cytec Engineered Materials (CEM) as an attempt to enhance the properties of the SOTA ablative, MX-4926, by introducing small amounts of MWNT [158, 159]. Test panels were manufactured using processes and materials similar to those used by CEM. MX-4926 MC (Molding Compound) is rayon precursor-based carbon fabric with a MIL-R-9299 phenolic resin. Composite panels were manufactured using rayon-based carbon fabric supplied by CEM, which is an eight-harness satin weave. The fabric has a weight of 261 g/m^2, a specific gravity of 1.84, and a thickness of 0.48 mm. Break strength is 0.496 MPa in the warp direction and 0.599 MPa in the fill direction, with 1.96 picks/mm warp and 2 picks/mm fill. SC-1008 phenolic resin meets the MIL-R-9299 specification, and was supplied by Hexion Specialty Chemicals. The viscosity is about 180–300 cps depending on storage conditions. SC-1008 contains about 20–25% IPA as a solvent. Graphistrength™C100 MWNTs were supplied by Arkema, Inc. Typical diameter of these MWNTs is 10–15 nm with a length of 1–10 μm. Graphistrength™C100 MWNTs appear to be agglomerated with dimensions of approximately 50–900 μm. High-shear mixing and sonication have been used in previous studies to provide good dispersion and exfoliation of nanoparticles [160–162]. Tate et al. investigated the dispersion characteristics using sonication, high-shear mixing, and a combination of the two in previous studies [158]. TGA results and SEM analysis suggested that the dispersion of MWNTs within IPA using sonication followed by high-shear mixing produces optimal dispersion characteristics for use in ablative panels. This method was adapted to produce composite panels with MWNT loading of 0.5, 1, and 2 wt%.

Oxyacetylene test bed (OTB) was developed to perform a small-scale testing of newly developed ablatives. Samples size of 12.7 mm × 12.7 mm × 10 mm were prepared and two holes of approximately 1.59 mm (1/16 in.) diameter drilled at depth of 2.7 mm and 7.7 mm for thermocouple placement. Samples were exposed to a heat flux of approximately 1,000 W/cm^2 (6 mm distance from the OTB) for 45 seconds [159]. Six samples were tested in each category. Ablative performance of the material was analyzed based on percentage mass loss, recession, peak temperatures at different depth in ablative samples, and SEM to analyze dispersion of MWNTs in phenolic resin. There was gradual decrease in percentage mass loss as weight percentage of MWNTs increased. For control samples, average mass loss was 26%; for 2 wt% MWNT, average mass loss was 23%; and for 0.5 wt% and 1 wt% MWNT, average mass loss was 25.5% and 25%, respectively. There is considerable reduction in mass loss as weight percentage of MWNTs increased. Typically mass loss is a combined pyrolysis of material and char erosion. Similarly, recession was also high for low wt% formulations and decreased gradually as wt% of MWNTs increased. Average recession was 0.83 mm for control samples, while it was 0.38 mm for 2wt% samples. Average peak temperatures were reduced considerably with the addition of MWNT to the composite (Figures 10.73 and 10.74) [158, 159].

In summary, the combination of sonication and high-shear mixing was used for uniform dispersion and exfoliation of MWNT. Ablative samples were tested on Oxyacetylene Test Bed for 45 seconds and at a heat flux of 1,000 W/cm^2. The samples were compared on the basis of percentage mass loss, recession, and peak in situ temperatures at depths of 10 mm and 5 mm from the flame front. The percentage

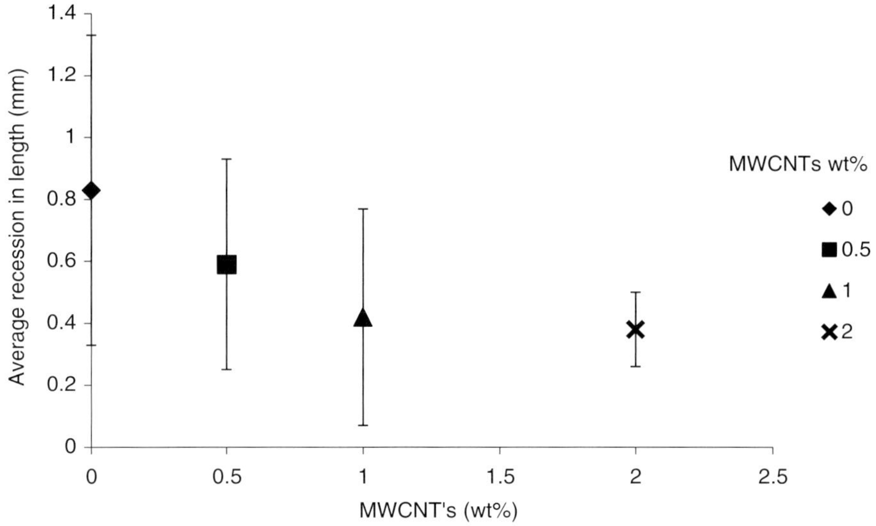

Figure 10.73. Decrease in recession in length with increase in MWNT weight percentage.

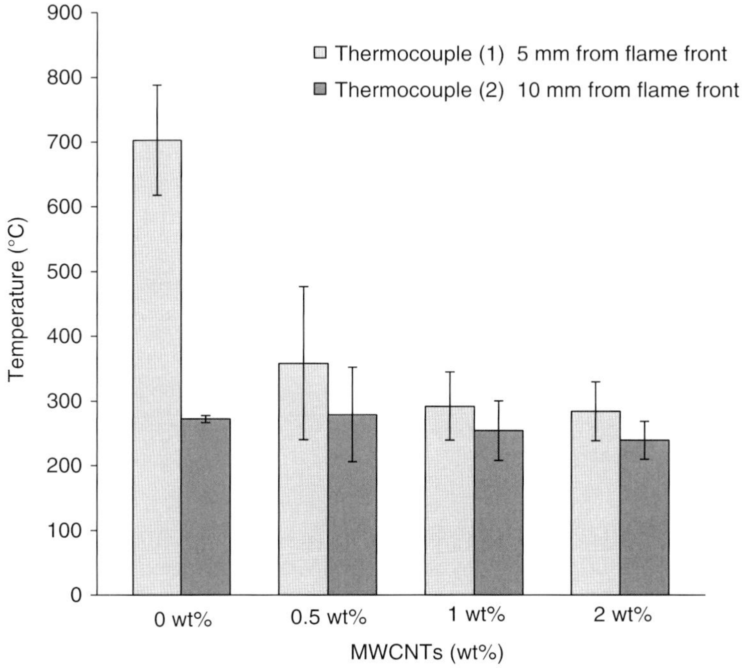

Figure 10.74. Peak in situ temperatures at 5 mm and 10 mm depths from the flame front surface.

mass loss for control sample was 26%, and it was 23% for the 2 wt% MCNT sample. Average recession was 0.83 mm for control samples while it was reduced to 0.39 mm for the 2 wt% MWNT samples. The peak temperatures at depths of 10 mm and 5 mm from the flame front showed decreasing trends as the MWNT weight percentage increased. Based on these results, it was concluded that the increase in weight

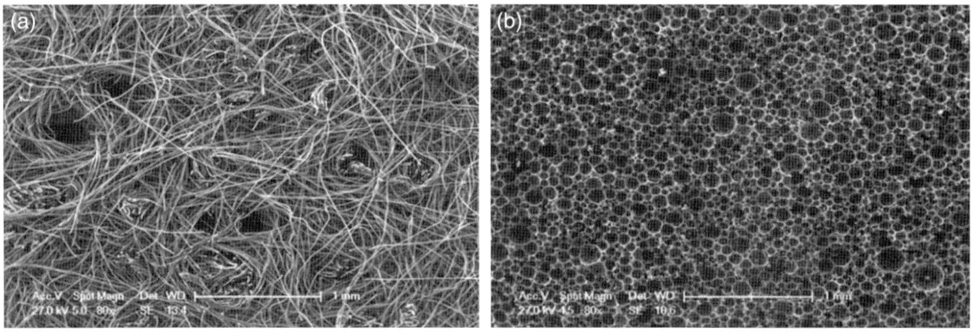

Figure 10.75. Microstructures of carbon-based reinforcements: (a) Sigratherm felt and (b) Grafoam.

percentage of MWNT improved ablation and insulation performance of phenolic-MWNT nanocomposite.

10.3.6.13 Phenolic-Graphitic Felt and Phenolic–Graphitic Foam Studies by University of Rome, Italy

Sapienza University of Rome group developed, manufactured, and tested a family of carbon-phenolic ablative TPS materials with the aim of fulfilling the thermal and mechanical requirements corresponding to the actual loads experienced by reentry vehicles [163–165]. Experimental investigations were conducted on two different composite systems: a resole resin coupled with a graphitic felt, and graphitic foam. Their aim was to optimize the manufacturing procedure and to characterize the mechanical behavior and the insulation performance of the fabricated composites.

A commercially available resole phenolic resin was employed and two carbon-based reinforcements were selected for impregnation with the phenolic resin. Graphitic foam (Grafoam® manufactured by Graftech International, Parma, OH; www.graftech.com) and a rigid graphitic felt (Sigratherm® manufactured by SGL Carbon SE, Wiesbaden, Germany) were used. Grafoam carbon foam is an open-cell, rigid, carbon-based material that can be produced in densities ranging from 0.03 g/cm^3 to 0.56 g/cm^3 and in a variety of sizes. Sigratherm felt has high thermal stability in vacuum up to 2,500°C and up to 3,000°C in inert atmospheres. It is well known that above 2,000°C physical properties of the ablative can change. Figure 10.75 shows the microstructures of carbon-based reinforcements.

Infiltration process was performed by the total immersion procedure: both reinforcements were totally immersed for 30 minutes in a phenolic solution with an isopropanol/resin ratio tailored to obtain the specific density (0.5 g/cm^3) after the cure process. After resin cure, a resin/composite volume ratio of 0.61 and 0.59 was calculated for the Sigratherm and Grafoam composites, respectively. Resin preheating (50°C) and vacuum infiltration (3 minutes at 10^{-3} bar) after soaking were mandatory for removing excess solvent and efficient impregnation. All tests were carried out under these conditions.

Preliminary ablation tests were performed on the two fabricated systems according to ASTM E285-80 standard [27]. The size of samples tested with oxyacetylene torch was 100 mm x 100 mm x 20 mm or 50 mm x 50 mm x 20 mm; the distance between nozzle and samples was 20 mm, and the temperature of the front and the back surface of the tiles was monitored by a two-color digital fiber-optics non-contact pyrometer (Impac Infrared, Frankfurt, Germany) and by a type-K thermocouple, respectively. Heat flux, as measured by a calorimeter (HFM 1000-Vatell Corp., Christiansburg, VA) during experimental tests, was 1.5 MW/m^2. Four exposure times were investigated: 1, 5, 10, and 20 min. Exposure tests were performed on five impregnated and two un-impregnated samples. Maximum temperature of about 4,000K was registered on the front face of the exposed samples.

Higher hardness values were noted on the surface rather than in the cross-sectional areas. Impregnated samples showed a homogeneous resin distribution throughout the thickness where the cross section of a 3 cm thick Sigratherm composite is observed [163]. Magnified views of impregnated samples' cross sections, showing suitable degree of resin impregnation, were achieved for both reinforcements [163]. All specimens exhibited a linear elastic behavior; Grafoam and Sigratherm composites show similar values of elastic modulus (about 2 GPa), but the felt composite showed a much higher (+144%) flexural strength than did the foam composite. Pure resin has an elastic modulus of 4 GPa, whereas un-impregnated materials have a modulus of 0.25 GPa and 0.5 GPa for the felt and the foam, respectively. It is evident that the resin provides most of the composite stiffness. Results of four-point bending tests indicated a higher flexural strength of impregnated Sigratherm with respect to Grafoam.

Thermophysical properties (thermal diffusivity, thermal conductivity, and specific heat) of the composite samples have been reported by Pulci et al. [163]. Similar behaviors were observed for the two types of materials. Foam composites exhibited a slightly lower thermal conductivity at high temperature (1,500–1,900°C), while impregnated felts exhibited lower thermal conductivity at temperatures lower than 800°C. Un-impregnated Grafoam and Sigratherm reinforcements showed a room temperature thermal conductivity of 0.12 and 0.22 W/(m-K), respectively. The results of a typical oxyacetylene exposure test on Sigratherm and Grafoam composite samples, in terms of front and back surface temperature, are reported by Pulci et al. [163]. The two materials show the same temperature characteristics with time, indicating comparable insulation properties. However, the analysis of samples after tests showed that marcroporous cells (Grafoam) were fractured during the flame exposure, whereas all felt-based samples (Sigratherm) passed undamaged during the longest oxyacetylene exposure tests.

There appears to be a clear indication that a carbon/phenolic composite material made of rigid non-woven graphitic felt impregnated with a resole phenolic resin can act as a very effective insulator in a very wide range of temperatures (from room temperature to about 2,000°C), making it suitable for forming/machining operations. A very accurate optimization of the main steps of the manufacturing procedures is required to achieve the above. The research multi-step curing cycle of 12 days at different temperatures is required to achieve optimal microstructure and highest mechanical properties of the composite. The outstanding performance of the manufactured composite system exposed to the intense heat flux of the oxyacetylene flame was achieved. Sigratherm-based ablators were selected for arc-jet testing in a Plasma Wind Tunnel (Scirocco, CIRA ScoA, Italy) to simulate reentry conditions.

In order to support the material/component design and to offer new and differential tools for the screening of TPS materials, this research group has developed a CFD (computational fluid dynamics) model of the oxyacetylene test apparatus per ASTM E285-80 standard. The CFD model simulates the combustion of the oxygen-acetylene mixture occurring in a conventional welding torch and calculates the heat flux incident on the material sample and the related temperature fields. The validation of the CFD model was obtained by performing suitable instrumented tests with direct measurements of both incident heat flux and temperatures of the front and back surfaces of an ablative sample. Two different numerical models using Fluent® software have been developed for the simulation: the first model is intended for the evaluation of the heat flux actually generated by oxyacetylene torch; the second model simulates a real test on an ablative material. Two different experimental setups were prepared in order to validate the results obtained by the numerical simulations. Good agreement was found between numerical results and experimental measurements [163, 164]. For the configuration corresponding to the real test of an ablative material, the comparison between simulation and test applies to the temperatures of the exposed face, as measured by the pyrometer, and of the sample back face. The range of the measured front face temperature is 1,970–2,070K [164]. The results of CFD simulation of this configuration are shown in [164]. The temperature value calculated at the center of the irradiated sample is about 1,900K; the correspondence between measured and simulations is acceptable.

Recently, additional lightweight ablative materials development data were obtained from Valente [165]. Figure 10.76 shows the (a) rigid graphite felt (long fibers) and (b) rigid graphite felt (short fibers) [165]. Innovative lightweight ablative materials using carbon-based reinforcement (felts) infiltrated with different volumetric fractions of phenolic resin to obtain different densities were investigated (Table 10.11) [165]. Mechanical properties (elastic modulus and flexural strength) were characterized using the four-point bend test. Microstructural characterization was investigated using the SEM analysis (Figure 10.76) [165]. Thermal performance by oxyacetylene exposure tests were conducted with two-color pyrometer to measure front face temperature and thermocouples for back face temperature, while heat flux sensor HFM 1000 was used to calibrate heat flux (Figure 10.77) [165]. Test conditions of heat flux 2 MW/m^2 and exposure time of 10 minutes were used. All samples exhibited good insulation performance with a surface temperature of about 1,700°C and back face temperature always below 500°C after 10 minutes of exposure using a sample thickness of 30 mm. Samples exhibited somewhat little recession of about 1.8–3.4 mm. Multilayer long fibers felt shows damage at the interface between joined layers, probably due to oxidation. Density of 0.25 g/cc samples appears to exhibit the worst thermal behavior in terms of insulation and recession.

Table 10.11. *Carbon-based reinforcements (felts) infiltrated with different volumetric fractions of phenolic resin to obtain different densities*

Felt	Matrix	Density (g/cc)	Additives
Carbon felt- long fibers	Phenolic	0.5	CNT mixture
Carbon felt- long fibers	Phenolic	0.25	no
Carbon felt- long fibers	Phenolic	0.5	Nano-Al_2O_3
Carbon felt- long fibers	Phenolic	0.5	Multilayer with graphitic adhesive
Carbon felt- short fibers	Phenolic	0.5	no
Carbon felt- short fibers	Phenolic	0.25	Nano-Al_2O_3

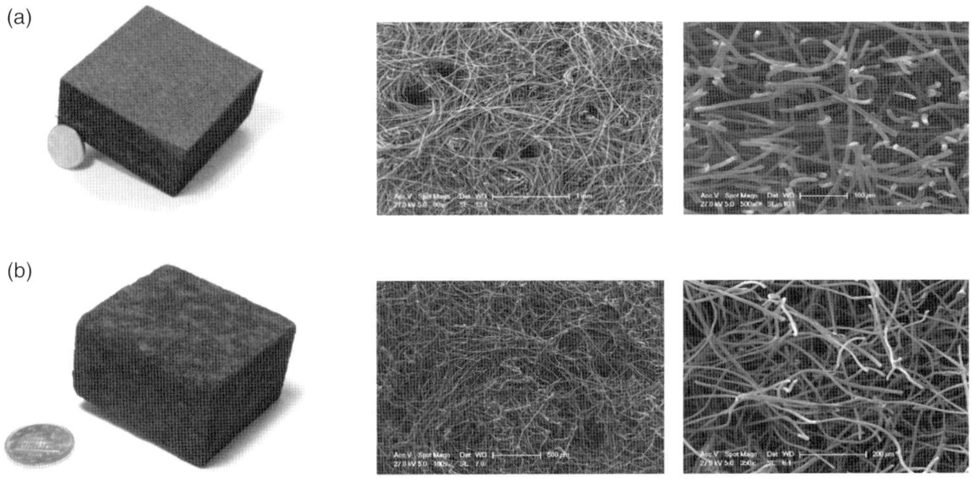

Figure 10.76. Microstructural characterization of (a) long fiber felt and (b) short fiber felt (Courtesy of Prof. Valente).

Figure 10.77. Ablation test setup at University of Rome: (a) during ablation testing and (b) posttest ablative (Courtesy of Prof. Valente).

MWNT dispersion in the phenolic resin is a critical point, but their addition seems to improve the thermal performance. Posttest observation of surfaces reveals nano-alumina forms alumina-like layer on the samples surface. This phenomenon could be effective to improve ablation resistant of charred layer.

10.4 Summary and Conclusions

A comprehensive review of polymer nanocomposites designed for advanced thermal protection materials was conducted. It has been demonstrated that the improvement in ablative performance of these polymer nanocomposites relative to traditional filled system is associated with their char-forming characteristics. The spatially uniform distribution of nano-scaled inorganic filler results in the formation of a uniform inorganic char at a relatively low fraction of inorganic. The nanostructured ablative materials will provide multifunctional properties, such as ablation resistance, thermal stability, insulation, and lighter weight. Ablation mechanisms were also studied and proposed by the researchers on their nanostructured ablatives.

Recently, Duffa published an excellent book on *Ablative Thermal Protection Systems Modeling* [166]. This volume discussed the board physical knowledge of modeling of phenomena, including (from microscopic to macroscopic) atomic physics, thermodynamics, gas kinetics, radiative transfer, physical and chemical reactions (both homogenous and heterogeneous), fluid mechanics, and turbulence applied to previously studied areas, such as rough walls. The author commented that many books were devoted to each of these research areas, but very little literature is devoted to the synthesis of ablation problem. Ablation research is undoubtedly strongly coupled to various physical domains. An overview of this particular research is given at the end of the first chapter, titled "Thermal Protection System Conception." The objective of this text is to develop physical skills in the research applied to the ablation modeling of thermal protection systems. It consists of thirteen chapters and seven appendixes that provide comprehensive technical discussions on ablation modeling phenomena, testing and specific test facilities, and selective thermal protection materials.

10.5 Questions and Problems

1. Discuss the thermochemical behavior of thermal protection materials under severe thermal attack in high heat flux and high velocity extreme environments.
2. In the thermoplastic polyurethane nanocomposites studies (Section 10.3.5), compare the ablation performance of (a) MMT clay, (b) carbon nanofiber (CNF), and (c) multi-walled carbon nanotubes (MWNT) and explain their respective ablation mechanisms. Postulate the ablation performance when a combination of (a) MMT clay-CNF, (b) MMT-MWNT, and (c) CNF-MWNT were used in the TPUNs. What percentages of MMT clay, CNF, and MWNT should be used in (a) to (c), and what is the sequence of polymer blends compounding?
3. As an extension to the thermoplastic polyurethane nanocomposites studies (Section 10.3.5), what would be the mechanical properties of the TPUN-clay,

TPUN-CNF, TPNU-MWNT, TPUN-clay-CNF, TPUN-clay-MWNT, and TPUN-CNF-MWNT nanocomposite systems? What are the important mechanical properties you should characterize?

4. What other nanofillers you would propose in the thermoplastic polyurethane nanocomposite studies to enhance the ablation performance, insulation, and mechanical properties?
5. In the phenolic nanocomposites studies (Section 10.3.6), (a) CNF, (b) MMT clay, (c) POSS, (d) nano-silica, (e) carbon black, and (f) MWNT were used as fillers in the phenolic resin and reinforced with carbon or glass fiber reinforcements. Discuss the protective mechanisms for each of fillers in their composites. Discuss the important mechanical properties of these phenolic nanocomposites and which mechanical properties you should characterize.
6. What other nanofillers you would propose in the phenolic nanocomposite studies to enhance its ablation performance and insulative and mechanical properties?
7. In the poly(phenylene oxide)/polystyrene (PPO/PS) studies (Section 10.3.1.2), what polymer-processing technique and sequence you would propose to better incorporate these ceramic particles into the polymer matrix? What dispersion technique you would propose to examine the degree of particle dispersions in the polymer matrix? What ablation testing you would propose to compare the ablation performance and heat transfer characteristics of these nanocomposites?
8. As an extension to the poly(phenylene oxide)/polystyrene (PPO/PS) studies (Section 10.3.1.2), characterize the important mechanical properties of these PPO/PS nanocomposites.
9. Discuss what other ablation test facilities, such as arc-jet or inductively coupled plasma (ICP) that are available in the industry to study thermal protection materials as well as the locations of these test facilities, and the major test parameters used in these ablation testing facilities?

REFERENCES

1. J. H. Koo, W. K. Ho, and O. A. Ezekoye (2006). A Review of Numerical and Experimental Characterization of Thermal Protection Materials – Part I. Numerical Modeling, AIAA-2006-4936. Presented at the *42nd AIAA/ASME/SAE/ASEE Joint Propulsion Conference*, Sacramento, CA, July 9–12.
2. J. H. Koo, W. K. Ho, M. Bruns, and O. A. Ezekoye (2007). A Review of Numerical and Experimental Characterization of Thermal Protection Materials – Part II. Material Properties Characterization, AIAA-2007-2131. Presented at the *48th AIAA/ASME/ASCE/AHS Structures, Structural Dynamics, and Materials Conference*, Honolulu, HI, April 23–26.
3. J. H. Koo, W. K. Ho, M. Bruns, and O. A. Ezekoye (2007). A Review of Numerical and Experimental Characterization of Thermal Protection Materials – Part III. Experimental Testing, AIAA-2007-5773. Presented at the *43rd AIAA/ASME/SAE/ASEE Joint Propulsion Conference*, Cincinnati, OH, July 8–11.
4. J. H. Koo, M. Natali, J. Tate, and E. Allcorn (2013). Polymer nanocomposites as advanced ablatives – a comprehensive review. *International Journal of Energetic Materials and Chemical Propulsion* 12(2), 119-162.

5. T. J. Pinnavaia and G. W. Beall (Eds.) (2000). *Polymer-Clay Nanocomposites*. New York: Wiley & Sons.
6. J. H. Koo (2006). *Polymer Nanocomposites: Processing, Characterization, and Applications*. New York: McGraw-Hill.
7. A. B. Morgan and C. A. Wilkie (Eds.) (2007). *Flame Retardant Polymer Nanocomposites*. Hoboken, NJ: Wiley & Sons.
8. R. A. Gupta, E. Kennel, and K. J. Kim (Eds.) (2010). *Polymer Nanocomposites Handbook*. Boca Raton, FL: CRC Press.
9. V. Mittal (Ed.) (2010). *Polymer Nanotube Nanocomposites: Synthesis, Properties, and Applications*. Hoboken, NJ: Wiley & Sons.
10. V. Mittal (Ed.) (2010). *Optimization of Polymer Nanocomposites Properties*. Weinheim, Germany: Wiley-VCH.
11. V. Mittal (Ed.) (2011). *Thermally Stable and Flame Retardant Polymer Nanocomposites*. Cambridge: Cambridge University Press.
12. V. Mittal (Ed.) (2012). *Characterization Techniques for Polymer Nanocomposites*. Weinhein, Germany: Wiley-VCH.
13. G. W. Beall and C. B. Powell (2011). *Polymer-Clay Nanocomposites*. Cambridge: Cambridge University Press.
14. R. A. Vaia, G. Price, P. N. Ruth, H. T. Nguyen, and J. L. Lichtenhan (1999). Polymer/layer silicate nanocomposites as high performance ablative materials. *Applied Clay Sciences* 15, 67–92.
15. B. Laub and E. Venkatathy (2003). Thermal Protection System Technology and Facility Needs for Demanding Future Planetary Missions. *Proceedings of the International Workshop on Planetary Probe Atmospheric Entry and Descent Trajectory Analysis and Science*, Lisbon, Portugal.
16. M. Lombardi, P. Fino, G. Malicelli, and L. Montanaro (2012). Exploring composites based on PPO blend as ablative thermal protection systems – Part I: The role of layered fillers. *Composite Structures* 94, 1067–1074.
17. M. Lombardi, P. Fino, and L. Montanaro (2014). Influence of ceramic particles features on the thermal behaviour of PPO-matrix composites. *Science and Engineering of Composite Materials De Gruyter*, 23-28, ISNN: 2191-0359.
18. P. Fino, M. Lombardi, A. Antonini, G. Malucelli, and L. Montanaro (2012). Exploring composites based on PPO blend as ablative thermal protection systems – Part II: The role of equiaxial fillers. *Composite Structures* 94, 1060–1066.
19. D. M. Lincoln, R. A. Vaia, J. M. Brown, and T. H. Benison Tolle (2000). Revolutionary Nanocomposite Materials to Enable Space Systems. *Proceedings of the 21st Century Aerospace Conference IEEE*, Big Sky, MT, vol. 4, pp. 183–192.
20. S. H. Philip, R. I. Gonzales, R. L. Blanski, and B. D. Viers (2002). Hybrid Inorganic/Organic Reactive Polymers for Severe Environment Protection. *Proceedings of the 47th SAMPE ISSE*, SAMPE, Covina, CA, May 12–16.
21. C. A. Yezzi and B. B. Moore (1986). Characterization of Kevlar/EPDM rubbers for use as rocket motor case insulators, AIAA-86-1489. Presented at the *AIAA/ASME/SAE/ASEE 22nd Joint Propulsion Conference*, Huntsville, AL, June 16-18.
22. E. C. Mathias and T. N. Johnson, (1999). *20th JANNAF Rocket Nozzle Subcommittee Meeting*, CPIA Pub. 694, pp. 237–266.
23. J. H. Koo, J. Polidan, et al. (2002). An Investigation of Polymer Nanocomposite Ablatives Characterization. Invited lecture at the *30th Annual Conference of the North American Thermal Analysis Society*, Pittsburgh, PA, September 23–25.
24. J. H. Koo, R. Blanski, et al. (2003). Nanostructured Ablatives for Rocket Propulsion System–Recent Progress, AIAA-2003-1769. Presented at the *44th AIAA/ASME/ASCE/AHS Structures, Structural Dynamics, and Materials Conference*, Norfolk, VA, April 7–10.

25. J. H. Koo and L. Pilato (2005). Polymer Nanostructured Materials for Propulsion Systems, AIAA-2005-3606. Presented at the *41st AIAA/ASME/SAE/ASEE Joint Propulsion Conference*, Tucson, AZ, July 10–12.
26. G. Gao, Z. Zhang, X. Li, X. Q. Meng, Y. Zheng, and Z. Jin (2010). Study on mechanical and ablative properties of EPDM/OMMT thermal insulating nanocomposites. *Journal of Nanoscience and Nanotechnology* 10, 7031–7035.
27. ASTM E-285-80 (2008). Standard test method for oxy-acetylene ablation testing of thermal insulation materials. Annual Book of ASTM Standards.
28. S. Singh, P. K. Guchhait, G. G. Bandyopadhyay, and T. K. Chaki (2013). Development of polyimide-nanosilica filled EPDM based light rocket motor insulator compound: Influence of polyimide-nanosilica loading on thermal, ablation, and mechanical properties. *Composites: Part A* 44, 8–15.
29. M. Natali, M. Rallini, D. Puglia, J. Kenny, and L. Torre (2013). EPDM based heat shielding materials for solid rocket motors: A comparative study of different fibrous reinforcements. *Polymer Degradation and Stability* 98, 2131–2139.
30. N. Iqbal, S. Sagar, M. B. Khan, and H. M. Rafique (2014). Elastomeric ablative nanocomposites used in hyperthermal environments. *Polymer Engineering and Science* 54, 255–263.
31. S. Khanlart and M. Kokabi (2010). Thermal stability, aging properties, and flame resistance of NR-based nanocomposites. *Journal of Applied Polymer Science* 119, 855–862.
32. Y. Guan, L. X. Zhang, L. Q. Zhang, and Y. L. Lu (2011). Study on ablative properties and mechanisms of hydrogenated nitrile butadiene rubber (HNBR) composites containing different fillers. *Polymer Degradation and Stability* 96, 808–817.
33. J. H. Koo (2006). *Polymer Nanocomposites: Processing, Characterization, and Applications*, New York: McGraw-Hill.
34. J. H. Koo, L. Pilato, and G. Wissler (2007). Polymer nanostructured materials for propulsion systems. *Journal of Spacecraft and Rockets* 44(6), 1250–1262.
35. J. H. Koo, O. A. Ezekoye, et al. (2009). Characterization of Polymer Nanocomposites for Solid Rocket Motor – Recent Progress. *Proceedings of the SAMPE 2009 ISSE*, SAMPE, Covina, CA.
36. J. H. Koo, et al. (2010). Flammability studies of a novel class of thermoplastic elastomer nanocomposites. *Journal of Fire Sciences* 28(1), 49–85.
37. J. H. Koo, O. A. Ezekoye, J. C. Lee, W. K. Ho, and M. C. Bruns (2011). Rubber-Clay Nanocomposites Based on Thermoplastic Elastomers. In *Rubber-Clay Nanocomposites*, M. Galimberti (Ed.). Hoboken, NJ: Wiley and Sons, pp. 489–521.
38. W. K. Ho, J. H. Koo, and O. A. Ezekoye (2009). Kinetics and thermophysical properties of polymer nanocomposites for solid rocket motor insulation. *Journal of Spacecraft and Rockets* 46(3), 526–544.
39. W. K. Ho, J. H. Koo, and O. A. Ezekoye (2010). Thermoplastic polyurethane elastomer nanocomposites: Morphology, thermophysical, and flammability properties. *Journal of Nanomaterials 2010*, Article ID 583234 (11pp). doi: 10.1155/2010/583234.
40. M. C. Bruns, J. H. Koo, and O. A. Ezekoye (2009). Population-based models of thermoplastic degradation: Using optimization to determine model parameters. *Polymers Degradation and Stability* 94, 1013–1022.
41. J. C. Lee, J. H. Koo, and O. A. Ezekoye (2009). Flammability Studies of Thermoplastic Polyurethane Elastomer Nanocomposites, AIAA-2009-2544. Presented at the *50th AIAA/ASME/ASCE/AHS/ASC Structures, Structural Dynamics, and Materials Conference*, Palm Spring, CA, May 4–7.
42. J. C. Lee, J. H. Koo, O. A. Ezekoye, et al. (2009). Heating Rate and Nanoparticle Loading Effects on Thermoplastic Polyurethane Elastomer Nanocomposite

Kinetics, AIAA-2009-4096. Presented at the *AIAA Thermophysics Conference*, San Antonio, TX, June 22–25.
43. J. C. Lee, J. H. Koo, and O. A. Ezekoye (2009). Thermoplastic Polyurethane Elastomer Nanocomposites: Density, Hardness, and Flammability Properties Correlations, AIAA-2009-5273. Presented at the *AIAA Joint Propulsion Conference*, Denver, CO, August 2–5.
44. J. C. Lee (2010). *Characterization of Ablative Properties of Thermoplastic Polyurethane Elastomer Nanocomposites*. Ph.D. Dissertation, The University of Texas at Austin, Department of Mechanical Engineering, Austin, TX.
45. J. C. Lee, J. H. Koo, and O. A. Ezekoye (2011). Thermoplastic Polyurethane Elastomer Nanocomposite Ablatives: Characterization and Performance, AIAA-2011-6051. Presented at the *47th AIAA/ASME/SAE Joint Propulsion Conference*, San Diego, CA, August 1–4.
46. E. Allcorn, M. Natali, and J. H. Koo (2011). Ablation Performance and Characterization of Thermoplastic Elastomer Nanocomposites. *Proceedings of the SAMPE 2011 ISTC*, Fort Worth, TX, October 17–20.
47. E. Allcorn, M. Natali, and J. H. Koo (2013). Ablation performance and characterization of thermoplastic elastomer nanocomposites. *Composites: Part A* 45, 109–118.
48. W. A. De Heer (2004). Nanotubes and the pursuit of applications. *Materials Research Society Bulletin* 29(4), 281–285.
49. M. S. Bell and W. Tam (1992). ASRM Case Insulation Design and Development, NASA-CR-191947.
50. C. M. Bhuvaneswari, S. D. Kakade, V. D. Deuskar, A. B. Dange, and M. Gupta (2008). Filled ethylene-propylene diene terpolymer elastomer as thermal insulator for case-bonded solid rocket motors. *Defence Science Journal* 58(1), 94–102.
51. C. M. Bhuvaneswari, M. S. Sureshkumar, S. D. Kakade, and M. Gupta (2006). Ethylene-propylene diene rubber as a futuristic elastomer for insulation of solid rocket motors. *Defence Science Journal* 56(3), 309–320.
52. M. Jaramillo, J. H. Koo, A. E. Edd, and D. M. Wells (2011). An Experimental Investigation of Char Strength of Polymer Nanocomposites for Propulsion Applications. *Proceedings of the International SAMPE Technical Conference*, Dallas, TX, October.
53. M. Moniruzzaman and K. I. Winey (2006). Polymer nanocomposites containing carbon nanotubes. *Macromolecules* 39(16), 5194–5205.
54. J. J. George and A. K. Bhowmick (2008). Fabrication and properties of ethylene vinyl acetate-carbon nanofiber nanocomposites. *Nanoscale Research Letters* 3, 508–515.
55. S. H. Lee et al. (2004). Thermal properties of maleated polyethylene/layered silicates nanocomposites. *International Journal of Thermophysics* 25, 1585–1595.
56. B. Schartel, A. Weib, H. Sturm, M. Kleemeier, A. Hartwig, et al. (2011). Layered silicate epoxy nanocomposites: Formation of the inorganic-carbonaceous fire protection layer. *Polymers for Advanced Technologies* 22(12), 1581-1592, doi: 10.1002/pat.1644.
57. M. Natali, M. Monti, D. Puglia, J. M. Kenny, and L. Torre (2011). Ablative properties of carbon black and MWNT/phenolic composites: A comparative study. *Composites: Part A* 43, 174–182.
58. P. Ambuken, H. Stretz, J. H. Koo, J. Lee, and R. Trejo (2012). High Temperature Flammability and Mechanical Properties of Thermoplastic Polyurethane Nanocomposites. In *Fire and Polymers VI: New Advances in Flame Retardant Chemistry and Science*, A. Morgan and C. Wilkie (Eds.). Washington, DC: ACS Books Series, pp. 344-360. doi: 10.1021/bk-2012-1118.ch023.

59. D. Yang, W. Zhang, B. Z. Jiang, and Y. Guo (2013). Silicone rubber ablative composites improved with zirconium carbide or zirconia. *Composites: Part A* 44, 70–77.
60. R. D. Patton, C. U. Pittman, Jr., L. Wang, and J. R. Hill (1999). Ablation, mechanical and thermal conductivity properties of vapor grown carbon fiber/phenolic matrix composites. *Composites: Part A* 30(9), 1081–1091.
61. R. D. Patton, C. U. Pittman, Jr., L. Wang, and J. R. Hill (1999). Vapor grown carbon fiber composites with epoxy and poly (phenylene fulfide) matrices. *Composites: Part A* 30, 1081–1091.
62. J. H. Koo, M. Kneer, et al. (1992). A cost-effective approach to evaluate high-temperature ablatives for military applications. *Naval Engineers Journal* 104(3), 166–177.
63. J. H. Koo, S, Lin, et al. (1992). Performance of High-Temperature Polymer Composite Ablatives under a Hostile Environment. *Science of Advanced Materials and Process Engineering Series*, 37, SAMPE, Covina, CA, pp. 506–520.
64. J. H. Koo, M. Miller, et al. (1993). Evaluation of Fiber-Reinforced Composites Ablatives for Thermal Protection. *Science of Advanced Materials and Process Engineering Series*, 38, SAMPE, Covina, CA, pp. 1085–1098.
65. F. B. Cheung, J. H. Koo, et al. (1993). Modeling of one-dimensional thermo-mechanical erosion of high-temperature ablatives. *Journal of Applied Mechanics* 60, 1027–1032.
66. D. Wilson, D. Beckley, and J. H. Koo (1994). Development of silicone matrix-based advanced composites for thermal protection. *High Performance Polymer* 6(2), 165–181.
67. Y. C. Shih, F. B. Cheung, and J. H. Koo (2003). Numerical study of transient thermal ablation of high-temperature insulation materials. *Journal of Thermophysics and Heat Transfer* 17(1), 53–61.
68. J. H. Koo, M. J. Miller, J. Weispfenning, and C. Blackmon (2011). Silicone polymer composites for thermal protection system: Fiber reinforcements and microstructures. *Journal of Composite Materials* 45(13), 1363–1380.
69. J. H. Koo, M. J. Miller, J. Weispfenning, and C. Blackmon (2011). Silicone polymer composite for thermal protection of naval launching system. *Journal of Spacecraft and Rockets* 48(6), 904–919.
70. J. H. Koo, H. Stretz, A. Bray, and W. Wootan (2001). Next Generation Nanostructured Ablatives for Rocket Propulsion System, AFOSR Contract No. F49620-00-C-0045, STTR Phase I Final Report, Submitted to AFOSR, Arlington, VA, September.
71. J. H. Koo, H. Stretz, A. Bray, W. Wootan, S. Mulich, et al. (2002). Phenolic-clay nanocomposite for rocket propulsion systems. *Proceedings of the 2002 SAMPE ISSE*, 47, SAMPE, Covina, CA, pp. 1085–1099.
72. J. H. Koo, H. Stretz, and A. Bray (2002). Nanocomposite Rocket Ablative Materials, AFOSR Contract No. F49620-00-C-0045, STTR Phase II Annual Report, Submitted AFOSR, Arlington, VA, September.
73. J. H. Koo, H. Stretz, A. Bray, J. Weispfenning, Z. P. Luo, and W. Wootan (2003). Nanocomposites Rocket Ablative Materials: Processing, Characterization, and Performance. *Proceedings of the 2003 SAMPE ISSE*, 48, SAMPE, Covina, CA, pp. 1156–1170.
74. J. H. Koo, H. Stretz, A. Bray, J. Weispfenning, Z. P. Luo, and W. Wootan (2004). Nanocomposite Rocket Ablative Materials: Processing, Microstructure, and Performance, AIAA-2004-1996. Presented at the *44th AIAA/ASME/ASCE/AHS Structures, Structural Dynamics, and Materials Conference*, Palms Springs, CA, April 19–22.
75. J. H. Koo, H. Stretz, J. Weispfenning, Z. P. Luo, and W. Wootan (2004). Nanocomposite rocket ablative materials: Subscale ablation test. *Proceedings of the 2004 SAMPE ISSE*, 49, SAMPE, Covina, CA, pp. 1000–1014.

76. J. H. Koo, W. K. Chow, H. Stretz, A. C.-K. Cheng, A. Bray, and J. Weispfenning (2003). Flammability properties of polymer nanostructured materials. *Proceedings of the 2003 SAMPE ISSE*, 48, SAMPE, Covina, CA, pp. 954–963.
77. J. H. Koo, L. Pilato, and G. E. Wissler (2005). Polymer nanostructured materials for high-temperature applications. *SAMPE Journal* 41(2), 7–19.
78. J. H. Koo and L. A. Pilato (2006). Thermal Properties and Microstructures of Polymer Nanostructured Materials. In *Nanoengineering of Structural, Functional, and Smart Materials*, M. J. Schulz, A. Kelkar, and M. J. Sundaresan (Eds.). Boca Raton, FL: CCR Press, pp. 409–441.
79. M. J. Miller, J. H. Koo, et al. (1993). Evaluation of Different Categories of Composite Ablative for Thermal Protection, AIAA-93-0839. Presented at the *31st AIAA Aerospace Sciences Meeting*, Reno, NV, January.
80. F. B. Cheung, J. H. Koo, et al. (1995). Prediction of Thermo-Mechanical Erosion of High-Temperature Ablatives in the SSRM Facility, AIAA-95-0254. Presented at the *33rd Aerospace Sciences Meeting*, Reno, NV, January.
81. M. VanMeter, J. H. Koo, et al. (1995). Mechanical Properties and Material Behavior of a Glass Silicone Polymer Composite. *Proceedings of the 40th International SAMPE Symposium*, SAMPE, Covina, CA.
82. J. H. Koo, et al. (1998). Effect of Major Constituents on the Performance of Silicone Polymer Composites. *Proceedings of the 30th International SAMPE Technical Conference*, SAMPE, Covina, CA.
83. J. H. Koo, et al. (1999). Thermal Protection of a Class of Polymer Composites. *Proceedings of the 44th International SAMPE Symposium*, SAMPE, Covina, CA.
84. MX-4926 Technical Data Sheet, Cytec Engineered Materials, Winona, MN.
85. W. Luehmann, Pratt & Whitney Space Propulsion/Chemical Systems Division, San Jose, CA, personal communication.
86. SC-1008 Technical Data Sheet, Borden Chemical, Louisville, KY.
87. J. H. Koo, C. U. Pittman, Jr., K. Liang, H., Cho, L. A. Pilato, et al. (2003). Nanomodified carbon/carbon composites for intermediate temperature: processing and characterization. *Proceedings of the International SAMPE Technical Conference* 35, 521–534.
88. J. H. Koo, L. A. Pilato, C. U. Pittman, and P. Winzek (2004). Nanomodified Carbon/Carbon Composites for Intermediate Temperature, AFOSR Contract No. F49620-02-C-0086, STTR Phase I Final Report, submitted to AFOSR, Arlington, VA, January.
89. J. H. Koo, L. A. Pilato, P. Winzek, K. Shivakumar, C. U. Pittman, Jr., and Z. P. Luo (2004). Thermo-oxidative studies of nanomodified carbon/carbon composites. *Proceedings of the International SAMPE Symposium and Exhibition* 49, 1214–1228.
90. R. Blanski, J. H. Koo, et al. (2004). Polymer Nanostructured Materials for Solid Rocket Motor Insulation – Ablation Performance. *Proceedings of the 52nd JANNAF Propulsion Meeting*, CPIAC, Columbia, MD, May.
91. J. H. Koo, D. Marchant, et al. (2004). Polymer Nanostructured Materials for Solid Rocket Motor Insulation – Processing, Microstructure, and Mechanical Properties. *Proceedings of the 52nd JANNAF Propulsion Meeting*, CPIAC, Columbia, MD, May.
92. P. Ruth, R. Blanski, and J. H. Koo (2004). Preparation of Polymer Nanostructured Materials for Solid Rocket Motor Insulation. *Proceedings of the 52nd JANNAF Propulsion Meeting*, CPIAC, Columbia, MD, May.
93. J. H. Koo, L. Pilato, et al. (2005). Epoxy Nanocomposites for Carbon Fiber-Reinforced Composites. *Proceedings of the SAMPE 2005 International Symposium*, SAMPE, Covina, CA, May 1–5.
94. J. H. Koo, L. Pilato, et al. (2005). Nanocomposites for Carbon Fiber-Reinforced Polymer Matrix Composites, AIAA-2005-1928. Presented at the *46th AIAA/ASME/ASCE/AHS Structures, Structural Dynamics, and Materials Conference*, Austin, TX, April 18–21.

95. J. H. Koo, L. Pilato, et al. (2004). Nanocomposite for Carbon Fiber Reinforced Polymer Matrix Composites, AFOSR STTR Phase I Final Report, submitted to AFOSR, Arlington, VA, October.
96. Koo, J.H., Pilato, L., Wissler, G.E. and Luo, Z.P. (2005) Flammability and Mechanical Properties of Nylon 11 Nanocomposites, *Proceedings of the International SAMPE 2005 Symposium and Exhibition (ISSE)*, SAMPE, Covina, CA, May.
97. J. H. Koo, L. A. Pilato, and G. E. Wissler (2005). Fire Retardant Polymer Nanocomposites for Selective Laser Sintering Processing, submitted USPO patent application on July 27.
98. J. Cheng, S. Lao, K. Nguyen, W. Ho, A. Cummings, and J. H. Koo (2005). SLS Processing of Nylon 11 Nanocomposites. *Proceedings of the 17th Solid Freeform Fabrication Symposium*, The University of Texas at Austin, Austin, TX, August.
99. J. H. Koo, L. Pilato, et al. (2005). Innovative Selective Laser Sintering Rapid Manufacturing Using Nanotechnology. *Proceedings of the 2005 Solid Freeform Fabrication Symposium*, The University of Texas at Austin, Austin, TX, August.
100. S. Lao, W. Ho, K. Nguyen, J. Cheng, and J. H. Koo (2005). Microstructural Analyses of Nylon 11 Nanocomposites. *Proceedings of the 37th International SAMPE Technical Conference (ISTC)*, Seattle, WA, October 31 – November 3.
101. A. Cummings, L. Shi, and J. H. Koo (2005). Thermal Conductivity Measurements of Nylon 11-Carbon Nanofiber Nanocomposites. *Proceedings of the IMECE2005 (2005 ASME International Mechanical Engineering Congress and Exposition)*, Orlando, FL, November 5–11.
102. S. C. Lao, T. Moon, J. H. Koo, et al. (2009). Flame-retardant polyamide 11 and 12 nanocomposites: Thermal and flammability properties. *Journal of Composite Materials* 43(17), 1803–1816.
103. S. C. Lao, J. H. Koo, et al. (2010). Flame-retardant polyamide 11 and 12 nanocomposites: Processing, morphology, and mechanical properties. *Journal of Composite Materials* 44(25), 2933–2951.
104. S. C. Lao, J. H. Koo, et al. (2011). Flame-retardant polyamide 11 nanocomposites: Further thermal and flammability studies, *Journal of Fire Sciences* 29(6), 479–498.
105. A. Bray, G. Beal, and H. Stretz (2004). Nanocomposite Rocket Ablative Materials, AFOSR STTR Phase II Contract F49620-02-0013, Final Report, submitted to AFOSR, Arlington, VA.
106. Y. Liu, Z. Lu, X. Chen, D. Wang, J. Liu, and L. Hu (2009). Study on phenolic-resin/carbon-fiber ablation composites modified with polyhedral oligomeric silsesquioxanes. *Proceedings of the 4th IEEE International Conference on Nano/Micro Engineered and Molecular Systems*, Shenzhen, China, January 5–8.
107. R. M. Laine (2005). Nanobuilding blocks based on the [OSiO] (x=6, 8, 10) octasilsesquioxanes. *Journal of Materials Chemistry* 15, 3725–3744.
108. L. Hu, X. Zhang, and Y. Sun (2005). Hardness and elastic modulus profiles of hybrid coating. *Journal of Sol-Gel Science and Technology* 34, 41–46.
109. K. Tanaka, S. Adachi, and Y. Chujo (2009). Structure-property relationship of octa-substituted POSS in thermal and mechanical reinforcements of conventional polymers. *Journal of Polymer Science: Part A: Polymer Chemistry* 47, 5690–5697.
110. E. Franchini, J. Galy, J. F. Gèrard, D. Tabuani, and A. Medici (2009). Influence of POSS structure on the fire retardant properties of epoxy hybrid networks. *Polymer Degradation and Stability* 94, 1728–1736.
111. Q-C. Yu and H. Wan (2012). Ablation capability of flake graphite reinforced barium-phenolic resin composite under long pulse laser irradiation. *Journal of Inorganic Materials* 27(2), 157–161 (in Chinese).
112. J. Si, J. Li, S. Wang, Y. Li, and X. Jing (2013). Enhanced thermal resistance of phenolic resin composites at low loading of graphene oxide. *Composites: Part A* 54, 166–172.

113. I. Srikanth, A. Daniel, S. Kumar, N. Padmavathi, V. Singh, et al. (2010). Nano silica modified carbon–phenolic composites for enhanced ablation resistance. *Scripta Materialia* 63, 200–203.
114. ASTM E1225-09 (2008). Standard test method for thermal conductivity of solids by means of the guarded-comparative-longitudinal heat flow technique, Annual Book of ASTM Standards.
115. S. Kumar, A. Kumar, A. Shukla, G. R. Devi, and A. K. Gupta (2005). Thermal-diffusivity measurement of 3D-stitched C–SiC composites. *Journal of the European Ceramics Society* 29(3), 489–495.
116. O. Knacke, O. Kubaschewski, and K. Hesselman (1991). *Thermo-Chemical Properties of Inorganic Substances*, 2nd ed. Berlin: Springer-Verlag.
117. J. Xiao, J.-M. Chen, H.-D. Zhou, and Q. Zhang (2007). Study of several organic resin coatings as anti-ablation coatings for supersonic craft control actuator. *Materials Science and Engineering: Part A* 452–453, 23–30.
118. Y. J. Lee and H. Joo (2004). Ablation characteristics of carbon fiber reinforced carbon (CFRC) composites in the presence of silicon carbide. *Surface Coating Technologies* 180–181, 286–289.
119. I. Srikanth, N. Padmavathi, S. Kumar, P. Ghosal, A. Kumar, and C. Subrahmanyam (2013). Mechanical, thermal and ablative properties of zirconia, CNT modified carbon/phenolic composites. *Composites Science and Technology* 80, 1–7.
120. A. R. Bahramian, M. Kokabi, M. H. Navid Famili, and M. H. Beheshty (2006). Ablation and thermal degradation behaviour of a composite based on resol type phenolic resin: Process modelling and experiment. *Polymer* 47, 3661–3673.
121. A. R. Bahramian, M. Kokabi, M. H. Navid Famili, and M. H. Beheshty (2007). Thermal degradation process of resol type phenolic resin matrix/kaolinite layered silicate nanocomposite. *Iranian Polymer Journal* 16(6), 375–387.
122. A. R. Bahramian, M. Kokabi, M. H. Navid Famili, and M. H. Beheshty (2008). High temperature ablation of kaolinite layered silicate/phenolic resin/asbestos cloth nanocomposite. *Journal of Hazardous Materials* 150, 136–145.
123. A. R. Bahramian and M. Kokabi (2009). Ablation mechanism of polymer layered silicate nanocomposite heat shield. *Journal of Hazardous Materials* 166, 445–454.
124. A. R. Bahramian and M. Kokabi (2011). Numerical and experimental evaluations of the flammability and pyrolysis of a resole-based nanocomposite by cone calorimeter. *Iranian Polymer Journal* 20(5), 399–411.
125. A. Paydeyesh, M. Kokabi, and A. R. Bahramian (2013). High temperature ablation of highly filled polymer-layered silicate nanocomposites. *Journal of Applied Polymer Science* 127, 2776-2785, doi: 10.1002/app.377588.
126. ASTM E1269-11 (2008). Standard test method for determining specific heat capacity by differential scanning calorimetry. Annual Book of ASTM Standards.
127. M. Bartholmai and B. Schartel (2004). Layered silicate polymer nanocomposites: new approach or illusion for fire retardancy? Investigations of the potentials and the tasks using a model system. *Polymer for Advanced Technologies* 15, 355–364.
128. M. Zanetti, G. Camino, and R. Mülhaupt (2001). Combustion behavior of EVA/flourohectorite nanocomposites. *Polymer Degradation and Stability* 74, 413–417.
129. S. Duquesne, C. Jama, M. L. Bras, R. Delobel, P. Recourt, and J. M. Gloaguen (2003). Elaboration of EVA-nanoclay systems-characterization, thermal behaviour and fire performance. *Composites Science and Technologies* 63, 1141–1148.
130. J. W. Gilman, C. L. Jackson, A. B. Morgan, R. Harris, E. Manias, et al. (2000). Flammability properties of polymer layered silicate nanocomposites polypropylene and polystyrene nanocomposites. *Chemistry of Materials* 12, 1866–1873.

131. A. R. Bahramian (2013). Pyrolysis and flammability properties of novolac/graphite nanocomposites. *Fire Safety Journal* 61, 265–273.
132. A. R. Bahramian and R. A. Astaneh (2014). Improvement of ablation and heat shielding performance of carbon fiber reinforced composite using graphite and kaolinite nanopowders. *Iranian Polymer Journal* 23, 979–985.
133. A. Mirzapour, M. H. Asadollahi, S. Baghshaei, and M. Akbari (2014). Effect of nanosilica on the microstructure, thermal properties and bending strength of nanosilica modified carbon fiber/phenolic nanocomposite. *Composites: Part A* 63, 159–167.
134. L. Torre, J. M. Kenny, and A. M. Maffezzoli (1998). Degradation behaviour of a composite material for thermal protection systems, Part I–Experimental characterization. *Journal of Materials Science* 33(12), 3137–3143.
135. L. Torre, J. M. Kenny, and A. M. Maffezzoli, (1998) Degradation behaviour of a composite material for thermal protection systems, Part II – Process simulation. *Journal of Materials Science* 33(12), 3145–3149.
136. L. Torre, J. M. Kenny, G. Boghetich, and A. M. Maffezzoli (2000). Degradation behaviour of a composite material for thermal protection systems, Part III – Char characterization. *Journal of Materials Science* 35(18), 4563–4566.
137. M. Natali, M. Monti, J. Kenny, and L. Torre (2011). Synthesis and thermal characterization of phenolic resin/silica nanocomposites prepared with high shear rate-mixing technique. *Journal of Applied Polymer Science* 120, 2632–2640.
138. M. Natali, M. Monti, J. Kenny, and L. Torre (2011). A nanostructured ablative bulk moulding compound: Development and characterization. *Composite: Part A* 42, 1197-1204. doi:10.1016/j.compositesa.2011.04.022.
139. M. Natali, M. Monti, D. Puglia, J. Kenny, and L. Torre (2012). Ablative properties of carbon black and MWNT/phenolic composites: a comparative study. *Composites: Part A* 43(1), 174–182.
140. A. J. Pavli (1968). Experimental evaluation of several advanced ablative materials as nozzle sections of a storable propellant rocket engine, NASA TM X-1559.
141. J. J. Warga (1979). Low Cost Fabrication Techniques for Solid Rocket Nozzles. *Proceedings of the National Aeronautics and Space Engineering and Manufacturing Meeting*, Los Angeles, CA, October 5–9, pp. 700–796.
142. LR1504 Prepreg datasheet. http://www.lewcott.com/ablatives.html.
143. Cytec Engineered Materials. http://www.cytec.com/engineered-materials/selectorguide.htm.
144. G. F. D'Aelio and J. A. Parker (1971). *Ablative Plastics*. New York: Marcel Dekker.
145. P. Sutton and O. Biblarz (2000). *Rocket Propulsion Elements*. New York: Wiley-IEEE.
146. D. A. Peterson, J. M. Winter, and A. M. Shinn, Jr. (1969). Rocket engine evaluation of erosion and char as functions of fabric orientation for silica-reinforced nozzle materials, NASA TM X-1721.
147. ASTM E457-08 (2008). Standard test method for measuring heat-transfer rate using a thermal capacitance (slug) calorimeter. Annual Book of ASTM Standards.
148. C. S. Wu, Y. L. Liu, and Y. S. Chiu (2002). Epoxy resins possessing flame retardant elements from silicon incorporated epoxy compounds cured with phosphorus or nitrogen containing curing agents. *Polymer* 43(15), 4277–4284.
149. W. J. Wang, L. H. Perng, G. H. Hsiue, and F. C. Chang (2000). Characterization and properties of new silicone-containing epoxy resin. *Polymer* 41(16), 6113–6122.
150. S. Zheng, H. Wang, Q. Dai, X. Kuo, D. Ma, and K. Wang (1995). Morphology and structure of organosilicon polymer-modified epoxy resins. *Macromolecular Chemistry and Physics* 196(1), 269–278.

151. M. Monti, M. Natali, R. Petrucci, D. Puglia, A. Terenzi, et al. (2011). Advanced Fiber Reinforced Composites Based on Nanocomposite Matrices. In *Wiley Encyclopaedia of Composites*, 2nd ed., L. Nicolais and A. Borzacchiello (Eds.). doi:10.1002/9781118097298.weoc025.
152. T. Kashiwagi, F. Du, K. I. Winey, K. M. Groth, J. R. Shields, et al. (2005). Flammability properties of polymer nanocomposites with single-walled carbon nanotubes: Effects of nanotube dispersion and concentration. *Polymer* 46, 471–481.
153. T. Kashiwagi, F. Du, J. F. Douglas, K. I. Winey, R. H. Harris, and J. R. Shields (2005). Nanoparticle networks reduced the flammability of polymer nanocomposites. *Nature Materials* 4, 928–933.
154. B. H. Cipiriano, T. Kashiwagi, S. R. Raghavan, Y. Yang, E. A. Grulke, et al. (2007). Effects of aspect ratio of MWNT on the flammability properties of polymer nanocomposites. *Polymer* 48, 6086–6096.
155. Z. Zhao and J. Gou (2009). Improved fire retardancy of thermoset composites modified with carbon nanofibers. *Science and Technologies of Advanced Materials* 10 (1), 015005 (6pp). doi: 10.1088/1468-6996/10/1/015005.
156. S. S. Rahatekara, M. Zammarano, S. Matko, K. K. Koziol, A. H. Windle, et al. (2010). Effect of carbon nanotubes and montmorillonite on the flammability of epoxy nanocomposites. *Polymers Degradation and Stability* 95, 870–879.
157. J. M. Park, D. J. Kwon, Z. J. Wang, J. U. Roh, W. I. Lee, et al. (2014). Effects of carbon nanotubes and carbon fiber reinforcements on thermal conductivity and ablation properties of carbon/phenolic composites. *Composites: Part B* 67, 22–29.
158. J. S. Tate, C. J. Jacobs, and J. H. Koo (2011). Dispersion of MWCNT in phenolic resin using different dispersion techniques and evaluation of thermal properties. *Proceedings of the 2011 SAMPE ISSE*, Long Beach, CA, May 23–26.
159. J. S. Tate, S. Gaikwad, N. Theodoropoulou, E. Trevino, and J. H. Koo (2013). Carbon/phenolic nanocomposites as advanced thermal protection material in aerospace applications. *Journal of Composites* 2013, ID 403656 (9pp). doi: 10.1155/2013/403656.
160. E. T. Thostenson, C. Li, and T. W. Chou (2005). Nanocomposites in context. *Composites Science and Technology* 65, 491–516.
161. J. Cheng (2006). *Polysyanate ester/small diameter carbon nanotubes nanocomposite*. Master's thesis, The University of Texas at Austin, Dept. of Mechanical Engineering, Austin, TX.
162. R. A. Safadi (2002). Multiwalled carbon nanotube polymer composites: Synethis and characterization of thin films. *Journal of Applied Polymer Science* 84, 2660–2669.
163. G. Pulci, J. Tirillo, F. Marra, F. Fossati, C. Bartuli, and T. Valente (2010). Carbon-phenolic ablative materials for re-entry space vehicles: Manufacturing and properties. *Composites: Part A* 41, 1483–1490.
164. F. Marra, G. Pulci, J. Tirillo, C. Bartuli, and T. Valente (2011). Numerical simulation of oxy-acetylene testing procedure of ablative materials for re-entry space vehicles. *Proceedings of the Institution of Mechanical Engineers, Part L: Journal of Materials: Design and Applications 225*, pp. 32–40. doi: 10.1177/14644207JMDA335.
165. G. Pulci, J. Tirillo, and T. Valente (2012). Ablative Materials for Thermal Protection Systems. Unpublished data, University of Rome, Rome, Italy.
166. G. Duffa (2013). *Ablative Thermal Protection System Modeling*. Reston, VA: AIAA.

11 Electrical Properties of Polymer Nanocomposites

11.1 Introduction

This chapter provides an overview for the electrical properties of polymer nanocomposites. The nano-sized fillers for the conductive polymer nanocomposites are introduced in two groups: inorganic nanomaterials, used mainly with intrinsically conductive polymers, which are not included in this chapter; and organic nanomaterials including carbon nanotubes, carbon black, carbon nanofibers, and graphite. The commonly used techniques for the characterization of electrical conductivity are presented in Chapter 6, as well as the mechanism of conductive network formation and some theories that explain the phenomenon.

The different processing and characterization of electrical properties of polymer nanocomposites are discussed in Chapters 5 and 6, respectively. Based on the different kinds of polymers used, the electrical properties of nanocomposites are discussed within three categories: thermoplastic-based nanocomposites, thermoset-based nanocomposites, and thermoplastic elastomer-based nanocomposites.

11.2 Electrical Properties of Thermoplastic-Based Nanocomposites

In this section, electrical properties of thermoplastic-based nanocomposites are discussed in groups of carbon nanotubes-, carbon nanofibers-, and graphene-based reinforced thermoplastic nanocomposites.

11.2.1 Carbon Nanotubes Reinforced Thermoplastic-Based Nanocomposites

One of the key challenges to fabricate CNT thermoplastic nanocomposites that possess good electrical and mechanical properties lies in the good processing to obtain excellent interconnective network of CNT [1]. Lew et al. [1] investigated the effect of the parameters of twin-screw extruders, such as mixing residence time, molding temperature, and screw speed, on the volume and surface resistivity of MWNT/PP nanocomposites. They also explained the possible reason for the

Table 11.1. *Properties of the CNTs used in Socher et al.'s study [7]*

	Diameter (nm)	Length (μm)	Carbon Purity (%)	Bulk Density (kg/m³)
Baytubes® C150P	13–16	1->10	>95	120–170
Nanocyl™ NC7000	9.5 (average)	1.5 (average)	>90	66*
FutureCarbon CNT-MW-K	15	–	90	28*
Printex® XE2	30–35	n. a.	>99	100–400

decrease in resistivity with higher value of some of these parameters from the perspective of exfoliation of MWNTs [1]. The agglomerate's delaminating into primary aggregates caused the decrease in electrical resistivity. With high shear, the primary aggregates further go into secondary aggregates that are loosely entangled and finally exfoliated CNTs.

With the understanding of the aggregation of CNTs that facilitate the formation of conductive network, there are more examples about the application of CNT reinforced nanocomposites that use thermoplastic, such as polyamide (PA). The influence of mixing speed, melt temperature, or residence time on the electrical conductivity has been investigated for MWNTs/PA nanocomposites [2–4]. Since most literature dealt with the PA6 [2, 5, 6] or PA6.6 [2] as the matrix for MWNTs, it was found that percolation threshold for PA6.6 was 1 wt%, and 2.5 wt% for PA6, which was slightly higher. However, Socher et al. [7] studied the electrical properties of various PA12 nanocomposites with different types of MWNTs. The PA12 can be categorized into low and high viscosity with two kinds of amine groups. Four different types of MWNTs and one type of CB (Pritx® XE2) were used for comparison (Table 11.1) [7].

The electrical volume conductivity of three different MWNTs and CB are shown in Figure 11.1 [7]. Obviously, the percolation thresholds of MWNTs are all lower than that of CB because of their higher aspect ratio. Among different types of MWNTs, the Nanocyl™ has the lowest percolation of 0.7 wt%, followed by Future Carbon MW-K at 1.3 wt% and 2.1 wt% for Baytubes®. The electrical percolation threshold for CB (Pritx® XE2) is determined at 4.3 wt%. Therefore, the influence of different types of carbon-based nanofillers on the electrical property of PA12 nanocomposites can be decided.

The electrical percolation behavior of different PA12 composites containing Baytubes® C150P can be seen in Figure 11.2 [7]. The composites with low viscous PA12 and acid termination has the lowest electrical percolation threshold with 2.1 wt%, and the one with low viscosity and amine excess has significantly higher percolation of 3.7 wt%, which indicates the influence of end group termination on the electrical percolation behavior, also reflected in a high-viscosity system. On the other hand, with the same end group, the ones with low viscosity present lower electrical percolation, suggesting the influence of viscosity on the electrical percolation behavior.

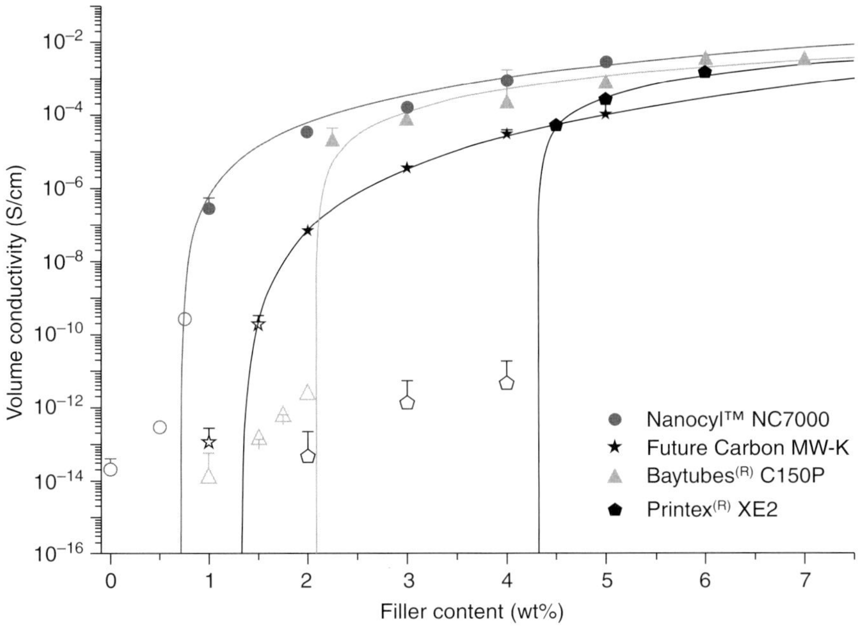

Figure 11.1. Electrical volume conductivity of polyamide 12 composites containing different MWNTs and CB [7].

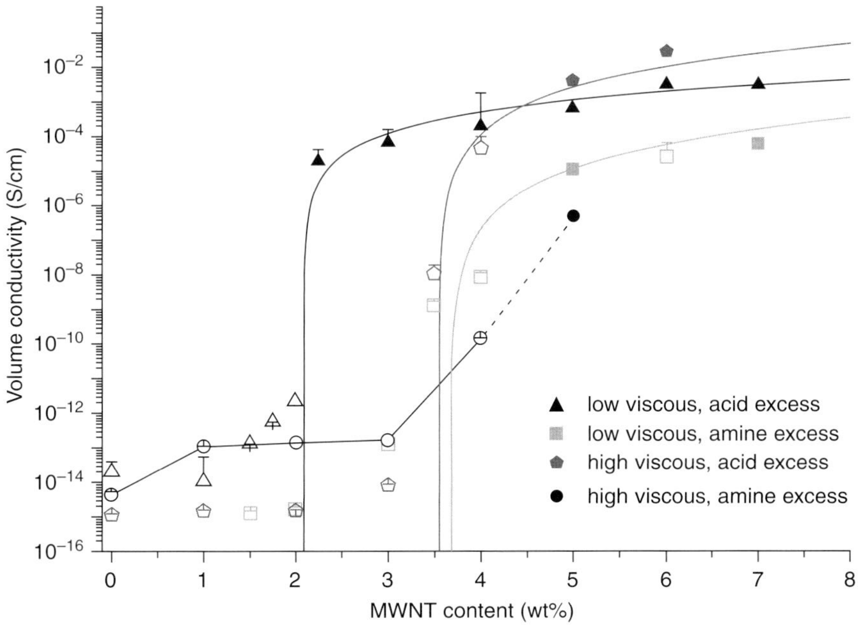

Figure 11.2. Electrical percolation behavior of different PA12 composites containing Baytubes® C150P [7].

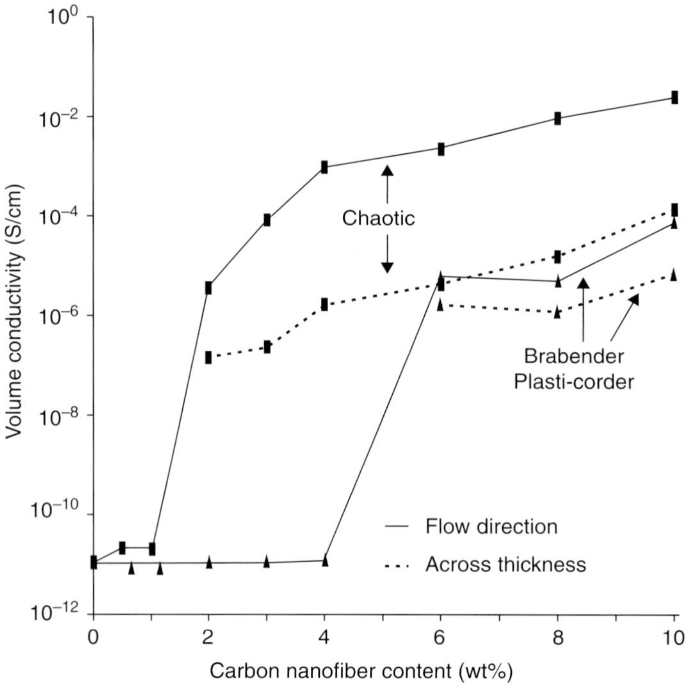

Figure 11.3. Volume electrical conductivity of PMMA-CNF composites. Results from materials mixed in Chaotic mixer and Brabender Plasticorder are identified. Measurements reflect a maximum standard deviation of 10% [11].

11.2.2 Carbon Nanofibers Reinforced Thermoplastic-Based Nanocomposites

One of the issues for CNF polymer nanocomposites is the poor interaction of fiber and polymer interface. Although shear mixing is an efficient way to achieve better dispersion, it causes severe damage to nanofibers and leads to low aspect ratios [8, 9]. Another way to disperse CNFs in a polymer matrix is to make use of functionalized nanofibers to improve the dispersion of nanofibers in matrix [10]. However, the usage of oxidized carbon nanofibers with PMMA of chaotic mixing leads to the decreasing of electrical conductivity with the improvement of dispersion, partially attributed to the low intrinsic conductivity of the functionalized carbon nanofibers.

Again, with the same mixing method, Jimenez and Jana [11] showed that the electrical conductivity of polymethylmethacrylate (PMMA)-CNF nanocomposites is dependent on dispersion of CNF and mixing time at low shear, especially at nanofibers loading (2 wt%) at percolation threshold. Figures 11.3 and 11.4 show that with better dispersion of nanofibers, the electrical conductivity of nanocomposites made of both methods increase significantly after reaching the percolation threshold and nanocomposites mixed by chaotic method have superior conductivity [11].

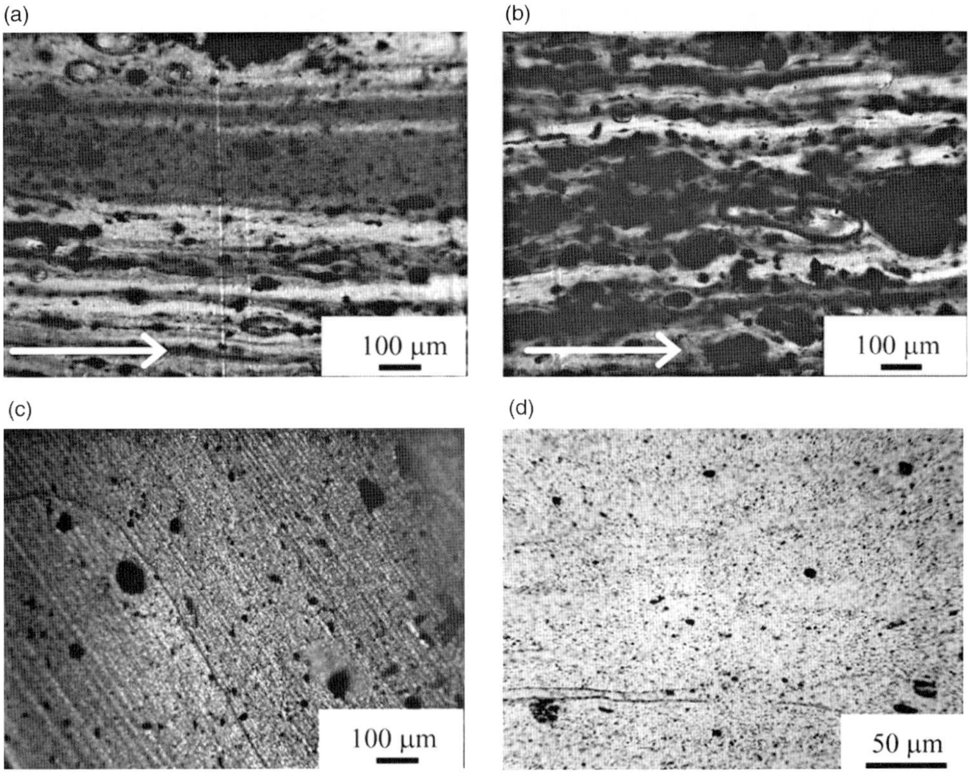

Figure 11.4. Optical micrographs of PMMA-CNF composites. Film thickness for (a), (b), and (c) was about 50 lm, and 1 lm for (d). Arrow indicates flow direction. Traces of lines seen in (c) are due to microtoming. (a) Chaotic 1 wt%, (b) Chaotic 2 wt%, (c) Brabender Plasticorder 4 wt%, and (d) Brabender Plasticorder 6 wt% [11].

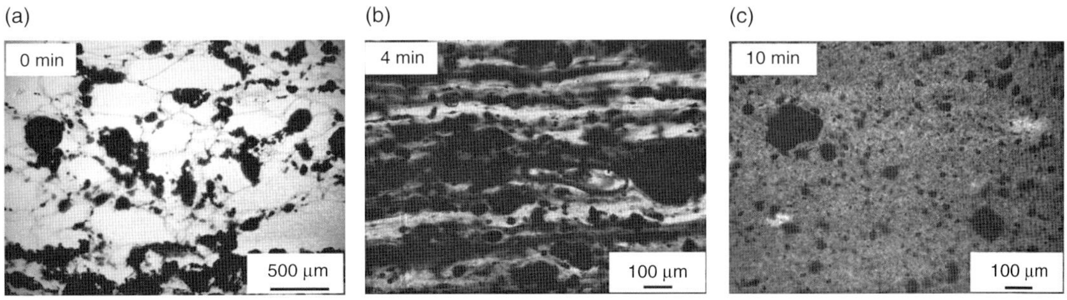

Figure 11.5. Effect of mixing time on nanofiber dispersion in 2 wt% CNF composite prepared in Chaotic mixer. Mixing times are indicated in the above figures [11].

Figure 11.5 illustrates that longer mixing time lowers the electrical conductivity despite of better dispersion; for example, for 2 wt% formulation, the materials is conductive up to 4 minutes but becomes insulated at 10 minutes [11]. One possible explanation is that the nanostructures of CNFs were broken with long time mixing, which leads to the reduction of conductivity.

11.2.3 Graphite Reinforced Thermoplastic-Based Nanocomposites

The electrical property of graphite reinforced thermoplastic-based polymer, such as polypropylene [12] nanocomposites, were reported. Li [13] added different loadings of expanded graphite (EG) into polyvinylidene fluoride (PVDF) with direct melt mixing and interestingly introduced the alternating current (AC) electrical measurement at various frequencies and the biased random walk [14] approach to explain the formation of conductive network at such a condition. As a result, the electrical conductivity of EG/PVDF system was found to be frequency and temperature dependent, which was known as tunnel effect in quantum mechanics [15]. Besides the factors like filler aspect ratio, content, and morphology that can affect the conductivity of nanocomposites, Goyal et al. [16] claimed that even for the same conductive filler, such as EG, the percolation threshold also depends on the types of polymer used in the system. This is because the good interaction between matrix and fillers is necessary to form excellent conductive network. For instance, the percolation for PMMA/expanded graphite was claimed to be 3.5 wt% [17], and with electrical conductivity at 25°C, applied voltage of 10V can be as high as 60 S/cm with 8 wt% [15]. The electrical conductivity of nylon 6/EG reached 10^{-4} S/cm when the graphite content was 2.0 vol% [18], worse conductivity than poly(phenylene sulfide) (PPS)/EG, which had an electrical conductivity percolation threshold of 1 wt% (0.6 vol%) [16].

Via et al. [19] compared the electrical property of different carbon-based fillers of carbon black (CB), multi-wall carbon nanotube, and exfoliated graphite nanoplatelets (GNT) in polycarbonate (PC). The main application of the above nanocomposites are for electrostatic dissipative and semiconductors that require relatively low electrical conductivity; for example, for fuel gauges, electrical conductivity is from 10^{-2} S/cm to 10^{-1} S/cm. According to the result, all the fillers exhibit low percolation threshold, for which CNT/PC was lowest at 1.2 vol%, 2.4 vol% for CB/PC, and highest for GNT/PC at 4.6 vol%. Additionally, for the first time in the literature, the authors applied the general effective media (GEM) models for these composites systems and found out that the modeling results agreed with the experimental result that with higher concentration of fillers the conductivity increases.

The synergistic effect of conductive networks formed by various carbon nanofillers in thermoplastic-based nanocomposites was investigated. For instance, Zhang et al. [20] produced conductive polypropylene-based nanocomposites with MWNT/CB mixtures in melt blending. The nanocomposites with only CB had higher percolation threshold that of nanocomposites with only MWNT, because of the higher aspect ratio of MWNT [20]. With 0.25 wt% of MWNT and 0.25 wt% of CB hybrid nanofillers, the nanocomposites became conductive, but it took 2.5 wt% of MWNT for the nanocomposites with only MWNT to become conductive. $MWNT_1$-CB_1 nanocomposites were more conductive than $MWNT_1$-CB_4 at the same total filler content. The resistivity of the composites containing hybrid nanofillers is higher than MWNT composites at the same filler content. However, the resistivity of $MWNT_4$-CB_1 and $MWNT_1$-CB_1 was similar to that of MWNT nanocomposites. As a result, synergistic effect between CB and MWNTs makes it possible of the partial replacement of high-aspect-ratio and high-priced nanofillers (MWNTs) with low-aspect-ratio, low-priced CB. The result is in correspondence with the theoretical modeling of electrical percolation of mixed MWNTs/CB and MWNTs/graphene reinforced polypropylene or polyoxymethylene nanocomposites [21].

As for the mechanism of formation of conductive network, it is easy to understand that for pure MWNT or CB reinforced nanocomposites, it was the nanofillers that were responsible for the formation of conductive network. In the hybrid system, there are three mechanisms for the formation of conductive networks: Mechanism 1: shortcut of network branches caused by new active branches of networks co-formed by CB; Mechanism 2: CB are introduced into MWCT networks by bridging MWNT local networks with CB; and Mechanism 3: bridging local contacts between CB clusters through MWNTs. For nanocomposites containing low filler content [20], conductive networks are formed in $MWNT_1$-CB_1 because there are bridges formed by CB to MWNTs local networks. For $MWNT_4$-CB_1, the MWNTs are more entangled, but not being efficiently used to build up conductive network. For $MWNT_1$-CB_4, the amount of MWNTs that have high-aspect-ratio filler is too low to build conductive networks. As a result, $MWNT_1$-CB_1 has much lower percolation threshold than the rest of the composites.

11.3 Electrical Properties of Thermoset-Based Nanocomposites

In the category of thermoset polymers, epoxy is widely used for advanced composites, because they display characteristics, such as good stiffness, specific strength, dimensional stability, chemical resistance, and strong adhesion with reinforcement [22]. Numerous research results on the electrical conductivity of carbon nanofibers reinforced epoxy nanocomposites are available. Cyanate ester (CE), a class of high-performance thermosetting resins that has received considerable attention due to its good mechanical properties, thermal stability, flammability properties, ease of process, and volatile-free curing process, is also included in this section [23].

11.3.1 Carbon Nanotubes Reinforced Thermoset-Based Nanocomposites

Battisti et al. [24] used the high-shear mixing method to process MWNT/unsaturated polyesters with CNT loading ranging from 0.5 wt% to 3 wt%. The electrical measurement was performed with AC impedance spectroscopy and DC conductivity measurement. Figure 11.6a shows the microstructure of the cured nanocomposites with 0.15 wt%, at which concentration of the reaggregation at microscale and nanoscale can be compared [24]. The low bright regions, which contain high content of CNTs, covered a large area of the sample, and could indicate that an aggregation had been formed that was different from those originally mixed in the matrix, because the new aggregations were packed as tightly as the original clusters. Figure 11.6b shows the charge contrast imaging used to demonstrate the morphology of the secondary agglomerate in matrix [24]. It is obvious that nanotubes were entangled and loosely

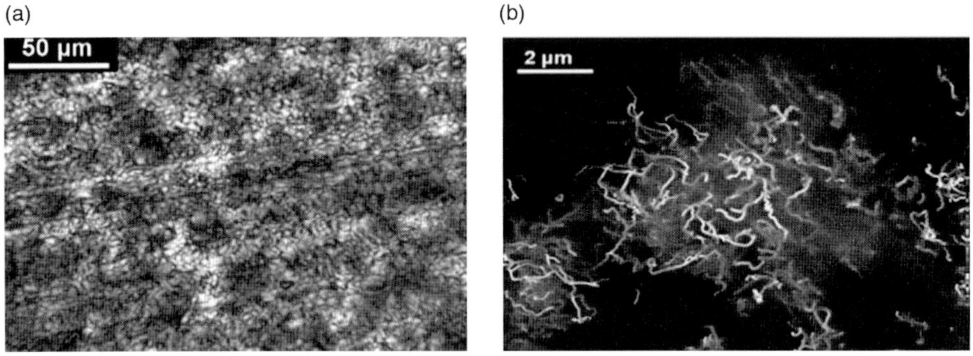

Figure 11.6. Nanocomposite containing 0.15 wt% of carbon nanotubes: (a) transmission light micrograph and (b) charge contrast imaging [24].

packed. So the reaggregation of CNTs during the cure process can explain well the low percolation threshold of 0.026 wt%, which was lower than theoretical value of about 0.96 wt%. Compared with Vera et al.'s study [25] of the electrical property of untangled MWNTs/unsaturated polyester nanocomposites processed without solvent, which has a percolation threshold of 0.1 wt%, the electrical property of nanocomposites in Ballisti et al.'s [24] study improves about four times.

Other similar reports about the CNT-reinforced thermoset polymer nanocomposites are available. For instance, Martin et al. [26] reported the effect of aspect ratio of fillers on the percolation threshold and analyzed the cluster formation that was responsible for the conductive network formation. Thostenson et al. [27] manufactured the MWNT/vinyl ester nanocomposites using three-roll milling technique and discovered that the conductive network of MWNTs formed in the matrix at the loading of >1 wt%. Sandler et al. [28] obtained similar percolation threshold as Battisti et al. [24] for the MWNT/epoxy nanocomposites produced by direct high-shear mixing with aligned MWNT at 0.025 wt%. Other researchers [29, 30] investigated the effect of treatment on the electrical property of CNT/epoxy nanocomposites and concluded that SWNTs needed more concentration than MWNTs to attain percolation threshold, and the treatment or functionalization did improve the mechanical properties of nanocomposites because of better interconnection between fillers and matrix [31, 32], but it decreased the electrical conductivity by making the aspect ratio smaller (Figure 11.7) [31].

Cyanate ester (CE), a class of high-performance thermoset, has good mechanical properties, thermal stability, flammability properties, ease of process, and volatile-free curing process [23]. The relatively high glass transition temperature (>250°C) of CE falls between that of epoxy and polyimide. With proper nano-modification and cure, CE has great potential to replace the hard-to-process polyimide resin. Multi-wall carbon nanotubes (MWNTs) are often selected to create polymer nanocomposites due to their unique combination of excellent mechanical, electrical, and thermal properties [33]. It is anticipated that the incorporation of MWNTs will enhance the mechanical and thermal properties as well as reduce electrical resistivity of the resulting CE-MWNT nanocomposites. Ultimately, the high-performance carbon fiber–reinforced

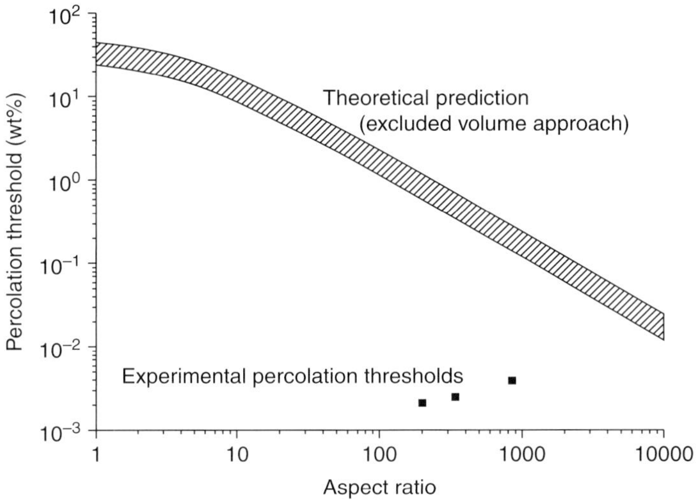

Figure 11.7. The electrical percolation threshold with the function of aspect ratios [31].

CE-MWNT nanocomposites will be fabricated for structural applications in the aerospace, aviation, and defense industries, which require excellent mechanical, thermal, and flammability properties as well as good electrical conductivity.

Recently, Lao, Moon, and Koo conducted CE nanocomposite research to determine the extent to which several standard processing techniques would affect/improve MWNT dispersion and attendant nanocomposite properties, such as the thermal, mechanical, and electrical properties as well as flammability [34]. A *processing-structure-property* relationship, as well as performance of this class of carbon-based CE nanocomposite, is established by the authors. Effects of different standard functionalizations of MWNTs (i.e., -OH, -COOH, and -NH_2) on MWNT dispersion in the CE resin are studied. In addition to aiding dispersion, these functionalizations have the potential to differentially affect/improve nanocomposite properties (such as mechanical properties, flammability properties, electrical conductivity, and thermal conductivity) due to the resulting improved interfacial bonding or adhesion between the CE resin and MWNTs. Preliminary results of carbon fiber–reinforced CE-MWNT composites are also reported [34].

Most processing techniques for nano-modifications aim to disperse nanomaterials uniformly in the polymer resin. Dispersion techniques vary for different polymer resins and different nanomaterials. Common methods include emulsion polymerization, in situ polymerization, melt compounding, solvent blending, ultrasonication, mechanical mixing, and high-shear mixing. For example, polyhedral oligomeric silsesquioxane (POSS®) is usually functionalized, specifically tailored to the chosen polymer resin, and processed by in situ polymerization. On the other hand, nanoclay usually requires shearing force, so that layers of clay platelets can be sheared apart. In the case of carbon nanotubes, vigorous stirring or high-shear mixing is believed to de-bundle the nanotubes aggregates. More details regarding these processing techniques and properties enhancement of the resulting CE-MWNT nanocomposites are discussed by the authors.

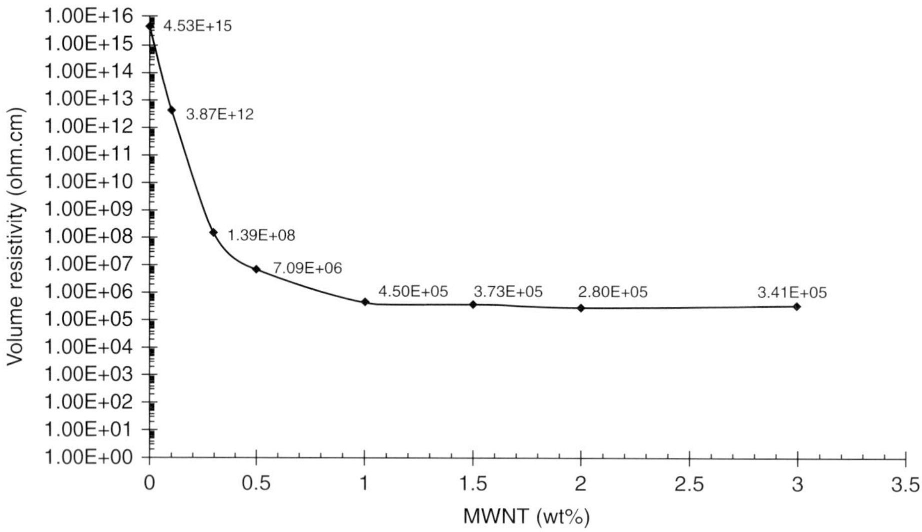

Figure 11.8. Electrical resistivity of CE-MWNT-5 (with 200 ppm Fe^{3+} catalyst) polymer nanocomposites processed by stand mixer with various MWNT loadings (wt%) (*Number of samples per MWNT wt%, N = 1)* [34].

In their study, several processing instruments – including ultrasonicator, planetary centrifugal mixer, high-shear mixer, three-roll mill, and stand mixer – are used to disperse the carbon nanotubes into the CE resin. The PNC morphology and MWNT dispersion are characterized by transmission electron microscopy (TEM) and scanning electron microscopy (SEM). Further, thermal stability is assessed by thermogravimetric analysis (TGA), while combustion characteristics are examined by microscale combustion calorimetry (MCC). The glass transition temperature is determined by dynamic mechanical analysis (DMA), while the electrical conductivity is measured by megohmmeter and plate electrodes. This section presents only the electrical results of the authors' study. Electrical conductivity/resistivity was calculated from the voltage drop measured by the Hioki Super Megohmmeter and plate sample electrode. Plate specimens of 1 mm thickness were used in the electrical volume resistivity measurement.

Neat MWNT-5 was mixed with CE and cured with 200ppm the Fe^{3+} catalyst or coupling agents, CA-1 and CA-2, followed by a post-cure at 250°C for 2 hours [34–37]. Specimens of 1 mm thickness were made for the electrical resistivity measurement. Figures 11.8 and 11.9 show that the addition via stand mixer of less than 0.5 wt% or 0.3 vol% MWNT lowers the neat CE's electrical resistivity. These measurements reveal that the percolation threshold is around 0.4 wt% or 0.24 vol% and the resistivity reaches a plateau beyond 1 wt% or 0.6 vol%. Furthermore, the electrostatic dissipation (ESD) requirement ($<10^{11}$ohm.cm) [38] is fulfilled by incorporating as little as 0.3 wt% or 0.18 vol% MWNT into the CE matrix. The resulting percolation threshold matches Hu et al.'s review [39], which reported percolation thresholds of 1 wt% multi- or single-wall carbon nanotubes (or below) in other resin systems. Similar percolation threshold of 1 wt% was reported by Ma et al.'s review [40]. In

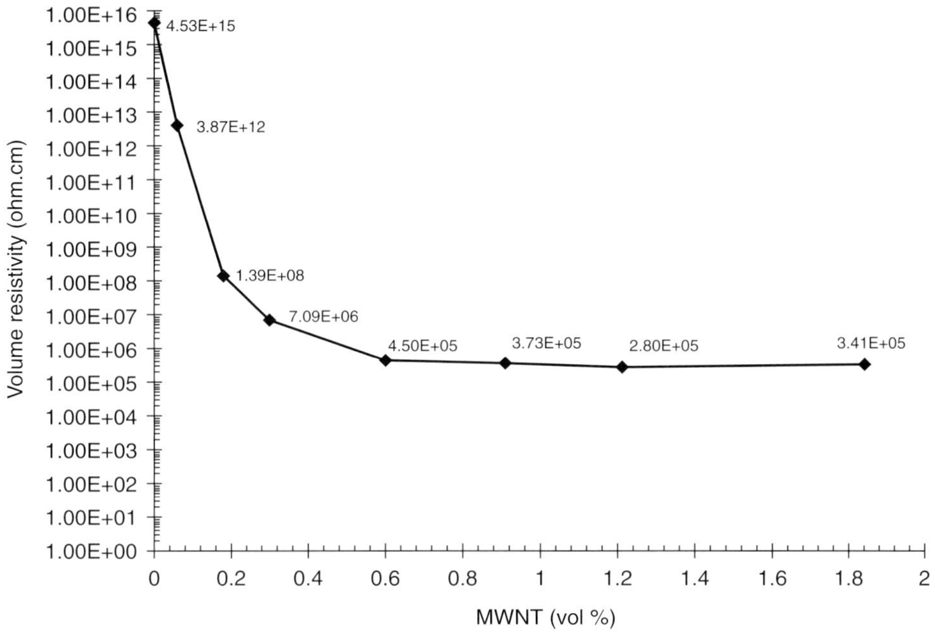

Figure 11.9. Electrical resistivity of CE-MWNT-5 (with 200 ppm Fe^{3+} catalyst) polymer nanocomposites processed by stand mixer with various MWNT loadings (vol%) (*Number of samples per MWNT vol%, N = 1)* [34].

fact, the viscosity of the polymer becomes too high to fabricate void-free polymer nanocomposites when the content of nanotubes is higher than 1 wt%. Therefore, it is crucial to use processing techniques, which can produce polymer-MWNT nanocomposites with a percolation threshold below 1 wt%. Among the processing techniques used in their study, stand mixer is able to achieve that.

Interestingly, similar electrical resistivity drops were not seen in samples processed by other techniques, even at the same MWNT weight loadings. For instance, the electrical resistivity of the CE-MWNT nanocomposites is still higher than the ESD requirement at a loading of 1 wt% or 0.6 vol% (Figures 11.10 and 11.11), although the addition of 1.5 wt% or 0.91 vol% of MWNT lowers the electrical resistivity by about four orders of magnitudes relative to the 1 wt% or 0.6 vol% sample. While it is likely that the percolation threshold is between 1 wt% and 1.5 wt% of MWNTs, additional data are needed to determine this value. Although the percolation threshold is above 1 wt%, void-free polymer nanocomposites can still be fabricated, since air bubbles are removed during the three-roll milling process because of the small gap distance between rollers.

Figure 11.12 compares the electrical resistivity of a CE-0.5%MWNT-5 polymer nanocomposite processed by five different techniques. The electrical resistivity values of samples prepared by three-roll mill, ultrasonication, and planetary centrifugal mixer are two orders of magnitude lower than that of neat CE, but still higher than the ESD requirement. Perplexingly, their readings are about six orders of magnitudes higher than the one processed by the stand mixer. This indicates that the MWNTs in the sample processed by the stand mixer were dispersed and de-bundled enough

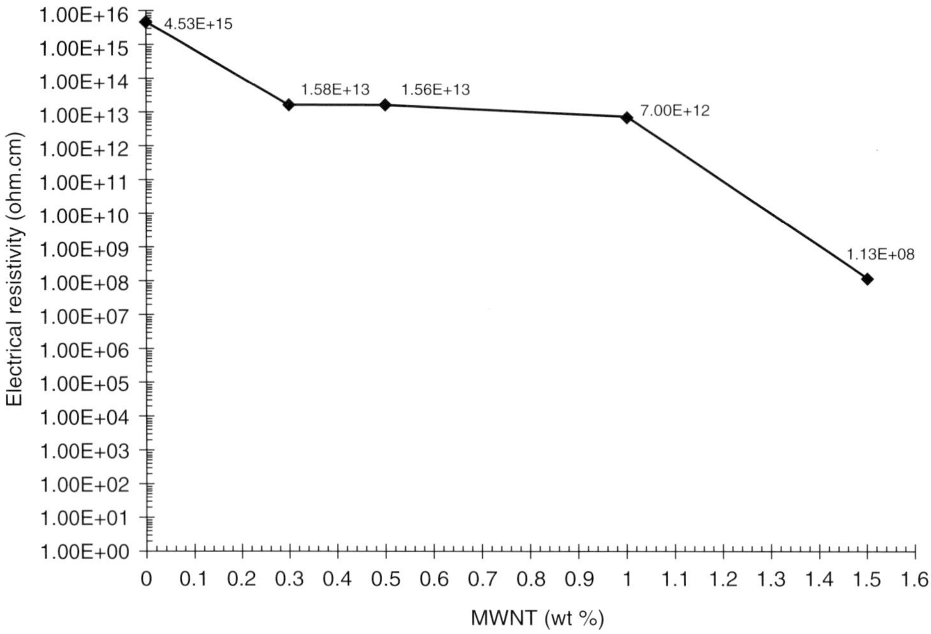

Figure 11.10. Electrical resistivity of CE-MWNT-5 (with 200 ppm Fe^{3+} catalyst) polymer nanocomposites processed by three-roll mill with various MWNT loadings (wt%) (*Number of samples per MWNT wt%, N = 1)* [34].

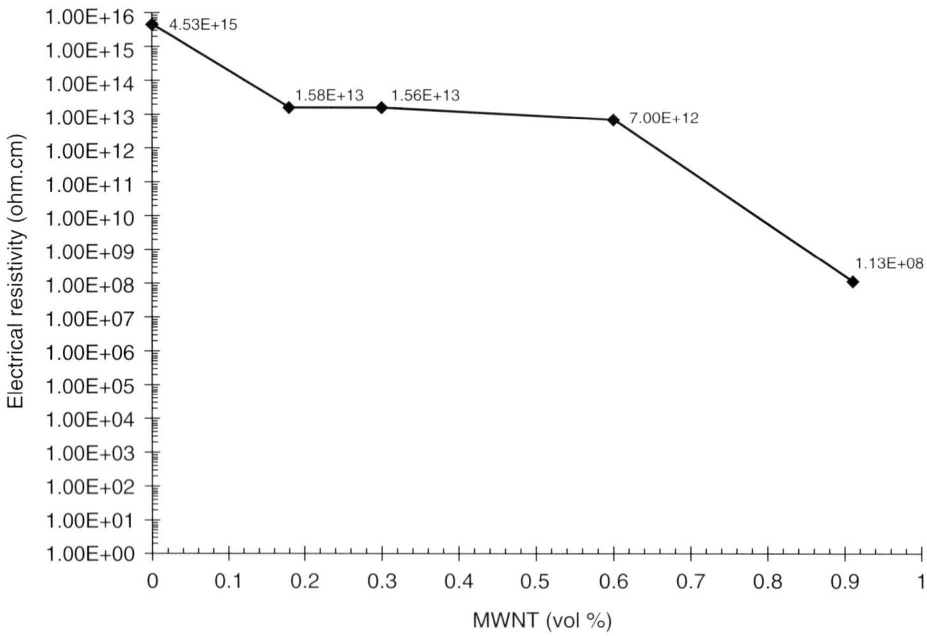

Figure 11.11. Electrical resistivity of CE-MWNT-5 (with 200 ppm Fe^{3+} catalyst) polymer nanocomposites processed by three-roll mill with various MWNT loadings (vol%) (*Number of samples per MWNT vol%, N = 1)* [34].

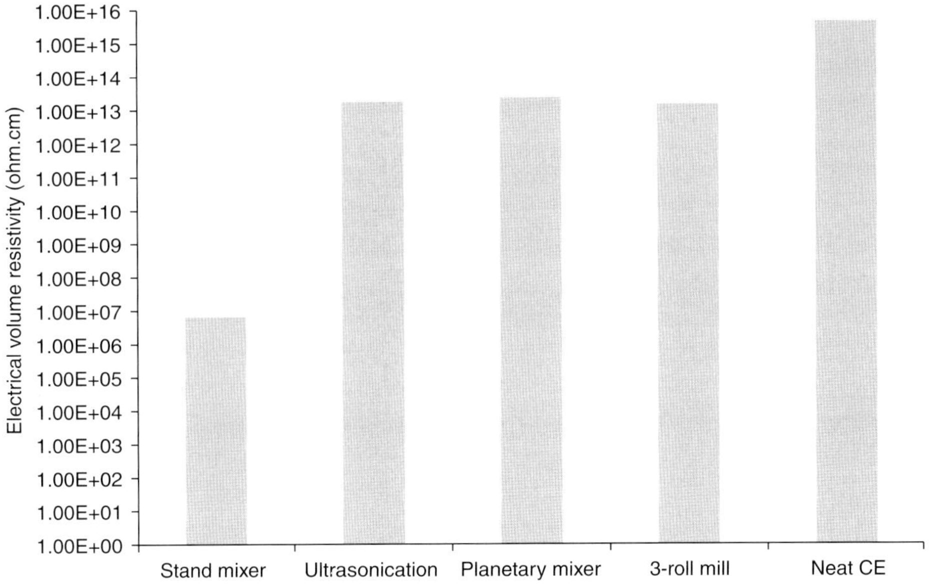

Figure 11.12. Electrical resistivity of CE-0.5%MWNT-5 (with 200 ppm Fe^{3+} catalyst) polymer nanocomposites processed by five different processing techniques (*Number of samples per processing technique, N = 1)* [34].

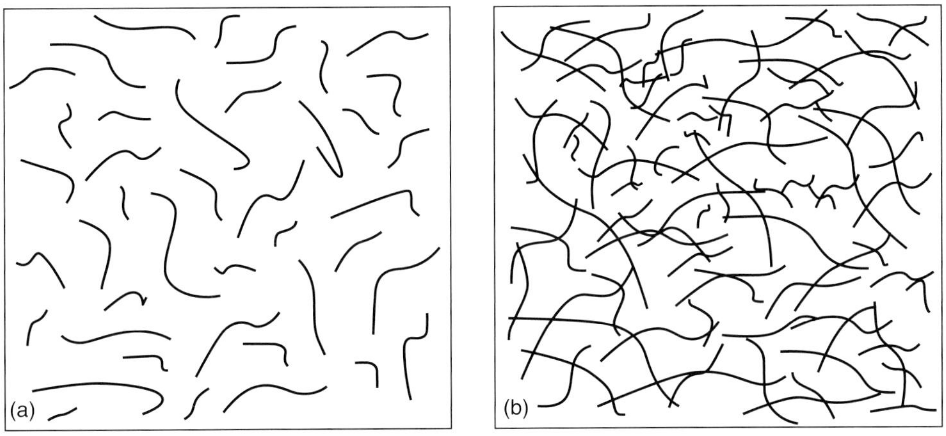

Figure 11.13. Illustration showing correlation of degree of dispersion and electrical conductivity: (a) more de-bundled and better dispersed nanotubes may not form continuous conductive path, and (b) conductive path may form in polymer matrix with a lower degree of nanotubes dispersion [34].

to form a much better conductive path for the electron to pass through the sample than the other samples, just as Ma et al. commented on the conducting behavior of composites consisting of conducting fillers and insulating matrices using percolation theory [40].

Figure 11.13 shows the relationship between nanotubes dispersion and electrical conductivity. In a sample with better nanotubes dispersion as shown in Figure 11.13a,

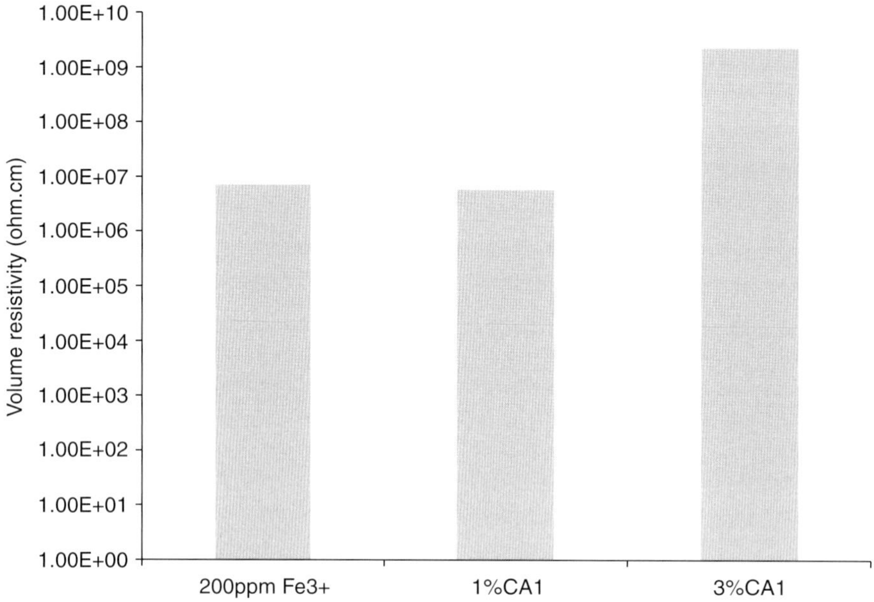

Figure 11.14. Electrical resistivity of CE-0.5%MWNT-5 polymer nanocomposites, cured with Fe^{3+} catalyst (200 ppm) or coupling agent CA-1 (1 or 3%)), and processed by stand mixer (*Number of samples per resin modifier, N = 1)* [34].

individual de-bundled nanotubes may appear in the polymer matrix uniformly. However, they may be separated so that electrons may not have a continuous conductive path to travel through the polymer matrix. On the contrary, as shown in Figure 11.13b, connected paths may be formed in a sample with lower degree of nanotubes dispersion, resulting in higher electrical conductivity.

Figure 11.14 shows that when 1 wt% of CA-1 is used instead of the Fe^{3+} catalyst (200 ppm), the electrical resistivity is the same. However, when higher weight loading (3 wt%) of CA-1 is used, the electrical resistivity is two orders of magnitudes higher.

In summary [34], morphology of the CE-MWNT-5 nanocomposites was characterized by SEM and TEM. Quality of MWNTs dispersion was also examined using both visual observation and quantification. Samples processed by different processing techniques show different aggregates sizes and various degrees of de-bundling with aggregates. Inconsistency exists between the visual observation and dispersion quantification methods, and further work is needed to eliminate this inconsistency. Dynamic mechanical analysis shows that both the catalyst and coupling agents had detrimental effects on the mechanical properties and glass transition temperatures of the CE-MWNT nanocomposites. However, the glass transition temperature is not affected by the mixing methods or weight loadings of MWNTs. The thermal stability of the nanocomposites was not significantly affected by the processing techniques or the weight loadings of the MWNTs. Microscale combustion calorimetry shows that the onset temperature of heat release rate remains unaffected by mixing techniques, modifiers, or weight loadings of MWNTs. Electrical resistivity of CE was reduced by

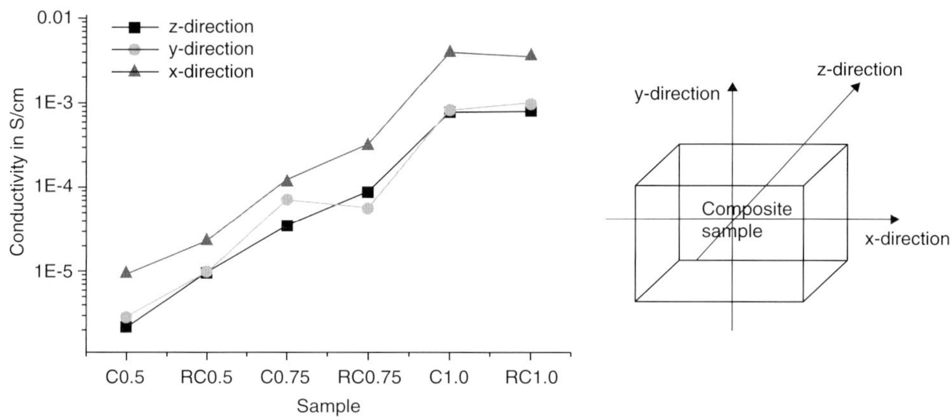

Figure 11.15. Electrical conductivity of CNF/ epoxy nanocomposites in three different directions along with a schematic of the sample indicating directions [43].

the incorporation of MWNTs. Sample processed by the stand mixer had the lowest electrical resistivity among those processed by other techniques.

11.3.2 Carbon Nanofibers Reinforced Thermoset-Based Nanocomposites

Smrutiskha [41] focused on the study of the critical content (0.5–1 wt%) of carbon nanofibers in epoxy that makes the composites conductive. The study's results showed that the nanocomposites with different contents of CNFs have the conductivity in the range of 2×10^{-6} S/cm to 4×10^{-3} S/cm (Figure 11.15) – that is, three to six orders-of-magnitude greater than that of neat epoxy (2.8×10^{-9} S/cm). The samples were tested in three directions, and Figure 11.15 shows that most fibers aligned in the x-direction, since the conductivity in the x-direction is higher than other directions, these nanocomposites can be used in electrostatic dissipation application. Ardanuy et al. [42] obtained similar result of surface electrical conductivity of CNF/epoxy nanocomposites around $1{\times}10^{-6}$ S/cm at 2 wt% loading, which indicates eight orders-of-magnitude increase over neat epoxy.

So far, as presented earlier in the chapter, most processing methods involve adding conductive fillers into polymer melts, which is convenient, but the conductivity of these materials would be lost during the melting process [43]. Cipriano et al. [43] proposed a new way, which is melt-annealing to achieve higher conductivity of MWNT/polystyrene and CNF/polystyrene nanocomposites. Figure 11.16 shows the effect of annealing temperature on the conductivity of two kinds of nanocomposites. From Figure 11.16a we can see that at 170°C, the annealing effect of MWNT at 4 wt% is moderate but not suitable for other loadings. At annealing temperatures of 230°C, the conductivity increased significantly for most loadings, especially at 4 wt%, which approached 1 S/cm, with about two orders-of-magnitude increase over the conductivity of composites at the annealing temperature of 170°C. Similar trend was observed in CNT-reinforced nanocomposites in Figure 11.6b. The reason for this result is that during the melt processing, such as twin-screw extruder

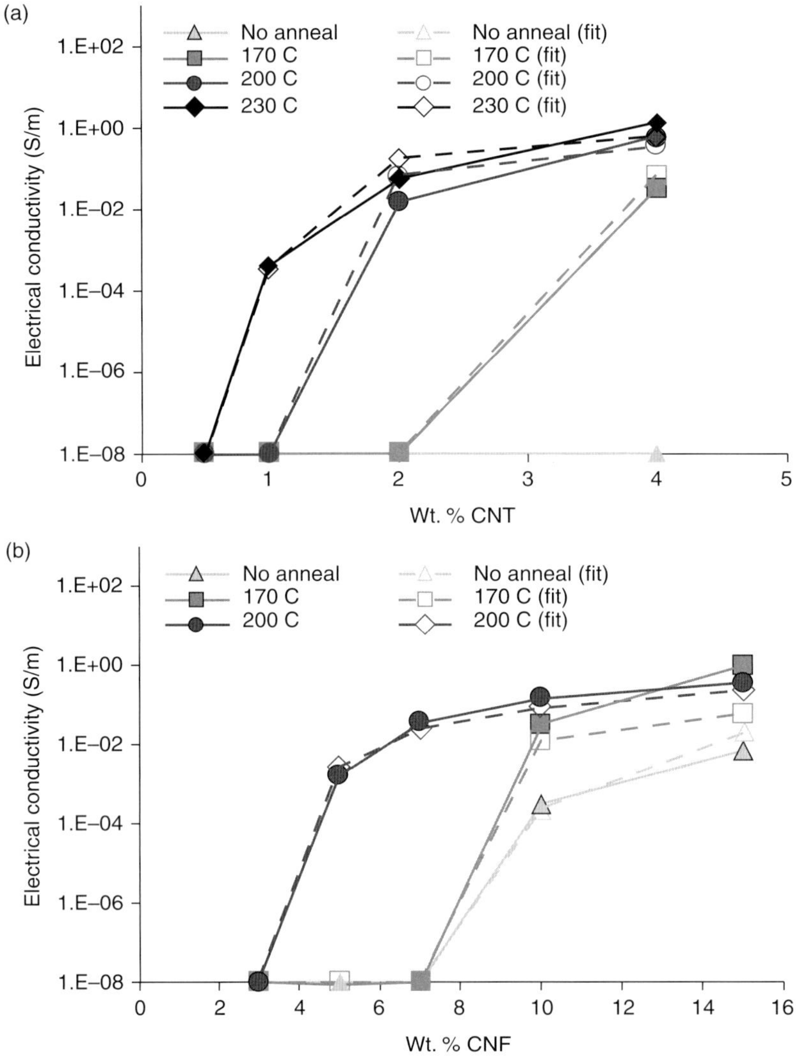

Figure 11.16. Effect of annealing temperature on the conductivity of (a) PS/CNT and (b) PS/CNF nanocomposites. Samples were annealed for 30 minutes [45].

and compressing molding, the particles become aligned in the flow direction, which results in poorer formation of the conductive network and decreased conductivity. However, the annealing process was able to re-establish the conductive network, leading to recovered electrical conductivity.

11.3.3 Carbon Black Reinforced Thermoset-Based Nanocomposites

The concept of tensoresistive effect was introduced by Knite et al. [44]: when the composites are stretched, the resistivity increases because the electro-conductive network is disrupted and the whole stretching process becomes irreversible. To solve this problem, Knite and colleagues made polyisopropene (PI)/carbon black nanocomposites

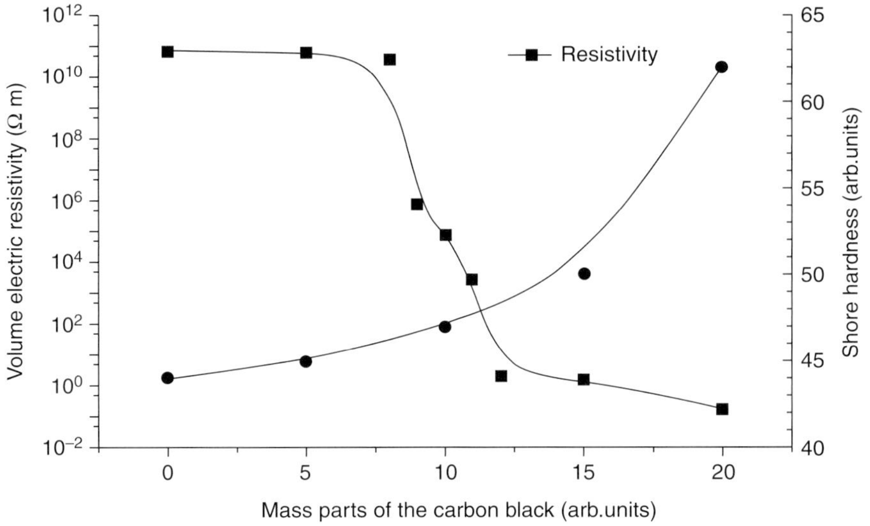

Figure 11.17. Bulk resistivity of the composite and Shore A hardness as functions of mass parts of the extraconductive carbon black added to 100 mp of polyisoprene [46].

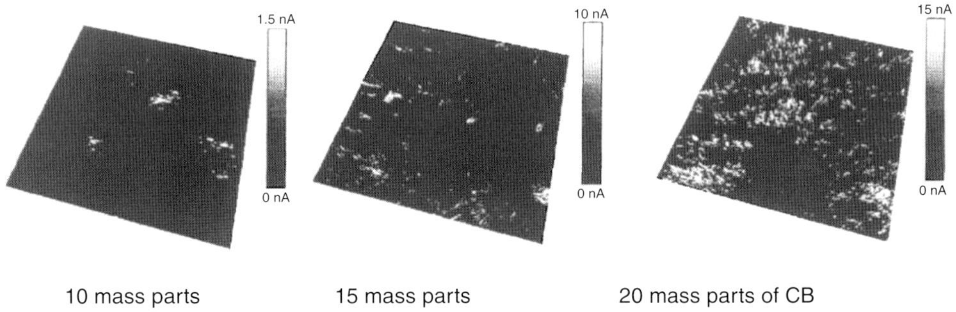

Figure 11.18. Nanoscale maps of electro-conductive channels of carbon black in the polyisoprene matrix of composites with 10, 15, and 20 mp of carbon black obtained by an atomic force microscope with a conductive tip [46].

on micro- and nanoscales to observe the relationship between electrical and elastic properties. To estimate the percolation threshold, the researchers prepared composites of 5, 8, 9, 10, 11, 12, 15, 20, and 30 mass parts (mp) of CB with 100 mp of polyisoprene. At micro level, Figure 11.17 shows that at 10 mp, the volume electric resistivity decreased drastically [44], which indicated the percolation around the value. As we can see, upon a 40% stretch, resistivity changed more than fourfold. When the samples were released, the resistivity went back to its previous value. This proved that the tensoresistive effect was reversible. However, at 5, 8, and 20 mp, there was no evident variation in electrical resistance, which agrees with the prediction that at low and high concentration of filler there will be weak or no tensoresistive effect.

At the nano level, the authors examined the tensoresistive effect by looking at the AFM images (Figure 11.18) [44]. The black spots of CB formed protrusions on the surface of the polymer matrix, which can be related to the electro-conductive channels that formed in the composites. The 10 mp of CB formulation had three

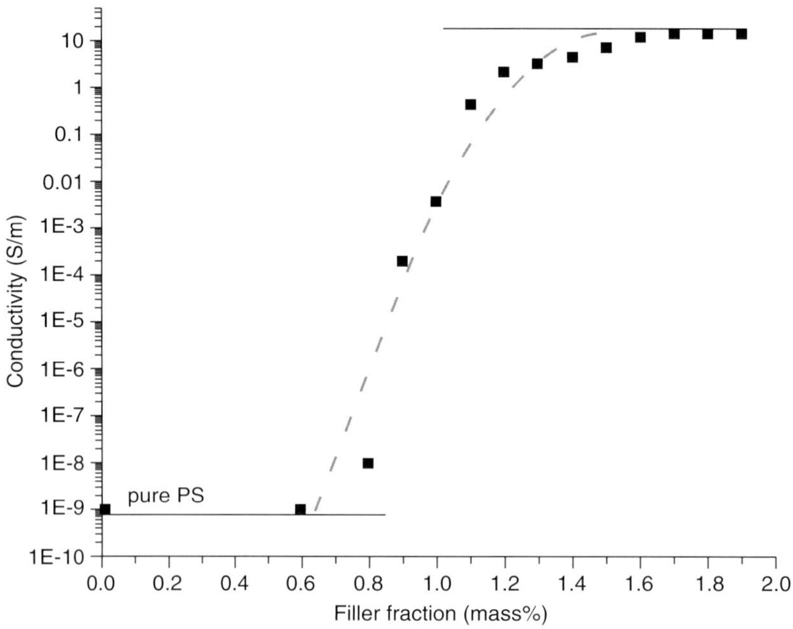

Figure 11.19. The conductivity of graphene/PS nanocomposites with different filler fractions [47].

electro-conductive channels in the nanoscale map, and with higher concentration of CB there were more and more channels, which agrees with the theory that higher concentration of CB leads to low resistivity.

11.3.4 Graphite Reinforced Thermoset-Based Nanocomposites

Tkalya et al. [45] used the latex technology to prepare graphene/polystyrene nanocomposites. The percolation threshold was very low, which was less than 1 wt% of graphene, and a maximum conductivity of about 15 S/m was achieved for 1.6 wt% to 2 wt% of the nanofiller. As has been shown before, with higher loading of nanofillers, the conductivity goes up (Figure 11.19) [45].

The expandable graphite (EG) used to make nanocomposites with polystyrene has the most advantage of improving the electrical property of thermoset matrix because of the nanostructure of the conductive filler [46]. By referring to the Chen et al.'s paper [46], it is not hard to find out that the percolation threshold was 1.8 wt%, which was higher than graphite powder with micron size (50 wt%), and the highest value achieved was 10^{-2} S/cm at 5 wt%, about 14 orders-of-magnitude improvement over the conductivity of the pure polystyrene, which was 10^{-16} S/cm. Similar results can be observed in EG reinforced poly(phenylene sulfide) (PPS) nanocomposites by melt-blending [47]. Compared with other fillers, graphite has the worm forms or filaments made of a large number of nanosheets, which is advantageous to form a good conductive network in the thermoset matrix.

Furthermore, Kim and Drzal [48] tried to compare the different properties of exfoliated graphite nanoplatelets (xGnP)/ethylene vinyl acetate (EVA) and CNT/

EVA nanocomposites. Despite the fact that CNT–EVA was electrically conductive with only 5 wt% CNT loading, and the percolation threshold for xGnP/EVA was about 14–16 wt%, the authors still recommended to use xGnP to make polymer nanocomposites because of other properties improvement [49]. With the same thermoset matrix of EVA, expanded graphite-reinforced nanocomposites had lower percolation threshold of 6 vol% than that of unexpanded graphite of 15 vol% [50].

Similar to the thermoplastic system mentioned earlier, the synergetic effects of hybrid filler were also investigated by researchers. Sumfleth [51] obtained the synergistic effect by looking at the conductivity values of CB/MWNT reinforced epoxy system compared with MWNT and CB system and explained the phenomenon based on the conductive network formation mechanism that were similar to what had been showed in [52]. Ma et al. [53] developed epoxy-based nanocomposites containing carbon nanotubes and carbon black in order to gain good electrical property and mechanical property at lower cost. And it turned out that in addition to the improvement of electrical conductivity at low total loading, CB effectively contributed to the enhancement of fracture toughness and ductility of nanocomposites and maintenance of high flexural modulus and strength, which indicates the synergistic effect of CB.

11.4 Electrical Properties of Thermoplastic Elastomer-Based Nanocomposites

11.4.1 Inorganic Filler in Thermoplastic Elastomer-Based Nanocomposites

Most electrically conductive polymer nanocomposites are made by using thermoplastic or thermosetting polymers, with thermoplastic elastomers being the less commonly used polymer type. According to Karttunen et al. [54], it was the very first time in the literature that the electrical conductivity of nanocomposites consisting of silver particles and thermoplastic elastomers was reported. In this paper, the agglomerated (average size from 1 μm to 10 μm) and non-agglomerated silver particles (average size ≤100 nm) were added to thermoplastic elastomer styrene-ethylene-propylene-styrene copolymer (SEPS) to form nanocomposites at different filling rates ranging from 5 vol% to 23.6 vol%. The percolation result is plotted in Figure 11.20 [54]. For the well-distributed silver particles, the minimum resistivity was 0.18 ohm-cm at 16 vol% and it remains stable at higher filling rates; for the agglomerated silver particles, the minimum resistivity was higher, 2.9×10^3 ohm-cm. As the size of the agglomerates goes up, the resistivity increases and the percolation threshold shifts to higher filling rates.

11.4.2 Organic Fillers in Thermoplastic Elastomer-Based Nanocomposites

Since an extremely low loading of CNT can efficiently achieve electrical conductivity that similar to conventionally filled system while maintaining the other properties of matrix polymers, CNTs are considered as the best filler for an elastomer

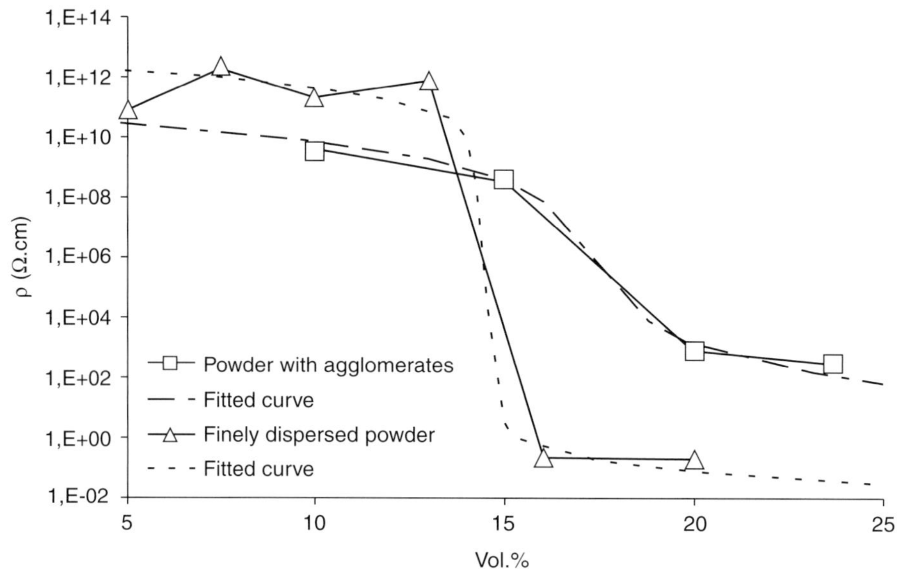

Figure 11.20. Resistivity versus degree of filling rate for nanocomposites containing uniformly distributed and agglomerated silver particles [56].

matrix to manufacture conductive nanocomposites where strecthability should be well maintained at low filler loading [55]. Koerner et al. [56, 57] successfully fabricated MWNT/thermoplastic polyurethane (PU) elastomer nanocomposites, and discovered that the elastomer nanocomposites possessed high electrical conductivity, about 1–10 S/cm, with good mechanical properties (Figure 11.21) with only a small loading of CNTs (0.5–10 vol%) [56]. Figure 11.21 shows that the initial modulus E(a) and engineering stress at the apparent yield point, σ_Y, increase as the volume fraction of CNTs goes up in the system. The strain hardening, λ_H, decreases with CNT loading after an initial increase with respect to the neat resin. The stress, σ_H, also increases. The elongation at rupture decreases slightly with CNT loading; however, even at 17 wt% (10.2 vol%) CNT, it still results in a thermoplastic elastomer with 1,200% elongation before failure. This is because the mechanical property of thermoplastic elastomer relies on the soft-segment crystalline content, and the addition of small amount of CNTs greatly increases that content, which leads to better mechanical property.

To enhance electrical conductivity, Sekitani [58] fabricated SWCNT/fluorine rubber nanocomposites with ionic liquid by solution blending. The resulted nanocomposites showed very good stretchability of more than 100%. Specifically, the conductivity of elastomer conductor made of fluorinated copolymer G801 and SWCNT was as high as 30 S/cm, but the elasticity and flexibility were not good, so it was suggested to add some ionic liquid to the nanocomposites. It is shown that when the content of SWCNT is equivalent to that of ionic liquid, the film has the highest conductivity about 50 S/cm at the bucky gel content of 30 wt% [58]. But with further increase of bucky gel content, the electrical conductivity goes down because the films became fragile and porous membranes are resulted.

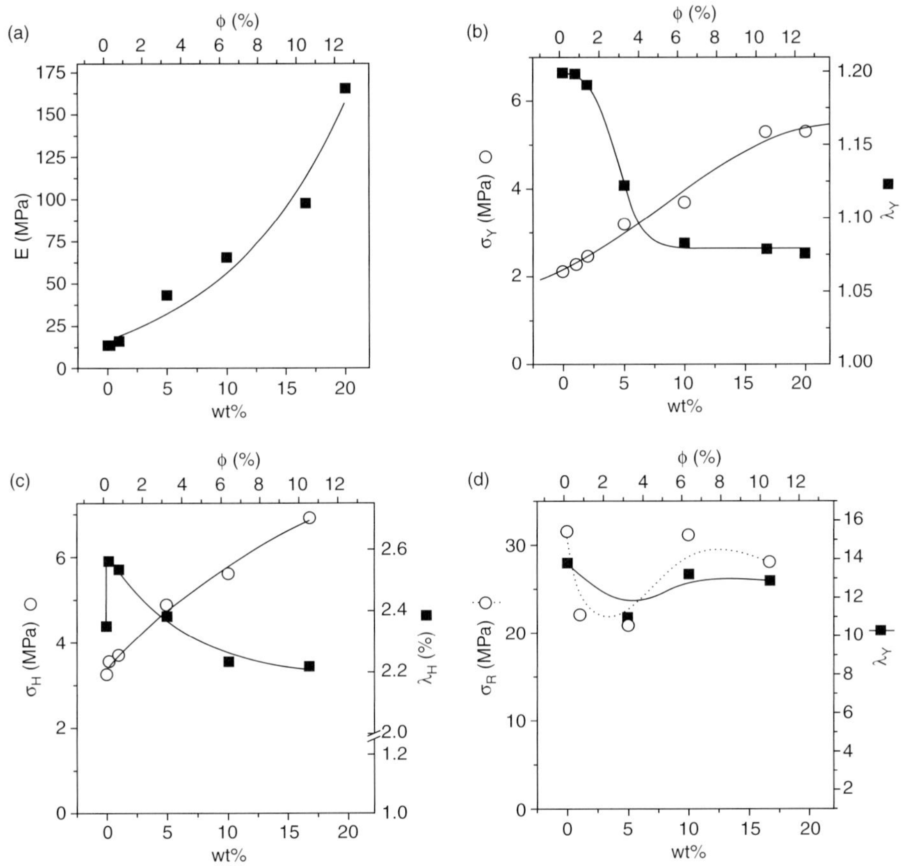

Figure 11.21. Mechanical characteristics of the CNT/PU nanocomposites: (a) initial Young's moduli, E; (b) stress and elongation at yield (σ_Y, λ_Y); (c) stress and elongation at hardening ($\sigma_, \lambda_H$); (d) stress and elongation at rupture (σ_R, λ_R). Lines express qualitative data trends. Open circles are stress (s) data and filled squares are elongation (l) values [58].

In terms of processing technique, most of the thermoplastic elastomer-based nanocomposites were prepared by solution blending or other methods that may have negative impact on the environment.

Some researcher sought to mix MWNTs with matrix with high-shear blending at heating temperature [59]. In this case, the polymer poly(vinylidene fluoride) (PVDF) was used, and the electrical property was compared between high-shear and low-shear mixing methods. As shown in curve (a) of volume conductivity in Figure 11.22, there is a stepwise increase in the conductivity of the nanocomposites [59]. Conductivity gradually increases with increasing nanotube content (≤1.2 wt%). The conductivity of the composite with 1.9 wt% MWNTs increases to 2.5×10^{-2} S/cm from that of the composite containing 1.2 wt% MWNT, which has the conductivity of 4.9×10^{-4} S/cm. When MWNT content is larger than 1.9 wt%, no further increase can be observed in PVDF/MWNT high-shear processed composites. The same trend was applied for surface conductivity of composites at low-shear processing, and the jump of surface conductivity at low-shear processing was found to be 1.5 wt% of MWNT

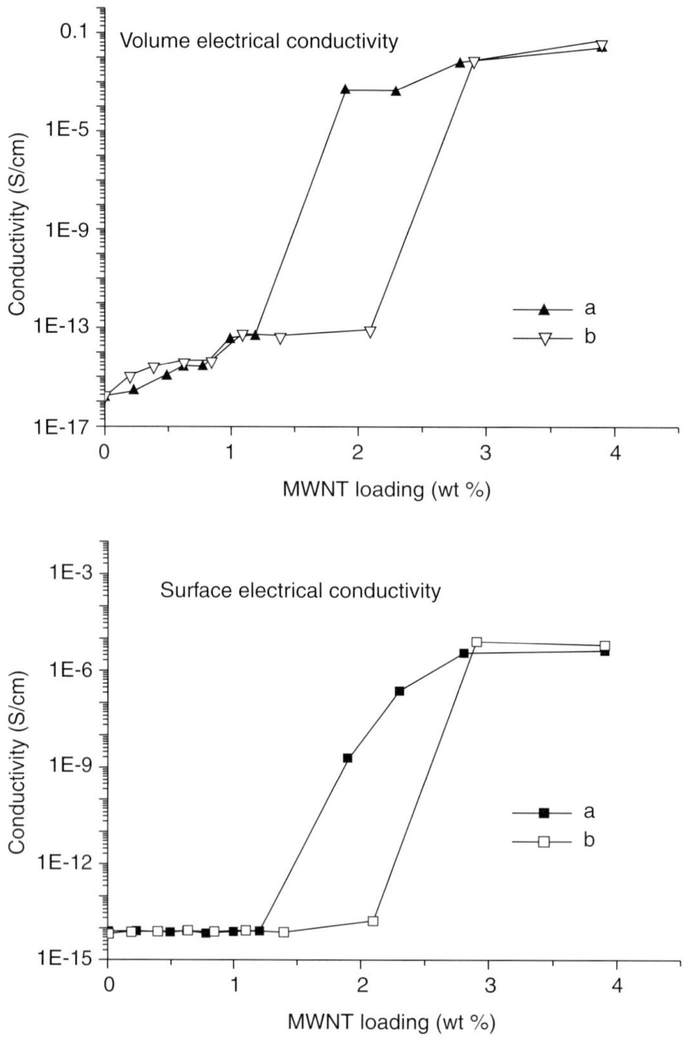

Figure 11.22. Electrical conductivity of the PVDF/MWNT composites as a function of the nanotube loading: (a) high-shearing composites and (b) low-shearing composites [59] (Courtesy of ACS).

loading. The generally accepted explanation of this stepwise change in conductivity is that the nanotube loading reached the conductivity threshold and a conductive network formed through hopping and tunneling processes at certain concentration of conductive fillers.

So the high-shear mixing process, which leads to better dispersion of MWNTs in matrix, made the electrical percolation threshold (1.5 wt%) lower than nanocomposites made via low-shear mixing process (2.5 wt%).

To achieve high stretchability and elasticity, Li and Shimizu [55] used a thermoplastic elastomer, poly[styrene-b-(ethylene-co-butylene)-b-styrene](SEBS), to melt-blend with MWNTs at high shear to fabricate nanocomposites, and studied the effect of MWNT loading (Figure 11.23) [55] and uniaxial strain on the electrical

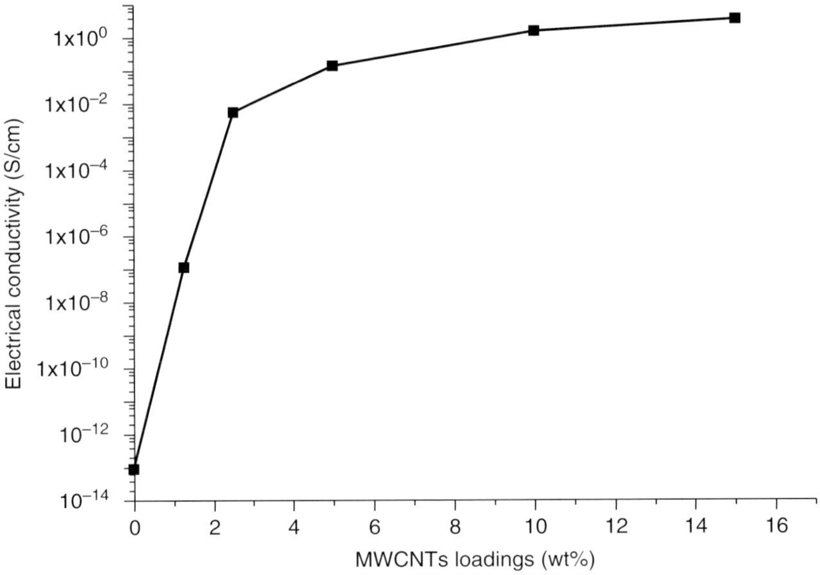

Figure 11.23. Electrical conductivity of SEBS/MWNT nanocomposites as a function of CNT loading [55] (Courtesy of ACS).

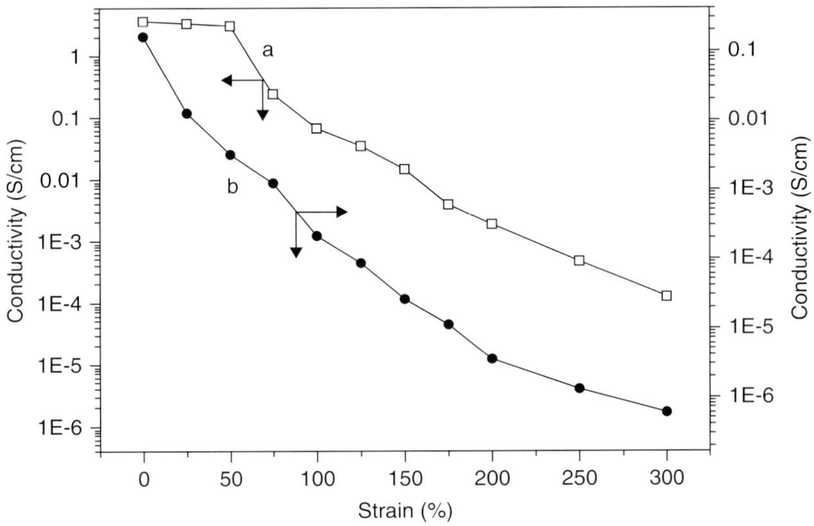

Figure 11.24. Electrical conductivity of SEBS/MWNT nanocomposites as a function of uniaxial tensile strain: (a) with 15 wt% MWNTs and (b) with 5 wt% MWNTs [55] (Courtesy of ACS).

conductivity (Figure 11.24) [55]. As MWNT loading content goes up, the electrical conductivity increases and the conductive network begins to form, which is indicated by the sudden increase of conductivity from 1.25 wt% to 2.5 wt% and the maximum conductivity of 5.16 S/cm at 15 wt%. This result corresponds with the SEM image of nanocomposites (Figure 11.25) [55]. Clearly there was no complete network between MWNTs (Figure 11.25b), but at 2.5 wt% a good conductive network was formed.

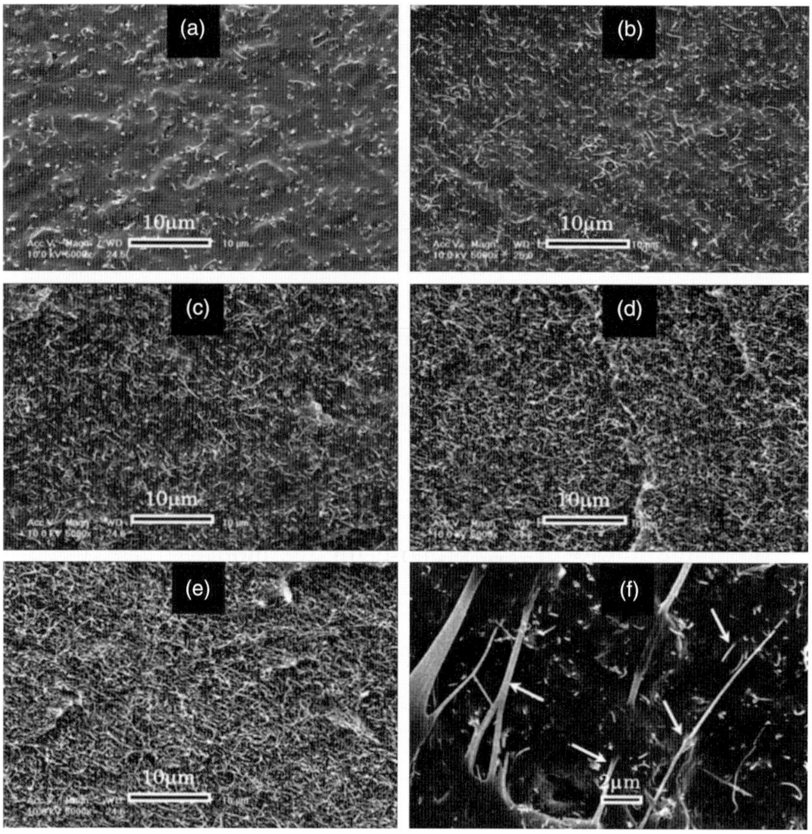

Figure 11.25. SEM images of SEBS/MWNT nanocomposites with different nanotube loadings: (a) 1.25 wt%, (b) 2.5 wt%, (c) 5.0 wt%, (d) 10 wt%, (e) 15 wt%, and (f) 1.25 wt% at higher magnification [55] (Courtesy of ACS).

Further increase in loading enhanced the entanglement of CNTs and thus the conductivity. Figure 11.25 can explain well the strain on the decrease in electrical conductivity, which can be explained by the fact that the electrical pathway was damaged when the CNT network had to align in the stretching direction, and that for the 15 wt% MWNT nanocomposites, small strain (50%) cannot easily destroy the network, resulting the stable electrical pathway.

11.5 Summary

This chapter briefly reviews the electrical properties of different nonconductive polymer-based nanocomposites reinforced by numerous types of conductive nanofillers. The characterization methods of electrical properties of nanocomposites have been demonstrated in details in Chapter 6. The electrical properties of polymer nanocomposites based on three categories of polymers are showed and some factors, such as polymer, types of nanofillers, and processing conditions that can influence the electrical properties of nanocomposites are introduced. The basic mechanism of formation of conductive networks that contributes to the conductivity of nanocomposites is examined and should be thoroughly understood to explain this phenomenon.

11.6 Questions and Problems

1. How does the nanomaterial dispersion affect the electrical properties of polymer nanocomposites in general? Provide examples to illustrate your claims.
2. This chapter looked at the enhancement of electrical properties of thermoplastic-based nanocomposites using carbon-based nanomaterials, such as MWNT, carbon black (CB), CNF, graphite, and mixed MWNT-CB. Briefly discuss the mechanism of the formation of conductive network of each type of nanofiller. Would surface treatment of these carbon-based nanomaterials enhance electrical conductivity?
3. Which types of nanomaterials are more effective for thermoplastic-based nanocomposites? Provide examples. Besides the aforementioned nanomaterials, what other nanomaterials would you propose for thermoplastic-based nanocomposites? Discuss the mechanical properties of these electrically conductive thermoplastic-based nanocomposites?
4. This chapter also explored the enhancement of electrical properties of thermoset-based nanocomposites using MWNT, CNF, CB, graphite, and CB/MWNT nanomaterials. Which types of nanomaterials are more effective for epoxy-based nanocomposites? Provide examples. Discuss the mechanical properties of these electrically conductive thermoset-based nanocomposites?
5. This chapter also covered the enhancement of electrical properties of thermoplastic elastomer-based nanocomposites using silver particles, MWNT, and SWNT. Which types of nanomaterials are more effective for thermoplastic elastomer-based nanocomposites? Provide examples. Besides these nanomaterials, what other nanomaterials can be used in thermoplastic elastomer to enhance its electrical conductivity? Discuss the mechanical properties of these electrically conductive thermoplastic elastomer-based nanocomposites?
6. What will be the effects of the multicomponent systems with two or more nanomaterials and conventional conductive polymer systems? What will be a good strategy to optimize the electrical properties in these polymer nanocomposites?
7. What are the electrical property mechanisms for these different types of nanomaterials in the polymer nanocomposites, as well as the hybrid material systems?

REFERENCES

1. C. Y. Lew and C. M. Luizi (2013). The Influence of Processing Conditions on the Electrical Properties of Polypropylene Nanocomposites Incorporating Multiwall Carbon Nanotube. Published on November 18 at http://www.nanocyl.com.
2. B. Krause, P. Pötschke, and L. Häußler (2009). Influence of small scale melt mixing conditions on electrical resistivity of carbon nanotube-polyamide composites. *Composites Science and Technology* 69, 1505–1515.

3. S. Pujari, T. Ramanathan, K. Kasimatis, J. I. Masuda, R. Andrews, et al. (2009). Preparation and characterization of multiwalled carbon nanotube dispersions in polypropylene: Melt mixing versus solid-state shear pulverization. *Journal of Polymer Science Part B: Polymer Physic* 47, 1426–1436.
4. T. Villmow, P. Pötschke, S. Pegel, L. Häussler, and B. Kretzschmar (2008). Influence of twin-screw extrusion conditions on the dispersion of multi-walled carbon nanotubes in a poly(lactic acid) matrix. *Polymer* 49, 3500–3509.
5. E. Logakis, C. Pandis, V. Peoglos, P. Pissis, J. Pionteck (2009). Electrical/dielectric properties and conduction mechanism in melt processed polyamide/multi-walled carbon nanotubes composites. *Polymer* 50, 5103–5111.
6. O. Meincke, D. Kaempfer, H. Weickmann, C. Friedrich, M. Vathauer, and H. Warth (2004). Mechanical properties and electrical conductivity of carbon-nanotube filled polyamide-6 and its blends with acrylonitrile/butadiene/styrene. *Polymer* 45, 739–748.
7. R. Socher, B. Krause, R. Boldt, S. Hermasch, R. Wursche, and P. Pötschke (2011). Melt mixed nano composites of PA12 with MWNTs: Influence of MWNT and matrix properties on macrodispersion and electrical properties. *Composites Science and Technology* 71, 306–314.
8. O. S. Carneiro, J. A. Covas, C. A. Bernardo, G. Caldeira, F. W. J. Van Hattum, et al. (1998). Production and assessment of polycarbonate composites reinforced with vapour-grown carbon fibres. *Composites Science and Technology* 58, 401–407.
9. G. G. Tibbetts, M. L. Lake, K. L. Strong, and B. P. Rice (2007). A review of the fabrication and properties of vapor-grown carbon nanofiber/polymer composites. *Composites Science and Technology* 67, 1709–1718.
10. G. A. Jimenez and S. C. Jana (2007). Oxidized carbon nanofiber/polymer composites prepared by chaotic mixing. *Carbon* 45, 2079–2091.
11. G. A. Jimenez and S. C. Jana (2007). Electrically conductive polymer nanocomposites of polymethylmethacrylate and carbon nanofibers prepared by chaotic mixing. *Composites Part A: Applied Science and Manufacturing* 38, 983–993.
12. K. Kalaitzidou, H. Fukushima, and L. T. Drzal (2007). A new compounding method for exfoliated graphite–polypropylene nanocomposites with enhanced flexural properties and lower percolation threshold. *Composites Science and Technology* 67, 2045–2051.
13. Y. Li and H. Shimizu (2008). Conductive PVDF/PA6/CNTs nanocomposites fabricated by dual formation of co-continuous and nanodispersion structures. *Macromolecules* 41, 5339–5344.
14. B. E. Kilbride, J. N. Coleman, J. Fraysse, P. Fournet, M. Cadek, et al. (2002). Experimental observation of scaling laws for alternating current and direct current conductivity in polymer-carbon nanotube composite thin films. *Journal of Applied Physics* 92, 4024–4030.
15. W.-P. Wang, Y. Liu, X.-X. Li, and Y.-Z. You (2006). Synthesis and characteristics of poly(methyl methacrylate)/expanded graphite nanocomposites. *Journal of Applied Polymer Science* 100, 1427–1431.
16. R. K. Goyal, S. D. Samant, A. K. Thakar, and A. Kadam (2010). Electrical properties of polymer/expanded graphite nanocomposites with low percolation. *Journal of Physics D-Applied Physics*, 43 (36), 365404(7 pp.).
17. W. Zheng and S.-C. Wong (2003). Electrical conductivity and dielectric properties of PMMA/expanded graphite composites. *Composites Science and Technology* 63, 225–235.
18. Y.-X. Pan, Z.-Z. Yu, Y.-C. Ou, and G.-H. Hu (2000). A new process of fabricating electrically conducting nylon 6/graphite nanocomposites via intercalation polymerization. *Journal of Polymer Science Part B: Polymer Physics* 38, 1626–1633.

19. M. D. Via, J. A. King, J. M. Keith, and G. R. Bogucki (2012). Electrical conductivity modeling of carbon black/polycarbonate, carbon nanotube/polycarbonate, and exfoliated graphite nanoplatelet/polycarbonate composites. *Journal of Applied Polymer Science* 124, 182–189.
20. S. M. Zhang, H. Lin, H. Deng, X. Gao, E. Bilotti, et al. (2012). Synergistic effect in conductive networks constructed with carbon nanofillers in different dimensions. *eXPRESS Polymer Letters* 6, 159–168.
21. Y. Sun, H.-D. Bao, Z.-X. Guo, and J. Yu (2008). Modeling of the electrical percolation of mixed carbon fillers in polymer-based composites. *Macromolecules* 42, 459–463.
22. D. Puglia, L. Valentini, and J. M. Kenny (2003). Analysis of the cure reaction of carbon nanotubes/epoxy resin composites through thermal analysis and Raman spectroscopy. *Journal of Applied Polymer Science* 88, 452–458.
23. L. Christodoulou and J. D. Venables (2006). Multifunctional material systems: The first generation. *Journal of Materials* 55, 39–45.
24. A. Battisti, A. A. Skordos, and I. K. Partridge (2010). Percolation threshold of carbon nanotubes filled unsaturated polyesters. *Composites Science and Technology* 70, 633–637.
25. J. Vera-Agullo, A. Glória-Pereira, H. Varela-Rizo, J. L. Gonzalez, and I. Martin-Gullon (2009). Comparative study of the dispersion and functional properties of multiwall carbon nanotubes and helical-ribbon carbon nanofibers in polyester nanocomposites. *Composites Science and Technology* 69, 1521–1532.
26. C. A. Martin, J. K. W. Sandler, M. S. P. Shaffer, M. K. Schwarz, W. Bauhofer, et al. (2004). Formation of percolating networks in multi-wall carbon nanotube–epoxy composites. *Composites Science and Technology* 64, 2309–2316.
27. E. T. Thostenson, S. Ziaee, and T.-W. Chou (2009). Processing and electrical properties of carbon nanotube/vinyl ester nanocomposites. *Composites Science and Technology* 69, 801–804.
28. J. K. W. Sandler, J. E. Kirk, I. A. Kinloch, M. S. P. Shaffer, and A. H. Windle (2003). Ultra-low electrical percolation threshold in carbon nanotube–epoxy composites. *Polymer* 44, 5893–5899.
29. F. H. Gojny, M. H. G. Wichmann, B. Fiedler, I. A. Kinloch, W. Bauhofer, et al. (2006). Evaluation and identification of electrical and thermal conduction mechanisms in carbon nanotube/epoxy composites. *Polymer* 47, 2036–2045.
30. A. Moisala, Q. Li, I. A. Kinloch, and A. H. Windle (2006). Thermal and electrical conductivity of single- and multi-walled carbon nanotube–epoxy composites. *Composites Science and Technology* 66, 1285–1288.
31. F. H. Gojny, M. H. G. Wichmann, U. Kopke, B. Fiedler, & K. Schulte (2004). Carbon nanotube-reinforced epoxy-composites: enhanced stiffness and fracture toughness at low nanotube content. *Composites Science and Technology* 64, 2363–2371.
32. F. H. Gojny, M. H. G. Wichmann, U. Kopke, B. Fiedler, & K. Schulte (2005). Influence of different carbon nanotubes on the mechanical properties of epoxy matrix composites – A comparative study. *Composites Science and Technology* 65, 2300–2313.
33. M. Moniruzzaman and K. I. Winey (2006). Polymer nanocomposites containing carbon nanotubes. *Macromolecules* 39, 5194–5205.
34. S. C. Lao (2013). *Multifunctional Cyanate Ester/MWNT Nanocomposites: Processing and Characterization.* Ph.D. dissertation, The University of Texas at Austin, Austin, TX, Dec.
35. K. Liang, G. Li, H. Toghiani, J. H. Koo, C. U. Pittman Jr., and C. Dave (2006). Cyanate ester/polyhedral oligomeric silsesquioxane (POSS) nanocomposites: Synthesis and characterization. *Chemistry of Materials* 18(2), 301–312.

36. H. S. Cho, K. Liang, S. Chatterjee, and C. U. Pittman Jr. (2005). Synthesis, morphology, and viscoelastic properties of polyhedral oligomeric silsesquioxane nanocomposites with epoxy and cyanate ester matrices. *Journal of Inorganic and Organometallic Polymers and Materials* 15(4), 541–553.
37. K. Liang, H. Toghiani, G. Li, C. U. Pittman Jr., and C. Dave (2005). Synthesis, morphology, and viscoelastic properties of cyanate ester/polyhedral oligomeric silsesquioxane nanocomposites. *Journal of Polymer Science, Part A: Polymer Chemistry* 43(17), 3887–3898.
38. *Fundamentals of Electrostatic Discharge, Part One – An Introduction to ESD* (2010), ESD Association, Rome, NY.
39. N. Hu, Z. Masuda, and H. Fukunaga (2009). *Carbon Nanotubes: New Research*, Nova Science Publishers, Inc., New York, NY, pp. 175–222.
40. P.-C. Ma, N. A. Siddiqui, G. Marom, and J.-K. Kim (2010). Dispersion and Functionalization of Carbon Nanotubes for Polymer-based Nanocomposites: A Review. *Composites: Part A* 41, 1345–1367.
41. B. Smrutisikha (2010). Experimental study of mechanical and electrical properties of carbon nanofiber/epoxy composites. *Materials & Design* 31, 2406–2413.
42. M. Ardanuy, M. A. Rodríguez-Perez, and I. Algaba (2011). Electrical conductivity and mechanical properties of vapor-grown carbon nanofibers/trifunctional epoxy composites prepared by direct mixing. *Composites Part B: Engineering* 42, 675–681.
43. B. H. Cipriano, A. K. Kota, A. L. Gershon, C. J. Laskowski, T. Kashiwagi, et al. (2008). Conductivity enhancement of carbon nanotube and nanofiber-based polymer nanocomposites by melt annealing. *Polymer* 49, 4846–4851.
44. M. Knite, V. Teteris, B. Polyakov, and D. Erts (2002). Electric and elastic properties of conductive polymeric nanocomposites on macro- and nanoscales. *Materials Science & Engineering C-Biomimetic and Supramolecular Systems* 19, 15–19.
45. E. Tkalya, M. Ghislandi, A. Alekseev, C. Koning, and J. Loos (2010). Latex-based concept for the preparation of graphene-based polymer nanocomposites. *Journal of Materials Chemistry* 20, 3035–3039.
46. G.-H. Chen, D.-J. Wu, W.-G. Weng, B. He, and W.-L. Yan (2001). Preparation of polystyrene-graphite conducting nanocomposites via intercalation polymerization. *Polymer International* 50, 980–985.
47. Y. F. Zhao, M. Xiao, S. J. Wang, X. C. Ge, and Y. Z. Meng (2007). Preparation and properties of electrically conductive PPS/expanded graphite nanocomposites. *Composites Science and Technology* 67, 2528–2534.
48. S. Kim and L. T. Drzal (2009). Comparison of exfoliated graphite nanoplatelets (xGnP) and CNTs for reinforcement of EVA nanocomposites fabricated by solution compounding method and three screw rotating systems. *Journal of Adhesion Science and Technology* 23, 1623–1638.
49. K. Kalaitzidou, H. Fukushima, and L. T. Drzal (2007). Mechanical properties and morphological characterization of exfoliated graphite-polypropylene nanocomposites. *Composites: Part A* 38, 1675–1682.
50. R. Tlili, A. Boudenne, V. Cecen, L. Ibos, I. Krupa, and Y. Candau (2010). Thermophysical and electrical properties of nanocomposites based on ethylene-vinylacetate copolymer (EVA) filled with expanded and unexpanded graphite. *International Journal of Thermophysics* 31, 936–948.
51. J. Sumfleth, X. Adroher, and K. Schulte (2009). Synergistic effects in network formation and electrical properties of hybrid epoxy nanocomposites containing multi-wall carbon nanotubes and carbon black. *Journal of Materials Science* 44, 3241–3247.

52. S. M. Zhang, L. Lin, H. Deng, X. Gao, E. Bilotti, et al. (2012). Synergistic effect in conductive networks constructed with carbon nanofillers in different dimensions. *Express Polymer Letters* 6, 159–168.
53. P.-C. Ma, M.-Y. Liu, H. Zhang, S.-Q. Wang, R. Wang, et al. (2009). Enhanced electrical conductivity of nanocomposites containing hybrid fillers of carbon nanotubes and carbon black. *ACS Applied Materials & Interfaces* 1, 1090–1096.
54. M. Karttunen, P. Ruuskanen, V. Pitkanen, and W. M. Albers, (2008). Electrically conductive metal polymer nanocomposites for electronics applications. *Journal of Electronic Materials* 37, 951–954.
55. Y. J. Li and H. Shimizu (2009). Toward a stretchable, elastic, and electrically conductive nanocomposite: Morphology and properties of poly[styrene-b-(ethylene-co-butylene)-b-styrene]/multiwalled carbon nanotube composites fabricated by high-shear processing. *Macromolecules* 42, 2587–2593.
56. H. Koerner, G. Price, N. A. Pearce, M. Alexander, and R. A. Vaia (2004). Remotely actuated polymer nanocomposites – stress-recovery of carbon-nanotube-filled thermoplastic elastomers. *Nature Materials* 3, 115–120.
57. H. Koerner, W. D. Liu, M. Alexander, P. Mirau, H. Dowty, and R. A. Vaia (2005). Deformation-morphology correlations in electrically conductive carbon nanotube thermoplastic polyurethane nanocomposites. *Polymer* 46, 4405–4420.
58. T. Sekitani, Y. Noguchi, K. Hata, T. Fukushima, T. Aida, and T. Someya (2008). A rubberlike stretchable active matrix using elastic conductors. *Science* 321, 1468–1472.
59. G.-X. Chen, Y. Li, and H. Shimizu (2007). Ultrahigh-shear processing for the preparation of polymer/carbon nanotube composites. *Carbon* 45, 2334–2340.

12 Optical Properties of Polymer Nanocomposites

12.1 Introduction

The impressive mechanical, thermal, flammability, ablation, and electrical properties of polymer nanocomposite materials are well documented. These properties have been discussed in previous chapters. Optical properties are intimately related to other properties and functionalities (e.g., electronic, magnetic, and thermal) that are of fundamental importance to many applications, such as energy conversion, chemical analysis, biomedicine, optoelectronics, communication, and radiation detection [1]. The interactions between various nanoscale particles and their host polymers allow for the combination and in some cases synergistic creation of interesting optical material properties that can be exploited to produce refractive devices, polarizing filters, optical switches, LEDs, optical lenses, sensors, actuators, and solar cells [1–5]. Zhang's text on *Optical Properties and Spectroscopy of Nanomaterials* would provide readers on more details of optical properties on nanomaterials [1]. This text covers several fundamental aspects of spectroscopic techniques for studying optical properties of nanomaterials. The different processing and characterization of optical properties of polymer nanocomposites are discussed in Chapters 5 and 6, respectively. The various optical properties and applications of thermoplastic-, thermoplastic elastomer- and thermoset-based nanocomposites are discussed in this chapter.

12.2 Optical Properties of Thermoplastic-Based Nanocomposites

Unfortunately, many nanoparticles commonly used to enhance mechanical properties impart optical properties that are undesirable for most optical applications. Adding carbon-based particles or structures, such as graphene imparts a black coloration (depending on the weight percentage used) to the polymer nanocomposite, making it useless for applications where transparency is desired. The addition of graphene and carbon nanotubes does provide some benefit to other useful optical properties, however, which will be discussed later in the thermoset portion of this chapter. Another difficulty preventing many particles from being used in optical applications is scattering. The vast relative surface area of nanoparticles means that even a small

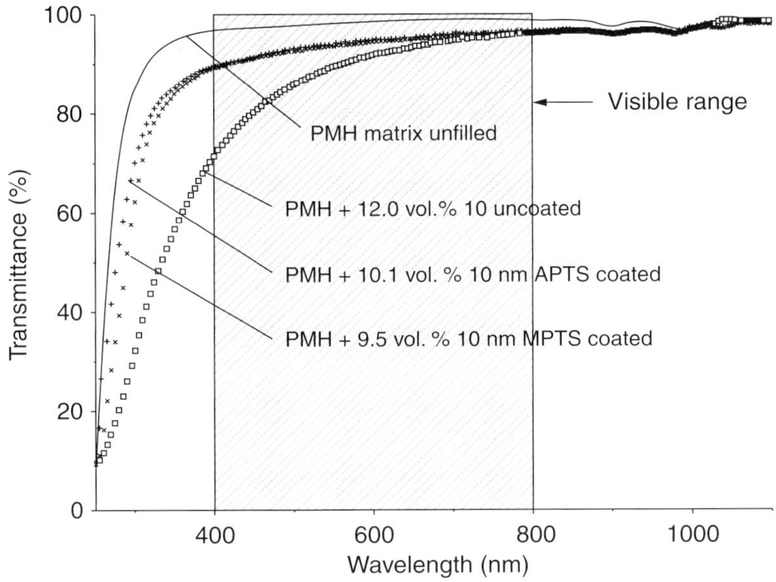

Figure 12.1. UV/VIS transmittance spectra for PMH nanocomposites containing approximately 10 vol.% 10 nm SiO_2 particles with different surface coating in comparison to the pure unfilled PMH copolymer matrix; sample thickness of 2 mm [7].

weight percentage addition of particles can cause scattering, which greatly decreases light transmitted through the material and results in opacity or "cloudiness." In general, scattering can be avoided by keeping particle size under 25 nm, or by matching the refractive index of nanoparticles to the polymer [3, 6]. If particle size is kept even smaller (under 10 nm), some materials can even remain transparent at high-volume fill rates, at which point processing may be the limiting factor [6]. However, controlling particle size in thermoplastic polymer nanocomposites can be quite difficult, and is the target of much research activities [3].

The advantage of using thermoplastic nanocomposites as lenses is that thermoplastic can be easily and inexpensively processed by injection molding. If thermoplastic-based nanocomposites are to replace glass lenses, they must be able to withstand long-term exposure to heat and light without degrading [3]. Many neat thermoplastics experience degraded mechanical properties at relatively low temperatures, but Figures 12.1 and 12.2 demonstrate that methacryloxypropyltrimethoxysilane (MPTS) coated SiO_2 particles added to polymethylmethacrylate-co-HEMA (PMH) at 9.5 vol% greatly increases the material modulus over neat PMH at elevated temperatures, while maintaining ~90% transmittance over the visible light spectrum [7]. The uncoated filled SiO_2 material and the acetoxypropyltrimethoxysilane (APTS) coated SiO_2 performed much worse with respect to strength, which can be attributed to poor dispersion as reported by Becker et al. [7]. These conclusions cannot necessarily be extended to other polymers; for example, Pandey et al. have shown that some PP and PE nanocomposites show increased degradation over neat materials when exposed to UV rays [5].

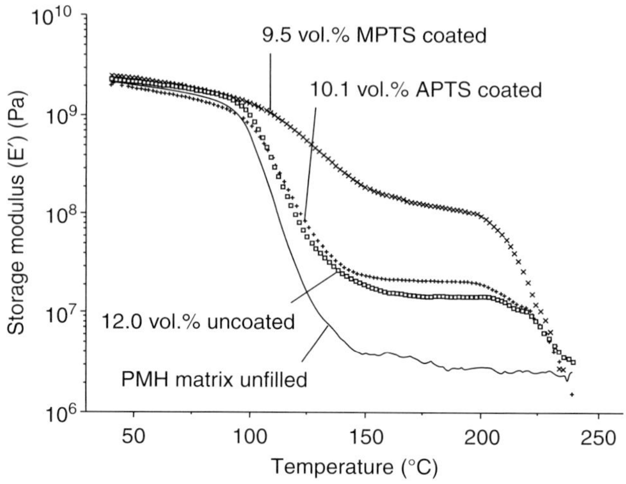

Figure 12.2. Temperature dependence of the storage modulus E' from DMTA measurements for PMH nanocomposites containing vol% of 10 nm SiO_2 particles with different surface coatings in comparison to the pure unfilled PMH copolymer [7].

The addition of inorganic nanoparticles into thermoplastics is a widely studied topic, with two of the main applications being the creation of refractive devices and use in polarizing filters. Neat polymers have refractive indices that range from 1.25 to 2.5, which is small compared to the indices achieved by inorganic materials [6]. The refractive index range of inorganic materials is much wider, as can be seen in Table 12.1.

Combining selected materials as nanoparticles with thermoplastics results in a refractive index that can be controlled depending on particle size, as demonstrated in Figure 12.3, particle refractive index, and the wt% added to the material [6]. Impressively, this refractive index can greatly exceed the values possible in neat polymers. In this specific case, PbS particles of various diameters have been added to polyethylene oxide to achieve a wide range of refractive indices.

Inorganic materials can also be added to reduce the refractive index of thermoplastic polymer nanocomposites. Table 12.1 shows that gold, silver, and copper have refractive indices much lower than those seen in organic polymers at the high end of the visible spectrum. Both gold and silver are commonly added to various polymers to decrease their refractive index, although gold is more commonly studied [6]. Gold also has the interesting property of color dependence based on particle size, which results in films appearing visibly tinted [6], as seen in Figure 12.4.

The important factors that determine optical properties of polymer nanocomposites with gold particles are particle diameter, weight percentage of gold particles, gold particle morphology, and polymer thickness [4, 8]. Two gold nanocomposites with equal volume percentages of gold particles can have vastly different properties if the gold particle size is different [8]. Initially, most studies focused on spherical gold particles, but now many groups employ colloidal chemistry to produce rod-shaped gold nanoparticles of controlled length and diameter. Homogenous dispersion, as seen in Figure 12.5, is also important but obtainable, since the gold is deposited onto or grown in thin films [8]. These films can be made in solution or by spinning [4, 6]. Common polymers for the film include PVA, PMMVA, and polystyrene [9, 10].

Table 12.1. *Refractive index of selected inorganic materials with extreme refractive indices at different wavelengths [5]*

Substance	Refractive index		
	413.3 nm	619.9 nm	826.6 nm
Os	4.05	3.98	2.84
W	3.35	3.60	3.48
Si crystalline	5.22	3.91	3.67
Si amorphous	4.38	4.23	3.86
Ge	4.08	5.59–5.64	4.65
GaP	4.08	3.33	3.18
GaAs	4.51	3.88	3.67
InP	4.40	3.55	3.46
InAs	3.20	4.00	3.71
InSb	3.37	4.19	4.42
PbS	3.88	4.29	4.50[a)]
PbSe	1.25–3.00	3.65–3.90	4.64
PbTe	1.0–1.8	6.4	3.8
Ag	0.173	0.131	0.145
Au	1.64	0.194	0.188
Cu	1.18	0.272	0.260

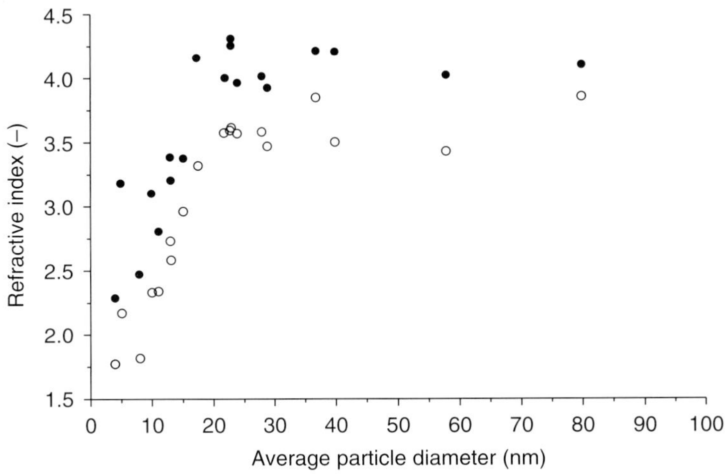

Figure 12.3. Refractive index of nanocomposites containing poly-(ethylene oxide) and PbS particles of different diameters (open circles) and the corresponding refractive indices extrapolated for the PbS colloids (filled circles) [6].

The addition of gold nanorods into polymers, such as PVA creates a new parameter to control – alignment. The difficult task of aligning the nanorods is made simple by heating and stretching the PVA film. As the film stretches, the PVA molecules elongate and orient the nanorods in the same direction [9, 11], as demonstrated in Figure 12.6.

Figure 12.7 shows the end result of aligning the gold nanorods: a polarizing filter has been produced [9].

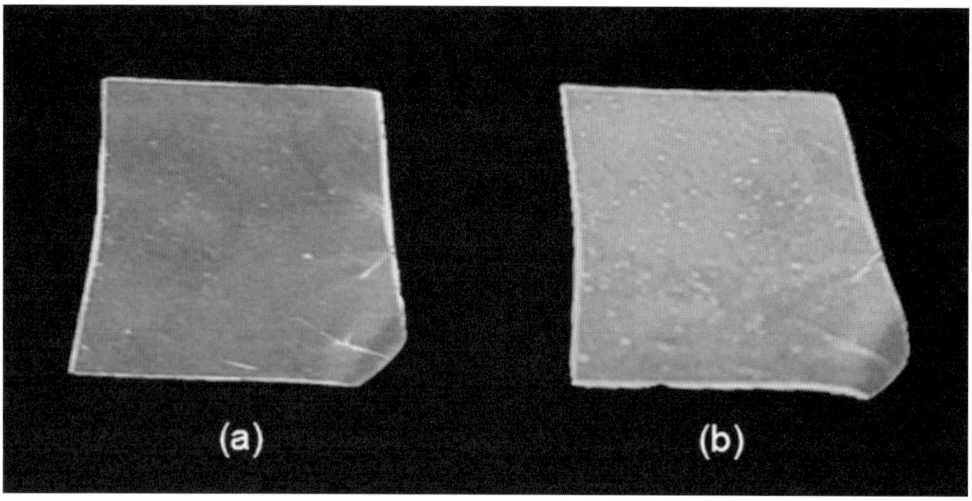

Figure 12.4. Typical image of colorless (a) and red color (b) transition of polystyrene-embedded gold quantum dot composites under UV-light source (λ = 365 nm) excitation [8].

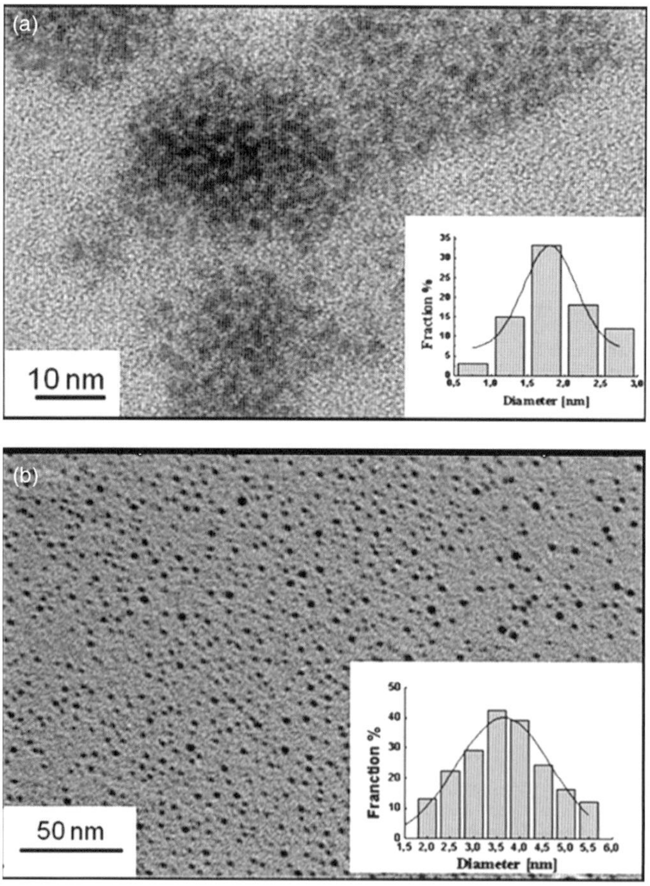

Figure 12.5. Representative TEM micrographs of gold nanoparticles into polystyrene matrix obtained by gold mercaptide-polystyrene blend thermolysis at (a) 200°C for 3 minutes and (b) 180°C for 5 minutes, with the corresponding size histograms [8].

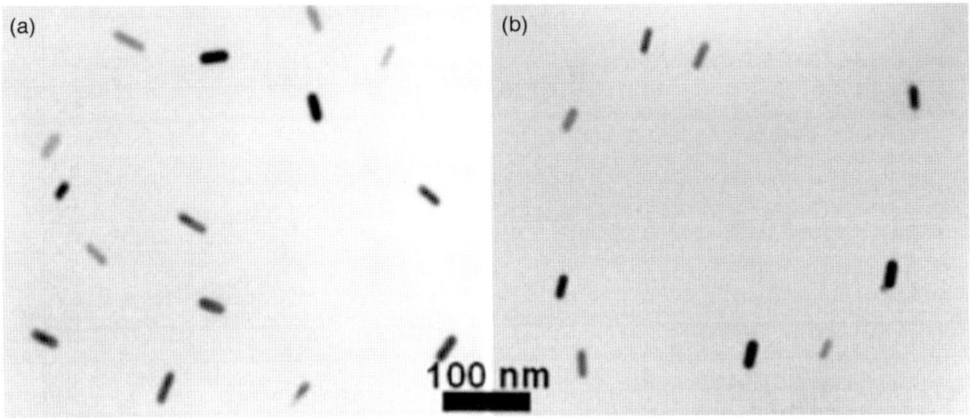

Figure 12.6. TEM images of PVA films containing gold nanorods (a) before and (b) after stretching [9].

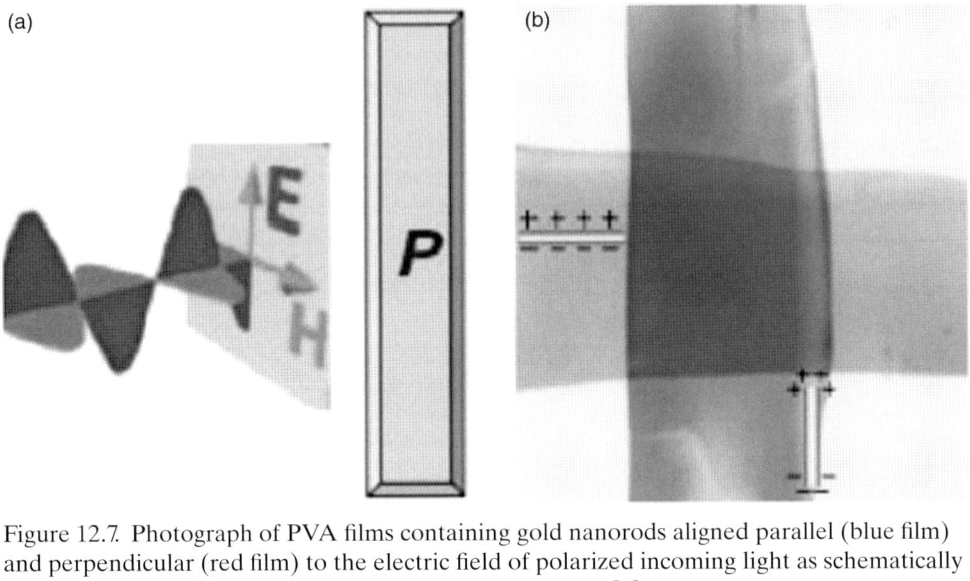

Figure 12.7. Photograph of PVA films containing gold nanorods aligned parallel (blue film) and perpendicular (red film) to the electric field of polarized incoming light as schematically shown to the left. The bar labeled P represents a polarizer [9].

As can be seen in Figure 12.8, increasing the aspect ratio of the gold nanorods from 2.23 to 2.94 shifted the wavelengths of absorbed light higher, as well as widening the bandwidth absorbed [9]. The aspect ratio can be controlled either in situ, which is difficult, or beforehand by pre-synthesizing particles, which makes dispersion difficult [9, 11]. Stretching the material slightly decreased the absorbed bandwidth, but more interestingly, it eliminated the first peak for both aspect ratios. This elimination is due to the realignment of nanorods that were previously perpendicular or near-perpendicular when they were randomly oriented [9]. Their contribution can be seen in the dotted lines.

Barber et al. made an interesting discovery that drives home the importance of particle morphology while they were attempting to amplify laser light with

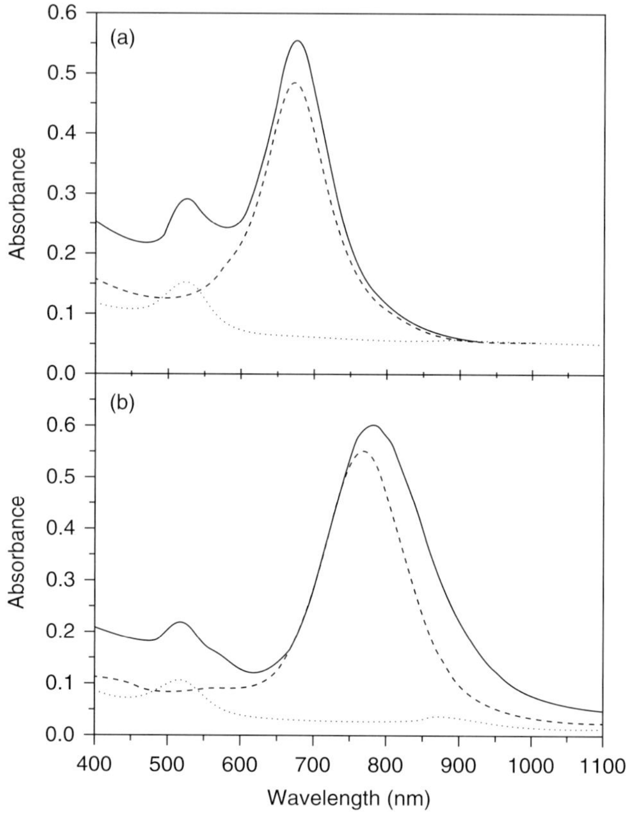

Figure 12.8. UV-vis-NIR spectra of PVA films containing gold nanorods: non-stretched (unbroken lines), stretched with parallel (dashed lines), and perpendicular polarization (dotted lines), for aspect ratios of (a) 2.23 and (b) 2.94 [9].

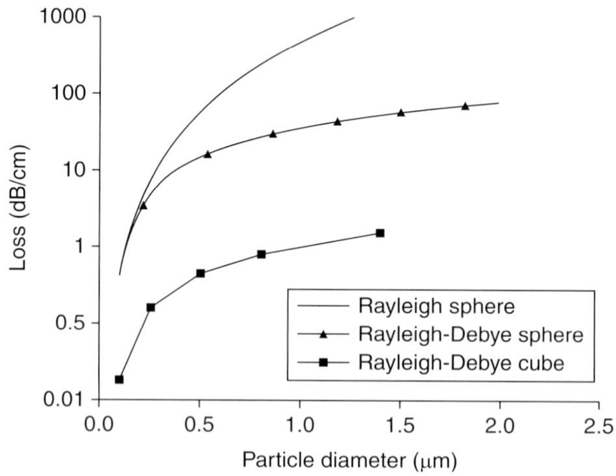

Figure 12.9. Rayleigh and Rayleigh-Debye scattering for 10% particle loading and Δn = 0.015 [12] (Courtesy of Optical Society).

Cr: forsterite and laser light with Cr: diopside crystals in polymer [12]. Although only minor amplification was achieved, they found that particle morphology strongly affected loss due to scattering. Figure 12.9 shows the Rayleigh-Debye cube shape greatly reduced losses due to scattering [12]. This cube shape might be useful in creating new polymer nanocomposites for optical use if other cubic nanoparticles can be generated.

12.3 Optical Properties of Thermoplastic Elastomer-Based Nanocomposites

Thermoplastic elastomers have a symbiotic relationship with the development of nanocomposites. On one hand, the addition of various nanoparticles can enhance (or have a neutral effect on) the optical properties of thermoplastic elastomers. On the other hand, the elastic properties of thermoplastic elastomers have been exploited to experimentally determine and improve nanoparticle selection for polymer nanocomposites.

As discussed earlier, polymer nanocomposites that contain MMT nanoclay generally have poor transmissivity, as the clay platelets contribute to scattering. However, Wang et al. reported that high transparency could be obtained in polyurethane loaded with up to 10 wt% MMT nanoclay [13]. In this instance, opacity was likely avoided because the refractive indices of polyurethane and the MMT clay matched well. The ability to achieve clarity in nanocomposites containing clay is very desirable, since clay is an extremely inexpensive additive and also enhances mechanical properties. Of further benefit is the fact that relatively high material strength was obtained at short cure times [13]. According to the authors, the main application of such a material would be in packing materials or sealants, where strength, elasticity, clarity, and low cost are all desired [13].

Choudhury et al. have demonstrated that solvent choice is also important when it comes to achieving clarity in thermoplastic elastomer-based clay nanocomposites. To test compatibility and look for synergistic effects between various solvent combinations, Choudhury et al. dissolved hydrogenated nitrile butadiene rubber (NHBR) and sepiolite clay in different solvents [14]. In total, five different solvents were used for the rubber and six different solvents were used for the clay. The goal was to find the solvent-solvent combination that gave the best dispersion and exfoliation of clay within the rubber, which in turn resulted in the highest material transparency. Table 12.2 shows that the Ch/MEK solvent combination achieved the best overall results, and that transmittance ranged from 78% for the best solvent combination down to 22% for the worst.

The solvent system with the highest transmittance percentage also possessed the most desirable mechanical properties, as can be seen in Tables 12.3 and 12.4. This relationship reaffirms the idea that transmittance percentage increases as clay dispersion and exfoliation increase [14] (this relationship has already been established for mechanical properties).

Gold and silver nanoparticles can also be combined with thermoplastic elastomers to form nanocomposites. Giesfeldt et al. deposited gold nanoparticles onto polydimethylsiloxane (PDMS) using physical vapor deposition (PVD). The use of PVD to impregnate polymers with gold or silver is common, as it has been shown to reliably disperse noble metals into polymers [15]. As with thermoplastic nanocomposites, the optical properties of thermoplastic elastomer nanocomposites containing gold or silver nanoparticles depend on the metal used as well as the morphology, orientation, and weight percentage of said particles [15]. Sometimes

Table 12.2. *Transparency of the cast film of HNBR-sepiolite nanocomposite for various solvent combinations*

SI no.	Solvent system	$\chi_{AB} - \chi_{CD}$	Transmittance (%) from UV
1	Ch/MEK	0.00	78
2	THF/THF	0.01	62
3	Ch/Ch	0.02	55
4	MEK/MEK	0.02	51
5	IAAc/IAAc	0.07	48
6	Ac/Ac	0.12	33
7	THF/Et	7.01	30
8	MEK/Et	7.11	28
9	Ch/Et	7.15	22

Table 12.3. *Mechanical properties of HNBR-sepiolite nanocomposite for various solvent combinations*

SI no.	System	$\chi_{AB} - \chi_{CD}$	Modulus at 100% elongation (MPa)	Tensile strength (MPa)	EAB[a] (%)
1	Neat rubber	–	0.50	2.00	2400
2	Ch/MEK	0.00	1.50	5.90	1600
3	THF/THF	0.01	1.20	4.80	1800
4	Ch/Ch	0.02	1.00	4.50	1700
5	MEK/MEK	0.02	1.00	4.40	1600
6	IAAc/IAAc	0.07	0.85	3.80	1800
7	Ac/Ac	0.12	0.80	3.30	1800
8	THF/Et	7.01	0.70	2.80	2100
9	MEK/Et	7.11	0.65	2.30	1900
10	Ch/Et	7.15	0.63	2.10	2000

[a] EAB = elongation at break.

Table 12.4. *Thermal properties of HNBR-sepiolite nanocomposite for various solvent combinations of oxygen atmosphere at 20°C/min*

SI no.	System	$\chi_{AB} - \chi_{CD}$	T_i (°C)	T_{max1} (°C)	T_{max2} (°C)	T_{max3} (°C)
1	Neat rubber	–	400	404	421	437
2	Ch/MEK	0	446	–	486	–
3	THF/THF	0.01	436	–	478	–
4	Ch/Ch	0.02	433	–	476	489
5	MEK/MEK	0.02	430	–	475	481
6	IAAc/IAAc	0.07	427	–	462	–
7	Ac/Ac	0.12	426	–	455	–
8	THF/Et	7.01	425	452	452	479
9	MEK/Et	7.11	423	–	451	470
10	Ch/Et	7.15	421	–	451	473

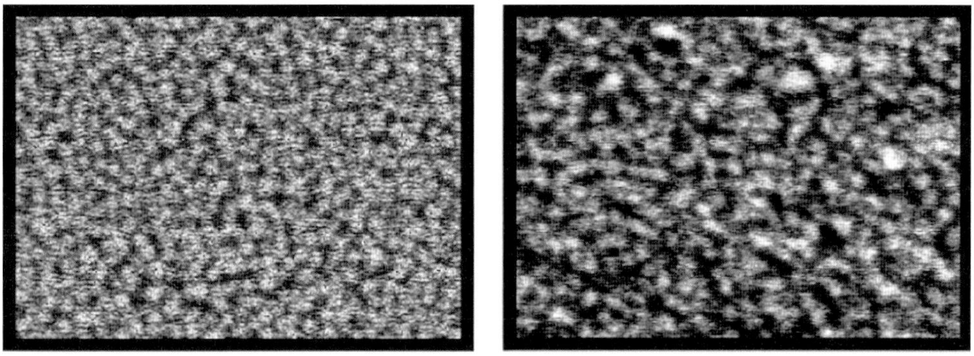

Figure 12.10. SEM micrographs of Ag-PDMS surface at 90,000-fold magnification for deposition rates of 1.0 Å/s (left) and 0.2 Å/s (right) [16] (Courtesy of Optical Society).

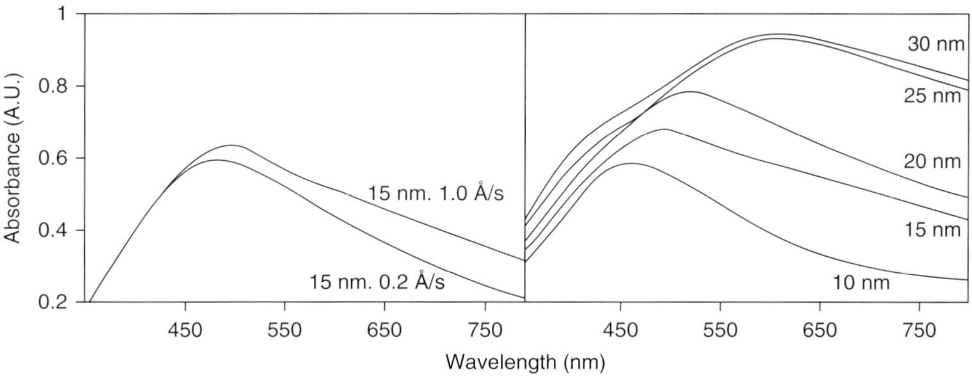

Figure 12.11. Optical extinction curves for 15 nm average Ag thickness Ag-PDMS substrates at the two different deposition rates (left). Optical extinction curves for Ag-PDMS obtained at a deposition rate of 1.0 Å/s for several different average thicknesses of Ag (right) [16] (Courtesy of Optical Society).

other factors besides optical properties play a role in the metal selected, as gold is often used instead of silver in biological applications where silver could trigger cell death. PDMS is used because as it is diffused with gold or silver, the metals form three-dimensional structures that absorb electromagnetic radiation and then scatter it, resulting in the resonance of surface electrons [15]. This behavior is the basis for surface-enhanced Raman spectroscopy (SERS) as well as a variety of biosensors, and can be controlled by variations in the gold deposition rate and thickness.

In a different set of experiments, Giesfeldt et al. used PVD to impregnate the PDMS with silver nanoparticles [16]. Although less biologically compatible, silver generally produces better results than gold for SERS. As can be seen in Figure 12.10, a slower deposition rate led to less uniform distribution of silver particles, as well as particle clumping [16]. The low deposition rate also increased particle size and decreased diffusivity.

The effects of deposition rate and total deposition thickness can be seen in Figure 12.11 [16]. The faster deposition rate of 1.0 Å/s had a higher maximum

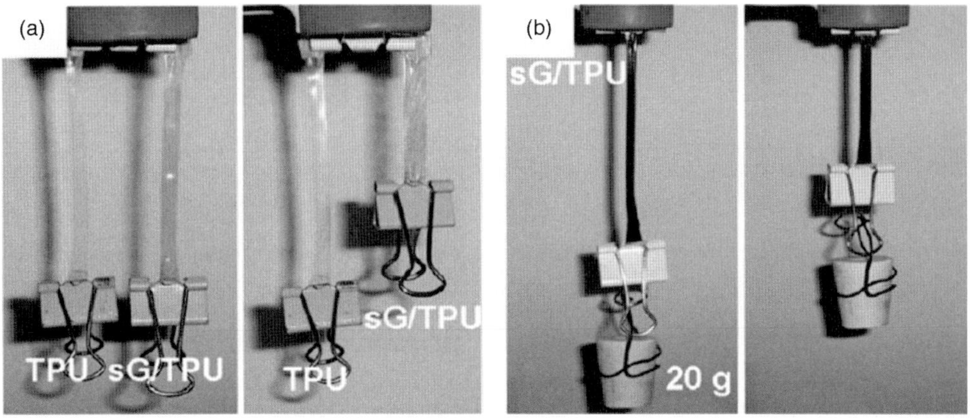

Figure 12.12. Optical images of IR actuation for graphene-based nanocomposites. (a) Comparison of shape recovery before (left) and after (right) remotely actuated by infrared light. Pure TPU did not recover; in contrast, 0.1 wt% sulfonated-graphene/TPU nanocomposites (sG/TPU) contracted after actuated by infrared irradiation. (b) The 1 wt% sulfonated-graphene/TPU film contracted and lifted a 21.6 g weight 3.1 cm with 0.211 N of force on exposure to infrared light and demonstrated an estimated energy density of 0.33 J/g [17] (Courtesy of ACS).

absorbance than the slower rate of 0.2 Å/s. The right half of Figure 12.11 shows an increase in peak absorbance and an increase in the wavelength of the peak as the average thickness of silver increases.

The effects of silver particle spacing in PDMS were also studied. In theory, PDMS is an excellent material for this as it allows for stretching and unstretching to change particle spacing. In practice, however, Giesfeldt et al. found it difficult to image the same set of particles as they moved when stretching the material. Furthermore, the number of particles in view as well as their alignment changed as the material was stretched, but did not necessarily return to their old orientation when unstretched, causing hysteresis and changes in reflectivity and absorption [16].

Thermoplastic elastomer-based nanocomposites have also been used to form other functional devices. Liang et al. showed that adding 1 wt% graphene sheets to thermoplastic polyurethane (TPU) allowed the material to serve as an actuator, as well as improving mechanical properties of the material [17]. Exposure to infrared light caused the material to contract, giving it the ability to lift significant weight. TPU can be used in this way without added nanographene, but addition of sulfonated graphene sheets increased the number of cycles and the recovery rate [17]. Figure 12.12 illustrates the lifting potential of this particular nanocomposite [17].

12.4 Optical Properties of Thermoset-Based Nanocomposites

Scattering occurs in thermoset-based nanocomposites much the same as in the materials discussed earlier. Figure 12.13 demonstrates that different types of clay result in visibly different levels of opacity from the neat material and from one another [18].

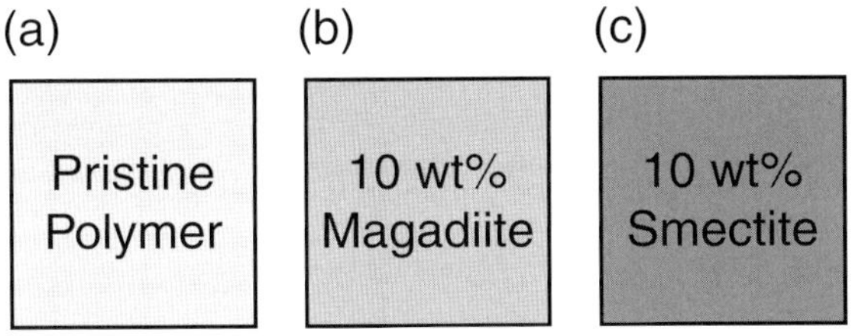

Figure 12.13. Comparison of the optical properties among (a) a pristine epoxy polymer, (b) an epoxy-exfoliated magadiite nanocomposite prepared from C18-magdaiite-PF, and (c) an epoxy-exfoliated smectite clay nanocomposite prepared from $CH_3(CH_2)(CH_2)_{17}NH_3^+$ ion-exchanged montomorillonite. The thickness of each sample is ~1 mm [18] (Courtesy of ACS).

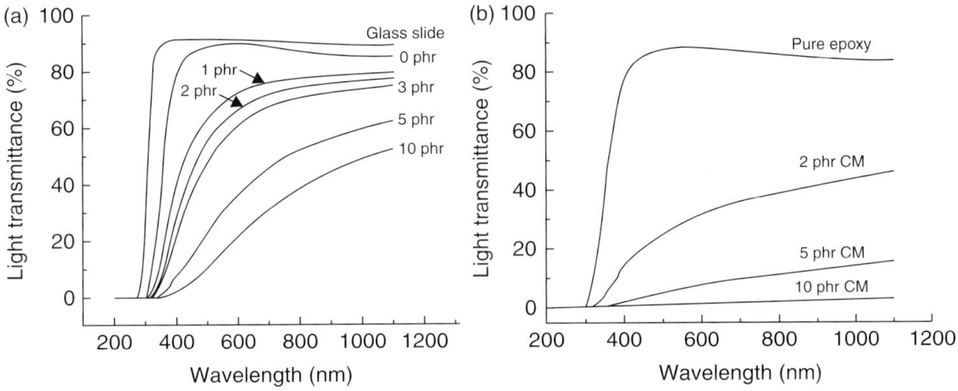

Figure 12.14. Light transmission spectra of pure epoxy, Ep/C18 (a) and Ep/CM (b) nanocomposites with different AMT contents [19].

Since the magadiite clay has a refractive index closer to that of the polymer, it appears clearer than the smectite, and not much darker than the pristine polymer as reported by Wang et al. [18]. Dispersion and exfoliation also have a positive correlation on material clarity and could have been partially responsible for the difference in opacity.

Deng et al. investigated the effects of MMT morphology on epoxy/MMT composite light transmittance [19]. Figure 12.14 shows that the Ep/C18 MMT variety had higher transmittance across the range of clay concentrations [19]. SEM and TEM images support the idea that the Ep/C18 variety of clay has higher light transmittance because it was the most evenly dispersed and was more exfoliated than the Ep/CM nanocomposite [19]. As expected, higher clay loading reduced transmittance in both cases. Interestingly, less curing led to more AMT exfoliation and transparency, which indicates that it is important to monitor and control the cure state of thermoset based nanocomposites.

So far, the literature has focused on designing materials to create the best optical properties. Conversely, optical properties of thermoset-based polymer nanocomposites can be monitored to reveal epoxy cure state. Monitoring cure state is challenging depending on the nanocomposite, since a small weight percentage of any

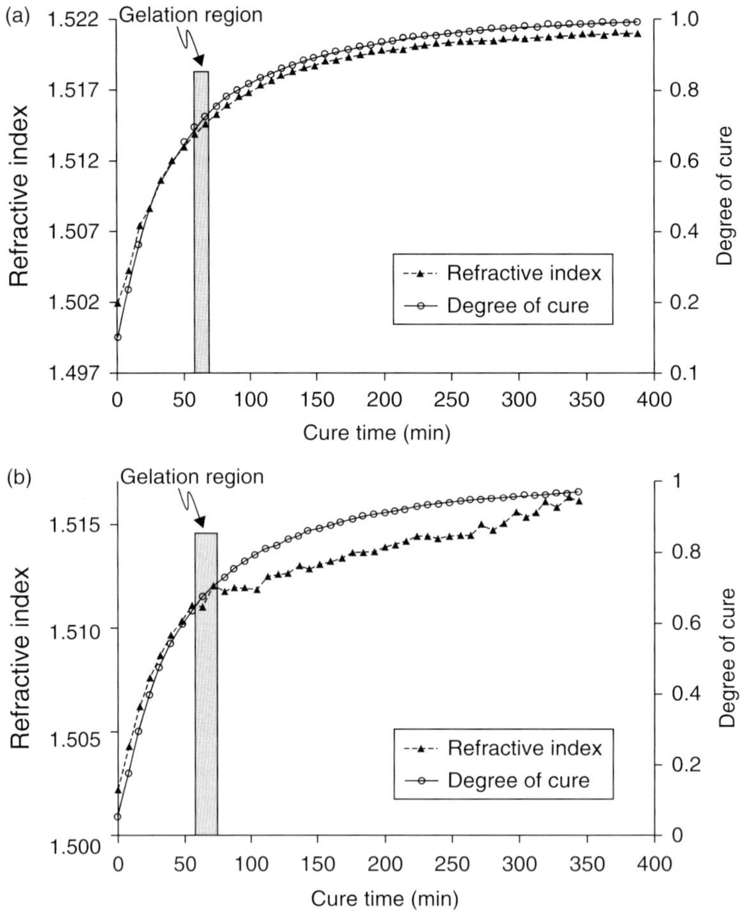

Figure 12.15. Comparison of the refractive index and degree of cure for the isothermal cure of (a) the MY750/HY5922 system and (b) the MY750/HY5922 system containing 0.06 wt% Hyperion CNTs at 80°C for 6 hours [20].

carbon-based material will make the material opaque. Dimopolous et al. tracked the refractive index of epoxies containing CNTs using an optical-fiber Fresnel refractometer [20]. This technique employed an embedded fiber, which allowed to obtain accurate readings despite the presence of CNTs. Figure 12.15 shows a close match between the measured refractive index and cure, making this technique an option when determination of cure state is important [20].

12.5 Summary

Many polymer nanocomposites achieve impressive material properties not possible with traditional fillers through the addition of small weight percentages of nanoparticles [21]. These nano-sized particles have a large surface area to interact with the surrounding material. Unfortunately, when it comes to optical properties that often means the polymer nanocomposites suffer from severe scattering, which renders them opaque and of limited use in optical applications. However, careful selection of nanoparticles as well as their size, morphology, and volume added to a given material has led to the development of useful optical properties in thermoplastic-, thermoplastic elastomer-, and thermoset-based nanocomposites.

Scattering problems can be somewhat mitigated by choosing nanoparticles that are below 25 nm in diameter or have a refractive index that is close to that of the polymer being used. Polymer-clay nanocomposites in all three categories were able to achieve good optical clarity through attention to refractive index, weight percentage, and the parameters necessary to achieve proper dispersion and exfoliation. Clay morphology and solvent-solvent interactions were examples of such parameters. Gold nanoparticles were also widely used, making polymer nanocomposites capable of everything from polarizing filters to biosensors. Control of gold nanoparticle morphology, orientation, and weight percentage was key to producing the desired optical properties, such as controlled refractive index or surface electron resonance.

There is a fairly major divide between applications of the polymer nanocomposites involving MMT clay and those involving gold nanoparticles. Most of the materials containing MMT clay would be useful in packaging applications, as they are optically clear, strong, inexpensive, and have the potential to scale up to production in industry. Most of the materials utilizing gold nanoparticles can only be made into thin films and through fairly intricate and time-intensive processes. The ability to scale the production of these materials may not be a huge issue; however, as most potential applications are fairly high end and low volume.

12.6 Questions and Problems

1. How does the nanomaterial dispersion affect the optical properties of polymer nanocomposites in general? Provide examples to illustrate your claims.
2. Discuss the major challenges to enhance the optical properties of polymer nanocomposites.
3. Briefly discuss the common nanomaterials used in thermoplastic-based nanocomposites, processing techniques used, and the different optical properties that can enhance the host polymers. Besides the several nanomaterials discussed in this chapter, what other nanomaterials can be used to enhance the desired optical properties of the polymers?
4. This chapter discussed the enhancement of optical properties of thermoplastic elastomer-based nanocomposites using MMT clay, sepiolite clay, gold, silver nanoparticles, and graphene sheets. What other types of nanoparticles can be used besides the above mentioned nanomaterials can be used to enhance optical properties of thermoplastic elastomer-based nanocomposites? Provide examples. How about the mechanical properties of these thermoplastic elastomer-based nanocomposites?
5. This chapter covered the enhancement of optical properties of thermoset-based nanocomposites using Magadiite clay, MMT clay, and CNT nanomaterials. What other types of nanoparticles besides the above mentioned nanomaterials can

be used to enhance optical properties of thermoset-based nanocomposites? Provide examples. How about the mechanical properties of these thermoset-based nanocomposites?

6. Is it feasible to use two or more nanomaterials in polymer nanocomposites to enhance optical properties? What will be a good strategy to optimize the optical properties of these polymer nanocomposites?

REFERENCES

1. J. Z. Zhang (2009). *Optical Properties and Spectroscopy of Nanomaterials*. Singapore: World Scientific Publishing.
2. M. Tamborra, M. Striccoli, R. Comparelli, M. L. Curri, A. Petrella, and A. Agostiano (2004). Optical properties of hybrid composites based on highly luminescent CdS nanocrystals in polymer. *Nanotechnology* 15, S240–S244.
3. L. L. Beecroft and C. K. Ober (1997). Nanocomposite materials for optical applications. *Chemistry of Materials* 9, 1302–1317.
4. S. Srivasta, M. Haridas, and J. K. Basu (2008). Optical properties of polymer nanocomposites. *Bulletin of Material Science* 31, 213–217.
5. J. K. Pandey, K. R. Reddy, A. P. Kumar, and R. P. Singh (2005). An overview on the degradability of polymer nanocomposites. *Polymer Degradation and Stability* 88, 234–250.
6. W. Caseri (2000). Nanocomposites of polymers and metals or semiconductors: Historical background and optical properties. *Macromolecular Rapid Communications* 21, 705–722.
7. C. Becker, P. Mueller, and H. Schmidt (1998). Optical and thermomechanical investigations on thermoplastic nanocomposites with surface modified silica nanoparticles. *Proceedings of SPIE* 3469, 88–98.
8. G. Carotenuto, A. Longo, and C. L. Hison (2009). Tuned linear optical properties of gold-polymer nanocomposites. *Journal of Materials Chemistry* 19, 5744–5750.
9. J. Pérez-Juste, B. Rodríguez-Gonzáles, P. Mulvaney, and L. M. Liz-Marzán (2005). Optical control and patterning of gold-nanorod-poly(vinyl alcohol) nanocomposite films. *Advanced Functional Materials* 15, 1065–1071.
10. M. R. Bockstaller and E. L. Thomas (2003). Optical properties of polymer-based photonic nanocomposite materials. *Journal of Physics and Chemistry* 107, 10017–10024.
11. J. Pérez-Juste, I. Pastoriza-Santos, L. M. Liz-Marzán, and P. Mulvaney (2005). Gold nanorods: Synthesis, characterization and applications. *Coordination Chemistry Reviews* 249, 1870–1901.
12. D. B. Barber, C. R. Pollock, L. L. Beecroft, and C. K. Ober (1997). Amplification by optical composites. *Optics Letters* 22, 1247–1249.
13. Z. Wang and T. J. Pinnavaia (1998). Nanolayer reinforcement of elastomeric polyurethane. *Chemistry of Materials* 10, 3769–3771.
14. A. Choudhury, A. K. Bhowmick, and C. Ong (2009). Novel role of polymer-solvent and clay-solvent interaction parameters on the thermal, mechanical and optical properties of nanocomposites. *Polymer* 50, 201–210.
15. K. S. Giesfeldt, R. M. Connatser, M. A. De Jesús, P. Dutta, and M. J. Sepaniak (2005). Gold-polymer nanocomposites: studies of their optical properties and their potential as SERS substrates. *Journal of Raman Spectroscopy* 36, 1134–1142.
16. K. S. Giesfeldt, R. M. Connatser, M. A. De Jesús, N. V. Lavrik, P. Dutta, and M. J. Sepaniak (2003). Studies of the optical properties of metal-pliable polymer composite materials. *Applied Spectroscopy* 57, 1346–1352.

17. J. Liang, Y. Xu, Y. Huang, L. Zhang, Y. Wang, et al. (2009). Infrared-triggered actuators from graphene-based nanocomposites. *Journal of Physics and Chemistry* 113, 9921–9927.
18. Z. Wang, T. Lan, and T. J. Pinnavaia (1996). Hybrid organic-inorganic nanocomposites formed from an epoxy polymer and a layered silicic acid (Magadiite). *Chemistry of Materials* 8, 2200–2204.
19. Y. Deng, A. Gu, and Z. Fang (2004). The effect of morphology on the optical properties of transparent epoxy/montmorillonite composites. *Polymer International* 53, 85–91.
20. A. Dimopoulos, S. J. Buggy, A. A. Skordos, S. W. James, R. P. Tatam, and I. K. Partridge (2009). Monitoring cure in epoxies containing carbon nanotubes with an optical-fiber Fresnel refractometer. *Journal of Applied Polymer Science* 113, 730–735.
21. K. I. Winey and R. A. Vaia (Eds.) (2007). Polymer nanocomposites. *MRS Bulletin* 32, 314–322.

13 Other Properties of Polymer Nanocomposites

13.1 Introduction

This chapter discusses several properties of the polymer nanocomposites that were not mentioned in the previous chapters. This chapter covers the (a) tribological properties of thermoplastic polymer materials containing several different nanoparticles including nanoclay, single and multi-walled carbon nanotubes, and graphene platelets; (b) permeability properties of polymer nanocomposites; and (c) oil field usage of nanotechnology. Tribological (abrasion, wear, and scratch resistance) and permeability characterization techniques are included in this chapter since these are more specialized techniques.

13.2 Tribological Properties of Polymer Nanocomposites

13.2.1 Abrasion, Wear, and Scratch Resistance Characterization Techniques

13.2.1.1 Taber Abraser

The abrasion resistance test is detailed in the ASTM D4060 standard [1]. This test is performed using a Taber Abraser [1, 2]. Two abrading wheels and the specimen rotate simultaneously to produce abrasion paths at a variety of angles. Figure 13.1 shows a typical Taber Abraser [2]. Loads of 1,000 g are set for each abrasion wheel, which is composed of rubber and aluminum oxide abrasive particles. The specimen is then subjected to a specified number of cycles. For tests covered in this section, the abrasion resistance is quantified by the weight loss during the process [1–4]. A lower weight loss indicates more wear/abrasion resistance.

13.2.1.2 Pin-on-Disc Tribometer

Several variations on the pin-on-disc tribometer were used in some of the following experiments to measure abrasion resistance. The tribometer consists of a stationary pin under a load in contact with a rotating disc similar to the diagram in Figure 13.2 [5]. The pin is generally fabricated from the polymer to be studied, and an abrasive

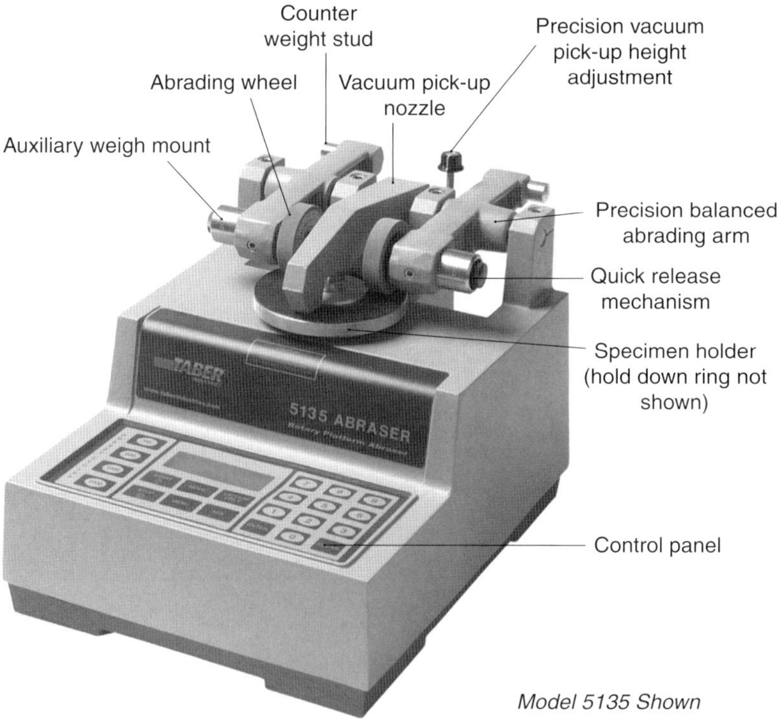

Figure 13.1. Taber Abraser [2].

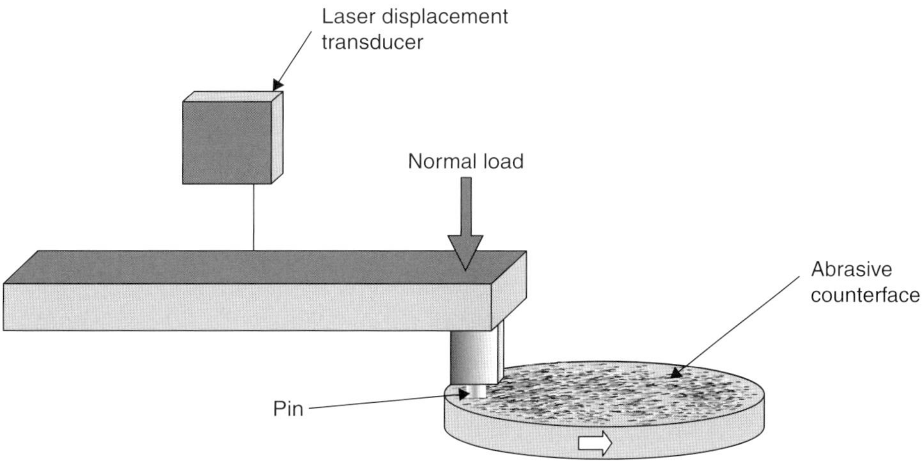

Figure 13.2. Pin-on-Disc Tribometer [5].

paper will be mounted to the disc. The frictional force and the change in dimensions of the pin can be measured to quantify the abrasion resistance of the material [5, 6].

13.2.1.3 Scratch Test

"The Standard Test Method for Evaluation of Scratch Resistance of Polymeric Coatings and Plastics Using an Instrumented Scratch Machine" as described by the

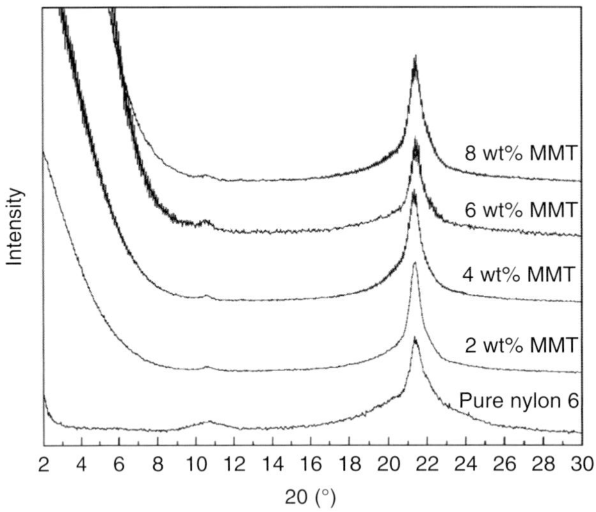

Figure 13.3. WAXS patterns of pure nylon 6 and nylon 6/nanoclay nanocomposites [3].

ASTM D7027-05 standard is used to quantify the scratch resistance of polymers [7]. Variations of this test method are common, but in general, the method involves using a conical indenter (usually diamond) under a load to scratch a specified length into the polymer. One scratch, or pass, of the indenter is used to quantify scratch resistance, but multiple passes can be used to quantify wear resistance. The scratch width and depth, frictional force, and normal force during the test are all parameters that can be analyzed and used to find the scratch hardness [7].

13.2.2 Wear and Abrasion Resistance of Polymer-Clay Nanocomposites

Nylon 6 is a polymer utilized in a wide array of industries. Nylon 6/nanoclay nanocomposites have been used since the Toyota Research Group first utilized them in the early 1990s. The thermal, mechanical, and permeability properties were much improved by the addition of nanoclay, but the abrasion resistance suffered [3]. Since abrasion resistance is an important property of nylon 6, understanding the abrasion resistance properties of nylon 6/nanoclay nanocomposites is important.

The abrasion resistance of nylon 6/nanoclay nanocomposites was studied by Zhou using the ASTM D4060 standard abrasion test [3]. Before the test, wide-angle X-ray (WAXS) scattering was performed on the nanocomposite to determine the structure. Nylon 6 is polymorphic, and two crystalline modifications can exist at room temperature, the α phase and γ phase. The WAXS data is shown in Figure 13.3. Diffraction peaks are shown at $2\theta=11°$ and $21.5°$, which correspond to the diffraction peaks characteristic of the γ crystal. No diffraction peaks corresponding to the α phase exist, indicating that the dominant crystalline structure for the neat nylon 6 as well as the nylon 6/nanoclay nanocomposites is the γ phase. Furthermore, diffraction peaks indicating the presence of nanoclay are generally seen at $2\theta < 10°$, so the absence of peaks in this region indicate that the nanoclay is fully exfoliated. The

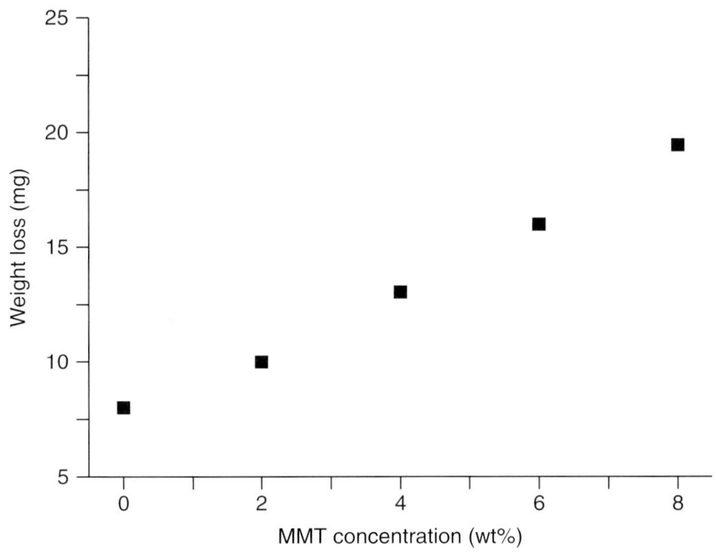

Figure 13.4. Abrasion weight loss of nylon 6/nanoclay nanocomposites versus nanoclay concentrations up to 8% [3].

abrasion results are shown in Figure 13.4. The weight loss increases almost linearly with the percentage of nanoclay ranging from 8.03 mg for the neat nylon 6 to 19.4 mg for the 8 wt% nanoclay nanocomposite [3]. The abrasion resistance decreases with increasing clay percentage.

The scanning electron microscopy (SEM) results shown in Figure 13.5 reveal the abrasion mechanisms for the neat nylon 6 and the 4 wt% and 8 wt% nanoclay after the abrasion test. Grooves seen specifically in the 8 wt% nanoclay, but also in the 4 wt% nanoclay, suggest that the cutting function is the main abrasion mechanism at work. The fractured particles evident in the neat nylon 6 and some in the 4 wt% nanoclay are due to the abrasive tractive stress. This pattern shows that as the nanoclay content increases, less fractured particles and more grooves will be seen on the abraded surface [3].

After the abrasion test, FTIR-ATR measurements were performed to understand the crystalline transformation of the nylon 6 during abrasion. The results showed that after abrasion, on the abraded surface of the neat nylon 6 as well as the nylon 6/nanoclay nanocomposites, there was an increase in α crystals and a decrease in γ crystals. The magnitude of the increase in α crystals was also highest for the neat nylon 6 and lowest for the 8 wt% nanoclay. Due to the tensile forces during abrasion causing stretching of the crystal chains, the γ crystals were transformed into α crystals. From this data it can be concluded that nanoclay additives decrease the ductility of nylon 6 and cause more cutting than tensile tearing [3].

X-ray photoelectron spectroscopy (XPS) was also used to characterize the composition of the sample surface before and after abrasion [3]. Before abrasion, the ratio of silicon to nitrogen atoms is relatively the same for every nanoclay concentration. After abrasion, the ratio increases overall and also increases with the increasing

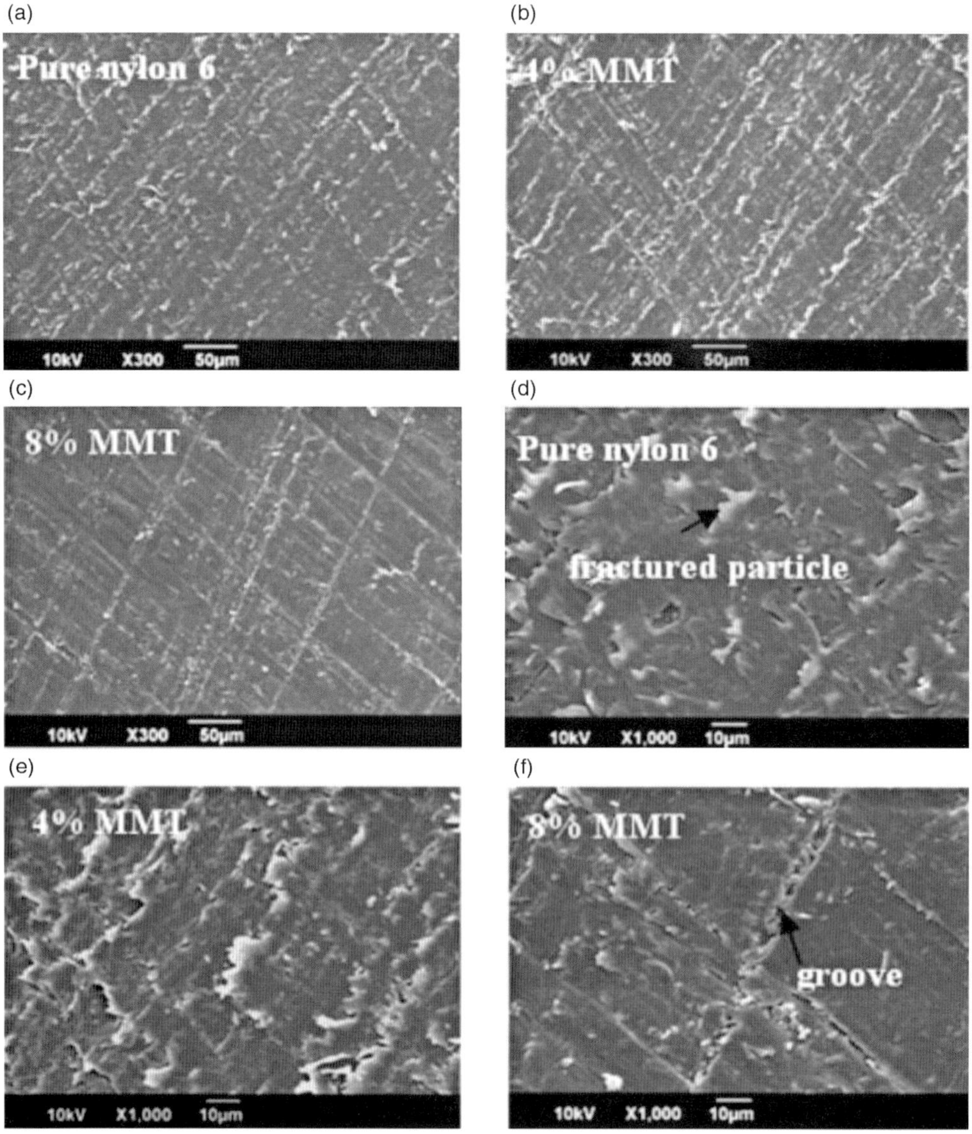

Figure 13.5. SEM micrographs of the abrasion surfaces of nylon 6 and the nylon 6/nanoclay nanocomposites [3].

nanoclay concentration. This leads to the conclusion that the presence of nanoclay particles makes it easier for the polymer to be removed from the surface. Nanoclay would increase the density of surface defects, making the nylon 6/nanoclay nanocomposite less abrasion resistant [3].

Liu et al. also performed abrasion tests on nylon 6/nanoclay nanocomposites using the ASTM D4060 standard, but their results were different [4]. The WAXD method of characterization in Figure 13.6 as well as the TEM images in Figure 13.7 showed the dispersion to be exfoliated for nanoclay weight percentage of up to 3%, but the 5 wt% nanoclay is somewhat intercalated or unmixed [4].

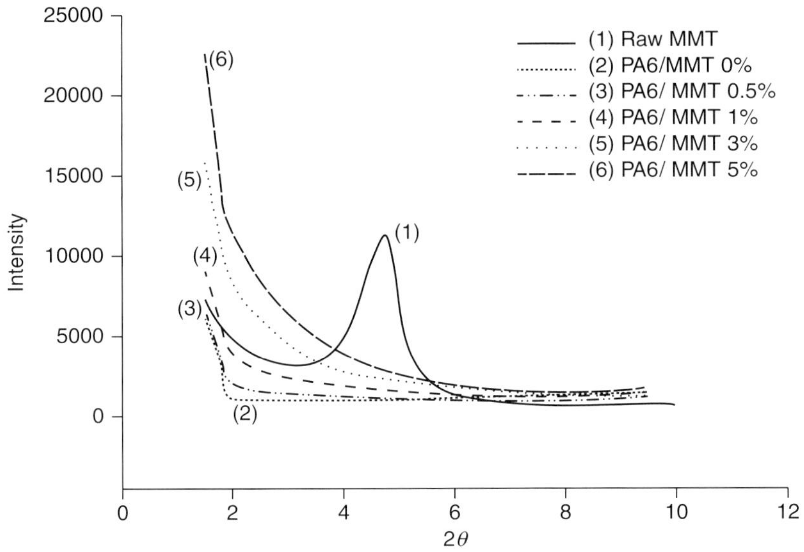

Figure 13.6. WAXD data for raw nanoclay, neat nylon 6, and nylon 6/nanoclay nanocomposites [4].

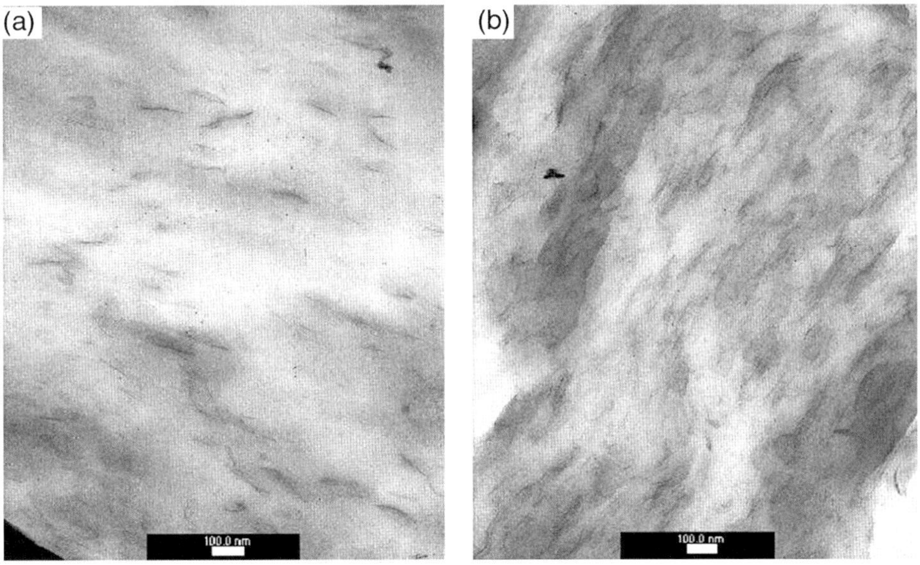

Figure 13.7. TEM images of (a) 3 wt% nanoclay and (b) 5 wt% nanoclay [4].

In Liu et al.'s wear resistance test, the wear mass loss decreased with increasing clay content up to 3 wt% nanoclay, and the change was more significant for the higher wearing speed rate [4]. On the other hand, the wear mass loss increased for 5 wt% nanoclay for both wearing speed rates. This increase was probably due to the aggregated (intercalated or unmixed) dispersion at the higher clay concentration. The SEM images show a rough abrasion surface for the neat nylon 6 as compared to the nylon 6/nanoclay nanocomposites (except for the 5 wt% nanoclay, which is

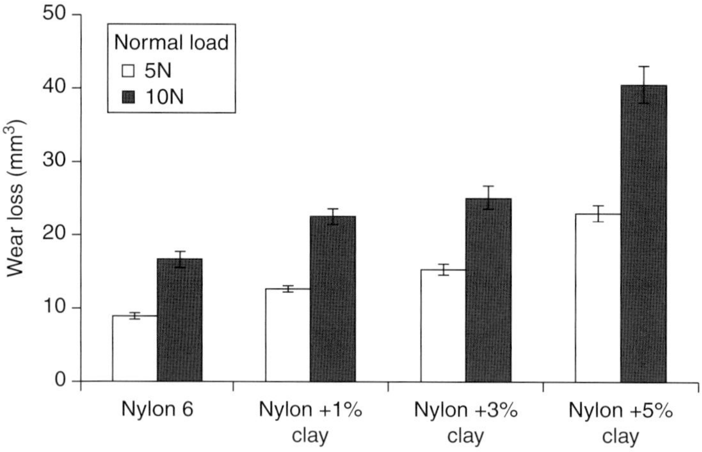

Figure 13.8. The effect of nanoclay concentration on wear loss for loads of 5 N and 10 N (grit #180, velocity 0.2 m/s) [5].

relatively rough). This would indicate that the inclusion of exfoliated nanoclay in nylon 6 enhances the wear resistance of nylon 6, but once the nanoclay concentration gets above a certain point, the wear resistance is lessened. The exact cause of this optimum clay content is not definitively known, but it is hypothesized that the phenomenon results from the competitive effects of transfer film development and the formulation of abrasive aggregates. Adding nanoclay to neat nylon 6 leads to increased hardness, so the corresponding increase in surface wear resistance could also be attributed to this increased hardness [4].

Srinath and Gnanamorthy conducted abrasive wear tests on nylon 6/nanoclay composites using the pin-on-disc test method [5]. The specimen pins were injection-molded, and various grades of abrasive paper were mounted on the steel disc. The friction force and dimensional change of the pin were continuously measured during the test. The effect of grit size, load, and sliding velocity on the wear resistance of the nylon 6/nanoclay composites was documented and can be reviewed in [5]. For this section, only the effect of nanoclay concentration on wear resistance is reviewed. WAXD was carried out on the specimens prior to testing and showed good exfoliation of the nanoclay in the nylon 6 polymer matrix. The wear loss versus nanoclay concentration ranging from neat nylon 6 to 5 wt% nanoclay was shown [5]. The neat nylon 6 showed less wear loss than any of the nylon 6/nanoclay nanocomposites, and the wear loss increases as more nanoclay is added for both the 5 N and 10 N normal loads as shown in Figure 13.8 [5].

The SEM images shown in Figure 13.9 show the abrasion surface of neat nylon 6 and 1, 3, and 5 wt% nanoclay. Neat nylon 6 is ductile, and grooves are seen resulting from the dominant ploughing mechanism. As nanoclay is added to the neat polymer, the material becomes harder and more brittle, and the cutting function dominates, leading to the grooves and fractured particles seen in Figure 13.9 (b, c, and d) [5].

In a modified pin-on-disc experiment, Mu et al. showed that the abrasion resistance of nylon 6 is heavily affected by the addition of nanoclay particles [6]. In their

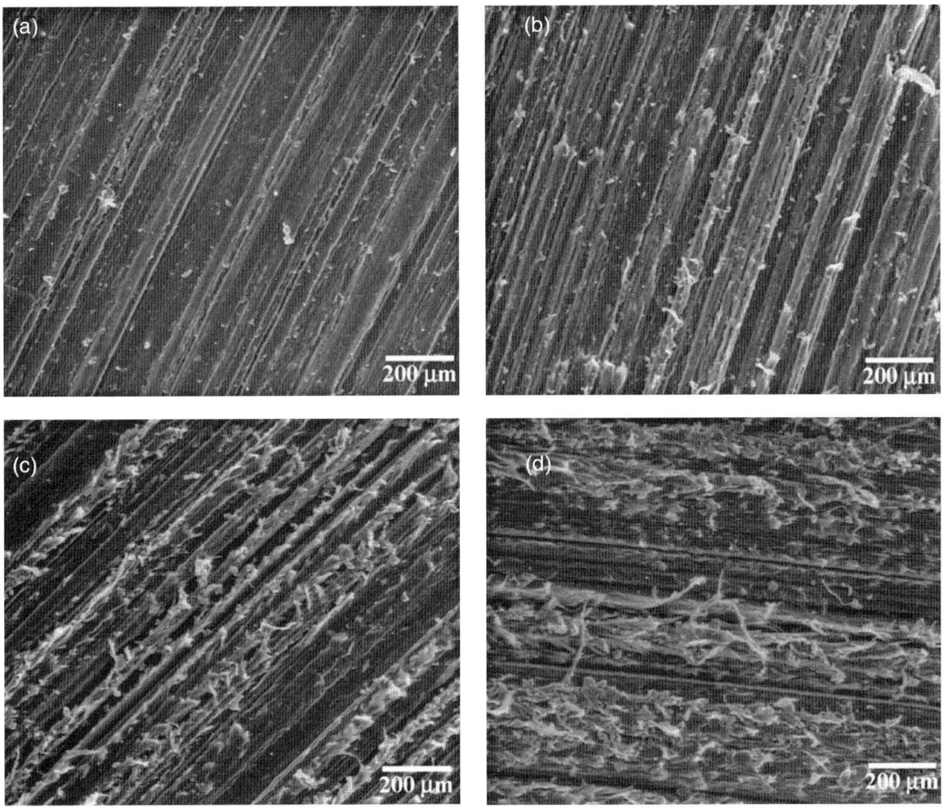

Figure 13.9. SEM of the worn pins (grit #180, 5N, .2 m/s): (a) neat nylon 6, (b) 1% nanoclay, (c) 3% nanoclay, and (d) 5% nanoclay [5].

test, the slider was a steel spherical ball, and the polymer specimens were injection-molded into small blocks in place of the disc. Before the test, WAXD was performed, indicating that the samples were a mix between exfoliated and intercalated nanoclay platelets. The TEM micrograph shown in Figure 13.10 confirms the dispersion [6].

The wear rate initially decreases only slightly from that of the neat polymer with the inclusion of nanoclay particles. Above 5 wt% nanoclay, the wear rate increases drastically and is much higher than the wear rate of the neat polymer (Figure 13.11) [6]. A similar trend is seen for the experiment run with a 10 N normal load, but the wear rate begins increasing at only 3 wt% nanoclay. The interfacial strengthening of the nanoclay in the nanocomposites acts to carry load and resist wear at lower nanoclay concentrations, but at larger concentrations, the abrasion action of the nanoclay aggregates damages the transfer film and causes additional wear on the abrasion surface [6]. Since the dispersion of the nanoclay platelets was a mix of intercalated and exfoliated particles, the trend in wear rate versus nanoclay content is likely related to the type of dispersion in each sample. Additional experiments in which the nanoclay is fully exfoliated could prove useful.

SEM images of the nanocomposites' worn surfaces and the steel ball pins showed the mechanisms of wear. The wear mechanism for neat nylon 6 is severe

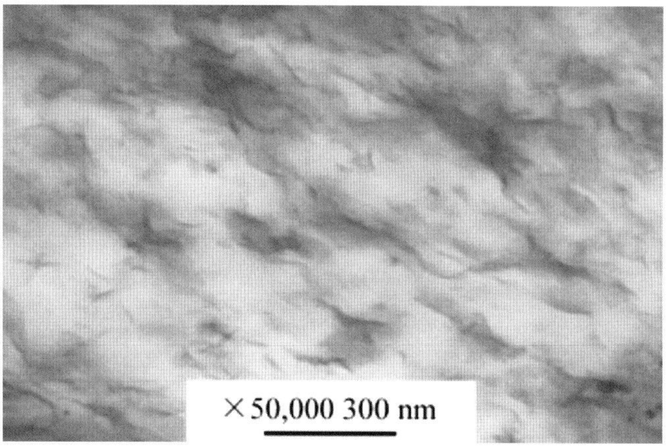

Figure 13.10. TEM micrograph of nylon 6 / 5 wt% nanoclay nanocomposite [6].

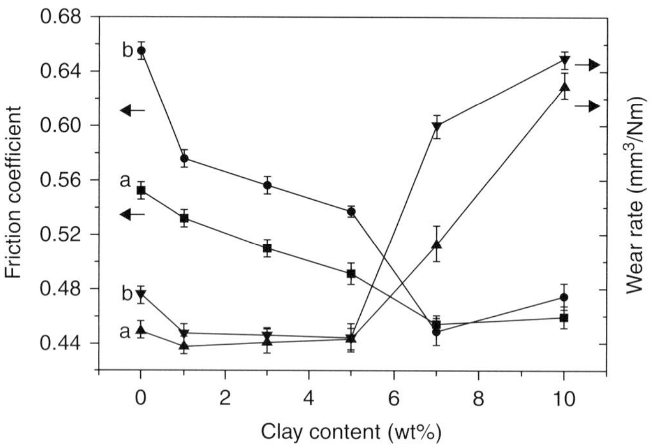

Figure 13.11. Friction coefficient and wear rate for nylon 6/nanoclay nanocomposites at different temperatures. Circles indicate friction coefficient while triangles indicate wear rate (5 N, 0.6 m/s): at (a) 25°C and (b) 60°C [6].

scuff and adhesion. With low concentrations of nanoclay, the adhesion of the nylon 6 appears lower, and at high nanoclay concentrations, fatigue crack is the main wear mechanism [6].

13.2.3 Wear and Scratch Resistance of Polymer-Carbon Nanotube Nanocomposites

Carbon nanotube-polymer nanocomposites have been studied quite extensively since 1994 when Ajayan first published his CNT research [8]. As previously mentioned, CNTs are known for their thermal, mechanical, and electrical properties, but studies focusing on their wear and scratch resistance are harder to find. Some experiments also yield contradicting evidence on whether or not they improve tribological

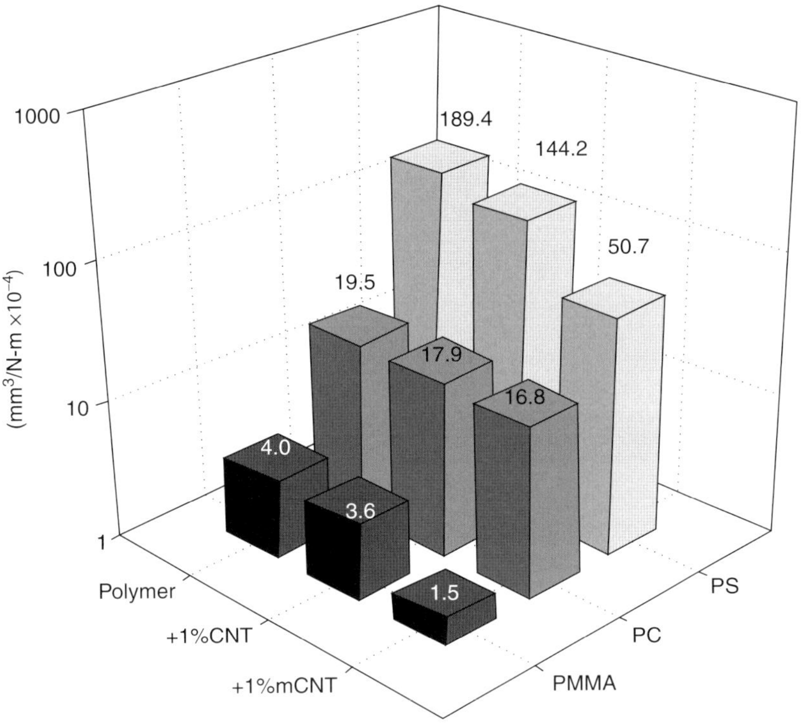

Figure 13.12. Specific wear rates for polymers and nanocomposites [8].

performance of polymers [9]. Regardless, understanding the tribological effects of CNTs in polymers is important, since CNTs are widely used to improve other properties in polymers.

Carrion et al. performed tribology experiments on multiple types of polymers to study the antiwear properties of thermoplastics with ionic liquid modified carbon nanotube additives [8]. The study tested the wear resistance of polystyrene (PS), polymethylmethacrylate (PMMA), and polycarbonate (PC). The properties of each of the neat polymers were compared to those of the polymer with SWNT additives. Some of the SWNT were added as-received from NanoAmor Inc. (referred to as CNT in this section), and some were modified with the ionic liquid, [OMIM]BF_4 (mCNT). The processing and preparation are detailed in the article by Carrion et al. [8]. Pin-on-disc tests were performed using polymer discs and stainless steel spherical pins. Raman spectroscopy was carried out to assess the degree of nanoparticle dispersion in the polymers, and the PS and PMMA show a more uniform dispersion than that of PC [8].

Figure 13.12 shows that the specific wear rates for all polymers decrease with the addition of SWNTs [8]. Furthermore, the nanocomposites containing the SWNTs treated with the ionic liquid have the lowest specific wear rates, thus confirming that the ionic liquid has an effective antiwear effect. PS has the highest specific wear rates; the addition of CNTs to the neat PS reduces the specific wear rate by 24% while the addition of mCNTs reduces the specific wear rate by 74% with respect to the neat

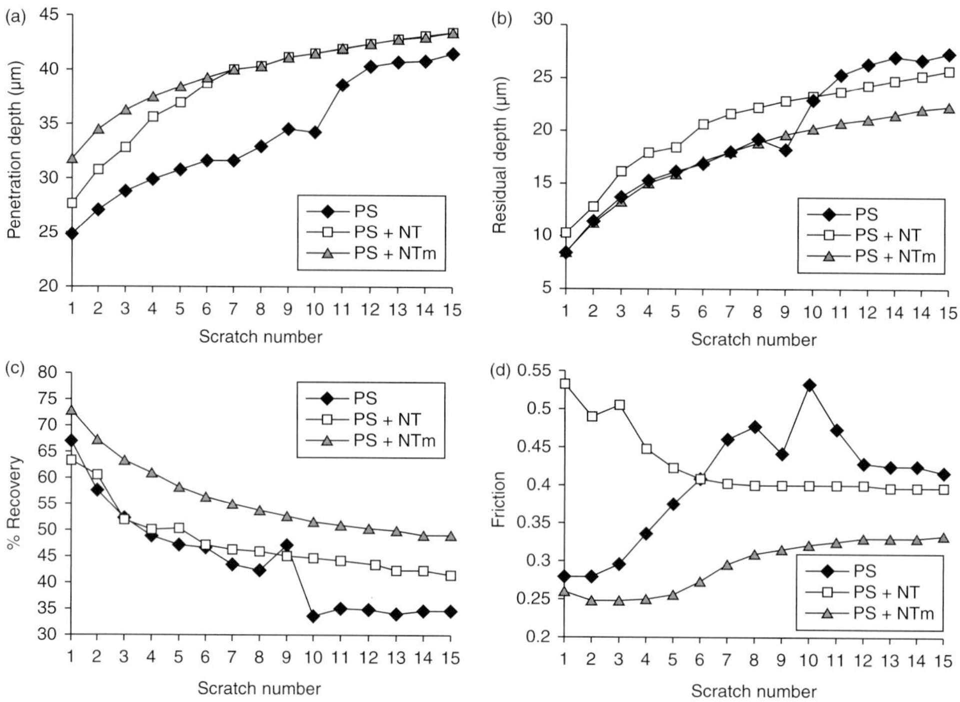

Figure 13.13. Multiple scratch behavior of compression molded discs [9].

polymer. The addition of CNTs to neat PC reduces the specific wear rate by 8%, while the addition of mCNTs reduces the specific wear rate by 14% with respect to the neat polymer. The addition of CNTs to neat PMMS reduces the specific wear rate by 35%, while the addition of mCNTs reduces the specific wear rate by 63% with respect to the neat polymer. PMMA has significantly lower specific wear rates overall than PC or PS. Overall, PS + mCNTs and PMMA + mCNTs show the maximum reduction in specific wear rates with respect to the neat polymer. The PC likely does not show as high of a reduction because of the less uniform dispersion of nanoparticles. The wear reduction is caused by the uniform dispersion of SWNTs in the polymer, which increases resistance to crack propagation and fracture, and the ionic liquid surface modification improves the lubricating ability of the nanoparticle additives [8].

The research group from above performed another set of experiments focusing only on polystyrene to determine the abrasive wear under multi-scratching of polystyrene. The effect of sliding friction and the same ionic liquid modification ([OMIM] BF_4) were studied. The previous study analyzed the effect of friction and wear under sliding conditions, while this study uses the ASTM D7027-05 scratch test method to determine the abrasive wear behavior of the nanocomposites. The specifics of sample preparation and test parameters are detailed in the article by Bermúdez et al. [9].

Figure 13.13 shows the multiple scratch behaviors of neat PS, PS+NT, and PS+NTm. After nine scratches, the neat PS starts showing more severe surface damage, indicating that strain hardening does not occur with increasing scratches.

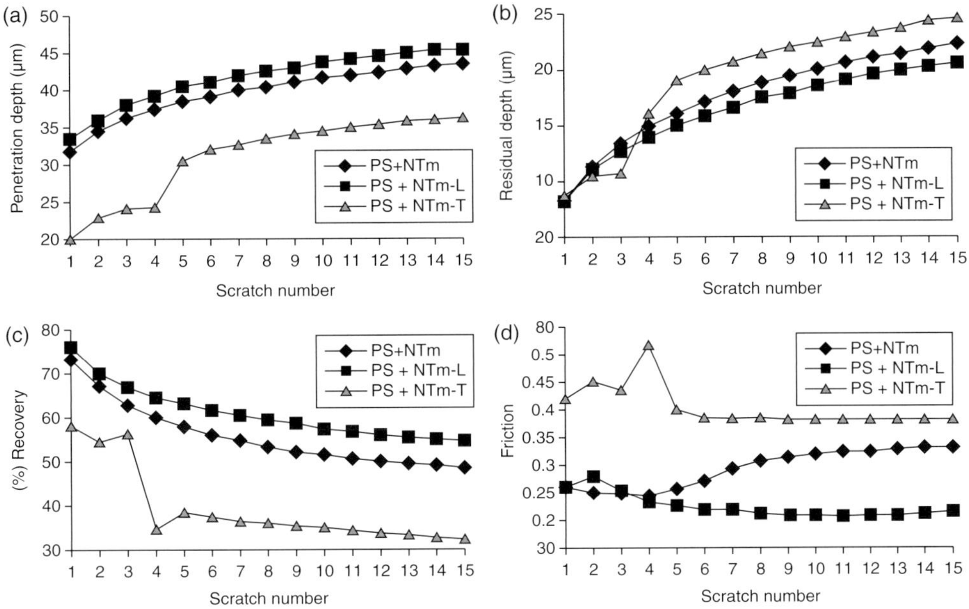

Figure 13.14. Multiple scratch behavior of compression-molded discs (PS+NTm) and injection-molded coupons in the longitudinal (L) and transverse (T) direction with respect to injection flow [9].

Conversely, the nanocomposites display an asymptotic response to increasing number of scratches indicating that the CNTs induce strain hardening. Neat PS is quite brittle, and the addition of CNTs acts to reduce this brittleness and increase instantaneous penetration. SEM micrographs, surface topography and cross section profiles show a rough surface with cracks and debris formation for the neat PS whereas both nanocomposites showed a lower surface roughness [9].

In Figure 13.14, the nanocomposite compression-molded scratch response is compared with the longitudinal and transverse (with respect to injection flow) scratch response of injection molded coupons. The result shows that the injection-molded coupons exhibit an anisotropic response, since the nanocomposite scratch behavior is dependent on the scratch direction. A scratch direction parallel to injection flow yields the highest scratch resistance and lowest friction coefficients, while the lowest scratch resistance and highest friction values are found for a scratch direction perpendicular to injection flow. The scratch resistance of the compression-molded discs is always intermediate to the longitudinal and transverse data. The polymer chains and nanoparticles align themselves in the longitudinal direction when they are injection-molded; thus they have a higher mobility in this direction and a hindered mobility in the transverse direction leading to the absence of strain hardening [9].

Giraldo et al. also studied scratch and wear resistance using the scratch test, but they used multi-wall carbon nanotubes dispersed in a nylon 6 polymer matrix [10]. Figure 13.14 shows the instantaneous penetration depth (R_p) and residual healing

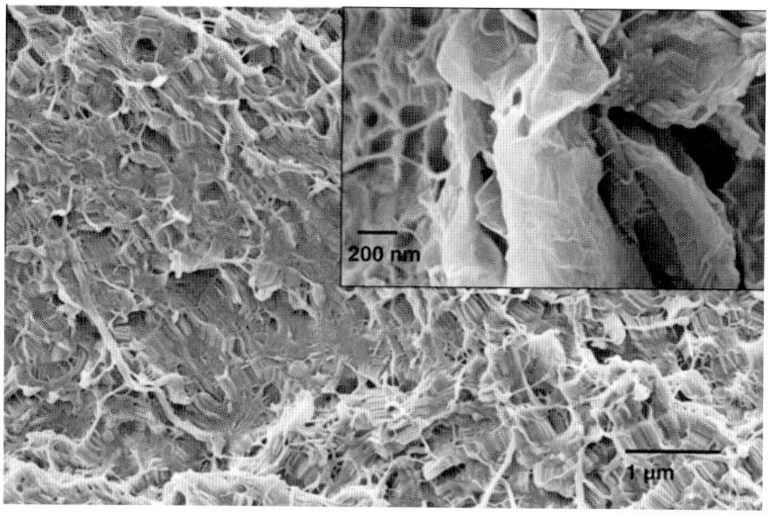

Figure 13.15. SEM image of 2 wt% graphene-PTFE composite (inset shows the wavy edges of graphene platelets) [11].

depth (R_h) as a function of scratch number for the neat nylon 6 and nanocomposites. The instantaneous depth is deepest for the neat nylon 6 and decreases with increasing MWNT concentration. Conversely, the residual depth is deepest for the 1 wt% nanocomposite and decreases with decreasing MWNT concentration. Both the instantaneous depth and the residual depth increase asymptotically with increasing scratches due to strain hardening. This phenomenon may be attributed to the high aspect ratio of MWNTs, since they act as reinforcing particles but may also induce more nucleation sites, thus hampering viscoelastic recovery [10].

SEM images confirm that the neat nylon 6 deforms in a relatively soft manner, and adding MWNTs increases the stiffness to better resist the deformation from the indenter. Overall, the addition of MWNTs increases abrasion resistance [10].

13.2.4 Wear Resistance of PTFE-Graphene Nanocomposites

Polytetrafluoroethylene (PTFE) is a commonly used thermoplastic solid lubricant with a wide range of industrial applications where using fluid lubricants is not feasible. PTFE provides low friction in dry sliding conditions, but is limited in its use by a very high wear rate. Micro-sized particles have been shown to reduce the wear rates of PTFE by several orders of magnitude, but the nano-sized graphene, with such large interfacial areas, is expected to reduce wear rates even more dramatically [11].

Kandanur et al. reported on the wear resistant properties of PTFE-graphene nanocomposites [11]. The authors used a tribometer to perform pin-on-disc experiments, and the test and preparations are detailed in the article. SEM was used to look at the dispersion of the graphene platelets in the PTFE matrix [11]. While TEM would give a better idea of dispersion, there is no indication of agglomeration of graphene platelets in the SEM image (Figure 13.15) [11].

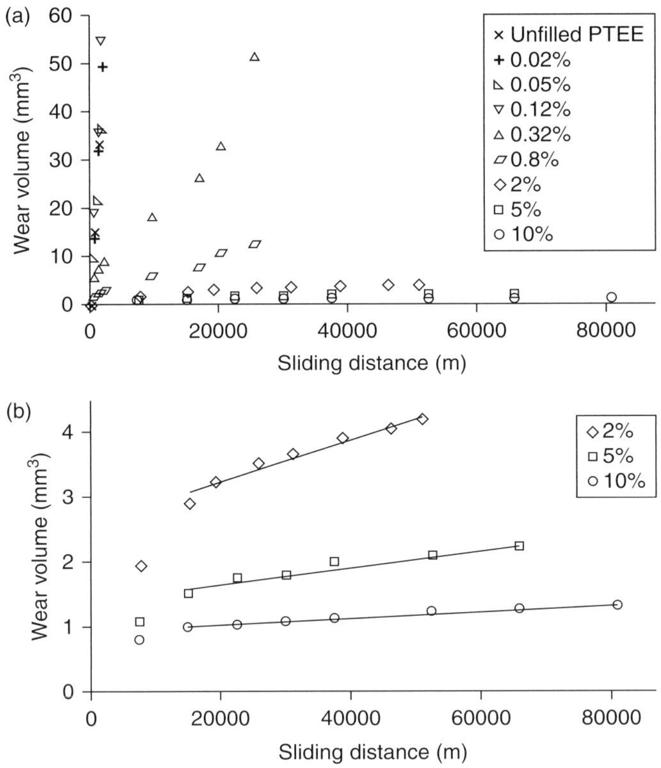

Figure 13.16. Wear volume of neat PTFE and graphene platelet-PTFE nanocomposites versus sliding distance [11].

Figure 13.16 shows that the wear volume versus sliding distance decreases as the concentration of graphene increases. The neat PTFE and nanocomposites with graphene concentrations up to 0.12 wt% show relatively the same wear rate as their data crowd near the vertical axis. However, once the graphene concentration reaches 0.32 wt%, the wear resistance increases dramatically with the 10 wt%, showing only about 1 mm^3 of wear volume even after sliding 80 km. The second graph also shows that the wear volume for 2, 5, and 10 wt% graphene seems to increase mainly during the first 7–15 km of sliding. Here, once the wear rate starts to decrease after the concentration reaches 0.12 wt%, the wear rate is found to decrease with approximately the square of the graphene platelet content. Furthermore, the wear rate does not appear to reach a minimum, so it is feasible that lower wear rates could be achieved with higher graphene platelet concentration [11].

Figure 13.17 shows the SEM images of the surface and counter-surface of neat PTFE and 10 wt% graphene-PTFE nanocomposite. The wear debris on the counter-surface of the neat PTFE is hundreds of micrometers in in-plane dimensions, while the wear debris on the counter-surface of the 10 wt% graphene-PTFE nanocomposite is much finer. Similarly, the wear surface of neat PTFE shows large, plate-like debris, whereas the 10 wt% graphene-PTFE nanocomposite shows only "mudflat" features. Neat PTFE is known to wear by ejecting large, plate-like debris likely due

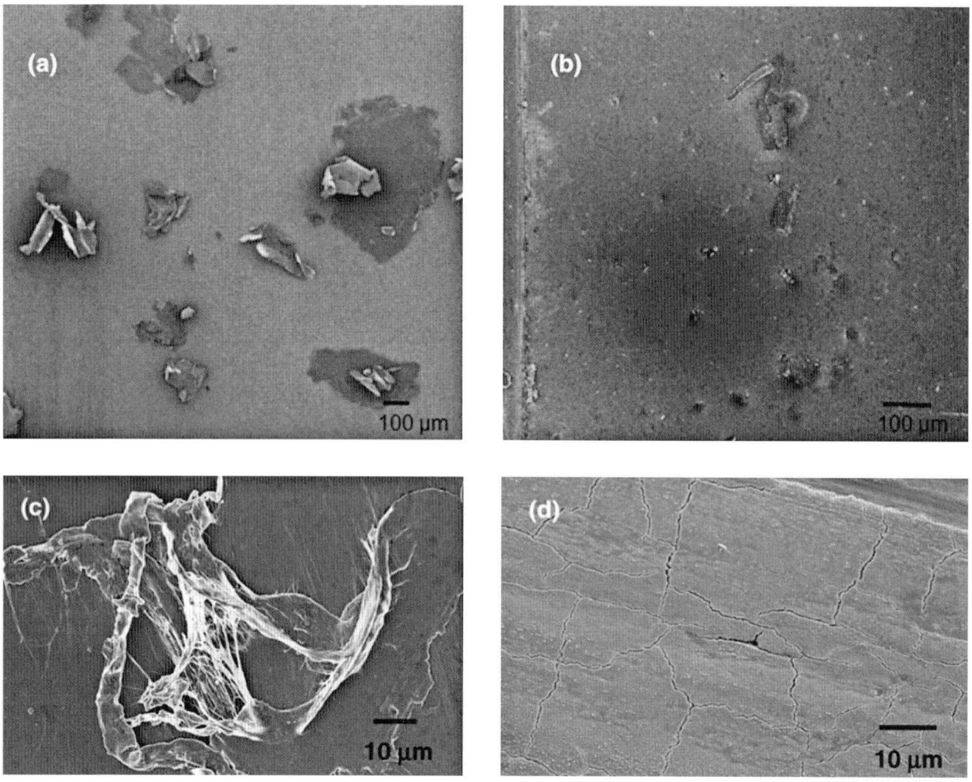

Figure 13.17. SEM images of: (a) counter-surface of neat PTFE, (b) counter-surface of 10 wt% graphene-platelet PTFE, (c) wear surface of neat PTFE, and (d) wear surface of 10 wt% graphene platelet-PTFE [11].

to propagation of subsurface cracks parallel to the sliding surface. The addition of graphene platelets could inhibit these subsurface cracks by deflecting them through a path of high-aspect-ratio graphene platelets. Graphene platelets show a definite increase in wear resistance when added to PTFE [11].

13.2.5 Summary of Tribological Properties of Polymer Nanocomposites

There are generally fewer studies on tribological properties of polymer nanocomposites than other, more sought-after properties, such as thermal, mechanical, or electrical. However, thermoplastic polymer nanocomposites generally exhibit better abrasion, wear, and scratch resistance than their unfilled polymer counterparts. These properties are often highly dependent on the degree of nanoparticle dispersion achieved in a polymer, since enhanced properties are often attributed to the extremely high aspect ratio of nanoparticles. In the case of nanoclay, more research is needed to determine exactly how to achieve better abrasion resistance, since different studies seem to produce contradictory results. The formation of nanoclay aggregates almost certainly decreases abrasion resistance, so an exfoliated dispersion within the polymer is required. Both CNTs and graphene platelets show a positive effect on the tribological properties. The mechanisms of wear in polymers is often crack propagation or debris formation on the abraded surface of the polymer, and well-dispersed nanoparticles often act to inhibit these mechanisms through better polymer-nanoparticle adhesion. Abrasion, wear, and scratch resistance are important properties for most polymers, and the inclusion of nanoparticles can create robust thermoplastic nanocomposites with enhanced tribological properties [12].

13.3 Permeability Properties of Polymer Nanocomposites

Permeability in polymers is of great interest due to the increasing demands of better membranes with decreased fouling and degradation. Separations, barriers, and even combustion prevention are dependent on good polymer films. A way to improve on the transport properties of polymers is with the addition of nano-sized particles for the production of polymer nanocomposites. Although organic nanofillers have been always added in polymers, such as rubber, studies of gas transport in nanocomposites are fairly recent [13]. It is the purpose of this section to highlight the use of nanomaterials in polymers for the enhancement of permeability properties. Furthermore, some polymer nanocomposites can even help by reducing or preventing the release of combustible gases from polymers and separate them from oxygen to prevent combustion from being sustained [14].

13.3.1 Permeability Properties Characterization Techniques

Permeability of gases and liquids can be measured in a variety of manners. This section explains the most commonly used methods. In all methods, a membrane of the polymer nanocomposite is prepared with a given thickness and placed in a permeation cell [15]. It should be noted that although these systems can be put together after obtaining all parts, commercial apparatuses exist that will perform said measurements. For a more detail understanding of these steady-state permeation measurement methods, see [13] (all images and captions in this section were taken from the aforementioned reference).

13.3.1.1 Constant-Pressure, Variable-Volume Method

Also known as continuous flow, this method applies constant gas feed pressure in the upstream side and measures the flux of gas in the downstream side using a flow meter. Before measurements, the downstream side will be purged with gas to remove any impurities. This method is preferred for high gas flux membranes.

13.3.1.2 Constant-Volume, Variable-Pressure Method

This system measures the permeate flux by measuring the pressure drop or increase of gas in a closed volume with the aid of a pressure transducer. Before starting measurements, the system will be subjected to vacuum to remove any impurities. This method is capable of measuring a wide range of gas flux values.

Table 13.1. *Effect of nanofiller incorporation on permeability of PLA films toward O_2 and N_2 [17]*

Sample	O_2 ($\times 10^{-18}$ m^2 s^{-1} pa^{-1})		N_2 ($\times 10^{-18}$ m^2 s^{-1} pa^{-1})	
	Permeability	Standard deviation (n = 3)	Permeability	Standard deviation (n = 3)
PLA	3.76	0.1980	1.10	0.1334
GO 0.2 wt%	1.34	0.0212	0.271	–[a]
GO 0.4 wt%	1.23	0.0141	0.306	–[a]
GO 0.6 wt%	1.49	0.0033	0.502	–[a]
GO 0.2 wt%	1.34	0.0265	0.270	–[a]
GO 0.4 wt%	1.20	0.1556	0.250	–[a]
GO 0.6 wt%	1.30	0.0555	0.237	0.1533

[a] Standard deviation not computed since only a single measurement was performed.

13.3.1.3 Mixed-Gas Permeability Measurements

This system is mainly used when studying the permselectivity of a membrane to a given gas. This method is similar to the ones mentioned beforehand with the additional requirement of a gas sensor both in the upstream and downstream sides to detect the types and concentrations of gases present.

13.3.2 Permeability Properties of Different Nanoscale Dimensions Nanomaterial-Based Nanocomposites

Addition of nanomaterials in polymers can have a drastic effect on the permeability of the composites to penetrant gas and liquid molecules. It is known that permeability of polymer membrane and coatings is affected by defects, kinetic diameter of penetrating molecules, physical properties, loading and particle size of nanomaterials, properties of the polymer matrix, feed pressure, plasticization by penetrating molecules, and the affinity between gas molecules and the dispersed phase [16]. It is the purpose of this section to show examples of how these factors can affect the permeability of polymer nanocomposites.

13.3.2.1 One Nanoscale Dimension Nanomaterial-Based Nanocomposites

Thermoplastic poly(lactic acid) (PLA) has been studied with graphene oxide (GO) and graphene nanoplatelets (GNP) as nanoadditives to show the effect of particle shape. The membranes were solvent-casted with different nanofiller loading raging between 0.2 wt% and 0.6 wt%. The oxygen and nitrogen gas permeabilities were studied using the constant-pressure variable-pressure method at room temperature with a feed pressure of 1 bar. The results of these studies are summarized in Table 13.1. As seen, the oxygen permeability of the PLA decreased the most (threefold) at a 0.4 wt% loading of GO or GNP. Nitrogen permeability of the membranes also showed a decrease with increasing GNP content (up to fourfold decrease). The nitrogen permeability, however, decreased as more GO was added to the PLA, yet it was still higher than for the pure PLA membrane [17].

Table 13.2. *Gas permeability of PLA/organoclay hybrid films [18]*

Clay (wt %)	O_2 (cc/m^2/day)		
	C_{16}-MMT	DTA-MMT	Cloisite 25A
0 (pure PLA)	777	777	777
4	449	455	–
6	340	353	430
10	327	330	340

Table 13.3. *Gas transport properties of PC-PU nanocomposites [19]*

Composition	Permeability (barrer)[a]				
	CH_4	N_2	O_2	CO_2	H_2
4692/1/0/18	1.3	0.8	2	21	7.1
4672/1/0/18	1.6	0.7	2.3	26	8.4
5652/1/0/19	2.6	1.1	3.1	35	11.1
5652/1/1/19	2.7	1.1	3.1	35	11.1

Only the last sample contains BO (1 wt%). The first number in the composition is the PC diol content while the last number in the composition is the percent of hard segment chains present [19].

Loadings of 10 wt% of nanoclays in PLA studied by Chang and co-workers [18], by contrast, have shown decreases in the permeability of PLA of about two times, as shown in Table 13.2. Given their shape, GNP in theory should stack better to make a more torturous path for penetrant gas molecules than GO does, yet this is not observed if the GNP is not aligned in the membrane.

Nanoclay bentonite (BO) has been added to thermoplastic elastomer copolymers of polyurethane and polycarbonate (PU-PC) by in situ polymerization to create nanocomposites. The gas transport properties were then tested using the constant-volume variable-pressure method. In this case, however, the gas transport properties were almost unaffected when 1 wt% of BO was added, as seen in Table 13.3, since the permeability of these copolymers is greatly determined by the order and content of the hard (glassy) segments [19].

The importance of dispersion and orientation can be seen when glass flakes were hand-mixed into epoxy with loadings ranging from 1 wt% to 30 wt%. It was observed that these nanofillers can experience settling due to gravity and the low viscosity of the resin. Samples were obtained, however, which suffered minimal settling and good nanofiller dispersion with the platelets being oriented and dispersed parallel to the membrane surface. It was then seen that the helium permeability is decreased by about one order of magnitude with higher loadings as seen in Figure 13.18 [20]. Nevertheless, the study was only conducted with up to 20 wt% loading due to the increase of epoxy resin viscosity as the filler was added.

Disc samples were prepared in order to allow filler loadings of 30 wt%. These disc samples showed a three order-of-magnitude decrease in permeability as seen

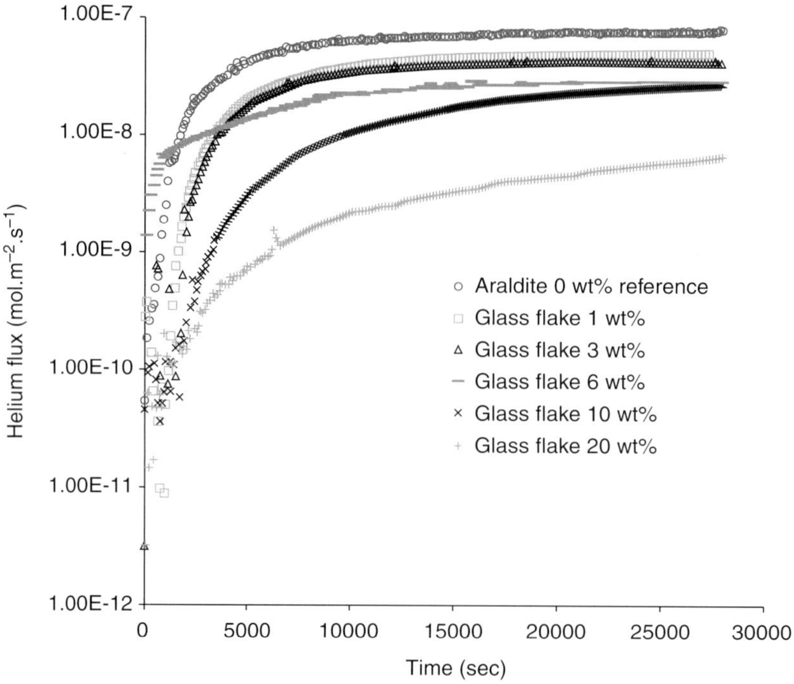

Figure 13.18. Helium flux measurements of the doctor blade sheets with up to 20 wt% glass flakes [20].

in Figure 13.19, and when the discs are multilayered, one can obtain more improvement due to the thinner layers, which causes the improved dispersion and orientation of the glass flakes [20].

13.3.2.2 Two Nanoscale Dimensions Nanomaterial-Based Nanocomposites

Nanofillers can be used to improve the permeability of otherwise porous polymers as seen when MWCNTs are added to poly(acrylonitrile-co-starch) (PAN-co-starch) by in situ polymerization. When MWCNTs are well dispersed, they can block the pores of the PAN-co-starch, as shown in the field emission scanning electron microscope (FESEM) image in Figure 13.20, leading to lower permeabilities. Figure 13.21 shows the permeability is then seen to decrease [21].

The dimensionality (particle size essentially) of the nanofiller can have a direct effect on the permeability. When Cloisite nanoclay (montmorillonite) and carbon fibers (CFs) were added by melt-mixing to styrene butadiene rubber (SBR), it was shown that Cloisite nanoclay provided a better improvement in oxygen permeability than the carbon fibers did, even though the relative permeability lowered with increasing nanofiller loading, as shown in Figure 13.22.

Yet the permeation and diffusion activation energies increased with filler loading. The nano-size of the fillers enables good polymer-filler interaction at the interface and decreases permeability. Thus Cloisite performs better due to its organic modifier, low surface, and cleavage energies. The decrease in permeation for this system can be attributed to the increased tortuosity as well as the reduced polymer mobility

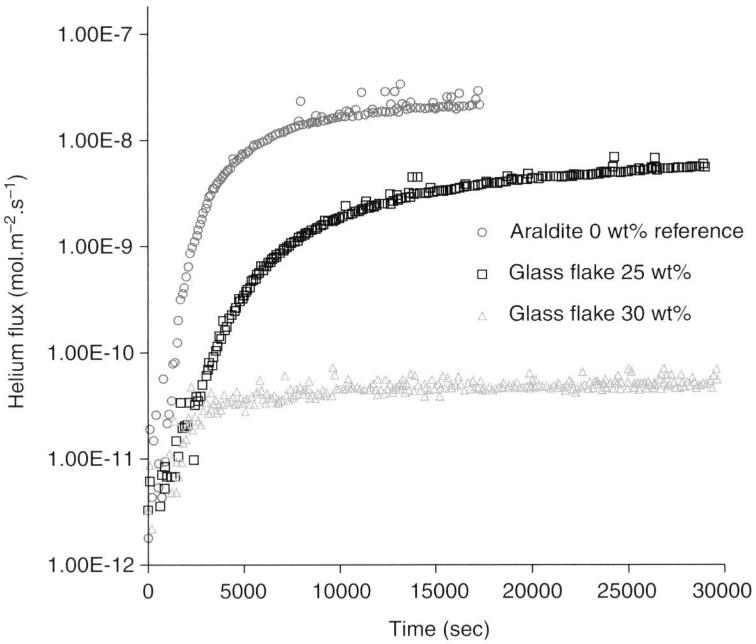

Figure 13.19. Helium flux measurements of the molded discs with up to 30 wt% glass flakes [20].

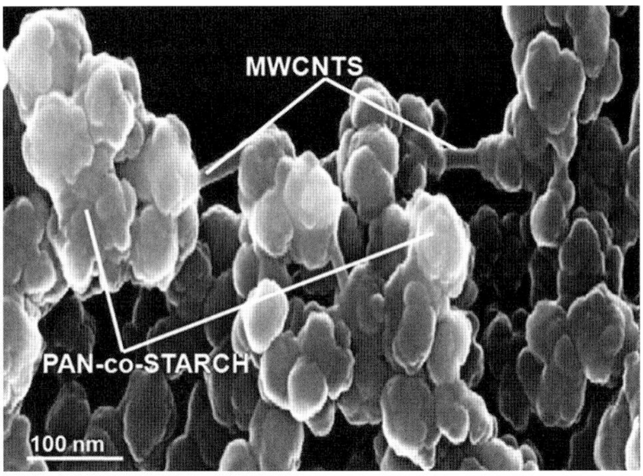

Figure 13.20. FESEM image of PAN-co-starch/MWCNT nanocomposite at 2 wt% MWNCT loading showing the MWCNT blocking the pores in the polymer [21].

and free volume. CFs showed a near-linear drop in permeability with increased loadings and were more effective in bringing down the diffusivities, signaling great control over the kinetic aspects of oxygen transport in SBR. Lastly, it was observed that the rates of diffusivity were enhanced in direct proportion to temperature, as seen in Figure 13.23 [13]. It should be noted that Cloisite is a one-dimensional nanomaterial, but unlike the case mentioned in the previous section with BO and PU-PC copolymers, the interaction of Cloisite and SBR allowed for a beneficial change in permeability.

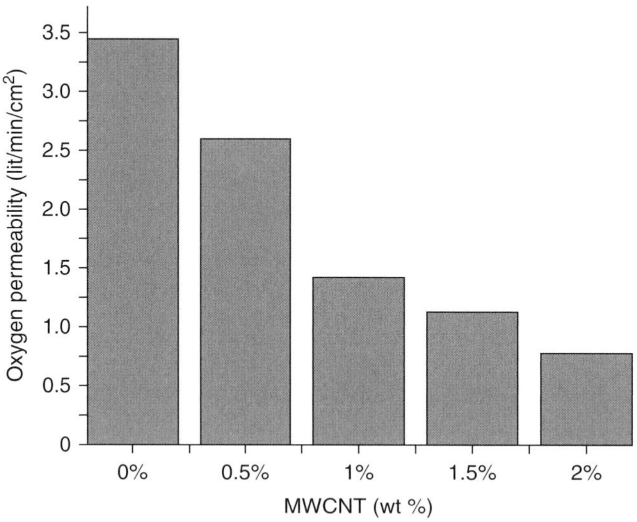

Figure 13.21. Oxygen permeabilities of the PAN-co-starch/MWCNT nanocomposites as a function of MWCNT content at constant pressure [21].

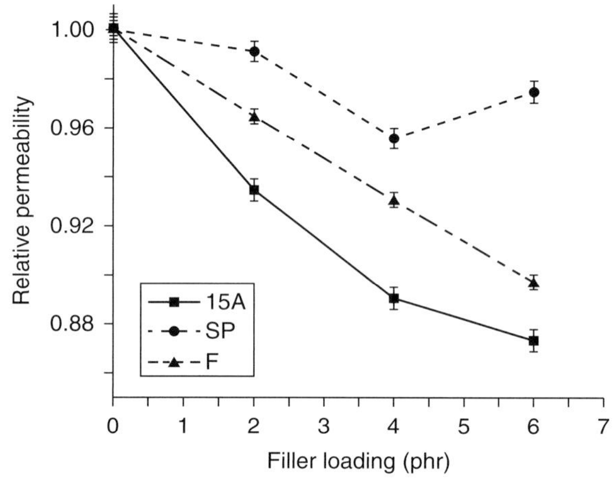

Figure 13.22. Relative O_2 permeabilities of nano-filled SBR nanocomposites at 30°C [13].

In recent years there has been an increasing interest using epoxy nanocomposites with buckypapers and carbon nanofiber papers. These buckypapers (BP) and carbon nanofiber papers (CNFP) are thin sheets made from the aggregation of carbon nanotubes and carbon fiber, respectively. Wu and co-workers studied the permeability of epoxy nanocomposites with SWNT buckypapers, MWNT buckypapers, and CNF papers. It was found that the gas flow through the nanocomposites decreases with increasing filler (BP or CNFP) thickness and density, as observed in Figure 13.24 [14]. However, CNFP showed a fast pressure drop in the system, even though it was the thickest paper due to its large pore size.

A direct correlation was also observed between the decrease in gas flow and the diameter of the CFs or CNTs used in the papers. MWNT BPs showed a slower drop in pressure than did SWNT BPs because the MWNT are thicker. Thus, any

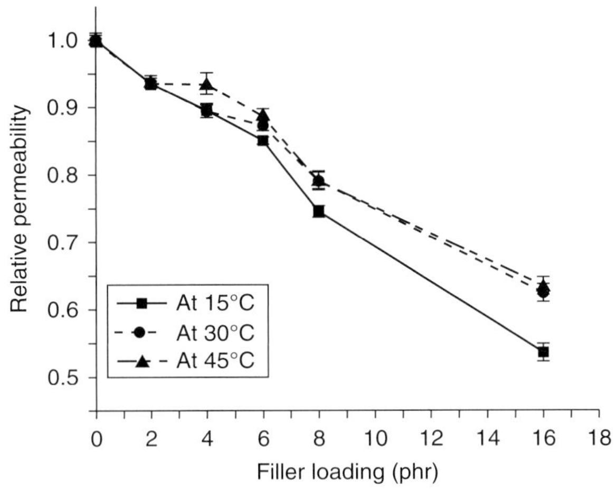

Figure 13.23. Relative permeabilities of nanocomposites at different loadings at different temperatures [13].

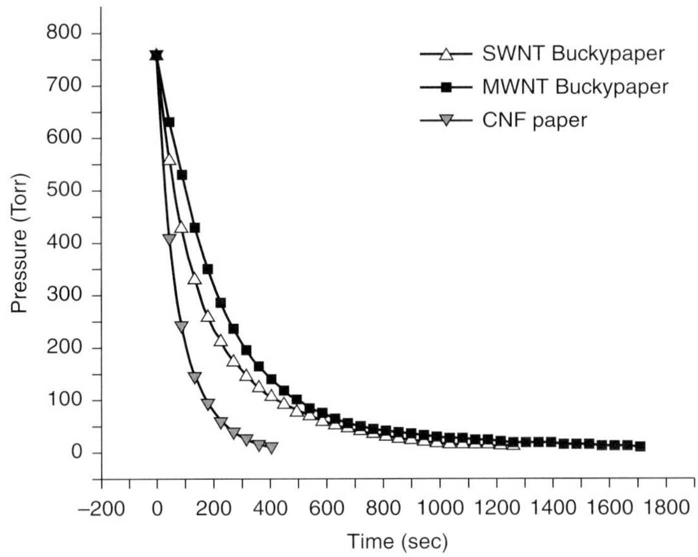

Figure 13.24. Pressure drop curves of gas flow through SWNT buckypaper, MWNT buckypaper, and CNF paper [14].

differences in the gas flow were directly attributed to the differences in the networks of the papers [14].

13.3.2.3 Three Nanoscale Dimensions Nanomaterial-Based Nanocomposites

Variable-pressure constant-volume measurements of the permeability of polyimide (PI) with zeolite (125 nm and 1105 nm average particle size) have been conducted with several gases to show the importance of pore size of the nanofiller. Two zeolite fillers were used with pore sizes approximately 13 Å and 4 Å. These studies showed that even though the membranes were free of voids and flaws, nitrogen, methane, and carbon dioxide permeabilities of the polymer nanocomposites still decreased

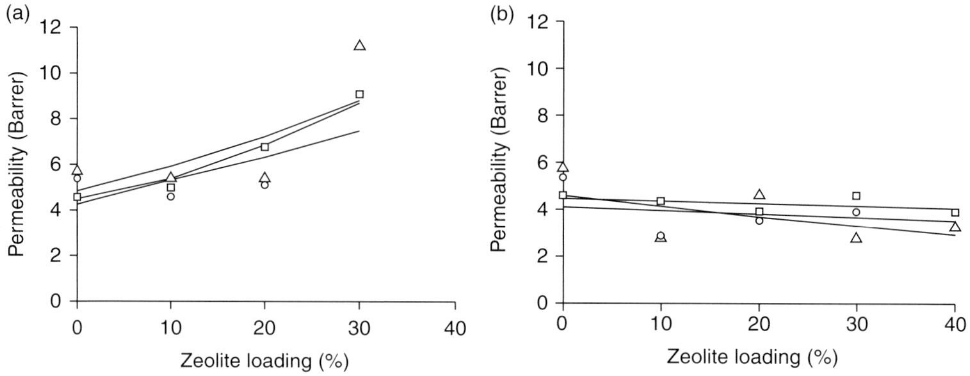

Figure 13.25. CO_2 permeability of the nanocomposites fabricated with (a) zeolite 13X and (b) zeolite 4A with different zeolite loading amounts at various feed pressures: (△) 4 atm, (○) 8atm, (□) 12 atm [16].

compared to that of the virgin PI. Figure 13.25 shows the CO_2 permeability of several composites studied [16].

The selectivity of carbon dioxide over methane and nitrogen also was greater for the nanocomposites due to the molecular sieving effect of zeolite, especially seen at 30% loading. This sieving effect allows the easy passage of carbon dioxide due to the gas molecule polarity and kinetic diameter. Sieving effects, however, were not observed in zeolite filler with the smaller pore size. In samples with large pore sizes, nitrogen and methane permeability was proportional to feed pressure because, even though they cannot easily penetrate the PI membranes, they can still find the pores in the nanofiller and pass through. The same was observed with methane (average size 3.8 Å) in the samples with the 4 Å pore size. Selectivity in these samples was the best when a 20% loading was used. Increasing the feed pressure caused a decrease in separation rate due to the presence of unrelaxed volume of glassy polymer. As before, there was no improvement in the selectivity when using the 4A zeolite due to its small pore size. Plasticization of membranes is undesired because of the lower carbon dioxide selectivity, yet these membranes show that no plasticization occurred at several filler loadings and feed pressures [16].

Another example highlighting the importance of pore size in nanomaterials is the polysulfone with titanosilicate nanocomposites. The pore size in the titanosilicate is smaller than the kinetic diameters of hydrogen and methane, yet it is so close to hydrogen that it should allow selective permeability of this gas molecule. Permeability results are shown in Figure 13.26. It was indeed observed that higher loadings showed a higher hydrogen selectivity (from 5 wt% to 10 wt %). However, once too much nanofiller is added, it cannot be embedded properly, and thus the selectivity decreases. If the exfoliation and layer structure preservation is improved in higher loading, an improvement in selectivity is still possible [22].

Commercial epoxy nanocomposites exist that are used as coatings for mortar exposed to corrosive environments. It is usually recommended that the more

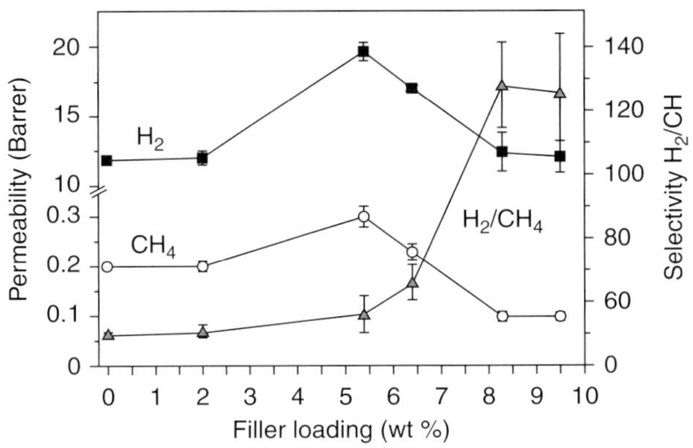

Figure 13.26. H_2/CH_4 separation experiments at 35°C with nanocomposites prepared with met compounding materials from swollen layered titanosilicate [22].

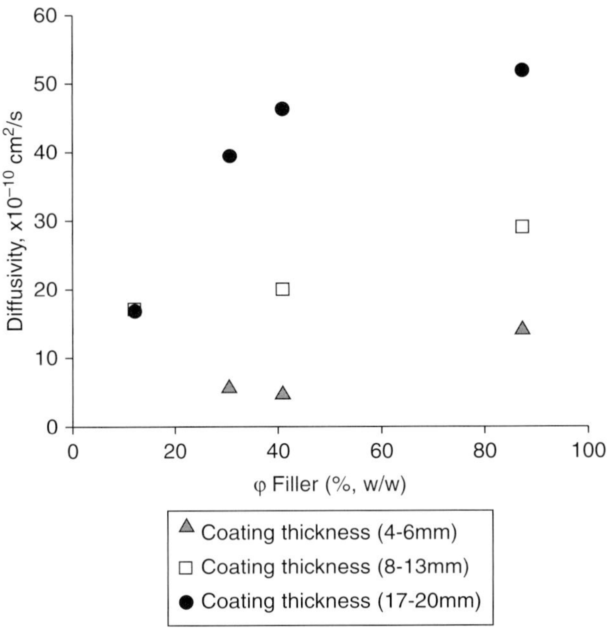

Figure 13.27. Effect of filler concentration on sulfuric acid diffusivity in various coating thicknesses [23].

corrosive an environment is, the thicker the epoxy nanocomposite coating should be. This, however, is not always the case. The nanofillers with lower loadings showed lower diffusivities to sulfuric acid regardless of film thickness, as seen in Figure 13.27 [23]. The increase in diffusivity was actually minor when the coating thickness increased and could be attributed to the presence of paracrystalline aggregates whose degree of order and orientation vary with film thickness.

When using nanofillers, the degree of chain alignment is maintained with change in thickness; thus, a thinner film will theoretically show a higher chain alignment and higher tortuosity for penetrant molecules. Lastly, it was noted that higher loadings could induce the clustering of filler, which would in turn increase the overall particle size and reduce the tortuosity [23].

13.3.3 Concluding Remarks

As shown, defects, kinetic diameter of penetrating molecules, physical properties, loading and particle size of nanomaterials, properties of the polymer matrix, feed pressure, plasticization by penetrating molecules, and the affinity between gas molecules and the dispersed phase all have strong effects on the permeability of nanocomposites to penetrant gas and liquid molecules. The effect can be a direct result of the interactions between nanoparticles and the polymer, as it could simply be the clogging of polymer pores by the nanofiller, as seen with the PAN-co-starch/MWCNT nanocomposites. The ability to improve the barrier properties of polymers is of great interest due to the increasing demand for better membranes with decreased fouling and degradation. Until this demand is fulfilled, polymer nanocomposites will continue to be explored to find an optimal material for these applications.

13.4 Application of Nanotechnology and Nanomaterials in the Oil Field

Nanotechnology is one of the fastest growing, most promising research fields. Specifically, nanotechnology has seen significant advances in the electronics, biomedical, materials and manufacturing, pharmaceutical, aerospace, photography, and energy industries [24]. Relatively few applications of nanotechnology are currently seen in the oil field. The oil and gas industry could greatly benefit from nanotechnology innovations in the oil field, since strong, stable materials are needed in the high-temperature, high-pressure environment of an oil well. Additionally, with fossil fuels growing scarcer, there is a greater need for effective equipment to extract oil from unconventional oil wells [25]. This section explores the benefits of nanocomposite materials, nanofluids, and nanosensors, with an emphasis on their potential applications in the oil field.

13.4.1 Areas of Application in the Oil Field

The upstream oil and gas industry is in a position to begin utilizing nanotechnology. Stronger, lighter, chemically resistant nanomaterials are perhaps the most obvious application of nanotechnology [26]. Nano-sized sensors could also be utilized to obtain more accurate measurements of temperature and pressure deep inside oil wells during the drilling or production phases. Nanosensors could also help engineers understand the nature of deep oil reserves where oil and gas could be trapped, and more advanced methods of oil recovery could be developed. By adding nanoparticles to fluids in order to optimize properties, such as density, viscosity, thermal conductivity, and specific heat, nanofluids or "smart fluids" could be developed to aid in drilling and production [27]. Table 13.4 provides an overview of the oil and gas industry needs and the nanotechnology solution that could be developed to address each need [24]. The following sections identify more specific applications of nanotechnology in three divisions of the oil and gas industry: exploration, drilling, and enhanced oil recovery.

Table 13.4. *Oil and gas industry needs and the nanotechnology solutions [24]*

Area	Industry Needs	Nanotech Solutions
Exploration	Less invasive methods of exploration, remote sensing Methods to "sniff" for new pockets of oil Enhanced resolution for subsurface imaging and computational techniques Improved temperature and pressure ratings in deep wells and hostile environments Improved instrumentation for gas adsorption Improved 1D, 2D, 3D, and 4D seismic resolution	Nanosensors and imaging
Reservoir management	Enhanced remote imaging, real-time continuous monitoring of flow rate, pressure and other parameters during production, wireless telemetry, in situ chemical sensing Accurate early warning detection and location of leaks (preventing environmental hazards) Improved reservoir illumination and characterization, including improved signal-to-noise ratio of subsalt events, improved velocity modeling accounting for anisotropy	Nanosensors
	Improved sand exclusion and mobility of inject ant Controlled agglomeration of particles Ability to capture and store CO_2	Nanomembranes
	Improved stability and pressure integrity and heat transfer efficiency Ability to minimize damage to formation of offshore platforms, reduce their weight requirements, and increase their sturdiness	Nanomaterials, fluids, and coatings
Drilling	Increased effectiveness and longevity of drilling components, making cheaper, lighter, and stronger pipes and drill bits Extended lifetime of equipment with corrosion resistance, adhesion enhancement, and wear resistance Improved strength-to-weight ratio for an expanding range of geological settings Expandable tubulars for deeper wells without needing to telescope the well, or casingless wells Improved cement integrity – light density and high strength, hole quality and well placement, hermetic seals Innovative drill engines that can be sent deep into the shaft; improved elastomers Ability to prevent bio-fouling	Nanomaterials and coatings
	Improved drilling fluids and thermal conductivity Removal of toxic metals (mercury, cadmium, lead) Ability to prevent drilling mud invasion, separating mud filtrate and formation water	Nanofluids and nanomembranes

13.4.2 Exploration

Today, the number of reservoirs easily accessible using current technology is dwindling. Approximately 67% of all U.S. oil is located in places that current technology cannot reach. Industry professionals believe that an improved understanding of the chemical and physical properties of existing reservoirs is crucial to extracting more oil and gas, so the ability to obtain detailed information about properties deep in reservoirs is becoming much more important. Current sensor technology lacks the needed resolution to ascertain these properties or the ability to deeply penetrate reservoir lithologies [24]. Before drilling can occur, the interaction between the rock and fluid needs to be understood to create new technologies designed to extract this trapped oil and gas. Sensors made from nanomaterials are ideally suited to this task due to alterations in their optical, magnetic, and electrical properties (compared to their bulk counterparts). Many nanoparticles are also anisotropic, which, when combined with smart fluids, could create sensors that have high resolution, are extremely sensitive, and are capable of working at extremely high temperatures and pressures [27].

A consortium of major oil and gas companies is currently working on creating subsurface nanosensors that can be injected into the tiny oil and gas well bores. The pores throats in typical oil-bearing sandstone are on the order of 1–15 μm [24], while pore throats in typical shale range from 10 nm to 30 nm [28]. These tiny nanosensors could travel out of wellbores through the pores of the rock to gather information about the surrounding geological structures [24].

13.4.3 Drilling

Extracting oil from far beneath the earth's surface is not a simple task, and there are frequent operational problems that can lead to lost productivity and additional costs. The drilling process is primarily dependent on the drilling fluid, oil well cement, and drilling bit. Stronger, lighter, more durable nanomaterials with desirable thermo-mechanical properties are likely to be the widest area of nanotechnology application [26]. Some applications of polymer nanocomposites include pipes, fiber-reinforced plastic tanks, valves, and elastomer seals. Thermoset pipes that absorb less water are very desirable as they can retain their mechanical properties. Fire performance is another important characteristic of oil field equipment. Nano-reinforcement of polysiloxane-phenolic resin pipes can help increase impact resistance while maintaining or improving the fire resistance capabilities [29].

Strong, lightweight equipment is desired for use on the surface of an oil rig due to safety concerns, since personnel injuries are common when operating heavy machinery. The Canadian company Katch Kan has developed an alternative mud bucket and drilling fluids tray composed of "super polymers." These parts weight approximately 1/10th of their steel counterparts and are designed to be operated by one person. These types of oil field equipment can increase safety while decreasing manpower on an oil rig [30].

Oil well cement is another vital element of the drilling and production processes. Without a stable casing in the borehole, the borehole is likely to have stability problems. Nanoparticles have been utilized to increase the compressive and flexural strengths of Portland and Belite cements [26].

Several applications of nanotechnology in the drilling stage are discussed in detail later in this chapter. Essentially, there are numerous issues affecting drilling operations that can be solved using nanomaterials with specially tuned properties.

13.4.4 Enhanced Oil Recovery (EOR)

Since the rate of new oil field discoveries is declining, the only way to meet the growing worldwide demand for oil is to create more effective methods of collecting the oil in known reserves, a process known as enhanced oil recovery (EOR). Approximately two-thirds of the oil in the world's reservoirs cannot be recovered using conventional production methods. Three major categories of EOR have been found to be commercially successful to varying degrees [24]:

- Thermal recovery, which involves the introduction of heat, such as the injection of steam, to reduce the viscosity of the heavy viscous oil and to improve its ability to flow through the reservoir.
- Gas injection, which uses gases, such as natural gas, nitrogen, or carbon dioxide that expand in a reservoir to push additional oil to a production wellbore, or other gases that dissolve in the oil to lower its viscosity and improve its flow rate.
- Chemical injection, which can involve the use of long-chained molecules called polymers to increase the effectiveness of waterflood, or the use of detergent-like surfactants to help lower the surface tension that often prevents oil droplets from moving through a reservoir.

Nanofluids are ideally suited for implementation in EOR since their density, viscosity, thermal conductivity, and specific heat can be tuned for each specific application. Dispersing CuO nanoparticles in CO_2 driving fluids can drastically increase the viscosity, creating a better mobility ratio and high sweep efficiency, which leads to higher oil recovery [24].

13.4.5 Examples of Nanotechnology Applications in the Oil Field

While there are many ideas about how nanotechnology and nanomaterials can be employed in drilling activities, relatively few are currently being researched in detail or implemented in the field. As oil reserves become scarcer, serious interest in pursuing nanotechnology solutions will increase. The following are several specific examples of how nanotechnology can be utilized in drilling operations.

13.4.5.1 Multiwall Carbon Nanotube Reinforced Rubber Seals for Extreme Downhole Drilling

Ito et al. have developed a multi-wall carbon nanotube (MWNT) reinforced fluoroelastomer rubber for use in extreme downhole drilling environments where the pressure and temperature are very high [31]. The current seals used in most drilling applications are made from fluoroelastomer rubber (FKM), which has a maximum

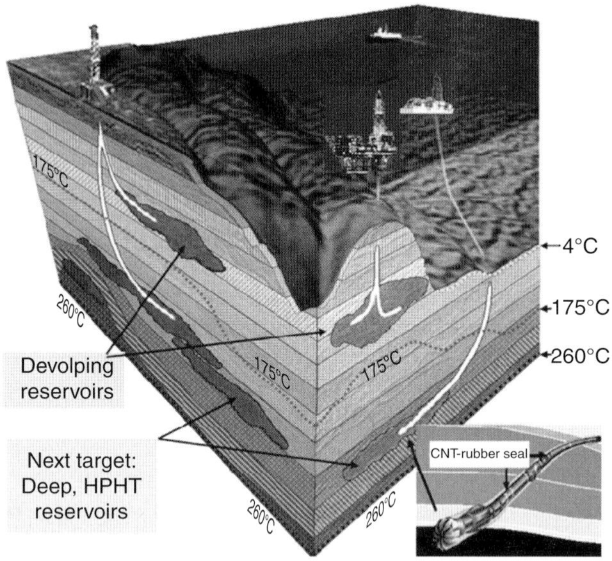

Figure 13.28. Diagram of typical oil reservoirs [31].

operating temperature of 175°C. Figure 13.28 shows a diagram of oil reservoirs and their corresponding temperatures and pressures at different depths [31]. Essentially, due to the temperature limitations of the neat fluoroelastomer rubber, current drilling operations are only able to access around 35% of the worldwide oil reservoirs. If technology can be developed to withstand temperatures and pressures exceeding 260°C and 240 MPa, respectively, oil and gas recovery could be increased to 70% of available reservoirs [31].

Improved sealing solutions will also have to account for the wide range of operating temperatures present when completing deepwater drilling operations. In this case, seals must perform well in cold seabeds where temperatures average 4°C, as well as in deep, high-pressure, high-temperature wells. Another type of rubber seals made out of perfluoroelastomer (FFKM) is available for high-temperature applications. However, FFKMs do not perform well at low temperatures like the ones encountered on the sea floor, and FFKM seals are more expensive than FKM seals. Hence, MWNT-reinforced FFKM has been developed as a cost-effective solution to access deeper, high-temperature, high-pressure oil reserves [31].

In a paper previously published by the same research group, the method of obtaining appropriate MWNT dispersion is described in detail. When MWNTs are dispersed in the rubber matrix, a "cellulation matrix" is formed when the MWNTs are firmly connected to each other and bonded to the rubber in a three-dimensional matrix. Figure 13.29 shows a diagram and TEM micrographs of complete cellulation [32]. In the cellulation model, partial cellulation is expected to take place at 11 wt% MWNTs while complete cellulation takes place at 16 wt% MWNTs. The complete development of the cellulation concept is described in more detail in the paper by Endo et al. [32].

The storage modulus of neat FKM and MWNT-FKM composites is shown ranging from 2 wt% to 26 wt% [31]. The storage modulus is shown as a function of

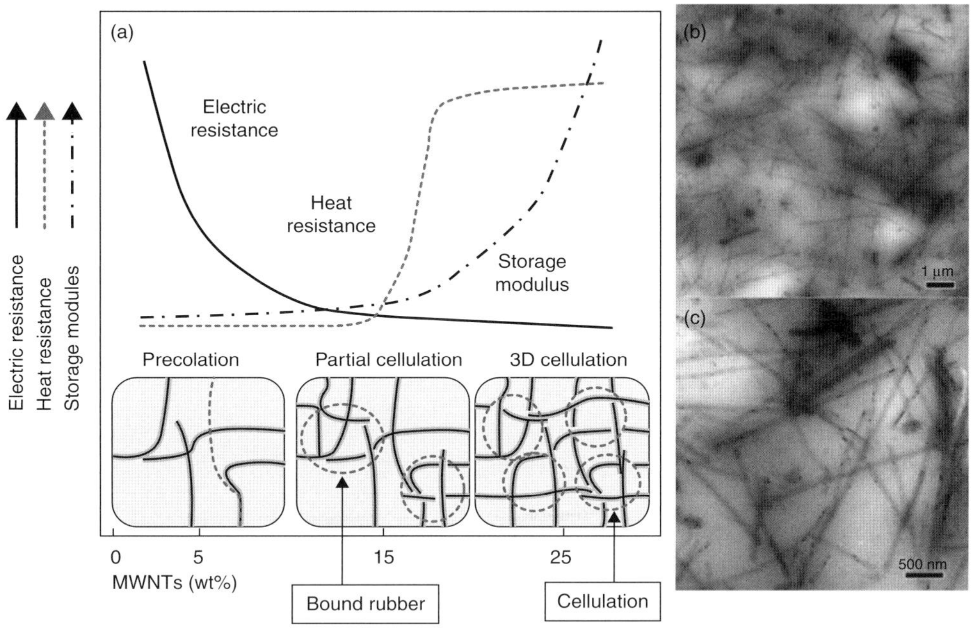

Figure 13.29. Diagram and TEM images of cellulation [32].

temperature for the varying compositions. The composites with a MWNT weight percentage above 9% maintain their storage modulus at temperatures above around 380°C, while the storage modulus of FKM and the other composites drops off dramatically. Additionally, the 26 wt% composite had the highest overall storage modulus and the neat FKM had the lowest. The storage modulus is shown as a function of MWNT wt% for various temperatures [31]. A critical point is reached at 9 wt% when the storage modulus increases for all temperatures. This indicates that the cellulation phenomenon has occurred. The storage modulus also does not decrease for temperatures above 200°C [31].

Table 13.5 shows the strength, elongation, glass transition temperature, maximum temperature rating, and compression-set percentage for the MWNT-FKM composite and the control groups of FKM and FFKM.

The strength and elongation are relatively the same for the three different materials at 25°C, but they differ slightly at 260°C, with the composite performing slightly better. The T_g is lower for the composite, which is desirable. The most significant benefit of the composite is seen in the compression-set data where the composite compression-set percentage is nearly 40% lower than for the two controls. This data indicates that the new MWNT-FKM material is a more durable sealing compound with application possibilities over a wider range of temperatures [31].

O-ring seals fabricated out of the three different materials were tested in a high-pressure, high-temperature fixture with silicon oil as the pressurizing medium. The seals were tested at different temperatures and pressurized to the point where they began leaking (if they began leaking). An example of the data obtained during testing is shown in Figures 13.30 (b and c) [31]. The overall results in Figure 13.30 show

Table 13.5. *Properties of composite and control seals [31]*

Properties	Test method	CNT-FKM present	Control-1 existing FKM	Control-2 existing FFKM
Strength @25°C, MPa	ASTM D1414, D412	15.8	11.9	15
Elongation @25°C, %	ASTM D1414, D412	100%	90%	90%
Strength @260°C, MPa	ASTM D1414, D412	6.1	1.8	4.8
Elongation @260°C, %	ASTM D1414, D412	40%	10%	40%
T_g by DMA, °C	ASTM D5026	−10.6	−3.2	12.7
Max temp rating, °C	Real seal test	260	175	260
Compression-Set, % @260°C/70h	ASTM D1414, D395B	39%	70%	83%

Properties of present compound and reference compounds were measured from AS 2–223 O-rings using ASTM methods and real seal life test.

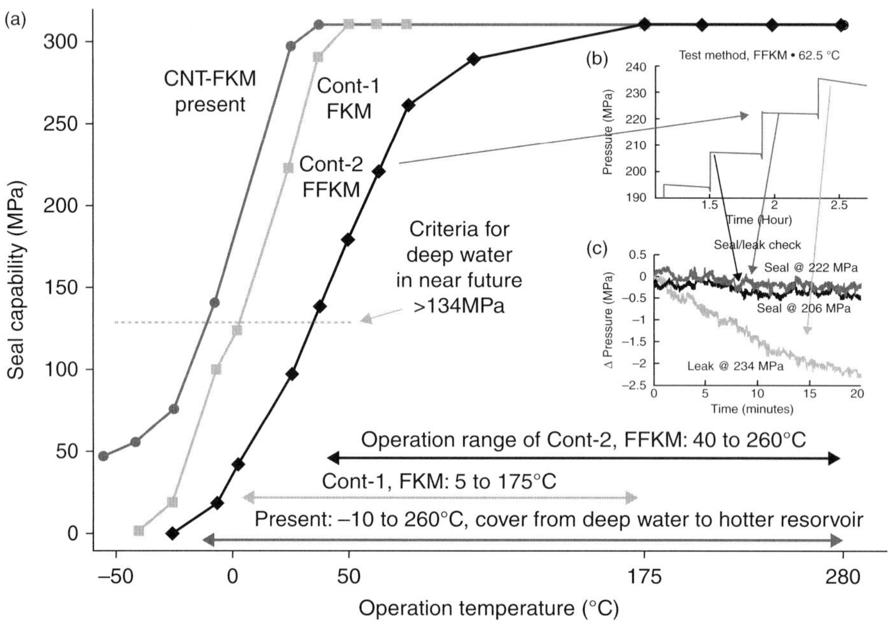

Figure 13.30. Seal capability versus operation temperature [31].

that the composite material has a higher seal capacity at each temperature, including the low temperatures. The resulting operating range of the composite at the required deepwater sealing pressure of 134 MPa is –10–260°C, whereas the FKM can only operate from 5°C to 175°C, and the FFKM can only operate from 40°C to 260°C. These results conclude that the MWNT-FKM composite can effectively seal in deepwater drilling operations where temperatures can range from 4°C to 260°C [31].

13.4.5.2 Controlled Electrolytic Metallic for Multi-Stage Shale Fracturing

In order to recover oil and gas reserves located deep beneath the earth's surface in shale rock, hydraulic fracturing has become necessary and commonplace in drilling operations to permeate the shale and allow the oil and gas to flow freely to the surface. For multi-stage shale fracturing, flow control devices, such as setting balls are used to seal each section of pipe for the frac. These devices, termed "frac balls," must withstand extremely high pressures and temperatures while in use. After fracturing is completed, these frac balls are intended to flow to the surface due to the backpressure from the well when the applied pressure is released. However, these frac balls often get stuck in the well due to deformations from the high pressure or simply because some of the frac balls are very small. They must then be drilled out to regain the full flowpath area, which is a costly and time-consuming process. Xu and his team have developed a new nanostructured material technology called controlled electrolytic metallic (CEM) to address this issue [33]. CEM balls are composites designed to be high strength, lightweight, and most importantly rapidly corrode at a predictable and controlled rate. The desired behavior of the CEM balls is illustrated dissolvable where CST is the critical service time at which the high-pressure pumping is completed [33]. These CEM balls can be used for multi-stage fracturing, and when left in the well, they will completely corrode, leaving the flowpath completely open. This technology has been tested in the Bakken shale field and performed as expected [33].

The structure of the CEM material consists of a cellular nanomatrix with metallic grains dispersed in the nanomatrix [33]. The specific metals and nanoparticles used were not specified in the article by Xu et. al., but from an article in Connexus, the metal material is specified as a magnesium-based alloy [34]. The nanomatrix is comprised of nanoscale metallic and/or ceramic layers. The nanomatrix reinforces the CEM balls, but also acts as the activation point for dissolution and controls the overall corrosion rate. The material is designed to rapidly corrode with typical downhole temperatures of 120°F or higher when immersed in fluids containing ions, such as seawater, completion brines, formation fluids, or remediation acids [33].

The CEM balls were subjected to numerous tests to evaluate their strength, dissolution rates, and impact resistance. Pressure tests were performed by applying hydraulic pressure to the top of a ball while in the ball seat for the specified operating time at an elevated temperature. The pressure leakage was monitored throughout the test, and the balls were evaluated after the test for deformation, cracking, and weight loss. A dissolution test was performed by submerging CEM balls in 3% potassium chloride at 200°F and monitoring the change in diameter and weight of each ball. At a dissolution rate of 0.005 inches per hour, it took approximately 41 hours for the 3.5 inch CEM ball to pass through a 0.25 inch increment ball seat. This ball dismount time is ideal for drilling operations as it matches the well shut-in time before full flowback and production flows are established [33].

The high-flowrate impact tests conducted on the CEM balls showed that there was no significant damage to the balls in high-flowrate impact simulations. In the wellbore, fluids can flow up to 80 bbl/min, and in the simulations, the only damage was very minor seal marks on the CEM balls [33].

More than 200 CEM frac balls were successfully tested in multi-stage fracturing systems in the Bakken shale field. One 3.5 in. frac ball used in the last stage of a 21-stage fracture was recovered for evaluation. This specific frac ball operated under a peak pressure of 8,000 psi and was left in the wellbore for about 7 days at a temperature of 120–160°F. Images of the frac ball before use and after the 7 days are shown in [33]. The diameter of the frac ball was reduced to 2.93 inches, but the inner material is still completely intact and mechanically sound. All of the CEM frac balls tested in the field performed their pressure-regulating functions and then degraded in the oil well without additional drilling operations needed to clear the fluid path [33].

13.4.5.3 Decreasing Shale Permeability Using Silica Nanoparticles

When extracting oil and gas from shale rock formations, water penetration into shale formations is very common. This water penetration causes the shale to swell, creating shale failure, which could lead to wellbore instability. Oil-based muds are one solution to this problem, as there is no interaction between oil and shale, but water-based muds are preferred due to environmental contamination issues. Water enters the shale through tiny pores on the shale surface. The shale pore throat size ranges from 10 nm to 30 nm, and theoretically, if these pores can be plugged, water will be unable to enter the shale. The diameter of traditional drilling mud additives, such as bentonite or barite ranges from 100 nm to 10,000 nm, which is much too large to plug the shale pores. Nano-sized particles are small enough to fit in the pores of the shale and prevent water from entering the shale [28].

Cai et al. investigated using nanoparticles to reduce the permeability of shale [28]. Nanomodified, commercially available silica nanoparticles were used in this specific study, since these types of nanoparticles are relatively inexpensive and could be employed on a much larger scale in the oil field. The silica nanoparticles vary in size from 5 nm to 22 nm, so in relation to the shale pore throat size, they would fit inside the pores. Three different types of water-based muds were tested: a fresh-water mud (FWM), a thick bentonite mud, and a low-solids mud (LSM) [28].

The pressure-penetration (PP) test was used to measure the shale permeability. A three-step testing procedure was used on hard and preserved Atoka shale. The first step was to flow seawater through the shale sample to ensure that each shale sample had the same original permeability. The second step was to run a PP test using the base mud to obtain a reference permeability before the final PP test was run with the 10 wt% nanosilica-modified mud.

The results of the experiments showed that the addition of the silica nanoparticles greatly reduced the amount of water that penetrated the shale. The permeability reduction percentage for the bentonite mud and the LSM for six different silica nanoparticles (NP-A through NP-F) is shown in [28]. The permeability reduction percentage was obtained by dividing the permeability difference of the base mud with 10% nanoparticles by the permeability difference of just the base mud [28].

The nanoparticles with diameters ranging from 7 nm to 15 nm are most effective at reducing permeability. Therefore, it seems that the nanoparticles in this range were the only ones that could effectively fit inside and plug the shale pores. Once the nanoparticle diameter became larger than 15 nm, the permeability reduction percentage dropped off quite sharply. Additional experiments and testing will have to be done to determine the optimum nanoparticle size to reduce permeability in different types of shale [28].

13.5 Summary

In the near future, it is likely that nanotechnology will be employed in all aspects of the oil and gas industry. As fewer oil reserves are being discovered, techniques to produce oil and gas from unconditional oil reserves will have to be developed. Nanosensors can be employed to help characterize the geology and interactions between rock and oil. Nanomaterials are ideally suited for this task as they can be structured to better handle the higher pressures and temperatures downhole. Nanofluids could help increase drilling efficiency or help prevent borehole instability by plugging nano-sized pores in the rock. There are endless possibilities for applications of nanotechnology, and the oil and gas industry is just beginning to seriously research and apply these techniques. Once the financial incentives of nanotechnology become more apparent, the oil and gas industry will likely see a much faster growth of nanotechnology implementation.

13.6 Questions and Problems

1. How does the nanomaterial dispersion affect the tribological properties of polymer nanocomposites in general? Provide examples to illustrate your claims.
2. How are the tribological properties of polymer-clay nanocomposites influenced by the type of clay, surface treatment of the clay, and how clay is processed in the polymer matrix? Provide examples.
3. This chapter reviewed the enhancement of tribological properties of thermoplastic-based nanocomposites using MMT clay, MWNT, SWNT, and graphene nanomaterials. Which types of nanomaterials are more effective for thermoplastic-based nanocomposites? Provide examples. Discuss the protective mechanism of these nanomaterials used to provide better tribological properties. Besides the nanoparticles you already mentioned, what other types of nanomaterials are used to improve tribological properties of polymer nanocomposites?
4. Discuss other commonly used characterization techniques other (a) Taber abraser, (b) pin-on-disc tribometer, and (c) scratch test mentioned in Section 13.2.1 to evaluate tribological properties of polymer nanocomposites.
5. Conduct a literature search on how the tribological properties of thermoset-based and thermoplastic elastomer-based nanocomposites are improved.
6. How does the nanomaterial dispersion affect the permeability properties of polymer nanocomposites in general? Provide examples to illustrate your claims.

7. This chapter covered the enhancement of permeability properties of polymer nanocomposites using one-dimensional (graphene oxide, graphene nanoplatelet, and nanoclay), two-dimensional (MWNT, CNF, buckypapers, and carbon nanofiber papers), and three-dimensional (zeolite) scaled nanomaterials. What other types of nanomaterials can be used effectively to enhance permeability properties of polymer nanocomposites? Provide examples.
8. In addition to the nanocomposites mentioned in Question 7, what other types of nanoparticles can be used to enhance permeability properties of polymer nanocomposites?
9. Describe other oil field applications using polymer nanocomposites with one-, two-, and three-dimensional scaled nanomaterials that were not mentioned in this chapter. Provide some examples.

REFERENCES

1. International, ASTM D4060 (2010). Standard Test Method for Abrasion Resistance of Organic Coatings by the Taber Abraser.
2. Taber Rotary Abraser 5135/5155 (2008), edited by Taber Industries.
3. Q. Zhou, K. Wang, and L. S. Loo (2009). Abrasion studies of nylon 6/montmorillonite nanocomposites using scanning electron microscopy, Fourier transform infrared spectroscopy, and X-ray photoelectron spectroscopy. *Journal of Applied Polymer Science* 113(5), 3286–3293.
4. S.-P. Liu, S.-S. Hwang, J.-M. Yeh, and C.-C. Hung (2011). Mechanical properties of polyamide-6/montmorillonite nanocomposites – prepared by the twin-screw extruder mixed technique. *International Communications in Heat and Mass Transfer* 38(1), 37–43.
5. G. Srinath and R. Gnanamoorthy (2006). Two-body abrasive wear characteristics of nylon clay nanocomposites: Effect of grit size, load, and sliding velocity. *Materials Science and Engineering: Part A* 435–436, 181–186.
6. Mu, B., Q. Wang, T. Wang, H. Wang, and L. Jian (2008). The Friction and Wear Properties of Clay Filled PA66. *Polymer Engineering & Science* 48 (1), 203–09.
7. International, ASTM (2009). The Standard Test Method for Evaluation of Scratch Resistance of Polymeric Coatings and Plastics Using an Instrumented Scratch Machine.
8. F. J. Carrión, C. Espejo, J. Sanes, and M. D. Bermúdez (2010). Single-walled carbon nanotubes modified by ionic liquid as antiwear additives of thermoplastics. *Composites Science and Technology* 70(15), 2160–2167.
9. M. D. Bermúdez, F. J. Carrión, C. Espejo, E. Martínez-López, and J. Sanes (2011). Abrasive wear under multiscratching of polystyrene + single-walled carbon nanotube nanocomposites: Effect of sliding direction and modification by ionic liquid. *Applied Surface Science* 257(21), 9073–9081.
10. L. F. Giraldo, W. Brostow, E. Devaux, B. L. Lopez, and L. D. Perez (2008). Scratch and wear resistance of polyamide 6 reinforced with multiwall carbon nanotubes. *Journal of Nanoscience and Nanotechnology* 8(6), 3176–3183.
11. S. S. Kandanur, M. A. Rafiee, F. Yavari, M. Schrameyer, Z.-Z. Yu, et al. (2012). Suppression of wear in graphene polymer composites. *Carbon* 50(9), 3178–3183.
12. J. Paulo Davim (Ed.) (2013). *Tribology of Nanocomposites*. Berlin Heidelberg: Springer-Verlag.
13. M. Bhattacharya, S. Biswas, and A. K. Bhowmick (2011). Permeation characteristics and modeling of barrier properties of multifunctional rubber nanocomposites. *Polymer* 52, 1562–1576.

14. Q. Wu, W. Zhu, C. Zhang, Z. Liang, and B. Wang (2010). Study of fire retardant behavior of carbon nanotube membranes and carbon nanofiber paper in carbon fiber reinforced epoxy composites. *Carbon* 48, 1799–1806.
15. H. Czichos, T. Saito, and L. E. Smith (Eds.) (2006). Mechanical Properties. In *Springer Handbook of Materials Measurement Methods*. New York: Springer, pp. 283–397.
16. H. Karkhanechi, H. Kazemian, H. Nazockdast, M. R. Mozdianfard, and S. M. Bidoki (2012). Fabrication of homogenous polymer-zeolite nanocomposites as mixed-matrix membranes for gas separation. *Chemical Engineering & Technology* 35(5), 885–892.
17. A. M. Pinto, J. Cabral, D. A. Pacheco Tanaka, A. M. Mendes, and F. D. Magalhaes (2013). Effect of incorporation of graphene oxide and graphene nanoplatelets on mechanical and gas permeability properties of poly(lactic acid) films. *Polymer International* 62, 33–40.
18. J. H. Chang, Y. K. An, and G. S. Sur (2003). Poly(lactic acid) nanocomposites with various organoclays. I. Thermomechanical properties, morphology, and gas permeability. *Journal of Polymer Science Part B* 41(1), 94–103.
19. R. Poreba, M. Spirkova, L. Brozova, N. Lazic, J. Pavlicevic, and A. Strachota (2013). Aliphatic polycarbonate-based polyurethane elastomers and nanocomposites. II. Mechanical, thermal, and gas transport properties. *Journal of Applied Polymer Science* 127(1), 329–341.
20. L. J. van Rooyen, J. Karger-Kocsis, O. C. Vorster, and L. D. Kock (2013). Helium gas permeability reduction of epoxy composite coatings by incorporation of glass flakes. *Journal of Membrane Science* 430, 203–210.
21. G. Prusty and S. K. Swain (2013). Dispersion of multiwalled carbon nanotubes in polyacrylonitrile-co-starch copolymer matrix for enhancement of electrical, thermal, and gas barrier properties. *Polymer Composites* 34(3), 330–334.
22. S. Castarlenas, P. Gorgojo, C. Casado-Coterillo, S. Masheshwari, M. Tsapatsis, et al. (2013). Melt compounding of swollen titanosilicate JDF-L1 with polysulfone to obtain mixed matrix membranes for H_2/CH_4 separation. *Industrial & Engineering Chemistry Research* 52(5), 1901–1907.
23. M. Valix, H. Mineyama, C. Chen, W. H. Cheung, J. Shi, and H. Bustamante (2011). Effect of film thickness and filler properties on sulphuric acid permeation in various commercially available epoxy mortar coatings. *Water Science & Technology* 64(9), 1864–1869.
24. X. Kong and M. Ohadi (2010). Applications of Micro and Nano Technologies in the Oil and Gas Industry – Overview of the Recent Progress. Paper presented at *Abu Dhabi International Petroleum Exhibition and Conference*, Abu Dhabi, UAE, 1-4 November. Society of Petroleum Engineers, SPE-138241-MS. doi: 10.2118/138241-MS.
25. S. K. Singh, R. M. Ahmed, and F. Growcock (2010). Vital Role of Nanopolymers in Drilling and Stimulations Fluid Applications. Paper presented at *SPE Annual Technical Conference and Exhibition*, Florence, Italy, 20-22 Sept. ISBN: 978-1-55563-300-4.
26. P. Pourafshary, S. S. Azimipour, P. Motamedi, M. Samet, S. A. Taheri, et al. (2009). Priority Assessment of Investment in Development of Nanotechnology in Upstream Petroleum Industry. Paper presented at the *SPE Saudi Arabia Section Technical Symposium*, Al Khobar, Saudi Arabia, 9-11 May. SPIE No. 126101.
27. N. Nabhani, M. Emami, and A. B. Taghavi Moghadam (2011). Application of nanotechnology and nanomaterials in oil and gas industry. *AIP Conference Proceedings* 1415(1), 128–131.
28. J. Cai, M. E. Chenevert, M. M. Sharma, and J. E. Friedheim (2012). Decreasing water invasion into atoka shale using nanomodified silica nanoparticles. *Society of Petroleum Engineers* 27(1), 103–112.

29. V. Savino, G. M. Fallatah, and M. S. Mehdi (2010). Applications of nanocomposite materials in the oil and gas industry. *Advanced Materials Research* 83–86, 771–776.
30. M. Clark (2009). Minimizing Environmental Footprint by Utilizing Prevention Technology. Paper presented at the *SPE Annual Technical Conference and Exhibition*, Houston, TX, 26-29 Sept. Society of Petroleum Engineers, SPE 90242.
31. M. Ito, T. Noguchi, H. Ueki, K. Takeuchi, and M. Endo (2011). Carbon nanotube enables quantum leap in oil recovery. *Materials Research Bulletin* 46(9), 1480–1484.
32. M. Endo, T. Noguchi, M. Ito, K. Takeuchi, T. Hayashi, et al. (2008). Extreme-performance rubber nanocomposites for probing and excavating deep oil resources using multi-walled carbon nanotubes. *Advanced Functional Materials* 18(21), 3403–3409.
33. Z. Xu, G. Agrawal, and B. J. Salinas (2011). Smart Nanostructured Materials Deliver High Reliability Completion Tools for Gas Shale Fracturing. Paper presented at the *SPE Annual Technical Conference and Exhibition*, Denver, CO, 30 Oct-2 Nov. Society of Petroleum Engineers, SPE 146586.
34. One colossal discovery: Materials science breakthrough creates nanostructured material of immense proportions. (2011). *Connexus* 2(2), 8–13.
35. A. Dasari, S. Lim, Z. Yu, and Y. Mai (2007). Toughening, thermal stability, flame retardancy, and scratch–wear resistance of polymer–clay nanocomposites. *Australian Journal of Chemistry* 60(7), 496–518.
36. A. Dasari, Z.-Z. Yu, and Y.-W. Mai (2009). Fundamental aspects and recent progress on wear/scratch damage in polymer nanocomposites. *Materials Science and Engineering: R: Reports* 63(2), 31–80.
37. J. H. Koo (2006). *Polymer Nanocomposites: Processing, Characterization, and Applications*. New York: McGraw-Hill.
38. S. K. Sinha, T. Song, X. Wan, and Y. Tong (2009). Scratch and normal hardness characteristics of polyamide 6/nano-clay composite. *Wear* 266(7–8), 814–821.
39. R. Verdejo, M. Mar Bernal, L. J. Romasanta, and M. A. Lopez-Manchado (2011). Graphene filled polymer nanocomposites. *Journal of Materials Chemistry* 21(10), 3301–3310.

PART THREE

CONCERNS AND OUTLOOK

14 Environmental and Health Impacts for Nanomaterials and Polymer Nanocomposites

14.1 Introduction

The use of nanomaterials in various products like polymer nanocomposites (PNCs) has seen a significant rise in research and development on their potential benefits in the last 25 years. With the increased use, the safety and environmental effects of nanomaterials and products resulting from their addition have been of concern. Being a fairly new technology, there is not a lot of long-term data to support a definitive answer regarding the safety of these new materials. The variety and diversity of nanomaterials makes it even more difficult to generate a general answer about their safety. The research community has begun preliminary experimentation into this subject and has determined two classes of hazards: direct and secondary. Direct hazards include direct contact with nanomaterials, while secondary hazards include the effects of PNC products when they experience chemical degradation. Affecting lives of not just those who interact with bulk nanomaterial but also everyday people who use products that includes nanomaterials like sunscreen, the safety and environmental effects are very important. Nanomaterials can be absorbed into the body in many ways, including inhalation, ingestion, and even through the skin. The uses range widely, from medical implants to automotive parts. As the usage of nanomaterials grows, the waste from these materials also increases. This leads to the important question of whether the products from new nanotechnology, nanomaterials, and polymer nanocomposites are safe materials for now and for the future.

This chapter includes three related subtopics:

1. Direct hazards: Safety and environmental health aspects of nanomaterials and polymer nanocomposites; risk assessment of nanomaterials and nanocomposites; safety standards;
2. Secondary hazards: Fire toxicity of polymer nanocomposites; and
3. Current regulation of nanomaterials and future perspectives.

14.1.1 Health and Safety Aspects of Nanomaterials

Because of their growing use in more applications, big impact in relatively small amounts, and ease of transmittance due to their size, the nanomaterials and materials that include nanomaterials are questioned for their safety and environmental effects. It is important to understand the safety and environmental effects of nanomaterials in order to understand the safety of polymer nanocomposites because, the nanomaterial is only an additive to the polymer under scrutiny. Although the polymer nanocomposites will act much more differently than the polymer and nanomaterial separately, there is still a chance that the nanomaterial can detach from the polymer nanocomposites and cause harm.

The overall effects of nanomaterial on human health and environmental safety are still not fully understood. Even Material Safety Data Sheets (MSDS) for nanomaterials, intended for full disclosure of the properties and safety of a substance supplied by the manufacturer, are not sufficient in communicating the hazards of the nanomaterial. From a selection of 23 MSDS documents published in 2010–2011, almost half (47.8%) were found to be unsatisfactory and in need of improvement [1]. Furthermore, the use of vague phrases, such as "No known toxicological effects" or "We are not aware of any reported health hazards for this product" supports the fact that the knowledge on the safety of nanomaterials is still limited and in need of extensive and rigorous research. Some nanomaterials, such as titanium dioxide and nanocarbon, however, have been researched more than other nanomaterials, and carbon nanotubes, for example, showed pathogenicity similar to that of asbestos [2]. Also, because of the diversity of nanomaterials, it will be difficult, if not altogether practically impossible, to come up with a general conclusion on their safety. The discussion of the safety of nanomaterials can be separated into three sections: (a) safety of nanomaterials in the short term, (b) environmental impact of nanomaterials in the long term, and (c) nanotoxicity in biomedical/biological applications.

14.1.1.1 Safety of Nanomaterials

When considering short-term effects of nanomaterials, exposure to nanomaterials is the main concern. For most people, exposure to nanomaterials will be through goods that incorporate nanomaterials in the product. Everyday products that may include nanomaterials are personal care items, sunscreen, household paints, and even sporting goods. However, a bigger safety concern would be at the factories where these products are produced, where the concentration and exposure levels are much higher than those from handling consumer products [3]. A common nanomaterial found in many consumer products, such as sunscreen and paints is titanium dioxide [4]. Titanium dioxide effects, risks, and usage guidelines have been studied by the Department of Health and Human Services, Centers for Disease Control and Prevention, and National Institute for Occupational Safety and Health.

Like other nanomaterials, titanium dioxide can be a particle with a commercial size from 0.22 μm to 0.25 μm. With it being used in the increasing number of applications, it is common for workers to be exposed to much higher than regular levels of titanium dioxide. There were 1.45 million metric tons of titanium dioxide produced in 2007, with an estimated number of 1,100 workers nationwide who handled it on a regular basis [4]. The permissible exposure limit for titanium oxide dictated by the Occupational Safety and Health Administration (OSHA) is 15 mg/m^3 in dust form – the same limit as for the particulates "not otherwise regulated." In 1988, the National Institute for Occupational Safety and Health (NIOSH) classified titanium oxide as possibly carcinogenic; six years later, NIOSH classified it as carcinogenic for animals with sufficient evidence, but only possibly carcinogenic for humans without enough evidence. So in short, both the safe limit and carcinogenic effect of titanium dioxide are not completely unknown, and a defined amount at which point it is harmful is still lacking. This generally seems to hold true for many other nanomaterials [1].

Case reports and studies performed on the safety of nanomaterials have dealt mainly with the uptake of nanomaterials through the respiratory system. This is the most common form of exposure for factory workers, because nanomaterials are often generated as aerosols [5]. Many case reports for titanium dioxide exposure in factory workers reported traces of titanium, not titanium dioxide, in the subjects' lungs [4]. The traces of mainly titanium signal some sort of interaction between the body and the nanomaterial. Even though the levels of titanium found in autopsies of patients with cancer were higher in the workers who were exposed to titanium dioxide, it cannot be directly correlated to titanium dioxide's level of safety at any quantitative level of exposure. This was not only because the studies showed that other known cariogenic particles like nickel and silica where present in the autopsies, but also because tobacco use of the subjects was also not taken into account.

In 1975, a study by Chen and Fayerweather was conducted on factory workers comparing the rate and cause of death among workers exposed to titanium dioxide to those at the same company who were not exposed to it. Although the quantitative measurement of titanium dioxide in 1975 can be questioned, the results led to a conclusion that there was no statically significant positive relationship found between titanium dioxide exposure and the cause of chronic respiratory disease. Another four studies done by Fryzek et al. in 2003, Boffetta et al. in 2001 and 2004, and Ramanakumar et al. in 2008 also found no relationship between exposure to titanium dioxide and chronic respiratory diseases of factory workers with different levels of exposure [4]. The study by Boffetta et al. in 2001, on the other hand, did find a relationship between an increased chance of lung cancer and the exposure to asbestos and BAP, which shows that the studies did show positive relational results for substances with known properties.

Based on the results of the aforementioned studies, it may seem that titanium dioxide is a safe nanomaterial to humans, although the results of similar experiments involving animals as exposure subjects were different. Studies have shown that exposure of rats to different grades of titanium dioxide caused pulmonary inflammatory response, cell damage, and cancer; additionally, frequency of occurrence for these afflictions was higher when the animals were exposed to ultrafine nanomaterial, compared to fine nanomaterial [4]. Knowing this, it can be concluded that exposure to titanium dioxide can have negative and potentially deadly effects to living organisms, although these effects may not be as apparent in humans, as they are in rats, given that no statistical correlation has been found to date. One reason suggested for the difference in results is the different levels of particle concentration in rats and humans lungs. Rat models were chosen, however, because they are known to be more accurate than mice and hamster models.

Even with the uncertainty between the rat model and actual human response, there is not enough evidence to discount the rat model for predicting actual human response. Until a study proves a correlation between human exposure and death, or no correlation between rat exposure and death, a solid conclusion cannot be drawn.

When looking at nanocarbon, another more extensively explored nanomaterial, the studies performed with multi-walled carbon nanotube on mice showed that mice exposed to the nanomaterial were much more likely to develop cancer, with a 90% incidence level [6]. Even though some studies have shown a relationship in cellular response between carbon nanotubes and asbestos, a known carcinogen, there is still no solid evidence that nanocarbon is indeed carcinogenic or even deadly to humans [2]. There are several animal experiments that have shown negative biological response to nanocarbons. Expressing that "the existing information on the lung toxicity of carbon nanotubes is limited and remains inconclusive" for humans only shows that work remains to define the safety of nanocarbons, which is generally true for the many new nanomaterials [7]. Finding a method of testing nanomaterials that gauges human responses more accurately would be greatly beneficial for discovering the toxicity of nanomaterials.

14.1.1.2 Environmental Effects

Direct exposure is important in understanding the short-term safety issues associated with a nanomaterial; however, as the use of nanomaterials increases in everyday items, the chances of nanomaterials spreading throughout the environment greatly increases. Adding a material that is foreign to the original environment may lead to issues that previously were not thought of. So where in the environment would most of the concerns be? A fast way for nanomaterials to enter in the ecosystem is through the water supply. Because the water supply is very critical in sustaining life as a direct water source for both humans and other animals, it would be advantageous to understand how nanomaterials can travel through water sources.

Nanomaterials can come in both hydrophobic and hydrophilic varieties. Different surface treatments can, however, affect a nanomaterial's interaction with water. Water, being hydrophilic, would carry hydrophilic materials very well, so it may seem logical to surface-treat nanomaterials to be hydrophobic to prevent the dispersion of nanomaterials via water. This idea is flawed, however, because surface treatment is defined for the compatibility of the nanomaterial in its intended use, which may be required to be hydrophilic. In addition, because of nanomaterial's unique properties (i.e., being very small), even hydrophobic nanomaterials like fullerene have the ability to form a stable colloid in water without any surface treatment [8].

Without good evidence of short-term nanomaterial safety, it is hard to estimate the concern for their environmental effects. The amount of nanomaterials needed in a practical application may not be very much compared to their macromolecule counterpart. For example, to sustain a fullerene colloid in water, the amount of fullerenes must be about 10 ppm. This level of nanomaterial, however, is around 100 times the safe solubility limit for hydrocarbons. Therefore, relative to the amount used in nanomaterial applications, it is possible for nanomaterials to have a significant impact on the environment. Even if nanomaterials may not be harmful to humans, in the end nanomaterials can interact directly with cells, which can be potentially deadly to small cellular organisms.

14.1.1.3 Nanomaterial Impact in Biomedical and Biological Applications

Even though regular people who do not work in an industry that uses nanomaterials may not receive "substantial" exposure to nanomaterials, their contact with nanomaterials is not nonexistent, although it is generally minimal. For example, a common everyday item containing titanium dioxide is sunscreen, which helps absorb UV radiation. This may be of concern because titanium dioxide has the possibility of forming free radicals when excited with UV radiation, which can cause cancer if absorbed into the body. Research that has been done on the effects on titanium dioxide on DNA and RNA found that the interaction would cause damage to the DNA and RNA, but these effects do not outweigh the risk of sun exposure [5, 9]. The risk of nanomaterials in everyday goods does exist, but benefits for the users generally override concerns over their well-being.

Another sometimes forgotten area in which nanomaterials may adversely affect patients is in implants. The implants themselves do not cause the issues of biocompatibility as they are correctly surface-treated; however, the wear in certain implants like a hip replacement will create particulates on a nanoscale. These particulates can cause complications, such as bone loss or inflammatory responses [5]. Different material interfaces can cause these particulates, including metal on metal, metal on polymer, metal on ceramic, and ceramic on ceramic, so the nanomaterials released depend on the type of interface [10]. Because the surface properties only affect the outside of the implant, the particulate that comes off from wear will not necessarily have the same surface properties as the bulk implant. As a result, the nanomaterial particulate and protein interactions at the site of the wear can cause many complications.

There have been many studies that deal with the cellular interactions of nanomaterials. Cellular response studies cannot always gauge the overall toxicity of the nanomaterial, however. Because nanomaterials are so small, they interact with living tissue on a cellular level. Cells often uptake nanomaterials via endocytosis, sometimes for disposal, which should not be seen as a toxic response, and other times for use as antioxidants, as in the case of fullerenes [11]. Fullerenes can also attach to the surface of cells. Although the complete interactions of fullerenes are not known, fullerene and other carbon-related nanomaterials' unique behaviors have sparked interested in the medical field to use nanomaterials to help find cures in areas relating to phototherapy and HIV [12]. Overall, research related to human body's reaction to nanomaterials have shown that dermal and oral toxicity seem fairly low while intravenous administration of the nanomaterial produces acute toxicity responses.

When discussing the safety of nanomaterials, the emphasis is often on their negative effects; however, it is important to note that by understanding the interactions with the human body, as well as the environmental impact, new beneficial uses can be discovered. Overall, the full impact of nanomaterials on the environment is still uncertain, with some research results pointing at their potential toxicity, but no evidence of causing serious, life-threatening complications in humans. As the use of nanomaterials increases, it is important to have the safety and environmental concerns well defined; the lack of knowledge is equally detrimental to understanding the already existing nanomaterials and to the development of new ones, or new applications for the existing ones.

14.1.2 Health and Safety Aspects of Polymer Nanocomposites

A specific application for nanomaterials is the creation of polymer nanocomposites by incorporating them into existing materials to augment the latter's properties. For example, adding different types of nanomaterials to a polymer can improve the polymer's ablation properties, as reviewed by Koo et al. [13]. The appeal of this application is in its ability to make small, carefully targeted adjustments to provide the original material with the characteristics it currently lacks, ensuring that it will not have to be discarded for certain uses, and eliminating the costly and often inefficient process of mixing-and-matching different materials in an attempt to improve performance. Nanomaterials allow polymers to adjust their properties, such as ablative performance, without sacrificing much of the material's original properties. As the popularity of nanotechnology increases and polymer nanocomposites increasingly feature in products associated with people's daily lives, their safety and environmental impacts must be identified to the fullest degree possible.

The major unknown safety and environmental concerns have to do with nanomaterials used in polymer nanocomposites. When discussing nanomaterials, both the short-term safety aspects and the longer-term environmental impact should be examined. Given that polymer nanocomposites as a material are fundamentally a polymer, their safety can be analyzed using the same techniques as for regular polymers.

14.1.2.1 Safety of Polymer Nanocomposites

Polymer nanocomposites are used in a variety of applications, but the ones with the most immediate impact on safety and health are the materials with the most direct human contact – biomedical applications. Polymers are often used in tissue scaffolding as a 3D matrix for new tissue to grow on. Other times, polymers are used to supply support for existing tissue. The scaffolds are temporary structures used during recovery, yet the polymer used for structural support would be a bone replacement meant for permanent implantation. With scaffolding, upon the end of recovery, the temporary material is surgically removed; with a degrading polymer, however, the invasive removal procedure is not necessary. Instead, the polymer naturally degrades at a set time designated by its design, with the component elements either absorbed or expelled by the body. Polymers used in the medical field, like many other polymers, can be augmented with nanocomposites to improve certain properties. Unique to polymers used in medical practices is the importance of biocompatibility and biodegradability..

Common polymers used in the medical field include polyglycolide (PGA) and polylactide (PLA). Synthetic polymers are more highly valued than natural polymers are, for their flexibility in properties [14]. Poly(butylene adipate-co-terephthalate) (PBAT) is a common synthetic material used in industrial applications and has the ability to have different physical properties, including biodegradability [15, 16]. However, PBAT has not been considered for implanted medical applications like joint replacements because of its poor mechanical resistance. To improve its mechanical resistance, nanomaterials of montmorillonites, synthetic fluorine mica, and sepiolite have been used [15, 16].

In order to test the biocompatibility of the polymer nanocomposite, three different tests were done to evaluate the overall safety of the material [15, 16]. The first set of test measured the in vitro cytotoxicity of the material, which included standard cell culture procedures with both L929 fibroblasts and hFOb 1.19 osteoblasts cells. Films of the material were essentially grown with the cells. The second test measured the protein adsorption (the paper erroneously used the term "absorption," which is the entrance of a protein into the material, where adsorption is the attachment of the protein on the surface) of the material. The importance of this is to reduce negative immunoresponse by attaching more natural proteins to the foreign material. The lack of adsorption will lead to the body rejecting the material. Usually this can be solved by surface treatments, but in this case the polymer's own properties are of interest. The procedures to measure protein adsorption is to immerse the material into a solution with human serum albumin and human plasma fibrinogen (two of the more common proteins in the body) and then measure the total proteins attached to the polymer. The last test is to test blood biocompatibility, which is easily done by immersing the polymer into rat blood and comparing the resulting hemoglobin concentration, red and white blood cell counts, and platelet count.

The results from comparing PBAT and PBAT with 10% clay show significant improvement in the compressible strength and modulus and good biocompatibility, and even improved biocompatibility in some cases with the nanomaterial [15, 16]. The cell growth of PBAT and PBAT with nanomaterials showed similar growth. The cells also grew on both the nanomaterials and the polymer, furthering the notion that the nanomaterial did not adversely affected the cells. Cell detachment or fragmentation of the cell cytoplasm was not observed in PBAT or PBAT with nanomaterials. Further analysis showed that in fact not only were the growth on PBAT with nanomaterials about the same as PBAT alone but PBAT with 10% montmorillonite (specifically NANOFIL 116 from Southern Clay Products) actually grew better that plain PBAT (Figure 14.1) [16].

In protein adsorption tests, similar results were achieved in terms of cell growth. The PBAT with nanomaterials for the most part outperformed PBAT alone (Figure 14.2) [16]. Human serum albumin adsorption was higher with nanomaterials, which

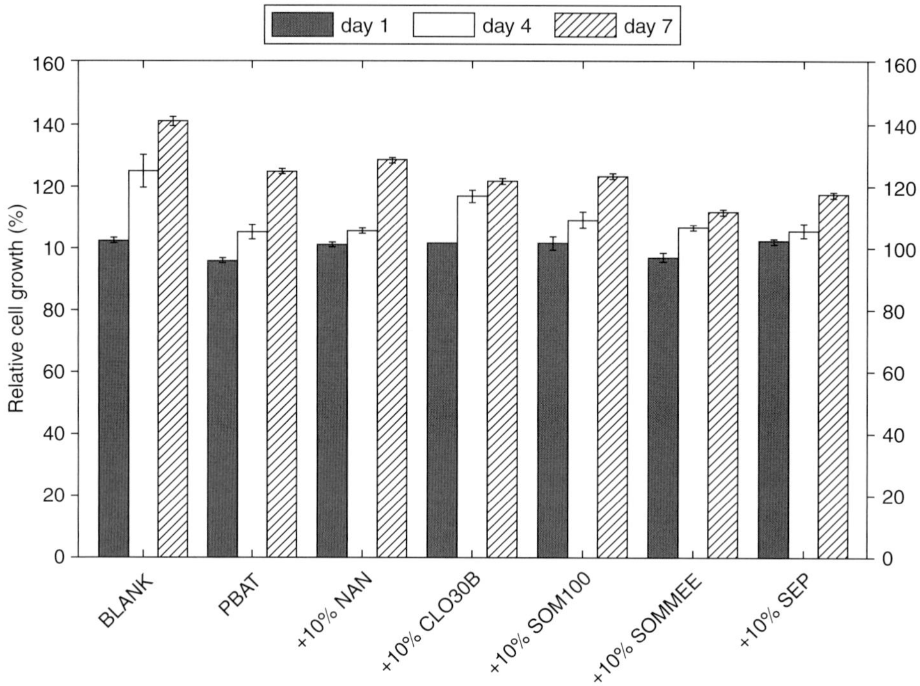

Figure 14.1. Growth of osteoblast cells in different PBAT membrane suspensions [16].

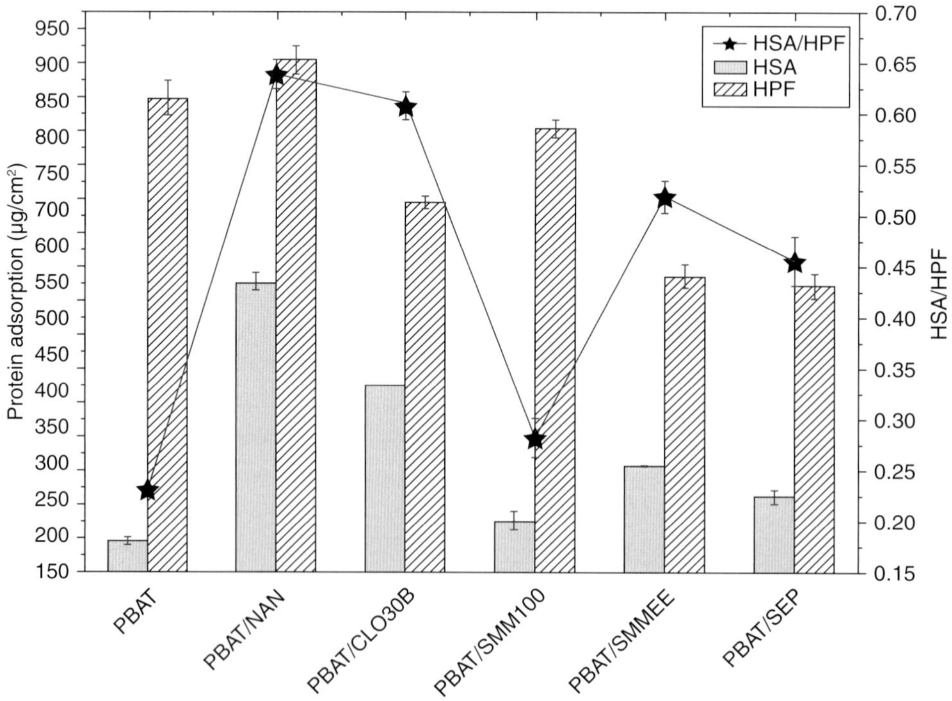

Figure 14.2. Protein adsorption of albumin and fibrinogen [16].

could be attributed to the electrical attraction between the nanomaterial and the protein. On the other hand, human plasma fibrinogen adsorption decreased, which is a desirable development, because plasma fibrinogen tends to increase the chances of thrombosis or thromboembolism. Overall, the ratio between albumin/fibrinogen adsorption is important to the biocompatibility of the material, which the PBAT with nanomaterial improved.

In blood compatibility tests, the white and red blood cell, platelet, and hemoglobin counts were similar to that of the standard control blood. This shows that both PBAT and PBAT with nanomaterial did produce negative hemostatic responses [15, 16].

This experiment on biocompatibility of a PBAT after being augmented with clay nanomaterial shows that polymer nanocomposites can indeed be safe enough for use inside the body. The paper did state, however, that clay's biocompatibility was unknown before this experiment, and also stated that this was one of the first experiments of this kind to test PBAT with nanomaterials. It is good to note that even though polymer has been used in the medical field for a while, the addition of nanomaterials to the polymer is still fairly new, and testing for the safety should be done before real use. Using nanomaterials not only can strengthen a polymer but can also increase biocompatibility, which is an ideal situation for the progression of nanotechnology.

14.1.2.2 Environmental Effects

Natural biodegradation of polymers is very useful with tissue scaffolding implants, but natural biodegradation is actually also very useful in minimizing the environmental impact a polymer can have. Polymers that cannot naturally biodegrade collect as permanent trash that affects the environment, often quite drastically. For example, a plastic that cannot be degraded can, over time, make its way into lakes and accumulate as surface scum. If enough trash floats on the surface of a lake, it begins to block sunlight, which leads to changes in the ecological balance of the underwater environment, causing damage to the ecosystem. Such accumulation may also affect water's drinkability, thus affecting public health and safety. A biodegradable polymer will break up into its component parts over time, which separately cannot do as much damage as an intact polymer can. Still, the only way even a biodegradable polymer will not detrimentally affect the environment is if its component parts are not biologically harmful. Nanomaterials alone, separated from the polymer, could be toxic, which would be more harmful to the environment than a combination of polymer and nanomaterials because of the nanomaterials' higher penetrability due to their smaller size [2]. The possibility of nanomaterials detaching from the polymer is especially likely if the polymer nanocomposites are incinerated with the rest of the waste materials. As a result, as more nanomaterials are used to augment polymers, the safety of the polymers – which should preferably be degradable – must be ensured.

Biodegradable polymers in common use include polylactic acid, PLA (corn-derived plastic), PHA (bacterial polyesters), and thermoplastic starch (TPS) [17]. Another type of polymer that is increasingly popular is cellulose plastics from trees. These polymers, such as cellulose acetate (CA) and cellulose acetate propionate

(CAP), would be suitable biodegradable materials; however, a major manufacturing difficulty is that the melt-processing temperature exceeds its decomposition temperature [17]. Using triethyl citrate as a plasticizer and nanoclay with the cellulose acetate has never been tested. Melt-processing through extrusion-injection molding was used to create the polymer nanocomposites, and several tests like X-ray diffraction and TEM were used to verify the material's strength. The results showed that it was possible to use nanoclay, triethyl citrate as a plasticizer in cellulose acetate, to make a biodegradable and safe polymer nanocomposite. A material of interest must fulfil the certain properties that can be tuned with nanomaterials, but the safety of the final polymer still needs to be considered.

Just like regular polymers, polymer nanocomposites can be treated similarly when evaluating their general safety. Whether the application is an internal implant or a polypropylene material substitution, a polymer nanocomposite's properties and external interactions change and can be treated as a new material. Another method of analyzing the safety polymer nanocomposites can be the analysis of its parts; however, the composite properties would be disregarded and this test will be of less total worth because the overall interactions may be different.

14.1.3 Risk Assessment of Nanomaterials and Nanocomposites and Safety Standards

The use of nanotechnology is still relatively new, and the standards to evaluate the safe usage handling and the evaluation of nanomaterial's safety and environmental impacts are still mostly under development. As described earlier, many MSDS are still very vague when it comes to the safety on the nanomaterial. For example, Sigma-Alrich manufactures carbon nanopowder with sizes less than >50 nm. The MSDS sheets that accompany the material explain that there are no "known Occupational Safety and Health Association (OSHA) Hazards" and that the potential health effects all start with "may be harmful if" inhaled, absorbed, and the like [18]. These descriptions are not very helpful and are most likely due to the lack of knowledge and testing procedures for nanomaterials. By understanding the safety of nanomaterials, it will be easier to identify the safety and environmental concerns of nanocomposites.

14.1.3.1 Titanium Dioxide Risk Assessment

The NIOSH did a quantitative risk assessment for titanium dioxide to find a recommended exposure limit. As explained earlier in this chapter, the data from human studies have revealed no correlation between exposure to titanium dioxide and increased chances of death, while experiments with rats have shown the animals' negative reaction to the nanoadditive. Therefore, to get data for this risk assessment, the data from the rat experiments were used. Figure 14.3 shows the flow chart of how the risk assessment process worked [4].

This process can be generalized, however, for any nanomaterial in need of a risk assessment. In short, an animal model is developed and tested. A measurement of

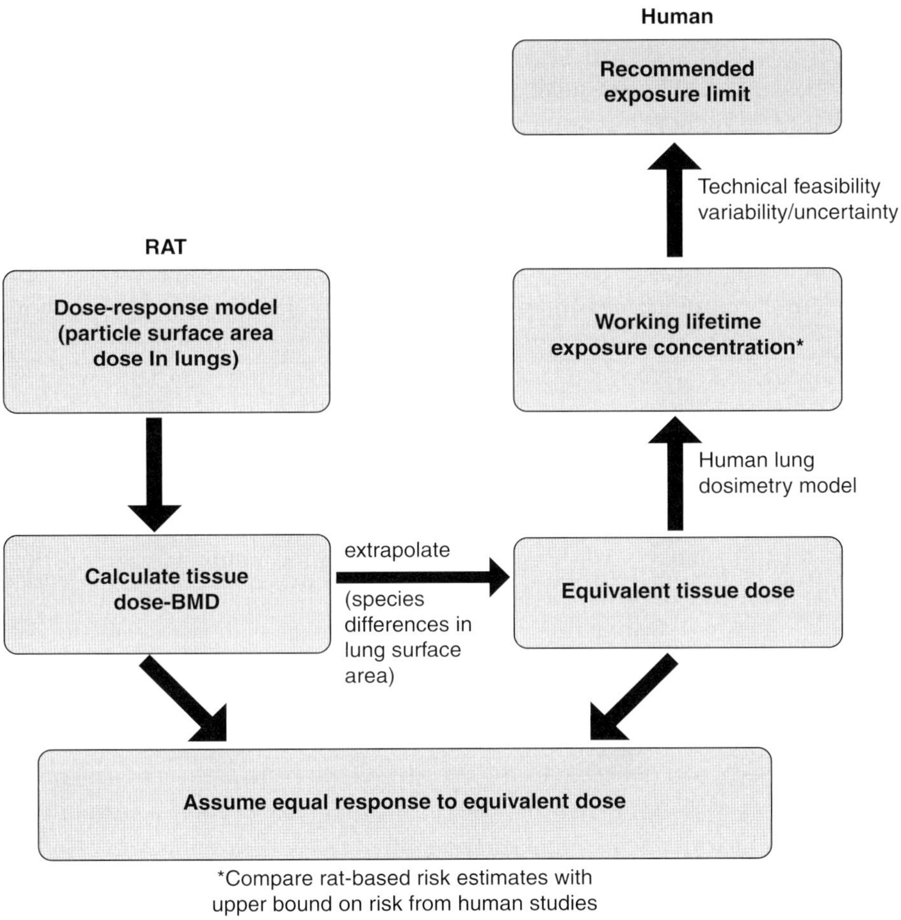

Figure 14.3. Risk assessment flow chart [4].

maximum exposure limit can be calculated by extrapolating from the animal model to correct for properties, such as differences in lung surface area, while taking into account the likelihood of exposure to a specific nanomaterial. These exposure limits are the quantitative measurements of the risk assessment. The response of the animal model, such as certain diseases or a descriptive cellular response, can also give insight into how the nanomaterial actually interacts with the cells within the body. This would be considered the qualitative measurements of the risk assessment. With these two components we can understand the probability of risk and the potential harm that may be caused by a nanomaterial.

Several properties that differ between rats and humans that must be taken into consideration for the calculation of the recommended exposure limit include lungs mass, lungs surface area, respiratory frequency, and tidal volume of the lungs. As a reference value, from the data of the rat experiments, the critical dosage per rat was 0.047 m^2 for fine titanium dioxide [4]. Critical dosage was defined as the dosage of nanomaterial retained in the lung that initiated inflammation or increased the risk of lung cancer. After extrapolation of the data, the maximum exposure limit of titanium dioxide is said to be 2.4 mg/m^3 for fine titanium dioxide and 0.29 mg/m^3 for ultrafine titanium dioxide [4]. The smaller amount needed for ultrafine titanium dioxide can possibly be explained by the increased surface-area-to-volume ratio, which may increase the absorption of the nanomaterial and increases its effects. These amounts, however, are most likely lower than the actual real maximum exposure amount because of the discrepancies in the results from human case studies compared to rat experiments, as concluded from a comparison of the final risk assessment data [4]. However, even though the limits are lower, NIOSH concludes that these values still "prudent to use" because there is no data to reasonably dismiss these values from being used to estimate risk of lung cancer from exposure to titanium dioxide.

Furthering the risk assessment to a recommended exposure limit to prevent an excess risk of lung cancer would be a lifetime mean of 13.2 mg/m^3 of fine titanium dioxide and 1.6 mg/m^3 of ultrafine titanium dioxide, both derived from rat experiments [4]. Consistent with the rest of the data, the ultrafine nanomaterial poses a higher risk than does its fine counterpart. An interesting conclusion that the NIOSH reached was that the higher risk associated with ultrafine nanomaterial was not because of the chemical composition of the compound, but rather its physical size. Taking this idea, it may be true that the smaller the nanomaterial, the higher the risk for that nanomaterial solely because of the nanoscale of the particle. The NIOSH also stated that this nanoscale effect was found to be true on several other poor-solubility and low-toxicity particles. This may be an interesting conjecture that can affect all different types of nanomaterial because it could be that the physical properties become more important than the chemical properties beyond a certain size. Future experiments for risk assessment can possibly identify the correlation between physical size and toxicity levels for general nanomaterials.

Like titanium dioxide, carbon nanotubes are classified by the International Agency for Research on Cancer as a Group B agent, which includes agents that are possibly carcinogenic to humans. A comparison with another risk assessment study done with carbon nanotubes showed that carbon nanotubes do produce negative biological responses, such as pulmonary fibrosis and inflammation [6]. What is more interesting is that single-walled carbon nanotubes caused a greater inflammatory response than multi-walled carbon nanotube [6]. This behavior seems to parallel the behavior of fine and ultrafine titanium dioxide. The recommended exposure limit for carbon nanotubes like titanium dioxide was derived from animal based experiments to model the human lung. However despite the differences, the recommended exposure limit of carbon nanotubes is 1 μg/m^3 which is far less compared to titanium dioxide's 1.6 mg/m^3 recommend exposure limit [6]. This may be due to the physical differences in a tube and a particle. To further support the importance of physical morphology, the recommended exposure limit for carbon black, which is like titanium dioxide in particle form, not tube form, is 2.5 to 5 mg/m^3, which is more similar to the limits for titanium dioxide.

Without solid evidence of the safety effect of titanium dioxide, it is hard to perform a risk assessment and taking the worst case scenario that there is a negative effect form the rat experiment, the value of recommend exposure limit may or may not be correct but is still a number with some valid reasoning. As with many other nanomaterials, without real solid evidence of the effects a nanomaterial has on humans, it is hard to achieve value of recommended exposure limit is hard to achieve even if we understand the impact on animals or even at a cellular level. Therefore, tests that attempt to bypass real-world animal and human studies by using microscopic cellular methods may be of interest. Also, assessments that can correlate the physical nature (size and shape) to different levels of toxicity would be beneficial as a general study on the risks of nanomaterials.

14.1.3.2 Standards for Testing Nanomaterial Safety

There are a few main associations that deal with the standardization of testing different materials. Even though nanomaterials are relatively new, there are already testing standardization for nanomaterials, which can help assess risks. However, many of these tests are for the physical or chemical characterization of the material rather than for the safety or toxicity of the material. From both the International Organization of Standardization (ISO) [19] and the American Safety for Testing and Materials (ASTM) [20] there are only a few that test the safety of the nanomaterial. No complete access to these documents has been given to the author, so the abstracts and supporting materials were used instead.

There are several ISO standards used to gauge the safety of a nanomaterial. ISO 13121:2011 evaluates the overall risk of a nanomaterial [19]. ISO 10801:2010 evaluates the toxicity of metal nanoparticles, such as silver or gold. It is intended to standardize the testing for the inhalation of nanomaterials in powder form with sizes smaller than 100 nm. A practical challenge stated in the standard is the difficulty dispersing the nanomaterial evenly in the air without aggregation that would create macromolecules. Although this standard seems groundbreaking, the title states that this procedure is for "metal" nanoparticles – presumably all of them – yet the standard makes an emphasis specifically on metal materials with melting temperatures and evaporation rates similar to silver and gold. For other nanomaterials, this safety-measuring method may not be applicable. This only furthers the notion that standardizing the safety testing for nanoparticles needs more work, but in order to create a useful standard, the basic understanding of the safety of nanomaterials must be established to validate the standards. Other ISO standards currently being developed include ISO 19007, which evaluates the effects of nanoparticles on cell viability, and ISO 19006, which evaluates the effects of nanoparticles on cell oxidative stress.

Among the ASTM standards, there are a few that evaluate the safety of nanomaterials [20]. The main one is E2526-08(2013), which evaluates with in vitro testing the cytotoxicity of nanomaterials in cells, specifically in porcine kidney cells and human hepatocarcinoma cells. Those cells are of interest because kidney cells are the primary blood filter cells of the body, getting rid of harmful waste, and hepatocarcinoma liver cells are commonly known to turn cancerous. Standards E2524-08(2013) and E2525-08(2013) both deal with cellular interactions between cells and nanomaterials. Most of these tests are derived from ASTM F1903, which is the standard in testing biological responses to particles in vitro. Performing this test, or a variation of it, can provide a better understanding of the safety of a nanomaterial on a microscopic scale. ASTM also includes the standard E2535-07(2013), which documents the appropriate method for handling airborne nanomaterials.

14.1.3.3 Risk Assessment for Polymer Nanocomposites

Because polymer nanocomposites are really polymers with a small amount of nanocomposites to augment the specific polymer's properties, the method of conducting risk assessments would be similar to that for regular polymers. There are many different standards for assessing the safety and environmental impact of polymers from both ISO and ASTM. The tests also happen to be very specific in what environment the polymer is being used. For example, ASTM F2820 is a group of specifications that include biological test methods to ensure the safety of polyetherketoneketone polymers used in implants.

Some polymer nanocomposites, however, have the nanomaterials that may cause complications in safe implementation or disposal. As stated earlier in the chapter, the main environmental danger of polymer nanocomposites is in their component parts, thus arising predominantly when the nanocomposite-infused polymer begins to disintegrate either by itself (biodegradation) or via an external force (incineration). Therefore, assessing what takes place during the disposal of polymer nanocomposites is necessary for a meaningful discussion of their overall safety issues.

Municipal solid waste disposal via incineration consists of burning the waste and, once cooled, processing the created solid residual and the off-gas. The process of identifying risks from polymer nanocomposites is based on whether or not the nanomaterial will detach from the original polymer during incineration (Figure 14.3) [2]. It is important to note that the toxicity of the polymer nanocomposites resides in the nanomaterials. The polymer should not be disregarded entirely, but for the sake of simplifying the risk assessment of polymer nanocomposites, an emphasis must be made on the nanomaterials.

Areas of interest in the municipal solid waste incineration include off-gas disposal and the incineration process [11]. Figure 14.4 shows that high temperatures associated with the incineration process are responsible for a variety of reducing and oxidizing reactions that can create new nanomaterials in addition to the ones already in the polymer nanocomposites. Because the newly create nanomaterials cannot be controlled, they pose additional risk that is hard to account for. Assuming the nanomaterial never leaves the polymer matrix or the polymer does not burn, the nanomaterial can be assumed to cause no harm. In the heat recovery and off-gas treatment phases, any solid residue filtered or extracted out may contain nanomaterials that may be harmful. Off-gas is the waste gas emitted in the process of incineration of waste products. There is a filtration system for off-gas; however, because nanomaterials are exceedingly small, around 20% of them are expected to pass through the filtration system [2]. Therefore, the level of nanomaterials in the off-gas from incinerated nanocomposites-containing waste can be from 10 to 10,000 times higher than in the off-gas from incinerated conventional waste [2]. Figure 14.4 identifies the risk of escaping as particle or gas for different nanomaterials [2]. The

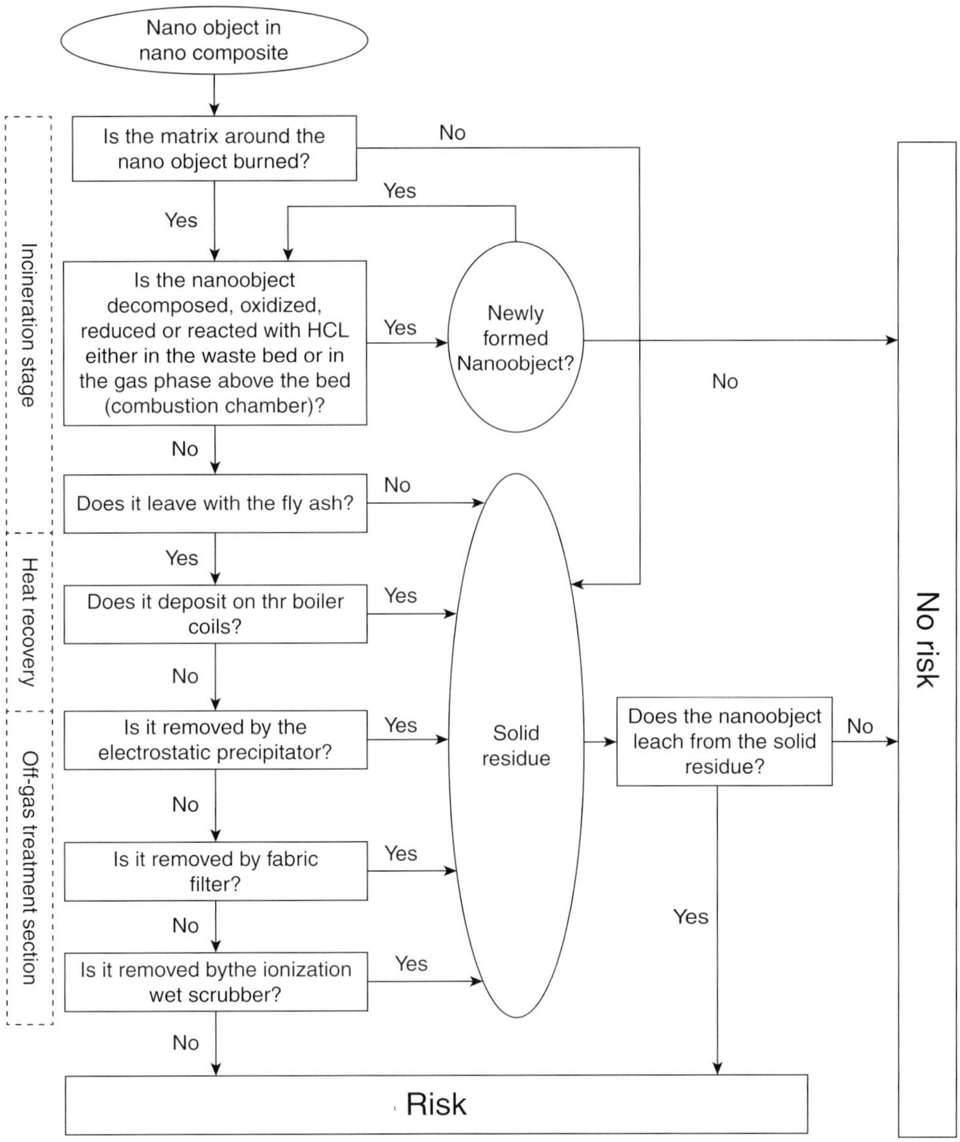

Figure 14.4. Decision tree for risk assessment [2].

increased risk shown by a specific nanomaterial in turn puts in jeopardy the general safety of the polymer nanocomposite containing it, since disposal is an important aspect of every material's lifetime.

The results of the risk analysis showed that, except for carbon nanotubes, there is no guarantee of complete removal from off-gas for any material below 100 nm in size (Table 14.1) [2]. The only way carbon nanotubes can adversely affect the environment is via leakage from solid residuals. Among the solid residual, all except magnesium hydroxide had nanomaterial particulates in the fly ash, which has the potential to leak into the environment [2]. Thus, it is fair to conclude for almost all polymer nanocomposites that the nanomaterials comprising them will eventually

Table 14.1. *The appearance of nanomaterial in difference areas [2]*

	Primary nano-object to fly ash?	Secondary nano-object(s) to fly ash?	Complete removal by off-gas treatment?	Risk of nano-object emission via stack?
Aluminum oxide	Yes	Yes	No	
Calcium Carbonate	Yes	Yes	No	
Magnesium hydroxide	No	Yes	No	
POSS	Yes	Yes	No	Yes
Silica	Yes	No	No	Yes
Titanium oxide	Yes	No	No	Yes
Zinc oxide	Yes	Yes	No	Yes
Zirconia	Yes	No	No	Yes
Mica	Yes	Yes	No	Yes
Montmorillonite	Yes	Yes	No	Yes
Talc	Yes	Yes	No	Yes
Cobalt	Yes	Yes	No	Yes
Gold	Yes	No	No	Yes
Silver	Yes	Yes	No	Yes
Carbon black	Yes	No	No	Yes
Carbon nanotubes	Yes	No	Yes	No
Fullerenes	Yes	No	No	Yes

leak into the environment from off-gas or solid residual. Types and levels of harm such leakage can cause depend on the type of nanomaterials being leaked. This is critical, because prior research into environmental safety of polymer nanocomposites up to date had covered only their intended use – not their disposal. It may be necessary to develop a disposal process – or processes – specifically for nanomaterials to ensure that they do no leak into the environment, given that municipal waste incineration is clearly ineffective in achieving this.

14.1.4 Summary and Conclusions

Nanotechnology has opened doors in many industries via a variety of applications. Because of their uniquely small size, nanomaterials have the ability to augment other materials by creating polymer nanocomposites. The safety-testing procedures for nanomaterials, however, are still under development. With contradicting results from studies involving animals and humans, the only solid conclusion that can be drawn is that nanomaterials do induce a reaction at the cellular level. Can this interaction cause cancer or other health complications? Perhaps, like carbon nanotubes, but it is still not proven, one way or another, for many other nanomaterials.

Looking beyond the basic toxicities of nanomaterials, other areas in which safety concerns are important can be affected. The safety of medical devices that incorporate nanomaterials depends on how the nanomaterials interact with proteins and living tissues. It is equally important to understand how various nanomaterials can potentially alter the environment for good or bad.

Using nanomaterials inside a polymer is gaining popularity because of their ability to augment the material with useful characteristics. Safety issues of the nanomaterial inside the polymer are generally disregarded because polymer nanocomposites interact with their environment in a totally different way than the nanomaterials alone. Therefore, the safety of polymer nanocomposites can be said to be less of a concern. However, during the disposal of these polymer nanocomposites via incineration, the nanomaterials can leak out as individual particles, which circles back to the original complications arising from interactions of individual nanomaterials with the environment.

Improvements in nanotechnology have the ability to bring new possibilities associated with new materials, but the safety of nanomaterials and polymer nanocomposites must not be forgotten. Public health and environmental safety cannot and must not be discounted for the sake of technological advancements. Research into environmental effects and safety of nanomaterials has been lagging far behind the material development of new polymer combinations and new nanoadditives. This situation is untenable and must not continue if we are to seek the full benefits of nanotechnology in our lives.

14.2 Fire Toxicity of Polymer Nanocomposites

Over the past years, significant research has been conducted on polymer nanocomposites (PNCs) to fulfill performance and safety requirements. PNCs have shown to improve mechanical, thermal, and flammability properties that make them an appealing fire-retardant (FR) material for a wide variety of applications. However, the fire toxicity of PNCs and the effluents that are released during chemical degradation have yet to be fully characterized. This section aims to explore the fire toxicity properties of FR thermoplastic PNCs using characterization methods to measure fire effluents.

14.2.1 Introduction

FR PNCs present a new alternative to conventional materials due to their proven ability to delay ignition time, reduce total mass loss, and decrease peak heat release rate during burning (see Chapter 9 for more details on this topic). However, the gaseous effluents and the hazardous risks they pose during thermal degradation of these materials have yet to be fully characterized. The main factors that determine the type and quantity of toxic gases include the material's chemical composition, flammability properties, and combustion conditions [21]. The most harmful and lethal products that result from combustion are asphyxiate gases, such as carbon monoxide (CO) and hydrogen cyanide (HCN) [21]. Respiratory problems, lingering injuries, and even death can occur when individuals are exposed to large quantities of these toxins. Other hazardous gases that are produced from combustion include nitrogen dioxide (NO_2), nitrogen oxide (NO), hypoxia, hydrocarbons, and organic species. As a result, thorough analysis of the effluents released during thermal degradation

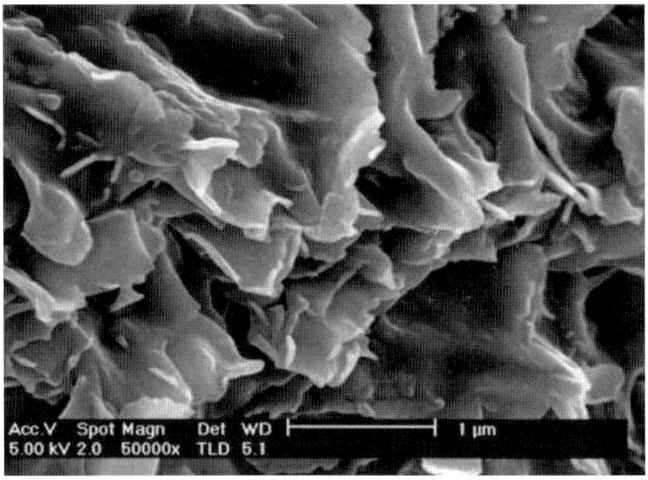

Figure 14.5. SEM images of EVA/nanoclay at low magnification [22].

of PNCs is pertinent in order to understand their potential risks. This section will explore the methods used to characterize fire toxicity properties and summarize recent findings on fire toxicity properties of some commonly used thermoplastic nanocomposites.

14.2.2 Materials

The materials that are the focus of discussion in this section are commonly used thermoplastic nanocomposite polymers, specifically FR ethylene-vinyl acetate (EVA), polyamide 6 (PA6), polypropylene (PP), polycarbonate (PC), polyethylene terephthalate (PET), and poly(methyl methacrylate) (PMMA).

14.2.2.1 Ethylene-Vinyl Acetate (EVA)

A study was conducted by Hull et al. [22] on the decomposition and burning behavior of EVA copolymer nanocomposite materials. The EVA copolymer consisted of 28% vinyl acetate (VA) [2]. The flame retardant selected for this study was aluminum trihydroxide (ATH), and the nanoparticle used was clay nanofiller. The formulations used for this study was EVA (28% VA), EVA (28% VA) with 5 parts per hundred (phr) nanoclay, EVA (28% VA) with 60 phr ATH, and EVA (28% VA) with 60 phr ATH and 5 phr nanoclay [22].

The copolymer nanocomposites were characterized by taking images of the dispersion of nanoclay into the copolymer using scanning electron microscopy (SEM), as shown in Figure 14.5. The low-magnification image shows that the nanoclay was not properly dispersed and instead formed clusters of about 0.01–0.05 mm in diameter [22]. The high-magnification image emphasizes the intercalated structure of the clustered nanoclay (Figure 14.5). Thus, the nanoparticles were dispersed as a microcomposite.

Table 14.2. *Formulations of PP and PA6 nanocomposites in wt% [23]*

Material	Polymer %	Fire Retardant %		Nanoclay %
		APP	AIP	
Polypropylene (PP) (all samples contain 5% PPgMA)	100			
PP + FR	70	30		
PP + NC	95			5
PP + FR + NC	65	30		5
Polyamide 6 (PA 6)	100			
PA 6 + FR	82		18	
PA 6 + NC	95			5
PA 6 + FR + NC	77		18	5

14.2.2.2 Polyamide 6 (PA6) and Polypropylene (PP)

Stec and Rhodes conducted a study on smoke and hydrocarbon yields of FR PA6 and PP polymer nanocomposites [23]. The fire retardant used for PA6 was OP 1311 (FR) and the nanoclay was Cloisite® 30A. The fire retardant and nanoclay used for PP were Exolit® APP 760 (FR) and Cloisite 20A (NC), respectively [23]. The formulations that were investigated are shown in Table 14.2.

Additional research into fire toxicity by Stec, Rhodes, and Smith also focused on PA6 and PP [24]. Both were processed with a twin-screw extruder with a length-to-diameter ratio of 40, screw speed of 200 rpm, and a dual feed (a main feed and a side gravimetric feed) [25]. For PP, polymer pellets were added in the main feed along with a fire retardant, 30 wt% of ammonium polyphosphate (AP 760). Further down the extruder, Cloisite 20A (5 wt%) was added in via the side feed. For PA6, polymer pellets were fed into the extruder and then 18 wt% of organic phosphinate (OP 1611) and 5 wt% of Cloisite 30B were added through the side feed. Temperatures along the extruder were varied depending on whether dispersion, conveying, or melting was targeted. The exact temperature distribution and screw layout are reported elsewhere [25]. Sample formation of films and slabs for both PP and PA6 were made by injection-molding.

Characterization of PA6 and PP nanocomposites were elucidated by TEM images of each polymer as visualized in [24]. The PP nanocomposite shows large agglomerates of nanoclay at the lower magnification that, when visualized at higher magnification, appear to be tactoids of tens of nanoclays that are partially intercalated with the polymer. The PA6 nanocomposite at both magnifications is well dispersed and exfoliated. At the higher magnification, there is a small agglomerate of 2–10 nanoclays showing the sample is not fully exfoliated but the vast majority is homogenous.

14.2.2.3 Polycarbonate (PC)

The thermal degradation of PC filled with solid bisphenol A bis(diphenyl phosphate) (S-BDP) and montmorillonite (OMMT) was investigated by Feng et al. [26].

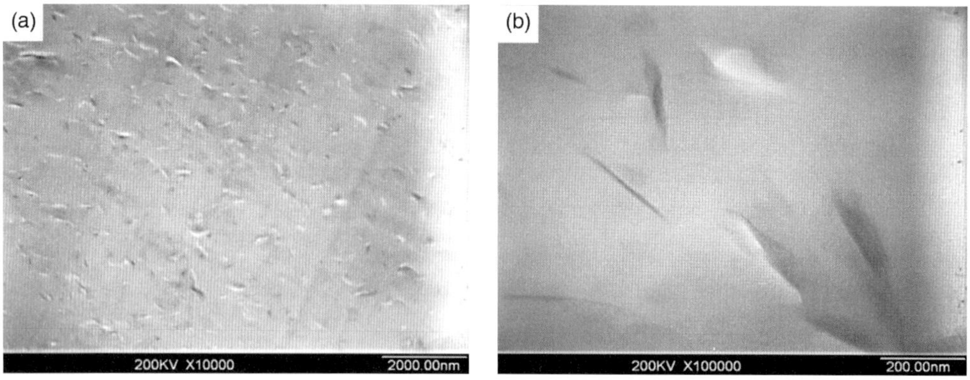

Figure 14.6. TEM images of (a) OMMT dispersed in PC at low magnification and (b) high magnification [26].

The material was processed by melt-blending the PC using a co-rotating twin-screw extrusion procedure at a rotary speed of 200 rpm and a temperature profile of 235, 240, 245, 250, 242, and 232°C from the hopper to the die [26]. Four formulations were used for the investigation: PC, PC with 8 wt% S-BDP, PC with 6 wt% S-BDP, and 2 wt% OMMT [26].

The dispersion of OMMT in PC was characterized using transmission electron microscopy (TEM), which is shown in Figure 14.6. The low-magnification image illustrates a uniform dispersion of the nanoclay platelets. However, the high-magnification image indicates a few clay platelets in an intercalated state.

14.2.2.4 Polyethylene Terephthalate (PET)

Didane et al. [27] characterized the fume opacity and toxicity of FR PET polymer nanocomposites. The fire retardant used was Exolit OP 950 and the nanofiller selected was octa-methyl POSS (OMPOSS). Three formulations were studied: PET, PET with 10 wt% OP (PET-OP 10), and PET with 9 wt% OP and 1 wt% POSS (PET-OP-OM 9-1) [27]. These samples were melt-blended using a twin-screw extruder at a rotary speed of 100 rpm and temperature profile of 260, 260, 280, 280, 280, 255, and 255°C in different zones of the twin-screw extruder [27].

SEM images of PET-OP 10 and PET-OP-OM 9-1 were taken to determine the morphology of the formulations. Figure 14.7a shows the elongated shape of the OP during the drawing process. Figures 14.7b and 14.7c illustrate raised white bumps on the surface and cross section of PET-OP-OM 9-1 [27]. Further investigation was carried out using energy dispersive X-ray analysis (EDX) to determine the composition of these lumped surfaces. The results indicated that they were silicon particulates that are only present in OMPOSS. Thus, the dispersion of the nanofiller was considered poor.

14.2.2.5 Poly(methyl methacrylate) (PMMA)

Calogine et al. conducted research on fire toxicity of PMMA mixed with various nanoparticles. The complete list of formulations is shown in Table 14.3 [28]. All

Table 14.3. *Formulations of PMMA nanocomposites [28]*

Sample	Formulation
1	PMMA
2	PMMA + 0.2 wt% Carbon Nanotubes
3	PMMA + 1 wt% Carbon Nanotubes
4	PMMA + 15 wt% Silica
5	PMMA + 15 wt% Alumina
6	PMMA + 15 wt% AP
7	PMMA + 5 wt% Alumina + 10 wt% AP

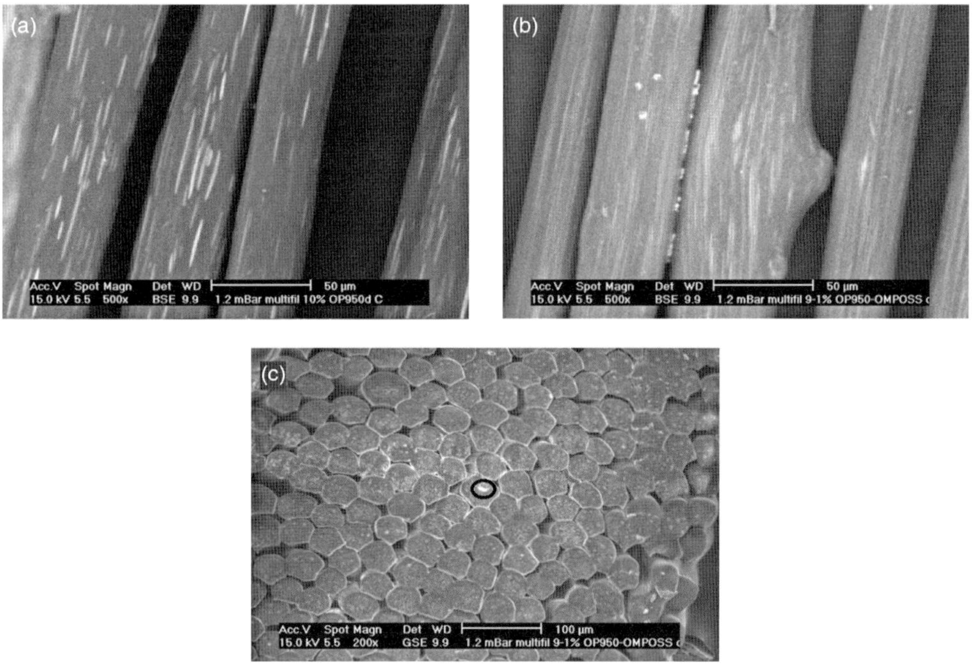

Figure 14.7. SEM images of (a) PET-OP 10, (b) PET-OP-OM 9-1, and (c) cross section of PET-OP-OM 9-1 [27].

PMMA composites were compounded with a twin-screw extruder with a L/D ratio of 48, a screw speed of 200 rpm, and a feed rate of 2.7 kg/h. CNTs were added to PMMA prior to the extruder and mechanically mixed before compounding; all other additives were processed within the extruder. No characterization data were reported for any of the formulations with PMMA.

14.2.3 Fire Toxicity Characterization Methods

The methods for measuring fire toxicity of FR PNCs include the steady-state tube furnace and thermogravimetric analysis (TGA) coupled with Fourier transform infrared spectroscopy (FTIR).

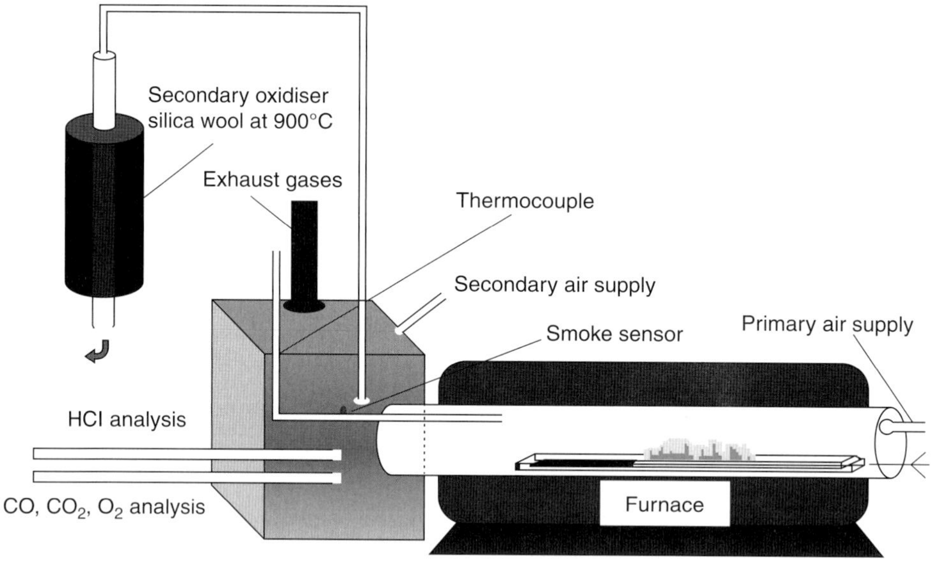

Figure 14.8. Steady-state tube furnace [29].

14.2.3.1 Steady-State Tube Furnace (ISO TS 19700)

The steady-state tube furnace is a bench-scale fire-testing apparatus capable of measuring the toxic products released during different fire conditions [29]. Fire conditions can be classified with an equivalence ratio (ϕ), which is the ratio of actual air/fuel ratio to the stoichiometric air/fuel ratio. The conditions of well-ventilated fires tend to be oxygen-rich and can be represented with an equivalence ratio of less than 0.75. Large vitiated fires, where the conditions are fuel-rich, have an equivalence ratio greater than 1.5 [30]. The steady-state tube furnace allows the control of the equivalence ratio in order to replicate any fire condition.

The steady-state tube furnace operates by driving a sample through the furnace at a fixed feeding rate. Meanwhile, a fixed air flow is passed through the furnace to simulate any ventilation associated with the fire condition. The configuration of the apparatus is shown in Figure 14.8 [29]. By controlling the temperature of the furnace, sample feed rate, and air flow, fire conditions, such as smoldering, well-ventilated flaming, under-ventilated flaming, small vitiated fires, and large vitiated fires can be replicated. After combustion occurs, a paramagnetic oxygen analyzer and electrochemical sensors are used to quantify the fire effluents, and a secondary oxidizer composed of silica wool at 900°C is used to fully oxidize some of the fire effluents. Toxic gases, such as CO and hydrocarbons are then determined by measuring the differences between effluent concentrations of the secondary oxidizer from the first oxygen analyzer [29].

Characterizations of fire toxicity properties for FR EVA, PA6, and PP nanocomposites were determined using the steady-state tube furnace. For the testing of EVA/ATH/Nano, the furnace was set at a nominal temperature of 750°C while the mass of the sample was varied to adjust to oxygen requirement [22]. The fire conditions simulated for this study were well-ventilated, under-ventilated, and small vitiated fires [22]. For the testing of PA6/OP/NC and PP/AP/NC, the fire conditions simulated were well-ventilated at 650°C ($\phi < 0.75$), small under-ventilated at 650°C ($\phi > 1.5$), and large under-ventilated at 825°C ($\phi > 1.5$) [23].

14.2.3.2 TGA/FTIR

For the characterization of fire toxicity properties, TGA can be coupled with FTIR to understand the thermal degradation of materials during combustion. TGA is a commonly used research method to measure the mass loss over time of materials under combustion. By coupling this method with FTIR, the gaseous effluents released during combustion travel directly through the FTIR device, which is then scanned to correlate an infrared absorption to a specific gas.

The studies that utilized TGA/FTIR to evaluate the thermal degradation of FR PNCs were EVA, PC, and PET. For EVA materials, approximately 5 mg was used for each sample and heated at a rate of 10°C/min [22]. About 10 mg was used for each PC sample and was heated at a rate of 20°C/min from 40°C to 800°C under a nitrogen flow rate of 60 mL/min [26]. For PET materials, roughly 7 mg was used for each sample and heated from 20°C to 700°C at a rate of 10°C/min under air and nitrogen [27].

14.2.3.3 Cascade Impactor

A cascade impactor operates as a set of inline filters allowing the separation of gaseous effluents by aerodynamic particle size, as shown in Figure 14.9 [24]. The upper filters have large orifices that decrease in size as the filters progress, capturing the effluents as they reach their cutoff size. The system is ideal because it separates effluents in the same way the lungs do, allowing a direct comparison as to which particles will interact with the lungs [24]. An impactor was utilized in the PA6 and PP nanocomposites experimentation by attachment to the gas outflow for 5 minutes during steady combustion; each test was performed three times [24].

14.2.4 Fire Toxicity Properties

Many studies have been conducted to evaluate the thermal properties of FR thermoplastic nanocomposites, but relatively little is known about the potential environmental, health, and safety consequences when these materials are chemically degraded. Thus, the assessment of fire gas toxicities is essential in order to analyze the fire hazards during the combustion of nanocomposites. The research presented below provides a brief summary of the FR properties and gaseous effluents produced from FR EVA, PA6, PP, PC, and PET polymers and nanocomposites.

14.2.4.1 EVA/ATH/Nano

TGA was performed to observe the thermal decomposition behavior of EVA/ATH/Nano in nitrogen and air. The results of percent residual mass in nitrogen and air are shown in Figure 14.10 [22]. For TGA conducted in nitrogen, the graph shows

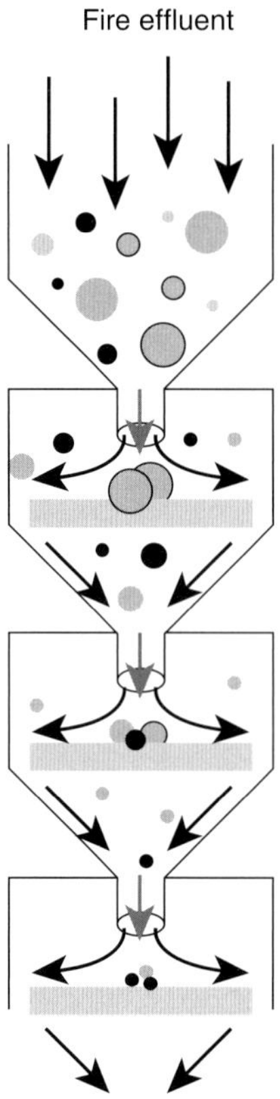

Figure 14.9. Cascade impactor [24].

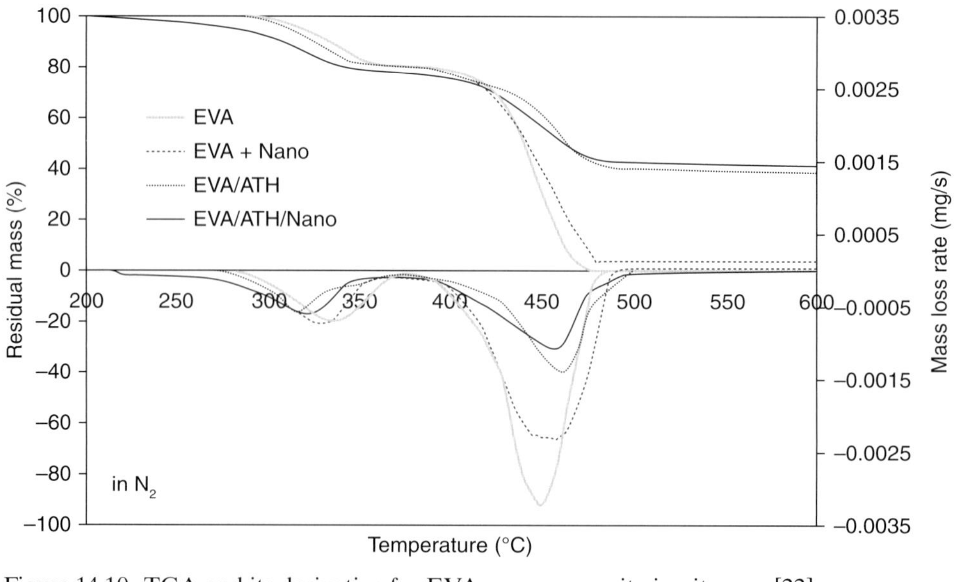

Figure 14.10. TGA and its derivative for EVA nanocomposite in nitrogen [22].

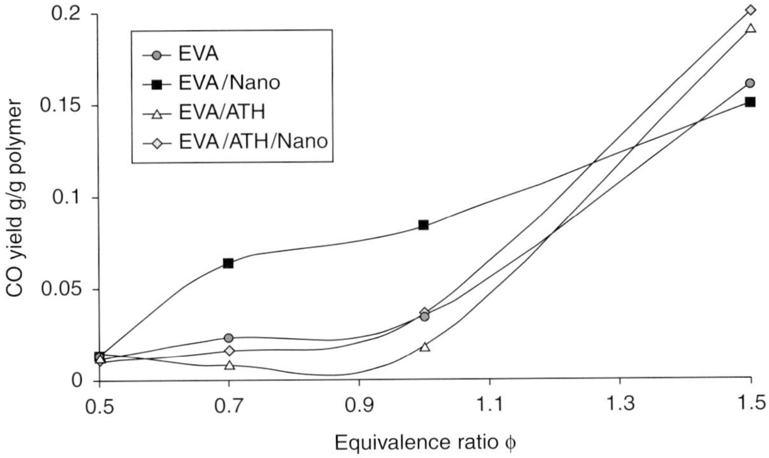

Figure 14.11. CO yield from EVA nanocomposite materials [22].

that the presence of ATH significantly increases the residual mass percentage. The derivative graph shows that the addition of clay also helps in reducing the rate of decomposition. For TGA conducted in air, the graph of the derivative illustrates that the neat polymer reaches a maxima around 420°C, whereas the clay composite does not reach its peak mass loss until about 470°C [22]. This indicates that the absence of clay greatly increases the decomposition rate. For ATH, similar trends were observed in air as in nitrogen. Overall, the data show that the addition of ATH increases the residual mass percentage and clay decreases the mass loss rate during thermal degradation.

To establish the fire toxicities of EVA and EVA clay composite under varying fire conditions, gaseous effluents were measured using the tube furnace. The yields of CO for the composite materials are shown in Figure 14.11 [22]. The results show a consistent trend of low CO levels under well-ventilated conditions ($\phi < 0.75$) that gradually increase as the combustion process approaches fuel-rich conditions ($\phi > 1.5$). However, the EVA clay composite shows the opposite trend of higher yields of CO under oxygen-rich conditions and lower yields under fuel-rich conditions compared to the other composites. The composites containing ATH show the highest yields of CO of about 0.2 g/g. Researchers state that the increase in the yield of CO under stoichiometric conditions can be attributed to the formation of the char layer, which reduces the materials' access to oxygen [22]. Based on these results, the yields of CO produced are consistent with the CO yields that are typically expected under varying fire conditions [22]. Thus, the addition of ATH or clay does not increase the amount of gaseous effluents.

14.2.4.2 PA6/OP/Nanoclay (NC) and PP/AP/Nanoclay (NC)

The steady-state tube furnace was used to simulate different fire conditions to investigate the fire toxicities of PA6 and PP with and without fire retardant and/or nanoclay. Figure 14.12 illustrates the main combustion effluents produced during the test: CO, HCN, NO_2, hypoxia, and organics [31]. These effluents were expressed as a

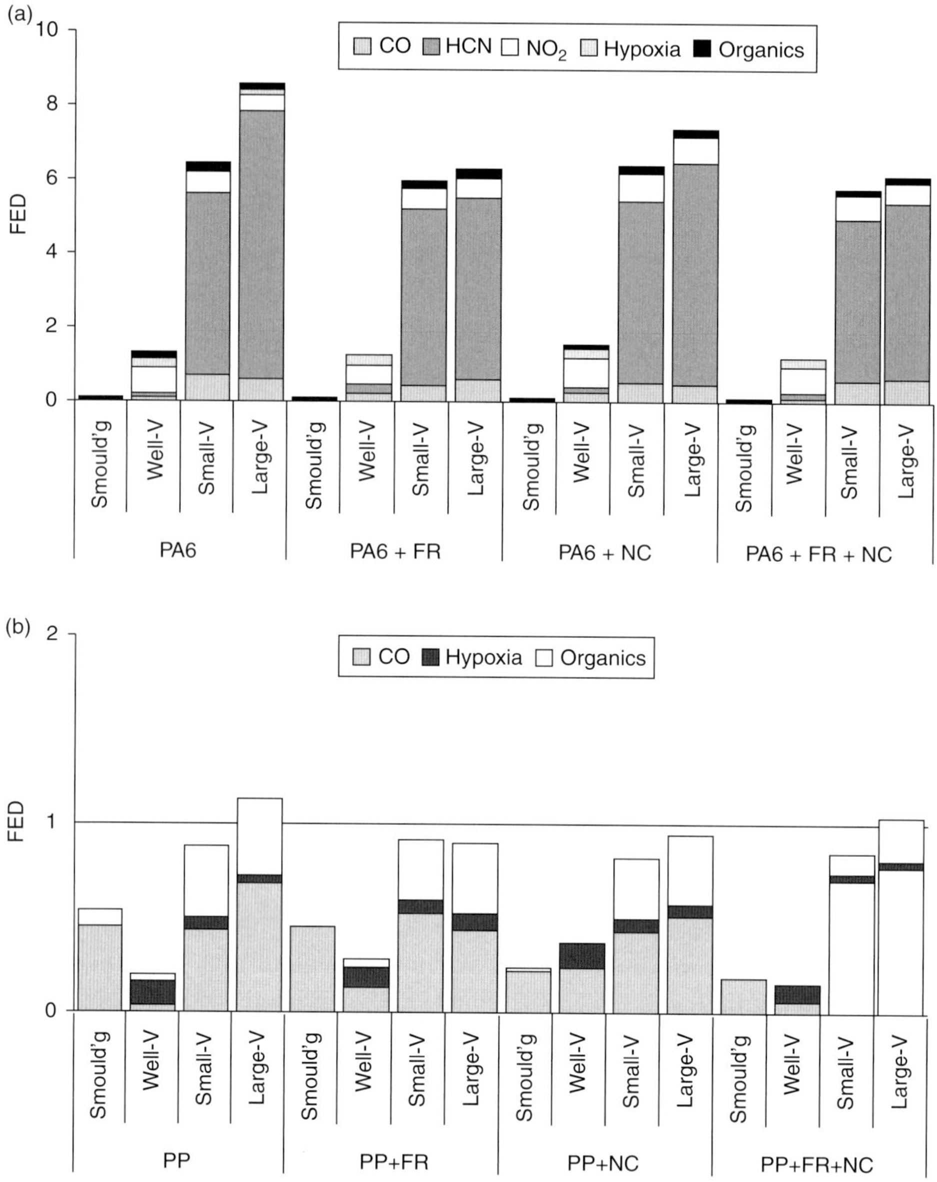

Figure 14.12. FED of toxic effluents for PA6 (left) and PP (right) [31] (Courtesy of ACS).

fractional effective dose (FED), which is an equation based on the tube furnace that relates the yield of each toxic product to its lethal amount [31].

For PA6, it can be seen that the growth of fire conditions increases the yield of CO and HCN. This is due to the decrease in ventilation as oxygen is quickly depleted in vitiated fire conditions. The presence of fire retardant and nanoclay decreases the overall FED compared to the neat PA6. However, the addition of nanoclay alone produces a higher toxic yield than if fire retardant was included for small and large vitiated fires. On the other hand, for PP, the main effluents produced were CO and

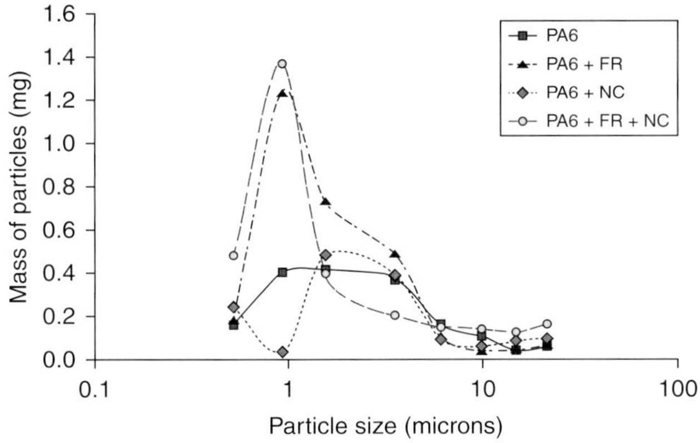

Figure 14.13. PA6 nanocomposite particle distribution after combustion for well-ventilated condition [24].

hydrocarbons. The yield of these toxins greatly increases under vitiated conditions. It should be noted that the temperature of the furnace did not have any correlations with the amount of toxins produced, and that the concentration of effluents from PA6 and PP samples increases as the oxygen is depleted. Based on these observations, the addition of fire retardant and nanoclay did not seem to have a significant effect on the overall yield of effluents for either PA6 or PP. Thus, they do not pose as a risk of increasing the toxic effluents.

The fire toxicity of the PA6 nanocomposites was further studied by measuring the particle distribution of fire effluents by a cascade impactor, as shown in Figure 14.13 [24]. For well-ventilated combustion, all materials showed a lower peak mass of particles produced with the fire retarded PA6 (PA6+FR) and fire-retarded PA6 with nanoclay (PA6+FR+NC), showing the highest peaks of 1.4 mg of 1μm size particles. In both under-ventilated cases, PA6+FR+NC shows a higher peak mass of particles of 3.5 mg of 1μm size particles while PA6+FR's is increased to 2.4 mg in only the small under-ventilated case. The PA6 nanocomposite (PA6+NC) exhibits a higher particle size for both the well-ventilated and large under-ventilated cases, equating to a lower fire toxicity and the mass of particles half that of PA6+FR+NC. Also, the two curves that are most similar in terms of mass of particles for each case are PA6 and PA6+NC, suggesting the NC alone does not affect the mass of fire effluents.

Thus, it seems the fire retardant, not the nanoparticle, plays a major role in producing ultra-fine particles that can be toxic. The increase in mass of particles for the PA6+FR+NC composites over PA6+FR composites is possibly caused by competition for interaction with the polymer as the weight percent increases. An interesting result of the experiment is the shift in particle size for PA6+NC from small to large under-ventilated. This signals a change in the NCs as the temperature increases, likely explained by agglomeration of the nanoclays to form shielding layers that, when burned, form larger particles, thus lowering the fire toxicity. Overall, the experiments show that fire toxicities' dependence on ventilation is inconclusive, as the

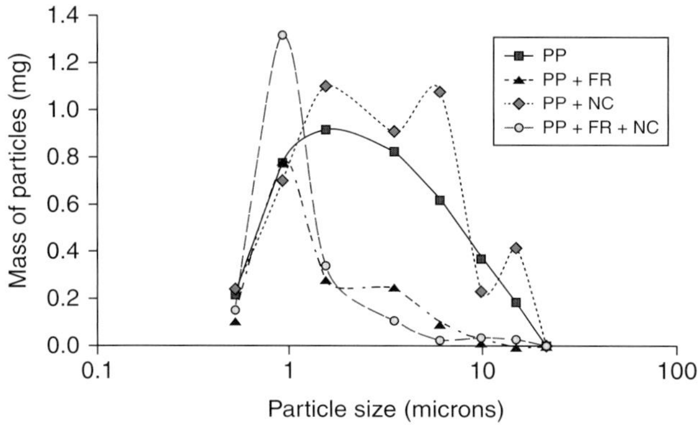

Figure 14.14. PP nanocomposite particle distribution after combustion for well-ventilated condition [24].

distribution of particles is similar for all cases but the mass of particles increases with under-ventilation. In addition, the PA6+FR+NC material does show a high level of fire toxicity by releasing high amounts of ultra-fine particles.

The PP NCs were tested in the same fashion as the PA6 NCs, and the results can be seen in Figure 14.14 [24]. The neat PP polymer exhibits a broad range of particle sizes released with under-ventilation, causing an increase in mass of particles as previously seen with PA6. In all ventilation cases, the PP+FR and PP+FR+NC exhibit the smallest particle size, near 1μm. This again shows the propensity of the FR to form ultra-fine particles that can be toxic regardless of the presence of the NC. The PP+NC composite exhibits larger particle size for large under-ventilated and well-ventilated conditions, similar to PA6+NC. The major difference between the PA6 and PP trials is the increase in the mass of particles for PP+NC. This is likely due to the difference in dispersion as the PA6 was exfoliated and PP showed large tactoids. Thus, when the NCs were released in PP, they were of similar size but consisted of many sheets, increasing the mass significantly. Overall, the results of the PP experiments were similar to that of PA6, exhibiting the fire toxicity of fire retardants by forming ultra-fine particles.

14.2.4.3 PC/S-BDP/OMMT

TGA coupled with FTIR was used to evaluate the thermal degradation behavior of PC nanocomposites. The thermal properties of the PC nanocomposites were measured first before evaluating their fire toxicities. The TGA results are shown in Figure 14.15 [26]. Out of the three formulations, neat PC had the highest onset temperature of about 478°C, whereas PC/S-BDP and PC/S-BDP/OMMT had lower onset temperatures of 465°C and 458°C, respectively [26]. This shows that S-BDP and OMMT do not enhance the onset temperature of the nanocomposite. However, despite having lower onset temperatures, the S-BDP and OMMT seemed to increase the peak temperature of the DTG curve compared to neat PC. This indicates that the addition of S-BDP and OMMT could promote char formation, which decelerates the degradation rate of the nanocomposite.

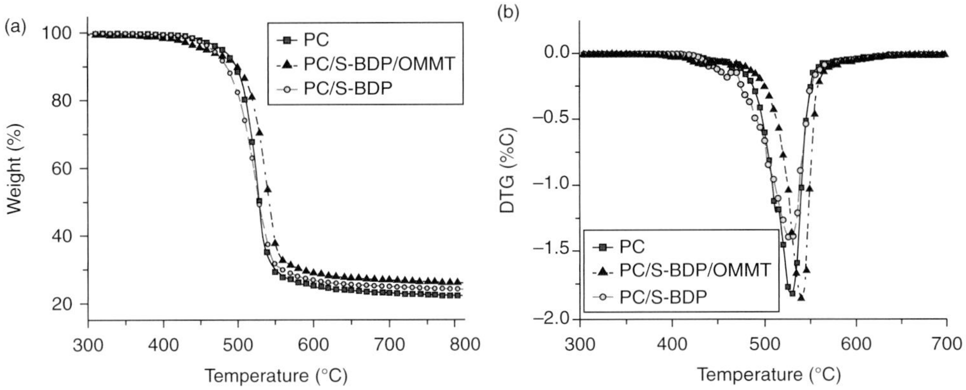

Figure 14.15. (a) TGA curve and (b) derivative TG for PC nanocomposite [26].

FTIR was then used to evaluate the CO_2 and CO production during combustion of the PC nanocomposites, as shown in Figure 14.16 [26]. The presence of S-BDP and OMMT reduced the maximum amount of CO_2 produced compared to neat PC. However, the amount of CO released is significantly increased when S-BDP alone is added to PC. The addition of OMMT to PC/S-BDP surpressed the amount of CO produced compared to PC/S-BDP, but did not seem to affect the maximum CO production compared to neat PC. These results imply that the combination of S-BDP and OMMT would maintain, if not enhance, the fire toxicity properties of neat PC and PC/S-BDP.

14.2.4.4 PET/OP/OM

Thermal analysis of FR PET nanocomposites was conducted using TGA in nitrogen and in air in Figure 14.17 [27]. The addition of OP and OMPOSS appears to lower slightly the onset temperature compared to neat PET. However, the DTG curve illustrates that PET/OP/OM produces a slightly higher maximum peak, which indicates that the presence of OMPOSS marginally decreases the rate of thermal degradation compared to neat PET and the addition of OP alone.

Figure 14.18 illustrates the density of the released gaseous effluents from PET nanocomposites using TGA coupled with FTIR [27]. From the graph it appears that none of the samples produced smoke at the same rate. PET/OP exhibited the fastest smoke production and reached maximum smoke density of about 339 at 250 seconds, whereas neat PET reached higher maximum smoke density of about 462 at about 450 seconds [27]. However, when combining OP and OMPOSS, the trend of the curve is completely altered. Smoke density gradually increases and appears to reach a maximum of only about 276, which is about a 40% reduction compared to PET [27]. Thus, the combination of OP and OMPOSS greatly improves the fire toxicity properties of PET.

14.2.4.5 PMMA/Various Nanoparticles

The various PMMA nanocomposites were all measured for their peak heat release rate (pHRR) to give a scale of their fire retardancy, as shown in [28]. For

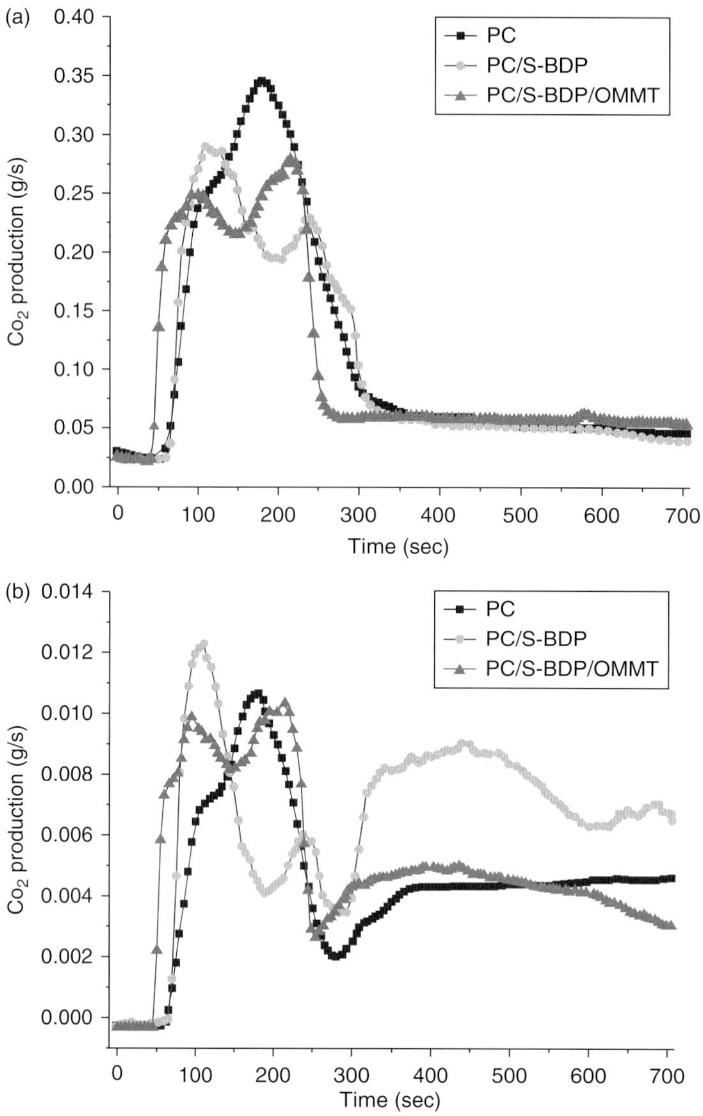

Figure 14.16. Production of (a) carbon dioxide and (b) carbon monoxide effluents from PC nanocomposites [26].

under-ventilated conditions the additive had little to no effect on the pHRR, but for well-ventilated conditions there were large differences. PMMA with 15 wt% silica showed the lowest pHRR, closely followed by 15 wt% alumina. CNT additives showed a pHRR most comparable with neat PMMA, with 0.2 wt% showing a slightly high pHRR. This is logical given that CNTs are excellent conductors of heat.

All PMMA composites were tested by FTIR for their emission of gaseous species when undergoing combustion in an under-ventilated and well-ventilated environments, as shown in [28]. As expected, gaseous emissions were much higher for the under-ventilated scenarios, especially in the case of CO and hydrocarbons (HCt).

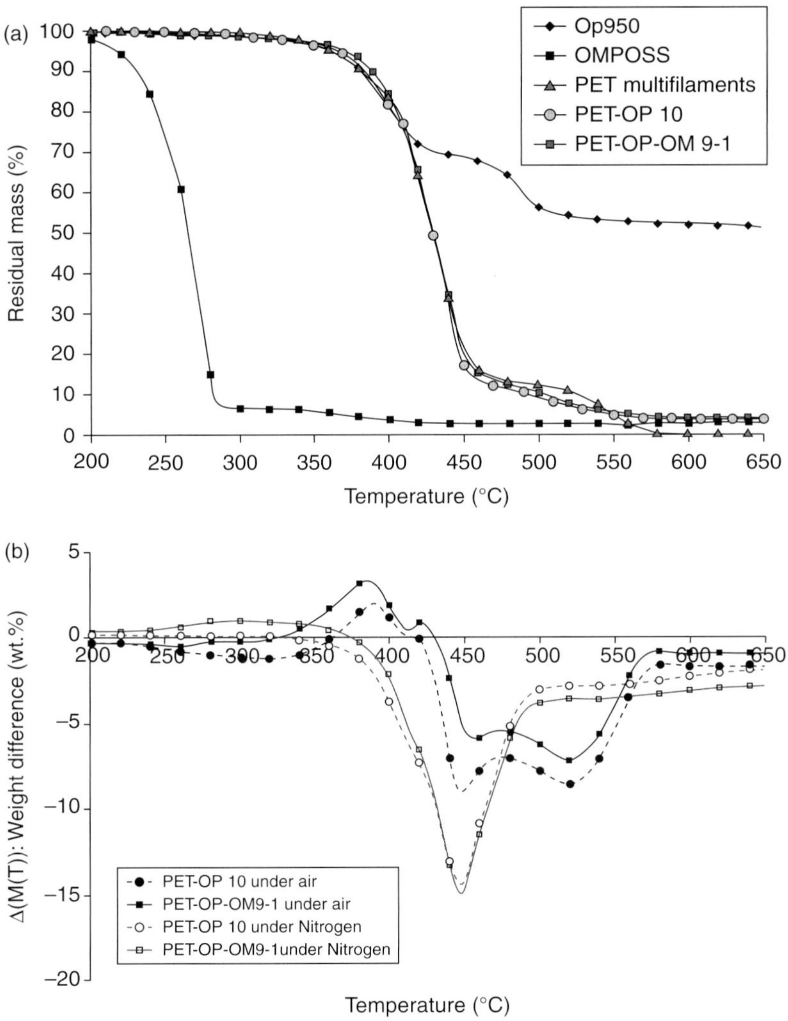

Figure 14.17. (a) TGA curve PET nanocomposites under air and (b) dTGA of PET nanocomposites under air and nitrogen [27].

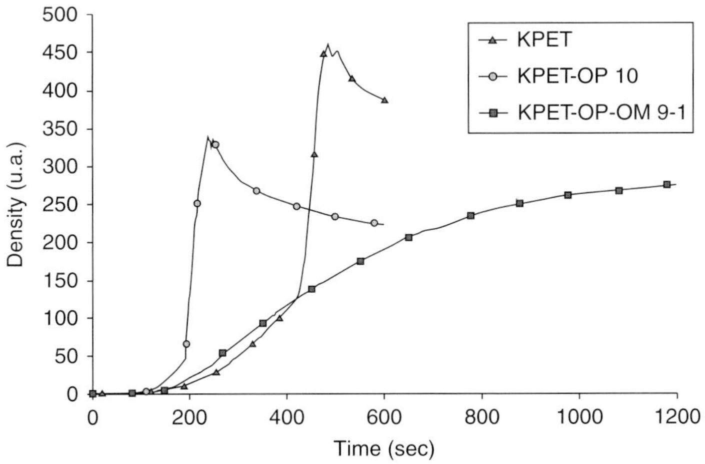

Figure 14.18. Density of smoke components produced during combustion of PET [27].

PMMA with CNT additives showed almost no change over the neat PMMA. With 0.2 wt% CNT, the gaseous emissions were almost identical, likely due to the minute presence of the CNT. With 1 wt% CNT, the gaseous emissions were around 20 mg/g lower than neat PMMA. The PMMA composite with 15 wt% silica shows a large decrease in the amount of gaseous emissions with an under-ventilated CO yield of 75 mg/g compared to 210 mg/g for neat PMMA. This reduction holds for all other gaseous emissions except for soot formation, which is greater than that of neat PMMA for both the well-ventilated and under-ventilated conditions. This can be explained by the char formation of silica used to retard fire.

The results for 15 wt% alumina are comparable to those for silica. PMMA with 15 wt% ammonium polyphosphate (APP) shows a reduction of all gaseous emissions by around half. It also shows an increase in soot over neat PMMA, again due to char formation for the purpose of fire retardancy. Finally, the PMMA composite with 5 wt% alumina and 10 wt% APP showed a decrease in CO emission for a well-ventilated environment but no change over neat PMMA in an under-ventilated environment. The hydrocarbon formation during well-ventilated combustion was greater than neat PMMA but was lower for under-ventilated conditions; for both cases the emission was around 25 mg/g, showing that hydrocarbon release for this combination may not be dependent on ventilation. The soot formation significantly increases in the under-ventilated environment but decreased over neat PMMA for the well-ventilated environment. The reasons for this and for the constant hydrocarbon formation are unknown and require further testing of this composite PMMA-Alumina-APP material. Overall, these experiments show gaseous emissions, especially CO, can be reduced by addition of nanoparticles; the extent of reduction depends on the type of nanoparticles and weight percentage.

14.2.5 Summary and Conclusions

Based on the results discussed, the main trend observed is that the combination of fire retardant and nanoparticles can reduce the amount of gaseous effluents produced during thermal degradation. Fire toxicity properties of FR EVA, PA6 and PP, PC, PET, and PMMA nanocomposites were all enhanced. However, using FR additives alone did not yield the same results as combining them, as demonstrated in PC and PET studies. Also, the morphology of the nanocomposite must be well dispersed and exfoliated in order to achieve the desired fire toxicity enhancements.

The understanding of fire toxicity properties of PNCs is still incomplete. The hazards and risks involved with the materials mentioned in this section cannot be completely disregarded, because each fire situation is dependent on the environment and surroundings. As the application of nanotechnology continues to grow, the general public should be aware of the health and environmental effects associated with PNCs. Future work on this subject will include greater efforts to reduce the yield of gaseous effluents produced from materials during combustion. There is still much to learn about material degradation behavior in order to prevent, control, and eliminate any hazards associated with materials throughout their usage life.

14.3 Current Regulation of Nanomaterials and Future Perspectives

14.3.1 Introduction

Nanomaterials are expected to play a profound role in society in the upcoming years. From applications in consumer electronics to medicine and energy, nanomaterials hold the promise of revolutionary changes that will benefit people all over the world. Because of their high potential, it is important to be educated about the environmental, health, and safety risks that nanomaterials pose. This section focuses on the implications and effects of governmental regulation on nanomaterial health and environmental hazards. First, an overview of the nanotechnology industry is provided. Next, the status of nanomaterial regulation is discussed, with insight on a possible risk assessment strategy. Lastly, a detailed investigation into the effects of nanomaterial safety inside the pesticides, food packaging, and medical industries is presented. A discussion about specific nanomaterial hazards is not included.

14.3.2 Current Trends in Nanomaterials

In the past few years, nanomaterials have become a pervasive force in the marketplace. Since 2001, the federal government has invested nearly $14 million in the National Nanotechnology Initiative [32]. In a study conducted by the Woodrow Wilson International Center for Scholars in 2009, there are more than 800 distinct products that incorporate nanomaterials produced by 484 companies located in 24 countries [32]. From 2005 to 2007, commercial and consumer products that used nanomaterials grew from a $30 billion industry into a $147 billion industry [32]. These numbers have significant implications to not only the staying power of nanomaterials but also their potential for health and environmental damage in the absence of effective regulation. Additional discussion of this topic can be found in Chapter 15.

14.3.3 Government Perspectives on Nanomaterial Safety

Since 2006, the government has increased funding for nanomaterial environmental and health safety research from $37.7 million to $123.5 million [33]. Despite this increase, funding toward nanomaterial safety has been well overshadowed by funding toward nanotechnology R&D, which has grown from $1.3 billion to $1.6 billion during that same time span [34]. This allocation of federal funding is reflective of the government's history of focusing on the business end of nanotechnologies over its safe production, usage, and distribution.

Currently, the legislative jurisdiction of regulating the safe usage and distribution of nanomaterials has been loosely divided between of the U.S.

Environmental Protection Agency, the U.S. Food and Drug Administration, and the Occupational Safety and Health Administration [33]. These agencies all share a common goal of protecting the environment and the public from nanomaterial hazards, which helps sustain consumer and business confidence. A key provision of their efforts, however, is to achieve this protection without stymieing business and innovation [33].

The federal government is currently responsible for creating regulation that controls the safety of a variety of products including food, consumer products, chemicals used in a work environment, medical products, and pesticides [33]. The government's strategy for regulating nanomaterials is similar to the regulation used for these mentioned products in that it is risk-based and science-based [33]. According to Principles for Regulation and Oversight of Emerging Technologies, federal agencies will have to abide by the following principles [33]:

- To ensure scientific integrity, base their decisions on the best available scientific evidence, separating purely scientific judgments from judgments of policy to the extent feasible.
- Seek and develop adequate information with respect to the potential effects of nanomaterials on human health and the environment and take into account new knowledge when it becomes available.
- To the extent feasible and subject to valid constraints (involving, for example, national security and confidential business information), develop relevant information in an open and transparent manner, with ample opportunities for stakeholder involvement and public participation.
- Actively communicate information to the public regarding the potential benefits and risks associated with specific uses of nanomaterials.
- Base their decisions on an awareness of the potential benefits and the potential costs of such regulation and oversight, including recognition of the role of limited information and risk in decision making.
- To the extent practicable, provide sufficient flexibility in their oversight and regulation to accommodate new evidence and learning on nanomaterials.
- Consistent with current statutes and regulations, strive to reach an appropriate level of consistency in risk assessment and risk management across the federal government, using standard oversight approaches to assess risks and benefits and manage risks, considering safety, health and environmental impacts, and exposure mitigation.
- Mandate risk management actions appropriate to, and commensurate with, the degree of risk identified in an assessment.
- Seek to coordinate with one another, with state authorities, and with stakeholders to address the breadth of issues, including health and safety, economic, environmental, and ethical issues (where applicable) associated with nanomaterials.
- Encourage coordinated and collaborative research across the international community and clearly communicate the regulatory approaches and understanding of the United States to other nations.

The federal government's strategy for nanomaterial regulation is to identify specific nanomaterial properties that are pertinent to health and environmental safety [33]. Some of these properties include exposure, biodistribution, persistence, bioaccumulation, toxicity, and pharmacokinetics [33]. Using these properties, agencies can then create regulations to effectively limit the hazards of nanomaterials and their applications [33].

One of the crucial elements of current nanomaterial regulation is that nanomaterial hazards are not categorized by their properties (i.e., size, shape, or charge) [33]. Instead, nanomaterial properties are used as a guide, rather than the driving determinant, for identifying hazards of nanomaterials [33]. Because different nanomaterials have different hazardous implications based on how they are produced and their applications, agencies avoid making generalizations about a nanomaterial's hazards outside the context of the nanomaterial's use and application [33].

14.3.4 DuPont's Nanomaterial Risk Assessment Strategy

The government's regulatory framework of investigating nanomaterial hazards on a case-by-case basis is supported by other research groups as well [33]. A joint study conducted by DuPont and the Environmental Defense in 2007 recommended using a regulatory framework that is driven by information and does not assume the risks and hazards of certain materials outside the context of their application [34]. The following section highlights the key components of each step in the framework.

14.3.4.1 Six Steps in DuPont's Framework [34]

In the **first step**, Describe Material and Application, information is gathered on the nanomaterial and its general usage by reviewing literature or consulting the manufacturer or developer of the nanomaterial's application. Information gaps can be filled in with similar materials and applications.

After completing the first step, the user is ready to enter the **second step**, Profile Lifecycles. During this stage, a complete study on the nanomaterial's properties, inherent hazards, and dosage and exposures metrics are conducted across the entirety of the nanomaterial's lifecycle from production and distribution to its application and disposal. The nanomaterial's properties include its physical and chemical properties, such as its shape, size, and charge. The nanomaterial's inherent hazards represent its potential to hurt the environment and human health. The associated dosage and exposure metrics identify how these nanomaterials can be introduced to the environment or humans in both accidental and non-accidental scenarios.

During the **third step**, Evaluate Risks, the information obtained in the previous steps is used to characterize the "nature, magnitude, and probability of risks" the nanomaterials and its application will pose to the environment and human health. Any gaps that have been left unfilled can be addressed by assuming a worst-case scenario. A high priority will then be given to fill these gaps will scientific evidence.

The **fourth step**, Assess Risk Management, assesses various options the user has to address the risks raised in the third step. Possible courses of action include engineering control, protective equipment, risk communication, and product and process modifications.

When an appropriate option has been considered, action to address the risks is taken in the **fifth step**, Decide, Document, and Act. The user can stop the production of the nanomaterial and its application or choose to wait and gather more information before making another decision. During this entire process, it is important that the user is completely transparent in his actions to stakeholders.

The **last step**, Review and Adapt, is about gathering new information and repeating the evaluation process. New information may include changes in the understanding of the nanomaterial's structure or acquired data of the nanomaterial's exposure levels. The user must continually update the information gathered about the nanomaterial and its application, share it with stakeholders, and apply the framework over again. Using these six steps, the framework provides a comprehensive and flexible risk management tool that can be iterated to evaluate nanomaterial risk.

14.4 Regulation by Industry

Current industry standards for the handling of nanoparticles and nanomaterials are controlled by specific government agencies, including the EPA and FDA. While the use of nanoparticles has become widespread across multiple industries, the markets in which such use is currently the most pervasive and potentially hazardous are the first to come under scrutiny and possible regulation. These markets are currently medical devices with nanomaterials, regulated by the FDA; food packaging comprised of nanomaterials, also regulated by the FDA; and pesticides with antimicrobial nanosilver (*n*-Ag), regulated by the EPA. These applications are discussed in the following subsections to elucidate the current trends for nanomaterial regulation; possible solutions are set forth as well.

14.4.1 Pesticide Industry

Nanopesticides were first introduced to the U.S. market in 2006 [35]. The nanoparticle of choice in all applications is *n*-Ag, which acts as an antimicrobial agent. Currently, the EPA has received four requests to register a nanopesticide with Ag. These requests prompted the EPA to start initiatives to explore the possible effects of nanoparticles.

14.4.1.1 Perceived Dangers

The foremost hazard for pesticides is the buildup of chemicals in runoff and its effects on the environment. This has classically caused many pesticides to be banned, including DDT. The difficulty is that the ban usually occurs after the pesticide has already been widely used. Nanoparticles are of major concern in pesticides because they will build up in wastewater, similar to chemicals, but their uniquely small size allows them to be more active and thus have more profound effects at lower concentrations [35].

The ability for n-Ag to become hazardous then depends on the amount of pesticide being sprayed, but given that most pesticides are used on multi-acre farms with few waterways, it seems likely, albeit unproven, that detectable – if not substantial – concentrations will build up. To examine this idea, a group of researchers modeled the increase of nanoparticles in wastewater due to use of cosmetics with nanoparticles n-Ag and n-TiO_2 for their case [36]. They found that the results depended heavily on the type of wastewater treatment available, but with conservative treatment, the waterways had significant levels of both nanoparticles. Given that this model assumes wastewater treatment occurs and the only source is runoff from washing cosmetics, pesticides would likely have a much larger effect.

14.4.1.2 Regulation

Regulation of nanopesticides by the EPA has slowly been evolving as more information about them has come to light. When the pesticides were first registered, the manufacturers were not required to disclose the nanoparticle unless it was an active ingredient. In 2009, the EPA was petitioned to regulate n-Ag due to safety and environmental concerns, but they declined the petition on reasoning that n-Ag works the same as larger Ag, and the small size of the particles only increases the activity [35]. Pesticides are now required, as of 2010, to list all nanoparticles under the Federal Insecticide, Fungicide, and Rodenticide Act (FIFRA) and update the EPA on any known adverse effects [35].

The EPA's current stance on nanotechnology is observation and investigation, but no regulation will be put in place until a nanoparticle application is conclusively shown to be harmful. To this end, the EPA currently has multiple studies occurring to determine any adverse effects of nanoparticles on the environment.

14.4.2 Food Packaging Industry

Food packaging was one of the first markets to formulate and manufacture nanocomposites for consumer use. The sustained growth of this market has led to more than 400 nano-enabled food packages on the market, and 25% of the market in the coming decade is projected to include some form of nanotechnology [37]. The attractiveness of these nanocomposites comes from their ability to act as a barrier to gas and UV light, thus extending the shelf life of foods. The nanocomposites can also be impregnated with nano-sized inorganic species to act as antimicrobial agents. For packaging containing nanocomposites, there is meant to be no contact between the food and the nanoparticles, but it inadvertently happens by migration.

A migration model by Simon et al. predicted that diffusion is highly possible but is dependent on the polymer or matrix the nanoparticle is held in and the dynamic size of the particle [38]. To further this study, researchers tested n-Ag-containing food packaging over varying time and temperatures for nanoparticle migration [39]. The results show that migration does occur and increases with the shelf life and temperature. This migration into food is alarming from a safety perspective, as ingestion bypasses almost all protective barriers, thus increasing the possibility that these nanoparticles can become hazardous.

14.4.2.1 Perceived Dangers

Current research into the dangers of ingesting nanoparticles has focused on their transmission and sequestration rather than the adverse effects of highly specific applications. This is largely due to the known differences in interaction of nanoparticles caused by differences in dynamic size, surface charge, processing, and so forth [40]. The transmission of nanoparticles through digestion is of concern, because the major clearance pathway is microvilli that brush food along the gastrointestinal (GI) tract. These microvilli are around 2 μm wide and thus have difficulty clearing nanometer-sized particles. Slow clearance allows nanoparticles to interact with capillaries and, again due to their small size, move within the capillary to different parts of the body. Overall this process largely increases the residence time and spreading of nanoparticles to sequestered organs for deposition [41].

A secondary clearance mechanism in the GI tract is the mucociliary elevator, which catches particulates in a mucus layer and translocates them out of the body. Many toxicologists believed this mechanism would prevent nanoparticle deposition, but recent research has shown otherwise. To test the fate of ingested nanoparticles, researchers delivered varying sizes (50 nm to 3 μm) of polystyrene beads to rats over a period of 10 days. The results showed that no particles larger than 300 nm were deposited but 26% of 100 nm and 34 % of 50 nm particles were deposited throughout the body, including in the blood, lung, and heart tissue [42]. While the explicit toxicity of nanoparticles used in food packaging is unknown and is the subject of current research, the extended residence time of nanoparticles within the human body could be hazardous; residence in the tissue alone can cause inflammation, leading to adverse effects.

14.4.2.2 Current Regulation

The major regulatory agency for food and food packaging, the FDA, has no explicit regulation in place covering nanoparticles or nanocomposites. The FDA's stance is that it regulates "products, not technologies," and thus nanotechnology as a whole requires no specific regulation. For the case of specific nanoparticles, the FDA holds that, unless research explicitly shows otherwise, materials at the macroscale "generally regarded as safe" (GRAS) are also safe at the nanoscale [43]. Thus, any material GRAS can be included in a nanoform and is not required to be labeled as said form.

Currently, the FDA is partnering with the NIH to establish viable assays to determine the toxicity of materials, and a Nanotechnology Task Force has been created to find the most efficient regulatory pathways for products incorporating nanotechnology [44]. While these last two steps are in the right direction, many experts believe the FDA is lagging the industry in applying regulations, and hence, the current safety of nano-enabled food packaging is unknown [45].

14.4.3 Medical Device Industry

The field of medicine sees significant potential in the application of nanotechnology. These applications can be segregated into three major markets: drug delivery, structural implants, and electronic implants. Drug delivery applications focus on the utilization of nanoparticles or nano-structured micelles to deliver molecules to targeted locations. Nanotechnology-enhanced structural implants are mostly orthopedic, with some dental applications where nanomaterials are used to increase mechanical properties or tissue ingrowth. Finally, electronic implants utilize miniaturized circuits, via a combination of nanoparticles and nanomaterials, to form sensors or impulse devices. For this discussion, drug delivery applications will be ignored because they are classified as pharmaceuticals, and there is a rigorous regulatory system in place for those.

14.4.3.1 Perceived Dangers

The hazards of implants with nanoparticles are thought to be similar to those of food packaging, namely that nanoparticles are being applied within the body bypassing the majority of the body's defenses. While migration is one hazard, implants also suffer from the propensity for nanoparticles to gather around the implant as it wears, causing a very high local concentration to the proximal cells. This is a common problem in joint orthopedic implants without nanoparticles, so their addition will likely only compound the issue. One example of such implants is a titanium core with a glass coating composed of a nanocomposite with TiO_2 and ZnO [46]. The glass coating provides better wear resistance, while the nanoparticles increase the bioactivity of the implant to enhance the ingrowth of cells. An example of an electrical implant utilizing nanotechnology is a small, neural electric impulse device for the treatment of Parkinson's disease, which has a carbon nanotube composite [47]. The carbon nanotubes provide enhanced conductivity for the interface of the device with the brain. While this application is empowering because it allows the creation of minute devices easily implantable in the brain, the close proximity of carbon nanotubes to the brain should not be overlooked. Carbon nanotubes are known to have adverse effects given the correct circumstances. In this light, are the advantages associated with enhanced conductivity worth the risk, especially with the limited knowledge of how carbon nanotubes may interact with cells of the brain?

14.4.3.2 Current Regulation

Regulation of medical devices by the FDA is the same regardless of whether or not nanomaterials are involved. Devices can be either Class I (low risk), Class II (medium risk), or Class III (high risk). Any device implanted within the body is Class II or higher, but if it does not sustain life, it is usually relegated to Class II. Thus, both of the devices mentioned are Class II. At this level the device must prove it will have no adverse reaction to cells but does not have to undergo clinical trials.

The glass nanocomposite coating was tested this way, with the coating applied to a petri dish and mouse cells applied over it [46]. The cells were monitored for any adverse events from death to cancerous growths, of which none of significance occurred. While this test is necessary, it is too simple considering the complexity of the nanoparticle's possible interaction. This testing assumes the nanoparticle undergoes no changes in the body and disregards any buildup or release due to wear of the coating. To fill this gap, the FDA is currently partnering with multiple government agencies to set up testing of nanomaterials and nanoparticles for adverse effects [48]. This testing is a step in the right direction, but without application testing it would seem difficult to assume any of these implants to be safe. The largest risk of these devices, which the FDA has no current framework to assuage, is the possibility of long-term effects caused by nanoparticle exposure and possible buildup.

14.4.4 Responsible Future Directions

Future safety testing of nanotechnology including nanomaterials needs to be more robust to ensure the safety of the consumer. Currently, nanomaterials for in vivo use must only undergo cell contact studies. To characterize the hazards of a nanomaterial more adequately without requiring clinical trials, many genetic assays, micronucleus or HPRT, can be used, but of importance is the style of testing [49]. The assays need to be performed with varying functional levels, such as high use to low use, to determine if the function increases the migration or deposition of nanoparticles. Depending on the cell response, the dose of nanoparticles at each functional level can be correlated to pure nanoparticle studies to give a better understanding of the hazard of the nanomaterial. This approach is ideal because it does not significantly increase the cost while providing a greatly enhanced understanding of the possible adverse effects of the nanomaterial.

14.5 Concluding Remarks

Exposure to nanomaterials presents a variety of health and environmental risks that must be controlled with federal regulation. At this point, however, there are few agencies in place that actively regulate the nanomaterial industry. This chapter introduced a risk assessment technique created by DuPont and reviewed the current status of regulation by the U.S. government, with a focus on the pesticide, food packaging, and medical industries. As the nanomaterial industry grows, federal agencies that execute efficient and effective risk management techniques must be used to protect those exposed to hazardous nanomaterials in every application.

14.6 Questions and Problems

1. The fire toxicity of polymer nanocomposites (Section 14.2) focused on thermoplastic-based polymers using FR ethylene-vinyl acetate (EVA), polyamide 6 (PA6), polypropylene (PP), polycarbonate (PC), polyethylene terephthalate (PET), and poly(methyl methacrylate) (PMMA) nanocomposites as examples. Conduct a literature search on the fire toxicity of other thermoset-based nanocomposites and thermoplastic elastomer-based nanocomposites that are not covered in this chapter.
2. The three fire toxicity characterization methods (Section 14.2.3) – steady-state tube furnace, TGA/FTIR, and cascade impactor – are widely used in the industry.

Conduct a literature search on other new ASTM or ISO standards-based methods available to the industry and provide brief descriptions of these new methods.
3. This chapter did not cover the handling and repairing, such as grinding reshaping, and end of life cycle, of polymer nanocomposites in a manufacturing plant with the focus on environmental and health impacts issues. Prepare a review paper on how the industry deals with this subject.

REFERENCES

1. A. Eastlake, L. Hodson, C. Geraci, and C. Crawford (2012). A critical evaluation of material safety data sheets (MSDSs) for engineered nanomaterials. *Journal of Chemical Health and Safety* 19(5), 1–8.
2. L. Roes, M. K. Patel, E. Worrell, and C. Ludwig (2012). Preliminary evaluation of risks related to waste incineration of polymer nanocomposites. *The Science of the Total Environment* 417–418, 76–86.
3. CDC – Nanotechnology – NIOSH Workplace Safety and Health Topic. (n.d.). Retrieved from http://www.cdc.gov/niosh/topics/nanotech/
4. AHHS, CDC, and NIOSH (2011). Occupational exposure to titanium dioxide. *Current Intelligence Bulletin 63*, NIOSH Publication No. 2011-160. Atlanta, GA: National Institute for Occupational Safety and Health (NIOSH).
5. V. L. Colvin (2003). The potential environmental impact of engineered nanomaterials. *Nature Biotechnology* 21(10), 1166–1170.
6. AHHS, CDC and NIOSH (2013). Occupational exposure to carbon nanotubes and nanofibers. *Current Intelligence Bulletin 65*, NIOSH Publication No. 2013-145. Atlanta, GA: National Institute for Occupational Safety and Health (NIOSH).
7. J. Muller, F. Huaux, N. Moreau, P. Misson, J.-F. Heilier, et al. (2005). Respiratory toxicity of multi-wall carbon nanotubes. *Toxicology and Applied Pharmacology* 207(3), 221–231.
8. S. Deguchi, R. G. Alargova, and K. Tsujii (2001). Stable dispersions of fullerenes C60 and C70 in water: Preparation and characterization. *American Chemical Society* 17(19), 6013–6017.
9. H. Hidaka, S. Horikoshi, N. Serpone, and J. Knowland (1997). Inorganic sunscreen agent on exposure to UVA and UVB radiation. *Journal of Photochemistry and Photobiology A: Chemistry* 111, 205–213.
10. J. D. Bolton (2002). Problems with wear in artificial orthopaedic joint replacements: A review. *Key Engineering Materials* 230–232, 447–454.
11. I. C. Wang, L. A. Tai, D. D. Lee, P. P. Kanakamma, C. K. Shen, et al. (1999). C(60) and water-soluble fullerene derivatives as antioxidants against radical-initiated lipid peroxidation. *Journal of Medicinal Chemistry* 42(22), 4614–4620.
12. T. D. Ros, M. Prato, S. Farmaceutiche, U. Trieste, and P. Europa (1999). Medicinal chemistry with fullerenes and fullerene derivatives. *Chemical Communications* 8, 663–669.
13. J. H. Koo, M. Natali, J. Tate, and E. Allcorn (2013). Polymer nanocomposites as advanced ablatives – A compressive review. *International Journal of Energetic Materials and Chemical Propulsion* 12(2), 119–162.
14. M. Okamoto and B. John (2013). Synthetic biopolymer nanocomposites for tissue engineering scaffolds. *Progress in Polymer Science* 38(10–11), 1487–1503.
15. K. Fukushima, A. Rasyida, and M.-C. Yang (2013). Characterization, degradation and biocompatibility of PBAT based nanocomposites. *Applied Clay Science* 80–81, 291–298.

16. K. Fukushima, M.-H. Wu, S. Bocchini, A. Rasyida, and M.-C. Yang (2012). PBAT based nanocomposites for medical and industrial applications. *Materials Science and Engineering, Part C* 32(6), 1331–1351.
17. H.-M. Park, M. Misra, L. T. Drzal, and A. K. Mohanty (2004). "Green" nanocomposites from cellulose acetate bioplastic and clay: effect of eco-friendly triethyl citrate plasticizer. *Biomacromolecules* 5(6), 2281–2288.
18. SIGMA-ALDRICH (2013). MSDS for Carbon Nanopowder, 1–6.
19. ISO/TC 299 – Nanotechnologies web link: http://www.iso.org/iso/home/store/catalogue_tc/catalogue_tc_browse.htm?commid=381983
20. ASTM E56.03 – Environment, Health and Safety: http://www.astm.org/COMMIT/SUBCOMMIT/E5603.htm
21. A. Stec and T. Hull (2007). Toxic combustion products from fire retarded nanocomposite polymers. *Proceedings of the 5th International Seminar on Fire and Explosion Hazards*, Edinburgh, UK, pp. 23–27.
22. T. Hull, D. Price, Y. Liu, C. Wills, and J. Brady (2003). An investigation into the decomposition and burning behaviour of ethylene-vinyl acetate copolymer nanocomposite materials. *Polymer Degradation and Stability* 82, 365–371.
23. A. Stec and J. Rhodes (2011). Smoke and hydrocarbon yields from fire retarded polymer nanocomposites. *Polymer Degradation and Stability* 96, 295–300.
24. J. Rhodes, C. Smith, and A. Stec (2011). Characterisation of soot particulates from fire retarded and nanocomposite materials and their toxicological impact. *Polymer Degradation and Stability* 96, 277–284.
25. F. Samyn, S. Bourbigot, C. Jama, S. Bellayer, S. Nazare, et al. (2008). Characterisation of the dispersion in polymer flame retarded nanocomposites. *European Polymer Journal* 44, 1631–1641.
26. J. Feng, J. Hao, J. Du, and R. Yang (2012). Using TGA/FTIR TGA/MS and cone calorimetry to understand thermal degradation and flame retardancy mechanism of polycarbonate filled with solid bisphenol A bis(diphenyl phosphate) and montmorillonite. *Polymer Degradation and Stability* 97, 605–614.
27. N. Didane, S. Giraud, E. Devaux, G. Lemort, and G. Capon (2012). Thermal and fire resistance of fibrous materials made by PET containing fiame retardant agents. *Polymer Degradation and Stability* 97, 2545–2551.
28. D. Calogine, G. Marlair, J.-P. Bertrand, S. Duplantier, J.-M. Lopez-Cuesta, et al. (2011). Gaseous effluents from the combustion of nanocomposites in controlled-ventilation conditions. *Journal of Physics: Conference Series* 304, 12–19.
29. A. Stec, R. Hull, and K. Lebek (2008). Characterization of the steady state tube furnace (ISO TS 19700) for fire toxicity assessment. *Polymer Degradation and Stability* 93, 2058–2065.
30. A. Stec and R. Hull (2009). Assessment of fire toxicity from polymer nanocomposites. In *Fire Retardancy of Polymers: New Strategies and Mechanisms*, T. R. Hull and B. K. Kandola (Eds.), Cambridge, UK: Royal Society of Chemistry Publishing, pp. 405–417.
31. A. Stec and R. Hull (2008). Effect of fire retardants and nanofillers on the fire toxicity of four common polymers. *American Chemical Society*, 342–366.
32. J. Valverde and I. Linkov (2011). Nanotechnology: Risk assessment and risk management perspective. *Nanotechnology Law and Business* 25, 1–4.
33. J. Holdren, C. Sunstein, and I. Siddiqui (2011). Policy principles for the U.S. decision-making concerning regulation and oversight of applications of nanotechnology and nanomaterials. *Memorandum for the Heads of Executive Departments and Agencies*, 1–4.
34. T. Medley and S. Walsh (2007). Defense – Nano risk framework. *DuPont Nano Partnership*, 23–75.
35. L. Bergeson (2011). Emerging nanomaterial governance systems: The state of play. *Molecular Imaging* 10, 17–27.

36. N. Musee (2010). Simulated environmental risk estimation of engineered nanomaterial: A case of cosmetics in Johannesburg City. *Human and Experimental Toxicology* 30, 1181–1195.
37. S. Alfadul and A. Elneshwy (2010). Use of nanotechnology in food processing, packaging, and safety – A review. *African Journal of Food, Agriculture, Nutrition, and Development* 10, 2719–2739.
38. P. Simon, Q. Chaudhry, and D. Bakos (2008). Migration of engineered nanoparticles from polymer packaging to food-a physicochemical view. *Journal of Food and Nutritional Research* 47(3), 105–113.
39. M. Cushen, J. Kerry, M. Morris, M. Cruz-Romero, and E. Cummins (2012). Nanotechnologies in the food industry – Recent developments, risks, and regulation. *Trends in Food Science & Technology* 24, 30–46.
40. Health and Safety Executive (HSE) (2006). Review of the adequacy of current regulatory regimes to secure effective regulation of nanoparticles created by nanotechnology.
41. H. Chen, J. Weiss, and F. Shahidi (2006). Nanotechnology in nutraceuticals and functional foods. *Food Technology* 3, 30–36.
42. P. Jani, G. W. Halbert, J. Langridge, and A. T. Florence (1990). Nanoparticle uptake by the rat gastrointestinal mucosa: quantitation and particle size dependency. *Journal of Pharmacy and Pharmacology* 42, 821–826.
43. U.S. FDA (2007). Nanotechnology: A Report of the U.S. Food and Drug Administration Nanotechnology Task Force. July 25. Available at: http://www.fda.gov/nanotechnology/taskforce/report2007.html (accessed January 15, 2008).
44. C. Chau, S. Wu, and G. Yen (2007). The development of regulations for food nanotechnology. *Trends in Food Science & Technology* 18, 269–280.
45. D. Bowman and G. Hodge (2008). "Governing" nanotechnology without government? *Science and Public Policy* 35, 475–487.
46. C. Soundrapandian, S. Bharati, D. Basu, and S. Datta (2011). Studies on novel bioactive glasses and bioactive glass-nano-hap composites suitable for coating on metallic implants. *Ceramics International* 37, 759–769.
47. F. Berger, S. Gevers, L. Siep, and K. Weltring (2008). Ethical, legal and social aspects of brain-implants using nano-scale materials and techniques. *Nanoethics* 2, 241–249.
48. B. Culliton (2008). Is special FDA regulation of nanomedicine needed? A conversation with Norris E. Alderson. *Health Affairs* 27, 315–317.
49. S. Doak, B. Manshian, G. Jenkins, and N. Singh (2012). In vitro genotoxicity testing strategy for nanomaterials and the adaptation of current OECD guidelines. *Mutation Research* 745, 104–111.

15 Opportunities, Trends, and Challenges for Polymer Nanocomposites

15.1 Introduction

Just because new technology works does not necessarily mean that it will be successful. There are many other aspects involved that people do not generally consider. This situation can be observed in many contemporary technologies that are being advocated for mass production and use, such as, for example, solar and wind energy. Currently, the startup cost is too high for many companies to consider switching over to these forms of energy. The pathway is similar for polymer nanocomposites in that although they do yield some exciting properties, they are difficult and expensive to produce on a mass scale. This chapter discusses the research funding and commercial market opportunities, developments currently being made to bring polymer nanocomposites to a larger consumer market, and the challenges of manufacturing nanomaterials and polymer nanocomposites.

15.2 Opportunities

15.2.1 U.S. Government Research Funding Opportunities

Since its inception in 2001, the National Nanotechnology Initiative (NNI; www.nano.gov) has received cumulatively more than $22 billion in government grants [1]. The president's 2016 budget provides $1.5 billion for the NNI, a continued investment in support of the president's priorities and innovation strategy [2]. This support reflects nanotechnology's potential to significantly improve our fundamental understanding and control of matter at the nanoscale and to translate that knowledge into solutions for critical national needs. In 2014, federal agencies invested a total of $1.57 billion in nanotechnology-related activities. The NNI investments in 2014 and 2015 and those proposed for 2016 reflect a renewed emphasis on accelerating the transition from basic R&D to innovations that support national priorities, while maintaining a strong foundation of broad, fundamental research in nanoscience that provides a continuing pipeline of new discoveries that will enable future revolutionary applications. Nearly half (43%) of this budget request is focused on applied R&D and support for the Nanotechnology Signature Initiatives (NSIs), reflecting an increased emphasis within the NNI on commercialization and technology transfer.

The NSIs are multi-agency initiatives designed to focus a spotlight on technology areas of national importance that may be advanced more rapidly through enhanced interagency coordination and collaboration. The following are the five current NSIs [2]:

- *Nanotechnology for Solar Energy Collection and Conversion: Contributing to Energy Solutions for the Future* spans efforts in fundamental and applied research to improve photovoltaic and thermophotovoltaic devices and advance the development of solar fuels.
- *Sustainable Nanomanufacturing: Creating the Industries of the Future* includes efforts focused on the development of robust nanomanufacturing methods for the cost-effective production of nanoscale materials and devices.
- *Nanoelectronics for 2020 and Beyond* is aimed at discovering and using novel nanoscale fabrication processes and innovative concepts to produce revolutionary materials, devices, systems, and architectures to advance the field of nanoelectronics.
- *Nanotechnology Knowledge Infrastructure (NKI): Enabling Leadership in Sustainable Design* is focused on providing a community-based, solutions-oriented knowledge infrastructure to accelerate nanotechnology discovery and innovation.
- *Nanotechnology for Sensors and Sensors for Nanotechnology: Improving and Protecting Health, Safety, and the Environment* is focused on the utilization of nanotechnology to develop physical, chemical, and biological sensors that have higher sensitivity and selectivity, are more portable, and have lower power demands than conventional sensors.

Federal agencies are committed to promoting the transition of nanotechnology-based discoveries from the laboratory to market. Although there is a renewed emphasis for 2014–2016 on activities aimed at promoting accelerated translation of nanotechnologies into commercial products, NNI agencies remain committed to strong and sustained support for fundamental research in nanoscience. Accordingly, the president's budget requested $512 million for foundational research in 2016, or 34% of the total NNI funding. Continued investment in this area is critical to maintaining the pipeline of basic discoveries in nanoscience that will enable future innovations in support of national priorities.

Table 15.1 shows NNI investments for 2014 through 2016 by federal agencies with budgets and investments for nanotechnology R&D [2]. Figure 15.1 shows NNI funding over time since the inception of the NNI in 2001 [2].

Figure 15.2 shows the 2016 NNI investments by Program Component Area (PCA), which are subdivided into: (a) Foundation Research (34%), (b) Applications, Devices, Systems (26%), (c) Signature Initiatives (17%), (d) Infrastructure & Instrumentation (16%), and (e) Environmental, Health, Safety (7%) [2].

Table 15.1. *NNI budget by agency, 2014–2016 ($ millions)*

Agency	2014 Actual	2015 Estimated[a]	2016 Proposed
CPSC	2.0	2.0	7.0
DHS[b]	25.0	28.4	17.0
DOC/NIST	97.8	83.6	86.3
DOD	189.6	141.7	130.4
DOE[c]	309.4	329.7	342.2
DOT/FHWA	2.1	1.5	1.5
EPA	15.5	15.1	15.3
DHHS (total)	430.3	437.8	448.6
FDA	9.3	15.8	14.6
NIH	410.0	411.0	423.0
NIOSH	11.0	11.0	11.0
NASA	22.4	19.1	15.8
NSF	464.5	413.4	416.4
USDA (total)	15.6	15.6	14.8
ARS	2.0	3.0	3.0
FS	5.6	4.6	4.5
NIFA	8.0	8.0	7.3
TOTAL[d]	**1574.3**	**1487.8**	**1495.3**

[a] 2015 numbers are based on 2015 enacted levels and may shift as operating plans are finalized.
[b] As of the date the 2016 budget was released, final 2015 appropriations for the Department of Homeland Security were not yet enacted. Therefore, the 2015 column of this table reflects amounts requested for DHS in the 2015 budget.
[c] Funding levels for DOE include the combined budgets of the Office of Science, the Office of Energy Efficiency and Renewable Energy (EERE), the Office of Fossil Energy, and the Advanced Research Projects Agency for Energy (ARPA-E).
[d] In columns 2–6, totals may not add, due to rounding.
Source: National Nanotechnology Initiative [2].

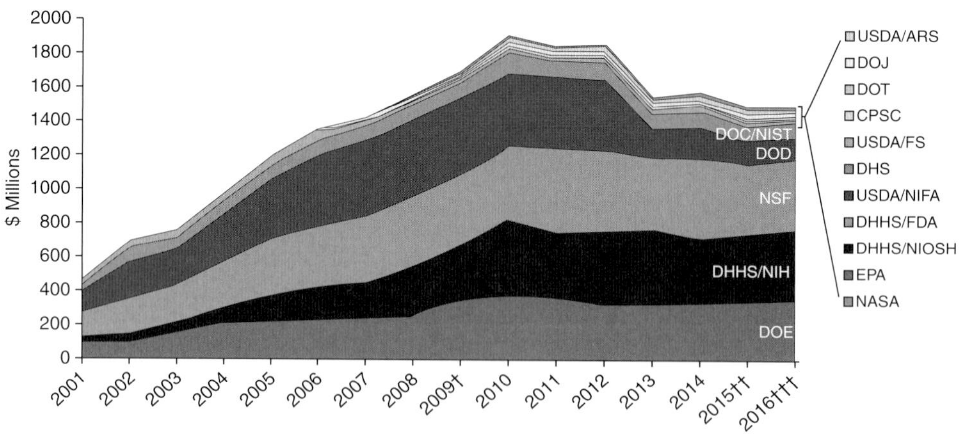

† 2009 figures do not include American Recovery and Reinvestment Act funds for DOE ($ 293 million), NSF ($ 101 million), NIST ($ 43 million), and NIH ($ 73 million).
†† 2015 estimated based on 2015 enacted levels and may shift as operating plans are finalized.
††† 2016 Budget.

Figure 15.1. NNI funding by agency, 2001–2016 (millions of dollars).
Source: NNI.

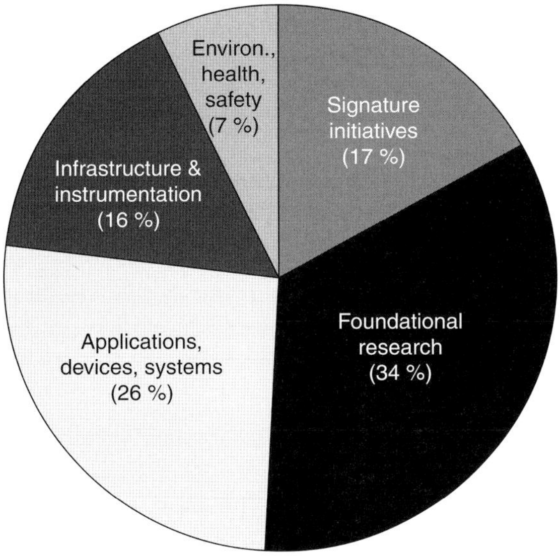

Figure 15.2. Breakout of NNI funding by Program Component Area (PCA) in the 2016 Budget.
Source: NNI.

15.2.2 Commercial Market Opportunities

Commercial market opportunities and trends data of nanomaterials and polymer nanocomposites are provided to the author by Lucintel [3] (consulting firm based in Irving, Texas; www.lucintel.com), Pyrograf Products, Inc., a division of Applied Sciences, Inc. [4] (carbon nanofibers manufacturer based in Cedarville, Ohio; www.pyrografproducts.com), and an excellent article published by Vance et al. [5]. According to Lucintel's executive summary of a recent nanomaterials/polymer nanocomposites market research report [3], the following statements were made:

- Global nanomaterials industrial reached \$1.7 billion in 2010, with an average annual growth rate of 10.4% over the last five years
  - Top 10 players are ~80% of global nanomaterial market
  - Nanoclays and carbon nanotubes are the dominant nanomaterials representing ~90% of the market value
- North American nanomaterials industrial accelerated by ~25% in 2010 and Europe by ~22%, while Asia and other regions grew by 32% in 2010, they were driven by
  - Active participation of government in nanotechnology R&D funding
  - Advancement in production processes
  - Heavy investment in R&D by major players
- Global nanomaterial market is forecast to become a \$6.1 billion business in 2016
  - Expected to grow at a CAGR of 23% during 2011 to 2016
  - Driven by health and personal care, energy, and electrical & electronics (E&E) applications
  - Emerging applications including field emission display (FED), drug delivery, and biofuels

- Health care and energy industries are expected to surpass the E&E industry in size over the coming 2 to 3 years, driven by nanotechnology advancements in biomedical and solar energy applications

The definition of "commercial product" is one that is both *offered for sale* and *used* in the regular production of a device or component in general commence. This section includes numerous examples of "commercial" and "potential commercial" nanomaterials. Table 15.2 shows an overview of currently available nanotechnology products, the developer/manufacturer, and the benefits of these nanomaterials applied to their prospective applications [6]. Table 15.3 illustrates some of nanotechnology's potential applications under development by various composite manufacturers, along with anticipated benefits [6].

An excellent article by Sanchez et al. summarizes the current and potential commercial applications of hybrid organic-inorganic nanocomposites [7]. They pointed out that organic-inorganic hybrid materials not only represent a creative way to design new materials and compounds for academic research, but the improved and unusual features of these methods will result in the development of innovative industrial applications. Most of the hybrid materials that are currently commercially available are synthesized and processed by using conventional soft-chemistry-based routes developed in the 1980s. The authors grouped all hybrid material processes into three routes:

- Copolymerization of functional organosilanes, marcomonomers, and metal alkoxides
- Encapsulation of organic components within sol-gel-derived silica or metallic oxides
- Organic functionalization of nanofillers with lamellar structures, and so forth.

The chemical strategies (self-assembly, nanobuilding block approaches, hybrid MOF [Metal Organic Frameworks], integrative synthesis, coupled processes, bio-inspired strategies, etc.) allow the development of new chemistry to direct the assembling of a large variety of structurally well-defined nano-objects into complex hybrid architectures hierarchically organized in terms of structure and functions. These new generations of hybrid materials will open very promising applications in: optics, electronics, ionics, mechanics, energy, environment, biology, medicine, functional smart coatings, fuel and solar cells, catalysts, sensors, and so on. The Sanchez et al.'s study [7] identifies selective examples of hybrid materials with an implied emerging potential commercialization as well as functional hybrids with real commercial opportunities.

Recently, selective applications of nanoclays and carbon nanotubes are also reported by Lucintel in Figure 15.3 [3]. Polymer-clay nanocomposites are used in: (a) bottling of O_2 and CO_2 sensitive products, such as beer and soft drink bottles; (b) food and packaging industries; (c) motor compartment of vehicle for casting and connectors; and (d) wire and cable applications. Polymer-CNT nanocomposites are used in: (a) timing belt covers in automotive; (b) scratch-resistant coatings in automotive and aircraft frame and body components; (c) EMI/RFI shielding and coatings in electronic devices to control radiation; (d) CMP slurries in electronic

Table 15.2. *An overview of current commercially available nanotechnology products*

Product	Developer/Manufacturer	Benefit/Function
Auto sideboards	GM/Basell/Southern Clay Products/Blackhawk	Nano TPO-clay provides stronger, lighter sideboards
Nylon film	Honeywell/Nancor/Ube/Unitika	Low oxygen barrier film for wrapping
Performance plastics masterbatch	Hyperion	EMI components for office and home appliances
Trousers	Nanotex/Burrlington Mills/Eddie Bauer	Stainless trousers
Mattress covers	Nanotex/Seally	Shield for stains
Floor coverings	Nanophase	Vinyl flooring with nanocoating abrasion resistant
Sunscreen	Nanophase	Nanocrystals allow sunscreen to be transparent
Tennis ball	InMat™/Wilson	Coating for butyl rubber core improves pressure seal for longer life
Red stain and grease removal	Eastman/P&G	Peroxide and nanoparticle to remove stain and grease
Fire retardant power cable	Kabelwerk Eupen	Combining ATH-based compound with organocaly improves on fire performance and smoke density
Basketball Shoe Pouch	Converse/Triton System	Nanocomposite pouch filled with helium inserts that fits into basketball shoe
Medical tubing	Foster-Miller	Nylon nanocomposite tubing with improved modulus for medical application
Chemical, mechanical planarization	DuPont/Air Products/Others	Nanosilica, nanocerium oxide, smooth chips
Nano-wax particle	BASF	Self-cleaning surfaces
Polymer-clay masterbatch	Nanocor	Molded parts
Nanoclay additives	Sud Chemie	Flame-retardant additives
Polycarbonate/ABS	Bayer	Flame-retardant polymer blends
Optical coating	Nanofilm	Antistreaking windshield coating
Polycarbonate panels	GE/Cabot	Nanogel translucent aerogel
Rocket fuel	Mach I	Nanoferric oxide additive to rocket fuel

Table 15.3. *An overview of potential commercial nanotechnology products*

Product	Developer	Benefit/Function
Tires	In Mat	Keep tires from losing air pressure, up to 2X as current seals
Power cells	Ballard	Fullerenes for H cells, notebook, auto computer
Drug delivery	Microchip/ChipRx	Medical devices, biosensors for drug delivery
Robots	NYU/USC	Tiny machine to clean ocean, attack gems, space research

chips and wafers to enhance electrical conductivity; and (e) diagnostics and imaging in health care to enhance clarity in X-ray films.

Several current and potential applications shown in Figure 15.4 can be subdivided into market segments of [3]: (a) transportation, such as engine and fuel systems, scratch-resistant exterior parts and coatings, car interior, aircraft structure and

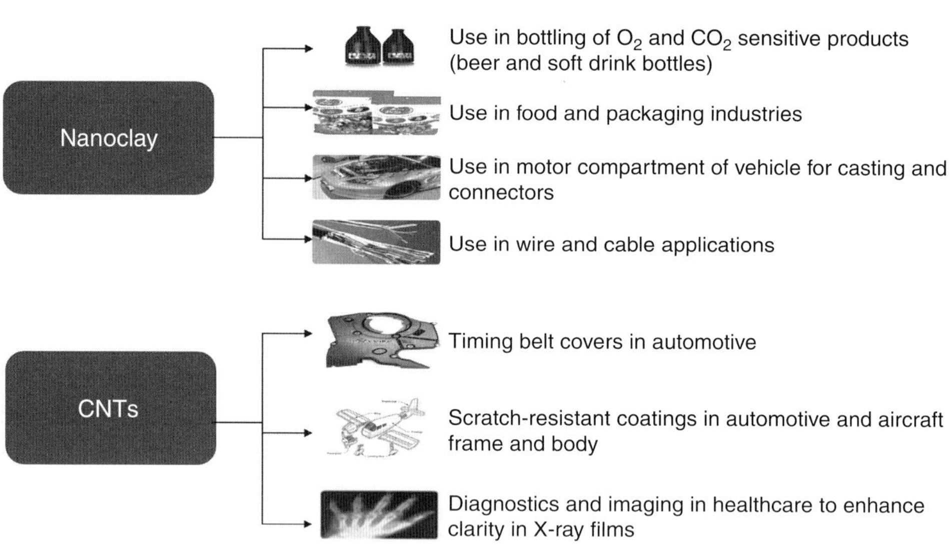

Figure 15.3. Selective applications for polymer-nanoclays and polymer–carbon nanotubes nanocomposites (Courtesy of Lucintel).

framing, and wear-resistant paints and coatings for defense vehicles; (b) construction, such as conductive flooring, pipes, insulating material for roofs and thatches, house and building siding, and self-cleaning windows; (c) packaging, such as meat and food packing, computers and electronics, medicines and pharmaceuticals, and soft drink and beer bottles; (d) consumer goods, such as home appliances, sporting goods and toys, furniture, and others; (e) energy, such as battery electrodes, fuel cell membranes, and super capacitors; (f) electrical and electronics, such as sensors, semiconductors, and hard disk storage in computers; (g) health care, such as body implants, medical devices, and dental filling materials; and (h) others, such as anti-foul coatings for marine ships, industrial equipment to increase strength, and fire-resistant clothes.

Polymer-carbon nanofiber nanocomposites applications in terms of TRL levels are shown in Figure 15.5 [4]. At the TRL 9 and 8 levels, applications, such as aerospace structural components, sporting goods, anti-corrosion/erosion coatings, audio equipment, conductive inks, thermal conductive grease, and ESD components are demonstrated. At the TRL 7 and TRL 6 levels, applications, such as ESD carpets/fabrics, electronic device foam, fire retardant foam, and automotive panels are used. At TRL 5 level, multi-functional/structural composites are used. At TRL 4 level, biosenors/smart composites are being considered. At TRL 3 level, solar cells and biomedical applications are investigated. At the lowest TRL levels of 1 and 2, continuous filaments for non-metallic cable and wiring and space elevator are being considered.

15.2.2.1 Hybrid Material for Sporting Goods

Due to their strength and light weight, several polymer nanocomposites are currently used in several sporting goods products as shown in Figure 15.6 [3]. Silicon dioxide (SiO_2) is used in badminton rackets. Carbon nanotubes (CNTs) are used in

Figure 15.4. Current and potential applications for polymer nanocomposites in different market segments, such as transportation, construction, package, consumer goods, energy, electrical and electronics, health care, and others (Courtesy of Lucintel).

tennis racquets, baseball bats, bicycle frames, and hockey sticks. Titanium dioxide (TiO_2) is used in fishing rods.

15.2.2.2 Hybrid Compounds for Cosmetics

New hybrid compounds are targeted for cosmetics and controlled release of "active ingredients" for applications concerning skin care and protection. They are based on the encapsulation of active organic components within porous silica micro-capsules. Undesirable skin alternations and melanomas created by sun-induced premature skin aging require the use of efficient but biocompatible sunscreens. Conventional commercial sunscreens are directly applied to the skin and usually contain an extremely high amount of active ingredients. These can

Figure 15.5. Polymer-carbon nanofiber nanocomposites are presented from TRL 9 to TRL 1 levels (Courtesy of Applied Sciences, Inc.).

Due to Strong and Lightweight Functions, Nanocomposites are Currently Used in Several Sporting Goods Applications

Sporting Goods Product	Nanomaterials
Badminton Racquets	SiO2 nanopowder
Tennis Racquets	Carbon nanotube
Baseball Bats	Carbon nanotube
Bicycle Frames	Carbon nanotube
Fishing Rods	Ti
Hockey Sticks	Carbon nanotube

Figure 15.6. Several nanomaterials are currently used in sporting goods products (Courtesy of Lucintel).

be detrimental to health, especially when they are absorbed by skin and are not photostable, thus generating free radicals that might cause damage to the DNA. Sol-Gel Technologies Ltd., based in Bet Shemesh, Israel (http://www.sol-gel.com) [8], provides a safe solution by encapsulating the active organic UV filters

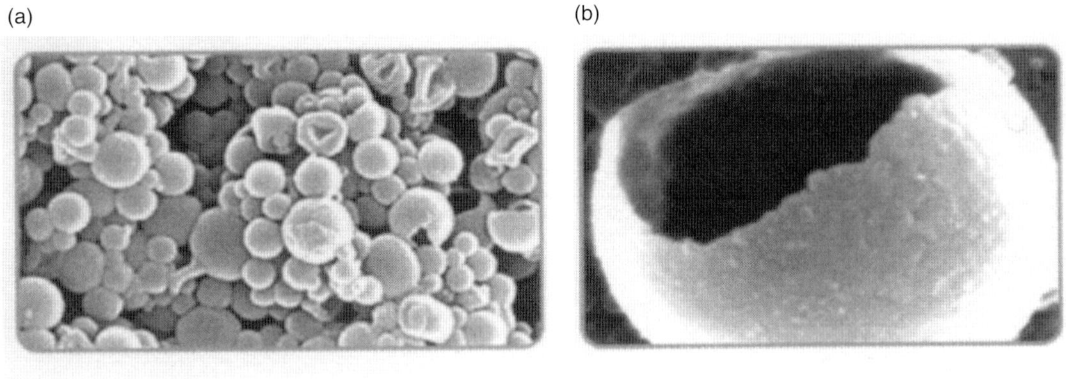

Figure 15.7. (a) Micro-capsules of UV pearls. (b) Micrograph of a core-shell capsule for carrying functional contents.

in silica micro-capsules. As a result, contact of these potent chemicals with skin is reduced and damage from free radicals within the porous host is prevented. These "UV-pearls" are incorporated into a suitable cosmetic vehicle to achieve high Sun Protection Factors (SPF), while providing an improved safety profile as the penetration of the absorbed UV is conveniently reduced. These newly developed products have already been marketed by companies for sunscreens and daily wear cosmetics. Figure 15.7a shows the "UV pearls" and Figure 15.7b shows the core-shell capsule for carrying functional contents [8].

Sol-Gel Technologies also developed silica rainbow pearls containing organic dyes for cosmetics applications. These silica active-pearls contain an effective acne medication, such as benzoyl peroxide (BP), which is as effective as antibiotics but does not cause bacterial resistance or stomach upset. Benzoyl peroxide in direct contact with skin causes skin irritation, dryness, and hyper-pigmentation in many patients. Encapsulating the benzoyl peroxide active ingredient in a silica shell prevents it from contacting the epidermis while gradually delivering BP into the follicular region where acne bacteria are found. Sol-gel active-pearls are presently a successful commercial development.

15.2.2.3 Hybrid Materials for Protective and Decorative Coatings

Many interesting and highly productive application of inorganic-organic hybrid materials have occurred in automotive coatings. Hybrid coatings provide coloration, scratch resistance, and protection from UV and chemical attacked for the substrate. DuPont with Generation 4® has developed a complex mixture of two hybrid polymers cross-linked simultaneously during curing to form a polymer network. This unique network is partially grafted and partially interpenetrated as a decorative coating for the automotive industry [9]. This hybrid material consists of a high-density acrylate polymer core with organically modified alkoxysilane groups and residual unsaturations. It confers a high modulus and scratch-resistant function, dispersed in a low cross-linked density polymer, which provides film-forming properties. The superior scratch and environmental-etch resistance of these hybrid coatings led to

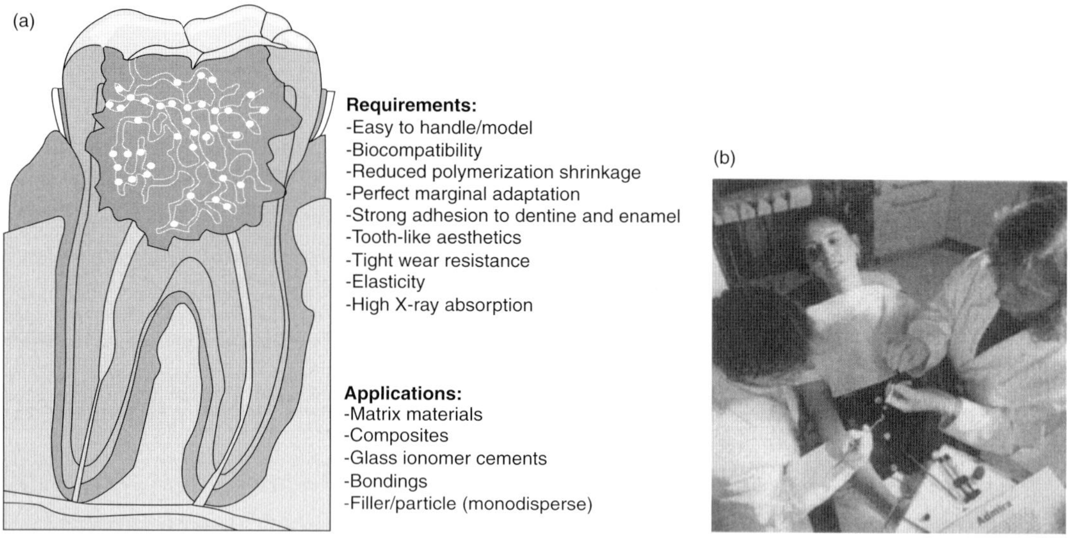

Figure 15.8. (a) Requirements and possibilities of dental applications of ORMOCER®s. (b) Application of an ORMOCER as dental filling material.

acceptance as topcoats for eight of the ten top-selling automobiles for 1997, including Ford Taurus, Toyota Camry, and Honda Civic/Del Sol.

Decorative hybrid coatings also encompass an industrial area of interest, such as packaging glass, glass sheets for furniture and sanitary appliances, and glass in architecture and the building industry. For example, a dishwasher-safe ORMOCER® color coating is used on glass (DEKOSIL®). The main advantages in terms of processing these coatings are the easy reproducibility of the colors and the finishing by conventional wet painting procedures (low curing temperatures <200°C). From an aesthetic point of view, some requirements, such as unlimited color palette, additional decorative effects like color gradients or partial coloration are also considered as key features. Another example is the titania siloxane-based hybrid organic-inorganic materials developed by several Japanese and European glass packaging industries [10]. These coatings provide a variety of colors, enhance the consumer appeal, and improve the mechanical properties of these glass bottles. These dyed colored bottles are recyclable as uncolored glass, because in contrast to conventional glass bottles, these glass bottles are color coated. The coloration does not originate from transition metals that are normally dissolved in glass and difficult to remove upon re-melting the glass for recycling. Conventional colored glass remains colored upon re-melting.

15.2.2.4 Hybrid Materials for Dental Applications

Inorganic-organic hybrid materials can be used as filling composites in dental applications [11–14]. As shown in Figure 15.8a these composites feature tooth-like properties (appropriate hardness, elasticity, and thermal expansion behavior) and are easy to use by the dentist as they can easily penetrate into the cavity and harden quickly under the effect of blue light [11]. These materials feature minimum shrinkage, are nontoxic, and sufficiently nontransparent to X-rays.

(a) (b)

Figure 15.9. Partners in development: (a) GM/Basell/Southern Clay Products/Blackhawk for the first commercial nano TPO-clay nanocomposites step assist on Astro and Safari vans in the automotive industry. (b) GM R&D manager stepping into a van with a step assist (Courtesy of SCP).

Traditional plastic filling composites had long-term adhesion problem and a high degree of polymerization shrinkage resulting in marginal fissures. The dual character of the ORMOCER®s as inorganic-organic copolymers is the key for improving the properties of dental filling composites. The organic, reactive monomers are bounded in the sol-gel process by the formation of an inorganic network. Thus, in the subsequent curing process, polymerization takes place with less shrinkage. Abrasion resistance is significantly enhanced by the existing inorganic Si-O-Si network.

An example of available commercial filling composites based on dental ORMOCER®s from Fraunhofer ISC, such as Definite® and Admira®, is shown in Figure 15.8b. In the case of Admira®, a specifically designed dentine-enamel bonding, an adhesive ORMOCER®, developed in cooperation with VOCO GmbH, is used to make this product especially attractive and brilliant in appearance. In the glass ionomer cement-based dental composites, blue light polymerized carboxyl functionalized ORMOCERs have been developed.

15.2.2.5 Hybrid Materials (Polymer-Clay Nanocomposites) with Structural Properties

The first successful development of polymer-clay nanocomposites (PCNs) was pioneered by Toyota's researchers [15–19] for high-performance reinforced plastics applications in automobiles. The earliest successes in the automotive industry are credited to the sonication process developed by Ford Corporation to better disperse the clay within the polymer. Related to these early disclosure, General Motors Corporation (GMC) developed a step assist component [20] (Figure 15.9) for 2002 GMC Safari and Chevrolet Astro vans, which are made of thermoplastic polyolefin-clay nanocomposite. This material is lighter and stiffer, more robust at low temperatures, and easily recyclable. GMC also used these PCNs in the lateral protection wire of the 2004 Chevrolet Impala because they are 7% lighter in weight and present a better surface appearance.

New applications of hybrid PCNs are envisaged by the automotive industry for commercialization: PP-clay for bodywork with anti-scratch properties (Dow Plastics/Magma); acetal-clay for ceiling lights (Showa Denko); PP-clay for panes of doors, consoles, and interiors decoration (Ford, Volvo) due to aesthetics, recyclability, and weight-saving properties; nylon-clay for bumpers with enhanced mechanical and weight-saving properties (Toyota); and nylon-clay for fuel reservoir with air tight properties (Ube America).

Beside the listed properties, the weight advantage significantly reduces the impact on the environment. It has been reported [21] that widespread use of PCNs by U.S. automotive industry could save 1.5 billion liters of gasoline in one year of automotive production, consequently reducing related carbon dioxide emissions by more than 10 billion pounds.

For environmental applications, nanoclays are used as effective reinforcing agents in "green nanocomposites" [22]. These renewable-resource-based biodegradable nanocomposites consist of cellulose plastics (plastics made from wood), polyactic acid (corn-derived plastic), or polyhydroxyalkanate (bacterial polyester). They are attractive materials as substitute for petroleum feedstock in making the manufacture of biodegradable plastic for the commercial market [23].

15.2.2.6 Hybrid Materials (PCNs) with Gas Barrier Properties

Through the incorporation of nanoclays in polymer matrices, it is possible to create a labyrinth within the structure, which physically delays the passage of molecules of gas. The excellent barrier properties against gas and vapor transmission have resulted in applications for food and beverage packaging, as well as barrier liners in storage tanks and fuel lines for cryogenic fuels in aerospace systems [24]. Researchers at Bayer Polymers have developed a hybrid material combining silicate nanoparticles with the features of polyamide 6 (PA6) and ethylene vinyl alcohol (EVOH), which is significantly better than neat PA6 [25].

Nanocor and Mitsubishi Gas Chemical Company developed the MXD6 high-barrier semi-aromatic nylon nanocomposites [25, 26]. Mitsubishi marketed this PCN under the Imperm® trade name. The new material features oxygen and carbon dioxide barrier properties that were previously unachievable. Imperm's barrier properties are superior to EVOH and its processing characteristics are ideal for multilayer films, bottles, and thermoformed containers (Figure 15.10).

Honeywell has developed a polyamide-clay nanocomposite with specific active and passive barrier properties against oxygen [25]. The passive barriers are the nanoclays, which render oxygen transmission inside the composite more difficult, but also conduct the oxygen molecules to specific oxygen captors. The clay incorporation limits the oxygen transmission up to 15–20% of the value for the pure polymer.

The beverage industry is not the only one that can benefit from implementing the aforementioned technologies. Food storage can also be enhanced through the same enhanced barrier properties used for cans and bottles. There is always demand for materials that would reduce the amount of food spoilage. Triton Systems and the U.S. Army collaborated on using PCNs as nonrefrigerated package materials, which

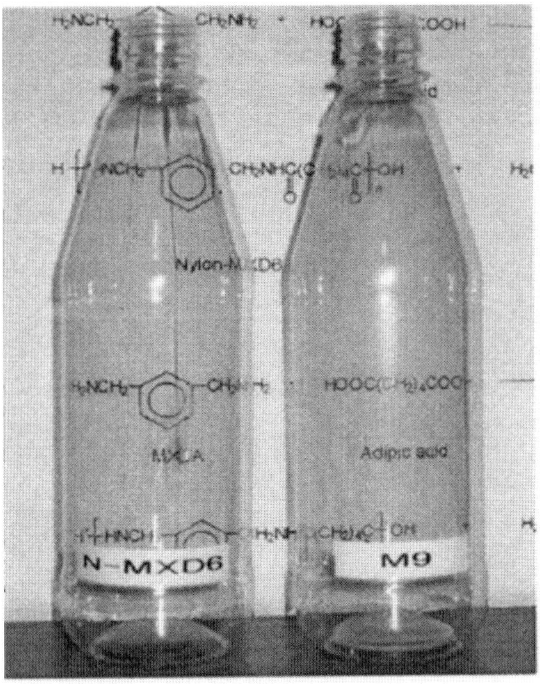

Figure 15.10. Nano N-MXD6 and N-MXD6 M9 are high-gas-barrier resins commercially available from Mitsubishi Gas Chemical Company.

can conserve food freshness for three years [25, 27]. Currently, the use of polymer nanocomposites is considered to enhance the preservation of military meal-ready-to-eat (MRE) packages, which currently use metal interfaces to preserve the food items (see Figure 15.11). The military projects that, if implemented correctly with the right filler and polymer combination, the cost of MREs could decrease by as much as 30 percent [28].

InMat Inc. has developed several applications for their nanocomposite coatings [29]. The nanocomposite coating of the Wilson tennis balls maintains the internal pressure for an extended period of time. Figure 15.12 shows the Wilson tennis balls containing a double core. This core is coated by a polymer-clay (vermiculite) nanocomposite that acts as a gas barrier, doubling the ball's shelf life. It is anticipated that this technology could be extended to the rubber industry and be incorporated into soccer balls or in automobile or bicycle tires.

The barrier properties of PCNs have also been extended to liquids (solvents) or molecules, such as water, since these liquids are responsible for polymer deterioration in some instances. De Bievre and Nakamura of Ube Industries reveal significant reduction in fuel transmission through polyamide-6/66 polymers by incorporation of nanoclay, resulting in reduced fuel emission for fuel tank and fuel line components [25, 30]. InMat Inc. collaborates with the U.S. Army with regard to the concept of liquid/hydrocarbon barrier [29]. Some envisaged applications are protective gloves and coatings for combustible and fuel tanks. Triton System Inc. and Converse created a polymer-clay nanocomposite pouch filled helium insert that fits into the basketball shoe (Figure 15.13).

Figure 15.11. An MRE food pouch sealed in a polymer bag.

Figure 15.12. The core of this Wilson tennis ball is covered with a polymer-clay nanocomposite coating that acts as a gas barrier, doubling its shelf life (Courtesy of SCP).

15.2.2.7 Hybrid Materials (PCNs) with Flame-Retardant Properties

Thermal stability and fire retardancy have motivated investigation of nanoclay as an FR additive for commodity polymers, as reported by Gilman et al. [25, 31]. Gitto (polymer manufacturer) and Nanocor (nanoclay manufacturer) are developing fire-retardant polyolefin-clay nanocomposites, since this polymer has a wide range of industrial applications [25, 32]. The heat distortion temperature (HDT) of PCNs

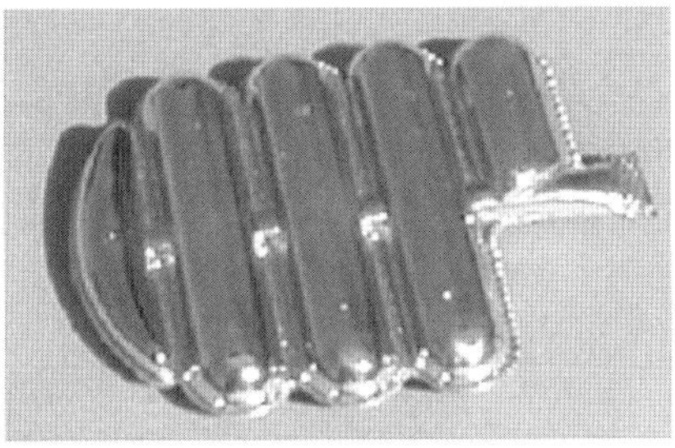

Nanocomposite Pouch filled with Helium

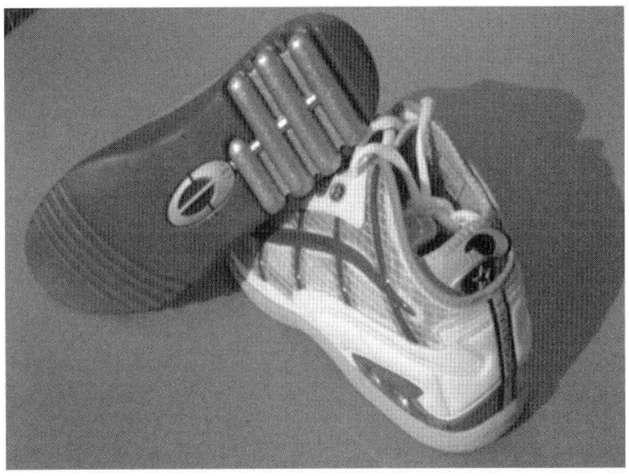

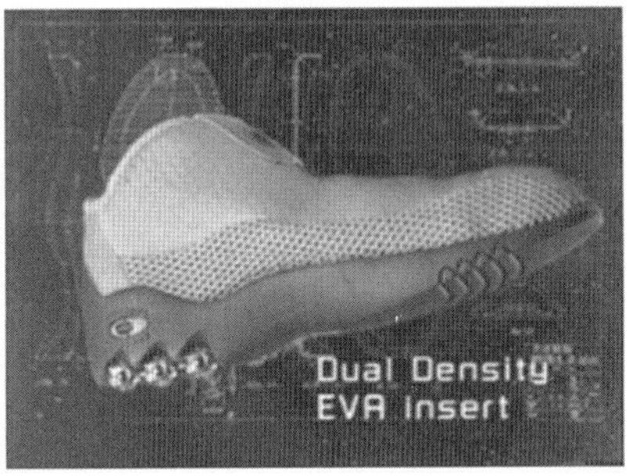

Insert that fits into the shoe

Figure 15.13. Converse basketball shoe pouch by Triton Systems Inc. (Courtesy of SCP).

can be increased up to 100°C, extending their usage to higher-temperature applications, such as automotive under-the-hood parts. Figure 15.14 shows a nanocomposite power cable manufactured by Kabelwerk Eupen with enhanced fire performance and smoke density. Additional flammability studies of polymer nanocomposites can be found in Chapter 9.

15.2.2.8 Hybrid Materials (PCNs) with Other Properties

In some cases, the value of PCNs technology relates to value-added performance of mechanical, barrier, or fire-retardant properties not present in the neat polymer but that occur via the low-volume introduction of these organically modified layered silicate (OLS) nanomaterials. For example, in contrast to conventional filled polymers that are opaque and exhibit compromised mechanical properties, the introduction of a low-volume fraction of OLS that is of nanoscopic dimensions and uniformly dispersed within the polymer results in PCNs with optimum transparency and clarity similar to the neat polymer. Further, these PCNs exhibit multifunctionality behavior that is unavailable in the neat polymer. Figure 15.15 shows a nylon 12 nanocomposite for medical applications with improved modulus.

In the area of fuel cells, polymer nanocomposites have been proposed as a viable material for producing electrical energy and reducing the solvent transmission to the fuel. This is an important area for any engine operating on fossil fuel. The implementation of polymer nanocomposites into the fuel lines will greatly lower the material costs. It may, in some cases, increase the fuel efficiency of the engine. This is

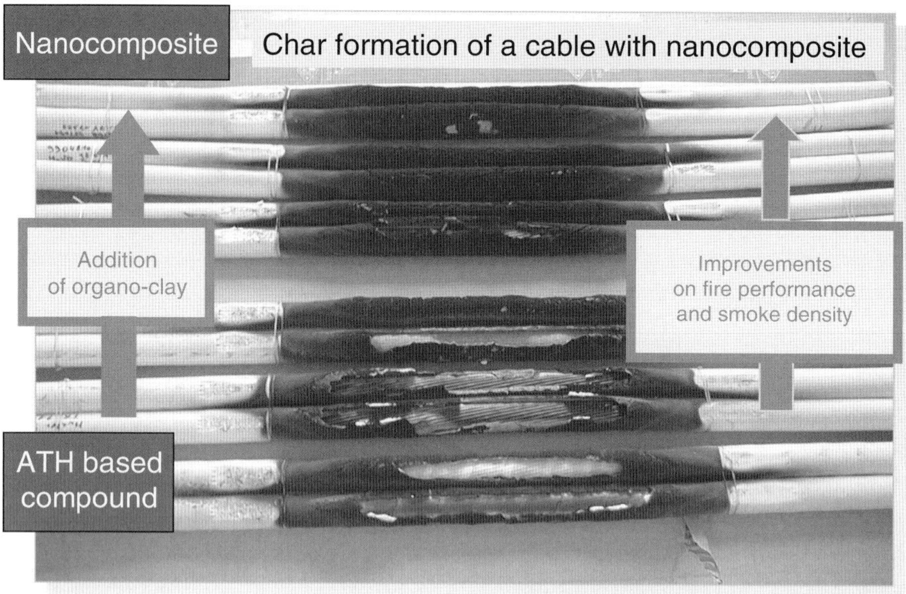

Figure 15.14. Nanocomposite power cable by Kabelwerk Eupen, showing char forming is more efficient, with enhanced improvement in fire performance and smoke density (Courtesy of SCP).

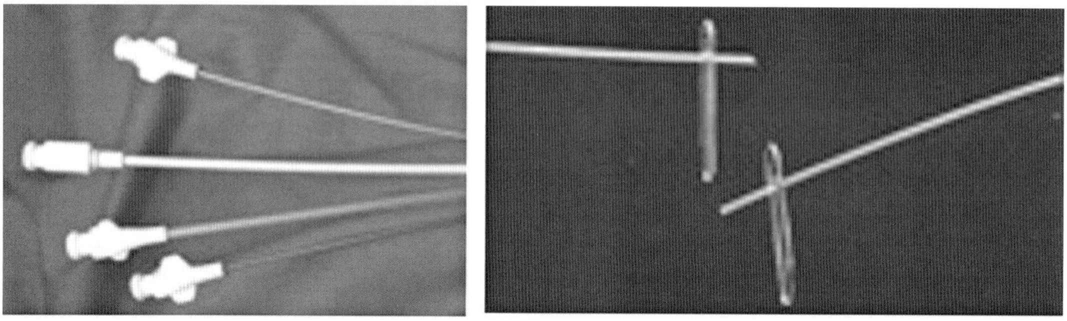

Tubing with improved modulus

Figure 15.15. Nylon 12 nanocomposite for medical application with improved modulus (Courtesy of SCP).

because the polymer nanocomposite will act against the fuel mixing with the solvent in the walls of the tube. Also, similar nanomaterials have been researched to inhibit the ability of certain polymers to absorb and transmit water. The same idea applies as in the case with the fuel cells. The nanofillers will act as a block and inhibit the water particles from permeating the matrix of the polymer. This technology is ideal in the area of environmental protection [33].

Protective coatings are a huge area of opportunity for nanostructured materials. First, ultraviolet (UV) light protection has recently become an important application. In unreinforced polymers, the high-energy UV beams can and will break the weak bonds holding monomers in a chain, resulting in the rapid degradation of polymers exposed to sunlight (i.e., photo-degradation). This results in loss of mechanical properties. Nanomaterials solve this problem by functioning as internal absorbers of UV light, preventing the damage that would normally take place [34, 35].

The use of nano-reinforcements in radar-absorbing materials (RAMs) has become another, new application. In practice, polymer nanocomposites can be tuned to absorb electromagnetic signals of specific frequency bands. As such, they can be used for shielding of sensitive electronic equipment or in stealth technology. RAMs were used in the development of the USAF F-117 and B-2 to reduce the radar signature of the aircraft. Other uses include antennas, car radios, cellular telephones, and other communications technologies [36].

Research into "hard yet tough" coatings is being driven by industry and will produce nanomaterial coatings that have a resistance to plastic deformation without being vulnerable to cracking. Focus for this application is on a dual-phase coating: a nanocrystalline phase along with a CNT-embedded matrix phase. The first provides the hardness and the second provides the toughness. A myriad of uses for this type of coating exist, including everything from automotive engine components to tools for machining [37].

Additionally, nanomaterials have been used for some time in the reinforcement of tires. This is particularly true of carbon black, which is used mainly for the production of tires. The nanomaterials enhance the mechanical properties and also help with heat dissipation. Likewise, it is used in other rubber products, such as belts, hoses, and pads. Although the industry has expanded into other applications, the focus for carbon black production will likely remain on tires [34, 38].

Because of their enhanced mechanical properties, polymer nanocomposites have also been researched for use in commercial vehicles. As shown in Figure 15.16, there are many areas in an automobile where these materials can be used [39].

The use of carbon black in tire technology has been mention earlier, but the auto industry also is constantly searching for ways to make the frames lighter without sacrificing durability and safety. This research rose partially from the need to increase the gas mileage of cars. Most car manufacturers decrease the weight of their cars by adding aluminum pieces to replace heavier steel pieces. However, this can also be accomplished using polymer nanocomposites with similar mechanical strength but significantly lower weight. Polymer nanocomposites were introduced into the car industry when the Toyota Research Group decided to use nylon 6-clay nanocomposites in their engine belt covers. Since then, GM has manufactured a one-piece, compression-molded rear floor assembly. Similar advancements can be made to systems like the suspension, using polymer nanocomposites to stabilize the frame. Carbon nanotubes are also starting to make their way into the market due to their ability to enhance mechanical strength in polymer matrices. The next step in the incorporation of nanocomposites into the automotive industry is to create biodegradable materials. The development of such materials could replace petroleum-based materials currently used in the assembly and therefore reduce the overall carbon footprint of the manufacturer [39].

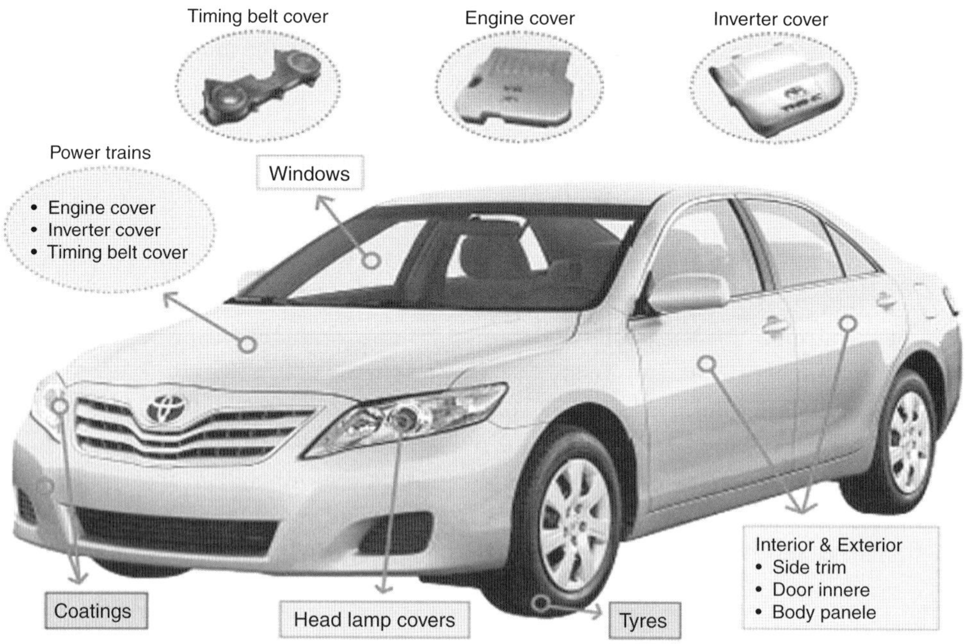

Figure 15.16. Nanocomposites for use in automobiles (Courtesy of Lucintel).

Corrosion plays a large part in the design of many products that deal with memory storage. For example, hard drives used to store massive amounts of data in a computer contain metal disks that are magnetically encoded with the data. The purity of this data and the ability to keep it depends highly on the condition of the disks over time. Since the disks are currently made of metals, they are prone to corrode, which affects the ability to encode the disks. Currently, there is research being conducted in order to find chemically stable polymer nanocomposites that may replace the metal disks in the hard drives. The same chemical instability occurs in many sensors, such as magnetic field sensors and explosive chemical sensors. A lightweight and highly electrically conductive polymer nanocomposite is needed to reduce the amount of oxidation that the metal sensors experience. Furthermore, similar magnetic and electrically conductive properties can be used to make electrochromism-based smart windows [40].

15.2.3 Cost and Property and Geographical Breakdown Analyses

The cost and property analysis of nanomaterials with respect to the other conventional materials, such as ceramic, Kevlar®, glass fibers, and carbon fibers, are shown in Figure 15.17 [3]. As one can see, the key features of nanomaterial's are higher strength, modulus, wear resistance, barrier properties, and electrical conductivity relative to other materials. In the areas of price competitiveness, weight, and surface area, nanomaterials score very low when compared with other conventional materials.

The geographical breakdown market of nanomaterials is shown in Figure 15.18 [3]. In 2010, global distribution of nanomaterials by region in 2010 was 38% for North America, 36% for Europe, and 25% for Asia and other regions. Some key

Materials	Strength	Elasticity	Price	Weight	Surface Area	Wear Resistance	Barrier Property	Electrical Conductivity
Nanomaterials	High	High	Least	Least	Least	High	High	High
Ceramic	Low	Medium	High	Medium	Medium	Low-Medium	Low-Medium	Low
Kevlar (Aramid)	Medium	Low-Medium	Medium	Low	Low	Medium	Low-Medium	Low
Glass Fibers	Low-Medium	Low-Medium	High	Low-Medium	Low	Medium	Medium	Low
Carbon Fibers	Medium	Medium	Low-Medium	Low	Low	Medium	Medium	High

High *Medium* *Low-Medium* *Low* *Least*

Nanomaterials have highest strength, modulus, wear resistance. barrier properties, and electrical conductivity relative to other materials

Figure 15.17. Cost and property analysis of nanomaterials with respect to other conventional materials (Courtesy of Lucintel).

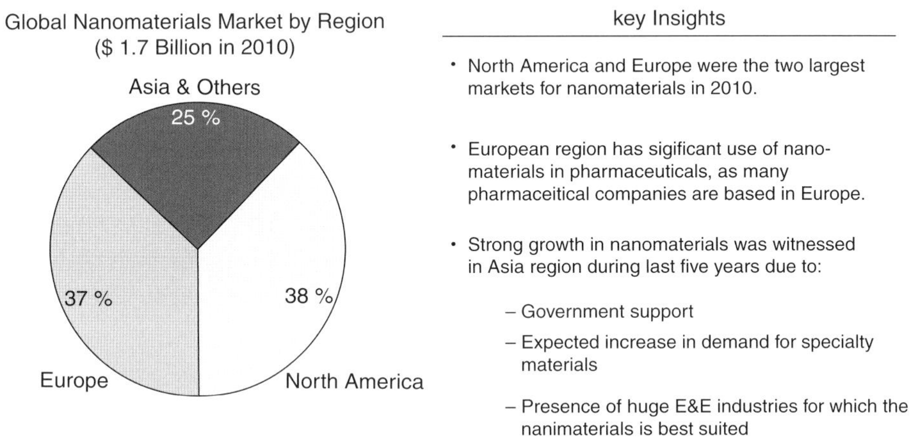

Figure 15.18. Geographical breakdown of nanomaterials markets (Courtesy of Lucintel).

insights are: (a) North America and Europe were the two largest markets for nanomaterials in 2010; (b) Europe showed significant usage of nanomaterials in pharmaceuticals, since many pharmaceutical companies are based there; (c) good growth of nanomaterials was witnessed in Asia during the last five years mainly due to (i) government support, (ii) continued advances in environmental consciousness, and (iii) expected increase in demand for specialty materials.

15.2.4 Technical and Funding Development

Figure 15.19 shows that the technical development in CT promises further capacity expansion and price reduction [3]. CNTs (MWNTs) prices have fallen dramatically,

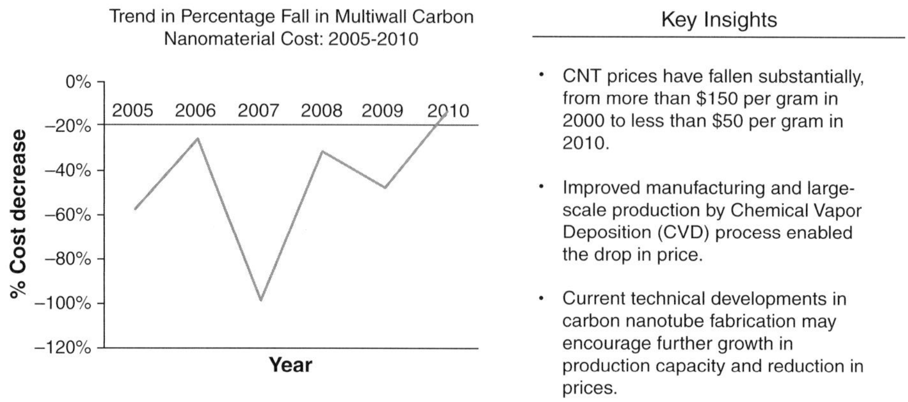

Figure 15.19. Technical development in CNT promise further capacity expansion and price reduction (Courtesy of Lucintel).

from more than $150 per gram in 2000 to about $0.125 per gram in 2013. Improved manufacturing and large-scale production by CVD process enabled the drop in price. Current technical developments in carbon nanotubes fabrication may encourage further increase in production capacity and further reduction in prices.

Figure 15.20 shows government funding for nanotechnology in millions of dollars between 2005 and 2009 [3]. The growing nanotechnology market promises to boost the growth of polymer nanocomposites. In the past five years a huge amount has been invested in nanotechnology R&D by U.S. and European countries. Japan has shown strong interest in nanotechnology development, spending approximately $1 billion in nanotechnology R&D in 2010. Global expenditures for nanotechnology R&D were expected to grow at approximately 23% annually until 2016 [3].

15.2.5 Nanomaterial Consumer Products Inventory

Vance and colleagues published an excellent research article in August 2015 documenting the marketing and distribution of nano-enabled products in the commercial marketplace [5]. Advancements in nanoscience and nanotechnology have resulted in myriad possibilities for consumer product applications, many of which have already migrated from laboratory to store shelves and e-commerce websites. Nanomaterials have been increasingly incorporated into consumer products, although research is still ongoing concerning their potential effects on the environment and human health. To document the penetration of nanotechnology in the consumer marketplace, the Woodrow Wilson International Center for Scholars and the Project on Emerging Nanotechnology created the Nanotechnology Consumer Product Inventory (CPI) in 2005, listing 54 products [41]. This first-of-its-kind inventory has become one of the most frequently cited resources showcasing the widespread applications of nanotechnology in consumer products. In 2010, the CPI listed 1,012 products from 409 companies in 24 countries. Even though it did not go through substantial updates in the period between 2010 and

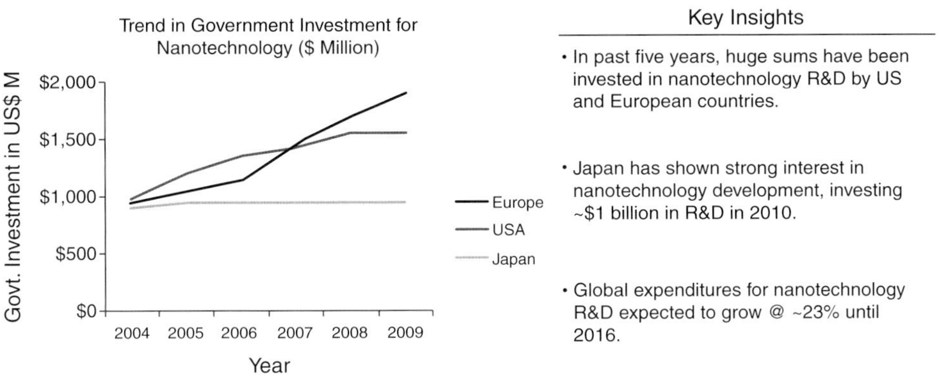

Figure 15.20. Government funding for nanotechnology in USD million $ from 2005 to 2009 (Courtesy of Lucintel).

2013, it continued being heavily cited in government reports and scientific literature – more than 2,580 times since its creation, according to Google Scholar [41].

The CPI has been criticized for lack of science-based data to support manufacturers' claims. Other long-standing suggestions for improvement included more frequent updates, indications when products were no longer available for purchase by consumers, and the inclusion of more product categories to improve the search ability of the CPI database [41]. Since the creation of the CPI, other nanotechnology-related inventories have been developed around the world by Germany, Japan, the European Consumers Organization (BEUC), the European Consumer Voice in Standardization (ANEC), the Danish Consumer Council and Ecological Council, and the Technical University of Denmark. The objective of Vance et al.'s research was to modify the CPI to improve its functionality, reliability, and utility to the diverse group of stakeholders who have come to depend on it as a critical resource for current information on nano-enabled consumer products.

15.2.5.1 CPI Growth Over Time

Table 15.4 lists the growth of the CPI since 2005 [5]. The new total of 1,814 products as of March 2015 represents a 30-fold increase over the 54 products listed in 2005, yet it is not a complete representation of the growth of this market. Based on Vance et al. 's review [41], the CPI is the largest online inventory of nanotechnology consumer products available. Products come from 622 companies located in 32 countries. The products listed on the CPI 2.0 satisfy three criteria: (1) they can be readily purchased by consumers; (2) they are claimed to contain nanomaterials by the manufacturer or another source; and (3) their claim to contain nanomaterials appears reasonable to CPI curatorial staff.

In the CPI, entries are grouped under eight generally accepted consumer goods categories that are loosely based on publicly available consumer product classification systems as shown in Figure 15.21. The health and fitness category includes the largest listing of products in the CPI, comprising 42% of all listed products. Within the health and fitness category, personal care products, such as toothbrushes, lotions,

Table 15.4. *Number of consumer products in the CPI over time [5]*

Year	Total products	Products added	Products archived	Data collection notes
2005	54	54	0	Beginning of CPI as a static pdf document.
2006	356	302	0	Launch of the online CPI.
2007	580	278	0	Nanoscale silver emerged as most cited nanomaterial.
2008	803	223	0	Health and fitness products represented 60% of the inventory.
2009	1015	212	107	Added archiving function to the CPI.
2010	1015	0	0	No data collected.
2011	1015	0	0	No data collected.
2012	1438	426	0	Beginning of CPI 2.0 project, focus on adding new products.
2013	1628	190	288	Launch of crowdsourcing component. Extensive effort put into adding and archiving products.
2014	1814[a]	238[a]	223[a]	Extensive effort put into adding and archiving products.

[a] The CPI now has crowdsourcing capabilities, so these numbers are a snapshot in time and will not represent the CPI at the time of reading.

and hairstyle tools comprise the largest subcategory (39%). Others categories include: home and garden, automotive, cross-cutting, electronics, food and beverage, appliances, and goods for children (Figure 15.21). Since 2012, a strong, continuous effort has been put into periodically checking products for their current availability and current claim to contain nanotechnology.

15.2.5.2 New Nanomaterial Descriptors

Eight new product descriptors were introduced in CPI 2.0 to facilitate the use of this database by a variety of stakeholders:

1. Main nanomaterial composition or type
2. Nanomaterial shape and size
3. Nanomaterial coating or stabilizing agent
4. Nanomaterial location within the product
5. Nanomaterial function in the product
6. Potential exposure pathways
7. "How much we know"
8. "Researchers say"

Several of these new quantitative descriptors are presented and discussed in the following subsections.

15.2.5.2.1 NANOMATERIAL COMPOSITION Of the 1,814 products listed in the CPI, 47% (846 products) advertise the composition of at least one nanomaterial composition and 62 of those products list more than one nanomaterial component (e.g., a product comprised of both silver and titanium dioxide nanomaterials). There are 39 different types of nanomaterial components listed in the inventory, which have been

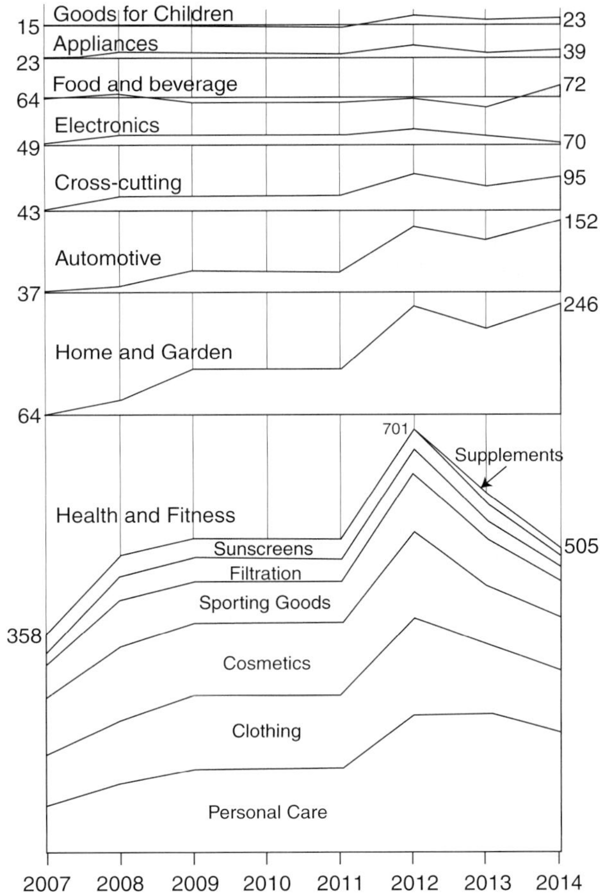

Figure 15.21. Number of available products over time (since 2007) in each major category and the Health and Fitness subcategories. Source: doi: 10.3762/bjnano.6.181.

grouped into five major categories as shown in Figures 15.22 and 15.23 to improve their legibility: metal, carbonaceous, silicon, not advertised, and other. Metals and metal oxides comprise the largest nanomaterial composition group advertised in the inventory, listed in 37% of products.

Titanium dioxide (TiO_2), silicon dioxide (SiO_2), and zinc oxide (ZnO) are the most produced nanomaterials worldwide (on a mass basis), and the global annual production of silver (Ag) nanoparticles is only 2% of that of TiO_2. However, silver nanoparticles are the most popular advertised nanomaterial in the CPI, present is 438 products (48%). Of carbonaceous nanomaterials (89 products), the majority of products listed contain carbon nanoparticles (sometimes described as carbon black – 39 products), and single- or multi-walled carbon nanotubes (CNT – 38 products). Unfortunately, 891 of the products (49%) included in the CPI do not present the composition or a detailed description of the nanomaterial used (Figure 15.24).

Figure 15.23 shows the availability of these major nanomaterial composition groups changed over time. Products advertised to contain metal and metal oxides nanomaterials, silicon-based nanomaterials (mostly SiO_2 nanoparticles), and a variety of other nanomaterial components (organics, ceramics, polymers, clays, nanocellulose, liposomes, nano micelles, carnauba wax, etc.) have been growing in popularity. Of the 846 products (46.5%) listed in the CPI for which the authors were able to

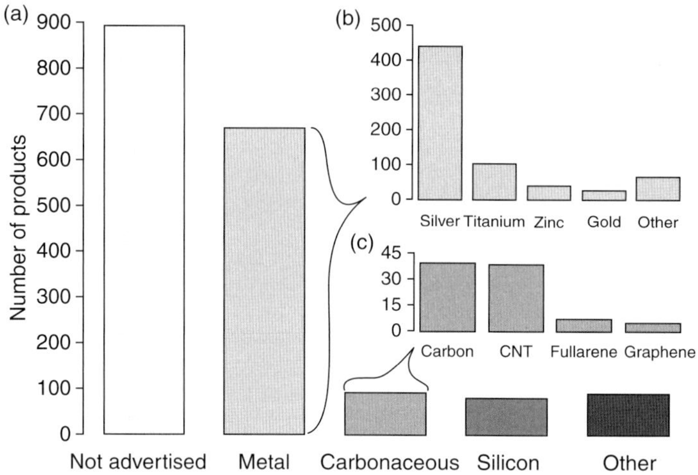

Figure 15.22. (a) Claimed composition of nanomaterial listed in the CPI, grouped in five major categories: not advertised, metal (including metals and metal oxides), carbonaceous nanomaterials (carbon black, carbon nanotubes, fullerenes, graphenes), silicon-based nanomaterials (silicon and silica), and other (organics, polymers, ceramics, etc.). (b) Claimed elemental composition of nanomaterials listed in the metals category: silver, titanium, zinc, gold, and other metals (magnesium, aluminum oxide, copper, platinum, iron, and iron oxides, etc.). (c) Claimed carbonaceous nanomaterials (CNT = carbon nanotubes). Silver nanoparticles are popular consumer product additives due to their well-documented antimicrobial properties. Source: doi: 10.3762/bjnano.6.181.

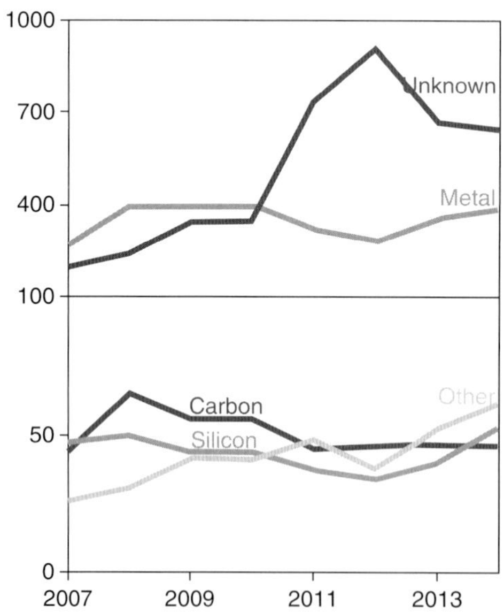

Figure 15.23. Major nanomaterial composition groups over time. Carbon = carbonaceous nanomaterials (carbon black, carbon nanotubes, fullerenes, graphenes). Other = organics, ceramics, polymers, clays, nanocellulose, liposomes, nano micelles, carnauba wax, etc. Note the difference scale between the top and bottom panels in this plot. *Source*: doi: 10.3762/bjnano.6.181.

	Gold	Carbon	Copper	Ceramics	Iron	Calcium	Magnesium	Silicon	Zinc oxide	Titanium	Silver
Gold	7	1		1				1		1	3
Carbon	1	10	1		2					2	3
Copper		1	10		1	1	1	2	1		1
Ceramics	1			11				1		2	7
Iron		2	1		11	1	1			2	1
Calcium			1		1	13	8	1			
Magnesium			1		1	8	13	1			
Silicon	1		2	1		1	1	14	1	2	4
Zinc oxide			1					1	14	10	2
Titanium	1	2		2	2			2	10	30	10
Silver	3	3	1	7	1			4	2	10	35

Figure 15.24. Major nanomaterial composition pairs in consumer products. Carbonaceous nanomaterials (carbon black, carbon nanotubes, fullerene, and graphene) were combined into the same category (carbon). Grey boxes in the diagonal represent the total times each nanomaterial composite has been listed with other composition in the same product. *Source:* doi: 10.3762/bjnano.6.181

determine a nanomaterial composition, 61 products (7%) advertised to contain more than one main nanomaterial components. Figure 15.24 presents 11 nanomaterial components that were most frequently listed with others in the same product.

Silver and titanium oxide are the nanomaterial components most likely to be combined with other nanomaterials in consumer products, with 35 and 30 product combinations, respectively. Silver and titanium oxide were paired in 10 products (cosmetics and electronics); titanium dioxide and zinc oxide were also paired in 10 products (sunscreens, cosmetics, and paints).

Calcium and magnesium were listed together in dietary supplements. Nanoceramics and silver are used in combination in water filtration products, cosmetics, and a humidifier. Vance et al. suggest the need to examine nanomaterial toxicity effects that could be synergistic, additive, or even antagonistic [41].

15.2.5.2.2 NANOMATERIAL LOCATION About 29% of consumer products in the CPI (528 products) contain nanomaterials suspended in a variety of fluids (e.g., water, skin lotion, oil, car lubricant). The second-largest group in this category, with 307 products, comprises solid products with surface-bound nanoparticles (e.g., hair curling and flat irons, textiles). Figure 15.25 shows the locations of nanomaterials for which a composition has been identified. The majority (64%) of carbonaceous

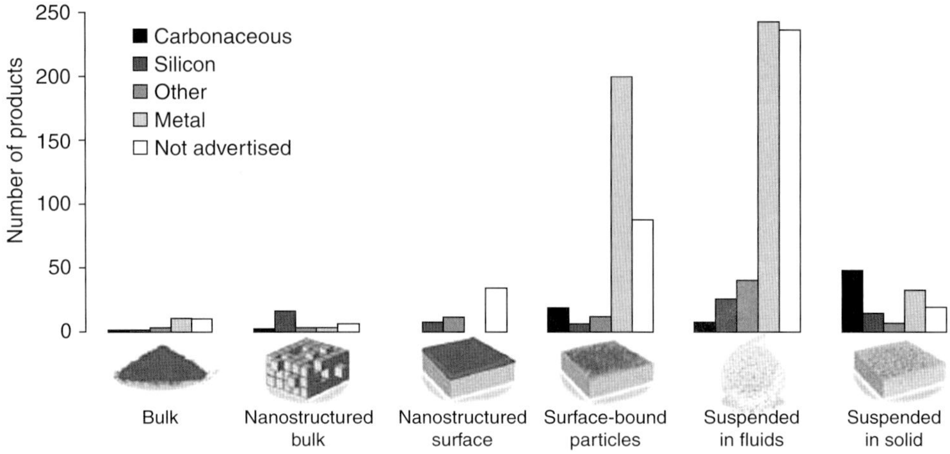

Figure 15.25. Locations of nanomaterials in consumer products for which a nanomaterial composition has been identified.
Source: doi: 10.3762/bjnano.6.18.

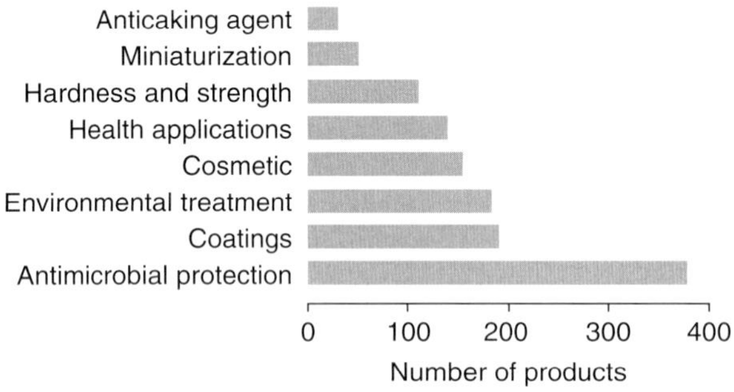

Figure 15.26. Expected benefits of incorporating nanomaterials additives in consumer products.
Source: doi: 10.3762/bjnano.6.181.

nanomaterials are embedded in solid products, whereas products of all other compositions are more commonly suspended in liquid. The largest group (42%) of consumer products available for purchase that contain bulk nanomaterials contains metals and metal oxide nanomaterials. Metals and metal oxides were also the largest composition for surface-bounded particles and those suspended in liquid products. However, most of these products do not advertise their use of nanomaterials, and Vance et al. believe that they are grossly underrepresented in the CPI [41].

15.2.5.2.3 NANOMATERIAL FUNCTION Of the 1,814 entries on CPI's list, 1,244 were grouped according to the expected benefits of adding such nanomaterials to the products, as shown in Figure 15.26. A significant portion of products in the CPI (31%) utilize nanomaterials, mostly silver nanoparticles as well as titanium dioxide and others for antimicrobial protection. Titanium and silicon dioxide nanoparticles are used to provide protective coatings (15%) and for environmental treatment (to protect product against environmental damage or to treat air and water in the

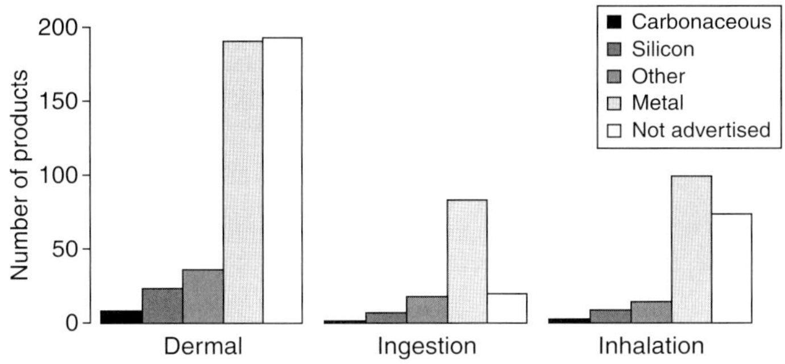

Figure 15.27 Potential exposure pathways form the exposure normal use of consumer products, group by major nanomaterial categories.
Source: doi: 10.3762/bjnano.6.181.

home – 15%). Cosmetic products (12%) are advertised to contain a variety of nanomaterials, such as silver nanoparticles, titanium oxide, nano-organics, gold, and others. A wide variety of nanomaterial compositions (silver, nano-organics, calcium, gold, silicon dioxide, magnesium, ceramics, etc.) were also advertised for health applications, such as dietary supplement (11%).

15.2.5.2.4 POTENTIAL EXPOSURE PATHWAYS Critical information, such as nanomaterial size and concentration, is not known for most products listed in the CPI; and so the actual health risks of these products remain largely unknown. The CPI can only be useful for inferring potential exposure pathways from the exposure stemming from normal use of the listed products. Vance et al. investigated a subset of 770 products from the CPI to determine their most likely routes to exposure (Figure 15.27). Vance et al. identified the skin as the primary route of exposure for nanomaterials from the use of consumer products (58% of the products evaluated) [41].

Of the products evaluated, 25% present nanomaterials that can be inhaled during normal use (e.g., sprays and hair driers) and 16% contain nanomaterials that may be ingested (e.g., supplements and throat sprays). Since metals and metal oxides are the most common nanomaterial composition in the CPI, they are also the most likely materials to which consumers will be exposed during the normal use of product via dermal, ingestion, and inhalation routes. Products containing nanomaterials of unknown composition are most likely to lead to exposure via the dermal route.

The Consumer Product Safety Commission recently requested $7 million to establish the Center for Consumer Product Applications and Safety Implications of Nanotechnology to help develop methods to identify nanomaterials in consumer products and to understand human exposure to those materials [41].

15.2.6 Conclusion of CPI Study

The modified version of the Wilson Center's nanotechnology consumer products inventory (CPI 2.0) has been released. Vance et al. improved the search ability and utility of the inventory by including new descriptors for both the consumer products and nanomaterial components of the products. The updated CPI 2.0 now links listed products to published scientific information and includes a metric to assess the reliability of the data associated with each entry. The CPI 2.0 has enabled crowdsourcing capabilities, which allows registered users to upload new findings, such as basic product composition information, human and environmental exposure data, and complete life cycle assessments. The CPI is a useful interactive database for educating consumers and legislators on the real-world applications of nanotechnology. It is useful for policy makers interested in regulating nanotechnology in consumer products by understanding their increasing numbers in the market and by the main nanomaterial components that are chosen by the manufacturers. Finally, the current lack of global standardized methods and metrics for nanomaterial characterization and labeling in the consumer products is an issue that can lead to greater understanding between all parties that are interested in nanotechnology development.

15.3 Nanotechnology Research Output

Typically, the research and production of polymer nanocomposites is focused on the enhanced barrier properties that they can provide. Advantages that polymer nanocomposites can provide for the food packaging industry and other air seal technologies are significant. The mechanical properties of the resulting composites are examined closely but are often not the main goal. Although adding nanofillers makes the polymer matrix stronger, it often can degrade the elongation property of the polymer; more research and development must be done to find ways to rectify this. As more research is done in the area of enhancing barrier properties, other ideas are generated from similar concepts. Recently, there has been interest in finding ways to use a sealing polymer composite in tissue bonding for materials such as skin grafts. Since skin acts as a protective layer for your body, the same barrier properties that help keep food fresh can help bond ends of tissue together with less risk of separation [38].

There are many physical constants in the incorporation of nanofillers into polymer matrices. These phenomena occur in almost every combination of matrix and filler used. First, the density of the polymer always increases with the incorporation of nanoparticles. This is an intuitive result considering that stronger materials are being packed into a weaker matrix. Secondly, the volume of polymer nanocomposites under mechanical strain increases faster than that of the neat polymer. As the particles used in the matrix get smaller, the volume increases under the same loading conditions. Also, as nanofillers are added, the polymer tends to become stiffer, and the elongation properties degrade [42]. This may be a problem, as brittle materials are seldom useful. Other mechanical properties, such as tensile strength and modulus, tend to increase as well in most polymer matrices. A study also found that in crystalline materials the T_g decreased while it actually increases in most amorphous structures. Finally, a property that is detrimental to the processing stage is the viscosity, which tends to increase as you add more nanomaterials. The increase

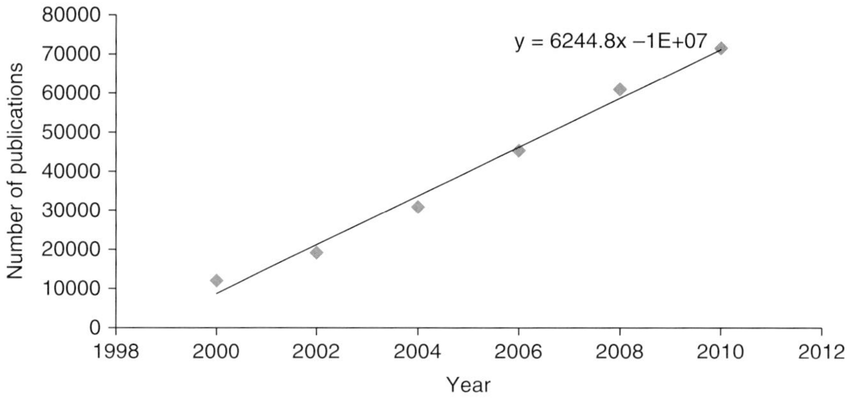

Figure 15.28. Publications containing "nano" from 2000 to 2010 with a trend line.

in polymer viscosity could cause the batch to mix improperly, or in some cases the mixer may not be powerful enough to overcome this significant increase of polymer viscosity during manufacturing [42].

Over the past decade, the field of nanomaterials studies has been on the rise. In 2000, there were only about 12,000 publications resulting from a search for "nano." This number has grown significantly, at a rate of about 6,200 additional papers per year. In 2010, there were over 71,000 papers containing "nano" in the title, as shown in Figure 15.28 [43].

According to Google Patents, the number of patents filed in the United States for "polymer nano" far exceeds Europe or international patent filing numbers, shown in Table 15.5. For the most part, filed patents appear to be way down from 2010. Google patents showed only 1,071 U.S. patents filed in 2015, down from more than 25,488 in 2010. The same trend was observed in Europe and internationally.

15.4 Trend and Forecast

The trend and forecast for the nanomaterials market are described in Figure 15.29 [3]. From 2005 to 2010, nanomaterial shipment grew from $1 billion to $1.7 billion, a healthy 11% CAGR growth. The forecast for shipments of nanomaterials from 2010 to 2016 will be from $1.7 billion to $6.1 billion, an extraordinary 23% CAGR growth! The electrical and electronics (E & E) market witness a robust growth rate due to large price decrease of carbon nanotubes and increase in mass production of nanomaterials. The drivers for nanomaterials investment can be summarized in the three factors: (a) emerging application – expected that nanomaterials will be used extensively in emerging applications in water treatment, drug delivery, and nanomedicines; (b) price – decrease in prices of nanomaterials may take place due to expanded production capacity; and (c) innovation – number of patents regarding manufacturing of nanomaterials are increasing extraordinarily upto 2010 (Table 15.5). The top three market segments can be subdivided regionally and are quite different (Figure 15.29): (a) North America – electrical and electronics,

Table 15.5. *Number of patents filed for by year according to Google Patents using keyword "polymer nano"*

Filed	Google Patents	"polymer nano"	
Year	US	Europe	International
2010	25,488	3,938	10,446
2011	19,497	2,458	6,710
2012	14,851	1,813	4,872
2013	9,091	1,043	2,850
2014	3,523	349	1,205
2015	1,071	135	391

Source: Google Patents

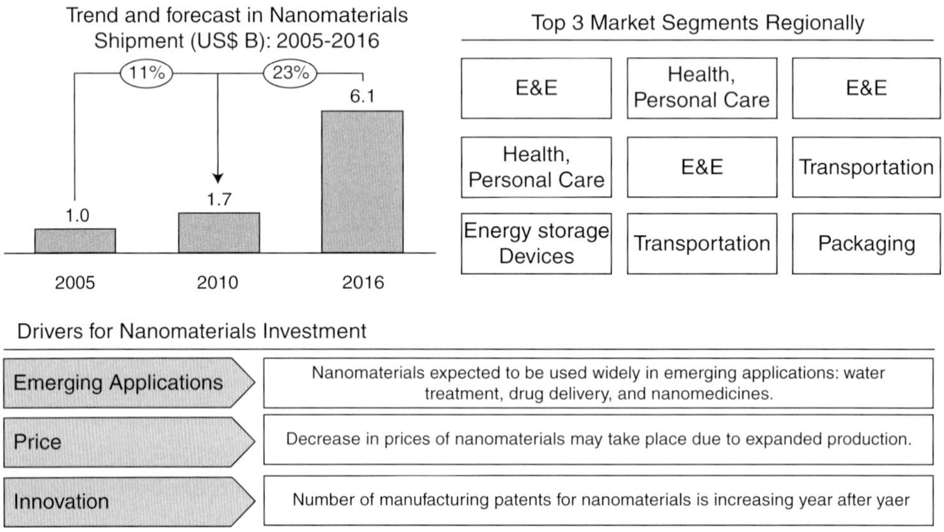

Figure 15.29. The trend and forecast for the nanomaterial market (Courtesy of Lucintel).

health and personal care, and energy storage devices; (b) Europe – health and personal care, electrical and electronics, and transportation; and (c) rest of the world (RoW) – electrical and electronics, transportation, and packaging.

Growth opportunities for nanomaterials in various applications are shown in Figure 15.30 [3]. The health care industry is forecast to have the most growth, followed by energy storage, then electric and electronics, transportation, and construction. Health care industry is expected to gain market share and overtake electric and electronics segment in size over the next five years. Several key insights are forecast: (a) nanomaterials have great potential in electric and electronics applications because of their extraordinary electrical conductivity; (b) global recession induced cost/price sensitivity in the electric and electronics segment, creating incentives to reduce manufacturing costs and increase product competitiveness; (c) packing is another important market segment, mainly flourishing in North America and Western Europe; and (d) the energy segment is expected to pick up a nice pace in the coming years at a double-digit growth rate.

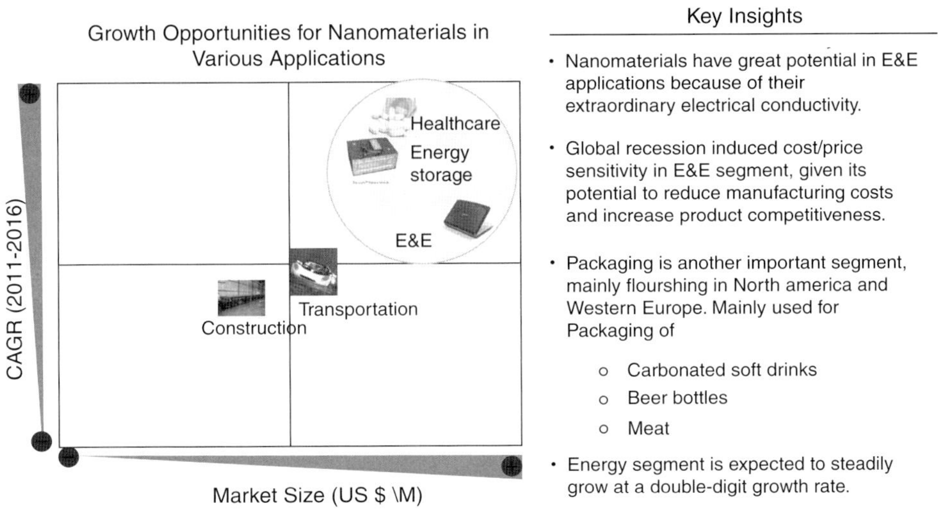

Figure 15.30. Grow opportunities for nanomaterials in various applications (Courtesy of Lucintel).

Markets	Nano-materials	Applications	Benefits
Aerospace & Defense	Carbon nanotubes	Fuselage and wings	• High strength-to-weight ratio • High impact resistance • Radiation protection
Wind Energy	Carbon nanotubes	Wind Turbine blades	• Increase mechanical efficiency Lightning protection
Automotive	Carbon nanotubes	External body	• Light weight and conductivity
Marine	Carbon nanotubes	Boat hull	• Light weight and increasing payload

Figure 15.31. Several potential applications for nanomaterials in composites for different market segments (Courtesy of Lucintel).

Figure 15.31 shows the various potential applications for nanomaterials in composites [3]. The different market segments using carbon nanotubes include: (a) aerospace and defense, for fuselage and wings, (b) wind energy, for wing blades, (c) automotive, for external body, and (d) marine for boat hulls. Wind energy, automotive, and marine industries are also using carbon nanotubes for various applications, mostly due to the light weight and improved mechanical, thermal, and electrical properties with which they provide various composites.

In modern aircraft, nanomaterials are expected to show healthy growth in the next 20 years due to high strength, light weight, and electrical conductivity properties,

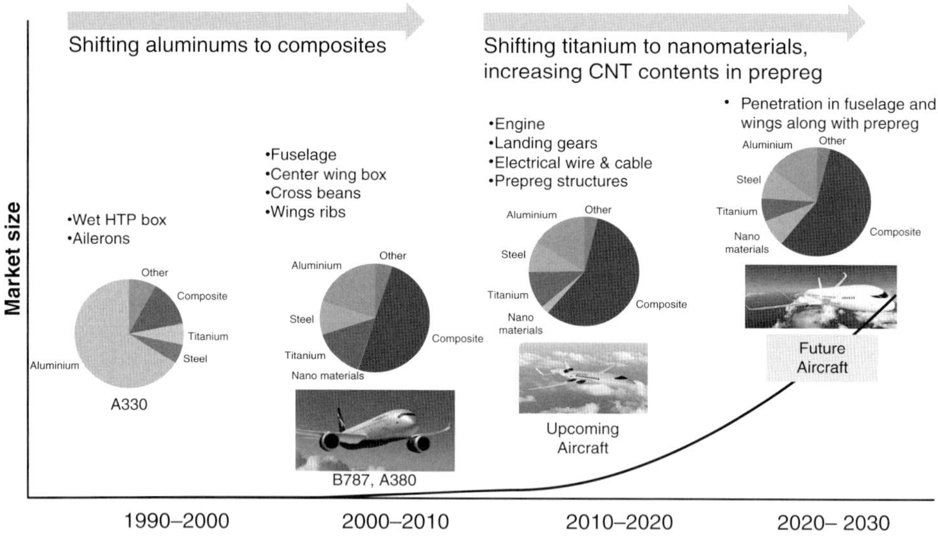

Figure 15.32. Past, present, and future usage of materials in the aircraft industry from 1990 to 2030 (Courtesy of Lucintel).

as shown in Figure 15.32 [3]. From 1990 to 2010, aircraft industry has been shifting from aluminums to composites. Both composites and polymer nanocomposites are currently integral components in such airlines as Boeing B787 and Airbus A380. For the upcoming aircraft (from 2010 to 2020) and future aircraft (from 2020 to 2030), the aircraft industry is shifting from titanium to nanomaterials and increasing carbon nanotube contents in prepreg that will be used to fabricate composites.

Figure 15.33 shows the different emerging trends in global nanomaterials and nanocomposites industry [3]. The trends are: (a) growing demand for high-strength, durable structural materials; (b) new material development and product design; (c) new and emerging market applications; (d) falling nanocomposites prices; (e) huge expansion by existing suppliers and new entrants; and (f) government support and R&D funding.

Figure 15.34 illustrates how the growing demand for high-strength and durable materials will create good future opportunities for polymer nanocomposites [3]. In 2010, demand for carbon fiber reached 80 million pounds, with a small amount replaced by carbon nanofiber and carbon nanotubes; the forecast for carbon fibers in 2015 was 150 million pounds. The strength and impact properties of carbon nanotubes/graphene nanoplatelets are superior to carbon/aramid/glass fibers. Other new and exciting uses for carbon nanotubes include aircraft wings and fuselages, boat hulls, sports cars, baseball bats, hockey sticks, and skis.

Figure 15.35 shows new material development and product design [3]. These new development include: (a) Nanoledge Inc.'s prepreg containing carbon nanotubes; (b) New Piranha using Zyvex nanomaterials for marine application; (c) Harbor Composites designing nanotube components for aircraft; and (d) Velozzi's Supercar using nanotube composites for structural parts.

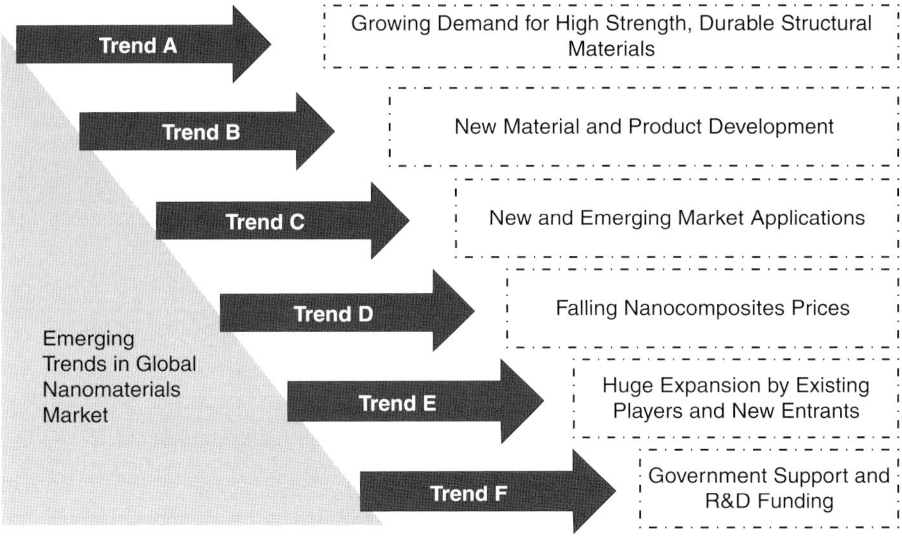

Figure 15.33. Emerging trends in global nanomaterials and nanocomposite industry (Courtesy of Lucintel).

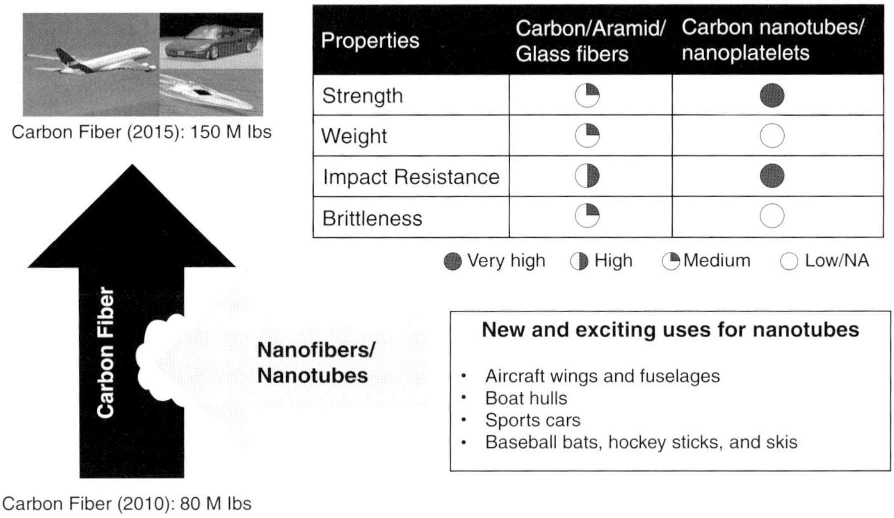

Properties	Carbon/Aramid/ Glass fibers	Carbon nanotubes/ nanoplatelets
Strength	Medium	Very high
Weight	Medium	Low/NA
Impact Resistance	High	Very high
Brittleness	Medium	Low/NA

Figure 15.34. Growing demand for high-strength, durable materials will create good future opportunities for nanocomposites (Courtesy of Lucintel).

Figure 15.36 shows how growing uses of nanocomposites in various areas in the transportation industry will create huge demand in future [3]. Figure 15.37 shows potential applications of nanocomposites in aircraft, though market remains nascent due to certification hurdles and overall slow adoption [3].

Figure 15.38 shows additional emerging applications for nanomaterials, including field emission display (FED), body armor, water purification, drug delivery, biofuel cell, and wind turbine blades [3]. The size of opportunities and the key drivers of these applications are described in Figure 15.38 [3].

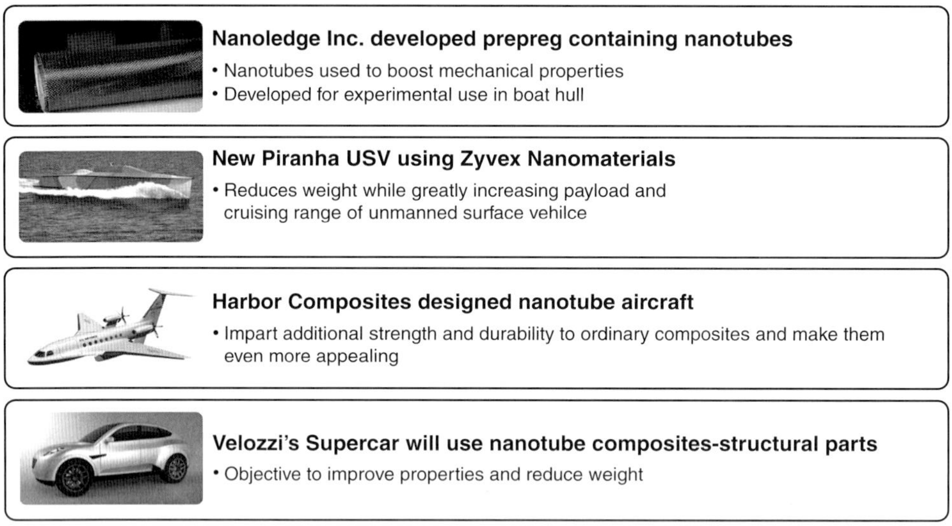

Figure 15.35. New material development and product design (Courtesy of Lucintel).

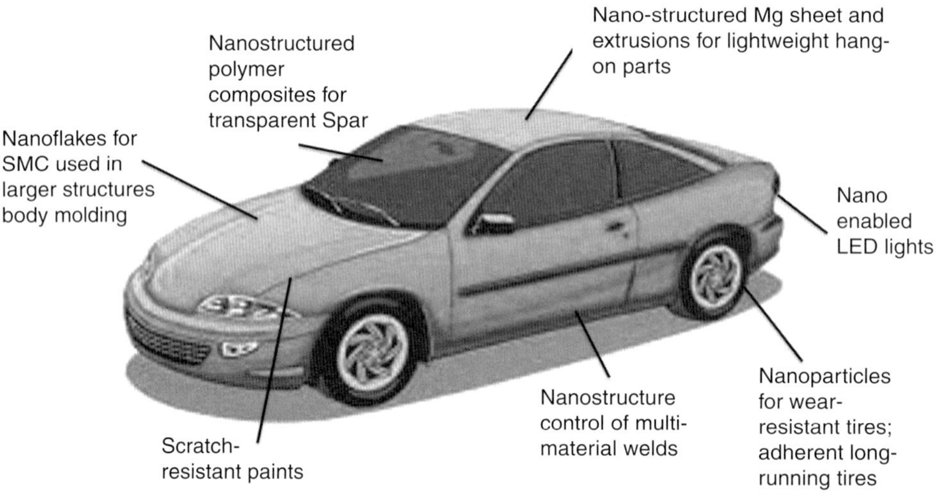

Figure 15.36. Growing uses of nanocomposites in transportation industry will create huge demand in future (Courtesy of Lucintel).

In addition, several polymer-CNF applications are provided by Applied Sciences [4]. Figure 15.39 shows ESD/wear improved pulleys of polymer-CNF nanocomposites and the benefits of using CNFs for this application [4]. Figure 15.40 shows electrostatic painting using CNFs and its benefits [4]. Figure 15.41 shows how CNFs are used in nanofluids applications to provide unique transport properties, such as no settling under gravity, no clogging, and large surface area. Figure 15.42 shows nanotechnology can be used for thermal management for top technical challenges facing transportation, defense, microelectronics, and energy industries [4].

Another emerging field for nanotechnology is human health and medicine. BASF Chemical Company is researching various ways that nanotechnology can be

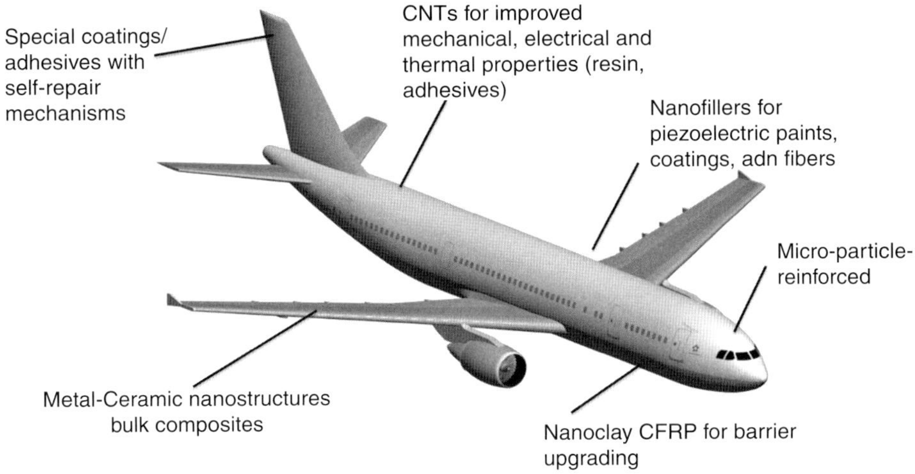

Figure 15.37. Potential applications of nanocomposites in aircraft, though market remains nascent due to certification hurdles and overall slow adoption (Courtesy of Lucintel).

used to improve peoples' standard of living [44]. BASF is working on developing nano-membranes to filter bacteria, germs, and viruses from water. They have also been developing vitamins with nanoparticles that allow better adsorption into the body. Using zinc oxide nanoparticles, BASF has been a global leader in UV-absorbing products as well.

NASA has been developing a shape-shifting aircraft that changes shape according to different manoeuvres and pressure variations along the wings [45], as shown in Figure 15.43. The technology uses pressure sensors along the length of the wing and actuators to change shape, similar to bird flight. The actuators are said to use carbon nanotube polymer composites and electrostriction to change shape by varying the stresses in the material from longitudinal to bending.

In summary, several market trend and forecast conclusions can be deduced from the Lucintel market research survey [3]:

- Top six manufacturers have secured two-thirds of the global nanomaterials market shipments; high opportunity for new manufacturers to enter this growth market.
- Decline in demand in 2009 for nano-enabled products in the automotive industry, such as, among others, automotive lubricants, catalytic converters, sensors, and filters, drove the heavy downturn in market opportunities for nanomaterials, such as multi-wall carbon nanotubes (MWNTs) and ceramic nanoparticles.
- Health care and energy markets were the two main application areas that helped global nanomaterial industry recover from the downturn in 2009.
- Health care industry is expected to surpass the electrical and electronics industry in size over the coming 2–3 years, driven by nanotechnology advancements in the biomedical field.
- North America will remain the market-leading region for several years, as it has significant ongoing R&D activities in nanomaterials.

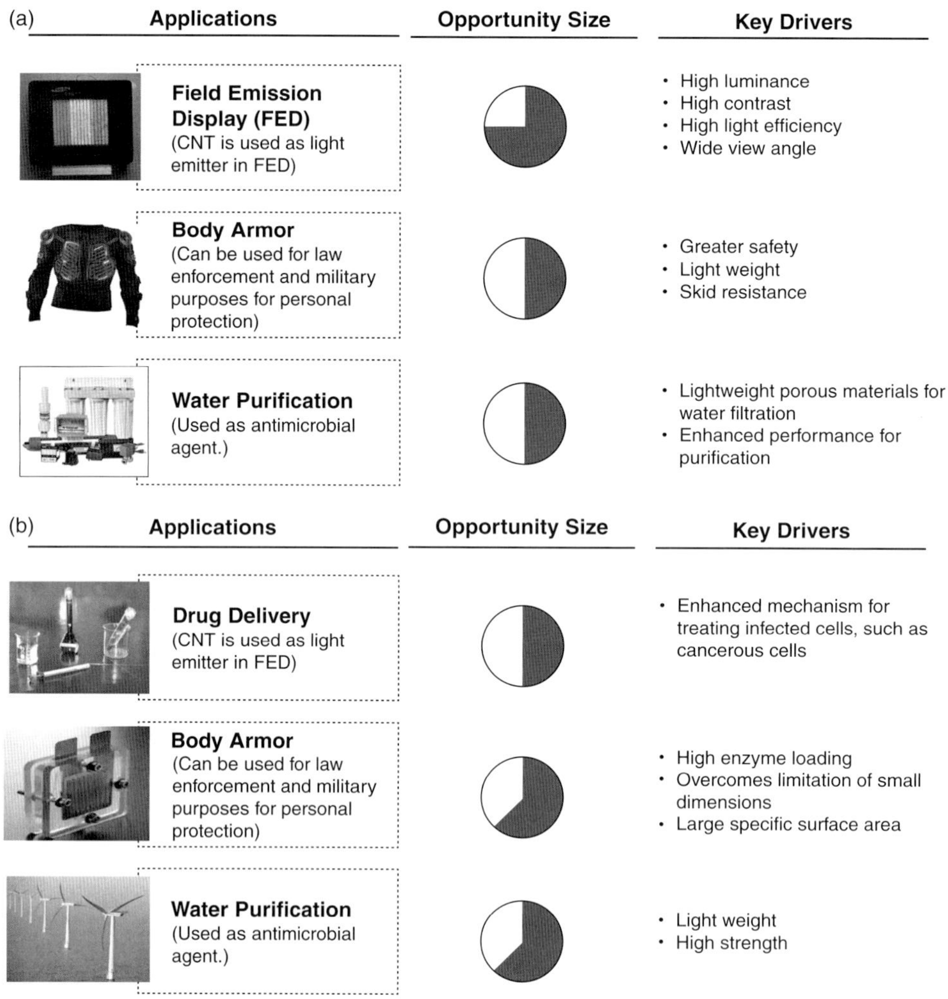

Figure 15.38. Emerging applications for nanomaterials (Courtesy of Lucintel).

- It is expected that nanomaterials will broaden their markets as they enter into more applications, such as drug delivery, armor, defense equipment, weapons, nanomedicines, and others.

15.5 Challenges

15.5.1 Manufacturability of Nanoparticles

The starting point of any polymer nanocomposite is choosing the appropriate nanomaterial, of which there are several. Many times, the cost of certain nanomaterial exceeds the gain from the property improvement, causing many combinations and weight percentages to be out of reach for many researchers. Finding inexpensive ways to mass-produce useful nanomaterials is an area that is highly important for the further development of all polymer nanocomposites. Current methods are

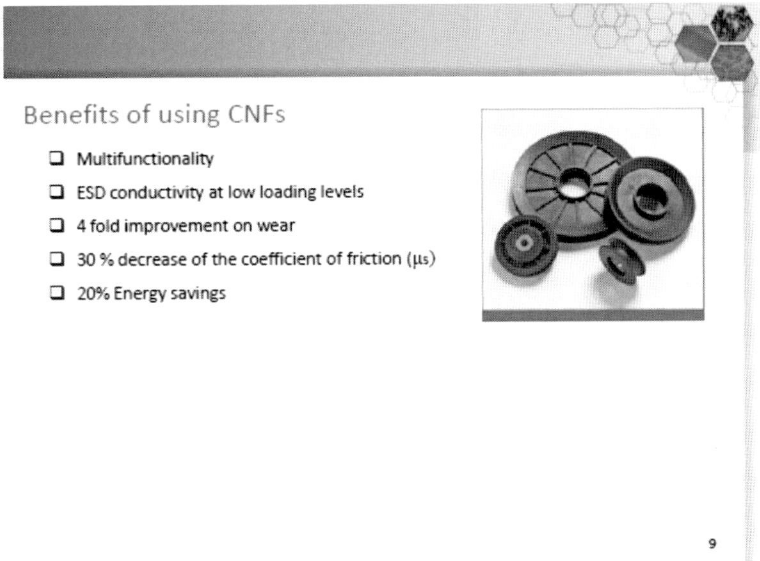

Figure 15.39. ESD/wear improved pulley using polymer-CNF nanocomposites and their benefits (Courtesy of Applied Sciences, Inc.).

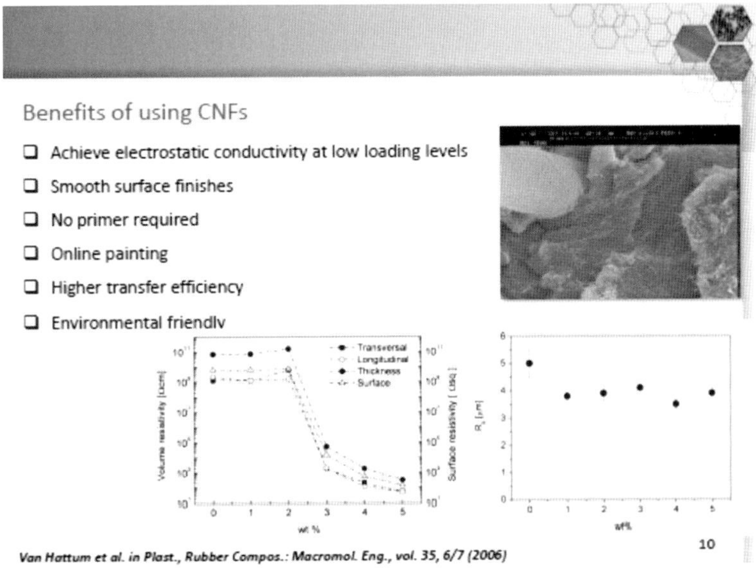

Figure 15.40. Electrostatic painting using CNFs and its benefits (Courtesy of Applied Sciences, Inc.).

successful and provide the industry with a steady supply of various nanomaterials, but other methods have potential to decrease cost and increase volume of nanomaterial production.

Currently, there are three commonly used ways to create carbon-based nanomaterials: arc discharge, laser ablation, and chemical vapor deposition. U.S. patent

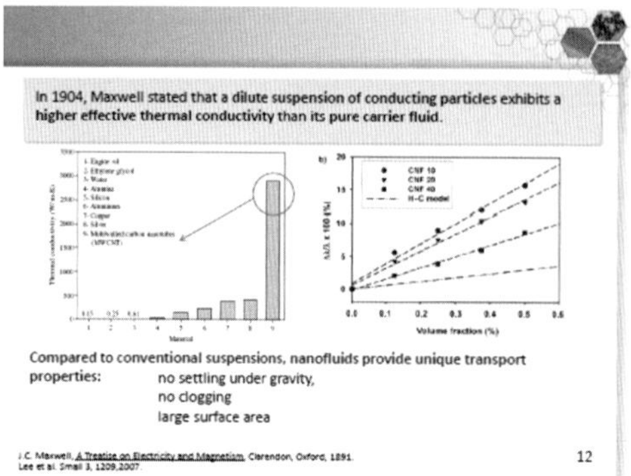

(a)

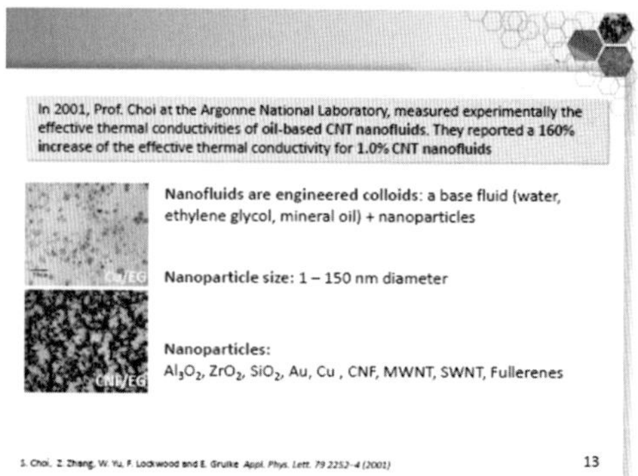

(b)

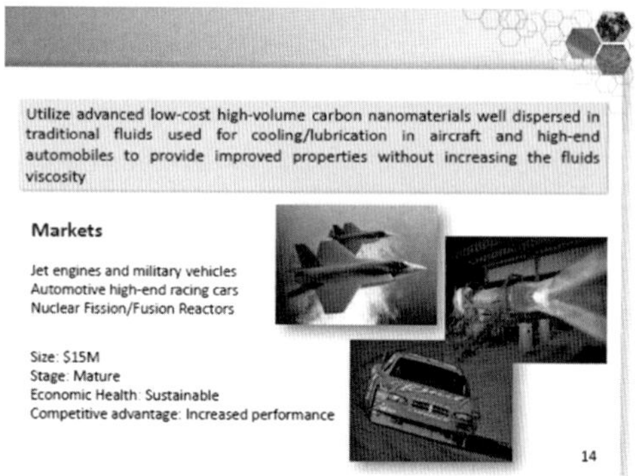

(c)

Figure 15.41. CNFs are used in (a, b) nanofluids application and (c) their opportunity and impact are shown (Courtesy of Applied Sciences, Inc.).

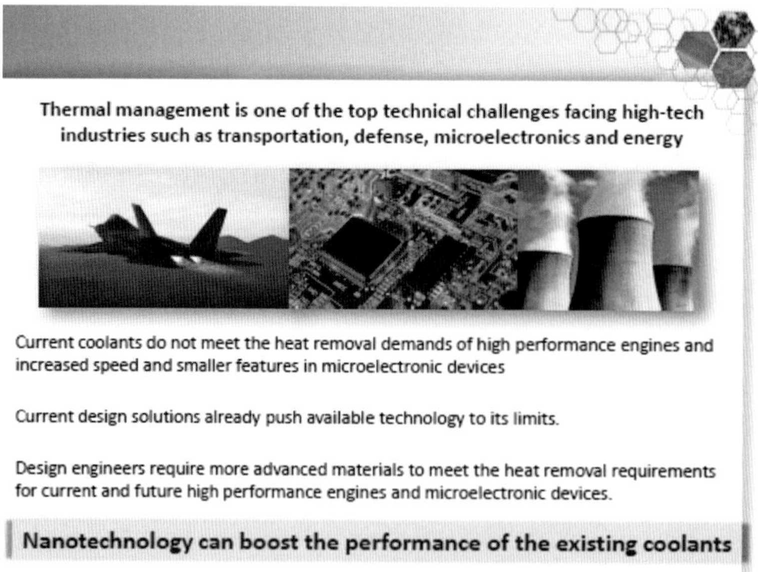

Figure 15.42. Nanotechnology can be used for thermal management of different high-tech industries (Courtesy of Applied Sciences, Inc.).

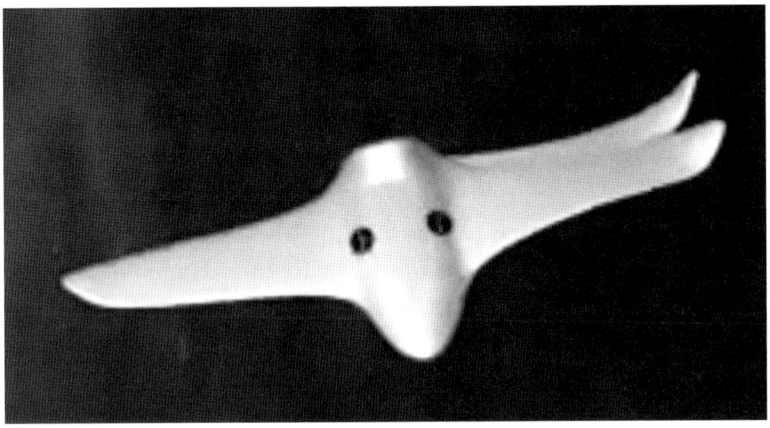

Figure 15.43. NASA shape-shifting plane simulation demonstrating how changing shape affects mobility.
Source: NASA.

number 20130089738 discusses a way to create activated nanocarbon from food waste using a new method that has the potential to reduce cost and reduce food wastes [46]. This method requires crushing food waste and sieving it to create a powder. Then, the powder is impregnated with an agent and activated using an activation gas at high temperatures, between 500°C and 800°C. Finally, the activated food waste is cryo-grinded to produce nanocarbon ranging from 10 nm to 100 nm.

For making other nanoparticles, grinding methods are sometimes used. According to Kim and Park, grinding methods still need refinement to obtain particles smaller than the micro-scale when using more traditional dry grinding methods [47]. Typically, wet grinding processes present problems, such as water growth and size expansion of the nanoparticles during water removal. To overcome problems associated with traditional grinding methods, the patent describes a method that first involves dissolving sucrose in water over several hours of stirring. Then, a fatty acid and surfactant are added and stirred at slightly elevated temperatures (about 60°C). Finally, an active ingredient is added, followed by drying, grinding many times, and stirring the resulting powder. Using this technique, mean particle sizes are expected to be smaller and more consistent than the ones obtained using traditional methods.

There is a wide variety of CNTs and CNFs provided by different manufacturers. The purity and consistency of these products are of great concern to the end-users. These parameters will affect the final properties of the manufactured polymer nanocomposites.

15.5.2 Manufacturability of Polymer Nanocomposites

The main challenges in implementing nanocomposites lie in the processing stage. Polymer nanocomposites require very precise measurements when it comes to the size of the filler materials. If the particles are too large, they may not disperse well in the matrix. Also, the orientation of the particles must be parallel to the surface of the object being constructed. It is very difficult to control the direction of the particles during the mixing process, but ignoring the orientation could have an adverse effect on the properties of the polymer nanocomposite material. It is also important to have nanofiller compatible with the polymer matrix. If the two are not compatible, the nanofiller may not mix evenly within the matrix, and the synthesis of a nanocomposite will have failed [48]. The interphase properties between the nanomaterials and the host polymer have two aspects: weak interfacial interaction favors network formation and strong interfacial affinity favors mechanical reinforcement.

The amount of capital required to distribute the amount of nanocomposite that is needed to even make a difference, as well as investing in new equipment and technologies, had been one of the main reasons why this technology has not already boomed. The process of making nanomaterials is expensive, which consequently makes the resulting nanocomposite expensive, in the long run much more so than the standard polymer. Put simply, someone with capital needs to take a risk on the polymer nanocomposite market. If a successful market can be established, then the prices to produce these products will decrease and the market will have a much more viable alternative financially than the one that exists now [49].

Along with individual product challenges, using polymer nanocomposites on a large scale requires more knowledge in the field than currently available. Research in this field is growing, but the awareness of the technology among the public is low. Because of the lack of interest in the market currently, companies have been reluctant to provide more funding to this field, fearing that it may take more time to produce results than they feel comfortable to invest in [38].

It is established that these materials are either already being used or could be used in the making of automobiles. There are certain challenges, however, that prevent these materials from becoming commonplace within the industry and setting new standards. First of all, with the number of automobiles produced by larger manufacturers, there is no guarantee that the conditions of the composites will stay consistent in mass production [50]. Also, the use of these new materials would require a large amount of government oversight before the company may proceed with their implementation (see Chapter 14). The start date for these projects could be pushed back if there are delays in the application process. While nanoclays provide some great additive properties, they are also thermally unstable and can degrade at temperatures as low as 170°C. This is undesirable considering that most parts are fabricated using some sort of melt-processing technique. Figure 15.44 is useful in showing how these nanostructured materials will affect the industry over time and relate them to market interests [51].

Another major obstacle to the continued use and growth of nanomaterials is safety. No all-encompassing safety regulations or procedures have been established to date, either by industry, government, or academia (see Chapter 14). Further complicating this is the plethora of competing safety strategies emerging from a variety of sources [52].

The effects of exposure to these materials are serious, and exposure can occur through many different means, such as inhalation, contact with the skin, and

	Short term impact (2011–2013)	Medium term impact (2013–2016)	Long term impact (2016–2021)
Growth Drivers Growth in high performance and super-size vehicles	Demand for high performance nanocomposites will follow the growth trends of high performance vehicles	As market for high end vehicles mature, demand for nanocomposites will stabilize	Innovations in new high performance nanocomposites with low cost will drive fater growth
Challenges Automotive markets facing global contraction	Global recessionary trends 2009-2012: delay in demand for new vehicles manufacturing	Slow growth, as vehicles manufacturing picks pace	Market for nanocomposites in automotive applications will mature, decline in market share

Figure 15.44. Incorporating polymer nanocomposites into automobiles.

ingestion. Animal studies have shown that smaller particles are more toxic than larger ones because they can penetrate farther into the lungs and may be able to move to other organs through barriers considered impermeable to larger particles [53] (see Chapter 14).

Six years ago, the EPA awarded $4 million in grants to study the risks associated with the proliferation of nanomaterials – an amount many considered insufficient. Funded studies included absorption and toxicity of nanoparticles on skin, nanoparticles in drinking water, nanoparticles' effect on lung tissue and airways, and nanoparticles' effect on marine life. To date, however, little effort has gone into establishing manufacturing and handling codes [54].

This lack of guidance has led to the emergence of many different suggestions, and navigating through them and choosing an appropriate procedure for one's application can be difficult. Also, some companies may hesitate to invest in technologies that do not have guidelines in place out of concern over how the government will eventually react. In other words, fears that the regulations eventually created will result in unfavorable market conditions may stifle current and future investment in nanomaterials research, development, production, and implementation. Strong leadership in this area is necessary if the projected growth in the industry is to be maintained.

These challenges can be solved by committing to further research in this field to develop better processing techniques. When the proper methods of fabrication are determined, what type of safety precautions can be determined to overcome the concerns? Research is growing in this area, but the demand for and awareness of polymer nanocomposites are still low. Educating the public and promoting awareness of this technology are key to overcoming these challenges discussed.

15.6 Concluding Remarks

Nanocomposites continue to play an ever increasing role in making products more lightweight, strong, flame retardant, conductive, and improving many other physical and chemical properties. This chapter discussed just a few of the ways nanomaterials are already changing the world. In recent years, research has been increasing heavily both in the United States and the rest of the world. As technologies develop and costs continue to decrease, the world will see more applications where nanomaterials will be used.

Polymer nanocomposites are highly promising for use in a vast number of applications because of their low weights and high strengths. Other emerging properties, such as electrical and thermal conductivity and increased thermal stability, make them highly sought after in extreme environments where currently only metals and ceramics are used.

Another view is that the nanomaterials industry is at a crossroads. With opportunities and applications abound, it could become a large and profitable sector with healthy research and development. On the other hand, safety, lack of guidance, and underfunded/underpursued R&D could cause the industry to flounder for many years. It has already shown itself capable of both disappointing results and incredible growth. It will fall to everyone involved to help direct the fledgling sector – those in academia, government, and industry.

REFERENCES

1. U.S. National Nanotechnology Initiative official website. http://www.nano.gov (accessed December 23, 2015).
2. "The National Nanotechnology Initiative – Supplement to the President's Budget for Fiscal Year 2016," National Science and Technology Council (NSTC), Committee on Technology (CoT), Subcommittee on Nanoscale Science, Engineering, and Technology (NSET), Washington, DC, March 2015.
3. Lucintel, Irving, TX, website: www.lucintel.com.
4. Applied Sciences, Inc., Cedarville, OH, website: www.apsci.com.
5. M. E. Vance, T. Kuiken, E. P. Vejerano, S. P. McGinnis, M. F. Hochella, Jr., et al. (2015). Nanotechnology in the real world: Redeveloping the nanomaterial consumer products inventory. *Beilstein Journal of Nanotechnology* 6, 1769–1780.
6. J. H. Koo (2006). *Polymer Nanocomposites – Characterization, Properties, and Applications*. New York: McGraw-Hill.
7. C. Sanchez, B. Julian, P. Belleville, and M. Popall (2005). Applications of hybrid organic-inorganic nanocomposites. *Journal of Materials Chemistry* 15, 3559–3592.
8. Sol-Gel Technologies Ltd., Bet Shemesh, Israel. http://www.sol-gel.com.
9. I. Hazan and M. Rummel, US Patent No. 5,162, 426, 1992.
10. Y. Akamatsu, A. Takamatus, K. Makita, H. Inaba, and T. Minami. (2001). Preparation of black-colored thin films from silica sol containing metal nitrates by sol-gel method. *Journal of Materials Science* 36(18), 4503–4510.
11. H. Wolter and W. Storch (1994). A new silane precursor with reduced polymerization shrinkage. *Journal of Sol-Gel Science and Technology* 2(1), 93-96.
12. H. Wolter, W. Storch, and H. Ott (1994). Better ceramics through chemistry VI. *Materials Research Society Symposium Proceedings* 346, 143-149.
13. H. Wolter, W. Glaubitt, and K. Rose (1992). Better ceramics through chemistry V. *Materials Research Society Symposium Proceeding* 271, 719-724.
14. C. Gellerman, W. Storch, and H. Wolter. (1997). Synthesis and characterization of the organic surface modifications of monodisperse colloidal silica. *Journal of Sol-Gel Science and Technology* 8(1), 173-176.
15. S. Fujiwara and T. Sakamoto (1976). Japanese Kokai Patent Application No. 109998 (assigned to Unichika K.K., Japan).

16. Y. Fukushima and S. Inagaki (1987). Synthesis of an intercalated compound of montmorillonite and 6-polyamide. *Journal of Inclusion Phenomena* 5, 473–482.
17. A. Okada, Y. Fukushima, M. Kawasumi, S. Inagaki, A. Usuki, et al. (1988). United States Patent No. 4739007 (assigned to Toyota Motor Co., Japan).
18. M. Kawasumi, M. Kohzaki, Y. Kojima, A. Okada, and O. Kamigaito (1989). United States Patent No. 4810734 (assigned to Toyota Motor Co., Japan).
19. A. Usuki, Y. Kojima, M. Kawasumi, A. Okada, Y. Fukusima, et al. (1993). Synthesis of nylon 6-clay hybrid. *Journal of Materials Research* 8, 1179–1184.
20. Southern Clay Products, Gonzales, TX. http://www.scprod.com/gm.html.
21. Nanocomposites New Low-Cost, High-Strength Materials for Automotive Parts, ATP Project 97-02-0047, National Institute of Standards and Technology, Gaithersburg, MD, 1997.
22. A. K. Mohanty, L. T. Drzal, and M. Misra (2003). Nano reinforcements of bio-based polymers – The hope and the Reality. *Polymeric Materials Science & Engineering* 88, 60-61.
23. S. L. Wilkinson (2001). All in good taste. *Chemical Engineering News*, January 22, 61.
24. G. W. Beall (2001). New Conceptual Model for Interpreting Nanocomposite Behavior. In *Polymer-Clay Nanocomposites*, T. J. Pinnavaia and G. W. Beall (Eds.). New York: Wiley & Sons, pp. 267–279.
25. J. N. Hay and S. J. Shaw, "A Review of Nanocomposites 2000," abstracted version from http://www.azom.com/details.asp?ArticleID-921.
26. Multilayer Containers Featuring Nano-Nylon MXD6 Barrier Layers with Superior Performance and Clarity, Mitsubishi Gas Co., Inc. and Nanocor, presented at Nava-Pac 2003. http://www.nanocor.com/techpapers.asp:
27. Triton Systems, Inc., Chelmsford, MA. http://www.tritonsys.com.
28. A. Downing-Perrault (2013). Polymer Nanocomposites Are the Future. University of Wisconsin-Stout. Retrieved from http://www.iopp.org/files/public/DowningPerraultAlyssaUWStroutNanoStructures.pdf.
29. InMat Inc., Hillsborough, NJ. http://www.InMat.com.
30. D. Bievre and K. Nakamura (2004). Laminate structure excelling in fuel permeation preventing performance. *PCT International Applications*, 26. Patent no. WO20004054802.
31. J. W. Gilman (1999). Flammability and thermal stability studies of polymer layered-silicate (clay) nanocomposites. *Applied Clay Science* 15, 31-49.
32. Nanocor, Inc., Arlington Heights, IL. http://www.nanocor.com.
33. S. Baksi, P. R. Basak, and S. Biswas (2013). Nanocomposites – Technology Trends & Application Potential. Technology Information, Forecasting and Assessment Council. Retrieved from http://www.tifac.org.in/index.php?option=com_content&view=article&id=523:nanocomposites--technology-trends-a-application-potential&catid=85:publications&Itemid=952.
34. Carbon Black Uses (2013). In *International Carbon Black Association*. Retrieved from http://www.carbon-black.org/uses.html.
35. Team Innovation (2007). Why Some Rotomolded Tanks are Black in Color – Carbon Black and Plastics. In *PolyProcessing Company*. Retrieved from http://www.polyprocessing.com/pdf/technical/CarbonBlackandPlastics.pdf.
36. Radar Absorbing Materials (2012). In *GlobalSpec*. Retrieved from http://beta.globalspec.com/learnmore/materials_chemicals_adhesives/electrical_optical_specialty_materials/radar_absorbing_materials_structures_ram_ras.
37. S. Zhang, H. L. Wang, S. E. Ong, D. Sun, and X. L. Bui (2007). Hard yet tough nanocomposite coatings – Present status and future trends. *Plasma Processes and Polymers* 4, 219–228.

38. J. Simonsen (2009). Bio-Based Nanocomposites: Challenges and Opportunities. Department of Wood Science & Engineering, Oregon State University, Corvallis, OR. Retrieved from http://www.cof.orst.edu/cof/wse/faculty/simonsen/Nanocomposites.pdf.
39. Polymer nanocomposites drive opportunities in the automotive sector. (2012). In *nanowerk*. Retrieved from http://www.nanowerk.com/spotlight/spotid=23934.php.
40. K. Bonsor (2012). How Smart Windows Work. In *howstuffworks*. Retrieved from http://home.howstuffworks.com/home-improvement/construction/green/smart-window4.htm.
41. The Project on Emerging Nanotechnologies. Consumer Products Inventory. http://www.nanotechproject.org/cpi (accessed December 30, 2015).
42. J. Jordan, K. I. Jacob, R. Tannenbaum, M. A. Sharaf, and I. Jasiuk (2005). Experimental trends in polymer nanocomposites – a review. *Materials Science and Engineering, Part A* 393, 1–11.
43. B. Karn (2012). NNI Research in NanoEHS, NSF Grantees Meeting Presentation, December 4.
44. BASF Chemical Company Website. http://www.basf.com
45. M. Berger (2008). NASA Nanotechnology Research into Shape-Shifting Airplanes. June 16. Retrieved from http://www.nanowerk.com/spotlight/spotid=6067.php
46. S. Al-Zahrani, A. Rahman, I. Ali, and R. Elliethy, U.S. Patent No. 2013 0089738.
47. K. Kim and J. Park, U.S. Patent No. 2013 005643.
48. A. Downing-Perrault (2013). Polymer Nanocomposites Are the Future. University of Wisconsin-Stout. Retrieved from http://www.iopp.org/files/public/DowningPerraultAlyssaUWStroutNanoStructures.pdf
49. S. Baksi, P. R. Basak, and S. Biswas (2013). Nanocomposites – Technology Trends & Application Potential. Technology Information, Forecasting and Assessment Council. Retrieved from http://www.tifac.org.in/index.php?option=com_content&view=article&id=523:nanocomposites--technology-trends-a-application-potential&catid=85:publications&Itemid=952
50. H. Presting and U. König (2003). Future nanotechnology developments for automotive applications. *Materials Science and Engineering, Part C* 23, 737–741.
51. Polymer nanocomposites drive opportunities in the automotive sector. (2012). In *nanowerk*. Retrieved from http://www.nanowerk.com/spotlight/spotid=23934.php
52. Nanomaterial Safety Plan (2010). In SLAC National Accelerator Laboratory – Stanford University. Retrieved from http://www-group.slac.stanford.edu/esh/eshmanual/references/hazmatplannano.pdf
53. R. Weiss (2004). EPA Backs Nanomaterial Safety Research. *The Washington Post*. Retrieved from http://www.washingtonpost.com/wp-dyn/articles/A43763-2004Nov11.html.
54. Conductive polymers and electronics: Tackling corrosion with nanotechnology (2011). In *Enginuity 2011*, Lamar University. Retrieved from http://www.lamar.edu/research/research-magazines/enginuity-2011/tackling-corrosion.html

Index